# Environmental Science and Engineering

The ultimate goal of this series is to contribute to the protection of our environment, which calls for both profound research and the ongoing development of solutions and measurements by experts in the field. Accordingly, the series promotes not only a deeper understanding of environmental processes and the evaluation of management strategies, but also design and technology aimed at improving environmental quality. Books focusing on the former are published in the subseries Environmental Science, those focusing on the latter in the subseries Environmental Engineering.

Nídia S. Caetano · Manuel Carlos Felgueiras
Editors

# The 9th International Conference on Energy and Environment Research

## Greening Energy to Shape a Sustainable Future

Springer

*Editors*
Nídia S. Caetano
School of Engineering (ISEP)
Polytechnic Institute of Porto
Porto, Portugal

Manuel Carlos Felgueiras
School of Engineering (ISEP)
Polytechnic Institute of Porto
Porto, Portugal

ISSN 1863-5520 ISSN 1863-5539 (electronic)
Environmental Science and Engineering
ISBN 978-3-031-43558-4 ISBN 978-3-031-43559-1 (eBook)
https://doi.org/10.1007/978-3-031-43559-1



This Springer imprint is published by the registered company Springer Nature Switzerland AG
The registered company address is: Gewerbestrasse 11, 6330 Cham, Switzerland

Paper in this product is recyclable.

# Organization

## Conference Chair

Nídia S. Caetano, LEPABE/FEUP/U. Porto and CIETI/ISEP/P.Porto, PT

## Program Chair

Manuel Carlos Felgueiras, CIETI/ISEP/P.Porto, PT

## Scientific and Technical Committee Chairs

Nídia S. Caetano, LEPABE/FEUP and CIETI/ISEP, PT
Carlos Felgueiras, CIETI/ISEP, PT

## Scientific and Technical Committee Members

Adriano Péres, Universidade Federal de Santa Catarina—UFSC Blumenau, BR
Alírio Rodrigues, LSRE-LCM/FEUP/U. Porto, PT
Ambra Giovanelli, University of Roma Tre, IT
Ana C. Meira Castro, ISEP/P. Porto and CERENA—Polo FEUP/U. Porto, PT
Ana I. Palmero-Marrero, CIENER/INEGI/FEUP/U. Porto, PT
Anabela A. Leitão, LESRA/FEUAN/Agostinho Neto University, AO
António Augusto Areosa Martins, University of Porto, PT
Antonio Casimiro Caputo, Roma Tre University, Roma, IT

António Curado, ProMetheus, Instituto Politécnico de Viana do Castelo/Porto University, PT
Antonio Zuorro, Sapienza University of Rome, IT
Aytun Onay, Academic Studies for Innovative Products, TR
Bachir Achour, University of Biskra, DZ
Barry A. Benedict, University of Texas at El Paso, US
Carlos Borrego, University of Aveiro, PT
Carlos Costa, UPorto LEPABE/FEUP/UPorto, PT
Carlos Pinho, CEFT-DEMEC-FEUP, PT
Catalin Popescu, Petroleum-Gas University from Ploiesti, RO
Catalina Rus Casas, University of Jaén, PT
Coriolano Salvini, Roma Tre University, IT
Eduardo Vivas, Managing Partner and Consultant at H2OPT, PT
Eugénio C. Ferreira, Universidade do Minho, PT
Fernando Neto, Departamento de Engenharia Mecânica, Universidade de Aveiro, PT
Florinda F. Martins, REQUIMTE/ISEP/P. Porto, PT
Franz Gassner, University of Saint Joseph (USJ), CN
Galyna Tabunshchyk, Zaporizhzhia National Technical University, UA
Gisela Marta Oliveira, Universidade Fernando Pessoa, Porto, PT
Gustavo R. Alves, Polytechnic of Porto, PT
Héctor A. Ruiz, Autonomous University of Coahuila, MX
Helder Santos, Polytechnic Institute of Leiria, PT
Helena I. Monteiro, Instituto de Soldadura e Qualidade, PT
Hugo Romero B., Technical University of Machala, EC
Isabel Paula Lopes Brás, Polytechnic of Viseu, PT
Isabel Soares, University of Porto, PT
José Beleza Carvalho, ISEP/P. Porto, PT
Laura Piedra-Muñoz, University of Almería (CIMEDES; ceiA3), ES
Lígia Pinto, University of Minho, PT
Luis Castro, Coimbra Institute of Engineering/ Polytechnic Institute of Coimbra, PT
Magdalena Ligus, Wroclaw University of Economics and Business, PL
Manuel Arlindo Amador Matos, DAO/University of Aveiro, PT
Mara Madaleno, University of Aveiro and GOVCOPP, PT
Margarita Robaina, University of Aveiro, PT
Maria Isabel Nunes, University of Aveiro, PT
Marta Ferreira Dias, University of Aveiro, PT
Meisam Tabatabaei, Universiti Malaysia Terengganu and Henan Agricultural University, MY/CN
Melih Onay, Van Yuzuncu Yil University, TR
Michael Hartnett, University of Galway, IE
Monique Branco, UFRJ, UPorto, IPK, DE
Nelson Amadeu Dias Martins, TEMA/DEM/University of Aveiro, PT
Nelson Soares, University of Coimbra, PT
O. Parthiba Karthikeyan, University of Houston, US
Orhan Ekren, Solar Energy Institute-Ege University, TR

Ornella Chiavola, Roma Tre University, IT
Paola Marrone, Roma Tre University, IT
Paulo Ramísio, University of Minho, PT
Pedro Faria, GECAD/ISEP/P. Porto, PT
Ramiro S. Barbosa, GECAD/ISEP/P. Porto, PT
Ricardo Jorge Costa, Polytechnic of Porto, PT
Rosa M. B. R. Pilão, CIETI/ISEP/P.Porto, PT
Rosa M. Quinta-Ferreira, University of Coimbra, PT
Rui A. R. Boaventura, Associate Laboratory LSRE-LCM/FEUP/U. Porto, PT
Rui Calejo Rodrigues, CONSTRUCT—FEUP, PT
Sérgio Ivan Lopes, Instituto Politécnico de Viana do Castelo, PT
Sergio Morales-Torres, University of Granada, ES
Sérgio Ramos, GECAD/ISEP/P.Porto, PT
Simona Varvara, "1 Decembrie 1918" University of Alba Iulia, RO
Sónia A. Figueiredo, REQUIMTE/LAQV-ISEP, PT
Teresa Fidélis, University of Aveiro, PT
Teresa Mata, UPorto LEPABE/FEUP/UPorto, PT
Vítor António Ferreira da Costa, University of Aveiro, PT
Vitor Manuel Ferreira Moutinho, University of Beira Interior, PT
Weihao Hu, University of Electronics Science and Technology of China (UESTC), CN
Wilson Galvão de Morais Júnior, Millhouse International (Pty) LTD, PT
Zita Vale, GECAD/ISEP/IPP—Instituto Superior de Engenharia do Instituto Politécnico do Porto, PT

# List of Scientific and Technical Committee Members Who Acted as Reviewers

Adriano Péres
Alírio Rodrigues
Ana C. Meira Castro
Ana I. Palmero-Marrero
Anabela A. Leitão
António Areosa Martins
Antonio Casimiro Caputo
António Curado
Aytun Onay
Bachir Achour
Barry A. Benedict
Carlos Borrego
Carlos Costa
Carlos Felgueiras
Carlos Pinho
Catalin Popescu
Catalina Rus Casas
Coriolano Salvini
Eduardo Vivas
Fernando José Neto da Silva
Florinda F. Martins
Franz Gassner
Galyna Tabunshchyk
Gisela Marta Oliveira
Gustavo R. Alves
Héctor A. Ruiz
Helder Santos
Helena Monteiro
Hugo Romero B.
Isabel Paula Lopes Brás
Isabel Soares

José Beleza Carvalho
Lígia Pinto
Luís Miguel Moura Neves de Castro
Magdalena Ligus
Manuel Arlindo Amador Matos
Mara Madaleno
Margarita Robaina
Maria Isabel Nunes
Marta Ferreira Dias
Melih Onay
Michael Hartnett
Monique Branco-Vieira
Nelson Martins
Nelson Miguel Lopes Soares
Nídia S. Caetano
O. Parthiba Karthikeyan
Orhan Ekren
Paola Marrone
Paulo Ramísio
Pedro Faria
Ramiro S. Barbosa
Ricardo Jorge Costa
Rosa M. B. R. Pilão
Rosa M. Quinta-Ferreira
Rui A. R. Boaventura
Rui Calejo Rodrigues
Sérgio Ivan Lopes
Sérgio Ramos
Simona Varvara
Sónia A. Figueiredo
Teresa Fidélis
Teresa M. Mata
Vítor António Ferreira da Costa
Vitor Manuel Ferreira Moutinho
Wilson Galvão de Morais Júnior
Zita Vale

# Preface

The International Conference on Energy and Environment Research, ICEER conference series, has reached maturity, after nine successful editions, that started in 2014, in Madrid, Spain. ICEER conference series has gained increased attraction, not only because of the interest and quality of the research that has been presented there, but also because of the informal, multicultural, and free atmosphere that we could create and experience. This informal discussion environment allowed vivid and fruitful discussions on the topics presented both by the keynote speakers, the authors of research mostly coming from academia but also from the professional field, and all participants, independent of their career stage.

The 9th edition of this conference, **ICEER 2022**, was organized for the first time in a hybrid mode and took place from September 13 to 16, 2022. **ICEER 2022** recovered the personal contact, with social activities, and combined it with the video interface communication, linking participants from thirty-five countries in five continents.

Due to the pandemic uncertainty in the World, with strong restrictions to travel in some countries, and a rather normal behavior in many others, the **ICEER 2022** was planned as a two-days live event in Porto@ISEP, Portugal, followed by two days online for all those that unfortunately could not travel. Aiming at contributing to the development of a sustainable World, where energy is the engine of development and where development has been causing dramatic impacts on the environment, of which climate changes are the biggest consequence, the theme for this edition of **ICEER 2022** was *Greening Energy to Shape a Sustainable Future.*

The unquestionable urgency for energy alternatives that could compensate the decrease of supply of fossil derived energy, while simultaneously contributing to energy decarbonization and to Sustainable Development, has driven researchers attention to renewable energy sources, integrated with energy storage systems and a more efficient distribution and use of energy. The use of renewable energy sources poses problems due to their intrinsic uncontrollability, making it urgent the development of efficient control systems for distributed energy sources and storage systems. Energy is being used not only for the production in factories and in the agrifood sector, but especially in transportation and buildings, where recently working from home has become much more frequent. Thus, cities represent huge energy consumers, not only for mobility but also for buildings, making it urgent to redesign cities and their mobility, food and water supply systems, using integrated approaches that increase circularity and make the whole cities self-sustained. Therefore, the adopted theme *Greening Energy to Shape a Sustainable Future* of **ICEER 2022**, allowed to address the issues associated with the need of sources of energy that are clean, easily available and that continuously supply the needs of a World under Sustainable Development conditions.

Linking together energy and environment research is not an easy task. However, it is now understood that these fields are interconnected and that the answer to the challenge of a sustainable future depends enormously on the willingness and capability of problem thinking in an integrated manner. Thus, upon a thorough revision process where 78 members of the Scientific and Technical Committee reviewed more than 155 full papers, almost 100 papers with at least two positive reviews were finally selected for publication. These papers report the research related to energy and environment, presented and discussed in **ICEER 2022**. In the end, 89 oral and 11 oral flash presentations were selected. Invited talks included two keynote lectures.

More than 100 delegates attended the conference from 35 countries including Qatar, Saudi Arabia, USA, Korea, Germany, France, Portugal, China, Australia, Canada, Ecuador, etc. Three best paper awards were assigned in the categories *Energy*, *Environment*, and *Sustainable Development*, respectively.

The research on renewable energy production, provision, and management, associated with the integrated analysis of the whole chain impacts in a life cycle perspective, together with the environmental concerns and education of the professionals of tomorrow, are the foundations of capacity building and of Sustainable Development focused in the edition of **ICEER 2022**.

Our small contribution to the dissemination of knowledge and of research results is the publication of this volume of the *Environmental Science and Engineering* book series. Thus, the proceedings of the ICEER 2002 collects one invited speech and 78 technical papers that were presented orally.

These papers are divided into nine following parts:

1: Advanced Energy Technologies
2: Sustainable Buildings
3: Life Cycle Analysis Methodologies
4: Modeling, Simulation, and Forecasting of Energy and Carbon Markets
5: Energy and Environment
6: Energy Efficiency
7: Renewable Energy
8: Energy Policy, Economics, Planning, and Regulation
9: Education for Sustainable Development.

This book can serve as a reference and hopefully inspire scientists, engineers, graduate students, and all other professionals who are working on the energy and environment-related fields.

Finally, we would like to acknowledge our members of the International Scientific and Technical Committee who borrowed their expertise and generously used their time and energy to review the manuscripts. At last, our special appreciation to the ICEER staff by their efforts collecting the revised manuscripts and all the efforts in publishing these proceedings.

Porto, Portugal

Prof. Nídia S. Caetano
Prof. Manuel Carlos Felgueiras

# Acknowledgements

The editors thank the Scientific Committee members of ICEER 2022, who kindly reviewed the papers included in this volume. Their valuable comments and recommendations contributed to substantial improvements of the quality of the reports provided by the authors.

Professor Nídia S. Caetano, Editor of this volume, is an integrated member of LEPABE—Laboratory for Process Engineering, Environment, Biotechnology and Energy, financially supported by LA/P/0045/2020 (ALiCE), UIDB/00511/2020 and UIDP/00511/2020 (LEPABE), funded by national funds through FCT/MCTES (PIDDAC).

Professor Manuel Carlos Felgueiras, Editor of this volume, is an integrated member of CIETI—Center for Innovation in Engineering and Industrial Technology, financially supported by UIDB/04730/2020 (CIETI), funded by national funds through FCT/MCTES (PIDDAC).

fct Fundação para a Ciência e a Tecnologia

REPÚBLICA PORTUGUESA

# Contents

# Contributors

**A. D. Aarthi** LKAB, Malmberget, Sweden

**A. E. M. van den Oever** Mobility, Logistics and Automotive Research Centre, Department of Electric Engineering and Energy Technology, Vrije Universiteit Brussel, Brussels, Belgium

**A. F. Teran-Alvarado** Universidad Politécnica Salesiana, Guayaquil, Ecuador

**A. G. M. B. Mustayen** Fuel and Energy Research Group, School of Engineering and Technology, Central Queensland University, Rockhampton, QLD, Australia; School of Engineering, University of Tasmania, Hobart, TAS, Australia

**A. O. Haikina** Novosibirsk State University, Novosibirsk, Russia

**A. V. Komarova** Novosibirsk State University, Novosibirsk, Russia;
Trofimuk Institute of Petroleum Geology and Geophysics, Novosibirsk, Russia

**A. Y. Novikov** Novosibirsk State University, Novosibirsk, Russia;
Trofimuk Institute of Petroleum Geology and Geophysics, Novosibirsk, Russia

**Abdul Rauf** United Arab Emirates University, Al Ain, UAE

**Abdulaziz I. Almulhim** Department of Urban and Regional Planning, College of Architecture and Planning, Imam Abdulrahman Bin Faisal University, Dammam, Saudi Arabia

**Adam Kubín** Faculty of Electrical Engineering, Czech Technical University in Prague, Prague, Czech Republic

**Adriano A. Santos** CIDEM, School of Engineering (ISEP), Polytechnic of Porto (P.Porto), Porto, Portugal;
INEGI—Institute of Science and Innovation in Mechanical Engineering and Industrial Engineering, Porto, Portugal

**Albina Ribeiro** CIETI, ISEP, Instituto Politécnico do Porto, Porto, Portugal

**Alejandro Hernández Serrano** CARTIF Technology Centre, Parque Tecnológico de Boecillo, Valladolid, Spain

**Alex Ranieri J. Lima** Center of Scientific Development, Butantan Institute, São Paulo, Brazil

**Alexandra Monteiro** CESAM and Department of Environment and Planning, University of Aveiro, Aveiro, Portugal

**Alexandre Soares dos Reis** Research Unit on Governance, Competitiveness and Public Policies (GOVCOPP), Department of Economics, Management, Industrial Engineering and Tourism (DEGEIT), University of Aveiro, Aveiro, Portugal

**Ali Cilesiz** Kastamonu University, Kastamonu, Turkey

**Ali Mahmud** Faculty of Agriculture and Animal Science, University of Muhammadiyah Malang, Malang, Indonesia

**Alice Tavares** Centre for Research in Ceramics and Composite Materials (CICECO), Department of Materials and Ceramic Engineering, University of Aveiro, Aveiro, Portugal

**Ammar M. Khourchid** Division of Sustainable Development, College of Science and Engineering, Hamad Bin Khalifa University, Qatar Foundation, Doha, Qatar

**Ana Almeida** CIETI, ISEP, Instituto Politécnico do Porto, Porto, Portugal

**Ana P. Marques** CBQF—Centro de Biotecnologia e Química Fina—Laboratório Associado, Escola Superior de Biotecnologia, Universidade Católica Portuguesa, Porto, Portugal

**Anabela M. Moreira** Instituto Politécnico de Coimbra, Instituto Superior de Engenharia de Coimbra, Coimbra, Portugal

**Andrea Molocchi** RSE Ricerca sul Sistema Energetico, Milano, Italy

**Andreia D. Santos** Department of Chemical Engineering, Faculty of Sciences and Technology, CIEPQPF—Chemical Engineering Processes and Forest Products Research Center, University of Coimbra, Coimbra, Portugal;
Department of Chemical and Biological Engineering, Polytechnic of Coimbra, Coimbra Institute of Engineering, Coimbra, Portugal

**Andrés Velastegui-Montoya** Facultad de Ingeniería en Ciencias de la Tierra, ESPOL Polytechnic University, Guayaquil, Ecuador;
Centro de Investigaciones y Proyectos Aplicados a las Ciencias de la Tierra, ESPOL Polytechnic University, Guayaqui, Ecuador

**Annie Feba Varghese** Department of Education, University of Kerala, Thiruvananthapuram, India

**António Curado** proMetheus, Instituto Politécnico de Viana Do Castelo, Viana Do Castelo, Portugal

**António A. Martins** LEPABE, Faculty of Engineering, University of Porto (FEUP), Porto, Portugal;
ALiCE, Faculty of Engineering, University of Porto, Porto, Portugal

**António A. Vicente** CEB, Centre of Biological Engineering, University of Minho Campus de Gualtar, Braga, Portugal

**António Ferreira da Silva** CIDEM, School of Engineering (ISEP), Polytechnic of Porto (P.Porto), Porto, Portugal;
INEGI—Institute of Science and Innovation in Mechanical Engineering and Industrial Engineering, Porto, Portugal

**Aris Winaya** Faculty of Agriculture and Animal Science, University of Muhammadiyah Malang, Malang, Indonesia

**Aybüke Taşer** Université Catholique de Louvain, Louvain-la-Neuve, Belgium

**B. Mainali** Department of Built Environment and Energy Technology, Linnaeus University, Växjö, Sweden

**Benedetta Marmiroli** RSE Ricerca sul Sistema Energetico, Milano, Italy

**Benigno Antonio Rodriguez Gomez** Universidade da Coruña, Galicia, Spain

**Bonifacio Doma** Mapua University, Manila, Philippines

**Bruna Moura** Low Carbon and Resource Efficiency, R&Di, Instituto de Soldadura e Qualidade, Grijó, Portugal

**C. M. Moreira** ESPOL Polytechnic University, Escuela Superior Politécnica del Litoral, ESPOL, Guayaquil, Ecuador

**C. M. O. Martinho** Department of Chemical Engineering, University of Coimbra, CIEPQPF, Coimbra, Portugal

**Carla Gama** CESAM and Department of Environment and Planning, University of Aveiro, Aveiro, Portugal

**Carlos Felgueiras** CIETI—School of Engineering (ISEP), Polytechnic of Porto (P.Porto), Porto, Portugal;
CIETI—Centre of Innovation on Engineering and Industrial Technology/IPP-ISEP, School of Engineering, Porto, Portugal

**Carlos Páscoa** Portuguese Air Force, Amadora, Portugal

**Carlos Pinho** CEFT-DEMEC—Faculdade de Engenharia, Universidade Do Porto, Porto, Portugal;
CEFT/FEUP, Faculdade de Engenharia da Universidade do Porto, Porto, Portugal

**Carolina Vicente** Instituto Politécnico de Coimbra, Instituto Superior de Engenharia de Coimbra, Coimbra, Portugal

**Cecilia V. Romero** Facultad de Medicina, Universidad Nacional de Trujillo, Trujillo, Peru

**Celeste Eusébio** GOVCOPP and Department of Economics, Management, Industrial Engineering and Tourism, University of Aveiro, Aveiro, Portugal

**Cesar Alvarez-Mendoza** Grupo de Investigación Ambiental en el Desarrollo Sustentable GIADES, Carrera de Ingeniería Ambiental, Universidad Politécnica Salesiana, Quito, Ecuador

**Cristina M. R. Rocha** CEB-Centre of Biological Engineering, University of Minho, Braga, Portugal;
LABBELS-Associated Laboratory, Braga, Guimarães, Portugal

**D. Delfín-Narciso** Grupo de Investigación en Ciencias Aplicadas Y Nuevas Tecnologías, Universidad Privada del Norte, Trujillo, Peru

**D. Khatiwada** KTH Royal Institute of Technology, Stockholm, Sweden

**Dale Bartolo** Institute of Applied Sciences MCAST, Raħal Ġdid, Malta

**Dânia S. Ascenção** IKEA Industry Portugal, SA, Penamaior, Portugal

**Daniel Efurosibina Attoye** United Arab Emirates University, Al Ain, UAE

**Daniele Costa** VITO/EnergyVille, Mol, Belgium;
Electric Vehicle and Energy Research Group (EVERGI), Department of Electrical Engineering and Energy Technology, Mobility Logistics and Automotive Technology Research Centre (MOBI), Vrije Universiteit Brussel, Brussels, Belgium;
VITO—EnergyVille, Unit Smart Energy and Built Environment (SEB), Genk, Belgium

**Débora de São José** Research Group on Intelligent Engineering and Computing for Advanced Innovation and Development (GECAD), Intelligent Systems Associated Laboratory (LASI), Polytechnic of Porto (P.Porto), Porto, Portugal

**Divya Chandrasenan** Department of Education, University of Kerala, Thiruvananthapuram, India

**Dominik Huber** Electric Vehicle and Energy Research Group (EVERGI), Department of Electrical Engineering and Energy Technology, Mobility Logistics and Automotive Technology Research Centre (MOBI), Vrije Universiteit Brussel, Brussels, Belgium

**E. A. Kuznetsova** Novosibirsk State University, Novosibirsk, Russia;
Trofimuk Institute of Petroleum Geology and Geophysics, Novosibirsk, Russia

**Edmundo Marques** Escola Superior de Tecnologia e Gestão, Instituto Politécnico de Viseu, Viseu, Portugal

**Elisa Ramalho** CIETI, ISEP, Instituto Politécnico do Porto, Porto, Portugal

**Elisabete Vieira** GOVCOPP, ISCA—Higher Institute for Accountancy and Administration of Aveiro, University of Aveiro, Aveiro, Portugal

**F. Diaz** Universidad Tecnológica del Perú, Lima, Peru

**F. Golzar** KTH Royal Institute of Technology, Stockholm, Sweden

**F. A. Nuno** Quinta das Bágeiras—Mário Sérgio Alves Nuno, Sangalhos, Portugal

**F. S. Henriques** Department of Chemical Engineering, University of Coimbra, CIEPQPF, Coimbra, Portugal

**Faruk Dayi** Kastamonu University, Kastamonu, Turkey

**Fernando Anuno** Faculty of Economics and Management, National University of Timor Lorosa'e (UNTL), Avenida Cidade de Lisboa, Díli, Timor-Leste;
Research Unit on Governance, Competitiveness and Public Policies (GOVCOPP), Department of Economics, Management, Industrial Engineering and Tourism (DEGEIT), University of Aveiro, Aveiro, Portugal

**Fernando Lezama** Research Group on Intelligent Engineering and Computing for Advanced Innovation and Development (GECAD), Intelligent Systems Associated Laboratory (LASI), Polytechnic of Porto (P.PORTO), Porto, Portugal

**Filipe Maciel** CEB, Centre of Biological Engineering, University of Minho Campus de Gualtar, Braga, Portugal

**Filipe Pereira** CIETI—School of Engineering (ISEP), Polytechnic of Porto (P.Porto), Porto, Portugal;
CIETI—Centre of Innovation on Engineering and Industrial Technology/IPP-ISEP, School of Engineering, Porto, Portugal;
CIDEM, School of Engineering (ISEP), Polytechnic of Porto (P.Porto), Porto, Portugal

**Florinda F. Martins** School of Engineering (ISEP), Polytechnic of Porto (P.Porto), Porto, Portugal

**Francesca Spagnol Gravino** Institute of Applied Sciences MCAST, Raħal Ġdid, Malta

**Ginga Sakanoue** Transdisciplinary Science and Engineering Program, Graduate School of Advanced Science and Engineering, Hiroshima, Japan

**Gisela Mello** Research Unit on Governance, Competitiveness and Public Policies (GOVCOPP), Department of Economics, Management, Industrial Engineering and Tourism (DEGEIT), University of Aveiro, Aveiro, Portugal

**Giulio Mela** RSE Ricerca sul Sistema Energetico, Milano, Italy

**Han Soo Lee** Transdisciplinary Science and Engineering, Graduate School of Advanced Science and Engineering, Hiroshima University, Hiroshima, Japan;

Center for Planetary Health and Innovation Science (PHIS), The IDEC Institute, Hiroshima University, Hiroshima, Japan

**Hassan Shokry** Environmental Engineering Department, Egypt-Japan University of Science and Technology, New Borg El-Arab City, Alexandria, Egypt;
Electronic Materials Research Department, Advanced Technology and New Materials Research Institute, City of Scientific Research and Technological Applications (SRTA-City), Alexandria, Egypt

**Hatem Mahmoud** Environmental Engineering Department, Egypt-Japan University for Science and Technology, Alexandria, Egypt;
Department of Architecture Engineering, Faculty of Engineering, Aswan University, Aswan, Egypt

**Helena Monteiro** Low Carbon and Resource Efficiency, R&Di, Instituto de Soldadura e Qualidade, Grijó, Portugal

**Hena Oak** Miranda House College, University of Delhi, New Delhi, India

**Humberto Varum** CONSTRUCT-LESE, Faculty of Engineering, University of Porto (FEUP), Porto, Portugal

**I. V. Filimonova** Novosibirsk State University, Novosibirsk, Russia;
Trofimuk Institute of Petroleum Geology and Geophysics, Novosibirsk, Russia

**I. V. Provornaya** Novosibirsk State University, Novosibirsk, Russia;
Trofimuk Institute of Petroleum Geology and Geophysics, Novosibirsk, Russia

**Isabel Soares** Faculdade de Economia do Porto and CEFUP, Porto, Portugal

**Isares Dhuchakallaya** Center of Excellence in Computational Mechanics and Medical Engineering, Faculty of Engineering, Thammasat School of Engineering, Thammasat University, Klong-Luang, Pathumthani, Thailand

**J. Almeida** Low Carbon and Resource Efficiency, R&Di, Instituto de Soldadura e Qualidade, Grijó, Portugal

**J. S. Cardoso** CIEPQPF, Department of Chemical Engineering, Faculty of Sciences and Technology, University of Coimbra, Coimbra, Portugal

**Jae-Soon Jeong** Transdisciplinary Science and Engineering Program, Graduate School of Advanced Science and Engineering, Hiroshima, Japan

**Jelle Smekens** Electric Vehicle and Energy Research Group (EVERGI), Department of Electrical Engineering and Energy Technology, Mobility Logistics and Automotive Technology Research Centre (MOBI), Vrije Universiteit Brussel, Brussels, Belgium

**Jiri Bim** Faculty of Electrical Engineering, Czech Technical University in Prague, Prague, Czech Republic

**Joana Pinto** Portuguese Air Force, Amadora, Portugal

**João Claro** INESC TEC and Faculdade de Engenharia, Universidade do Porto, Porto, Portugal

**João Moura** Faculdade de Economia do Porto and CEFUP, Porto, Portugal

**João F. C. Silva** Instituto Politécnico de Coimbra, Instituto Superior de Engenharia de Coimbra, Coimbra, Portugal

**João Monney Paiva** Escola Superior de Tecnologia e Gestão, Instituto Politécnico de Viseu, Viseu, Portugal

**João R. Silva** Instituto Politécnico de Coimbra, Instituto Superior de Engenharia de Coimbra, Coimbra, Portugal;
Department of Chemical Engineering, Faculty of Sciences and Technology, CIEPQPF—Chemical Engineering Processes and Forest Products Research Center, University of Coimbra, Coimbra, Portugal;
Department of Chemical and Biological Engineering, Polytechnic of Coimbra, Coimbra Institute of Engineering, Coimbra, Portugal

**Joaquim C. G. Esteves da Silva** Chemistry Research Unit (CIQUP), Institute of Molecular Sciences (IMS), DGAOT, Faculty of Sciences of University of Porto (FCUP), Porto, Portugal

**Joaquim P. Silva** 2Ai, School of Technology, IPCA, Barcelos, Portugal

**Jonathan Salar Cabrera** Transdisciplinary Science and Engineering Program, Graduate School of Advanced Science and Engineering, Hiroshima, Japan

**Jorge Luis Amaya-Rivas** ESPOL Polytechnic University, Escuela Superior Politécnica del Litoral, ESPOL, Guayaquil, Ecuador;
Facultad de Ingeniería Mecánica y Ciencias de la Producción, ESPOL Polytechnic University, Guayaquil, Ecuador

**José Chen-Xu** Public Health Research Centre, National School of Public Health, NOVA University of Lisbon, Lisbon, Portugal;
FEP—School of Economics and Management, University of Porto, Porto, Portugal

**José A. Teixeira** CEB-Centre of Biological Engineering, University of Minho, Braga, Portugal;
LABBELS-Associated Laboratory, Braga, Guimarães, Portugal

**Jose Armando Hidalgo Crespo** Carrera de Ingeniería Industrial, Universidad Politécnica Salesiana, Guayaquil, Ecuador;
Universidade da Coruña, Coruña, Spain;
G-Scop Laboratory, Polytechnic Institute of Grenoble, Grenoble, France;
Carrera de Ingenieria Industrial, Universidad de Guayaquil, Guayaquil, Ecuador;
Universidad de Guayaquil, Carrera de Ingeniería Industrial, Guayaquil, Ecuador;
Facultad de Ciencias, Universidade da Coruña, Coruña, Spain;
G-Scop Laboratory, University of Grenoble Alpes, Grenoble, France;
Facultad de Ingeniería Industrial, Universidad de Guayaquil, Guayaquil, Ecuador

**José N. F. G. Rodrigues** Chemistry Research Unit (CIQUP), Institute of Molecular Sciences (IMS), DGAOT, Faculty of Sciences of University of Porto (FCUP), Porto, Portugal

**Juan José Bonello** Institute of Applied Sciences MCAST, Raħal Ġdid, Malta

**K. Mahapatra** Department of Built Environment and Energy Technology, Linnaeus University, Växjö, Sweden

**K. D. Gladkikh** Novosibirsk State University, Novosibirsk, Russia;
Trofimuk Institute of Petroleum Geology and Geophysics, Novosibirsk, Russia

**Kenneth Mensah** Environmental Engineering Department, Egypt-Japan University for Science and Technology, Alexandria, Egypt

**Khalid Mohammed Almatar** Department of Urban and Regional Planning, College of Architecture and Planning, Imam Abdulrahman Bin Faisal University, Dammam, Saudi Arabia

**L. Borja-Mora** Carrera de Ingenieria Industrial, Universidad de Guayaquil, Guayaquil, Ecuador

**L. M. Gando-Ferreira** CIEPQPF, Department of Chemical Engineering, Faculty of Sciences and Technology, University of Coimbra, Coimbra, Portugal

**Leandro Madureira** CEB, Centre of Biological Engineering, University of Minho Campus de Gualtar, Braga, Portugal

**Leonardo Alvaro Banguera Arroyo** Facultad de Ingeniería Industrial, Universidad de Guayaquil, Guayaquil, Ecuador;
Carrera de Ingenieria Industrial, Universidad de Guayaquil, Guayaquil, Ecuador;
Universidad de Guayaquil, Carrera de Ingeniería Industrial, Guayaquil, Ecuador

**Leonel J. R. Nunes** proMetheus, Instituto Politécnico de Viana Do Castelo, Viana Do Castelo, Portugal

**Luis Gomes** GECAD—Research Group on Intelligent Engineering and Computing for Advanced Innovation and Development, LASI—Intelligent Systems Associate Laboratory, Polytechnic of Porto, Porto, Portugal

**Luís Soares** Chemistry Research Unit (CIQUP), Institute of Molecular Sciences (IMS), DGAOT, Faculty of Sciences of University of Porto (FCUP), Porto, Portugal

**Luís M. Castro** Instituto Politécnico de Coimbra, Instituto Superior de Engenharia de Coimbra, Coimbra, Portugal;
Department of Chemical Engineering, Faculty of Sciences and Technology, CIEPQPF—Chemical Engineering Processes and Forest Products Research Center, University of Coimbra, Coimbra, Portugal;
Instituto Politécnico de Coimbra, Instituto de Investigação Aplicada, Laboratório SiSus, Coimbra, Portugal;

Laboratório SiSus, Instituto Politécnico de Coimbra, Instituto de Investigação Aplicada, Coimbra, Portugal;
Department of Chemical and Biological Engineering, Polytechnic of Coimbra, Coimbra Institute of Engineering, Coimbra, Portugal

**Lukáš Janota** Faculty of Electrical Engineering, Czech Technical University in Prague, Prague, Czech Republic

**M. De La Cruz-Noriega** Vicerrectorado de Investigación, Universidad Autónoma del Perú, Lima, Peru

**M. Gonçalves** Low Carbon and Resource Efficiency, R&Di, Instituto de Soldadura e Qualidade, Grijó, Portugal

**M. A. Hazrat** Centre for Regional Economies and Supply Chain (CRESC), Central Queensland University, Rockhampton, QLD, Australia

**M. A. Sattar** Engineering School, Chisholm Institute, Dandenong, VIC, Australia

**M. G. Rasul** Fuel and Energy Research Group, School of Engineering and Technology, Central Queensland University, Rockhampton, QLD, Australia;
School of Engineering, University of Tasmania, Hobart, TAS, Australia

**M. I. Jahirul** Engineering School, Chisholm Institute, Dandenong, VIC, Australia;
Fuel and Energy Research Group, School of Engineering and Technology, Central Queensland University, Rockhampton, QLD, Australia

**M. J. Quina** Department of Chemical Engineering, University of Coimbra, CIEPQPF, Coimbra, Portugal

**M. M. Hasan** Fuel and Energy Research Group, School of Engineering and Technology, Central Queensland University, Rockhampton, QLD, Australia

**Maarten Messagie** Department of Electric Engineering and Energy Technology, Mobility, Logistics and Automotive Research Centre, Vrije Universiteit Brussel, Brussels, Belgium;
Electric Vehicle and Energy Research Group (EVERGI), Department of Electrical Engineering and Energy Technology, Mobility Logistics and Automotive Technology Research Centre (MOBI), Vrije Universiteit Brussel, Brussels, Belgium

**Maeva Lavigne Philippot** Electric Vehicle and Energy Research Group (EVERGI), Department of Electrical Engineering and Energy Technology, Mobility Logistics and Automotive Technology Research Centre (MOBI), Vrije Universiteit Brussel, Brussels, Belgium

**Mahmoud Samy** Environmental Engineering Department, Egypt-Japan University of Science and Technology, Alexandria, Egypt;
Public Works Engineering Department, Faculty of Engineering, Mansoura University, Mansoura, Egypt

**Manabu Fujii** Environmental Engineering Department, Egypt-Japan University of Science and Technology, Alexandria, Egypt;
Department of Civil and Environmental Engineering, Tokyo Institute of Technology, Tokyo, Japan

**Manuel Soto** Universidade da Coruña, Coruña, Spain;
Facultad de Ciencias, Universidade da Coruña, Coruña, Spain

**Mara Madaleno** Research Unit On Governance, Competitiveness and Public Policies (GOVCOPP), Department of Economics, Management, Industrial Engineering and Tourism (DEGEIT), University of Aveiro, Aveiro, Portugal

**Marcos Santos-Méndez** Universidad de Guayaquil, Carrera de Ingeniería Industrial, Guayaquil, Ecuador

**Margarita Robaina** Research Unit on Governance, Competitiveness and Public Policies (GOVCOPP), Department of Economics, Management, Industrial Engineering and Tourism (DEGEIT), University of Aveiro, Aveiro, Portugal

**Maria João Carneiro** GOVCOPP and Department of Economics, Management, Industrial Engineering and Tourism, University of Aveiro, Aveiro, Portugal

**Maria Leonor Carvalho** RSE Ricerca sul Sistema Energetico, Milano, Italy

**Mariana Cardoso** CONSTRUCT-LESE, Faculty of Engineering, University of Porto (FEUP), Porto, Portugal

**Mariana Santos** CIETI, School of Engineering (ISEP), Polytechnic Institute of Porto (P.Porto), Porto, Portugal

**Marianne Balat-Pichelin** PROMES, UPR CNRS 8521, Font-Romeu-Odeillo, France

**Marta Ferreira Dias** Research Unit on Governance, Competitiveness and Public Policies (GOVCOPP), Department of Economics, Management, Industrial Engineering and Tourism (DEGEIT), University of Aveiro, Aveiro, Portugal

**Marwa Elkady** Chemical and Petrochemical Engineering Department, Egypt-Japan University of Science and Technology, New Borg El-Arab City, Alexandria, Egypt

**Mehzabeen Mannan** Division of Sustainable Development, College of Science and Engineering, Hamad Bin Khalifa University, Qatar Foundation, Doha, Qatar

**Michaela Valentová** Faculty of Electrical Engineering, Czech Technical University in Prague, Prague, Czech Republic

**Michelle Almendrala** Mapua University, Manila, Philippines

**Mohamed Gar Alalm** Public Works Engineering Department, Faculty of Engineering, Mansoura University, Mansoura, Egypt

**Mona G. Ibrahim** Environmental Engineering Department, Egypt-Japan University of Science and Technology, Alexandria, Egypt;
Environmental Health Department, High Institute of Public Health, Alexandria University, Alexandria, Egypt

**Monique Branco-Vieira** Julius Kühn-Institute, Federal Research Centre for Cultivated Plants, Institute for Strategies and Technology Assessment, Kleinmachnow, Germany;
Faculty of Engineering, LEPABE-Laboratory for Process Engineering, Environment, Biotechnology and Energy, University of Porto (FEUP), Porto, Portugal;
Faculty of Engineering, ALiCE-Associate Laboratory in Chemical Engineering, University of Porto, Porto, Portugal

**Morhaf Aljber** Transdisciplinary Science and Engineering Program, Graduate School of Advanced Science and Engineering, Hiroshima, Japan

**Muhyiddine Jradi** Center for Energy Informatics, University of Southern Denmark, Odense, Denmark

**N. M. Berrones-Rivera** Universidad Politécnica Salesiana, Guayaquil, Ecuador

**Nacef Tazi** Circular Services, Toulouse, France;
Interdisciplinary Research on Society-Technology-Environment Interactions, University of Technology of Troyes, Troyes, France

**Nadine Töpfer** Institute for Plant Sciences, University of Cologne, Köln, Germany

**Nelson Soares** ADAI, Department of Mechanical Engineering, University of Coimbra, Coimbra, Portugal

**Nídia S. Caetano** Faculty of Engineering, LEPABE-Laboratory for Process Engineering, Environment, Biotechnology and Energy, University of Porto (FEUP), Porto, Portugal;
Faculty of Engineering, ALiCE-Associate Laboratory in Chemical Engineering, University of Porto, Porto, Portugal;
CIETI—Centre of Innovation on Engineering and Industrial Technology/IPP-ISEP, School of Engineering, Porto, Portugal;
LEPABE—Laboratory for Process Engineering, Environment, Biotechnology and Energy, Faculty of Engineering, University of Porto (FEUP), Porto, Portugal;
ALiCE—Associate Laboratory in Chemical Engineering, Faculty of Engineering, University of Porto, Porto, Portugal;
CIETI, School of Engineering (ISEP), Polytechnic Institute of Porto (P.Porto), Porto, Portugal

**Nuno Lopes** 2Ai, School of Technology, IPCA, Barcelos, Portugal

**Nuno Teixeira** GECAD—Research Group on Intelligent Engineering and Computing for Advanced Innovation and Development, LASI—Intelligent Systems Associate Laboratory, Polytechnic of Porto, Porto, Portugal

**P. Brito** Centro de Investigação de Montanha (CIMO), Instituto Politécnico de Bragança, Brgança, Portugal

**P. V. Almeida** Department of Chemical Engineering, University of Coimbra, CIEPQPF, Coimbra, Portugal

**Patcharin Saechan** Faculty of Engineering, King Mongkut's University of Technology North Bangkok, Bangsue, Bangkok, Thailand

**Paulo Silva** CIETI—Centre of Innovation on Engineering and Industrial Technology/IPP-ISEP, School of Engineering, Porto, Portugal

**Pedro Faria** Research Group on Intelligent Engineering and Computing for Advanced Innovation and Development (GECAD), Intelligent Systems Associated Laboratory (LASI), Polytechnic of Porto (P.Porto), Porto, Portugal

**Pierpaolo Girardi** RSE Ricerca sul Sistema Energetico, Milano, Italy

**R. Godina** UNIDEMI—Research and Development Unit for Mechanical and Industrial Engineering, Department of Mechanical and Industrial Engineering, NOVA School of Science and Technology, Universidade NOVA de Lisboa, Caparica, Portugal;
Laboratório Associado de Sistemas Inteligentes, LASI, Guimarães, Portugal

**R. Nazario-Naveda** Vicerrectorado de Investigación, Universidad Autónoma del Perú, Lima, Peru

**R. Robles-Iglesias** Universidade da Coruña, Coruña, Spain

**R. Calejo Rodrigues** Porto University—FEUP—NI&DEA—GEQUALTEC—CONSTRUCT, Porto, Portugal

**R. P. Rodrigues** Department of Chemical Engineering, University of Coimbra, CIEPQPF, Coimbra, Portugal

**Ralph Carlo Evidente** POSTECH, Pohang, Gyeongbuk, South Korea

**Raymond Caruana** Aquaculture Directorate, Forti San Luċjan, Marsaxlokk, Malta

**Renata Pacheco** INESC TEC and Faculdade de Engenharia, Universidade do Porto, Porto, Portugal

**Ricardo N. Pereira** CEB-Centre of Biological Engineering, University of Minho, Braga, Portugal;
LABBELS-Associated Laboratory, Braga, Guimarães, Portugal

**Robert Crawford** The University of Melbourne, Melbourne, Australia

**Roberto Novais** CEB, Centre of Biological Engineering, University of Minho Campus de Gualtar, Braga, Portugal

**Roger Garcia** PROMES, UPR CNRS 8521, Font-Romeu-Odeillo, France

**Rolando Azevedo** Centro de Interface Tecnológico E Industrial, Arcos de Valdevez, Portugal

**Rosa Pilão** CIETI, ISEP, Instituto Politécnico do Porto, Porto, Portugal

**Rosa M. Quinta-Ferreira** Department of Chemical Engineering, Faculty of Sciences and Technology, CIEPQPF—Chemical Engineering Processes and Forest Products Research Center, University of Coimbra, Coimbra, Portugal

**S. Alves** Department of Mechanical and Industrial Engineering, School of Sciences and Technology, Universidade NOVA de Lisboa, Caparica, Portugal;
Low Carbon and Resource Efficiency, R&Di, Instituto de Soldadura e Qualidade, Grijó, Portugal

**S. Rojas-Flores** Vicerrectorado de Investigación, Universidad Autónoma del Perú, Lima, Peru

**Sami G. Al-Ghamdi** Division of Sustainable Development, College of Science and Engineering, Hamad Bin Khalifa University, Qatar Foundation, Doha, Qatar;
Environmental Science and Engineering Program, Biological and Environmental Science and Engineering Division, King Abdullah University of Science and Technology (KAUST), Thuwal, Saudi Arabia;
KAUST Climate and Livability Initiative, King Abdullah University of Science and Technology (KAUST), Thuwal, Saudi Arabia

**Santiago M. Benites** Vicerrectorado de Investigación, Universidad Autónoma del Perú, Lima, Peru

**Sara Ciarlantini** DIST—Interuniversity Department of Regional and Urban Studies and Planning, Polytechnic University of Turin and University of Turin, Turin, Italy

**Sara G. Pereira** CEB-Centre of Biological Engineering, University of Minho, Braga, Portugal;
LABBELS-Associated Laboratory, Braga, Guimarães, Portugal

**Sedef Uçarıylmaz** Izmir Institute of Technology, Izmir, Turkey

**Sérgio I. Lopes** Centro de Interface Tecnológico E Industrial, Arcos de Valdevez, Portugal;
ADiT-Lab, Instituto Politécnico de Viana Do Castelo, Viana Do Castelo, Portugal;
IT—Instituto de Telecomunicações, Campus Universitário de Santiago, Aveiro, Portugal

**Silvia Magdalena Coello Pisco** Facultad de Ingeniería Industrial, Universidad de Guayaquil, Guayaquil, Ecuador;
Universidade da Coruña, Galicia, Spain

**Sofía Mulero-Palencia** CARTIF Technology Centre, Parque Tecnológico de Boecillo, Valladolid, Spain

**Sylvester William Chisale** Transdisciplinary Science and Engineering, Graduate School of Advanced Science and Engineering, Hiroshima University, Hiroshima, Japan;
Department of Applied Studies, Malawi University of Science and Technology (MUST), Limbe, Malawi

**Takuji Matsumoto** Faculty of Transdisciplinary Sciences for Innovation, Kanazawa University, Ishikawa, Japan

**Tânia Ferreira** Escola Superior de Tecnologia e Gestão, Instituto Politécnico de Viseu, Viseu, Portugal

**Telys Kurlyana** Faculty of Agriculture and Animal Science, University of Muhammadiyah Malang, Malang, Indonesia

**Teresa M. Mata** LAETA-INEGI, Associated Laboratory for Energy and Aeronautics, Institute of Science and Innovation in Mechanical and Industrial Engineering, Porto, Portugal

**Tiago Ramos da Silva** Low Carbon & Resource Efficiency, R&Di, Instituto de Soldadura e Qualidade, Grijó, Portugal

**Victor Moutinho** Department of Management and Economics, NECE—Research Center for Business Sciences, University of Beira Interior, Covilhã, Portugal

**Yomar Alexander González Cañizales** Universidad de Guayaquil, Carrera de Ingeniería Industrial, Guayaquil, Ecuador

**Yomar Alexander González Cañizales** Facultad de Ingeniería Industrial, Universidad de Guayaquil, Guayaquil, Ecuador

**Youcef Bouzidi** Interdisciplinary Research on Society-Technology-Environment Interactions, University of Technology of Troyes, Troyes, France

**Youssef Berro** PROMES, UPR CNRS 8521, Font-Romeu-Odeillo, France

**Yuji Yamada** Faculty of Business Sciences, University of Tsukuba, Tokyo, Japan

**Z. Lin** Department of Chemistry, CICECO, University of Aveiro, Aveiro, Portugal

**Zeynep Durmuş Arsan** Izmir Institute of Technology, Izmir, Turkey

**Zhane Ann Tizon** Mapua University, Manila, Philippines

**Zita Vale** Research Group on Intelligent Engineering and Computing for Advanced Innovation and Development (GECAD), Intelligent Systems Associated Laboratory (LASI), Polytechnic of Porto (P.PORTO), Porto, Portugal

**Herwintono** Faculty of Agriculture and Animal Science, University of Muhammadiyah Malang, Malang, Indonesia

**Sutawi** Faculty of Agriculture and Animal Science, University of Muhammadiyah Malang, Malang, Indonesia

# Part I
# Advanced Energy Technologies

# Chapter 1
# The Future of Transportation: Recyclable Solar Metal Fuel

**Youssef Berro, Roger Garcia, and Marianne Balat-Pichelin**

**Abstract** The novel development of clean sustainable fuels became a necessity in the last two decades as a response to crude oil depletion and global warming issues. In this context, we proposed, under the STELLAR project, the use of metal fuels as zero-carbon recyclable substitutes for conventional transport fuels through combustion/reduction cycles. Further, we investigated the use of concentrated solar power for the regeneration of those fuels, through the carbothermal reduction of the combustion oxide products, thus making the process more economically and environmentally beneficial. A solar process was developed to operate in batch or semi-continuous modes allowing the production of highly-pure Mg and Al metal powders. Promising results of 96% Mg and 77% Al yields were reached, using commercial magnesia and alumina in batch mode, by optimizing the process and reaction conditions. Similar yields were obtained using oxides produced from the combustion of Mg and Al metal fuels. Furthermore, we were able to surpass different experimental difficulties, encountered during the semi-continuous solar processing of magnesia, and attain up to 75% Mg yield. The regenerated Mg and Al powders were micro-sized and stable to be stored in native air with no re-oxidation risks.

**Keywords** Carbothermal reduction · Circular economy · Concentrated solar power · Energy carrier · Metal-fueled combustor

## 1.1 Introduction

Since the beginning of the twenty-first century, researchers worked side by side to develop innovative clean processes facing the excessive need for energy associated with two critical issues, the depletion of fossil fuels and global warming. In this context, a strategy was proposed for the production of zero-carbon sustainable transport fuels as transportation is the most critical sector both from energetic and

Y. Berro · R. Garcia · M. Balat-Pichelin (✉)
PROMES, UPR CNRS 8521, 66120 Font-Romeu-Odeillo, France
e-mail: marianne.balat@promes.cnrs.fr



N. S. Caetano and M. C. Felgueiras (eds.), *The 9th International Conference on Energy and Environment Research*, Environmental Science and Engineering,
https://doi.org/10.1007/978-3-031-43559-1_1

environmental aspects (Capellán-Pérez et al. 2014). This strategy consists of the direct combustion of metal fuels, having high energy density, and their regeneration through the reduction of their oxide combustion products (Bergthorson et al. 2015). This combustion/reduction approach becomes even more attractive when those fuels are recycled using renewable energy sources thus promoting the circular economy concept (Bergthorson 2018). Based on this approach, the STELLAR project was launched, in collaboration with leading institutions in the fields of combustion and concentrated solar power (CSP), to study the production of recyclable solar metal fuels and the development of metal-fueled burners (Berro and Balat-Pichelin 2022).

The choice of possible candidates to be used as fuels must consider the properties of metals, their combustion behavior (kinetics and modes) and the ease of their regeneration from their oxide products. Considering those parameters, Mg/MgO and Al/$Al_2O_3$ were among the most prospective couples to be used for combustion/reduction cycles (Berro and Balat-Pichelin 2022). Combustion engineering allowed to stabilize the air/metal flame in the burner, control the size of the produced oxides and improve the burning rate that is dependent form the burning rate and flame thickness (Goroshin et al. 1996). The production of oxides with particle sizes bigger than several microns allows their efficient collection, for regeneration, through cyclone technology (Wang et al. 2012). Recently, an aluminum-air flame was stabilized in a Bunsen burner by researchers at the PSA group (now Stellantis), AVL group and ICARE-CNRS laboratories proving that 7 μm Al particles are reactive as much as conventional hydrocarbons with similar burning velocities (Lomba et al. 2019). Similarly, a magnesium-air flame was swirled-stabilized in an innovative system allowing to recover about 80% of the power released by the combustion (up to 11 kW) and to recover approximately 98% of the produced submicron MgO (Laraqui et al. 2020).

The regeneration of metal fuels, through the reduction of their oxides (magnesia and alumina), can be assisted using carbon as a reducing agent thus decreasing the equilibrium temperature of the reaction (Steinfeld and Palumbo 2003). A further decrease of both the equilibrium and onset temperatures of the reduction can be achieved by operating under low vacuum conditions (Puig and Balat-Pichelin 2016). Considering the magnesia carbothermal reduction, the CO partial pressure ($P_{CO}$) is reduced under those conditions, thus making the C/MgO solid–solid phase boundary kinetics more favorable over the CO/MgO gas–solid diffusion kinetics (Xiong et al. 2019). In this case, one must consider the effect of the C/MgO properties and magnesia sintering along the reaction (Coray and Jovanovic 2019). Similarly, $P_{CO}$ is reduced during the vacuum-assisted carbothermal reduction of alumina, thus limiting the formation of undesired by-products mainly $Al_2OC$, $Al_4C_3$ and $Al_4O_4C$ (Kruesi et al. 2011). Furthermore, using CSP for the reduction of oxides grants to reach high temperatures very quickly (heating rates up to 500 K $s^{-1}$), thus reducing the magnesia sintering and the formation of unwanted Al-oxycarbides (Vishnevetsky and Epstein 2015).

Therefore, this study aims to improve the efficiency, under optimized conditions, of a novel semi-continuous solar carbothermal reduction process for the regeneration of Mg and Al metal powders.

## 1.2 Process Description

Reduction experiments were performed in the Sol@rmet reactor, shown in Fig. 1.1, using a 1.5 kW solar furnace consisting of a solar-tracking heliostat and a parabolic mirror allowing to concentrate the solar radiation up to 15,000 times. The reactions occur under low-pressure conditions supplied by a dry pump and controlled by adjusting the flow of the argon carrier gas. During batch processing, one C/oxide pellet (255 mg) was placed on graphite support at the focus of the parabola, while during semi-continuous processing pellets were put in an elevator allowing to push them progressively to the focus (at 1.5 mm $min^{-1}$). The reaction was stoichiometric during alumina reduction, while during magnesia reduction an excess of carbon was used (C/MgO = 1.25 or 2). The temperature at the surface of the pellet was measured using an optical monochromatic pyrometer and controlled by opening the shutters placed between the heliostat and the parabola.

The produced metal powders were driven by the argon carrier gas to condense mainly on a porous metallic filter (placed ahead of the pump) and also on the reactor walls. Whereas the produced CO and $CO_2$ were sent to a gas analyzer to follow the reaction kinetics with time. Characterization methods (granulometry, X-ray diffraction, scanning electron microscopy) were performed to determine the structural (morphology and size) and chemical (purity and by-products) properties of the produced metal powders. The reduction efficiency was presented by the yields ($y_{Mg}$ and $y_{Al}$) and purity (%Mg and %Al) of the produced metal powders that can be

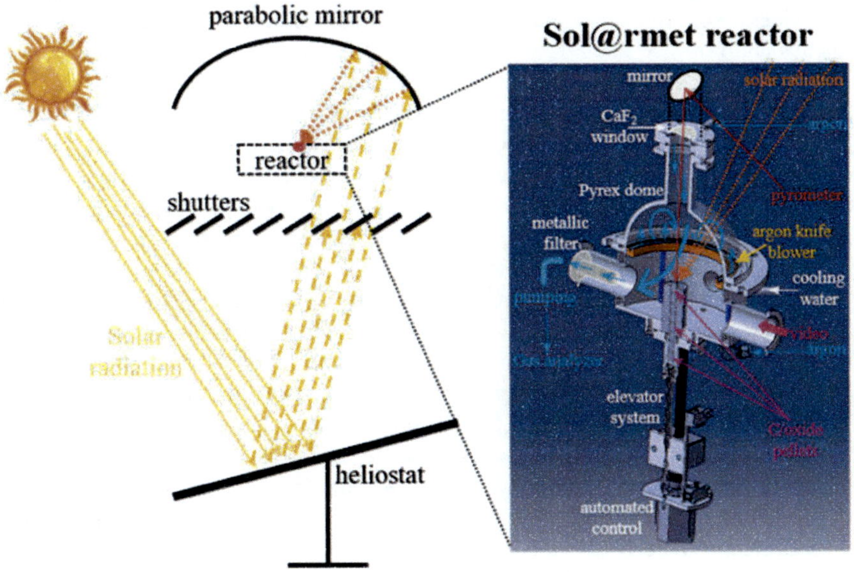


**Fig. 1.1** Schematic presentation of the 1.5 kW solar furnace including the Sol@rmet reactor

determined according to Eqs. 1.1 and 1.2 respectively.

$$y_{Mg}(\%) = 100 m_{Mg}\ \%\ Mg/m_{Mg_{max}} \quad \text{where}$$
$$m_{Mg_{max}} = m_{MgO_{initial}} M_{Mg}/M_{MgO} \tag{1.1}$$

$$y_{Al}(\%) = 100 m_{Al}\ \%\ Al/m_{Al_{max}} \quad \text{where}$$
$$m_{Al_{max}} = 2 m_{Al_2O_{3_{nitial}}} M_{Al}/M_{Al_2O_3} \tag{1.2}$$

## 1.3 Results and Discussion

Solar reduction experiments were performed through batch or semi-continuous mode as mentioned previously, using commercial oxides (magnesia and alumina) or those recovered from the combustion of metal fuels in burners developed by our collaborators under the STELLAR project.

### *1.3.1 Parameters Optimization Using Commercial Oxides*

We aimed through the following batch solar experiments, performed using commercial oxides, to optimize the operating conditions and parameters, control the reaction kinetics and improve the reduction efficiency. We proved through numerical simulations, using ANSYS-CFX software, and validated experimentally that creating a swirl circulation in the Sol@rmet reactor grants an efficient purging of the products (CO and metal powders) thus boosting the reduction extent (Berro et al. 2021a). Further, despite the favorable impact of using CSP to quickly achieve high temperatures, we showed that a gradual increase of the temperature, after reaching the onset temperature, allows dispersing the CO production along the reaction and thus increasing the metal production (Berro et al. 2021a).

More importantly, as previously discussed, the C/oxide properties are critical when operating under low pressure (low $P_{CO}$) which encourages investigating the effect of the charcoal reducing agent. Thus, we synthesized charcoals under various pyrolysis conditions (temperature, heating rate…) and using different biomass sources (wood-, starch-, cellulose-, sugar-, mushroom-, fungus-, okara-based). We confirmed through solar experiments that those with refined chemical (fixed carbon content, ash content…) and structural (porosity, specific surface area…) properties promote the magnesia reduction (Berro et al. 2021b). Furthermore, we studied the catalytic effect of metal catalysts (Fe, Ni, Fe–Ni) at different percentages demonstrating their adverse effect as carbon is consumed rapidly at the beginning of the reaction which promotes the magnesia sintering and the formation of undesired Al-oxycarbides. On the other hand, the mechanical milling of C/oxide powders allows reducing the particle size and raising the C/oxide contact which ameliorates the reduction kinetics and thus the metal production yields.

To sum up, highly pure Mg powders were produced, during magnesia reduction at 830 Pa, with a 96% yield when using 5% starch–5% bentonite binder that showed a catalytic-like behavior by improving the C/MgO contact and homogeneity (prevent sintering) at the beginning of the reaction. The magnesium powders consist of micro-sized agglomerates of particles and crystals with a $D_{90}$ size of around 100 μm (Berro et al. 2021c). Nevertheless, this favorable role of starch-bentonite binder becomes reversible during alumina reduction as more undesired by-products were formed at the beginning of the reaction. Herein, the aluminum yield was increased by around 5%, to reach up to 77%, when reducing the pressure from 830 to 285 Pa as $Al_4O_4C$ and $Al_2OC$ by-products were prevented. The produced powders were pure (91% Al–6% $Al_4C_3$–3% $Al_2O_3$) and consisted of micro-sized agglomerates of nano- and micro-particles with a $D_{50}$ and $D_{90}$ size of 50 nm and 3 μm respectively. The non-reacted remains of the C/$Al_2O_3$ pellet consisted mainly of $Al_2O_3$, $Al_2OC$, $Al_4C_3$, $Al_xO_y$, $Al_3O_{3.5}C_{0.5}$ with some traces of Al (< 1%). Similar yield results and particles size were obtained independently from the alumina initial size (100–200 or 325 mesh) making difficult to produce bigger Al particles and thus to prevent the partial oxidation or carbonization of those particles.

### 1.3.2 Batch Processing Using Combustion Oxide Products

During the solar reduction of magnesia powders recovered from the combustion of Mg fuels, similar results were obtained as for commercial magnesia with up to 91% Mg yield. This small difference was associated with the purity of the produced powders as those collected on the filter (around 85% of the total) were highly pure with 95% Mg, while those collected on the metallic walls had 50–60% Mg purity and contained remarkably around 10% of magnesium carbonate. The presence of $MgCO_3$ by-product was attributed to the existence of a small amount (up to 7%) of unburned Mg, in the oxides recovered from the combustion, that may have reacted with the produced CO (Stopic et al. 2018).

During the solar reduction of alumina powders recovered from the combustion of Al fuels, similar results were obtained at 285 Pa as for commercial alumina with a little improvement, up to 82% Al yield. This enhancement was ascribed to the existence of around 7% of unburned Al in those oxides that may have catalyzed the reaction. In light of this, we should mention that those oxides consisted of various phases (20% γ-, 44% θ-, 13% δ- and 16% α-$Al_2O_3$) unlike the commercial 100% α-$Al_2O_3$. This difference caused the explosion of pellets when heated quickly using CSP as those phases are unstable and expand in volume when heated, so we controlled very finely the heating at the beginning of the reaction to prevent the explosion of the pellet. Figure 1.2 shows the advance of the reaction along the reduction period in terms of CO and $CO_2$ formation as a function of the temperature (shutters opening).

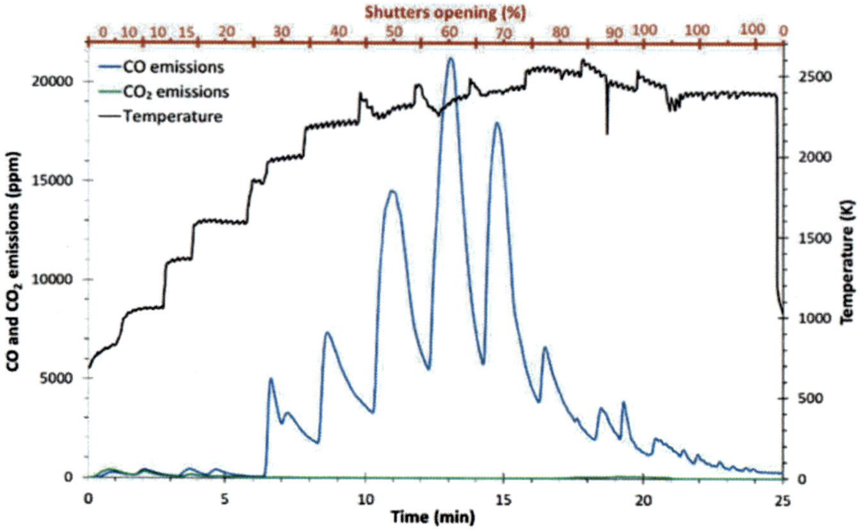


**Fig. 1.2** Gaseous emissions during alumina reduction as a function of time and temperature

During the alumina solar reduction, temperatures up to 2600 K were reached and the reaction was almost complete after 15 min at $T > 2400$ K. As seen, $CO_2$ emission occurred only at the beginning of the reaction and was minor compared to the CO emission.

### *1.3.3 Semi-continuous Processing of Magnesia*

A more realistic approach for the validation of the sustainable production of solar metal fuels requires the development of a semi-continuous process allowing the extrapolation on a larger industrial scale in the long term. Two major issues were encountered during the semi-continuous process, shown in Fig. 1.1, the clogging of the elevator and the condensation of the produced metal powders on the glass dome. The clogging was related to the coalescence of the melted C/MgO pellets with the surrounding graphite support and could be prevented by increasing the pushing speed of the elevator (see Fig. 1.3). However, this led to a reduced reaction time and thus decreased the reaction extent. The metal condensation on the glass dome was limited during batch processing by optimizing the gas circulation (swirl flow) in the Sol@rmet reactor (Berro et al. 2021a). Whilst, the cumulative metal condensation, on some parts of the glass dome, could not be prevented during the semi-continuous process despite increasing the gas flow or adding new gas inlets (knife blower shown in Fig. 1.1). This condensation phenomenon caused the obstruction of the solar radiation and the overheating of the glass dome. In that regard, a very

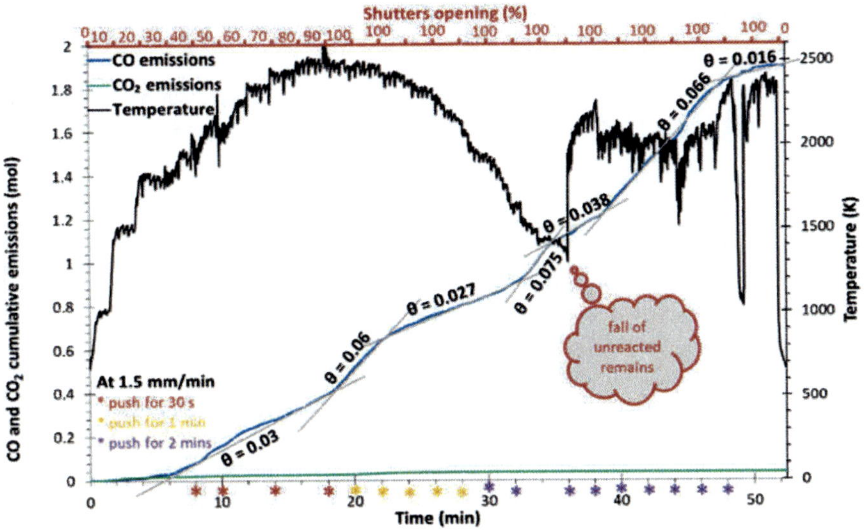


**Fig. 1.3** Reaction kinetics progress as a function of time, temperature and C/MgO pellets consumption

recent study was published discussing the aerodynamic-aided window protection of solar thermochemical reactors (Wang et al. 2022).

Despite the overheating of the dome, we were able to process 10 C/MgO pellets (2.55 g) yielding up to 75% Mg powders where ~ 0.67 g (82% purity) were collected on the filter while 0.21 g (51% purity) were collected on the reactor walls. The lower powder purity, compared to the batch processing, was attributed to the higher operating pressure (1240 Pa compared to 830 Pa during batch mode) that was raised when increasing the argon flow up to 10 L $min^{-1}$ (to reduce metal condensation on the glass dome).

Figure 1.3 reveals an idea about the reaction kinetics progress in terms of CO and $CO_2$ formation along the reduction period as a function of the temperature and the consumption of pellets. We observed that the higher measured temperature value reaches around 2600 K when shutters are fully opened (at t = 18 min), however, the measurement becomes less accurate as the agglomerated magnesia residues (colored white) accumulate over the unreacted C/MgO pellet. This uncertain measurement is manifested by the progressive decrease of the temperature to around 1300 K and then the sudden re-increase up to 2300 K when the white magnesia residues fall (at t = 36 min), as shown in Fig. 1.3. Furthermore, by interpreting the cumulative CO production along the reaction period, we can conclude that the reaction rate accelerates when shutters were fully opened (at t = 18 min) and when pellets were pushed more quickly to the most heated (concentrated radiation) region at the focus of the parabola (for t > 30 min). The reaction ended after around 53 min when the CO formation became negligible (< 100 ppm).

## 1.4 Conclusion

The attractiveness of the use of metal fuels, as sustainable and clean substitutes for conventional transport fuels, accrues when using renewable energy sources to recycle them. In this context, we developed a process for the regeneration of metal fuels through the vacuum-assisted carbothermal reduction of their corresponding oxides (magnesia and alumina) using concentrated solar power. Firstly, we used commercial oxides in batch processing to optimize the reaction parameters and to study the effect of the pressure, CO partial pressure, temperature, heating, gas circulation, C/oxide properties, charcoal reducing agent, milling, binders and catalysts. Following these investigations, we proved the ability to produce highly pure micro-sized Mg and Al powders, that are stable in native air, with yields up to 96% Mg (at 830 Pa) and 77% Al (at 285 Pa) respectively. Similar results were obtained using oxides recovered from the combustion of metal fuels in burners developed by our collaborators under the STELLAR project. However, few differences were observed due to the existence of a small amount of unburned Mg and Al in the oxides that caused the formation of $MgCO_3$ by-product during magnesia reduction (91% Mg yield) and catalyzed the reaction during alumina reduction (82% Al yield) respectively. The semi-continuous processing of 10 C/MgO pellets (2.55 g) at 1240 Pa allowed producing around 0.67 g (63% yield) of 82%-purity and 0.21 g (12% yield) of 51%-purity Mg powders, corresponding to a total of 75% Mg yield.

**Funding** This work was supported by the French National Research Agency [ANR-18-CE05-0040–02 and ANR-10-EQPX-49-SOCRATE].

## References

Bergthorson JM (2018) Recyclable metal fuels for clean and compact zero-carbon power. Prog Energy Combust Sci 68:169–196

Bergthorson JM, Goroshin S, Soo MJ, Julien P, Palecka J, Frost DL, Jarvis DJ (2015) Direct combustion of recyclable metal fuels for zero-carbon heat and power. Appl Energy 160:368–382

Berro Y, Balat-Pichelin M (2022) Metal fuels production for future long-distance transportation through the carbothermal reduction of MgO and $Al_2O_3$: a review of the solar processes. Energy Convers Manage 251:114951

Berro Y, Masse R, Puig J, Balat-Pichelin M (2021a) Improving the solar carbothermal reduction of magnesia for metallic fuels production through reactor designing, milling and binders. J Clean Prod 315:128142

Berro Y, Kehrli D, Brilhac J-F, Balat-Pichelin M (2021b) Metal fuel production through the solar carbothermal reduction of magnesia: effect of the reducing agent. Sustain Energy Fuels 5:6315–6327

Berro Y, Puig J, Balat-Pichelin M (2021c) Improving the solar carbothermal reduction of magnesia as a production process of metal fuels. IJMMME

Capellán-Pérez I, Mediavilla M, de Castro C, Carpintero Ó, Miguel LJ (2014) Fossil fuel depletion and socio-economic scenarios: an integrated approach. Energy 77:641–666

Coray A, Jovanovic ZR (2019) On the prevailing reaction pathways during magnesium production via carbothermic reduction of magnesium oxide under low pressures. React Chem Eng 4:939–953

Goroshin S, Fomenko I, Lee JHS (1996) Burning velocities in fuel-rich aluminum dust clouds. Symp (Int) Combust 26:1961–1967

Kruesi M, Galvez ME, Halmann M, Steinfeld A (2011) Solar aluminum production by vacuum carbothermal reduction of alumina—thermodynamic and experimental analyses. Metall and Materi Trans B. 42:254–260

Laraqui D, Leyssens G, Schonnenbeck C, Allgaier O, Lomba R, Dumand C, Brilhac J-F (2020) Heat recovery and metal oxide particles trapping in a power generation system using a swirl-stabilized metal-air burner. Appl Energy 264:114691

Lomba R, Laboureur P, Dumand C, Chauveau C, Halter F (2019) Determination of aluminum-air burning velocities using PIV and laser sheet tomography. Proc Combust Inst 37:3143–3150

Puig J, Balat-Pichelin M (2016) Production of metallic nanopowders (Mg, Al) by solar carbothermal reduction of their oxides at low pressure. J Magn Alloys 4:140–150

Steinfeld A, Palumbo R (2003) Solar thermochemical process technology. In: Encyclopedia of physical science and technology. Elsevier, pp 237–256

Stopic S, Dertmann C, Modolo G, Kegler P, Neumeier S, Kremer D, Wotruba H, Etzold S, Telle R, Rosani D, Knops P, Friedrich B (2018) Synthesis of magnesium carbonate via carbonation under high pressure in an autoclave. Metals 8:993

Vishnevetsky I, Epstein M (2015) Solar carbothermic reduction of alumina, magnesia and boria under vacuum. Sol Energy 111:236–251

Wang H, Zhang Y, Wang J, Liu H (2012) Cyclonic separation technology: researches and developments. Chin J Chem Eng 20:212–219

Wang B, Rahbari A, Hangi M, Li X, Wang C-H, Lipiński W (2022) Topological and hydrodynamic analyses of solar thermochemical reactors for aerodynamic-aided window protection. Eng Appl Comput Fluid Mech 16:1195–1210

Xiong N, Tian Y, Yang B, Xu B, Dai T, Dai Y (2019) Results of recent investigations of magnesia carbothermal reduction in vacuum. Vacuum 160:213–225

# Chapter 2
# Waste Plastics to Hydrogen ($H_2$) Through Thermochemical Conversion Processes

**M. G. Rasul, M. A. Sattar, M. I. Jahirul, and M. M. Hasan**

**Abstract** Plastic products are essential parts of modern life as they are used in household items, packaging, electronics, building, automotive and many more. Amongst about 350 Mt of waste plastics produced worldwide, only about 20% are recycled and 80% are landfilled. The landfilled waste plastic has a serious negative impact on the environment which causes land diversity, havoc in marine, etc. Pyrolysis of waste plastics into usable energy products such as syngas, char and liquid oil can alleviate the burden of plastic waste management. While pyrolyzed liquid oil can be converted to plastic diesel through distillation and hydrotreatment processes, and char can be used for agricultural purposes, the pyrolysis syngas can be processed further through different reforming processes to produce $H_2$. $H_2$ is expected to dominate in fuel sector by substituting fossil fuels. In this paper, the experimental findings of waste plastics pyrolysis into oil, char and syngas are reported which shows that waste plastic can be converted to liquid oil by about 80%, remaining are approximately 10% char and 10% syngas. The possibility of $H_2$ production from pyrolysis syngas through different reforming processes such as steam, partial oxidation, autothermal, plasma, aqueous phase, etc. are reviewed and critically analysed in this paper. The literature indicated that 530 Mt of $H_2$ is needed to achieve net zero by 2050 worldwide. It is envisaged that converting all waste plastic into $H_2$ will meet the demand of $H_2$ and support net zero goal by 2050. Amongst different reforming processes, steam reforming is better than others to produce $H_2$ from syngas. Therefore, the waste plastics can be a significant potential source of $H_2$ and will benefit the society and the environment from negative impact and support achieving net zero by 2050.

**Keywords** Waste plastic · Pyrolysis syngas · Reforming processes · $H_2$ production · Net zero by 2050

---

M. G. Rasul (✉) · M. M. Hasan
Fuel and Energy Research Group, School of Engineering and Technology, Central Queensland University, Rockhampton, QLD 4702, Australia
e-mail: m.rasul@cqu.edu.au

M. A. Sattar · M. I. Jahirul
Engineering School, Chisholm Institute, 121 Stud Road, Dandenong, VIC, Australia



N. S. Caetano and M. C. Felgueiras (eds.), *The 9th International Conference on Energy and Environment Research*, Environmental Science and Engineering,
https://doi.org/10.1007/978-3-031-43559-1_2

## 2.1 Introduction

Plastic is used in every part of daily life which includes food packaging, electrical devices, household utensils and many more because of their favorable properties. The production of plastic items is increasing day by day and the safe disposal of waste plastic has become an urgent need to all countries in the world. Global plastic production exceeds 350 million tons in 2018 (Kusenberg et al. 2022). The main types of waste plastic that are found in the municipal waste are high-density polythene (HDPE), polyvinyl chloride (PVC), low-density polythene (LDPE), polythene terephthalate (PET), polypropylene (PP), and polystyrene (PS) (Kusenberg et al. 2022). Most of the plastics are non-biodegradable and take longer time to degrade which are causing many problems in land and seas. Incineration of plastic produces noxious and toxic fumes which poses health hazard. Therefore, incineration is not a good solution for disposing the plastics. Pyrolysis can an alternative and comparatively safe process of utilising waste plastic. Pyrolysis is a process of decomposing solid wastes into oil, char, and syngas in the absence of oxygen. The oil can be used as a fuel for internal combustion engine after refinement to plastic diesel. The char can be used for soil abatement and other purposes. The Syngas are mostly neglected and released into the atmosphere.

Combustion of hydrocarbons produces $CO_2$ and other harmful GHG gases which has detrimental effects on the environment. Researchers are searching for environmentally friendly alternative fuel, for example $H_2$ as a clean energy which can be a suitable substitute for fossil fuel (Hazrat et al. 2022; Sarker et al. 2023). $H_2$ has many favorable properties such as the highest calorific value and does not produce any harmful gases during the combustion. The demand of $H_2$ in 2020 was around 88 Mt and expected to increase to 530 Mt in 2050 to achieve net zero goal (IEA 2019; PwC 2017).

The syngas, normally discarded into the atmosphere, can be converted into $H_2$ using different reforming processes. In this paper, the pyrolysis of waste plastic is discussed first, then different types of reforming processes are briefly described to produce $H_2$.

## 2.2 Pyrolysis of Waste Plastics

As introduced earlier, pyrolysis is a thermochemical process of decomposing the waste plastic into oil, biochar, and syngas in the absence of oxygen. There are different types of pyrolysis reactors used for pyrolysis as shown in Fig. 2.1 (Lewandowski et al. 2019; Papari and Hawboldt 2015). Authors used 20L vertical fixed bed reactor for their own research, the pictorial view of which is shown in Fig. 2.2.

In this reactor, three electrical heaters are used to heat the reactor and k-type thermocouples are used to measure the pyrolysis operational temperature. Before any experiment, nitrogen gas is purged through the system to make it inert (free of

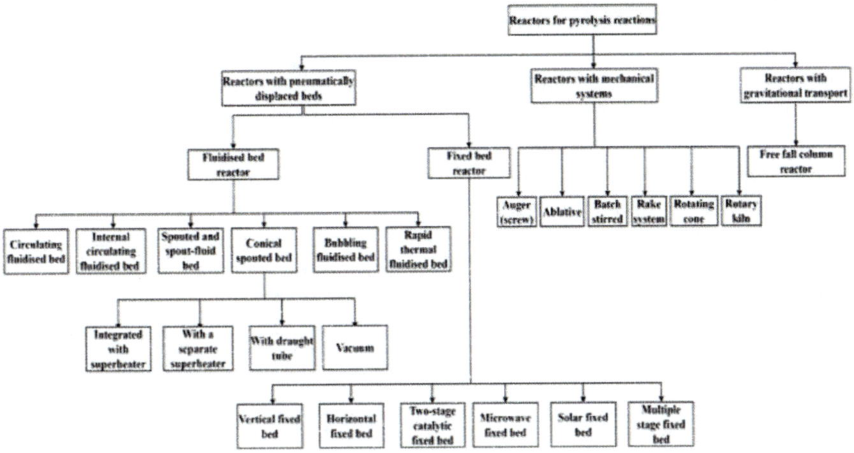


**Fig. 2.1** Different types of reactor for pyrolysis (Lewandowski et al. 2019; Papari and Hawboldt 2015)

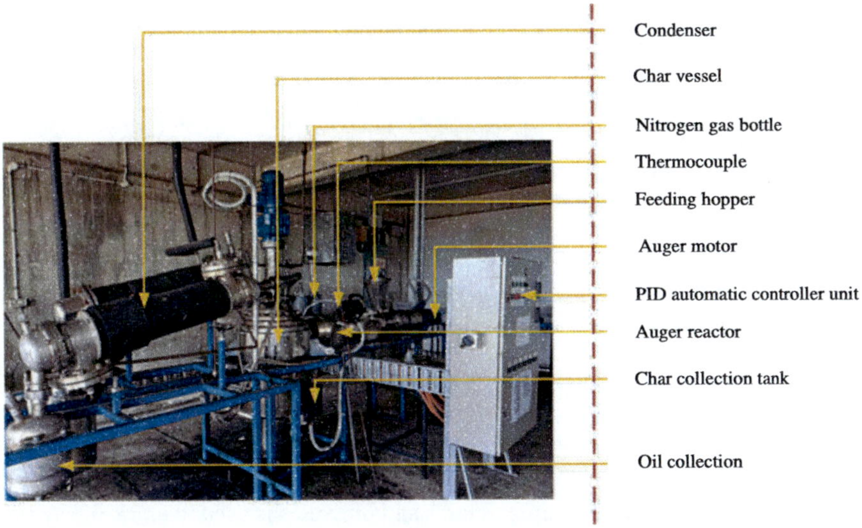


**Fig. 2.2** Pyrolysis set up at Fuel and Energy Research Laboratory, CQU Rockhampton campus

oxygen). A PID controller is used to operate the reactor at atmospheric pressure. A chiller unit is used to condense the pyrolysis vapors after decomposition of feed materials into crude oil. Polyethylene glycol solution circulated chiller unit temperature can be set − 5 to 20 °C. The mixed waste plastic was fed into the reactor through a feeding hopper and kept sealed after nitrogen was purged. Then, the data logging system and electrical heater were turned on to heat the reactor to the desired temperature and maintained at the desired time. During this period, the feedstock was converted into vapor and char. The vapor passed through a water-cooled condenser to produce crude oil which was collected from the oil collection tank. The system was turned off until the reactor temperature reduced to room temperature, then char was taken out from the bottom opening of the reactor. The crude oil can be refined through distillation process to produce oil equivalent to standard diesel to use in diesel engine. The ultimate and proximate analysis of different types of waste plastics can be found in Zhou et al. (2014). The yield from the pyrolysis of waste plastic is presented in Table 2.1. The results show that the oil yield can be found $\geq 80\%$ including findings of the authors, except yield of PET, PE, and few mixtures (Rasul et al. 2022).

## 2.3 Production from Syngas

Syngas produced from pyrolysis of waste plastic is a mixture of $H_2$, CO and small amount of $CO_2$ and methane. The general equation for syngas production from pyrolysis can be expressed by:

$$Waste\,platisctics \overset{Pyrolysis}{\Rightarrow} \mathrm{H_2} + \mathrm{CO} + \mathrm{CO_2} + Hydrcarbon\,gases \tag{2.1}$$

Hydrogen can be separated from the syngas. The CO can be transformed into $CO_2$ by producing $H_2$ using different reforming process as shown in Fig. 2.3 (Rasul et al. 2022). Reforming is the process of converting low quality hydrocarbon to high quality hydrocarbon which has been used in industry for quite long. Reforming process can be used to produce $H_2$ by cracking the natural gas, methane, gaseous hydrocarbon, liquid hydrocarbon, ethanol, methanol, coal, and naphtha. Brief description of different types of reforming processes are given below.

### *2.3.1 Catalytic Steam Reforming*

In this process, steam at high temperatures and pressures is used to crack the hydrocarbon to produce CO and $H_2$ rich syngas. Steam reforming process can be carried out with and without catalyst. After steam reforming process, water–gas-shift (WGS) process is carried out to convert CO into $CO_2$ and dissociate $H_2$ from water. This combined process is used to produce $H_2$ from the hydrocarbon. The highest efficiency of SR can reach up to 80–90%. The reforming reactions (2.2) to (2.4) are given below:

Oxygenated hydrocarbon:

$$\mathrm{C}_x\mathrm{H}_y\mathrm{O}_z + (x - z)\mathrm{H_2O}(+heat) \rightarrow x\mathrm{CO} + (x + y/2 - z)\mathrm{H_2} \tag{2.2}$$

**Table 2.1** Yield of pyrolysis of waste plastic

| Plastic types | Yield % | | | References |
|---|---|---|---|---|
| | Oil | Char | Syngas | |
| PP | 88.86 | 1.84 | 9.3 | Martínez et al. (2013) |
| | 83.6 | 0.4 | 16.0 | Williams and Slaney (2007) |
| PE | 92.3 | 0.1 | 7.6 | Williams (2006) |
| | 85.3 | 10.9 | 3.4 | Miskolczi et al. (2004) |
| PS | 80.8 | 6.2 | 13 | Rehan et al. (2016) |
| | 97 | 0.5 | 2.5 | Onwudili et al. (2009) |
| PET | 38.89 | 52.13 | 8.98 | Fakhr Hoseini and Dastanian (2013) |
| | 15 | 53 | 32 | Kunwar et al. (2016) |
| HDPE | 89.1 | 6.7 | 4.2 | Abbas-Abadi et al. (2014) |
| | 81.00 | 17.50 | 1.5 | Seo et al. (2003) |
| LDPE | 75.6 | 8.2 | 7.5 | Uddin et al. (1997) |
| | 84 | 14.25 | 1.75 | Supriyanto and Richards (2021) |
| HDPE[a] | 58 | 6.4 | 4.6 | Authors work |
| PS | 85 | 5.5 | 9.5 | Authors work |
| PP | 90 | 3.3 | 6.7 | Authors work |
| HDPE + PP + PS (1:1:1) | 81 | 8.5 | 10.5 | Authors work |
| HDPE + PP + PS (1:1:2.2) | 85 | 4.5 | 11.5 | Authors work |
| HDPE + PP + PS + PET(4:4:3:1) | 76 | 10.5 | 13.5 | Authors work |
| HDPE + PP + PS + PET + PVC (2.1:3.7:3:2.3:2.2) | 64 | 21.5 | 14.5 | Authors work |
| HDPE + PP + PS + PET (3:4:3:2) | 66 | 18.75 | 15.25 | Authors work |

[a]Rest is wax

Non-oxygenated hydrocarbon:

$$C_xH_y + (x - z)H_2O(+heat) \rightarrow xCO + (x + y/2)H_2 \tag{2.3}$$

WGS process.

$$CO + H_2O(-heat) \rightarrow CO_2 + H_2 \tag{2.4}$$

Catalyst is used to enhance the efficiency of the $H_2$ production. Nahar et al. (2015) found 94% yield efficiency at 650 °C in the presence of 10 wt.% Ni/Ce-Zr catalyst.

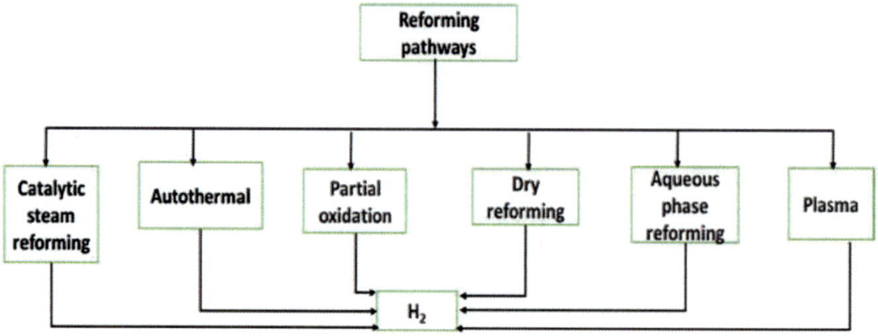


**Fig. 2.3** Different types of reforming process to produce hydrogen

Typical operating conditions for cracking natural gas is 3–25 bar and 700–1000 °C in the presence of a catalyst. Steam reform process requires high temperature if catalyst is not used for example $CH_4$ cracks into various radicals (e.g., $C_2H_4$, $C_2H_2$, and C) at 1000 °C, and at over 1500 °C to produce $H_2$ gas (Rostrup-Nielsen et al. 2011). The water–gas shift process is performed after the steam reforming process to decompose water with CO to increase $H_2$ production. Haber–Bosch process is being used for quite long to produce $H_2$. The combination of SR and WGS process has drawn the attention to produce $H_2$ in recent time. Globally about 48 and 30% of the total $H_2$ is produced using the SR and WGS methods (Ugurlu and Oztuna 2020).

### 2.3.2 *Autothermal Reforming*

Autothermal steam reforming (ATR) is a combination of SR and partial oxidation (POX) where fuel, steam, and water input into the reactor to produce $H_2$. The heat is produced through POX therefore no additional heat is required. This process requires oxygen, so oxygen separation plant is needed to supply and carry out this process. WGS process is carried out to enhance the $H_2$ concentration in the mixture. ATR process can produce $H_2$ at a cost of \$1.69–\$2.55 per kg and adding the carbon capture, utilization, and storage (CCUS) techniques can make it green process (Ahmed and Krumpelt 2001). The $CO_2$ emissions from this process is less than other processes. Catalysts can enhance the production of $H_2$ as for example the addition of Pd-Zn/γ-$Al_2O_3$ at 400 °C can produce about 45% (v/v) of $H_2$ efficiently (Oni et al. 2022). The typical reaction process of the ATR or OSR process is shown in Eq. 2.5.

$$Fuel\left(C_xH_yO_z\right) + air + \text{steam} \rightarrow CO_2 + H_2 + N_2(-\Delta H) \qquad (2.5)$$

Cortazar et al. (2022) reported the production of $H_2$ from several waste plastics (HDPE, PP, PS, PET), mixed plastics, biomass, and HDPE and found that the highest $H_2$ production was from PP (64.1%), and HDPE (64%).

### 2.3.3 Partial Oxidation Reforming (POX)

POX process generates heat during the reaction. This process does not require external heat like SR process. POX process produces less $H_2$ than that of SR process. POX reforming process is carried out at sub-stoichiometric quantity of oxygen as can be seen in Eqs. 2.6 and 2.7 (Rasul et al. 2022). POX is exothermic whereas SR is endothermic process.

$$CH_4 + 1/2O_2 \rightarrow CO + 2H_2(+\Delta H) \tag{2.6}$$

$$\text{WGS reaction}: CO + H_2O \rightarrow CO_2 + H_2(+\text{ small amount of heat}) \tag{2.7}$$

Hydrogen is separated using the pressure swing absorption process. Three-fourth of the total global $H_2$ is produced usually using this technology. The generic POX reaction of hydrocarbon fuels is presented in Eq. 2.8.

$$Hydrocarbon\ fuel: C_nH_m + (n/2)O_2 \rightarrow nCO + \left(\frac{m}{2}\right)H_2 + heat \tag{2.8}$$

The process can be carried out without catalyst therefore there is no drawback of degradation of catalyst effectiveness. The main challenge of POX is the requirement of high temperature and lower $H_2/CO$ ratio (Rasul et al. 2022). WGS is conducted after POX to convert CO to $CO_2$. POX reaction for methanol and ethanol can be expresses as in the Eqs. 2.9 and 2.10, respectively (Rasul et al. 2022).

$$POX of methanol: CH_3OH + \frac{1}{2}O_2 \rightarrow 2H_2 + CO_2 + \left(\Delta H \approx 192.2\frac{kJ}{mol}\right) \tag{2.9}$$

$$POX of ethanol: C_2H_5OH + \frac{3}{2}O_2 \rightarrow 3H_2 + 2CO_2 + \left(\Delta H \approx 620.3\frac{kJ}{mol}\right) \tag{2.10}$$

Catalyst can lower operating temperature. Agrell et al. (2001) used Cu (%40) Zn (%60) to conduct POX reaction of methanol at 185–215 °C and found that the $H_2$ production increases.

### 2.3.4 Dry Reforming (DR)

In drying reforming process, $CO_2$ and $CH_4$ react to produce CO and $H_2$ at 700–900 °C. The main challenge of this process is the deactivation of the catalyst and low $H_2/CO$ ratio (Uddin et al. 1997). In the dry reforming process $CO_2$ is used to crack the hydrocarbon to produce the syngas. The dry reforming process can be expressed in Eq. 2.11 below.

$$2C_xH_y + CO_2(+heat) \rightarrow 2(x + 1/2)CO + yH_2 \quad (2.11)$$

The Boudouard and reverse water–gas shift (RWGS) reaction can be written as:

$$Boudouard\,reaction : 2CO \leftrightarrow CO_2 + C_{(s)}; (\Delta H = -172\,kJ/mol) \quad (2.12)$$

$$RWGS\,reaction : CO_2 + H_2 \leftrightarrow H_2O + CO; (\Delta H = 41\,kJ/mol) \quad (2.13)$$

The efficiency of this process depends on the performance of the catalyst therefore a highly active catalyst is desired. Ru, Pt and Pd show higher catalytic activity for DR process (Medeiros et al. 2022). Ballarini et al. (2019) reported that K-L Zeolite, K-$Al_2O_3$, K-Mg/Al oxides, and MgO along with Pt-based catalyst has high stability to produce $H_2$. The maximum yield was obtained when MgO/Pt was used. Xie et al. (2018) found that PtCo/$CeO_2$ has high stability and effectiveness to produce $H_2$. Hajizadeh et al. (2022) found that 48.07 kg/h biogas produces 8.11 kg$H_2$/h in presence of Co-Ni-$Al_2O_3$ catalyst.

### 2.3.5 *Aqueous Phase Reforming*

In the aqueous phase reforming (APR), oxygenated/non-oxygenated hydrocarbon is cracked in aqueous solution at lower temperature to produce $H_2$. This process operates at 200–250 °C and at 60 bar in the presence of catalyst. Platinum (Pt), tin (Sn), cobalt (Co) or nickel (Ni)-based metallic, and alumina can be used as a catalyst support (Shabaker and Dumesic 2004). The reaction can be given by Eq. 2.14.

$$C_xH_y + 2xH_2O \rightarrow +\left(2x + \frac{y}{2}\right)H_2 + xCO_2 \quad (2.14)$$

This process consumes less energy than other processes and is termed as greener and therefore this process is an economical process to produce $H_2$ from organic compounds. There is complexity of producing $H_2$ directly from the biomass which can be overcome by converting the biomass into liquid as an intermediate material and then producing $H_2$ through reforming. The stoichiometric APR reaction of sugar–alcohol sorbitol ($C_6O_6H_{14}$) in the presence of Pt catalyst can be shown as in Eq. 2.15 (Shabaker and Dumesic 2004):

$$C_6O_6H_{14}(l) + 6H_2O(l) \leftrightarrow 13H_2(g) + 6CO_2(g) \quad (2.15)$$

$H_2$ and $CO_2$ react in the presence of catalyst therefore it is essential to capture $CO_2$ otherwise the overall efficiency of the process will decrease. In the APR process, bimetallic catalysts like PtNi, PdFe, PtFe show better performance than that of monometallic catalysts.

### *2.3.6 Plasma Reforming*

Electron at high temperature in the plasma supports the decomposition of organic material. At high temperature, waste plastic and other hydrocarbon decomposes into CO, $H_2$, and other hydrocarbons. There are different types of plasma reactors exists, such as dielectric barrier discharge (DBD) reactor, pulse plasma reactor, gliding arc plasma reactor, and microwave plasma reactor etc. (Budhraja et al. 2023). Comparative analysis of different types of plasma reactors can be found in Budhraja et al. (2023). Song et al. (2019) investigated the conversion of ammonia into methane and $H_2$ using DBD plasma reactor and found that catalyst enhances the conversion of methane to $H_2$. Morgan and ElSabbagh (2017) used a pulse plasma reactor to convert methane into $H_2$ and found that 92% of methane converted into $H_2$. Wang et al. (2019) used gliding arc reactor to convert n-heptane and found 50.1% $H_2$. Wang et al. (2021) used MW plasma reactor and found 94% methane conversion with 74% $H_2$ yield. The comparison of different reforming processes is given in Table 2.2.

## 2.4 Challenges and Conclusions

Waste plastic is posing a global threat for environment as only 20% of the total used plastic is disposed safely and the rest is thrown to landfill which takes more than hundred years to decompose. Incineration and other technique are not sustainable and environmentally friendly because it produces toxic gases. Literature review suggests that pyrolysis of waste plastic can produce more than 80% liquid oil which can be refined to diesel fuel for engine. The syngas produced pyrolysis of waste plastic can be used as feedstock to produce $H_2$ through reforming processes. It is found from the literature that steam reforming process is better than other reforming processes to produce $H_2$ which has efficiency of more than 90%. The conversion of waste plastic into $H_2$ through pyrolysis and reforming process can solve the issue of safe disposal of waste plastics. The production of $H_2$ can also help meet the demand of $H_2$ and achieve the net zero by 2050. The main challenges of producing hydrogen from waste plastic using reforming process are:

- Cost of producing hydrogen from waste plastic.
- Scale up the production process for mass production.
- Deactivation and reduction of catalyst effectiveness.

**Table 2.2** Comparison of different reforming process to produce $H_2$

| Reforming Process | Catalyst | Operating condition | Feedstock | Advantage | Disadvantage | References |
|---|---|---|---|---|---|---|
| Steam | $Cu/ZnO/Al_2O_3$, Cu/Zn/Mg, Pt, Ru, Rh/$SiO_2$ | 200–750 °C 0.05–37 bar | Organic compound | Highly efficient (65–75%) | Endothermic | Supriyanto and Richards (2021), Rasul et al. (2022) and Nahar et al. (2015) |
| Autothermal | Pd-Zn/$\gamma$-$Al_2O_3$ | 950–1050 °C 30–50 bar | Plastic, biomass | No external heat required | Need Oxygen supply | Rostrup-Nielsen and Hansen (2011), Ugurlu and Oztuna (2020) and Ahmed and Krumpelt (2001) |
| Partial oxidation | Au-CuO-ZnO, Cu40Zn60 | 1150 – 1315 °C, 6 bar | Heavy hydrocarbon, Petrol, Diesel | Produce heat, less $CO_2$ | lower $H_2/CO$ ratio, High temperature | Uddin et al. (1997) and Oni et al. (2022) |
| Dry | K-L Zeolite, K-$Al_2O_3$, K-Mg/Al oxides, MgO/Pt, PtCo/$CeO_2$ | 700 °C and 900 °C | Hydrocarbon, methane | Sustainably promising thermos-catalytical processes | Catalyst deactivation, Equilibrium, endothermic | Uddin et al. (1997), Cortazar et al. (2022), Agrell et al. (2001), Medeiros et al. (2022) and Ballarini et al. (2019) |
| Aqueous phase | Pt, Sn, Co or nickel (Ni)-based metallic catalysts | 200–250 °C, up to 60 bars | Oxygenated/ non-oxygenated hydrocarbons, biomass | Less energy consuming | $H_2$ and $CO_2$ react | Xie et al. (2018) |
| Plasma | $NiO/Al_2O_3$ Cu/ZnO | 300–400 °C | Hydrocarbon, Biomass, | Less costly and bulky | Unstable plasma discharge, nonuniform temperature | Hajizadeh et al. (2022), Shabaker and Dumesic (2004), Budhraja et al. (2023), Song et al. (2019) and Morgan and ElSabbagh (2017) |

**Acknowledgements** The authors would like to express their gratitude to Fuel and Energy Research Group of CQUniversity, Australia for providing technical support to conduct this research. The work was supported by CRC-P8 project on "Australian Standard Diesel from Mixed Waste Plastics Waste: Maximizing Recovery from Waste", Project No: CRCPEIGHT000194.

## References

Abbas-Abadi MS, Haghighi MN, Yeganeh H, McDonald AG (2014) Evaluation of pyrolysis process parameters on polypropylene degradation products. J Anal Appl Pyrol 109:272–277

Agrell J, Hasselbo K, Jansson K, Järås SG, Boutonnet M (2001) Production of $H_2$ by partial oxidation of methanol over Cu/ZnO catalysts prepared by microemulsion technique. Appl Catal A 211:239–250

Ahmed S, Krumpelt M (2001) $H_2$ from hydrocarbon fuels for fuel cells. Int J $H_2$ Energy 26:291–301

Ballarini AD, Virgens CF, Rangel MC, Miguel SRD, Grau JM (2019) Characterization and behaviour of Pt catalysts supported on basic materials in dry reforming of methane. Braz J Chem Eng 36:275–284

Budhraja N, Pal A, Mishra RS (2023) Plasma reforming for $H_2$ production: pathways, reactors and storage. Int J $H_2$ Energy 48(7):2467–2482

Cortazar M, Gao N, Quan C, Suarez MA, Lopez G, Orozco S et al (2022) Analysis of $H_2$ production potential from waste plastics by pyrolysis and in line oxidative steam reforming. Fuel Process Technol 225:107044

de Medeiros FGM, Lopes FWB, Rego de Vasconcelos B (2022) Prospects and technical challenges in $H_2$ production through dry reforming of methane. Catalysts 12:363

Fakhr Hoseini SM, Dastanian M (2013) Predicting pyrolysis products of PE, PP, and PET using NRTL activity coefficient model. J Chem 2013:1–5

Hajizadeh A, Mohamadi-Baghmolaei M, Cata Saady NM, Zendehboudi S (2022) $H_2$ production from biomass through integration of anaerobic digestion and biogas dry reforming. Appl Energy 309:118442

Hazrat MA, Rasul MG, Jahirul MI, Chowdhury AA, Hassan NMS (2022) Techno-economic analysis of recently improved hydrogen production pathway and infrastructure. Energy Rep 8:836-844

IEA (2019) The future of H2. International Energy Agency (IEA), Paris

Kunwar B, Cheng HN, Chandrashekaran SR, Sharma BK (2016) Plastics to fuel: a review. Renew Sustain Energy Rev 54:421–428

Kusenberg M, Roosen M, Zayoud A, Djokic MR, Thi HD, De Meester S, Ragaert K, Kresovic U, Van Geem KM (2022) Assessing the feasibility of chemical recycling via steam cracking of untreated plastic waste pyrolysis oils: feedstock impurities, product yields and coke formation. Waste Manag 141:104–114

Lewandowski WM, Januszewicz K, Kosakowski W (2019) Efficiency and proportions of waste tyre pyrolysis products depending on the reactor type—a review. J Anal Appl Pyrol 140:25–53

Martínez JD, Puy N, Murillo R, García T, Navarro MV, Mastral AM (2013) Waste tyre pyrolysis—a review. Renew Sustain Energy Rev 23:179-213

Miskolczi N, Bartha L, Deák G, Jóver B (2004) Thermal degradation of municipal plastic waste for production of fuel-like hydrocarbons. Polym Degrad Stab 86(2):357–366

Morgan NN, ElSabbagh M (2017) $H_2$ production from methane through pulsed DC plasma. Plasma Chem Plasma Process 37:1375–1392

Nahar G, Dupont V, Twigg MV, Dvininov E (2015) Feasibility of $H_2$ production from steam reforming of biodiesel (FAME) feedstock on Ni-supported catalysts. Appl Catal B Environ 168–169:228–242

Onwudili JA, Insura N, Williams PT (2009) Composition of products from the pyrolysis of polyethylene and polystyrene in a closed batch reactor: effects of temperature and residence time. J Anal Appl Pyrol 86(2):293–303

Oni AO, Anaya K, Giwa T, Di Lullo G, Kumar A (2022) Comparative assessment of blue H2 from steam methane reforming, autothermal reforming, and natural gas decomposition technologies for natural gas-producing regions. Energy Conv Manage 254:115245

Papari S, Hawboldt K (2015) A review on the pyrolysis of woody biomass to bio-oil: Focus on kinetic models. Renew Sustain Energy Rev 52:1580–1595

PwC (2017) The world in 2050: the long view—how will the global economic order change by 2050? The PwC network

Rasul MG, Sattar MA, Jahirul MI (2022) Experimental and numerical investigation of mixed waste plastics pyrolysis. ICEER 2022:12–16

Rasul MG, Hazrat MA, Sattar MA, Jahirul MI, Shearer MJ (2022) The future of $H_2$: challenges on production, storage and applications. Energy Convers Manage 272:116326

Rehan M, Nizami AS, Shahzad K, Ouda OKM, Ismail IMI, Almeelbi T, Iqbal T, Demirbas A (2016) Pyrolytic liquid fuel: a source of renewable electricity generation in Makkah. Energy Sour Part A Recov Utiliz Environ Effects 38(17):2598–2603

Rostrup-Nielsen JR, Hansen JB (2011) Chapter 4—Steam reforming for fuel cells. In: Shekhawat D, Spivey JJ, Berry DA (eds) Fuel cells: technologies for fuel processing. Elsevier, Amsterdam, pp 49–71

Sarker AK, Azad AK, Rasul MG, Doppalapudi AT (2023) Prospect of green hydrogen generation from hybrid renewable energy sources: a review. Energies 16:1556 (Q2, MDPI)

Seo Y-H, Lee K-H, Shin D-H (2003) Investigation of catalytic degradation of high-density polyethylene by hydrocarbon group type analysis. J Anal Appl Pyrol 70(2):383–398

Shabaker JW, Dumesic JA (2004) Kinetics of aqueous-phase reforming of oxygenated hydrocarbons: Pt/$Al_2O_3$ and Sn-modified Ni catalysts. Ind Eng Chem Res 43:3105–3112

Song L, Liang T, Liu C, Li X (2019) Experimental investigation of $H_2$ production by $CH_4$–$CO_2$ reforming using rotating gliding arc discharge plasma. Int J $H_2$ Energy 44(56):29450–29459

Supriyanto PY, Richards T (2021) Gaseous products from primary reactions of fast plastic pyrolysis. J Anal Appl Pyrol 158:105248

Uddin MA, Koizumi K, Murata K, Sakata Y (1997) Thermal and catalytic degradation of structurally different types of polyethylene into fuel oil. Polym Degrad Stab 56(1):37–44

Ugurlu A, Oztuna S (2020) How liquid $H_2$ production methods affect emissions in liquid $H_2$ powered vehicles? Int J $H_2$ Energy. Int J Hydrogen Energy 45(60). https://doi.org/10.1016/j.ijhydene.2020.01.250

Wang B, Peng Y, Yao S (2019) Oxidative reforming of n-heptane in gliding arc plasma reformer for $H_2$ production. Int J $H_2$ Energy 44(41):22831–22840

Wang Q, Wang J, Zhu T, Zhu X, Sun B (2021) Characteristics of methane wet reforming driven by microwave plasma in liquid phase for $H_2$ production. Int J $H_2$ Energy 46(69):3

Williams PT (2006) Yield and composition of gases and oils/waxes from the feedstock recycling of waste plastic. In: Scheirs J, Kaminsky W (eds) Feeds tock recycling and pyrolys is of waste plastics: converting waste plastics into diesel and other fuels. Wiley, West Sussex, pp 285–309

Williams PT, Slaney E (2007) Analysis of products from the pyrolysis and liquefaction of single plastics and waste plastic mixtures. Resour Conserv Recycl 51(4):754–769

Xie Z, Yan B, Kattel S, Lee JH, Yao S, Wu Q et al (2018) Dry reforming of methane over $CeO_2$-supported Pt-Co catalysts with enhanced activity. Appl Catal B 236:280–293

Zhou H et al (2014) Classification and comparison of municipal solid waste based on thermochemical characteristics. J Air Waste Manag Assoc 64(5):597–616

# Chapter 3
# Novel Nanocomposite Electrospun Polyaniline/Zirconium Vanadate for LPG Gas Detection

**Hassan Shokry and Marwa Elkady**

**Abstract** Novel polyaniline (PANI)/zirconium vanadate (Zr-V) hybrid nanomaterials were prepared successfully via sol–gel technique of Zr-V onto the polymerization process of PANI. The most proper synthetic nanohybrid was characterized using SEM, XRD, FTIR, and TGA. The optimum ratio was recorded using the first route to be 1 PANI: 1.5 Zr-V. The synthetic nanohybrid showed a crystalline structure with nanotube branching (like cauliflower) morphology. Electrospinning technique was used to fabricate a composite nanofiber matrix using polyvinyl alcohol (PVA) with the optimum prepared PANI/Zr-V nanocomposite. Electrospinning parameters including flow rate, collecting distance, and applied voltage were optimized to attain uniform composite nanofibers. The gas sensitivity of the synthetic nanofiber composite towards liquefied petroleum gas (LPG) as a function of temperature was specified by measuring the conductivity of two sputtered electrodes of the sensor's devices. The nanofiber matrix showed the highest sensing performance for LPG within 10 s at 200 °C.

**Keywords** Polyaniline · Nanofiber · Gas detector · Flammable gas · Electrospinning

---

H. Shokry (✉)
Environmental Engineering Department, Egypt-Japan University of Science and Technology, New Borg El-Arab City, Alexandria, Egypt
e-mail: hassan.shokry@ejust.edu.eg

M. Elkady
Chemical and Petrochemical Engineering Department, Egypt-Japan University of Science and Technology, New Borg El-Arab City, Alexandria, Egypt



N. S. Caetano and M. C. Felgueiras (eds.), *The 9th International Conference on Energy and Environment Research*, Environmental Science and Engineering,
https://doi.org/10.1007/978-3-031-43559-1_3

## 3.1 Introduction

Many industrial and commercial activities require the protection of the environment from hazardous and harmful gases. To detect these gases and define the quantities and hazardous levels; it's preferable to develop a modern detection tool that is empirically optimized. The properties of semiconductor gas sensing material change upon exposure to gas molecules, subsequently, this change transfers as a signal to the alarm system. The sophisticated detectors can measure changes in values of the impedance, current, frequency, capacitance, electromotive force and potential difference (Shokry et al. 2015).

Among the conducting polymers, PANI is often used as an organic part to prepare nanohybrids because of its easy preparation, low cost, controllable unique properties by protonation and oxidation state and excellent environmental stability. PANI is used in broad applications such as corrosion protection in organic coatings, batteries, supercapacitors, electronic devices and sensors (Fawal et al. 2019).

Zr-V is one of the inorganic nanomaterials that have received excessive attention due to their unparalleled electrical properties where they show good electrical conductivity which is attributed to semiconductor behavior that ranges from $10^{-5}$ to $10^{-6}$ $\Omega^{-1}$ $cm^{-1}$ (Elkady et al. 2015). Composites and alloys containing organic and inorganic nanomaterials present a totally new category of materials with new properties. It also leads to a marked change in electrical, thermal, and mechanical, properties compared to pure organic materials due to their presence of inorganic nanoparticles in the nanoscale (Ali et al. 2020).

Electrospinning is one of the simplest techniques for obtaining polymer/inorganic composite nanofiber materials with a diameter ranging from nanometers to microns which have a large surface area to volume ratio using electrostatic force. There are various parameters affected by the electrospinning process such as flow rate, applied voltage, needle diameter, collecting distance, and collector design (Elkady et al. 2020).

This work aims to investigate PANI/Zr-V nanocomposite through simple interfacial polymerization and then fabrication of a composite nanofiber matrix using PVA via electrospinning technique. The sensitivity of the fabricated composite nanofiber matrix towards LPG will be measured and examined.

## 3.2 Experimental Procedure

### *3.2.1 Preparation of PANI/ZrV Nanohybrid*

ZrV nanoparticles were prepared using sol–gel technique by adding sodium vanadate solution (0.4 M) drop wisely into a solution of 0.1 M of zirconium oxychloride octahydrate ($ZrOCl_2 \cdot 8H_2O$) with constant stirring. A fine yellow precipitate will appear after the addition completed. The reaction mixture was diluted to 1 l and left to settle for 24 h to complete precipitation. The products were filtered by suction and washed several times with distilled water and ethanol until the filtrate become colorless; finally, the washed precipitate was then dried by gentle heating at 40 °C.

The PANI nanotube was prepared using simple interfacial polymerization in which 0.2 M of aniline monomer was dissolved in distilled water in a flask conical of 50 ml solution under stirring. 50 ml solution of ammonium persulfate (0.25 M) was prepared. Both solutions were mixed in a beaker with slow agitation for 24 h at room temperature. The products were filtered by suction and washed several times with deionized water and finally, washed then dried by gentle heating at 40 °C.

### *3.2.2 Preparation and Characterization of PANI/ZrV Nanofiber Using Electrospinning Technique*

0.5 g from the optimum ratio 1 PANI: 1.5 ZrV was dissolved in 10 ml N, N-Dimethyl formamide (DMF) 99%. The solution was stirred for 24 h at room temperature, followed by 15 min of sonication. 1.8 g PVA was dissolved in 16.2 ml deionized water with stirring for 2 h at 70 °C until getting a viscose solution. Then, the PVA solution was added to the nanohybrid solution. The resultant solution was then moved to the electrospinning machine at 17.5 kV with an average distance between the needle and the collector of 16 cm and a flow rate of 0.5 ml/h. The prepared nanohybrid was characterized using SEM, XRD, FTIR, and TGA.

### *3.2.3 Devices Fabrication and Measurements*

Platinum contact electrodes were deposited by sputtering machine onto the surface of the prepared nanofiber composite layer. The gas sensitivity of the synthetic nanofiber composite was measured LPG and plotted as a function of temperature by measuring the resistance of two electrodes of the sensor's devices.

## 3.3 Results and Discussion

### *3.3.1 Characterization of Prepared PANI/ZrV and Hybrid Nanofiber*

XRD patterns of all prepared materials are shown in Fig. 3.1. Figure 3.1a shows two strongest peaks appeared at $2\theta = 22.8°$and 23.5°, revealing that PANI nanostructure has semi-crystalline properties. As shown in Fig. 3.1b three strongest peaks appeared

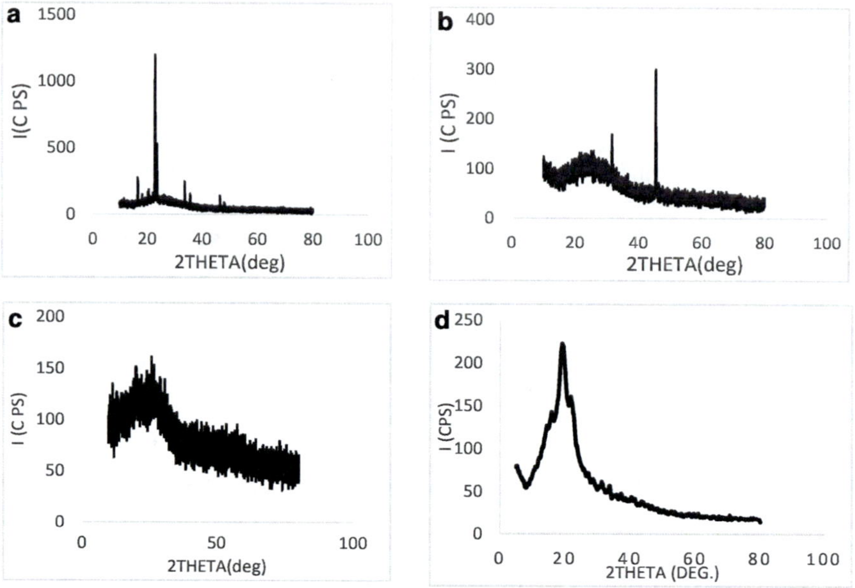


**Fig. 3.1** XRD pattern of **a** PANI, **b** ZrV, **c** PANI/ZrV Nano-hybrid 1:1.5 and **d** PANI/ZrV/PVA Nanofiber

at 2θ = 45.4°, 31.7° and 32.3°, revealing that some degree of crystalline structure of ZrV.

The average diameter of prepared zirconium vanadate was calculated between 14.7 and 54.8 nm, which was calculated from the full width at half-maximum of the peak using Debye-Scherer's equation

$$D = K\lambda/(\beta \cos\theta) \tag{3.1}$$

where D is the average particle size in, K is Scherer constant. (If β is FMHM, K = 0.89.) The XRD peaks of the optimum mixing ratio showed in Fig. 3.1c. The strongest two peaks at 2θ = 20° and 2θ = 25° are corresponds to the typical crystalline properties of PANI with reference to JCPDS file no. 53-1819 (Liu et al. 2014). Figure 3.1d shows the XRD patterns of the prepared PANI/Zr-V Nnanofiber where it showed that the different peaks appeared at 2θ = 19.3°, 21.7°, 23.5°, 15.8°, 14.2°, 33.6° and 31.28°. These peaks are corresponding to orthorhombic structure of PANI/PVA/ZrV nanofiber (Elkady and Shokry 2021).

Figure 3.2a. shows the SEM images of the PANI Nanostructure. The particles of the PANI have branched nanotube with a diameter of about 70–130 nm. This form allows further studies with Zr-V for obtaining a nanohybrid. On the other hand, the SEM image in Fig. 3.2b has shown spherical shape of Zr-V nanoparticles in the range of 45–95 nm. Also, these measurements are identical to XRD results.

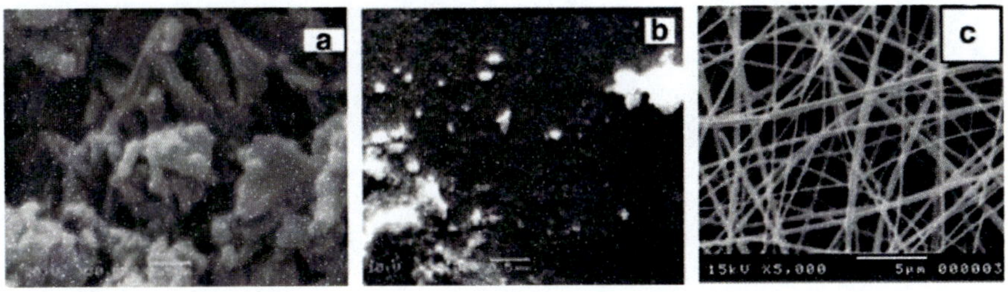


**Fig. 3.2** SEM images of the synthesized **a** PANI, **b** ZrV and **c** PANI/ZrV/PVA nanofibers

A very strong and broadband at 1130 cm$^{-1}$ has been assigned to the B–NH$^+$ indicating that the PANI is conductive and in the form of emeraldine salt. The peak at 816 cm$^{-1}$ is attributed to an aromatic C–H out-of-plane bending. The FTIR spectra for Zr-V-O spherical nanoparticles are shown in Fig. 3.3b. The peak at 3390 cm$^{-1}$ corresponds to the presence of interstitial water and hydroxyl group, whereas the band at 1620 cm$^{-1}$ can be assigned to the bending mode (H-O-H) of coordinated water. The characteristic peaks around 977 cm$^{-1}$ are assigned to the symmetric stretching vibration of the$VO_4$ tetrahedral. The peaks around 690 cm$^{-1}$ are assigned to $VO_4$ asymmetric bending in combination with ZrO6 octahedral stretching (Gowri et al. 2015). FT-IR spectrum of the optimum ratio (1 PANI: 1.5 Zr–V) is shown in Fig. 3.3c. A new peak at 1356 cm$^{-1}$ is attributed to the C–N= stretching vibration between benzenoid and quinoid units. The FTIR spectrum of PANI/PVA/Zr-V nanofiber shows in Fig. 3.3d. The peaks around 3444 cm$^{-1}$ arise from the stretching vibration of N-H group of PANI. The characteristic peaks around 2958 cm$^{-1}$ arising from the stretching vibration of the aromatic C–H group of PANI and aliphatic C–H group of PVA can be seen from this FTIR graph. The peak at 2353 cm$^{-1}$ corresponds to the presence of overtones from PVA–PANI. The characteristic peak of benzenoid rings at 1537 cm$^{-1}$ shifted to 1624 cm$^{-1}$ after the electrospinning process. A sharp peak at 796 cm$^{-1}$ shows the presence of para-substituted aromatic rings, indicating a linear structure of PANI (Mansour and Eslahi 2014).

Thermal stability of the prepared PANI was studied by TGA in Fig. 3.4a. The weight loss of 5.7% occurs within the temperature range of 23–84 °C is attributed to the evaporation of surface adsorbed water and other volatile matter associated with the polymerization. Slightly change in weight loss (4.7%) is observed from 84 to 235 °C. The TGA of Zr-V spherical nanoparticles are shown in Fig. 3.4b. The weight loss of 17% occurs within the temperature range of 18–137 °C is attributed to the evaporation of surface adsorbed water. The second step of the weight loss of 6.2% occurs within the temperature range of 137–350 °C may be caused by the decomposition of the condensation dehydration of the hydroxyls. There is no significant weight loss after 362–600 °C, which indicates that no structural changes can occur in the materials. Figure 3.4c. shows the TGA of PANI/Zr-V composite nanohybrid.

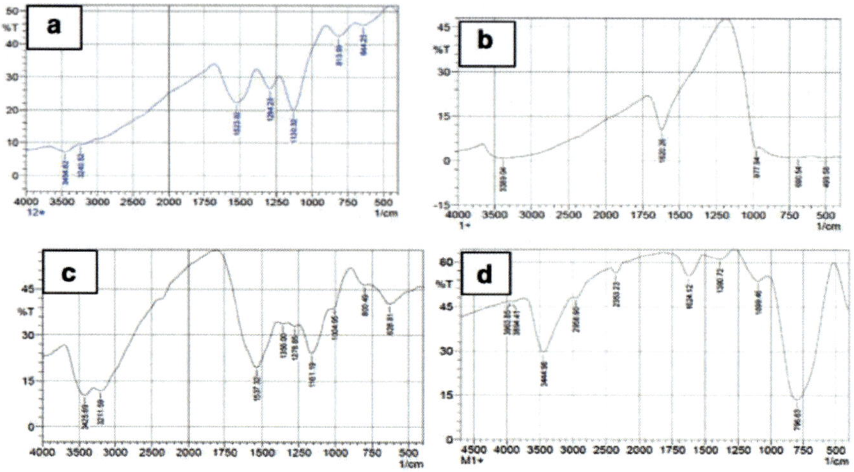


**Fig. 3.3** FTIR spectrum of **a** PANI, **b** Zr-V, **c** PANI/Zr-V nanohybrid and **d** PANI/Zr-V/PVA nanofiber

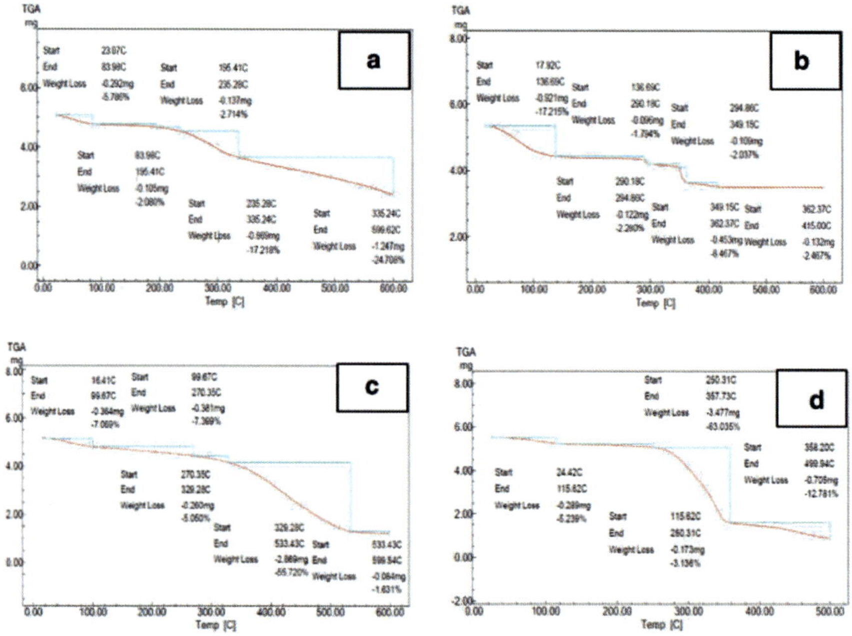


**Fig. 3.4** TGA analyses of **a** PANI, **b** Zr-V, **c** PANI/Zr-V nanohybrid and **d** PANI/Zr-V/PVA nanofiber

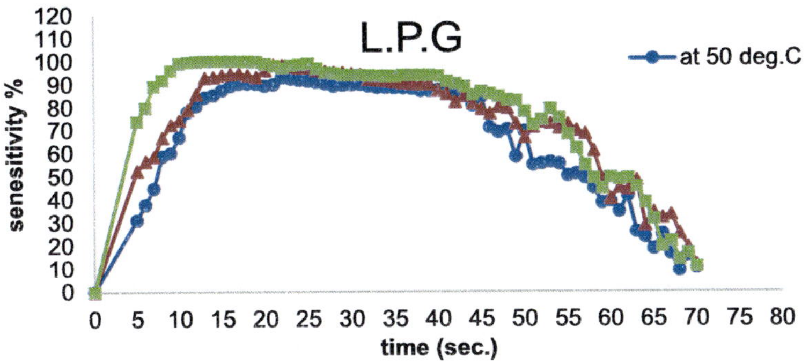


**Fig. 3.5** Sensing performance of PANI/Zr-V/PVA nanofibers for LPG at different temperatures

The average weight loss at this temperature range not exceeded 19% from the material weight until 329 °C which is attributed to wetness. This indicates the presence of Zr-V in the PANI chain where it's noted that the thermal stability of the hybrid is higher than that in the pure PANI. Thereafter a continuous loss is observed in the temperature range of 329–533 °C showing a maximum weight loss of 55%. This weight loss is attributed to the decomposition as well as degradation of PANI/Zr-V Nano-hybrid. The TGA of PANI/Zr-V/PVA composite nanofiber is shown in Fig. 3.4d. The weight loss of 8.2% occurs within the temperature range of 24–250 °C (Shokry 2019).

### *3.3.2 Gas Sensing Performance of LPG*

The characteristics of the fabricated nanofiber matrix gas sensor for LPG were measured as a function of time with a gas concentration of 800 ppm and at a working temperature of 50, 100, and 200 °C using DC resistance measurements as shown in Fig. 3.5. In this experiment, sensor's sensitivity sharply increased when exposed of LPG, where it reaches 100% force (Abozeid et al. 2019).

## 3.4 Conclusion

PANI/Zr-V/PVA nanofiber matrix has been synthesized successfully. SEM, XRD, FTIR, and TGA were used in the prepared materials characterization. It was indicated that the PANI has a highly crystalline structure with nanotube branching (like cauliflower) morphology. Gas sensitivity of the PANI/Zr–V/PVA composite nanofiber sensor depends on the variation of the resistivity by exposure to LPG at a working temperature of 50, 100, and 200 °C. The maximum sensitivity scored at 100% sensitivity.

## References


Abozeid MA, Shokry HH, Morsi I, Kashyout AB (2019) Development of nano-$WO_3$ doped with NiO for wireless gas sensors. Arab J Sci Eng 44:647–654

Ali I, Kashyout AB, Tayel M, Shokry HH, Rizk M (2020) Ruthenium (Ru) doped zinc oxide nanostructure-based radio frequency identification (RFID) gas sensors for $NH_3$ detection. J Market Res 9:15693–15704

El FawalGF, Shokry Hassan H, El-Aassar MR, Elkady MF (2019) Electrospun polyvinyl alcohol nanofibers containing titanium dioxide for gas sensor applications. Arab J Sci Eng 44:251–257

Elkady MF, Shokry Hassaan H (2021) Photocatalytic degradation of malachite green dye from aqueous solution using environmentally compatible Ag/ZnO polymeric nanofibers. Polymers 13:2033

Elkady MF, Shokry Hassan H, El-Sayed EM (2015) Basic violet decolourization using alginate immobilized nanozirconium tungestovanadate matrix as cation exchanger. J Chem 1–10:385741

Elkady M, Salama E, Amer WA, Ebeid EM, Ayad MM, Shokry H (2020) Novel eco-friendly electrospun nanomagnetic zinc oxide hybridized PVA/alginate/chitosan nanofibers for enhanced phenol decontamination. Environ Sci Pollut Res 27:43077–43092

Gowri S, Rajiv Gandhi R, Sundrarajan M (2015) Structural, optical, antibacterial and antifungal properties of zirconia nanoparticles by biobased protocol. J Mater Res Technol 30:782–790

Liu Q, Yang J, Rong X, Sun X, Cheng X, Tang H, Li H (2014) Structural, negative thermal expansion and photocatalytic properties of $ZrV_2O_7$: a comparative study between fibers and powders. Mater Charact 96:63–70

Mansour G-M, Eslahi H (2014) Synthesis, characterization and antibacterial properties of a novel nanocomposite based on polyaniline/polyvinyl alcohol/Ag. Arab J Chem 7:846–855

Shokry HH (2019) Role of preparation technique in the morphological structures of innovative nano-cation exchange. J Market Res 8:2854–2864

Shokry Hassan H, Kashyout AB, Morsi I, Nasser AAA, Abuklill H (2015) Development of polypyrrole coated copper nanowires for gas sensor application. Sens Biosens Res 5:50–54

# Chapter 4
# Engine Performance and Emission Characteristics of Diesel Produced from Pyrolysis of Mixed Waste Plastics

**M. A. Hazrat, M. G. Rasul, M. I. Jahirul, and A. G. M. B. Mustayen**

**Abstract** Globally accumulated waste plastics have become an alarming environmental hazard for rapid increase in production and lack of recycling opportunities. In this study, mixed waste plastics containing an equal proportion of high-density polyethylene, polypropylene, and polystyrene were converted into oil through a 20 l batch pyrolysis reactor at 540 °C. Further upgrading of plastic pyrolysis oil (PPO) via a vacuum distillation process resulted in the separation of PPO into petrol cut (gasoline), diesel cut (diesel), and naphtha. This study defines the PPO diesel cut as plastic made diesel (PMD). The PMD fuel was blended with commercial ultra-low-sulphur diesel (ULSD) at 5% (defined as PMD5) and 10% (v/v) (defined as PMD10) to assess the performance and emissions characteristics of a naturally aspirated direct injection diesel engine. The results are highly comparable with the results of commercial ULSD. The maximum brake power of ULSD was only 0.9% higher than PMD10 fuel. The PMD5 and PMD10 showed lower brake specific fuel consumption (BSFC) and $NO_x$ emissions and comparable CO and $CO_2$ emissions with ULSD in an unmodified diesel engine. Therefore, waste plastics can be considered an energy source by producing plastic diesel that can reduce environmental pollution and meet growing energy demand.

**Keywords** Diesel · Distillation of pyrolysis oil · Emission reduction · Waste plastics · Waste management · Pollution control

---

M. A. Hazrat
Centre for Regional Economies and Supply Chain (CRESC), Central Queensland University, Rockhampton, QLD 4701, Australia

M. G. Rasul (✉) · M. I. Jahirul · A. G. M. B. Mustayen
Fuel and Energy Research Group, School of Engineering and Technology, Central Queensland University, Rockhampton, QLD 4701, Australia
e-mail: m.rasul@cqu.edu.au

M. G. Rasul · A. G. M. B. Mustayen
School of Engineering, University of Tasmania, Hobart, TAS 7001, Australia

 
N. S. Caetano and M. C. Felgueiras (eds.), *The 9th International Conference on Energy and Environment Research*, Environmental Science and Engineering,
https://doi.org/10.1007/978-3-031-43559-1_4

## 4.1 Introduction

Alternative resourcing of fuel production for energy supply is one of the key focuses due to depleting fossil fuel reserves (BP 2021), the demand of reduced carbon intensity (CI) and the mitigating effect of climate change and global warming from the energy supply chain (Welsby et al. 2021). Not only the combustion of fossil fuels but also the increased nonrecycled plastic wastes are emitting a large amount of greenhouse gases to the environment. About 3.4% [i.e., 1.8 Gt $CO_2$-e (Giga tons of carbon dioxide equivalent)] of total global emissions engendered from waste plastics in the year 2019 (OECD 2022). Globally, about 9% of waste plastics are recycled, 19% are sent to incinerators for energy production, and the rest of the waste plastics end up in landfills and uncontrolled dumpsites (OECD 2022). Though overall collection of waste plastics is about 15%, the preliminary filtration processes prior to recycling processes discard about 40% of the collected plastics as unacceptable to be recycled (OECD 2022). There is a projection of the cumulative plastic production from 1950 to 2050 to be about 34 bn Mt (Billion Metric Tonnes) of which about 24.8 bn Mt will be left as garbage on earth (UNEP 2021). As plastics are made of hydrocarbons, there is a great potential for converting plastic wastes into usable energies, which is recognized as one of the effective plastic waste management processes to reduce wastes from the environment. Pyrolysis is one of the effective thermochemical processes that can convert waste plastics into usable hydrocarbons (Li et al. 2021), which can be further processed to produce fuel for energy production such as diesel, petrol, and jet fuel (Jahirul et al. 2022; Zhang et al. 2019; Papari et al. 2021). Pyrolysis oils can be used as fuel for boilers or as fuel blending with diesel (Papari et al. 2021), or the oils can be reformed to produce hydrogen fuels (Cortazar et al. 2022).

Arjharn et al. (2022) conducted pyrolysis of waste MSW plastics at 350–400 °C to obtain plastic pyrolysis oil (PPO), which was distilled within 176–360 °C to obtain 10% heavy oil, 60% diesel-like middle distillates (i.e., PMD), and 30% naphtha. Cetane index (CI) was 72.89 and 65.8, and calorific value (CV) was 44.98 MJ/kg and 46.04 MJ/kg for PPO and PMD fuels, respectively. The engine test showed that the brake specific fuel consumption (BSFC) for PMD and reference ULSD blended with 7% biodiesel (BD7) (CV = 45.39 MJ/kg, and CI = 57.01) were comparable. In the case of the emissions, nitrogen oxides ($NO_x$) were observed 10–25% higher than that of BD7. Carbon monoxide (CO) and unburned hydrocarbon (uHC) emissions were similar for both BD7 and PMD. Though this experimental study did not mention the plastic constituents in the MSW stream, this indicated a reliable transformation of fuel production from MSW pyrolysis and distillation that could reinforce the fuel supply chain while reducing the non-biodegradable plastic wastes.

Upgrading of the PPO can also be performed by hydroprocessing, which can yield a higher portion of transport-grade fuels. Mangesh et al. (2022) conducted catalytic pyrolysis of high-density polyethylene (HDPE), low-density polyethylene (LDPE), and polypropylene (PP) individually that showed a higher average yield (75%) of liquid than that with mixed plastics (60%). An equal amount of HDPE, LDPE, and PP were mixed to make mixed plastics pyrolysis oil, which was hydro-processed in the presence of a catalyst to obtain 90% diesel-like fuel (PMD). Both the CV and CI were reported as 43.825 MJ/kg, 43.225 MJ/kg, 43.92 MJ/kg, and 62, 60, 52 for the PMD, PPO, and ULSD fuels, respectively. At full load operation and with comparison to ULSD fuel, PMD demonstrated 2–5% lesser in-cylinder pressure, 2.8–6.8% lower heat rejection rate (HRR), 36% more CO, 2–9.3% more carbon dioxide ($CO_2$), 3–9% more $NO_x$, and 0.7–1.8% lesser UHC emissions.

In this study, mixed waste plastics were converted into oil via a pyrolytic process, and an efficient fractionating distillation technique was adopted to produce PMD. This fuel was blended with commercial ULSD to investigate performance and emissions characteristics in a direct injection diesel engine. An optimized thermochemical processing parameter was employed to obtain more PPO. Thahir et al. (2019) conducted higher temperature (500–650 °C) pyrolysis of PP that helped to separate the two grades of diesel fuel, i.e., kerosene and diesel fuel. However, in this study, pyrolysis was done at 550 °C. The key motivation of this study was to produce plastic diesel from waste plastics and tested in a diesel engine for its feasibility to use as a good automobile fuel.

## 4.2 Materials and Methodology

The stages of this work are pyrolysis, fractionating distillation, separating diesel grade fuels, blending with the commercial ULSD, and comparing the performance and emission characteristics against the reference ULSD fuel. Figure 4.1 shows the stages of whole methodologies. Their brief description is given below. The mixed plastic samples were prepared by mixing same proportion of HDPE, PP, and PS (i.e., HDPE: PP: PS = 1:1:1) after individually shredding them into fine granules (< 1 mm). The pyrolysis experiment was done in our 20 l batch reactor at about 540 °C. The temperature increase rate was 10 °C/min. Detailed operating procedure can be found in our article published in the Renewable Energy (Hasan 2022). The oil produced from waste plastic pyrolysis process has been defined as WPO. The fractionating distillation was done in a vacuum distillation setup (model: iFischer AutoDest 800/860 AC, vacuumed pressure up to 266 Pa) facilitated with automatic/pre-set programming for a highly reliable distillation process. The gasoline cut was collected from the distillation products under 170 °C, diesel fuels (PMD) were collected from the distilled liquids obtained within 170–380 °C, and the other residuals were obtained at a temperature higher than 380 °C.

### *4.2.1 Blending of PMD with ULSD*

The desired amount of ULSD and PMD was mixed with the help of magnetic hot plate stirrers to prepare PMD-ULSD blends. The mixing was done for 15 min for each of the blends to ensure proper diffusion and blending occurred. Two different

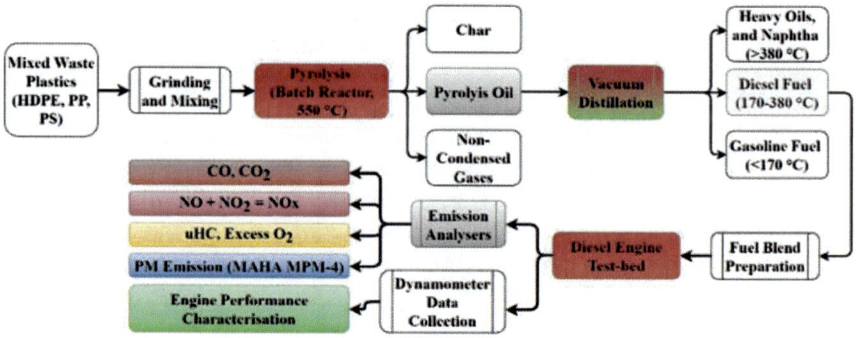


**Fig. 4.1** Flow diagram of methodologies

**Table 4.1** Physical properties of tested fuels

| Properties | PMD | ULSD |
|---|---|---|
| Calorific value (MJ/kg) | 45.01 | 42.3 |
| Cetane index (CI) | 48 | 46.2 |
| Density@15 °C (kg/m$^3$) | 847 | 850 |
| Lubricity (WSD), μm | 200 | 450 |
| Flash point (°C) | 63 | 61.7 |
| Kinematic viscosity @40 °C (mm$^2$/s) | 2.4 | 3.9 |
| Moisture content (g/kg) | 0.015 | 0.04 |

blends were prepared by mixing 5% (v/v) and 10% (v/v) PMD with 95% (v/v) and 90% (v/v) ULSD. The fuel blends were termed PMD5 and PMD10, respectively. Key fuel properties are presented in the Table 4.1.

### *4.2.2 Engine Test Bed Set-Up*

A naturally aspirated direct injection diesel engine (Kubota V3300) with Dyno Dynamics chassis dynamometer, particulate matter analyser (Model: MAHA MPM-4), and an emission analyser (Model: Infralyt N-Saxon Exhaust Gas Analyzer for Combustion Engines) set-up were used for measuring performances such as brake thermal efficiency (BTE), brake specific fuel consumption (BSFC) and brake specific energy consumption (BSEC) and emissions namely CO, $CO_2$, NO, $NO_2$, and uHC at various speeds and loads conditions. In this study three sets of tests were done under mentioned operating conditions. All these results were compared with the performance and emission parameters. Engine specifications is given in Table 4.2. Besides, Fig. 4.1 schematically represents the experimental process (mixed waste plastic to engine testing) with methodology of the current study.

**Table 4.2** Engine specifications and operating conditions

| Engine model | Kubota V3300 |
|---|---|
| Stroke (S) | 98 mm |
| Bore (B) | 110 mm |
| Connecting rod length | 170 mm |
| Rated power output (kW/rpm) | 50.7/2600 |
| Rated torque (Nm/rpm) | 230/1400 |
| Injection system | Direct |
| Compression ratio (CR) | 22.6:1 |
| Fuel injection timing | 16° before TDC |
| Injection pressure (MPa) | 13.73 |
| Inlet manifold temperature | 80 °C |
| EVO | 51° BBDC |
| EVC | 28° ATDC |
| IVO | 17° BTDC |
| IVC | 63° ABDC |

The following equations were used to determine BTE, BSFC, and BSEC.

$$\text{BSFC}\left(\frac{\text{g}}{\text{kW hr}}\right) = \frac{\text{m}_\text{f}\left(\frac{\text{kg}}{\text{h}}\right)}{\text{BP (kW)}} \times \left(\frac{1000\,\text{g}}{\text{kg}}\right) \tag{4.1}$$

$$\text{BSEC}\left(\frac{\text{J}}{\text{Wh}}\right) = \text{BSFC} \times \text{Q}_{\text{CV}} \tag{4.2}$$

$$\text{BTE}(\%) = \frac{3600}{\text{BSFC(kWh)} \times \text{Q}_{\text{HV}}(\text{MJ/kg})} \times 100 \tag{4.3}$$

where, $\text{m}_\text{f}$ is the fuel flow rate (kg/h), BP is the effective power output (kW), and QCV is the calorific/heating value of the fuel in kJ/kg, and QHV is the heating value of fuel in MJ/kg. The data acquisition system provides the values of BP and the fuel flow rate, whereas the calorific values were determined as per ASTM D975 standards (ASTM 2021).

## 4.3 Result and Discussion

The physical properties of both ULSD and the distilled diesel from WPO (PMD) are presented in Table 4.1. It can be seen from the table that the higher heating value and cetane index of the PMD are about 6.4 and 3.9% higher than the ULSD, respectively. The kinematic viscosity of the ULSD is about 60% higher than the PMD. Though no additives were mixed with the PMD, the commercial ULSD may contain various additives that could help prolong the quality of the diesel sold in fuel pumps in diverse environmental conditions in Australia.

### *4.3.1 Brake Power (BP) and Brake Thermal Efficiency (BTE)*

Figure 4.2 shows the BP and the BTE at full load condition. At 800 rpm, the idle condition could be observed for the given loading condition which shows that the ULSD, PMD5, and PMD10 fuels are producing almost same (only 0.5% variation) power and efficiency. The BP values are increasing gradually towards 46 kW power output as the speed increased to 2400 rpm and starts decreasing beyond that speed.

The ULSD produced about 0.9% more power than the PMD10 fuel. On the other hand, the higher thermal efficiency was observed at 2100 rpm for all the fuels. Here, both the PMD5 and PMD10 fuels showed 2.36 and 3.34% more efficiency than that of the ULSD. Both the higher cetane index and heating values of the blend fuels could be the key reasons to demonstrate higher efficiency. Mangesh et al. (2022) observed 5.7% lower efficiency with hydro-processed mixed plastics (equal proportion of HDPE, LDPE, and PP) fuel. Hydro-processing can convert the distilled fuel into commercially available diesel fuels. Padmanabhan et al. (2022) reported lower efficiency with the oil derived from pyrolysis of HDPE, but observed higher efficiency when mixed with oxygenated additives. Kizza et al. (2022) reported that the quality and quantity of the oils derived from PS are higher than the other plastics, which can also improve the thermal efficiency of the fuel blends. In this study, the

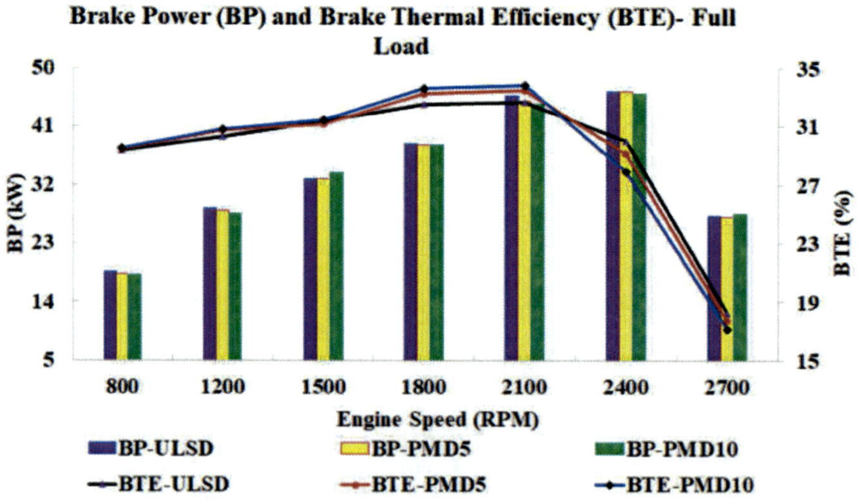


Fig. 4.2 BP and BTE at full load and various speeds

mixed plastics oil was produced from HDPE, PP, and PS mixtures. Also, the pyrolysis oil was distilled in a very efficient industry standard distillation system that helped to achieve higher heating values more than that of the ULSD used here. Hence there could be a small increase in the thermal efficiency from the combustion of the PMD5 and PMD10 fuel blends. Further reasons could be clarified from the BSFC and BSEC as presented in the Sect. 4.3.2.

### *4.3.2 Brake Specific Fuel Consumption (BSFC) and Brake Specific Energy Consumption (BSEC)*

Figure 4.3 shows both the BSFC and BSEC characteristics of the fuels (i.e., ULSD, PMD5, and PMD10) at full load and various speeds. From 800 to 2100 rpm, the BSFC values kept decreasing for all the blends at an order of ULSD > PMD5 > PMD10, but the values and the order changed for higher speeds at an order of PMD10 > PMD5 > ULSD. Since the blends fuels had higher heating values, variation observed until 2100 rpm is well understood which is also demonstrated by Ashok et al. (2017) with the help of biodiesel fuels. The BTE starts decreasing after 2100 rpm, but the fuel consumption increases to run the engine at full load and higher speeds. As a result, the fuel consumption increases at higher speeds beyond achieving maximum brake thermal efficiency.

BSEC is another parameter to explain this situation. It is the efficiency of energy recovery from the combustion of fuels to generate the unit amount of usable power. Since the BSEC is higher after the engine reaches to its maximum BTE speed, the

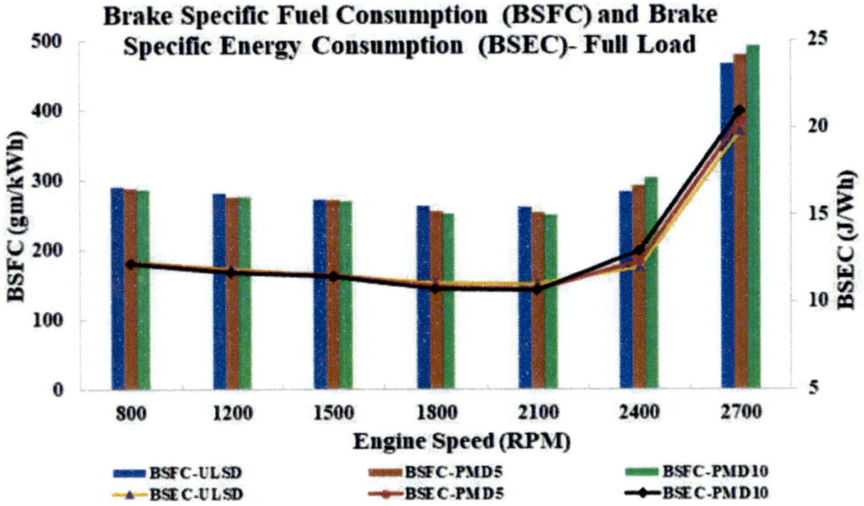


Fig. 4.3 BSFC and BSEC at full load and various speeds

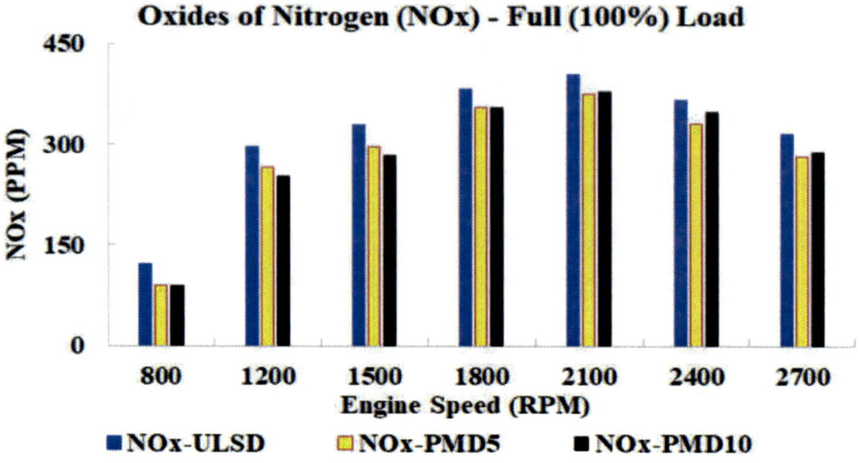


**Fig. 4.4** $NO_x$ at full load and various speeds

fuel consumption increases. Lower BSEC of the blend fuels is another reason for higher or comparable BTE, as observed in Fig. 4.2.

### 4.3.3 *Nitrogen Oxide ($NO_x$) Emissions*

Figure 4.4 shows the emissions of the nitrogen oxide ($NO_x$), which are the inherent yield of the combustion due to higher temperature (Zeldovich mechanism), the amount of air drawn into the combustion chamber, and the availability of oxygen. The result shows the $NO_x$ emission for diesel fuel is higher than the PMD5 and PMD10 under all operating conditions. In Fig. 4.4, PMD5 at 1200 and 1500 rpm shows more $NO_x$ emissions than PMD10. In contrast, PMD5 shows less $NO_x$ at 2100, 2400, and 2700 rpm operation compared to PMD10; however, 800 and 1800 rpm show equal value. The higher viscosity of the diesel fuel may have influenced more fuel to be injected compared to blends, which caused more $NO_x$ formation by the diesel fuel. At higher speeds than 2100 rpm causes, more PMD10 fuel blends are drawn into the combustion chamber that has more fuel to be burnt, which could be the reason for higher $NO_x$ than the PMD5.

### 4.3.4 *Unburnt Hydrocarbons (uHC) and Particulate Matters (PM) Emissions*

Figure 4.5 and 4.6 show the unburnt hydrocarbons (UHC) and particulate matters (PM) emissions, respectively. At 800 rpm, the idle combustion condition at full load

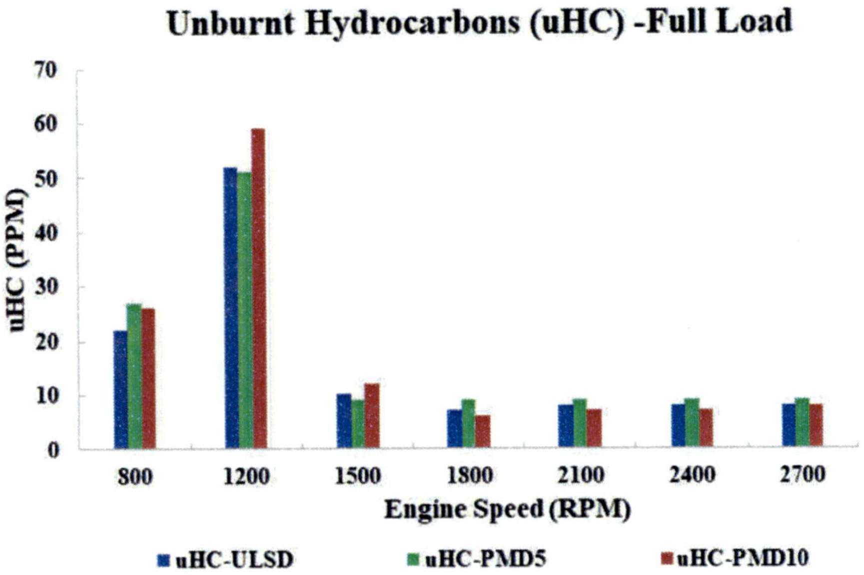


**Fig. 4.5** uHC at full load and various speeds

causes a higher air–fuel ratio, resulting in incomplete combustion. The higher uHC as well as PM emissions are the consequences of the idle combustion condition. The engine gains higher power and torque when the speed increases from idle to 1200 rpm at full load. It is results in more fuel intake that ends up releasing higher amount of uHC and PM for all the three fuels. Due to higher carbon content, the PMD10 emits the highest amount of uHC at 1200 rpm. But the PM is almost equal for ULSD, PMD5, and PMD10 fuels. Beyond 1200 rpm, ULSD emits lesser PM and uHC due to complete combustion. Beyond 2400 rpm, oxygen availability in the combustion chamber causes more emissions due to complete combustion in the chamber for all fuels. Very low PM emission by the ULSD indicates the commercially produced good quality fuel that went through hydro processing after the distillation. The PMD5 and PMD10 fuels had more carbon in their blends, which caused higher uHC than the ULSD. Padmanabhan et al. (2022) also observed lower uHC in waste plastic-derived pyrolytic oil than standard diesel fuel.

### *4.3.5 Carbon Monoxide (CO) and Carbon Dioxide ($CO_2$) Emissions*

Figures 4.7 and 4.8 show the emissions of CO and $CO_2$ for all ULSD and two PMD blends at full load under variable speed operation, respectively. The higher viscosity in the ULSD caused more inadequacy (Padmanabhan et al. 2022) of combustion due

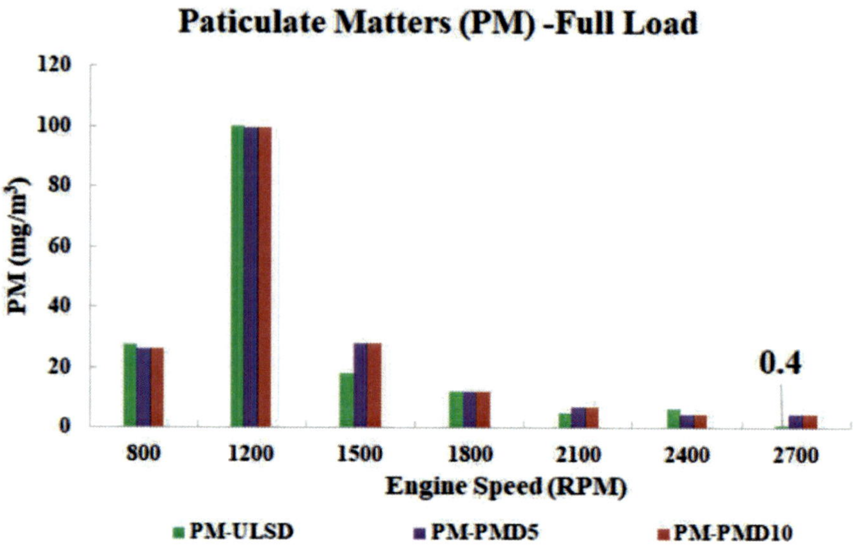


**Fig. 4.6** PM at full load and various speeds

to carbon contents at 800 and 1200 rpm, resultant all three fuels (ULSD, PMD5, and PMD10) show higher CO emissions. Because of higher carbon contents at 1200 and 1500 rpm conditions, PMD5 and PMD10 show slightly more CO emissions than the ULSD. When the engine speed increases from the idle condition, the torque and load requirement for overcoming the static inertia to gain desired speed for overdrawing fuel into the combustion chamber through the direct injection spraying system. This causes more $CO_2$, and CO emissions for all the fuels tested in this study. With increased speed and balancing of the dynamic conditioning of the engine power delivery requirements, a very small amount of CO emission can be observed (Fig. 4.8) for speeds beyond 1200 rpm for all the fuels. Between 1200 and 2400 rpm, the $CO_2$ emission looks almost uniformly decreasing for all the fuels. PMD5 and PMD10 fuels show just little increase in $CO_2$ than the ULSD which can be explained with the higher fuel content and resulting better combustion conditions. At 2700 rpm, the full load condition delivers lesser useful power by drawing lesser fuel and conducting moderately good combustion process.

## 4.4 Conclusions

The PMD fuel from an equal amount of waste HDPE, PP, and PS is of very high quality via its heating value, cetane index, and lower viscosity. Though there are many references to using pyrolysis oil blended with diesel fuels in specific applications, the distillation process increases the acceptability of the fuel as that of the

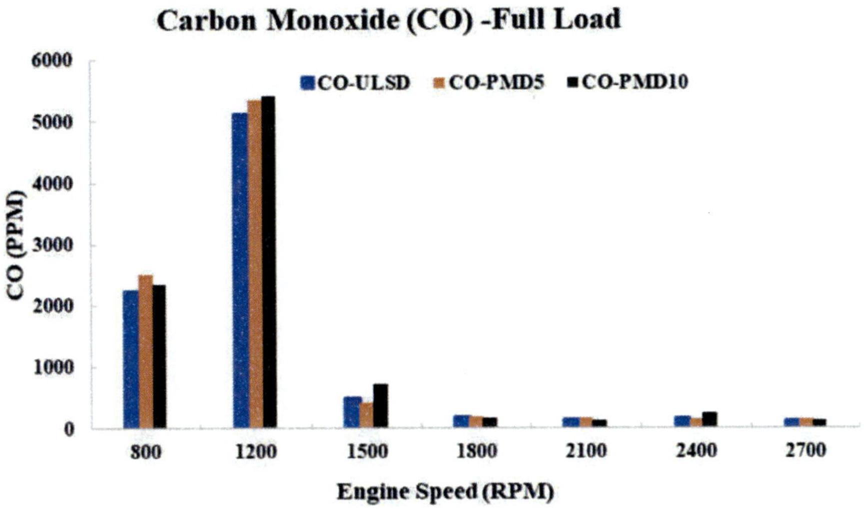


Fig. 4.7 CO at full load and various speeds

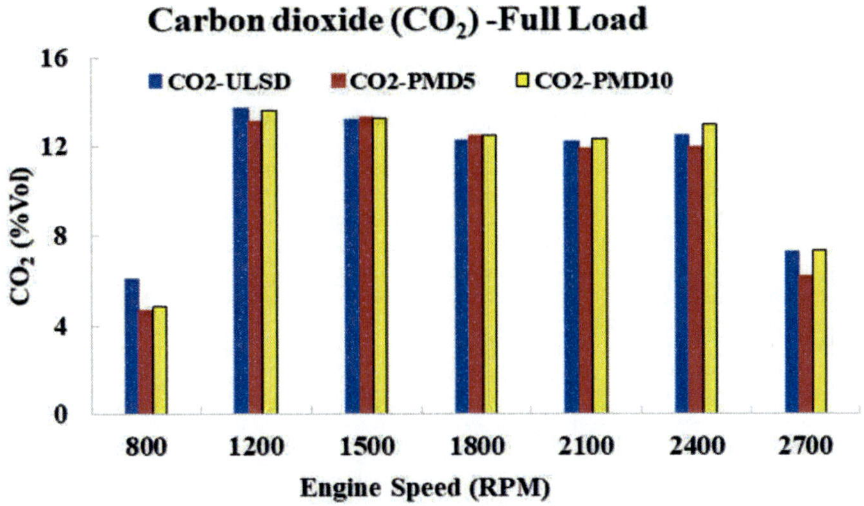


Fig. 4.8 $CO_2$ at full load and various speeds

commercially available fuels. The blends in this study do not show much difference in fuel consumption, power production, and critical emissions like $NO_x$, CO, and $CO_2$ compared to commercial ULSD fuel. Distillation is the first stage of upgrading the pyrolysis oil obtained from the plastics. Due to the separation of the hydrocarbons like gasoline, diesel fuel, and naphtha, the combustion of the derived PMD fuel blended with ULSD indicated more combustion uniformity. This application indicates a very reliable method of converting mixed waste plastics into energy producing fuels, which can offer better waste management options in Australia and worldwide. Further upgrading via hydro processing can be another option for PMD fuel to assess the performance and emission characteristics of the engine in the future study.

## References

Arjharn W et al (2022) Distilled waste plastic oil as fuel for a diesel engine: fuel production, combustion characteristics, and exhaust gas emissions. ACS Omega 7(11):9720–9729

Ashok B, Nanthagopal K, Vignesh D (2017) *Calophyllum inophyllum* methyl ester biodiesel blend as an alternate fuel for diesel engine applications. Alex Eng J 57:1239–1247

ASTM (2021) ASTM D975-21: standard specification for diesel fuel. ASTM International, p 28

BP (2021) bp's Statistical review of world energy 2021. British Petroleum

Cortazar M et al (2022) Analysis of hydrogen production potential from waste plastics by pyrolysis and in line oxidative steam reforming. Fuel Process Technol 225:107044

Hasan MM et al (2022) Fast pyrolysis of beauty leaf fruit husk (BLFH) in an auger reactor: effect of temperature on the yield and physicochemical properties of BLFH oil. Renew Energy 194:1098–1109

Jahirul MI et al (2022) Transport fuel from waste plastics pyrolysis—a review on technologies, challenges and opportunities. Energy Convers Manage 258:115451

Kizza R, Banadda N, Seay J (2022) Qualitative and energy recovery potential analysis: plastic-derived fuel oil versus conventional diesel oil. Clean Technol Environ Policy 24(3):789–800

Li D et al (2021) Study on the pyrolysis behaviors of mixed waste plastics. Renew Energy 173:662–674

Mangesh VL et al (2022) Hydroprocessing mixed waste plastics to obtain clean transport fuel. J Clean Prod 358:131952

OECD (2022) Global plastics outlook.OECD Publishing, Paris

Padmanabhan S et al (2022) Energy recovery of waste plastics into diesel fuel with ethanol and ethoxy ethyl acetate additives on circular economy strategy. Sci Rep 12(1):5330

Papari S, Bamdad H, Berruti F (2021) Pyrolytic conversion of plastic waste to value-added products and fuels: a review. Materials 14(10):2586

Thahir R et al (2019) Production of liquid fuel from plastic waste using integrated pyrolysis method with refinery distillation bubble cap plate column. Energy Rep 5:70–77

UNEP (2021) From pollution to solution: a global assessment of marine litter and plastic pollution. United Nations Environment Programme (UNEP), Nairobi

Welsby D et al (2021) Unextractable fossil fuels in a 1.5 °C world. Nature 597(7875):230–234

Zhang Y et al (2019) Jet fuel production from waste plastics via catalytic pyrolysis with activated carbons. Appl Energy 251:113337

# Part II
# Sustainable Buildings

# Chapter 5
# Design and Smartness Evaluation of Building Automation and Management Systems in Danish Case Studies

**Muhyiddine Jradi**

**Abstract** The building sector is a key component of energy efficiency and environmental strategies and initiatives worldwide due to its major contribution in the energy consumption and the corresponding share in carbon emissions. These goals along with the evolution towards digitalization of the building stock have called for smarter, flexible, and proactive buildings. In this context, a well-designed and properly installed and operated building automation and control system (BACS) is critical to achieving energy efficiency and environmental goals as well as providing high comfort levels and empowering occupants with information. In this study, an innovative tool 'IBACSA' for BACS assessment and building smartness evaluation is used as a basis to evaluate the design functionalities and services provided by the automation and control system. The tool assesses eight various technical domains in the building, covering 60 major building services, reporting the rating and performance of each domain, and considering multiple impact criteria. IBACSA is implemented in three Danish buildings of different type, size, and use. The BACS design and functionalities are evaluated and rated in each case, and actions for improvement are reported.

**Keywords** Smart buildings · Building automation · Control systems · EN15232 standard · Technical domains · Energy efficiency

## 5.1 Introduction

With its major contribution to the overall energy consumption and the corresponding carbon emissions, the building sector is at the center of the majority of energy and environmental goals and strategies. Thus, a large block of theoretical and practical applications and investigations were devoted to improving the performance

M. Jradi (✉)
Center for Energy Informatics, University of Southern Denmark, Campusvej 55, 5230 Odense, Denmark
e-mail: mjr@mmmi.sdu.dk



N. S. Caetano and M. C. Felgueiras (eds.), *The 9th International Conference on Energy and Environment Research*, Environmental Science and Engineering,
https://doi.org/10.1007/978-3-031-43559-1_5

of buildings and enhancing their efficiency quotient by targeting building constructions and envelope (Dascalaki et al. 2021), building services and systems, including heating, ventilation, and air conditioning (HVAC) units (Mirnaghi and Haghighat 2020), equipment and devices (Chauhan and Chauhan 2019) with an overall aim of enhancing the quality of design, improving the operation, and reducing energy consumption and cost. On the other hand, the building automation and control system (BACS) is the heart of the building, allowing monitoring of the overall performance, controlling, and managing different building services. It is the critical component for establishing effective interaction and integration between the different energy systems, controls, services, meters, and sensors. With its control and management capabilities and functionalities, it is the driver of every single decision-making process in the building on various levels (Engvang and Jradi 2021).

BACS are currently playing a key role in both newly built buildings and existing energy-efficient buildings. The initiatives towards building sector digitalization in recent decades and the increase in the number of installed sensors, IoT devices, and smart meters, will require building automation and control systems to play even a bigger and more important role in the future. In recent years, major developments and advancements have been made in the field of BACS, both to ensure a proper design and to establish proper operation and functionality (Domingues et al. 2016; Li and Wang 2022; O'Grady et al. 2021). However, only a very limited number of studies have been reported dealing with auditing and commissioning of the building automation and control systems to make sure that the design is complying with the national and international standards and that it is equipped with optimal functionalities and services. Considering demonstrated potential of improving the building performance and optimizing systems operation, a well-designed, installed, and operated BACS is critical to achieving energy and environmental goals.

In recent decades, the trend in building commissioning processes is that they are done on the whole building level, evaluating if the envelope is living up to the standards and that the predicted consumption is in line with the regulations. Thus, there is a reported lack of information and systematic frameworks to aid the evaluation of the BACS and to ensure that these systems are properly designed and installed. Considering the BACS impact and connection to every single system and functionality in the building, a proper BACS design and operation has a direct impact on the different associated services and their technical and economic efficiencies. Thus, there is a need for user-friendly and comprehensive tools to support decision-making on the energy-efficient and optimal design and operation of automation and control systems. In addition, a systematic and methodical approach for auditing and evaluating various building services and the overall building smartness is required.

## 5.2 IBACSA Interactive Tool

IBACSA 'Interactive Tool for Building Automation and Control Systems Auditing' was developed in a previous study (Engelsgaard et al. 2020) to address the missing link between the regulatory standards and legislation on one hand and the actual design of current and future BACS components and the corresponding functionalities on the other hand. Thus, IBACSA provides a first-of-its-kind instrument for BACS assessment and building smartness evaluation. IBACSA was developed and launched to serve as a basis for initial and retro commissioning of the BACS in a large set of different buildings and facilities. In its development, the tool is driven by the requirements set by the European Standard EN15232 (2012) for building automation and control, but with a more holistic approach in terms of the impact criteria considered. The tool was developed under the research project BuildCOM—Automated Auditing and Continuous Commissioning of Next Generation Building Management Systems (Jradi et al. 2021), supported by the Danish Energy Agency.

In terms of the auditing and evaluation methodology, IBACSA employs a hybrid qualitative-quantitative multi-criteria holistic framework, where it considers and evaluates eight major building domains: Heating, Hot Water, Cooling, Ventilation, Lighting, Dynamic Envelope, Electricity and Monitoring and Control. Each of these domains comprises a set of different services as defined by the European Standard EN15232. The sum of all services under all domains is 60, where each service is characterized by different control and functionality levels. Table 5.1 provides an example of the structure of one of the heating domain services, the 'Heat Supply Control' service. This service, for example, has five progressive levels of functionality and control as highlighted in the table, and a higher level of control will imply a higher degree of service intelligence and smartness. As highlighted in Table 5.2, a points-based grading score is employed to quantify the impact of each functionality. A service with a higher functionality level will lead to more points accrued.

**Table 5.1** Heat supply control service list of functionality levels

| Service | Functionality level | Description |
|---|---|---|
| Heat supply control | 0 | No automatic control |
| | 1 | Central automatic control |
| | 2 | Individual room control |
| | 3 | Level 2 + communication between controllers |
| | 4 | Level 3 + presence control |

Table 5.2 'Energy efficiency' and 'comfort' scores of the heat supply control

| Heat supply control functionality level | Energy efficiency | Comfort |
|---|---|---|
| Level 0 | 0 | 0 |
| Level 1 | 1 | 1 |
| Level 2 | 2 | 2 |
| Level 3 | 2 | 2 |
| Level 4 | 3 | 2 |

## 5.3 IBACSA BACS Evaluation Criteria

When auditing the BACS employing IBACSA, the assessor will evaluate 60 services by inserting information on the right functionality level corresponding to each service under all domains. IBACSA evaluates the BACS and quantifies the impact of the selected control levels in each of the eight domains considered against five impact criteria: (1) Energy efficiency, (2) Maintenance and fault prediction, (3) Energy flexibility, (4) Comfort and (5) Information to occupants. The selection of the five impact criteria is inspired by two previously developed BACS evaluation schemes, the euBAC scheme (Schönenberger 2015) and the Smart Readiness Indicator—SRI scheme (Verbeke et al. 2018). While euBAC only considered one impact criterion in the evaluation of the building, which is the 'Energy Efficiency' criterion, IBACSA moves a step further and considers four other impact criteria. On the other hand, while SRI scheme considers some of the other impact criteria as well in evaluating the BACS functionalities, it reports the results considering one weighted average score, which can't be used as a basis to evaluate the single technical domains. Thus, some of the poor functionalities are hidden under this overall score. IBACSA, however, assesses each of the eight technical domains individually and reports the evaluation and impact for each of the five criteria. link between the regulatory standards and legislation on one hand and the actual design of current and future BACS.

The IBACSA evaluation matrix is shown in Fig. 5.1, where the overall approach is to evaluate each domain by itself against each of the impact criteria, allowing identification of the poorly performing domains. In addition to the overall score provided for each domain, shown in the vertical line to the right, the matrix provides an assessment of the BACS with one overall score against each of the five impact criteria, shown in the horizontal line at the bottom. This provides the auditor with the flexibility to consider different evaluation impacts depending on the application and the interest. IBACSA is developed as a standalone executable application, with a user-friendly interface and clear graphical elements and tabs. Figure 5.2 depicts the 'Results Summary' tab showing the scores obtained in each of the single eight domains. The results tab also shows scores for each domain against each of the impact criteria, along with an overall score for each impact criteria by itself. In addition, IBACSA aids the user in the decision-making process in terms of retrofitting the current BACS, allowing simple and fast evaluation and comparison of the impacts of BACS functionalities selected.

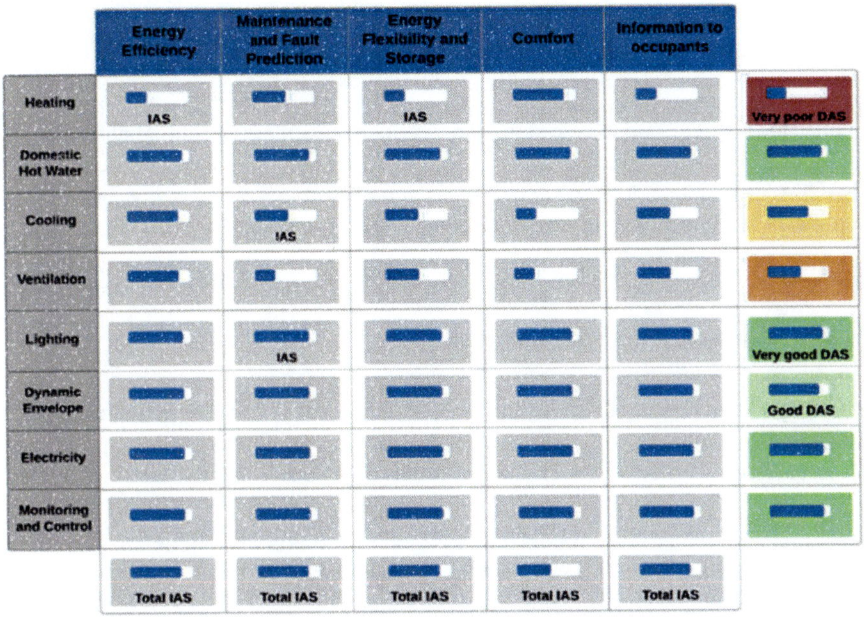


Fig. 5.1 IBACSA BACS evaluation matrix (Engelsgaard et al. 2020)

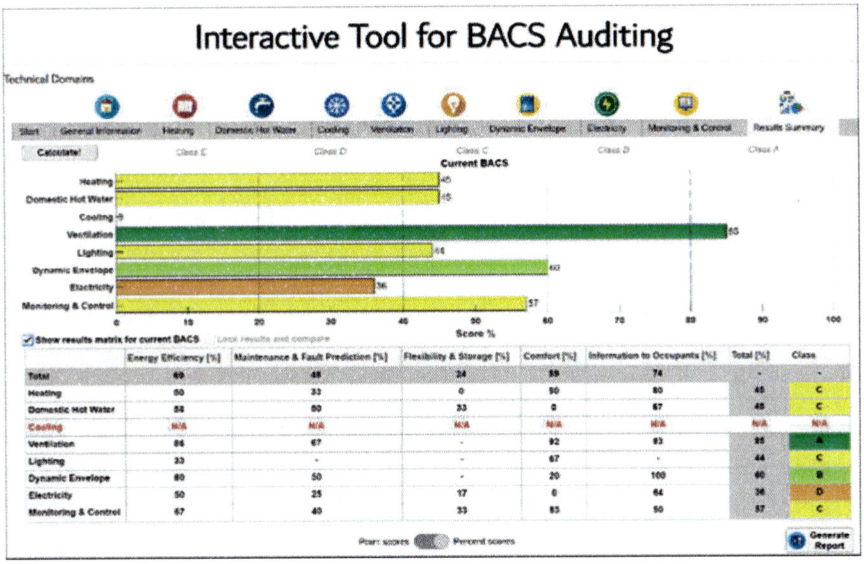


Fig. 5.2 An overview of the 'Results Summary' generated by IBACSA

## 5.4 Building Case Studies and Corresponding Technical Systems Details

In this study, three Danish case study buildings are considered to assess and evaluate the BACS design functionalities and smartness levels using the recently developed IBACSA tool. The portfolio of buildings considered covers an office building, a healthcare lab building, and a hospital patient ward building.

### *5.4.1 Building A Specifications*

Building A is an office university building, built and opened in 1995, comprising two floors and a small basement. It is mainly used now as an office building by staff and researchers on a daily basis. The building has 110 zones, mainly personal offices, in addition to some meeting rooms and laboratories. An in-direct district heating system is used to provide the demands of space heating and domestic hot water. In addition, radiators with thermostatic valves are used for water supply control. A central automatic control is used for heating supply temperature, and the distribution pumps follows an on/off control. In terms of ventilation, the building relies mainly on natural ventilation to provide fresh air to the different spaces through controlled operable windows. In addition, a main mechanical ventilation system is employed to serve some special rooms, kitchens, and laboratories. The ventilation system is not equipped with any overheating or icing protection. The supply flow rate is controlled based on a set time schedule. The building has no cooling system and relies only on the ventilation system to provide proper indoor thermal comfort in summer.

The building has a conventional lighting system with a mix of halogen lights and some LED tubes. The lighting system is controlled using daylight sensors and motion sensors on the level of most of the rooms. Blinds in the building are automatically controlled and integrated with the overall HVAC system control. In terms of the building metering and sensing infrastructure, the building is equipped with electricity and heating meters on the whole building level, but also on the level of each of the energy supply components, including domestic hot water, space heating, ventilation, and lighting. However, the electricity and heating metering are not deeply sub-metered beyond the large systems and cover less than 50% of the loads. The building has a brand-new building automation system for systems and services operational management. The building is also equipped with occupancy counters at the two entrances to track the number of people. It also has a weather station mounted on the roof.

### *5.4.2 Building B Specifications*

Building B is a healthcare public building that was initially constructed in 2000 and exhibited a deep energy retrofit process in 2021. The building is used primarily as a blood donor center, with multiple laboratories, offices, treatment units, and screening rooms. As part of the retrofitting process, the whole building interior envelope was upgraded. A brand-new cooling system along with four new mechanical ventilation units were installed. In addition, a new building management system is in place, allowing advanced monitoring, control, and management of building operation on various levels. The majority of the information on the building energy systems and technical services was provided by the technical manager. In addition, multiple interviews and field visits were carried out to collect additional information from building users. The space heating and domestic hot water requirements are met by an indirect district heating system. The heating supply temperature is controlled on the level of each room with direct communication between individual controllers on thermostatic valves and the central building automation system. There is also a thermally activated building system with advanced automatic control, self-regulating the room temperatures based on the principle of low heating demand and higher comfort level. The centralized heating setpoint follows the ambient air temperature change, and the associated pumps have variable speed operation based on demand. Submetering on the heating consumption within the building is provided for over 50% of the relevant heating loads, and capabilities for fault detection and preventive maintenance are also provided. There is no domestic hot water storage onsite.

The building has two electrical cooling chillers, and the cooling supply is controlled at a room level, considering the ambient air temperature. The cooling system operation is time-scheduled, and the distribution pumps have variable speed operation. In terms of ventilation needs, four new balanced mechanical ventilation systems are installed, controlled by occupancy detection sensors and the corresponding $CO_2$ level in each room. The supply flow rate is automatically controlled based on pressure, and the units are equipped with icing and overheating protection. The setpoint for supply air temperature depends on the outside temperature. A night ventilation mode is also included in the building management system. Indoor air quality and thermal comfort conditions are reported in real-time at a room level. In terms of lighting control, the building has both automatic motion detection and daylight sensors. The building also has motorized blind operation with window opening position detection sensors. On the other hand, the building has no onsite energy generation systems due to the limited roof space. The electricity consumption in the building is very well sub-metered, covering more than 50% of the installed capacity. The operation time for the HVAC system is based on dynamic demand in addition to a predictive control based on external grid signals. A fault detection and diagnostics system is integrated for preventive maintenance. The building has no smart grid integration.

### *5.4.3 Building C Specifications*

Building C is a newly built hospital building that opened in 2020 and since then has been in use to serve patients. The building has four floors and comprises mainly patient rooms, staff rooms, and technical rooms. It was built in compliance with the newest building regulations in Denmark, the low energy class 2020 with indoor climate class A. Most of the information on the technical systems was collected from documents and digital design files. The rest of the information was provided by technical managers and users through interviews and field visits. The building's heating demand is fulfilled by a district heating system with thermostatic radiators controlled at the level of each room. The hot water distribution pumps are multi-stage with automated on/off operation. The heat supply temperature follows the ambient air temperature. In addition, the building has multiple electric chillers providing cooling to the different rooms, and the supply temperature is demand-controlled. In terms of ventilation, the building has a balanced mechanical ventilation system with multiple units, with automatic pressure control in the air handling units, controlled on a room-level, with icing and overheating protection. Energy consumption and indoor air quality reporting is in real-time, and the operation monitoring is also equipped with fault capabilities. The lighting system is controlled by motion sensors in all the rooms and by daylight sensors in some specific rooms.

The building has automatic blind control combined with the HVAC control, with no automatic operable windows and no reporting on the opening position. The building is also equipped with a PV system situated on the roof for onsite power generation, and the excess production is fed into the grid. There is real-time monitoring of the generation/performance. The building also has EV charging possibilities. Electricity sub-metering is provided on the level of each building system and zone with over 50% coverage of the electric loads. Occupancy detection devices are used to control multiple building functions, including the lighting system and ventilation supply. The heating and cooling setpoints are managed from a central platform, taking into account the periods of no occupancy presence.

## 5.5 BACS Auditing in Case Study Buildings

### *5.5.1 Building a BACS Auditing*

Generally, the results reported by the IBACSA tool for the BACS auditing and evaluation for Building A are not very encouraging and highlight a poor BACS design and limited functionalities in the majority of the services, as shown in Fig. 5.3. In terms of the overall rating, the best performing domain is the lighting system domain, with the building having motion sensors and daylight sensors equipped on a room level. Thus, the lighting domain scores 67%. All the other technical domains score lower than 31%, with the ventilation system domain scoring only 11%, the domestic

| | Energy Efficiency [%] | Maintenance & Fault Prediction [%] | Flexibility & Storage [%] | Comfort [%] | Information to Occupants [%] | Total [%] | Class [%] |
|---|---|---|---|---|---|---|---|
| Total | 29 | 16 | 10 | 26 | 25 | - | - |
| Heating | 30 | 33 | 0 | 38 | 60 | 31 | D |
| Domestic Hot Water | 33 | 0 | 11 | 0 | 0 | 17 | E |
| Cooling | N/A | N/A | N/A | N/A | N/A | N/A | N/A |
| Ventilation | 14 | 0 | - | 15 | 0 | 11 | E |
| Lighting | 67 | - | - | 67 | - | 67 | B |
| Dynamic Envelope | 0 | 0 | - | 0 | 0 | 0 | E |
| Electricity | 29 | 0 | 0 | 0 | 38 | 21 | D |
| Monitoring & Control | 22 | 40 | 17 | 50 | 25 | 30 | D |

**Fig. 5.3** Building A automation and control system evaluation results

hot water system with 17% and the Electricity domain with a cumulative score of 21%. The reason for such a low score for the ventilation domain is that the current mechanical ventilation system has a very basic control pattern, with a scheduled operation time and constant air flow rates and operational settings. To improve the performance of this domain, a more dynamic control approach is needed for proper management of the air flow as well as air temperature. The impact criterion with the lowest score in evaluating the building technical domains is Flexibility and Storage, with only a 10% overall score. In addition, all the rest of the impact criteria score relatively low, with Energy Efficiency and Comfort being the two highest scoring impact criteria, with only 29% and 26%, respectively. Overall, it is very evident from the building automation and control system auditing in Building A that the design of this BACS doesn't comply with the lowest acceptable automation standard and there is an urgent need for a deep retrofit for the technical systems, in addition to the corresponding control functionalities services.

### *5.5.2 Building B BACS Auditing*

It is evident in Fig. 5.4 that Building B scores relatively high in most of the domains, with some exceptions. The domains with the highest scores were lighting with 100%, ventilation with 89% and monitoring and control with 89%, with all three being classified as Class A. The worst scoring technical domains are the electricity domain with 17%, domestic hot water with 38%, and dynamic envelope with 47%. As the building was recently energy retrofitted, including the installation of a new BACS platform, it is not a surprise to see that the majority of the domains score high, including heating, lighting, and ventilation. Furthermore, the monitoring and control domain also scored 89. On the other hand, the reason for the electricity domain's low score (17%) is the lack of onsite electricity generation and storage and the limitation of the connection to the grid. As shown in the performance matrix for Building B, the building domains lead to positive impacts with high scores in the majority of the criteria considered, with an average impact score of between 72 and 80%. Moreover, 4 out of the 8 building domains considered score the full mark (100%) with respect to Information to Occupants and Maintenance and Fault Prediction criteria, while the lighting and monitoring and control domains score 100% against the Comfort

| | Energy Efficiency [%] | Maintenance & Fault Prediction [%] | Flexibility & Storage [%] | Comfort [%] | Information to Occupants [%] | Total [%] | Class |
|---|---|---|---|---|---|---|---|
| Total | 73 | 80 | 17 | 72 | 73 | - | - |
| Heating | 78 | 100 | 0 | 80 | 100 | 73 | B |
| Domestic Hot Water | 42 | 100 | 0 | 33 | 100 | 38 | D |
| Cooling | 68 | 75 | 20 | 67 | 83 | 61 | B |
| Ventilation | 81 | 100 | - | 92 | 100 | 89 | A |
| Lighting | 100 | - | - | 100 | - | 100 | A |
| Dynamic Envelope | 88 | 0 | - | 33 | 0 | 47 | C |
| Electricity | 29 | 0 | 0 | 0 | 25 | 17 | E |
| Monitoring & Control | 93 | 100 | 57 | 100 | 100 | 89 | A |

**Fig. 5.4** Building B automation and control system evaluation results

criteria. Thus, it is noted that the evaluation results provided by IBASA for the five different impact criteria are very encouraging, with an acceptable good indoor air quality and thermal control. This comes except for the Flexibility and Storage criteria, with a reported score of 17%. The low score in this domain is a main demonstration that the absence of any connection to the grid and the absence of power generation onsite lead to low flexibility and storage capabilities.

### 5.5.3 *Building C BACS Auditing*

Based on the evaluation results reported in Fig. 5.5 for Building C, it is noted that the ratings and scores associated with the majority of the technical domains are acceptable, considering that the building is still relatively new, opened in 2020, and equipped with a brand-new building management system. According to IBACSA findings, the dynamic envelope (26%), domestic hot water (34%), and heating (47%), are the lowest-scoring technical domains in the building. Regarding the dynamic envelope, the low score could be enhanced by employing automated operable windows with sensors to monitor the window opening position. Employing smart windows could be another option. However, none of these options were feasible in the case of this specific building, as informed by the technical manager, mainly due to the solution's high cost and the corresponding economic impacts.

Nevertheless, the domestic hot water score is relatively low, mainly due to the absence of onsite storage, which was also claimed to be a no-go by the manager due to healthcare legislation related to legionella bacteria. The best performing domain is the ventilation system, scoring 91%, followed by the monitoring and control domain with

| | Energy Efficiency [%] | Maintenance & Fault Prediction [%] | Flexibility & Storage [%] | Comfort [%] | Information to Occupants [%] | Total [%] | Class |
|---|---|---|---|---|---|---|---|
| Total | 68 | 56 | 27 | 71 | 68 | - | - |
| Heating | 57 | 50 | 0 | 60 | 83 | 53 | C |
| Domestic Hot Water | 42 | 50 | 0 | 33 | 100 | 34 | D |
| Cooling | 61 | 75 | 40 | 67 | 83 | 61 | B |
| Ventilation | 100 | 83 | - | 92 | 67 | 91 | A |
| Lighting | 67 | - | - | 100 | - | 78 | B |
| Dynamic Envelope | 38 | 0 | - | 33 | 0 | 26 | D |
| Electricity | 80 | 50 | 42 | 80 | 64 | 62 | B |
| Monitoring & Control | 86 | 40 | 43 | 86 | 60 | 68 | B |

**Fig. 5.5** Building C automation and control system evaluation results

68%, and the electricity and the cooling domains, both scoring 62%. The performance matrix provided for Building C highlights that the domains have acceptable scores against the majority of the five impact criteria, with an average impact reported ranging between 55 and 71%. The lighting system attained a score of 100% for Comfort, but the standout performer is the ventilation domain, scoring high in all criteria, including 100% for Energy Efficiency and 92% for Comfort. The criteria scoring the lowest against the domains are the Flexibility and Storage, with 27% as a cumulative score. Saying that, some of the technical domains attain an 'acceptable' score relative to Flexibility and Storage, including the electricity domain (42%) and the monitoring and control (43%).

## 5.6 Conclusion

IBACSA, an innovative tool for building automation systems auditing and evaluation, was used in this study to evaluate and assess the building automation and control systems design and functionalities in three case study buildings in Denmark. The technical domains and the associated services are evaluated against five impact criteria, including energy efficiency, flexibility, and comfort. The results show that there are large differences in the control and functionality levels of the BACS in the considered buildings, mainly depending on the building age, type, services, and technical systems. It was shown that the ventilation system domain scored relatively high in the considered three buildings, highlighting the importance of this domain in Danish buildings to attain proper indoor air quality and thermal comfort. On the other hand, the worst scoring criterion in the considered buildings was Flexibility and Storage, with the absence of energy storage onsite and the lack of connection to the grid being two major reasons. With the provided capabilities, IBACSA serves as a potential instrument for BACS initial and retro commissioning processes, aiding in reducing performance gaps.

**Acknowledgements** This work was supported by the 'BuildCOM: Automated Auditing and Continuous Commissioning of Next Generation Building Management Systems' project, funded by the Danish Energy Agency under the Energy Technology Development and Demonstration Program (EUDP), ID no. (64019-0081).

## References

Chauhan RK, Chauhan K (2019) Building automation system for grid-connected home to optimize energy consumption and electricity bill. J Build Eng 21(2019):409–420

Dascalaki EG, Argiropoulou P, Balaras CA, Droutsa KG, Kontoyiannidis S (2021) Analysis of the embodied energy of construction materials in the life cycle assessment of Hellenic residential buildings. Energy Build 232(2021):110651

Domingues P, Carreira P, Vieira R, Kastner W (2016) Computer standards & interfaces building automation systems: concepts and technology review. Comput Stand Interf 45(2016):1–12

Engelsgaard S, Alexandersen EK, Dallaire J, Jradi M (2020) IBACSA: an interactive tool for building automation and control systems auditing and smartness evaluation. Build Environ 184(2020):107240

Engvang JA, Jradi M (2021) Auditing and design evaluation of building automation and control systems based on eu.bac system audit—Danish case study. Energy Built Environ 2(2021):34–44

European Technical Standard EN 15232 (2012) Energy performance of buildings-impact of building automation, control, and building management, 2nd edn. CEN, Brussels

Jradi M, Boel N, Madsen BE, Jacobsen J, Hooge JS, Kildelund L (2021) BuildCOM: automated auditing and continuous commissioning of next generation building management systems. Energy Inform 4(2021):2

Li W, Wang S (2022) A fully distributed optimal control approach for multi-zone dedicated outdoor air systems to be implemented in IoT-enabled building automation networks. Appl Energy 308(2022):118408

Mirnaghi MS, Haghighat F (2020) Fault detection and diagnosis of large-scale HVAC systems in buildings using data-driven methods: a comprehensive review. Energy Build 229(2020):110492

O'Grady T, Chong H-Y, Morrison GM (2021) A systematic review and meta-analysis of building automation systems. Build Environ 195(2021):107770

Schönenberger P (2015) eu.bac system. Energy Build 100(2015):16–19

Verbeke S, Ma Y, Van Tichelen P, Bogaert S, Waide P, Uslar M, Schulte J (2018) Support for setting up a smart readiness indicator for building and related impact assessment final report. VITO NV, Brussels

# Chapter 6
# Influence of Concrete Composition on the Carbon Footprint and Embodied Energy of a Frame Structure

**Mariana Cardoso, Teresa M. Mata, Helena Monteiro, Humberto Varum, and António A. Martins**

**Abstract** The construction sector is responsible for a significant part of global greenhouse gas emissions, with concrete being one of the most used building materials, which significantly affects the overall sustainability of the built environment. Thus, it is necessary to reduce its embodied energy and carbon emissions. For this purpose, a regular frame structure is analyzed in this work, for estimating the energy and carbon emissions of three different concrete formulations (C25/30, C25/30 with 30% recycled aggregates, and C35/45), following the guidelines defined in the EN 15084 and ISO 21930 standards, and also the Life Cycle Assessment methodology. To simplify the comparison between the concrete formulations, the chosen functional unit is the structural frame of the building, which is evaluated in this work following a "cradle-to-gate" approach. The inventory data was obtained by designing the frame structure,

M. Cardoso · H. Varum
CONSTRUCT-LESE, Faculty of Engineering, University of Porto (FEUP), R. Dr. Roberto Frias, s/n, 4200-465 Porto, Portugal

T. M. Mata (✉)
LAETA-INEGI, Associated Laboratory for Energy and Aeronautics, Institute of Science and Innovation in Mechanical and Industrial Engineering, R. Dr. Roberto Frias 400, 4200-465 Porto, Portugal
e-mail: tmata@inegi.up.pt

H. Monteiro
Low Carbon and Resource Efficiency, R&Di, Instituto de Soldadura e Qualidade, 4415-491 Grijó, Portugal
e-mail: himonteiro@isq.pt

A. A. Martins
LEPABE, Faculty of Engineering, University of Porto (FEUP), R. Dr. Roberto Frias, s/n, 4200-465 Porto, Portugal

ALiCE, Faculty of Engineering, University of Porto, R. Dr. Roberto Frias, 4200-465 Porto, Portugal

A. A. Martins
e-mail: aamartins@fe.up.pt

 
N. S. Caetano and M. C. Felgueiras (eds.), *The 9th International Conference on Energy and Environment Research*, Environmental Science and Engineering,
https://doi.org/10.1007/978-3-031-43559-1_6

ensuring that it fulfills the structural safety requirements according to Eurocodes. Results show that C35/45 concrete has a significant carbon footprint of 307.62 kg $CO_2$ eq per $m^3$ and the values for C35/45 concrete, with and without 30% RCA, are respectively 11.4 and 10.3% lower. The structure built using C25/30 concrete with 30% RCA contributes 4.35% less to carbon footprint, corresponding to a reduction of 356.09 kg $CO_2$ eq, than the structure built with C35/45 concrete. Concerning the embodied energy, results show that the C25/30 concrete has a lower embodied energy of 65,900 MJ-eq/FU than the C35/45 concrete with 68,134 MJ-eq/FU to which, reinforced steel contributes respectively, 67 and 64%.

**Keywords** Building structural frame · Concrete formulation · Life cycle assessment

## 6.1 Introduction

According to the International Energy Agency and the United Nations Environment Programme (IEA and UNEP 2019), buildings and construction accounted for 39% of energy and process-related greenhouse gas (GHG) emissions in 2018. Among them, production of construction materials, as for example steel, cement and glass, accounted for around 11% of the global GHG emissions, most of them as result of direct energy consumption. It is well known that concrete is the most widely used material in the construction industry and the second most used material in the world, after water (Watts 2019). Concrete is build-up of different components, of which aggregates constitute a large part. As concrete production increases, so do concerns about the associated environmental impacts and about the exploitation of natural resources needed for its production, in order to ensure access to all the necessary components. One of the most widely adopted methods for reducing the impacts on the environment is the use of construction and demolition waste as a recycled source of raw materials, thus making it possible to control and reduce some of the environmental impacts created by its disposal in landfill sites and the exploitation of new virgin materials. Production of concrete contributes about 4–8% to global GHG emissions. Half of this amount results from the production of clinker, the most energy-intensive part of the cement production process, which is one of the major constituents of concrete (Watts 2019). On the other hand, recycled aggregates (RA) extracted for example, from the construction and demolition waste, can be used for concrete production in the construction sector, with lower carbon footprint, when compared to cement production (Weil et al. 2006). Aggregates production accounts for about 13–20% of total $CO_2$ emissions from concrete. Also, the use of RA to replace natural aggregates (NA), such as fine (sand) and coarse (gravel), in concrete production reduces the consumption of primary resources by up to 44% (Pavlu et al. 2019). Yet, the reduction of a building carbon footprint due to the utilization of RA also depends on the construction itself, in particular the relative amounts of other materials.

Due to the high consumption of raw materials and energy, the construction sector has significant impacts (IEA and UNEP 2019). Thus, to increase its sustainability, it is fundamental to assess its environmental impact, in order to define measures and/or policies to improve its environmental performance. It is consensual that these assessments should be made following a Life Cycle Thinking (LCT) approach (Tártaro et al. 2017), considering all the activities and materials associated with the life cycle of a product or a process (Rodrigues et al. 2018). The evaluation itself of the environmental impacts is usually made using the Life Cycle Assessment (LCA) methodology, that allows to determinate the impacts of a process system (material, product, building, etc.) or service, throughout the various stages of its life (Rodrigues et al. 2018). The LCA methodology is established by ISO 14040 (2006a) and ISO 14044 (2006b) standards, defining a common framework that enables, for example, the comparison between different materials (Rodrigues et al. 2022). Also, it allows the identification of the process hotspots in terms of environmental impacts, and which improvement measures should be considered first. Hence, in this work the embodied energy and carbon footprint of a model building structure are evaluated, comparing different concrete formulations, and the potential inclusion of RA.

## 6.2 Materials and Methods

This study follows the LCA methodology that involves four sequential phases (ISO 2006a): (i) Definition of the objective and scope of the study, (ii) life cycle inventory (LCI) analysis, (iii) Life cycle impact assessment (LCIA) and (iv) Interpretation.

### *6.2.1 Goal and Scope Definition*

**Study Goal**. The goal of this study is to evaluate the embodied energy and carbon footprint of a model building structure (shown in Fig. 6.1a) consisting of linear structural elements (columns and beams) that ensure the building safety. In particular, the study focuses on the reinforced concrete frame structure (Fig. 6.1b), comparing the following three concrete formulations:

1. Lower resistance conventional (C25/30),
2. Lower resistance conventional concrete (C25/30), with incorporation of 30% of recycled concrete aggregates (RCA), and
3. Higher resistance conventional concrete (C35/45).

Even though the structure is simple, it is a good approximation of a typical five storey building, commonly found in cities around the globe for residential or office use. Thus, this LCA study results may be helpful to inform and support decision making on design and construction of future real buildings.

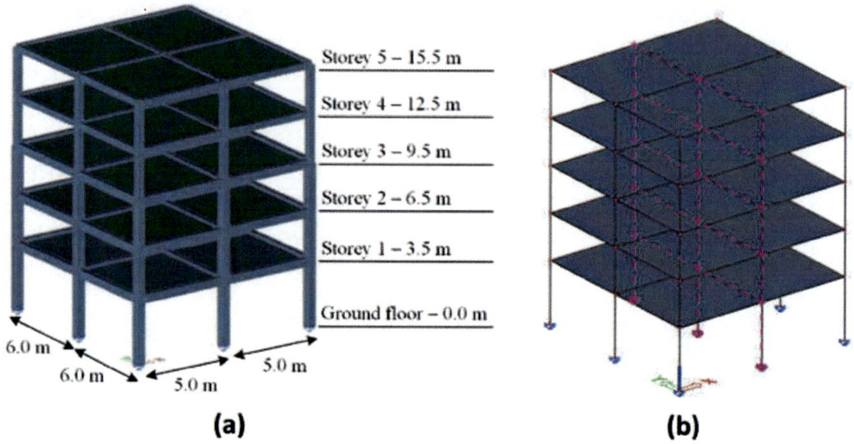


**Fig. 6.1** **a** Building structure and **b** central frame (in pink color), considered in this LCA study

**Functional Unit**. The functional unit (FU) chosen for this study is defined as the central frame structure of the building (as shown in Fig. 6.1b in pink color). The reason to select the central frame structure as FU is to facilitate the data management and the comparison between the carbon footprints of the three concrete formulations. Even though the FU is the dimensioned central frame structure, the quantities of materials (concrete blend and steel reinforcement) are considered in terms of mass or volume depending on the material type.

**Study Scope**. This LCA study follows a "cradle-to-gate" approach, i.e. it considers the life cycle steps from the extraction of raw materials, to the production of construction materials, which are mainly concrete and steel in this case. Since the study aims to compare three concrete formulations (C25/30, C25/30 with 30% recycled aggregates, and C35/45), the construction of the building structure was not considered, as the same environmental impacts are assumed, regardless of the concrete formulation used. Moreover, the materials transportation to the construction site, and the building use and demolition phases, are assumed to have similar impacts regardless of the concrete formulation used. This means that the analysis focuses on the raw materials extraction and concrete and steel production for constructing the central frame of the building structure, comparing the three concrete formulations described above. Thus, the system boundary can be represented as shown in Fig. 6.2.

This LCA study is attributive, as the environmental impacts are directly attributed to the functional unit. For the inventory data, European conditions were considered whenever possible, in particular for the energy supply and production of raw and construction materials. Whenever possible, information and data valid for regions with similar technologic and economic development were used. Regarding time coverage, current conditions and state-of-the-art technologies were considered, and the results are valid for a time horizon of at least 10 years. The target audience include

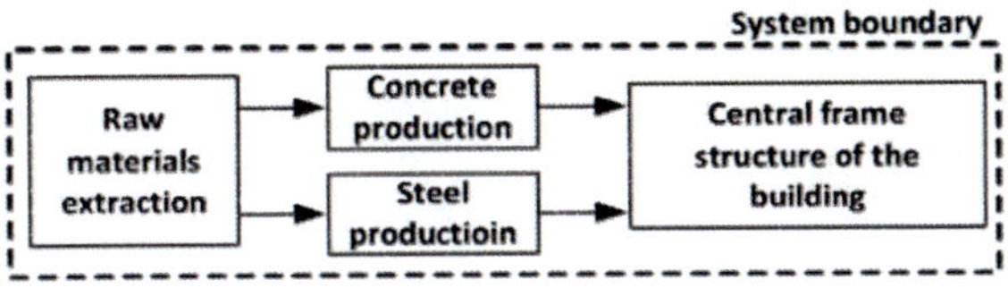


**Fig. 6.2** System boundary considered for the LCA study, including the raw materials extraction and the concrete and steel production, used for the construction of the central frame structure of the building

the construction professionals and companies, and the policy makers, interested to assess the potential relevance of recycled aggregates to reduce the carbon footprint and embodied energy in the building sector.

### 6.2.2 *Life Cycle Inventory Analysis*

Considering the study assumptions, the relevant inventory items are the concrete and steel, which are necessary to construct the central frame of the building structure. As the study considers a virtual simple model structure, yet representative of real buildings, it is not possible to use real/primary data. Thus, in order to obtain the inventory data, the structure was dimensioned for vertical loads, according to existing construction regulations, in particular the Eurocode, to ensure that the structure analyzed is close to a real building. Due to space limitations and the fact that the issues addressed in the dimensioning are outside of this study scope, no details of the procedure are given in this work. More information on how it was modelled can be found in the work of Cardoso (2022).

The life cycle inventory data for the production of steel and conventional concrete with higher resistance was obtained from the EcoInvent life cycle inventory database V3.8, integrated in Simapro V9.3. For the conventional concrete, with lower resistance and 30% substitution of natural aggregates by RCA, the process was not so straightforward. In this case, inventory data was obtained from the literature and Environmental Product Declaration (EPD) (Argent Materials Inc 2021), concerning the emission values related to the impacts of recycled concrete aggregates. For C25/30 with 30% recycled aggregates, the EPD chosen (Argent Materials Inc 2021) refers to an American product produced with 100% renewable energy mix, consisting of 50% solar and 50% wind power, with an emission factor of 0.0211 kg $CO_2$ eq/kWh.

### 6.2.3 *Life Cycle Impact Assessment (LCIA)*

The environmental impact assessment was based on the standards and regulations aimed specifically to the construction sector, in particular, the standards EN 15804

**Table 6.1** Inventory analysis for C35/45 and C25/30 concrete types in the central frame structure designed for vertical loads

| Concrete type | Total concrete volume of each structural member ($m^3$/FU) | | | Total weight of steel reinforcement in structural members (ton/FU) | | | | |
|---|---|---|---|---|---|---|---|---|
| | Columns | Beams | Total | Columns | | Beams | | Total |
| | | | | Longitudinal | Transversal | Longitudinal | Transversal | |
| C25/30 | 5.51 | 7.2 | 12.71 | 0.55 | 0.11 | 1.12 | 0.17 | 1.95 |
| C35/45 | 5.51 | 7.2 | 12.71 | 0.55 | 0.11 | 1.06 | 0.17 | 1.91 |

(EN 2013) and ISO 21930 (2007), which define the relevant potential environmental impact categories and the evaluation methodologies. In particular this study aims to focus on the carbon footprint and embodied energy. Regarding the carbon footprint, these standards consider three types of GHG emissions: fossil, biogenic, and land use/land use change, all expressed in kg $CO_2$ eq.

## 6.3 Results and Discussion

### *6.3.1 Inventory Analysis*

Table 6.1 shows the inventory data for the lower resistance concrete formulation (C25/30) and the higher resistance concrete formulation (C35/45) in the central frame structure designed for vertical loads. As expected, the inventory data for both concrete formulations are similar, as they need to meet certain criteria in order to be used in the construction of buildings. The inventory results for concrete C25/30 with 30% RCA (not shown in Table 6.1) are similar to those of concrete C25/30.

### *6.3.2 Carbon Footprint and Embodied Energy Evaluation*

Considering the utilization of higher resistance concrete (C35/45) as the reference scenario, Fig. 6.3a compares the contribution to carbon footprint of the three concrete formulations, normalized in relation to the reference scenario (C35/45). Considering the inventory analysis (in Table 6.1) and the cumulative energy demand of concrete (C25/30, C25/30) and of reinforced steel, obtained from the EcoInvent database, the embodied energy was estimated for the conventional concrete formulations and for the reinforced steel needed to be incorporated in each concrete, as shown in Fig. 6.3b.

Concerning the carbon footprint (Fig. 6.3a), results show that the lower resistance concrete (C25/30) has a lower contribution to carbon footprint than the higher resistance C35/45 concrete. The differences are respectively 10.3 and 11.4% for C25/

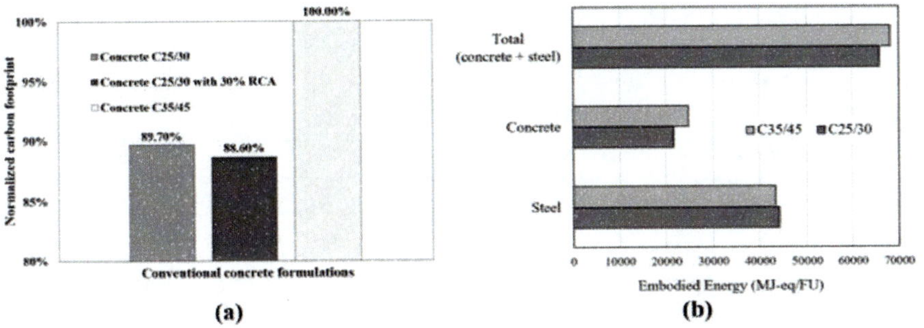


**Fig. 6.3** Comparison of concrete formulations in terms of: **a** Carbon footprint normalized in relation to the reference scenario (concrete C35/45), and **b** Embodied energy of concrete (C25/30 and C25/30) and of rein-forced steel needed for each concrete

30 concrete without and with incorporation of RCA in the formulation. Although the differences are small, as C35/45 concrete has a significant carbon footprint (of 307.62 kg $CO_2$ eq per $m^3$) even a small reduction can have a significant impact in reducing the overall carbon emissions. Also, although lower quantities of the conventional higher resistance C35/45 concrete are used, as the increase in the carbon footprint is much larger than the reduction in the quantity used, it can be concluded that using C35/45 leads to higher carbon emissions. Cement is the largest contributor (around 90%) to carbon footprint, either in C25/30 and C35/45 concretes, mainly due to energy consumption, followed by sand and gravel (around 7%) used as natural aggregates. Typically, to produce one ton of cement about 4 GJ of energy is required in a well-equipped cement plant (Cantini et al. 2021). Regarding the inclusion of 30% RCA in the lower resistance C25/30 concrete, it will reduce the individual impact of gravel. Concerning the embodied energy (Fig. 6.3b), results show that the lower resistance C25/30 concrete has a lower embodied energy (of about 65,900 and 21,689 MJ-eq/FU respectively, with and without the steel incorporated), than the higher resistance C35/45 concrete (of about 68,134 and 24,830 MJ-eq/FU respectively, with and without the steel incorporated. This is even considering that low strength concrete needs more steel than high strength concrete. The steel represents 67% of the embodied energy in the case of the lower resistance concrete and 64% of the energy in the case of the higher resistance concrete.

Figure 6.4a presents the carbon footprint of the building structure designed for vertical loads, normalized in relation to the reference concrete C35/45, comparing two concrete formulations: C25/30 lower resistance concrete with 30% RCA and C35/45 higher resistance concrete. Figure 6.4b presents the steel and concrete's relative contributions to carbon footprint, in the overall frame structure designed for vertical loads, comparing two concrete formulations (C25/30 lower resistance concrete with 30% RCA and C35/45 higher resistance concrete). Figure 6.4a shows that the structure built with C25/30 concrete with 30% RCA contributes 4.35% less to the carbon footprint (corresponding to a reduction of 356.09 kg $CO_2$ eq) than the frame structure built with C35/45 concrete. Although this is a small percentual reduction, as the

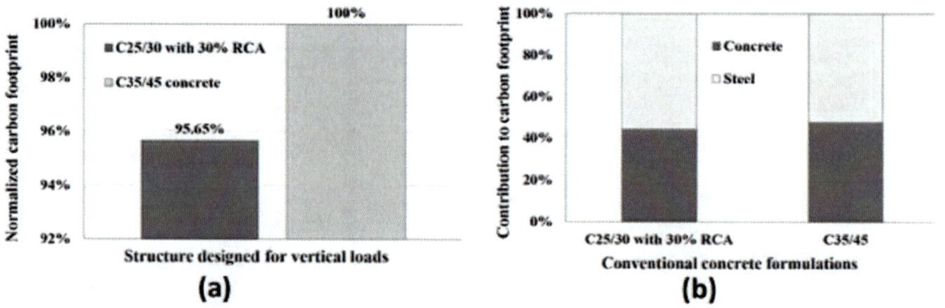


**Fig. 6.4** **a** Carbon footprint of the building structure designed for vertical loads, normalized in relation to the reference concrete C35/45; and **b** Steel and concrete's relative contributions to carbon footprint, in the structure designed for vertical loads, comparing concrete C25/30 with 30% RCA and C35/45 concrete

magnitude of the carbon footprint of the structure with C35/45 concrete is very large (8186.043 kg $CO_2$ eq), small improvements will have a significant impact.

The relative contribution to carbon footprint of concrete and steel used to build the structure is similar for both concrete formulations (C25/30 concrete with 30% RCA and C35/45 concrete). This result was expected, as the quantities of materials used for both cases are similar according to the inventory data presented in Table 6.1. For the higher resistance concrete C35/45, a smaller quantity of steel is needed, as it has better resistance characteristics when compared to concrete C25/30, either with or without 30% RCA.

## 6.4 Conclusions

This work analyzed the influence in the carbon footprint and embodied energy of a central frame structure of a building, of using different concrete formulations. The LCA methodology was followed and a combination of tools was used, in particular design software, for determining the sizes of sections and quantities of materials for the model structure. The results show that concrete C25/30 with 30% RCA can reduce about 4.35% the carbon footprint in relation to C35/45 concrete, thus contributing to minimize the building environmental impact. Although this is a small percentage reduction, as the magnitude of the structure's carbon footprint is very large, small improvements may have a significant impact. Concrete C25/30 has a lower embodied energy than concrete C35/45, for which the contribution of reinforced steel is significant (between 60 and 70%). For more sustainable buildings' construction, it is important to develop LCA studies during the design phase of a construction project, in order to evaluate and compare alternatives and decide on the best solutions.

**Acknowledgements** This work was financially supported by base Funding of the following projects: UIDB/04708/2020 and Programmatic Funding—UIDP/04708/2020 (CONSTRUCT—Instituto de I&D em Estruturas e Construções), LA/P/0045/2020 (ALiCE), UIDB/00511/2020 (LEPABE) and UIDB/50022/2020 (LAETA), funded by national funds through FCT/MCTES (PIDDAC). António Martins gratefully acknowledge the Portuguese national funding agency for science, research and technology (FCT) for the financial support through program DL 57/2016—Norma transitória. Teresa Mata gratefully acknowledge the funding of Project NORTE-06-3559-FSE-000107, cofinanced by Programa Operacional Regional do Norte (NORTE2020), through Fundo Social Europeu (FSE). All authors acknowledge Mr. Helder Maranhão for his support in the structural design of the structures studied in this research work.

## References

Argent Materials Inc. (2021) Recycled aggregate products from Argent Materials Inc. Environmental Product Declaration, EPD S-P-03101. The International EPD® System, Stockholm

Cantini A, Leoni L, De Carlo F, Salvio M, Martini C, Martini F (2021) Technological energy efficiency improvements in cement industries. Sustainability 13(7). https://doi.org/10.3390/su13073810

Cardoso M (2022) Design and comparative life cycle assessment of reinforced concrete frame structures made with natural and recycled aggregates. Dissertation submitted in partial fulfilment of the requirements for the degree of Master's in Civil Engineering: Specialization in Structures, Porto, p 79

EN 15804 (2013) Sustainability of construction works—environmental product declarations—core rules for the product category of construction products. British Standards Institution, p 70

IEA and UNEP (2019) Towards a zero-emissions, efficient and resilient buildings and construction sector. In: 2019 global status report for buildings and construction. Global Alliance for Building and Construction, London, p 41

ISO 14040 (2006a) Environmental management—life cycle assessment—principles and framework. International Organization for Standardization

ISO 14044 (2006b) Environmental management—life cycle assessment—requirements and guidelines. International Organization for Standardization

ISO 21930 (2007) Sustainability in building construction—environmental declaration of building products. International Organization for Standardization, Geneva

Pavlu T, Kocí V, Hájek P (2019) Environmental assessment of two use cycles of recycled aggregate concrete. Sustainability 11(21). https://doi.org/10.3390/su11216185

Rodrigues V, Martins AA, Nunes MI, Quintas A, Mata TM, Caetano NS (2018) LCA of constructing an industrial building: focus on embodied carbon and energy. Energy Proc 153. https://doi.org/10.1016/j.egypro.2018.10.018

Rodrigues I, Mata TM, Martins AA (2022) Environmental analysis of a bio-based coating material for automobile interiors. J Clean Prod 367. https://doi.org/10.1016/j.jclepro.2022.133011

Tártaro AS, Mata TM, Martins AA, Esteves da Silva JCG (2017) Carbon footprint of the insulation cork board. J Clean Prod 143. https://doi.org/10.1016/j.jclepro.2016.12.028

Watts J (2019) Concrete: the most destructive material on earth. The Guardian. https://www.theguardian.com/cities/2019/feb/25/concrete-the-most-destructive-material-on-earth. Accessed 22 May 2022

Weil M, Jeske U, Schebek L (2006) Closed-loop recycling of construction and demolition waste in Germany in view of stricter environmental threshold values. Waste Manag Res 24(3):197–206. https://doi.org/10.1177/0734242X06063686

# Chapter 7
# Effect of Building Envelope and Environmental Variables on Building Energy Performance: Case of a Residential Building in Mediterranean Climate

**Aybüke Taşer, Sedef Uçaryılmaz, and Zeynep Durmuş Arsan**

**Abstract** At least 30% of the World's energy consumption and greenhouse gas emissions originate from buildings. Thus, design decisions should be well studied during the design phase of buildings following energy efficiency approaches. Environmental variables and properties of the building envelope are significant for energy efficiency. Thus, this study aims to investigate the potential of a residential building in the Mediterranean climate of İzmir, Turkey, regarding decreasing energy use and understanding the significance of architectural decisions during the design stage of buildings. Eight design scenarios were created by defining seven variables affecting energy consumption for room electricity, heating, and cooling. The first three scenarios focused on environmental-related variables, i.e., surrounding buildings, ground surface materials, and building orientation, while the last five scenarios investigated building envelope-related variables, i.e., thermal transmittance of the wall, floor and roof, glass, window frame, and door types, shading elements, and natural ventilation. Then, energy modeling and simulation are applied to test their potential for minimizing energy consumption. Research findings proposed that early architectural design decisions significantly influenced the case building's energy performance. Thermal transmittance of the building components provided an annual energy saving of 22.4%, thus, was seen as the best-performed variable for the case building.

**Keywords** Building envelope · Environmental variables · Energy modeling · And simulation

A. Taşer (✉)
Université Catholique de Louvain, Louvain-la-Neuve, Belgium
e-mail: aybuke.taser@uclouvain.be

S. Uçaryılmaz · Z. D. Arsan
Izmir Institute of Technology, Izmir, Turkey



N. S. Caetano and M. C. Felgueiras (eds.), *The 9th International Conference on Energy and Environment Research*, Environmental Science and Engineering,
https://doi.org/10.1007/978-3-031-43559-1_7

## 7.1 Introduction

Buildings account for at least 30% of the World's energy consumption and greenhouse gas emissions; thus, their design decisions are significant for reaching sustainability goals and tackling with global climate crisis (Ascione et al. 2021). Excessive building energy consumption and greenhouse gas emissions accelerate global warming and prevent achieving sustainability targets (Giouri et al. 2020). Necessary interventions must be implemented in the buildings. New buildings must be designed energy-efficiently, while existing building stock should be renovated. Because residential buildings are responsible for 75% of energy consumption in the construction sector (Intergovernmental Panel on Climate Change 2015), they convey a high potential for decreasing energy use and its related greenhouse gas emissions (Vishwamitra and Somaroo 2020).

Space heating and cooling, cooking, electrical equipment use, water heating, and lighting contribute to a residential building's energy load (International Energy Agency (IEA) 2013). In the future, there will be greater demand for building energy use. The study of Yang et al. (2021) intended to thoroughly examine the effects of climate change on energy efficiency and thermal comfort in European residential buildings. The findings show that there will be more cooling demand and less heating demand in the future. Building energy performance is affected by many variables.

Environmental and envelope properties are some of the most significant and primarily investigated variables. Environmental variables include the building orientation and surrounding environment, i.e., buildings, ground surfaces, geography, and climate. The building envelope comprises the foundation, walls, windows, roof, and doors. Aydınol (2016) proved that the transmittance of the building envelope significantly influences buildings' energy performance. Different glazing types and transmittance values are applied on different facades of buildings in various climatic zones. It is noted that each façade responds differently to the scenarios. Saeed et al. (2021) proposed that its roof properties and orientation strategies highly influence a residential building's energy consumption in the warm climate of Kirkuk, Iraq. The building envelope components and orientation are directly associated with the building's total heat gain/loss and heat transfer, affecting its energy demand. Nasrollahzadeh (2021) developed an optimization methodology of building envelope components to understand their influence on a residential building's energy, daylight, and thermal performance in Iran's humid subtropical/Mediterranean climate. The variables in the study are wall and roof material and thicknesses, window-to-wall ratio, and glazing and shading device types. The study emphasized the importance of defining proper building components in the design phases of buildings. Studies have shown that environmental variables and building envelope components significantly affect a residential building's energy performance. Although some studies in the literature test these variables' effect on building energy consumption, more detailed analyses must be conducted for each variable that may influence a building's energy demand. Thus, this study aims to evaluate the potential of a residential building in the Mediterranean climate of İzmir, Turkey, whose design process is still ongoing, in terms of decreasing energy use, and explore the importance of architectural decisions in the design phase of buildings. The study demonstrates the relationships between architectural design decisions and building energy performance. It exemplifies using tools to simulate building performance and incorporate ambiguity and appropriate assessments into building performance evaluation throughout the design stage.

## 7.2 Material and Methods

### *7.2.1 Study Design*

Firstly, a residential building in the Mediterranean climate in İzmir, Turkey, whose design process is ongoing, is selected for the study. Then, the reference building is modeled, and its energy performance is simulated in the DesignBuilder dynamic simulation software. The study investigates several variables' influence on a residential building's energy performance. The variables' effects are tested individually in eight different scenarios. The variables in the study and respective scenarios are (1) surrounding buildings, (2) building orientation, (3) ground surface materials, (4) building envelope materials' (i.e., wall, roof, and floor) thermal transmittance values, (5) glass, opening frame, door types, (6) shading elements, and (7) natural ventilation. The simulation engine calculated the room's electricity, heating, cooling, and total kWh and kWh/m$^2$ energy consumption. The results are compared to the base case results in which no implementations and modifications are applied to the building. Scenario results are calculated by the simulation engine and analyzed in detail. Results are compared and presented both in tables and graphs.

### *7.2.2 Case Building*

A reference building is selected to assess the building's energy performance. The "A House" was designed as a residential building on the Aegean Sea coast in Narlıdere, İzmir, west of Turkey. The elevation from sea level is 2 m. The city has a Mediterranean climate with warm and rainy winters and hot and dry summers due to its proximity to the Aegean Sea (Çetinkaya et al. 2017). The surrounding environment is almost suburban, primarily consisting of 2–4 story residential buildings. The reference building comprises four stories with a total closed area of 373 m$^2$. Each floor has an area of 111 m$^2$, except the roof floor, with an area of 38 m$^2$. The orientation of the building to the North is 342.5 °C. The house has one living room, kitchen, and three bedrooms, and it is occupied by four people, one woman and three men.

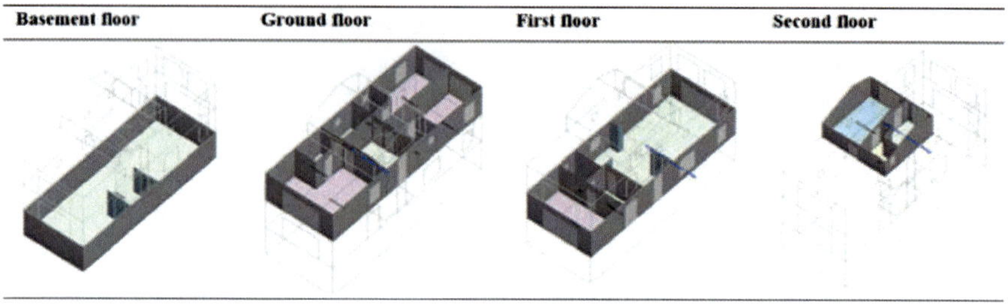


**Fig. 7.1** An axonometric view of each story representing thermal zones in DesignBuilder Software

### *7.2.3 Simulation Models*

The simulation model is generated on DesignBuilder v7.0.1.006 energy modeling and simulation software (see Fig. 7.1). DesignBuilder is a dynamic simulation software that uses EnergyPlus as the calculation engine (Download 2021; EnergyPlus 2021). It works with multi-zone approach modeling; each zone corresponds to its schedule and specification (Outgouga and Hicham 2022). An EnergyPlus weather file of İzmir-ADB is selected for climate-based simulations. HVAC systems and construction materials for windows, walls, roof, unoccupied roof, ground floor, and interior floors are assigned. The whole building is heated and cooled throughout the year except for the basement floor. Activity schedules and equipment are also introduced to the software. The base case situation is modeled in the DesignBuilder simulation software. All the inputs for the base case situation are presented to the simulation engine as in Table 7.1. All the rooms have their schedules for occupancy, computer and office equipment use, catering, and miscellaneous tools. Surrounding buildings, streets, and trees are modeled in DesignBuilder. Their materials and schedules are also defined for the sensitivity of the analysis. Energy consumption is calculated for seasonal heating and cooling, annual room electricity, and total energy consumption in kWh and kWh/m$^2$. Room electricity is identical for each scenario since no implementation is based on electrical equipment and tools.

## 7.3 Results and Discussion

### *7.3.1 Base Case Results*

In the base case modeling, surrounding buildings and ground surface materials, e.g., road types and green areas and their annual schedules, are included in the calculations for sensitive and proper analysis. According to the existing design decisions of the building, the total consumption for heating and cooling is calculated as 46,587 kWh, which equals 124.7 kWh/m$^2$. The total consumption of natural gas and electricity is

**Table 7.1** Data input details and schedules for the energy simulation model

| Base program | | Building area | | | |
|---|---|---|---|---|---|
| | | Living room + kitchen | Bedroom 1 | Bedroom 2 | Bedroom 3 |
| People | People/area | 0.055 | 0.058 | 0.05 | 0.051 |
| | Occupancy | Until: 07:00, 0, Until: 08:00, 1, Until: 09:00, 0.5 Until: 18:00, 0.25 Until: 20:00, 1 Until: 24:00, 0.5 | Until: 07:00, 1 Until: 24:00, 0 | Until: 07:00, 1 Until: 20:00, 0 Until: 24:00, 1 | Until: 07:00, 1 Until: 20:00, 0 Until: 24:00, 1 |
| Electricity equipment | Watts/area | 5.54 | – | – | – |
| Miscellaneous | Watts/area | 1.66 | | | 5.10 |
| Doors | Schedule | Closed | Closed | Closed | Closed |
| Natural ventilation | Min. indoor temperature (°C) | 21 | 21 | 21 | 21 |
| | Max. indoor temperature (°C) | 28 | 28 | 28 | 28 |
| HVAC condition | Energy source | Natural gas/ electric | Natural gas/ electric | Natural gas/ electric | Natural gas/ electric |
| Heating/ cooling setpoint | Degree (°C) | 21/19 26/28 | 21/19 26/28 | 21/19 26/28 | 21/19 26/28 |

50,054 kWh. The results convey that the highest annual energy consumption values share due to space heating and natural gas usage. Base case building's lighting, heating, and cooling energy consumption cover total energy consumption by 6.92%, 84.1%, and 8.98%, respectively.

### 7.3.2 *Scenario 1: Effect of Surrounding Context*

In this scenario, the model has no surrounding building, while ground surface materials are kept as default as assigned by the software. The aim is to understand the surrounding context's effect on the building's annual energy consumption. The simulation engine consists of all the schedules, materials, equipment, and systems as in the base case model. Total energy consumption (i.e., heating and cooling load) is calculated as 46,459 kWh, which equals 124.33 kWh/m$^2$. The electricity energy

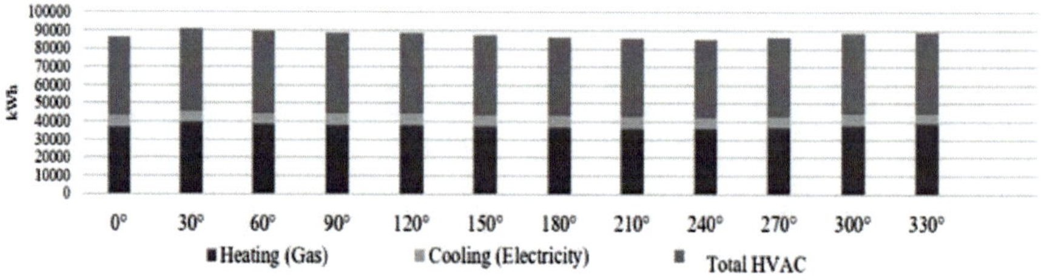


**Fig. 7.2** Distribution of energy consumption for heating, cooling, and total HVAC

consumption of the building is estimated as 9521 kWh the consumed gas is calculated as 36,936 kWh. Results show no significant difference in the total heating and cooling energy consumption compared to the base case. It may be due to the building being in a suburban area with low-rise buildings and a low-dense urban fabric. Thus, the effect of the surrounding context is found to be minor for the reference building.

### *7.3.3 Scenario 2: Effect of the Orientation*

In the second scenario, the case building's energy performance with different orientations is individually simulated and compared to determine the best orientation for the first scenario. The building is rotated along with the surroundings. An orientation of 240° is defined as the best to provide the least energy consumption (see Fig. 7.2). Results show that an orientation of 30° has the highest energy consumption and is thus defined as the worst orientation. Total energy consumption is calculated as 122.62 kWh/m$^2$ for the best orientation. It is 133.90 kWh/m$^2$ for the base case and 130.54 kWh/m$^2$ for the worst orientation. The total energy decrease of 9.24% is provided when the first scenario and orientation of 240° are compared.

### *7.3.4 Scenario 3: Effect of Surrounding Context with Default Materials and Schedules*

The difference between this scenario and the base case model is that the software default-defined materials and schedules of the surrounding context, e.g., trees, buildings, roads, etc. The software-defined materials for the surrounding buildings and ground surfaces as plastic. They had a transmittance of 0% and a schedule of 24/7.

Trees had 0% transmittance and 24/7 throughout the year. In this scenario, surrounding building and ground surface materials are defined as plaster, asphalt, pavement, and soil. The schedule of the trees is changed as summer (northern hemisphere), and their transmittance is defined as 50%. The aim is to see the materials and schedules' effect on the building's energy performance. According to the results, there could achieve no significant difference in the building's energy performance. It may be because the building is located in a suburban area with low-rise buildings, and natural materials are used in the surrounding context.

### 7.3.5 *Scenario 4: Effect of Thermal Resistance on the Components of Building Envelope'*

In this scenario, the building envelope components' thermal resistance is changed. These components are walls, floors, and roofs. U-values of walls are decreased to 0.2, external/ground floor to 0.3, and flat/pitched roof to 0.3 W/m$^2$ K. All the implementations and previous values can be seen in Table 7.2. The TS825, the national regulation of Turkey, defined existing U-values. According to the results, the total energy consumption from heating and cooling is calculated as 35,383 kWh, equal to 94.7 kWh/m$^2$. The total energy consumption from gas and electricity is calculated as 38,850 kWh. Consumption decreased by 24.1 and 22.4% compared to the base case situation. It is noted that building component U-values significantly influence the case building's energy consumption.

### 7.3.6 *Scenario 5: Effect of Glazing, Windows, and Doors*

In this scenario, the material properties of glass panes of windows, doors, and framings are changed. U-values of glass panes, window framing, and door framing are decreased to 1.49 W/m$^2$ K, 1 W/m$^2$ K, and 0.7 W/m$^2$ K, respectively, as seen in Table 7.3. This scenario aims to understand the effect of the properties of glass surfaces and their selection criteria on building energy consumption. Windowpane is selected as double glazing filled with argon, while the window frame is defined as plywood. This change shows that total energy consumption from heating and cooling is calculated as 43,689 kWh, which equals 116.92 kWh/m$^2$ K. The total consumption of electricity and gas is calculated as 47,156 kWh. Compared to the base case, these decreased by 6.63 and 6.14%.

### 7.3.7 *Scenario 6: Effect of the Shading Element*

In this scenario, different shading elements are implemented in the windows of the base case. Two sun-shading elements and one fixed window shading are applied to only south-facing windows. Local shadings are selected as drapes with medium weaves and blinds with medium reflectivity slats. They kept close when solar radiance exceeded 120 W/m$^2$. Fixed window shading element is defined as 0.5 m overhangs.

**Table 7.2** Proposed building materials for Scenario 4

| Component | Material | Specific heat (J/kg-k) | Density (kg/$m^3$) | Thickness (m) | U-value (W/$m^2$ K) New | U-value (W/$m^2$ K) Old |
|---|---|---|---|---|---|---|
| External wall | Gypsum plasterboard | 1000 | 900 | 0.01 | 0.20 | 0.70 |
| | Cement/plaster | 840 | 1860 | 0.03 | | |
| | EPS expanded polystyrene | 1400 | 10 | 0.12 | | |
| | Concrete block | 1000 | 600 | 0.20 | | |
| | Cement/plaster/ mortar -gypsum | 840 | 1200 | 0.03 | | |
| Below grade walls | Lean concrete | 1000 | 1000 | 0.02 | 0.20 | 0.70 |
| | XPS extruded polystyrene | 1400 | 35 | 0.07 | | |
| | Curtain wall | 1000 | 1400 | 0.25 | | |
| | Board insulation | 840 | 160 | 0.08 | | |
| | Cement/plaster/ mortar | 840 | 1680 | 0.10 | | |
| Basement floor | Clay underfloor | 2085 | 1500 | 0.25 | 0.47 | 0.47 |
| | Brick slips | 1000 | 1700 | 0.02 | | |
| | Curtain wall | 1000 | 2000 | 0.10 | | |
| | EPS expanded polystyrene | 1400 | 15 | 0.06 | | |
| | Flooring screed | 1000 | 1200 | 0.05 | | |
| | Ceramic, glazed | 840 | 2500 | 0.02 | | |
| Flat roof | Ceramic tile | 840 | 2500 | 0.02 | 0.30 | 0.44 |
| | Asphalt | 1000 | 2100 | 0.19 | | |
| | Concrete | 1000 | 300 | 0.10 | | |
| | XPS extruded polystyrene | 1400 | 35 | 0.04 | | |
| Sloped roof | Wooden board | 1000 | 2100 | 0.02 | 0.30 | 0.44 |
| | Air gap 15 mm | 0.01 | | | | |
| | Mineral fiber/ wool | 1000 | 99 | 0.10 | | |
| | Roofing felt | 834 | 960 | 0.01 | | |

The best-performed shading element is defined as blind with high reflectivity slats. In that scenario, the total energy consumption from heating and cooling is calculated as 46,520 kWh, which equals 124.5 kWh/$m^2$ K. Table 7.4 shows the distribution of energy for heating and cooling. The gas and electricity consumptions are calculated as 49,987 kWh, as seen in Table 7.5. Results are increased by 0.14 and 0.13% compared

Table 7.3 Proposed glazing, window, and door materials for Scenario 5

| Building component | | Material | U-value (W/m$^2$ K) |
|---|---|---|---|
| Glass pane | Former | Double glazing clear glass | 1.98 |
| | New | Double Low-E glazing filled with Argon | 1.49 |
| Glass frame | Former | Painted wooden window frame | 3.91 |
| | New | Plywood (lightweight) | 1.02 |
| Door | Former | Painted oak | 2.99 |
| | New | Opaque door | 0.70 |

Table 7.4 Distribution of energy for heating and cooling results for Scenario 6

| Shading type | Room electricity (kWh) | Natural gas heating load (kWh) | Electricity cooling load (kWh) | Total energy consumption (kWh) |
|---|---|---|---|---|
| Blind with high reflectivity slats | 3466.86 | 42,147.82 | 4372.73 | 46,520.55 |
| Drapes open wave dark | 3466.86 | 42,098.72 | 4492.51 | 46,591.23 |
| Louvre—0.50 m overhang | 3466.86 | 42,144.02 | 4492.13 | 46,636.15 |

Table 7.5 Fuel total results of Scenario 6

| Shading type | Electricity consumption (kWh) | Gas consumption (kWh) | Total energy consumption (kWh) |
|---|---|---|---|
| Blind with high reflectivity slats | 7839.59 | 42,147.82 | 49,987.40 |
| Drapes open wave dark | 7959.37 | 42,098.72 | 50,058.09 |
| Louvre—0.50 m overhang | 7958.98 | 42,144.02 | 50,103 |

to the base case situation. The effect of the shading element on energy consumption may be minor due to low solar radiation on the window surface.

### *7.3.8 Scenario 7: Effect of the Natural Ventilation*

This scenario aims to test the effect of natural ventilation on building energy consumption. A natural ventilation schedule is applied to the building. All above-ground rooms are ventilated naturally. The prevailing wind directions of İzmir are southeast and west. Thus, natural cross ventilation reduces cooling loads, especially

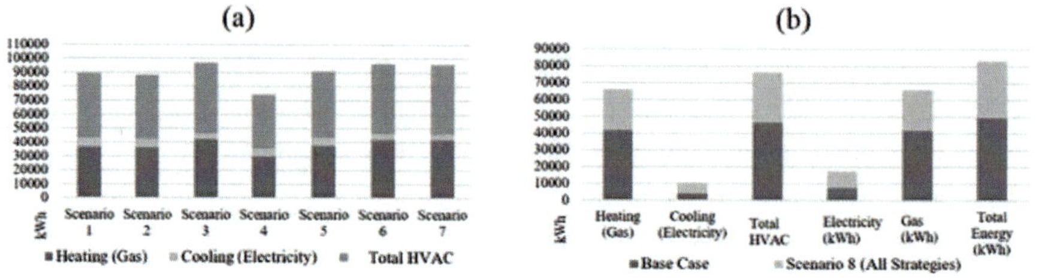


**Fig. 7.3** **a** Comparison of scenarios **b** Energy consumption for heating, cooling, and electricity for the base case condition and Scenario 8

in the east–west. In the summer period from March 31 to September 30, a natural ventilation schedule was applied from 9 am to 6 pm to reduce indoor air temperatures. The results provided that the total energy consumption from heating and cooling is calculated as 46,297 kWh, corresponding to 123.9 kWh/m$^2$ K. The total consumption from gas and electricity is calculated as 49,764 kWh. These are decreased by 0.62 and 0.58% compared with the base case situation. This decrease was observed to be 6.42% in only cooling energy consumption.

### *7.3.9 Scenario 8: Combining Scenarios 4, 5, 6, and 7*

A merged model is created by combining scenarios 4, 5, 6, and 7. These scenarios performed the best in reducing the energy consumed for the building. Thus, they are combined, and a new model is created. In this model, U values of building components, windows, and doors are decreased; shading elements are applied, and natural ventilation is enabled. According to the results, the total consumption from heating and cooling is calculated as 29,751 kWh, equal to 79.6 kWh/m$^2$. The total consumption from gas and electricity is calculated as 33,218 kWh. These are decreased by 36.1% and 33.6%, respectively. The implementations significantly affected the building's annual energy consumption (see Fig. 7.3).

## 7.4 Conclusion

This study displays a comprehensive analysis of the impact of environmental and envelope-related variables on the energy performance of a residential building during the design stage in the Mediterranean climate of İzmir, Turkey. The study aims to understand the effectiveness of the early design decision on the energy performance of residential buildings.

The scenarios are tested using the energy simulation models regarding room electricity, heating, and cooling loads. Among the applied scenarios, building envelope components' thermal resistance significantly affects the energy saving of heating the case building. A lower U-value provides better insulation, thus, can reduce the heating loads in primarily heated buildings. Thus, in such climates, it may be preferable to use low U-value building components to increase thermal resistance and reduce building heat losses. Regarding the energy-saving of cooling loads, the best results were obtained when natural ventilation was applied on the first floor. Natural ventilation can decrease indoor air temperatures and provide thermal comfort for occupants, especially during summer when the outdoor air temperature reaches extreme heat. Thus, natural ventilation can be a tool for such climate regions to reduce indoor air temperatures and decrease the cooling load of the building. The minor effective variables for this case building were the surrounding context, materials, and heating and cooling energy consumption schedules. It may be because the case building is located in a suburban area where the surrounding buildings are low-rise and constructed with more natural and urban-friendly materials. Thus, their effect may have become neglectable in this sense. It is equally essential to retrofit total energy consumption, especially in Mediterranean climates where heating and cooling energy consumption are similar. When the total annual energy consumption is analyzed and compared, envelope-related variables are performed better than the environment-related variables in terms of the energy saving of the building. In this respect, it was observed that energy-saving improvement with building envelope-related variables is more effective than environmental-based ones for the case building. Comments on the overall research can be gathered to discuss how climate change scenarios and the design process relate to one another, how to use tools for simulating building performance, and how to incorporate ambiguity and proper analyses into building performance evaluation during the design stage.

## References

Ascione F, De Masi RF, Mastellone M, Ruggiero S, Tariello F, Vanoli GP (2021) Energy performance of buildings: improvements, limits and future perspectives during the last twenty years of energy and sustainability policies. In: 6th international conference on smart and sustainable technologies (SpliTech), pp 1–6

Aydınol F, Azize Z (2016) Impact of different transparency ratios of building envelope on building energy performance in Various Climatic Zones (dissertation). Yıldız Technical University

Çetinkaya G, Selman A, Ozturk MZ (2017) Climate types of Turkey according to köppen-geiger climate classification. J Geogr 35:17–27

Download, DesignBuilder (2021) https://designbuilder.co.uk/. Accessed 20 May 2021

EnergyPlus (2021) https://energyplus.net/. Accessed 20 May 2021

Giouri ED, Tenpierik M, Turrin M (2020) Zero energy potential of a high-rise office building in a Mediterranean climate: using multi-objective optimization to understand the impact of design decisions towards zero-energy high-rise buildings. Energy Build 209:109666

Intergovernmental Panel on Climate Change (2015) Climate change 2014: mitigation of climate change: working group III contribution to the IPCC fifth assessment report. Cambridge University Press, Cambridge

International Energy Agency (IEA) (2013) Modernising building energy codes. IEA, France

Nasrollahzadeh N (2021) Comprehensive building envelope optimization: improving energy, daylight, and thermal comfort performance of the dwelling unit. J Build Eng 44
Outgouga F, Hicham J (2022) Improving the passive energy efficiency of an office building in Agadir Morocco using dynamic thermal simulations in Design Builder. In: International conference on decision aid sciences and applications (DASA), pp 1765–1769
Saeed M, Sadegh A, Zhaleh HZG (2021) Simulation of the roof shapes and building orientation on the energy performance of the buildings. J Build Pathol Rehabil 6(1)
Vishwamitra O, Somaroo N (2020) Multi-objective optimization of the energy performance of residential buildings in tropical climates. In: 3rd international conference on emerging trends in electrical, electronic and communications engineering (ELECOM), pp 240–243
Yang Y, Kavan J, Vahid N (2021) Climate change and energy performance of European residential building stocks: a comprehensive impact assessment using climate big data from the coordinated regional climate downscaling experiment. Appl Energy 298:117246

# Chapter 8
# Knowledge Retrieval Mechanism for Smart Buildings Based on IoT Devices Data

**Nuno Teixeira, Luis Gomes, and Zita Vale**

**Abstract** The use of smart building solutions can bring advantages to both a building and its users. The main motivation of this work is to propose a solution based on internet of things devices to retrieve knowledge from historic data and, using the knowledge gained, contribute to the comfort of the user and the building's sustainability. The use of internet of things devices enables the easy deployment of the proposed system in today's buildings, where this kind of devices are commonly found. The knowledge regarding the role of each device and how they interact or impact each other will be studied by the proposed solutions that will deploy a continuous mechanism. The designed software architecture was developed in python, to allow easy implementation of different features. To test the proposed solution, it was used real data from a research building where several types of sensors were correlated to assess relations among data. The results show significant correlations between celling light consumption and the room's light intensity sensor, and between the operation of air conditioning units and the room's temperature, for example, data from 2 months, regarding air conditioner unit in room N102, demonstrates a − 0.85 correlation with the room temperature, when operating in cooling mode.

**Keywords** Ambient intelligence · Internet of things device · Knowledge retrieval · Smart building

## 8.1 Introduction

The concept of smart buildings opens the possibility of turning our buildings in smart and even intelligent solutions that can support our activities while applying efficient management that can promote, among others, the sustainability of the building

N. Teixeira · L. Gomes (✉) · Z. Vale
GECAD—Research Group on Intelligent Engineering and Computing for Advanced Innovation and Development, LASI—Intelligent Systems Associate Laboratory, Polytechnic of Porto, Rua Dr. António Bernardino de Almeida 431, 4200-072 Porto, Portugal
e-mail: lufog@isep.ipp.pt; lfg@isep.ipp.pt

 
N. S. Caetano and M. C. Felgueiras (eds.), *The 9th International Conference on Energy and Environment Research*, Environmental Science and Engineering,
https://doi.org/10.1007/978-3-031-43559-1_8

(Panteli et al. 2020), the reduction of energy costs (Lezama et al. 2019), the support for maintenance actions (Kisel 2021), and the increase of persons comfort (Panchalingam and Chan 2021). The use of smart buildings can also contribute to the existence of nearly zero-energy buildings, that have been pushed by European recommendations (2016).

In today's market and buildings, the existence of internet of things (IoT) devices cannot be disregarded. Noticing the potentialities of IoT devices, some works adopt these systems to manage smart building. In Farahani et al. (2018), it is proposed a patient-centric healthcare solutions based on IoT devices. In Wang et al. (2020), IoTs devices are used to deploy and test heating, ventilating and air conditioning (HVAC) units' management models. To increase the efficiency of IoT devices, some works proposed a context-aware approach where the IoT device is able to know its context (Nweke et al. 2018; Gomes et al. 2019a).

The usage of heterogeneous devices in smart buildings might provide challenges because they do not share the same information, control, or methods of communication. In order to explain the gadgets in smart buildings, a semantic method might be utilized (Iddianozie and Palmes 2020). The use of semantics is usually done by applying ontologies to describe the system knowledge (Esnaola-Gonzalez et al. 2020). This can be a problem for volatile systems that tend to get worse over time when more devices are added, or the system becomes more user-friendly. Additionally, it is not currently feasible to describe all IoT devices on the market in an ontology.

Artificial Intelligence (AI) methodologies and strategies are recognised for interpreting the environment and representing the information and knowledge extracted from the environment (Ramos et al. 2008). Ambient Intelligent (Aml) system performance depends on the modeling of the various IoT devices present in a building. This makes it possible to identify the IoT devices that affect a particular attribute (Antón et al. 2019). Correlations can also be a useful tool for modeling without the need for ontologies by connecting the shared characteristics of many IoT devices (Francisti et al. 2020).

The primary goal of this research is to investigate the viability of a continuous mechanism that retrieves knowledge based on the operation of IoT devices without requiring users or a prior knowledge model. The suggested mechanism enables the development of intelligent systems without requiring the user to provide a knowledge base. To do this, the suggested approach will employ mathematical relationships to determine what a certain IoT device in the building effects. Future research made possible by this identification can teach the system relational rules to control the smart building according to a parameter. The suggested mechanism is a component of the Aml solution that is also presented in this paper. Real smart buildings with real data will be used to test and evaluate the knowledge retrieval method.

## 8.2 Building Ambient Intelligent System

Taking advantage of the use of IoT devices, this paper proposes the building ambient intelligent system (BAIS) supported by IoT devices and considering data security and privacy. BAIS is an improvement of the proposed architecture in Teixeira et al. (2022).

### *8.2.1 Software Architecture*

The software architecture proposed was designed to facilitate the development of an IoT-enabled ambient intelligence system while maintaining user privacy. This study suggests a framework that enables IoT devices to optimize energy consumption data and model various IoT devices as well as the parameters they will affect through correlations, allowing the learning of how new IoT devices introduced to the network will affect various parameters.

As can be seen in Fig. 8.1, the software architecture of the proposed solution is divided into three layers. The Connectivity layer allows the connection between the user and the system, where we have a REST-based API with different routes, which allows access to energy consumption according to user privacy and allows the generation of a JavaScript Object Notation (JSON) Web Token (Teixeira et al. 2021; Jones et al. 2015). This enables the creation of the token that gives access according to user permissions to energy consumption data, the token manager module allows to validate this token generated for the user, so that data is not viewed by unreliable third parties. To enable the users to generate tokens, a graphical user interface has been developed so that the user can request the token according to the permissions and can visualise the energy consumption data. The Middleware layer allows the integration of IoT devices so that it is possible to monitor and collect data from IoT devices and control them as well. This solution solves a very common problem which is the integration of heterogeneous IoT devices thus promoting interoperability and cooperation between them, implementing multiple connection drivers where all IoT devices present in the building can be connected. The data will be sent to a NoSQL database that will allow historical management. The IoT forecasting module aggregates the anticipated energy of each IoT to forecast the energy required to balance demand and load supply. Additionally, this module provides IoT usage forecasting, which enables users to see whether they are utilizing a certain IoT in a given time, and IoT energy flexibility forecasting, which may be used for energy resource balancing.

Also present in the middleware is the ambient intelligence module. This module is subdivided and contains the knowledge discovery module, which allows, through the parameter finder and the correlation finder, to find correlations between the parameters that the IoT devices have in common so that in this way it is known which parameters are the IoT devices influencing to be simpler to make the optimization.

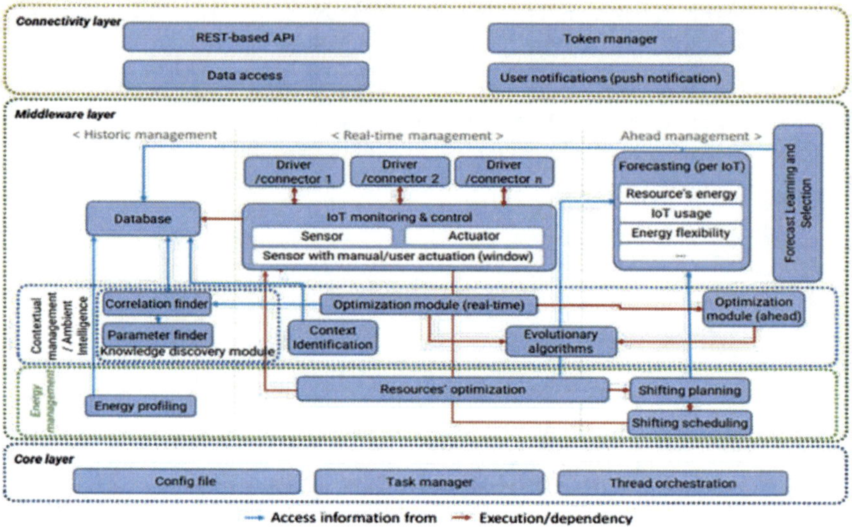


**Fig. 8.1** Software architecture

After this it is forwarded to the optimization module in real time and ahead time, which will allow the optimization of energy resources using evolutionary algorithms. There is also context identification, which will enable the identification of contexts so that it is possible to understand some patterns in energy consumption. The system's engine, as well as its foundation, is the Core layer. It includes a configuration file that uses a JSON structure, allowing the configuration of various IoT devices. This means that a new IoT device can be added without having to change the system's primary code. Threads are used to monitor IoT devices, making it easier and faster to do so, and the task manager allows management and execution of the tasks that will be executed by the system.

## *8.2.2 Knowledge Retrieval*

Modelling the different resources can be a difficult task. To do it the system will use correlations. For example, if a new $CO_2$ sensor is integrated, it will be identified that a new parameter will have to be monitored, and after that the system will start to understand which actuators/resources cause an impact on $CO_2$. The parameter can only be controlled/optimised if there is a correlation. This modelling is done in this way so that the system can model itself through its own intelligence whenever a new IoT device is added to the system, without having a previous knowledge model. It can be useful in data analysis and modelling to better understand the relationships between variables. The relationship between two variables is referred to as their correlation. Variables can be related by a linear relationship, which is called covariance. A covariance is not easy to interpret, but if the covariance has a value of zero it indicates that both variables are completely independent.

Due to the difficult interpretation of correlation in covariance, it is preferable to use the Pearson correlation coefficient. The coefficient will return a value between − 1 and 1 which represents the value of the correlation with 0 meaning no correlation. The value should be interpreted, where normally a perfect correlation is when the value is close to 1 or − 1, a perfect correlation, indicates that when one variable increases the other variable tends to increase as well if the correlation is positive or decrease if the correlation is negative. If coef ∈ [0.5…0.1] or coef ∈ [− 1… − 0.5], then it can be said to be a strong correlation. If coef ∈ [0.3…0.5[ or coef ∈ ]− 0.5… − 0.3], then it is considered as a medium correlation, and if the value is below 0.29 or above − 0.29, it is a small correlation.

## 8.3 Case Study

For this case study it will be used a research building where multiple IoT devices and a supervisory control and data acquisition (SCADA) system are deployed to allow the monitoring and control of the building. This building was part of multiple studies from forecasting algorithms (Ramos et al. 2020) to real-time load optimization (Gomes et al. 2019b). In this paper, an intelligent solution is promoted to create knowledge from data without the user's input, meaning that no semantic rules are inputted by the user of by the developer. In this case study the system will retrieve knowledge based on historical data that is monitored by the solution presented in Sect. 8.2.

To test the solution presented in this paper, two research offices (i.e., N102, and N103) will be used and one meeting room (i.e., N101). This are adjacent rooms that shared prefabricated walls, as can be seen in Fig. 8.2. All the rooms present in Fig. 8.2 have actuators and sensors of all types that will be described later.

These locations have a normal occupation between four and seven researchers. The locations have deployed two three-phase energy analyzer that separates, by phase, the consumptions of the energy sockets, the consumptions of the artificial lights, and the consumptions of the air conditioning (AC) units. The rooms also have access to the generation data that is measured by the photovoltaic (PV) inverter.

Regarding the deployment of sensors, each room has: a temperature sensor, a humidity sensor, a light sensor, an $CO_2$ sensor, a volatile organic compounds (VOC) sensor, and a door sensor that indicates if the door is open or not. Both researcher

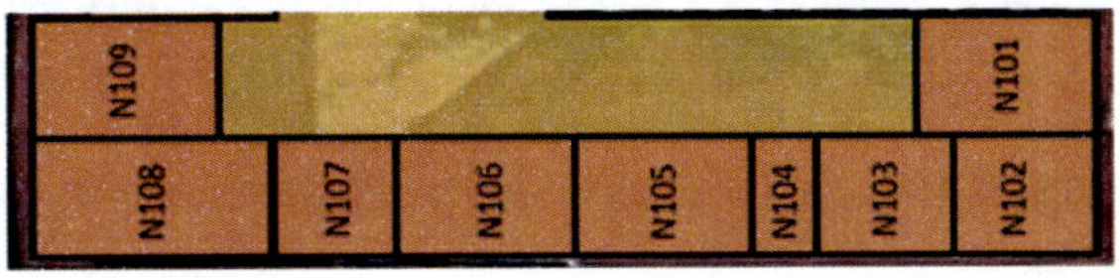


**Fig. 8.2** Research building

offices have passive infrared (PIR) movement sensors, one office has four PIR sensors and the other has two PIR sensors. The building also as an outdoor temperature sensor and a day/night sensor. Internally the system models IoT devices as being associated to a single room. The platform performs the knowledge retrieval mechanism, and it is periodically updated to consider new historical data. The stated data is recorded every 10 s in the database of the suggested solution.

To test and validate the knowledge retrieval mechanism proposed in sub Sect. 8.2.2, three sizes of historical data will be used: historical data of 1 week, historical data of 2 months, and historical data of 1 year. This enables the detection of knowledge that the system can produce will increasing the historical data.

## 8.4 Results

Actuators and sensors are the two types of IoT devices, depending on whether the IoT can be controlled. The correlations that will be made will only be made based on actuators, because what we want to know is how actuators affect the environment in which they are inserted, and the consequences produced by actuators will be recorded in the remaining sensors present in the environment. Table 8.1 demonstrates the results obtained by the Pearson correlation coefficient between the actuators and the respective sensors of the same rooms.

These results were obtained through Python using the pandas library, resorting to the use of the corr() method, to visualize the correlations between actuator and sensors. To make it easier to find the most relevant correlations using the seaborn library, a heatmap was created.

As can be seen in Table 8.1, the air conditioners in the different rooms were divided into two strands: when they are on to heat the room and when they are on to cool the room. In the table can be seen the grouped celling lighting of the three rooms considered in the case study.

In the AC of room N101 turned on to cold the room, there is no significant correlation, which means, there are no correlations greater than 0.50 or less than − 0.50. But there was an improvement in correlations from 1 week to 2 months, and the correlation with temperature is very close to − 0.50. The most obvious correlations during the time considered are temperature and humidity. When the AC was on to heat the room, it had better correlations, because there are significant correlations, after 2 months the temperature has a strong correlation with the AC, after 1 year it is still a significant correlation but decreases compared with the data of 2 months.

In the AC of room N102, there were no results from the correlations of the week studied, because the AC was not turned on during that week. When the AC was switched on for cooling, after 2 months, there were quite a few significant correlations in $CO_2$, VOC, temperature, and light intensity. But after a year, some of these correlations are lost and those that remain are no longer significant. For the AC on for heating the room, the same correlations found when AC was on for the cooling

**Table 8.1** Knowledge retrieval results considering the more relevant correlations

| Room | Resource operation | One week correlations | Two months correlations | One year correlations |
|---|---|---|---|---|
| N101 | Air conditioner—cold | Temperature (− 0.39)<br>Humidity (0.37) | Temperature (− 0.48)<br>Humidity (0.38) | Temperature (− 0.20)<br>Humidity (0.026) |
| N101 | Air conditioner—heat | Temperature (0.43)<br>Humidity (− 0.34) | Temperature (0.63)<br>Humidity (− 0.48) | Temperature (0.55)<br>Humidity (− 0.40)<br>VOC (− 0.41) |
| N102 | Air conditioner—cold | The AC was not used in the studied week | CO2 (0.92)<br>VOC (0.79)<br>Temperature (− 0.85)<br>Light intensity (0.83) | VOC (0.40)<br>Temperature (− 0.48) |
| N102 | Air conditioner—heat | The AC was not used in the studied week | CO2 (− 0.81)<br>VOC (− 0.89)<br>Temperature (0.65)<br>Light intensity (− 0.79) | VOC (− 0.56)<br>Temperature (0.64) |
| N103 | Air conditioner—cold | CO2 (0.48) | Light intensity (0.25) | Temperature (0.22) |
| N103 | Air conditioner—heat | VOC (0.35)<br>Temperature (0.32) | Temperature (0.31)<br>Humidity (− 0.30) | Light intensity (0.27) |
| N101<br>N102<br>N103 | Celling lighting | N103 light intensity (0.77)<br>N101 light intensity (0.38) | N103 light intensity (0.73)<br>N101 light intensity (0.41) | N103_light intensity (0.59)<br>N101_light intensity (0.46) |

are present after 2 months, but after 1 year the VOC and the temperature remained with significant correlations regarding the AC.

In the AC of room N103, there were no significant correlations either when it was on to heat the room or when it was on to cool the room, but the most significant correlations were temperature, humidity and $CO_2$.

The Celling lighting represents all the lighting in the different rooms, N101, N102 and N103, which were considered in this case study. After 1 week of correlations, we can see that the intensity of the lights in room N103 and room N101 were the strongest correlations found, and the intensity of the lights in room N103 was quite significant, passing 0.5. Over time we can see that the correlations for light intensity in room N101 went up, but never reached a significant correlation, and the light intensity of room N103 went down but never went below 0.5, remaining always a significant correlation with celling lighting.

## 8.5 Conclusion

In this paper it was proposed a knowledge retrieval mechanism for smart buildings based on IoT devices, using correlations. Our platform guarantees the modelling of the different IoT devices using correlations, this modelling is done so that the system can model itself through its own intelligence whenever a new IoT device is added, without previously having a knowledge model.

The results showed that the correlations indeed relate well the actuators with the different sensors, being that for air conditioners, the system found correlations with temperature, humidity and $CO_2$, for example, in 2 months of consumption data from the air conditioner in room N101 turned on for heating, there was a correlation of 0.63 with temperature sensor and $-$ 0.48 with humidity sensor, which show to be relevant correlations. It is possible to extract significant correlations after 2 months. A whole year of data is not necessary to extract significant correlations, because in some actuators certain correlations are lost or their coefficient decreases. Also, 1 week of correlations is not enough to have significant correlations.

To conclude, this study showed that correlations can be important and an asset to infer knowledge without needing a knowledge base or ontologies.

**Acknowledgements** The present work has received funding from European Regional Development Fund through COMPETE 2020—Operational Programme for Competitiveness and Internationalisation through the P2020 Project TIoCPS (ANI|P2020 POCI-01-0247-FEDER-046182), and has been developed under the EUREKA—ITEA3 Project TIoCPS (ITEA-18008), we also acknowledge the work facilities and equipment provided by GECAD research center (UIDB/00760/2020) to the project team.

## References

Antón MÁ, Ordieres-Meré J, Saralegui U, Sun S (2019) Non-invasive ambient intelligence in real life: dealing with noisy patterns to help older people. Sensors 19(14):3113. https://doi.org/10.3390/s19143113

Commission Recommendation (EU) (2016) Commission Recommendation (EU) 2016/1318 of 29 July 2016 on guidelines for the promotion of nearly zero-energy buildings and best practices to ensure that, by 2020, all new buildings are nearly zero-energy buildings (OJ L 208, CELEX, p 46. https://eur-lex.europa.eu/legal-content/EN/TXT/?uri=CELEX:32016H1318. Accessed 2 Aug 2016)

Esnaola-Gonzalez I, Bermudez J, Fernandez I, Arnaiz A (2020) Ontologies for observations and actuations in buildings: a survey. Seman Web 11:593–621. https://doi.org/10.3233/SW-200378

Farahani B, Firouzi F, Chang V, Badaroglu M, Constant N, Mankodiya K (2018) Towards fog-driven IoT eHealth: promises and challenges of IoT in medicine and healthcare. Futur Gener Comput Syst 78:659–676. https://doi.org/10.1016/j.future.2017.04.036

Francisti J, Balogh Z, Reichel J, Magdin M, Koprda Š, Molnár G (2020) Application experiences using IoT devices in education. Appl Sci 10(20):7286. https://doi.org/10.3390/app10207286

Gomes L, Ramos C, Jozi A, Serra B, Paiva L, Vale Z (2019a) IoH: a platform for the intelligence of home with a context awareness and ambient intelligence approach. Future Internet 11(3):58. https://doi.org/10.3390/fi11030058

Gomes L, Spínola J, Vale Z, Corchado JM (2019b) Agent-based architecture for demand side management using real-time resources' priorities and a deterministic optimization algorithm. J Clean Prod 241. https://doi.org/10.1016/j.jclepro.2019.118154

Iddianozie C, Palmes P (2020) Towards smart sustainable cities: addressing semantic heterogeneity in building management systems using discriminative models. Sustain Cities Soc 62. https://doi.org/10.1016/j.scs.2020.102367

Jones M, Bradley J, Sakimura N (2015) JSON web token (JWT), RFC 7519. https://doi.org/10.17487/RFC7519

Kisel (2021) Cost efficiency of building information modeling use at the stage of real estate object maintenance. Int J Technol 12:1468–1478. https://doi.org/10.14716/ijtech.v12i7.5382

Lezama F, Soares J, Hernandez-Leal P, Kaisers M, Pinto T, Vale Z (2019) Local energy markets: paving the path toward fully transactive energy systems. IEEE Trans Power Syst 34(5):4081–4088. https://doi.org/10.1109/TPWRS.2018.2833959

Nweke HF, Teh YW, Al-garadi MA, Alo UR (2018) Deep learning algorithms for human activity recognition using mobile and wearable sensor networks: state of the art and research challenges. Expert Syst Appl 105:233–261. https://doi.org/10.1016/j.eswa.2018.03.056

Panchalingam R, Chan KC (2021) A state-of-the-art review on artificial intelligence for smart buildings. Intell Build Int 13:203–226. https://doi.org/10.1080/17508975.2019.1613219

Panteli C, Kylili A, Fokaides PA (2020) Building information modelling applications in smart buildings: from design to commissioning and beyond A critical review. J Cleaner Prod 265. https://doi.org/10.1016/j.jclepro.2020.121766

Ramos C, Augusto JC, Shapiro D (2008) Ambient intelligence—the next step for artificial intelligence. IEEE Intell Syst 23(2):15–18. https://doi.org/10.1109/MIS.2008.19

Ramos D, Teixeira B, Faria P, Gomes L, Abrishambaf O, Vale Z (2020) Using diverse sensors in load forecasting in an office building to support energy management. Energy Rep 6:182–187. https://doi.org/10.1016/j.egyr.2020.11.100

Teixeira N, Barreto R, Gomes L, Faria P, Vale Z (2022) A trustworthy building energy management system to enable direct IoT devices' participation in demand response programs. Electronics 11(6):897. https://doi.org/10.3390/electronics11060897

Teixeira N, Gomes L, Vale Z (2021) Data access mechanism to allow multiple level permissions in energy management solutions supported by IoT devices. In: 2021 IEEE international conference on environment and electrical engineering and 2021 IEEE industrial and commercial power systems Europe (EEEIC/I&CPS Europe), pp 1–6. https://doi.org/10.1109/EEEIC/ICPSEurope51590.2021.9584750

Wang C, Pattawi K, Lee H (2020) Energy saving impact of occupancy-driven thermostat for residential buildings. Energy Build 211. https://doi.org/10.1016/j.enbuild.2020.109791

# Chapter 9
# Designing a Qualitative Pre-diagnosis Model for the Evaluation of Radon Potential in Indoor Environments

**Joaquim P. Silva, Nuno Lopes, António Curado, Leonel J. R. Nunes, and Sérgio I. Lopes**

**Abstract** In a very early stage of implementation of a comprehensive experimental campaign for indoor radon assessment, a pre-evaluation selection of the variables that play a leading role in influencing expected results must be insightfully assessed. Hence, a practical methodology for variable selection based on an analysis of historic data plays a key role concerning radon potential assessment. Given the circumstances, this work is focused on the design of a qualitative pre-diagnosis model for the evaluation of radon potential in indoor environments, for different energy efficiency scenarios, by considering a set of relevant variables carefully selected to characterize occupants' risk exposure. A prior survey was done to identify all relevant characteristics that most affect Indoor Air Quality (IAQ), mainly concerning local geology, built environment performance, and occupancy schedules. The selected parameters will be afterward weighted and combined into performance indicators through an evidence-based literature review. In the current early stage, the requirements to drive the software development are presented, together with a software architecture proposal. Finally, it is expected that this pre-diagnosis model will allow

J. P. Silva (✉) · N. Lopes
2Ai, School of Technology, IPCA, Barcelos, Portugal
e-mail: acurado@estg.ipvc.pt

A. Curado · L. J. R. Nunes
proMetheus, Instituto Politécnico de Viana do Castelo, Rua da Escola Industrial e Comercial Nun'Álvares, 4900-347 Viana do Castelo, Portugal
e-mail: acurado@estg.ipvc.pt

S. I. Lopes
Centro de Interface Tecnológico e Industrial, Largo da Feira, 5, 4970-786 Arcos de Valdevez, Portugal

ADiT-Lab, Instituto Politécnico de Viana do Castelo, Rua da Escola Industrial e Comercial Nun'Álvares, 4900-347 Viana do Castelo, Portugal

IT—Instituto de Telecomunicações, Campus Universitário de Santiago, 3810-193 Aveiro, Portugal



N. S. Caetano and M. C. Felgueiras (eds.), *The 9th International Conference on Energy and Environment Research*, Environmental Science and Engineering,
https://doi.org/10.1007/978-3-031-43559-1_9

a more refined sample selection for indoor radon assessment, by choosing the most susceptible variables that influence radon potential in a given scenario.

**Keywords** Indoor radon assessment · Radon performance indicator · Radon potential · Energy efficiency

## 9.1 Introduction

Radon is a noble radioactive gas found freely in the natural environment, odorless, colorless, and tasteless. It arises through the decay of uranium and is prominent on granite and schist soils and substrates, rocks, and even in borehole water. Though in smaller concentrations, radon is also present in some specific building materials such as concrete, brick, and aggregates (WHO 2009). In normal environmental conditions, such as those that exist on the earth's surface, radon exhalation from the soil occurs in the gas form, presenting in outdoor environments low concentrations of approximately 10 Bq $m^{-3}$ with reduced known impact on human health (Steck et al. 1999). However, the accumulation of radon gas in indoor environments poorly ventilated, which is approximately eight times denser than air (Soltani-Nabipour et al. 2019), represents a known public health problem, extensively reported by the World Health Organization (WHO), given the established relationship between high indoor radon concentration levels and the prevalence of respiratory diseases, mainly lung cancer (WHO 2009). Previous research works have been carried out to highlight the correlation between high indoor radon concentration, and the lack of air renovation (Curado et al. 2020). However, an extensive analysis including other influencing parameters is yet to be done.

Besides ventilation, the relationship between local geology (soil type, composition, and layers), built environment characteristics and features (floors below ground level versus floors above ground level, applied building materials, and surface finishing, the existence of technical galleries, culverts, pipes or other infrastructures under buildings' floors, etc.), as well as building occupancy schedules, are additional parameters very relevant for determining the radon potential of a given room part of a critical building located in a specific risk location.

The main objectives of this research work are the following: (1) propose a methodology for radon potential assessment—Rn potential—based on historic data analysis, by considering a set of influencing variables previously selected, systematized, and categorized; (2) design a pre-diagnosis model for sample selection in a preliminary stage of an experimental campaign for indoor radon assessment.

This pre-diagnosis model will incorporate the most influencing variables previously selected by the Rn potential methodology. A pre-diagnosis assessment is a form of pre-assessment where the radon potential of a certain site can evaluate a building's tendency concerning indoor radon exposure. With this form of assessment, a meaningful and efficient experimental campaign can be implemented, without spending unnecessary time and resources on measuring buildings with low radon indoor risk exposure.

The next section summarizes previous research works that identified the major factors contributing to high indoor radon concentration throughout several countries. Section 9.3 proposes a method to design a qualitative pre-diagnosis model for assessing indoor radon potential. Section 9.4 describes the expected outcomes of this model.

## 9.2 Related Works

According to the literature, a pre-diagnosis model to assess radon potential must include a preliminary task consisting of gathering systematic information collected from previous experimental campaigns, creating in an inventory, a set of radon measurements carried out for different types of buildings: residential, school, kindergarten, administrative offices, historical buildings, etc.

High radon potential is consensually attributed to geosites underlying constructions (Hahn et al. 2015). Most authors emphasize that the territory is endowed with high geodiversity, namely: residual; granitic; tectonic; fluvial; wind, and geocultural geoforms, therefore increasing the tendency for increasing radon exposure. Beyond geologic issues, most instrumented buildings are operated with natural ventilation, therefore indoor radon concentrations show daily and seasonal variations, floating throughout the day and over the year. Indoor radon concentration tends to be higher at night than during the day, and during winter when compared to other seasons, since buildings are densely occupied and airtight, with the heating systems turned on, and so the heat flow tends to increase air pressure difference, acting as a radon soil "vacuum cleaner". On the other hand, in summer due to increased natural ventilation, indoor concentration tends to reduce.

The following researches state the importance of site geology, building construction, and ventilation on indoor radon potential:

- Barros-Dios et al. (2007) conducted a cross-sectional study analyzing indoor radon concentration in a set of distinct 983 homes. The bivariate analysis found that several factors influence indoor radon concentration, mainly the building age, the measurement location, the building material, and finishing coverings. In general, dwellings made of granite stone show higher radon concentrations. Additionally, the study revealed a positive correlation between indoor radon concentration and the altitude above sea level. Multivariate analysis found four variables with significant influence on radon concentration: age of dwelling; the number of floors; building height, and building finishing coats. The model variability explained 10% of the studied cases. No significant differences between houses with and without cellars were detected.
- Regarding the same subject, Andersen et al. (2007) developed a linear regression model indoor radon prediction in Danish houses. The model was calibrated against

radon measurements in a set of 3116 single-family houses and apartments, by using an independent dataset with 788 house in-situ measurements concerning model performance assessment. Nine exploratory variables were tested, however, the most influencing variables concerning radon potential are the house type and the local geology.

To assess radon potential, Cortina et al. (2008) performed a year-round study concerning indoor radon assessment in Santiago de Compostela, Galicia, Spain. A total of 233 samples at 59 different locations in 16 buildings were analyzed, showing a direct correlation between the indoor radon concentration and outdoor air temperature. No other correlation could be established with other experimental variables (wind direction, humidity, etc.). According to the authors, room ventilation, including human activities, is the main factor influencing indoor radon concentration.

Martins et al. (2013) measured indoor radon concentration in a set of 73 dwellings in three different geological sites in Amarante, Portugal: soil in coarse-grained, porphyritic biotite granite (AT1), soil in medium-grained biotite granite (AT2), and soil in metasediments rocks of Paleozoic age (MTS). Average indoor radon concentrations in winter conditions are, respectively, 85, 220, and 430 Bq $m^{-3}$ for dwellings in MTS, AT1, and AT2 sites. Based on the results, the authors concluded that local geology is the most relevant factor concerning indoor radon concentration.

Groves-Kirkby et al. (2015), assessed indoor radon concentration and air temperature in an environmentally stable, rarely-visited basement in a public-service building, from June 2003 to March 2008. The authors detected a good correlation between indoor radon concentration and internal–external temperature difference, suggesting that the principal driver for indoor radon potential is the atmospheric moisture content rather than indoor relative humidity.

Martins et al. (2016) in Vila Pouca de Aguiar, North of Portugal, stated that 62.6% of analyzed dwellings (n = 57) present average indoor radon concentrations exceeding, by far, the national legal limit. The authors reported that the most problematic dwellings have less than 50 years old, are constructed with a basement, low ventilated, and well-insulated. Recent buildings are more critical than the old ones, due to the lack of ventilation.

Collignan et al. (2016) conducted an indoor radon monitoring campaign evolving 3400 dwellings in Brittany, France from 2011 to 2014. Based on the study, the authors concluded that indoor radon potential is related to the total floor number, the sampling level, the type of foundation, the construction period, the building material, the ventilation system, the use of wood heating, and the existence of thermal retrofit.

- Under the scope of R&D Project Fundação Ilídio Pinho (Curado et al. 2020), a set of nine buildings in Alto Minho region, Viana do Castelo, Portugal, was monitored continuously to assess indoor radon concentration. The authors state that all instrumented buildings share similar types of construction, identical characteristics of the foundation soil, but different occupancy regimes and ventilation rates. As expected, rooms' occupancy and ventilation actions performed by the occupants played an important role in indoor radon concentration, and consequently on radon potential assessment.

R&D Project RnMonitor (2021) monitored two sets of 15 public buildings made with granite blocks, in two different regions in the Northwest of Portugal. The monitored buildings diverged on construction period, occupancy, ventilation schemes, and devices. The results showed the influence of local geology and room ventilation on indoor radon potential. Rooms in basements and ground-floor levels, low ventilated, show high indoor radon concentration (Azevedo et al. 2021).

Based on the related research works, a pre-diagnosis model concerning radon potential assessment must include the analysis of variables like site geology, building materials, ventilation, occupancy, and heating systems. Previous research found a correlation between the indoor radon concentration and several factors, that can be external or internal to the building. The external factors are related to the building localization and the climate conditions: geology and lithology, region, elevation above sea level, location on a hill or slope, Air pressure difference, outdoor temperature, and total atmospheric moisture. The internal factors are composed of the foundation type, the existence of a basement, the building materials, the story (distance from the ground), ventilation system, heating system, thermal retrofit, and indoor relative humidity. Table 9.1 shows the list of factors and respective researches that found the correlation of each factor with the radon gas concentration.

## 9.3 Materials and Methods

Based on the evidence presented in the literature, the first step in the design of a qualitative pre-diagnosis model for indoor radon potential evaluation is the selection of a set of relevant variables correlated with indoor concentration level. These variables consist of a set of characteristics that will be selected from the work developed in the scope of previous Indoor Air Quality (IAQ) research, according to three categories: local geology, built environment performance, and occupancy schedules.

In parallel, the research team is gathering systematic data on radon concentration measurements and enriching this data with these three categories of characteristics. The inventory of our previous studies comprises radon measurements carried out in different types of buildings: dwellings, schools, and administrative offices. This list includes several heritage buildings in the historic center of cities of North of Portugal. Then, it will be assessed the correlation and relative weight of each variable concerning the radon to select the set of the most relevant characteristics or factors. Based on that set of variables, it will be developed a qualitative pre-diagnosis model for the evaluation of radon concentration potential in indoor environments through the possible use of classification and regression trees.

A pre-diagnosis model is an interactive tool where the user can verify the need to carry out a radon assessment study, starting from a set of variables. This approach allows the determination of the level of radon occurrence potential in a building, be it for housing or as the workplace. The pre-diagnosis model aims to measure the level of risk so that the necessary measures can be taken, after confirmation by assessment, to mitigate the problem and consequent monitoring. Thus, the pre-diagnosis model

**Table 9.1** List of factors correlated with indoor radon concentration

| Category | Factor | References |
|---|---|---|
| External factors | Geology and lithology | Andersen et al. (2007), Martins et al. (2013), Martins et al. (2016) and Otton (1992) |
| | Region | Andersen et al. (2007) |
| | Elevation above sea level | Barros-Dios et al. (2007) |
| | Location on a hill or a slope | Otton (1992) |
| | Air pressure difference | Otton (1992), IAEA (2019), Cerqueiro-Pequeño et al. (2021) and McGrath and Byrne (2021) |
| | Outdoor temperature | Cortina et al. (2008), IAEA (2019) and Cerqueiro-Pequeño et al. (2021) |
| | Total atmospheric moisture | Groves-Kirkby et al. (2015) |
| Internal factors | Foundation type | Andersen et al. (2007), Martins et al. (2013), Collignan et al. (2016) and Otton (1992) |
| | Existence of basement | Barros-Dios et al. (2007), Andersen et al. (2007), Martins et al. (2016), Otton (1992) and Barros (2016) |
| | Building materials | Andersen et al. (2007), Collignan et al. (2016) and Otton (1992) |
| | Storey (distance to the ground) | Barros-Dios et al. (2007), Andersen et al. (2007), Collignan et al. (2016) and Barros (2016) |
| | Ventilation system | Collignan et al. (2016) and Barros (2016) |
| | Heating system | Collignan et al. (2016), IAEA (2019) and Barros (2016) |
| | Thermal retrofit | Collignan et al. (2016) |
| | Indoor relative humidity | Cerqueiro-Pequeño et al. (2021) |

is based on the determination of the potential of radon concentration, assuming a set of variables, namely the location, the type of occupancy of the building, and the type of construction, as shown in the flowchart of Fig. 9.1.

In the specific case of the location, which is a critical variable because it is directly related to the geology of the place, is decisive for the occurrence of radon. However, also for the other variables, the user must provide indications so that the pre-diagnosis model can determine the potential occurrence of the radon. In this way, the user must enter information about the variable's location, occupation mode, and the built environment. For example, does the occupancy take place permanently in the building? Or is it occasional? Does the building only have a ground floor? Or does it develop in height with several floors? Or, does the occupation of the building take place in a basement? After assigning a weight to each one, these are the characteristics that will allow the pre-diagnosis model to assess the potential for radon occurrence inside the building. When the result points to a low potential, it is not necessary to implement

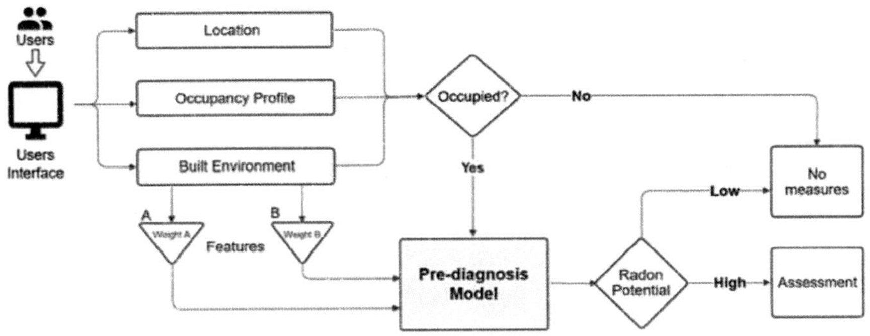


**Fig. 9.1** Pre-diagnosis model

any measure. However, if the pre-diagnosis model points towards the verification of high potential, assessment measures should be taken to quantify radon concentrations and to be able to calculate the doses to which the building's occupants may be exposed.

## 9.4 Expected Results and Preliminary Conclusions

The goal of this work was the design of a qualitative pre-diagnosis model for the evaluation of radon potential in indoor environments, by considering a set of relevant factors carefully selected to characterize occupants' risk exposure. In the scope of the work, we identified the relevant characteristics that most affect indoor radon concentration, concerning the building's external and internal factors. At the time this survey is conducted, we are preparing radon measurements data sets that integrate the selected factors.

The implementation of a pre-diagnosis model for indoor radon potential is a decisive tool for indoor radon potential evaluation. The model proposed in this paper will be laid upon a set of key factors which determine indoor radon concentration, mainly concerning building construction, its applied technologies, devices, and features, but also regarding the geological characteristics of the foundation soil and its type. Following work undertaken, radon measurements data sets that integrate the selected factors are already under preparation, and a web platform to give the direct public access to the pre-diagnosis model is an ongoing task.

The goal is to provide a software tool to support the first step of a broader approach to public health problems. After this stage is concluded, a short-term radon concentration measurement will be implemented whenever a user gets a high-risk estimate. And, if the short-term measurement confirms the high risk, a further long-term radon concentration measurement should be carried out. At last, after the long-term measurement confirms the high risk an in-situ remediation program will be implemented to reduce indoor radon exposure.

Recently retrofitted buildings play an important role in the current model development since well-insulated walls, windows, and roofs with high energy efficiency performance can lead to an air renovation reduction, designed to optimize winter thermal comfort. Nevertheless. the ventilation strategies are essential to improve Indoor Air Quality and reduce indoor radon potential. Radon mitigation strategies should accomplish a good trade-off between Indoor Air Quality and energy efficiency in order to improve indoor air quality with low impact on energy efficiency.

**Acknowledgements** This research is a result of the project TECH—Technology, Environment, Creativity and Health, Norte-01-0145-FEDER-000043, supported by Norte Portugal Regional Operational Program (NORTE 2020), under the PORTUGAL 2020 Partnership Agreement, through the European Regional Development Fund (ERDF). L.J.R.N. was supported by proMetheus, Research Unit on Energy, Materials and Environment for Sustainability—UIDP/05975/2020, funded by national funds through FCT—Fundação para a Ciência e Tecnologia. António Curado co-authored this work within the scope of the project proMetheus—Research Unit on Materials, Energy and Environment for Sustainability, FCT Ref. UID/05975/2020, financed by national funds through the FCT/MCTES.

## References

Andersen CE, Raaschou-Nielsen O, Andersen HP, Lind M, Gravesen P, Thomsen BL (2007) Prediction of 222Rn in Danish dwellings using geology and house construction information from central databases. Radiat Protect Dosim 123:83–94

Azevedo R, Silva JP, Lopes N, Curado A, Lopes SI (2021) Short-term indoor radon gas assessment in granitic public buildings: a multi-parameter approach. In: Advances in science, technology & innovation. Springer, Cham, pp 415-418

Barros N, Steck DJ, Field WR (2016) Utility of short-term basement screening radon measurements to predict year-long residential radon concentrations on upper floors. Radiat Protect Dosim 171(3):405–413

Barros-Dios JM, Ruano-Ravina A, Gastelu-Iturri J, Figueiras A (2007) Factors underlying residential radon concentration: results from Galicia, Spain. Environ Res 103(2):185–190

Cerqueiro-Pequeño J, Comesaña-Campos A, Casal-Guisande M, Bouza-Rodríguez J-B (2021) Design and development of a new methodology based on expert systems applied to the prevention of indoor radon gas exposition risks. Int J Environ Res Public Health 18(1):269

Collignan B, Le Ponner E, Mandin C (2016) Relationships between indoor radon concentrations, thermal retrofit and dwelling characteristics. J Environ Radioact 165:124–30

Cortina D, Durán I, Llerena JJ (2008) Measurements of indoor radon concentrations in the Santiago de Compostela Area. J Environ Radioact 99:1583–1588

Curado A, Lopes SI, Antão A (2020) On the relation of geology, natural ventilation and indoor radon concentration: the Northern Portugal case study. In: Comunicações Geológicas, vol 107. LNEG, São Mamede de Infesta, pp 31–41

Groves-Kirkby CJ, Crockett RGM, Denman AR, Phillips PS (2015) A critical analysis of climatic influences on indoor radon concentrations: implications for seasonal correction. J Environ Radioact 148:16–26

Hahn EJ, Gokun Y, Andrews WM, Overfield BL, Robertson H, Wiggins A (2015) Radon potential, geologic formations, and lung cancer risk. Prev Med Rep 2:342–346

International Atomic Energy Agency (IAEA) (2019) Design and conduct of indoor radon surveys. In: Safety reports series, no. 98. International Atomic Energy Agency Vienna International Centre, Vienna, pp 1–128

Martins LMO, Gomes MEP, Neves LJPF, Pereira AJSC (2013) The influence of geological factors on radon risk in groundwater and dwellings in the region of Amarante (Northern Portugal). Environ Earth Sci 68:733–740

Martins LMO, Gomes MEP, Teixeira RJS, Pereira AJSC, Neves LJPF (2016) Indoor radon risk associated to post-tectonic biotite granites from Vila Pouca de Aguiar Pluton, Northern Portugal. Ecotoxicol Environ Safety 133:164–175

McGrath JA, Byrne MA (2021) An approach to predicting indoor radon concentration based on depressurization measurements. Indoor Built Environ 30(8):1042–1050. https://doi.org/10.1177/1420326X20924747

Otton JK (1992) The geology of radon. In: General interest publication, USGS Publications Warehouse, United States

Soltani-Nabipour J, Khorshidi A, Sadeghi F (2019) Constructing environmental radon gas detector and measuring concentration in residential buildings. In: Physics of particles and nuclei letters, vol 16. Springer Nature, Berlin, pp 789–795

Steck DJ, Field RW, Lynch CF (1999) Exposure to atmospheric radon. Environ Health Perspect 107(2):123–127

World Health Organization (WHO) (2009) WHO handbook on indoor radon: a public health perspective. World Health Organization, Geneva

# Part III
# Life Cycle Analysis Methodologies

# Chapter 10
# Prospective Life Cycle Assessment of REDIFUEL, an Emerging Renewable Drop-in Fuel

**A. E. M. van den Oever, Daniele Costa, and Maarten Messagie**

**Abstract** A novel drop-in renewable fuel process was recently developed in project named 'Robust and Efficient processes and technologies for Drop In renewable FUELs for road transport' (REDIFUEL). The plant concept consists of a dual-fluidized bed gasifier, followed by a Fischer–Tropsch process and a hydroformylation process to produce a mixture of high-cetane hydrocarbons and C6-C11 alcohols. The Fischer–Tropsch process can also be preceded by synthetic gas production from renewable hydrogen and carbon capture. The environmental impacts of three production pathways for this fuel were compared with Life Cycle Assessment methodology. REDIFUEL has a climate change impact of 8.1–22.8 $gCO_2eq/MJ$ when produced from bark, 16.2–30.9 $gCO_2eq/MJ$ for short rotation coppice, and 11.1–19.8 $gCO_2eq/MJ$ for carbon capture. REDIFUEL production from short rotation coppice leads to the highest environmental impacts in 12–14 out of 16 impact categories assessed, especially in marine eutrophication. REDIFUEL from carbon capture has the lowest impacts in 9 out of 16 of the assessed environmental impact categories, but it leads to high material resource depletion impacts. REDIFUEL production from bark has the lowest impacts in 6 or 7 categories, and it was not associated with specific risks compared to the other scenarios.

**Keywords** Biofuel · Comparison · Gasification · Fischer–Tropsch · Life cycle assessment (LCA)

A. E. M. van den Oever (✉) · M. Messagie
Mobility, Logistics and Automotive Research Centre, Department of Electric Engineering and Energy Technology, Vrije Universiteit Brussel, Pleinlaan 2, 1050 Brussels, Belgium
e-mail: anne.van.den.oever@vub.be

D. Costa
VITO/EnergyVille, Boeretang 200, 2400 Mol, Belgium

 
N. S. Caetano and M. C. Felgueiras (eds.), *The 9th International Conference on Energy and Environment Research*, Environmental Science and Engineering,
https://doi.org/10.1007/978-3-031-43559-1_10

## 10.1 Introduction

In the context of the Green Deal, the European Commission has set a target to reduce the European greenhouse gas (GHG) emissions by 55% in 2030 (EC 2022). All sectors, transport included, need to ramp up renewable energy integration to reach this goal. Therefore, the the latest proposal for the revised Renewable Energy Directive, RED III, increases the minimum blending share of advanced biofuels to 2.2% and sets a new blending share for renewable fuels from non-biological origin (RFNBOs) of 2.6% (EC 2021). These fuels must comply with minimum GHG emission saving targets of 80% for advanced biofuel plants starting operation after 2026, and 70% for RFNBOs. RED III also acknowledges the need for applying the principles of circular economy to the bioenergy sector. Therefore it maintains the strict definition of 'advanced' feedstocks from the current renewable energy directive, RED II (Council Directive 2018). An emerging technology for renewable drop-in fuel production was recently developed in the REDIFUEL project (Heuser et al. 2020; REDIFUEL 2020). The proposed conversion technology produces a mixture of high-cetane hydrocarbons and C6-C11 alcohols, henceforward REDIFUEL. The fuel properties are almost fully compliant with the EN590 standard, allowing high blending rates with fossil diesel. In addition, the fuel has proven to reduce a vehicle's fuel consumption, which positively affects its environmental impacts (Mojtaba Lajevardi et al. 2019; Sacchi et al. 2021; Sen et al. 2017). REDIFUEL can be produced from diverse feedstocks, such as bark for advanced biofuel production as defined in RED II, short rotation coppice (SRC) for non-advanced second-generation biofuel production, and carbon capture and usage (CCU) for RFNBO production. Further technology improvements are still required, but the fuel is expected to be brought to the market at a competitive price by 2030. The main objective of this work was to compare the environmental performance of producing REDIFUEL from bark, SRC, and CCU in the medium term (2030).

## 10.2 Methodology

### *10.2.1 Goal and Scope*

Life Cycle Assessment (LCA) methodology (ISO 2006a, 2006b) was used to evaluate the environmental performance of REDIFUEL production in the European Union in 2030. The functional unit is 1 MJ of fuel, based on the lower heating value (LHV), delivered to a vehicle tank. An attributional modelling approach was taken and allocation was used for solving multifunctionality. The background model was the Ecoinvent cut-off system model (Wernet et al. 2016), which applies economic allocation. In the foreground model, fair use of economic allocation is complex since the economic values of the generated co-products are uncertain. As all products function as energy

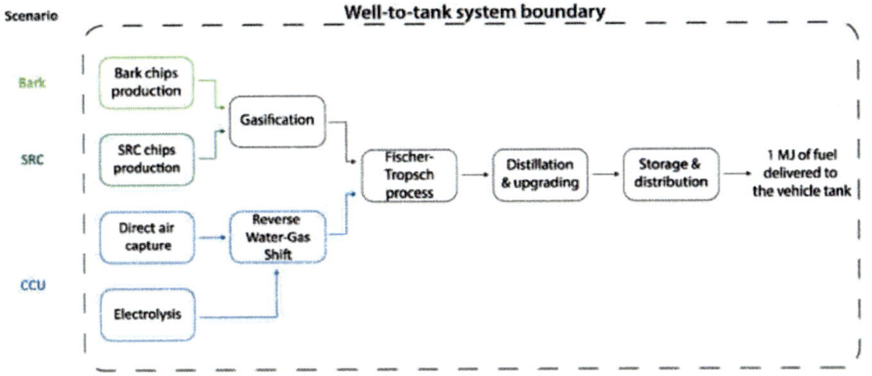


**Fig. 10.1** Well-to-tank system boundaries of three scenarios for REDIFUEL production. Grey processes are the same for all scenarios. Legend: *CCU* carbon capture and usage, *SRC* short-rotation coppice

carriers, allocation based on energy content (LHV) was adopted to solve the multi-functionality of the products from the REDIFUEL plant. The system boundaries of all scenarios are well-to-tank (Fig. 10.1).

Three scenarios with REDIFUFEL have been defined (Fig. 10.1) with different feedstocks: bark and SRC for biofuel production and carbon capture and usage (CCU) for RFNBO production. An explorative scenario approach was taken (Börjeson et al. 2006), i.e., a best-case and worst-case were defined for each situation. Scenario-specific assumptions are given in the life cycle inventory (Sect. 10.3).

### *10.2.2 Life Cycle Impact Assessment*

The most recent European impact assessment method, Environmental Footprint 3.0 (Fazio et al. 2018; Saouter et al. 2020), was used. The LCA was performed in the Activity Browser software (Steubing et al. 2020).

## 10.3 Life Cycle Inventory

A superstructure background database was created with the python library premise (Sacchi et al. 2022). This database is a prospective transformation of the Ecoinvent 3.8 cut-off database (Wernet et al. 2016), using output from the integrated assessment model IMAGE (PBL 2022). The selected worst-case and best-case scenarios for the superstructure database were SSPS2-base (no climate policy) and SSP2-RCP19 (1.5 °C climate target). The data for the foreground model of REDIFUEL production (Fig. 10.2) was generated via pilot-scale experiments and process simulation in Aspen

Plus® during the REDIFUEL project. Pathway A in Fig. 10.2 refers to the two biomass cases, Bark and SRC, while pathway B refers to the e-fuel case (CCU). No experimental data or process modelling was done for the blocks blending, direct air capture, CO production RWGS, and electrolysis (Fig. 10.2).

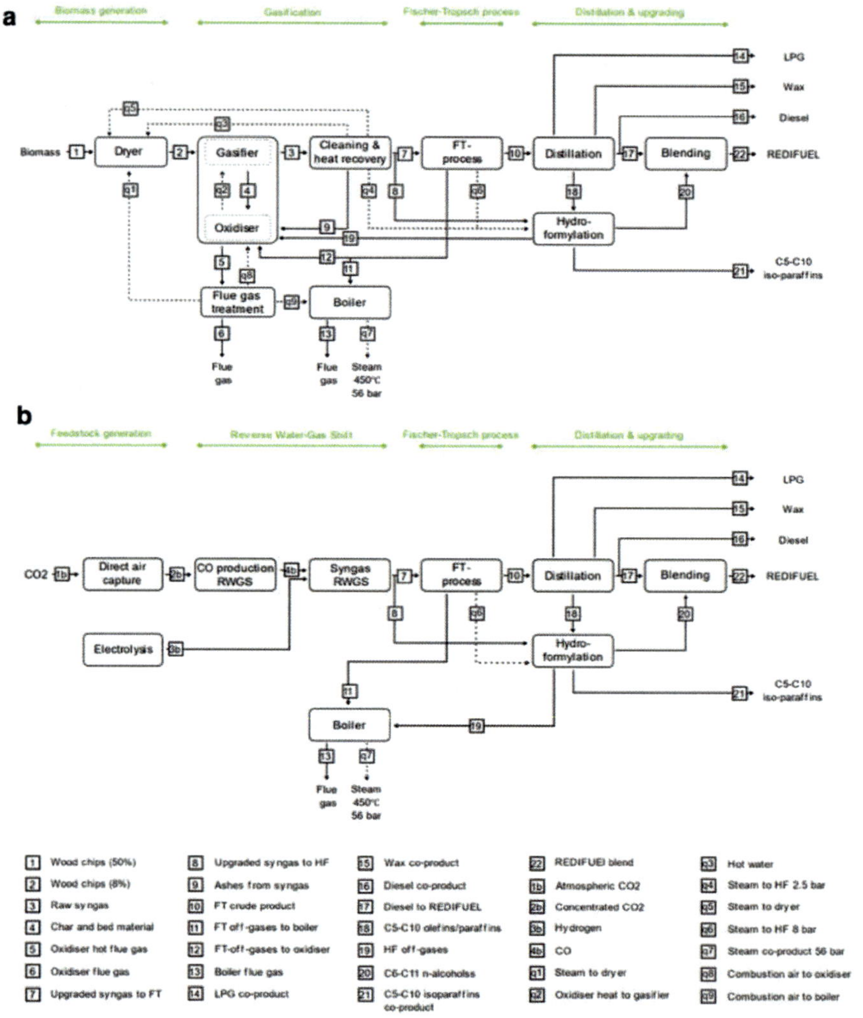


**Fig. 10.2** REDIFUEL process description. The green arrows indicate the corresponding life cycle phases in Fig. 10.1. Part **a** represents the REDIFUEL process with biomass gasification. **b** represents a hypothetical future adaptation of the REDIFUEL process, with direct $CO_2$ air capture and electrolysis. Legend: *FT* Fischer–Tropsch, *HF* hydroformylation, *LPG* liquified petroleum gas, *RWGS* reverse water gas shift

### 10.3.1 Biomass Generation

Two biomass types have been considered: the industrial residue bark and SRC, grown for energy generation. The REDIFUEL production process was demonstrated using Finnish bark. Therefore, its moisture content and chemical composition were considered. It was assumed that the average European bark has the same composition as the Finnish bark used in the project. The bark LHV was obtained experimentally from the project. The bark chips arrive at the plant with a moisture content of 50% and are dried to a moisture content of 12% with a belt dryer. The environmental impact of bark production was modelled with an average European market mix from the superstructure database.

In contrast to bark, the SRC wood is not a by-product and carries all the burdens of the plantation production and harvesting processes. A 3 years rotation period was assumed. For SRC, a German production process of willow chips was selected. For both biomass types, the transport distance from the sawmill to the REDIFUEL plant was assumed as 100 km, which corresponds to the economically feasible transport distance of wood chips (Trzciński et al. 2021).

### 10.3.2 Carbon Capture and Electrolysis

The syngas production process for CCU was taken from the superstructure database. The $CO/H_2$ ratio was adapted to match the REDIFUEL syngas composition. It was assumed that the electricity consumed in all processes (direct carbon capture, electrolysis, CO production via the reverse water gas shift (RWGS), and syngas production via RWGS in Fig. 10.2) is derived from renewable sources (Lebois et al. 2020).

### 10.3.3 Gasification

Gasification occurs in a dual fluidized-Bed gasifier, as Frilund et al. (2021) described. The flue gas from the oxidizer is filtered to separate the remaining fly ashes. In the worst-case scenario, the fly ashes are landfilled, while they are used for landfarming in the best-case scenario. The syngas also goes through a reformer with a nickel catalyst. In the worst-case, the lifetime of the nickel catalyst is 3 years, while, in the best-case, it is assumed that regular regeneration with excess steam from the plant can double the lifetime to 6 years.

### 10.3.4 Fischer–Tropsch (FT) Process and FT-Catalyst Production

The FT-process converts syngas into crude FT oil. The novel FT catalyst developed in the REDIFUEL project was modelled based on mass and energy balances from lab-scale experiments (Jeske et al. 2021), and the upscaling was done following the framework of Piccinno et al. (2016). The catalyst loading in the FT-reactor was calculated by dividing the syngas feed to the reactor by the weight hourly space velocity (WHSV), which expresses the feed syngas weight per unit weight of the catalyst per hour. The worst-case WHSV (5 $h^{-1}$) and the best-case WHSV (33 $h^{-1}$) were derived from the lab-scale experiments.

### 10.3.5 Distillation and Upgrading

The crude FT-product coming from the FT-process is distilled to separate the wax fraction (C22+), the diesel fraction (middle distillate C11-C21), the C5-C10 olefins and the paraffin fraction, and the light fuel gas fraction (C3/C4). Various industries could directly use waxes (cosmetics, adhesives, candle making, etc.) or it could be hydrocracked and sold as a fuel. The C5-C10 olefins/paraffins fraction goes to hydroformylation (Püschel et al. 2021; Rösler et al. 2021), producing C6-C11 n-alcohols for the REDIFUEL blend and C5-C10 iso-paraffins, which could be sold as a co-product as a renewable petrol substitute.

### 10.3.6 Storage, Blending and Distribution

For all scenarios, the storage, blending and distribution were modelled with data from the superstructure database. It was assumed that REDIFUEL would be produced and consumed in Europe and that the distribution impact would be similar to domestically produced biodiesel.

## 10.4 Results and Discussion

The climate change (CC) impact of REDIFUEL ranges from 8.1–22.8 $gCO_2eq/MJ$ for Bark, 16.2–30.9 $gCO_2eq/MJ$ for SRC, and 11.1–19.8 $gCO_2eq/MJ$ for CCU (Fig. 10.3). In the worst-case scenario, REDIFUEL production does not comply with the RED II GHG emission savings targets for Bark and SRC. Electricity consumption contributes to 62% of the total impact for Bark, and to 46% for SRC. The current

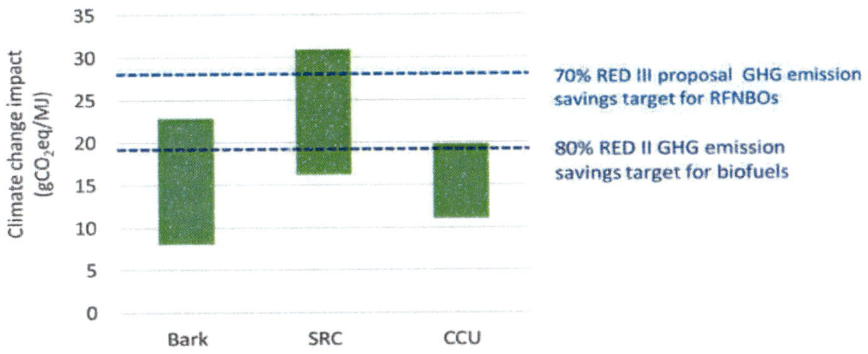


**Fig. 10.3** Climate change impact of REDIFUEL from bark, short rotation coppice (SRC), and carbon capture and usage (CCU) in $gCO_2eq/MJ$. The bars represent the range between the worst-case and the best-case. The Renewable Energy Directive (RED II) mandates a minimum of 80% greenhouse gas (GHG) emission savings for biofuel installations starting operation after 2026, while the RED III proposal sets a minimum of 70% GHG emission savings target for renewable fuels from non-biological origin (RFNBOs)

electricity consumption is based on process simulation, but future process optimization and industrial learning can decrease this consumption. When the electricity consumption of the REDIFUEL process is decreased by 30%, scenario Bark reaches the RED II target even in the worst-case. In the best-case Bark and SRC already comply, leading to 91 and 83% GHG emission savings compared to the fossil fuel reference. CCU leads to 79–88% GHG emission savings, and it complies with the proposed RED III savings target in both the worst-case and the best-case.

The results for all other impact categories are given relative to the scenario Bark (Fig. 10.4). In the worst-case, Bark has the lowest impact for 6 out of 16 impact categories assessed: acidification potential (AP), human toxicity, carcinogenic (HT-c), material resource depletion (MRD), ozone depletion potential (ODP), particulate matter formation (PMF), and photochemical ozone depletion (POF). SRC has the lowest impact for non-carcinogenic human toxicity (HT-nc), and CCU has the lowest impact in 9 out of 16 impact categories: CC, ecotoxicity (ET), energy resource depletion (ERD), freshwater eutrophication (EP-f), marine eutrophication potential (EP-m), terrestrial eutrophication potential (EP-t), ionizing radiation potential (IRP), land use (LU), and water use (WU). SRC has the highest impacts in 12 out of 16 environmental impact categories. In the best-case, scenario Bark had the best performance in 7 out of 16 impact categories: CC, AP, EP-t, HT-c, MRD, PMF, and POF. In the 9 remaining environmental impact categories, CCU had the lowest impacts, and in 14 out of 16 categories, SRC had the highest impacts.

Some of the results stand out, for example, the relatively high EP-m impacts of SRC in both cases, which are caused by fertilizer use during the production phase. The relatively high HT-c and HT-nc impacts of CCU in the worst-case compared to the best-case scenario stand out too, since the absolute HT-c and HT-nc impacts of CCU differ by only 3% between the worst-case and best-case. CCU itself does not

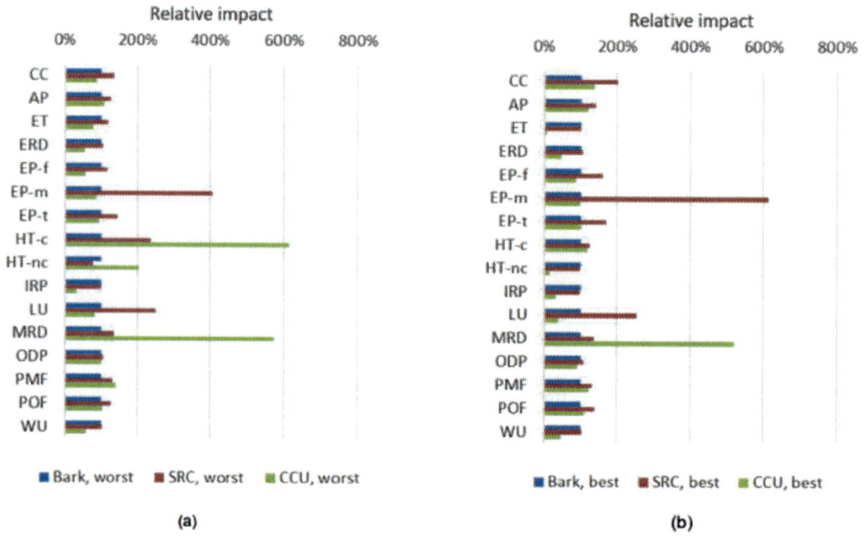


**Fig. 10.4** Results of the well-to-tank Life Cycle Assessment of 1 $MJ_{LHV}$ REDIFUEL in the worst-case (**a**) and the best-case (**b**) compared to bark (= 100%). Legend: *CC* climate change, *AP* acidification potential, *ET* ecotoxicity, *ERD* energy resource depletion, *EP-f* freshwater eutrophication potential, *EP-m* marine eutrophication potential, *EP-t* terrestrial eutrophication potential, *HT-c* ***human toxicity, carcinogenic, *HT-nc* human toxicity, non-carcinogenic, *IRP* ionizing radiation potential, *LU* land use, *MRD* material resource depletion, *ODP* ozone depletion potential, *PMF* particulate matter formation, *POF* photochemical ozone depletion, *WU* water use, *SRC* short rotation coppice, *CCU* carbon capture and usage

cause the relative high HT-c and HT-nc impacts, but by the lower absolute HT-c and HT-nc impacts of Bark and SRC in the worst-case due to different fly ash management practices. Landfilling (worst-case) encloses pollutants, leading to lower toxicity impacts than landfarming, where the pollutants are dispersed in the environment. At last, it can be observed that CCU leads to relatively high MRD impacts due to the mineral resource requirements of renewable energy technologies.

## 10.5 Conclusions

The LCA presented in this paper has shown that the climate change impact of renewable drop-in REDIFUEL production ranges from 8.1 to 30.9 $gCO_2eq/MJ$. From all feedstocks assessed, dedicated energy crops such as SRC are the least preferable feedstocks for REDIFUEL production, as they lead to lower GHG emission savings and the highest impacts in 75–88% of the impact categories. In particular, high EP-m risks were found. REDIFUEL production from an advanced feedstock such as bark has the potential to comply with the RED II GHG emission savings targets provided a clean electricity mix is used or if the future REDIFUEL plant's electricity consumption is decreased by 30% compared to the current simulation-based estimations. The scenario Bark lead to the lowest impacts in 6 or 7 out of the 16 assessed environmental impact categories, depending on the background scenario. REDIFUEL production from CCU complies with the RED III GHG emission savings target for RFNBOs, regardless of the background used. It also leads to the lowest impacts in 9 out of 16 environmental impact categories. However, it is associated with high MRD impacts. To conclude, the REDIFUEL technology has the potential to contribute to the Green Deal's climate mitigation goals by 2030. Future developments should consider advanced biofuel and RFNBOs production pathways, reduce electricity consumption, and address the depletion of mineral resources of renewable energy technologies.

**Acknowledgements** This work was supported by REDIFUEL, which has received funding from the European Union's Horizon 2020 research and innovation programme (Grant Agreement no. 817612). The authors thank all REDIFUEL project partners for their efforts and the productive collaboration. The authors would also like to express their gratitude to Romain Sacchi for his help with premise.

## References

Börjeson L, Höjer M, Dreborg KH, Ekvall T, Finnveden G (2006) Scenario types and techniques: towards a user's guide. Futures 38:723–739. https://doi.org/10.1016/j.futures.2005.12.002

Council Directive (2018) 2018/2001/EU on the promotion of the use of energy from renewable sources, vol L328

EC (2021) Proposal for a directive of the European parliament and of the council amending directive (EU) 2018/2001 of the European Parliament and of the Council, Regulation (EU) 2018/1999 of the European Parliament and of the Council and Directive 98/70/EC of the E 2021;0218

EC (2022) A European green deal 2022. https://ec.europa.eu/info/strategy/priorities-2019-2024/european-green-deal_en. Accessed 25 July 2022

Fazio S, Biganzioli F, De Laurentiis V, Zampori L, Sala S, Diaconu E (2018) Supporting information to the characterisation factors of recommended EF life cycle impact assessment methods, version 2, from ILCD to EF 3.0, EUR 29600 EN. Ispra. https://doi.org/10.2760/002447

Frilund C, Tuomi S, Kurkela E, Simell P (2021) Small- to medium-scale deep syngas purification: biomass-to-liquids multi-contaminant removal demonstration. Biomass Bioenerg 148:106031. https://doi.org/10.1016/j.biombioe.2021.106031

Heuser B, Vorholt A, Prieto G, Graziano B, Schönfeld S, Messagie M et al (2020) REDIFUEL: robust and efficient processes and technologies for drop-in renewable FUELs for road transport. Transp Res Arena 2020 (Helsinki, Finland)

ISO (2006a) ISO 14040:2006. Environmental management—life cycle assessment—principles and framework. International Organization for Standardization, Switzerland

ISO (2006b) ISO 14044:2006. Environmental management—life cycle assessment—requirements and guidelines. International Organization for Standardization, Switzerland

Jeske K, Kizilkaya AC, López-Luque I, Pfänder N, Bartsch M, Concepción P et al (2021) Design of cobalt Fischer–Tropsch catalysts for the combined production of liquid fuels and olefin chemicals from hydrogen-rich syngas. ACS Catal 11:4784–4798. https://doi.org/10.1021/acscatal.0c05027

Lebois O, Boersma P, McGowan D, Rzepczyk T, Sönmez C, Powell D et al (2020) TYNDP 2020—scenario report. ENTSO-E and ENTSOG, Brussels

Mojtaba Lajevardi S, Axsen J, Crawford C (2019) Comparing alternative heavy-duty drivetrains based on GHG emissions, ownership and abatement costs: simulations of freight routes in British Columbia. Transp Res Part D Transp Environ 76:19–55. https://doi.org/10.1016/j.trd.2019.08.031

PBL (2022) IMAGE—integrated model to assess the global environment. https://www.pbl.nl/en/image/about-image. Accessed 17 July 2022

Piccinno F, Hischier R, Seeger S, Som C (2016) From laboratory to industrial scale: a scale-up framework for chemical processes in life cycle assessment studies. J Clean Prod 135:1085–1097. https://doi.org/10.1016/j.jclepro.2016.06.164

Püschel S, Störtte S, Topphoff J, Vorholt AJ (2021) Green process design for reductive hydroformylation of renewable olefin cuts for drop-in diesel fuels 2021:5226–5234. https://doi.org/10.1002/cssc.202100929

REDIFUEL (2020) Project 2020. https://redifuel.eu/project/. Accessed 26 Nov 2020

Rösler T, Ehmann KR, Köhnke K, Leutzsch M, Wessel N, Vorholt AJ et al (2021) Reductive hydroformylation with a selective and highly active rhodium amine system 400:234–243. https://doi.org/10.1016/j.jcat.2021.06.001

Sacchi R, Bauer C, Cox BL (2021) Does size matter? The influence of size, load factor, range autonomy, and application type on the life cycle assessment of current and future medium- and heavy-duty vehicles. Environ Sci Technol 55:5224–5235. https://doi.org/10.1021/acs.est.0c07773

Sacchi R, Terlouw T, Siala K, Dirnaichner A, Bauer C, Cox B et al (2022) Prospective environmental impact assessment (premise): a streamlined approach to producing databases for prospective life cycle assessment using integrated assessment models. Renew Sustain Energy Rev 160:112311. https://doi.org/10.1016/j.rser.2022.112311

Saouter E, Biganzoli F, Ceriani L, Versteeg D, Crenna E, Zampori L et al (2020) Environmental Footprint: update of life cycle impact assessment methods—ecotoxicity freshwater, human toxicity cancer, and non-cancer, EUR 29495 EN. Luxembourg. https://doi.org/10.2760/300987

Sen B, Ercan T, Tatari O (2017) Does a battery-electric truck make a difference?—life cycle emissions, costs, and externality analysis of alternative fuel-powered Class 8 heavy-duty trucks in the United States. J Clean Prod 141:110–121. https://doi.org/10.1016/j.jclepro.2016.09.046

Steubing B, de Koning D, Haas A, Mutel CL (2020) The activity browser; an open source LCA software building on top of the brightway framework. Softw Impacts 3. https://doi.org/10.1016/j.simpa.2019.100012

Trzciński G, Tymendorf Ł, Kozakiewicz P (2021) Parameters of trucks and loads in the transport of scots pine wood biomass depending on the season and moisture content of the load. Forests 12:1–17. https://doi.org/10.3390/f12020223

Wernet G, Bauer C, Steubing B, Reinhard J, Moreno-Ruiz E, Weidema B (2016) The ecoinvent database version 3 (part I): overview and methodology. Int J Life Cycle Assess 21:1218–1230. https://doi.org/10.1007/s11367-016-1087-8

# Chapter 11
# Life Cycle Environmental Impacts of Water Use in Buildings: A Case Study in Qatar

**Mehzabeen Mannan and Sami G. Al-Ghamdi**

**Abstract** Water consumption in buildings is a significant contributor to global freshwater utilization, yet research on the consequences of water usage has been limited. This study aims to evaluate the environmental implications of life-cycle water consumption in a multi-family residential building located in Doha, Qatar, using a comprehensive life cycle assessment (LCA). The analysis of the building was conducted using Building Information Modelling (BIM) as the primary tool. The LCA results revealed substantial impacts during the raw water treatment phase in Doha, which is characterized by energy-intensive thermal desalination. The water usage phase accounted for nearly half of the total impact observed. Throughout the life cycle of the modeled building, the cumulative annual emission reached 59,440 kg of $CO_2$. This research provides valuable insights for water authorities and the building research community, facilitating the development of more sustainable water usage policies tailored to specific regions or countries. By understanding the environmental ramifications of water consumption in buildings, policymakers can make informed decisions to reduce the ecological footprint associated with water usage. Overall, this study underscores the importance of considering the life-cycle perspective in assessing the environmental impact of water consumption in buildings. It emphasizes the need for sustainable water management practices and encourages the adoption of efficient technologies and policies to mitigate the environmental consequences of water usage in residential structures.

M. Mannan · S. G. Al-Ghamdi (✉)
Division of Sustainable Development, College of Science and Engineering, Hamad Bin Khalifa University, Qatar Foundation, Doha, Qatar
e-mail: sami.alghamdi@kaust.edu.sa

S. G. Al-Ghamdi
Environmental Science and Engineering Program, Biological and Environmental Science and Engineering Division, King Abdullah University of Science and Technology (KAUST), Thuwal 23955-6900, Saudi Arabia

KAUST Climate and Livability Initiative, King Abdullah University of Science and Technology (KAUST), Thuwal 23955-6900, Saudi Arabia



N. S. Caetano and M. C. Felgueiras (eds.), *The 9th International Conference on Energy and Environment Research*, Environmental Science and Engineering,
https://doi.org/10.1007/978-3-031-43559-1_11

**Keywords** Life cycle assessment · Environmental impact · Urban water cycle

## 11.1 Introduction

Buildings are recognized as significant consumers of fresh water, and their water usage yields diverse environmental repercussions (Yoonus and Al-Ghamdi 2020). Despite growing concerns regarding freshwater availability, research in this domain has predominantly concentrated on ensuring a reliable supply and minimizing water consumption within buildings, overlooking the evaluation of the environmental implications associated with building-specific water use and their corresponding importance in both building and water research studies (Arpke and Hutzler 2006).

### *11.1.1 Water Use in Qatar*

Qatar, located in the arid region of the Arabian Peninsula, experiences a desert climate characterized by minimal rainfall, averaging just 80 mm annually. The region's high temperatures during the summer, exceeding 40 °C, coupled with evaporation rates of 2200 mm per year, exacerbate the scarcity of freshwater resources. Evapotranspiration rates range from less than 2 mm per day in December to a peak of 10 mm per day in June (Forecasts IBMIS 2011).

In spite of being considered a water-scarce nation, Qatar faces an intriguing predicament. This country has one of the highest domestic water consumption rates in the world, consuming approximately 430 L of water per person per day (Fig. 11.1), which is a mean estimate of municipal desalinated water production (approximately 600 L/capita/day) and treated wastewater (approximately 250 L/capita/day) (Lawler et al. 2023). Water use in Qatar has experienced a significant increase in recent years, with statistics highlighting a substantial rise in used water quantity across various sectors. This exceptionally high demand for water is attributed to rapid infrastructure development and rising standards of living, which have contributed to an increased need for water resources. Research findings for the period 2006–2016 emphasized the dominant roles of agriculture and domestic sectors in water allocation, while also highlighting notable growth rates in the government and commercial sectors (Water Statistics in the State of Qatar 2017).

To address its freshwater requirements, Qatar relies heavily on seawater desalination and groundwater abstraction, with treated wastewater playing a comparatively smaller role. Seawater desalination technologies have been harnessed to convert saline seawater into potable water, effectively augmenting the available freshwater supply. Groundwater abstraction serves as another vital source of water, extracted from underground aquifers. However, this practice raises concerns regarding long-term sustainability due to the finite nature of groundwater resources.

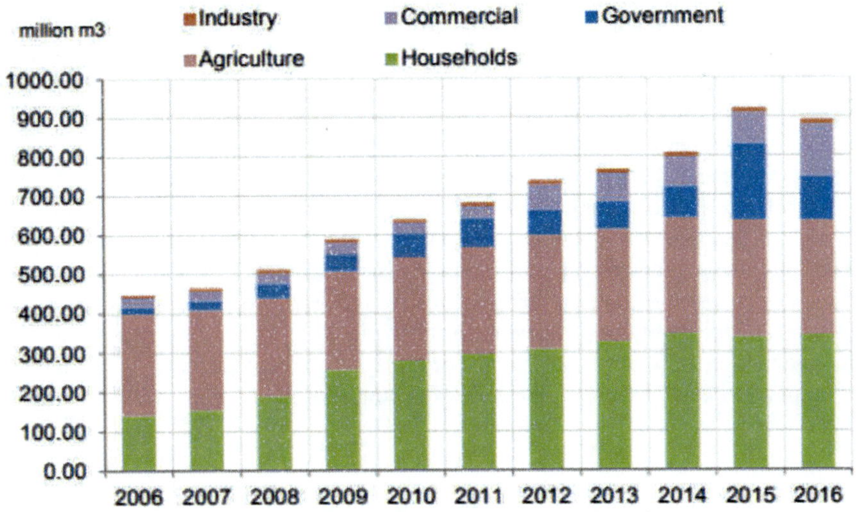


**Fig. 11.1** The volumetric measurement of water use by economic activity, excluding losses, from 2006 to 2016. Adapted from Water Statistics in the State of Qatar (2017)

## *11.1.2 Sustainability Assessment of Domestic Water Use*

Considering the unique water scarcity situation in Qatar, where water demand surpasses available freshwater resources, it becomes imperative to examine the water production systems in use. As Qatar predominantly relies on energy intensive seawater desalination, understanding the environmental implications of the entire water system for domestic water use is crucial for devising effective strategies to optimize water production processes and minimize their impact on the environment. Hence, the aim of this study is to conduct a comprehensive life cycle assessment (LCA) of water use in residential house in Qatar, providing insights into their sustainability performance.

The LCA methodology offers a systematic framework for evaluating the environmental impacts associated with the entire life cycle of water production systems, including raw water extraction, treatment, distribution, and end-use (Mannan and Al-Ghamdi 2020). By quantifying resource consumption, energy use, greenhouse gas emissions, and other relevant indicators, LCA enables a holistic assessment of the environmental footprint of water use system (Devkota et al. 2015; Machado et al. 2007; Raluy et al. 2004). This assessment is essential for identifying potential hotspots and areas where improvements can be made to enhance the sustainability of water resources in Qatar.

Typically, studies on water use impacts have predominantly focused on buildings located in the United States, Europe, or Australia, while insufficient attention has been given to top water-stressed countries where desalination plays a crucial role in water supply. The Gulf Cooperation Council (GCC) countries in the Middle East serve as a compelling case, characterized by extreme water scarcity and the implementation of energy-intensive water supply systems. Additionally, high GDPs and local customs contribute to some of the highest per capita water consumption rates worldwide. Qatar, specifically, exemplifies this situation, with more than 99% of its municipal water supply relying on desalination. By addressing the unique context of water scarcity and reliance on desalinated water in Qatar, this research contributes to filling the knowledge gap in water use impacts and sustainability considerations specific to the region. The findings serve as a crucial foundation for informed decision-making, enabling policymakers to design and implement effective measures to enhance water efficiency and promote sustainable practices in residential buildings. Moreover, the outcomes of this study offer insights for the adaptation of building rating systems, facilitating their alignment with local environmental goals and promoting sustainable development in Qatar and neighboring GCC countries.

## 11.2 Methods and Procedures

This study employed two primary steps to analyze the environmental impacts of water use in buildings. Firstly, a residential building was meticulously modeled to capture its characteristics and features accurately. Subsequently, a comprehensive life cycle assessment (LCA) model was developed to evaluate and quantify the associated environmental impacts throughout the building's water use stages. The following sections provide a detailed description of the methods employed in this research, outlining the building modeling process and the subsequent development of the LCA model.

### *11.2.1 Modeled Residential Unit in Doha, Qatar*

In this study, the modeling of a representative case study residential building was carried out utilizing Building Information Modeling (BIM) techniques. BIM, a promising 3D process, facilitated the virtual construction of a comprehensive building model, encompassing design and operational phase considerations (Al-Ghamdi and Bilec 2017). Figure 11.2 showcases the resulting model, which served as the basis for subsequent analysis.

To determine the baseline annual energy and water requirements for the modeled building, Autodesk Green Building Studio (GBS) was employed. This software tool played a crucial role in assessing and quantifying the initial energy and water consumption parameters for the building model. By utilizing GBS, accurate estimations were obtained, enabling a reliable foundation for further investigations into the environmental impacts of water use. The combination of BIM and GBS technologies provided a robust framework for modeling and evaluating the selected residential

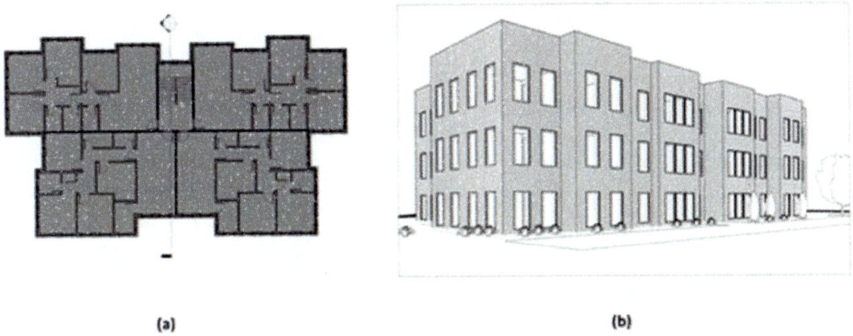


**Fig. 11.2** BIM-model building. The floor plan **a** shows the modelled building's details, while the 3D image **b** shows the completed construction

building. This approach ensures a comprehensive understanding of the building's energy and water demands, setting the stage for subsequent LCA analysis.

The case study building was specifically designed as a multi-family residential structure, accommodating a total of 30 residents, and spanning up to the third floor. The floor area of the model building was measured at 1189 square meters. During the design phase, strict adherence to the relevant codes and regulations of Qatar was ensured to meet all applicable standards.

The energy calculations were conducted in accordance with the guidelines specified in ASHRAE 90.1-2010, ensuring compliance with the relevant standards. On the other hand, the water use calculations drew upon the insights provided by the American Water Works Association (AWWA) Research Foundation 2000 Residential/Commercial and Institutional End Uses of Water report and the 2000 Uniform Plumbing Code of the International Association of Plumbing and Mechanical Officials (IAPMO). To estimate water consumption, the calculations were based on building codes that assume uniform water requirements irrespective of location or human behavior. Within the modeled building, residents were assumed to spend approximately 58% of their day indoors. This consideration played a significant role in determining the water consumption patterns.

Figure 11.3 illustrates the annual energy requirements obtained from the analysis, providing a comprehensive overview of the building's resource demands. The information derived from GBS serves as a vital basis for understanding the energy and water consumption characteristics of the model building. The annual water demand has been determined to be 3130 $m^3$/year.

### 11.2.2 LCA Framework

The LCA model employed in this study followed the four-step framework outlined in ISO 14040 (ISO 2006): goal and scope definition, life cycle inventory analysis

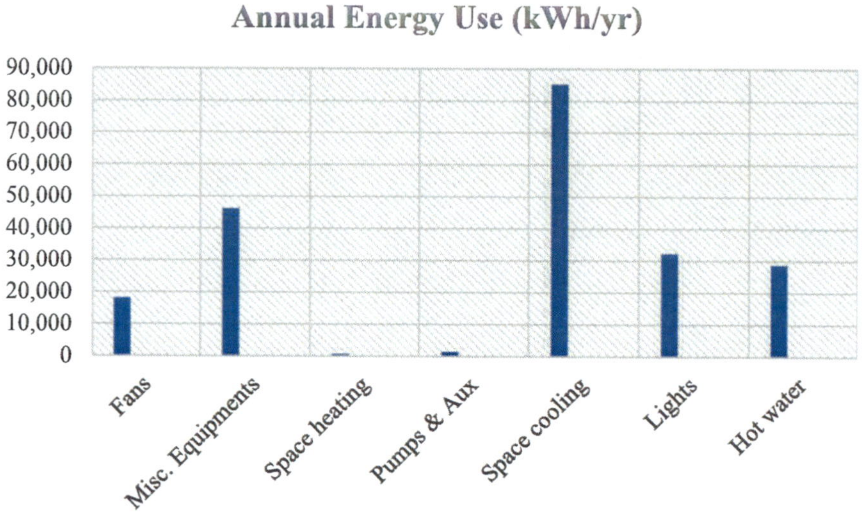


**Fig. 11.3** Annual energy requirements for Green Building Studio case study building

(LCI), life cycle impact assessment (LCIA), and interpretation (ISO 2006). This section provides a concise overview of the LCA framework utilized to evaluate the environmental impacts of water use in the case study buildings.

The primary objective of this study was to quantify the life cycle impacts associated with water use in residential buildings. The functional unit chosen was 3130 cubic meters ($m^3$) of high-quality potable water consumed annually per residential unit, determined as the basic water requirement by GBS. The system boundary encompassed various stages, including raw water treatment, conveyance, water use within the residential unit, wastewater transport, and wastewater treatment. However, the construction and dismantling scenarios for treatment infrastructure and the building itself were excluded from the system boundary. Due to data limitations, the analysis did not consider the impacts of brine from the desalination plant in Doha.

The LCI step involved the assessment of inputs and outputs for each process. Resource consumption, such as electricity input, was accounted for. Data collection relied on a combination of primary resources (real-life data) and secondary resources (published literature values and GaBi local databases). The specific details of the raw water treatment in Doha, which constituted the first part of this study, can be found in the literature by Mannan et al. (Mannan et al. 2019).

Household water use data were collected from GBS for the modeled residential unit, while wastewater treatment data were sourced from published literature, governmental websites, and the GaBi database. The electricity grid mix data for Qatar, considering the reference year 2013, was obtained from the GaBi database. It should be noted that electricity production in Qatar relies entirely on natural gas (NG), and the data for grid transmission losses were also obtained from GaBi.

In the LCIA step, the magnitude of environmental impacts associated with water use was evaluated across different impact categories. The GaBi 6 tool was utilized to convert inventory results into environmental impact scores, employing the ReCiPe midpoint assessment method. Impact categories such as climate change (kg $CO_2$-eq), fossil depletion (kg oil eq), terrestrial acidification (kg $SO_2$-eq), and marine eutrophication (kg N-eq) were investigated and further discussed in the results section.

### *11.2.3 Overall Water Use Life Cycle*

The comprehensive assessment of water use impacts in residential buildings considers five distinct stages: water treatment, water transportation, water use within households, wastewater transport, and wastewater treatment (Fig. 11.4). The analysis encompasses the entire cycle of domestic water, from its initial treatment to its discharge from buildings. The specific details of each stage, along with the corresponding LCA framework employed for assessment, are described below.

Water Treatment and Distribution: In Qatar, potable water demand is met solely through seawater desalination, with multi-stage flash desalination (MSF) being the predominant method. This process involves seawater extraction, physical and chemical pre-treatment, and the use of energy, primarily in the form of electricity. For the Qatar case study, an MSF plant was selected, with a total energy requirement of 4.05 kWh/$m^3$ and 127 kJ/kg for electricity and thermal energy, respectively. The study also accounts for water loss during transportation, with reported real losses in 2014 estimated at 30.5 million $m^3$ or approximately 6.3%. The LCA modeling details can be found in the literature by Mannan et al. (2019).

Water Use Within Homes: Within residential buildings, water use impacts primarily arise from house pumps and water heating. House pumps are responsible for transporting incoming water from the network to elevated tanks on building roofs, typically utilizing centrifugal pumps. Water heating within residential units predominantly relies on electricity.

Wastewater Transport and Treatment: Wastewater transportation energy requirements for Doha were collected through personal communication, with an estimated value of 0.08 kWh/$m^3$. The largest wastewater treatment plant in Doha, known as the Doha West Wastewater Treatment Plant, serves as the model system for the case study. The treatment process includes primary treatment involving screening and degritting, followed by activated sludge secondary treatment for organic, nitrogen, and phosphorus removal, and tertiary treatment comprising rapid sand filtration, ultrafiltration, and chlorination. Energy requirements and chemical consumption data were obtained from the Public Works Authority of Qatar (Ashghal) (Fig. 11.4).

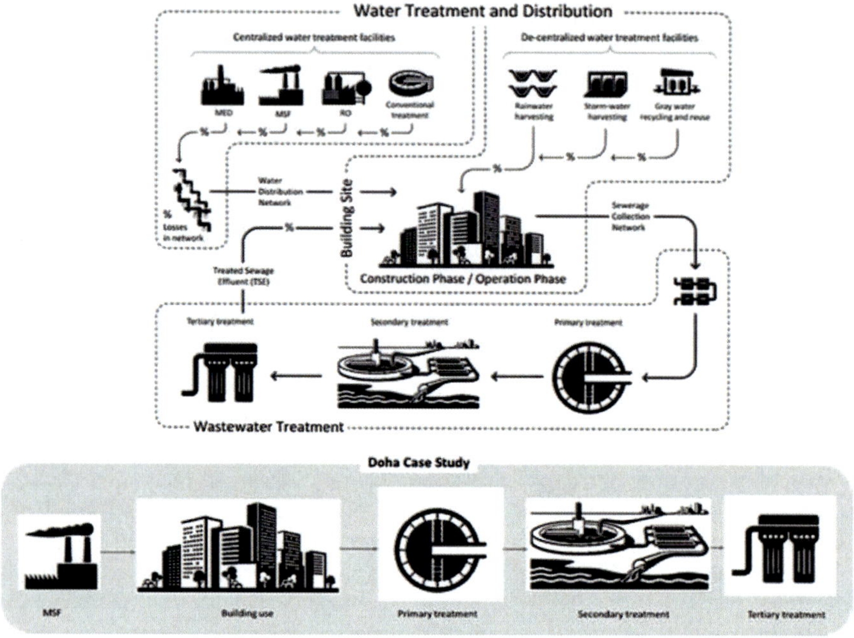


**Fig. 11.4** Doha domestic water use diagram. Top: built environment water-use phases and treatment options in different countries. Bottom: Doha case study water use cycle including water and wastewater treatment stages. Adapted from Mannan and Al-Ghamdi (2022)

## 11.3 Results and Discussions

Figure 11.5 illustrates the $CO_2$ emissions associated with water use impacts in the modelled building.

The analysis shows that the raw water treatment stage has the largest impact, primarily due to the heavy energy consumption of MSF thermal desalination. Steam generation for thermal energy supply and electricity usage for pumping during desalination contribute significantly to $CO_2$ emissions, while the chemical use in pre- and post-treatment has a relatively smaller impact. The second-highest impact arises from the building use stage, accounting for nearly half of the total impact compared to the water treatment stage. Specifically, the $CO_2$ emissions from raw water treatment and water use within the buildings amount to 38,070 kg/year and 18,249 kg/year, respectively. The $CO_2$ emissions from wastewater treatment are considerably lower at 2044 kg/year. Regarding transportation, the emissions associated with the transport of potable water are estimated at 907 kg/year, while wastewater transportation contributes 170 kg/year of $CO_2$ emissions.

Furthermore, the $CO_2$ emissions resulting from the annual electricity consumption in the modelled building were evaluated based on the specifications of Qatar's electricity grid mix. The analysis revealed that the building in Doha accounted

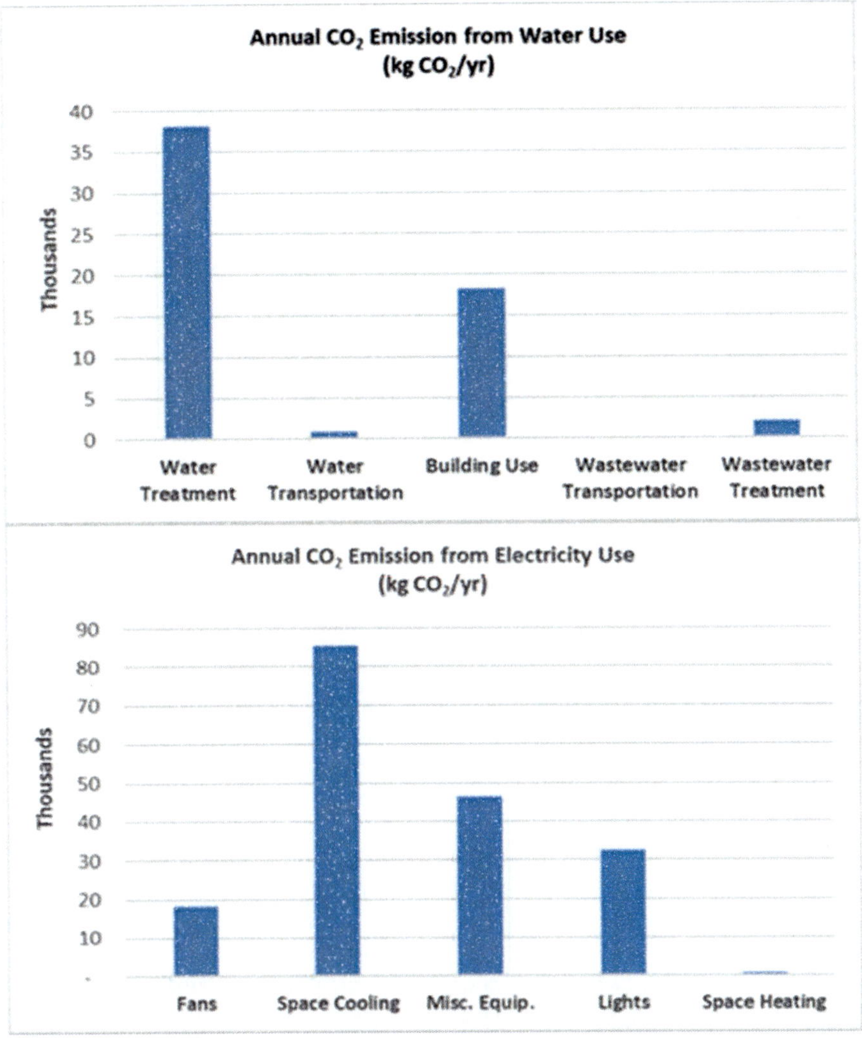


**Fig. 11.5** Case study building's annual $CO_2$ emissions from water usage and electricity consumption. The left side of the graph plainly demonstrates that the raw water treatment stage has the greatest impact on Doha. The right portion of the graph depicts the $CO_2$ emission potential of the case study building based on its annual electricity consumption

for 111,314 kg/year of $CO_2$ emissions from electricity use. Overall, the total $CO_2$ emissions for all water use stages reached 59,440 kg/year.

Figure 11.6 presents the results of the life-cycle impact assessment for four different impact categories. Consistent with previous findings, the water treatment stage in the Doha case study is the primary driver of higher impacts in each category. The excessive use of thermal energy during water treatment contributes to higher

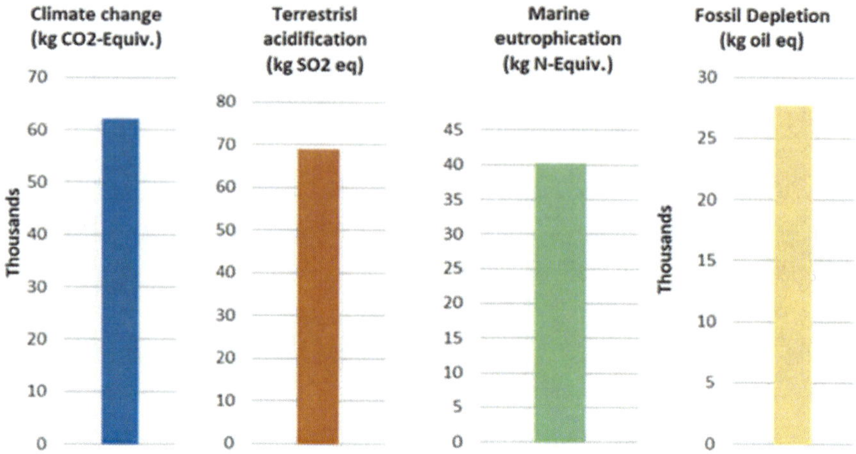


**Fig. 11.6** Impact assessment results for case study building

greenhouse gas (GHG) emissions, thereby impacting the global warming potential as a key climate change impact category. Additionally, the combustion of fossil fuels to meet thermal energy requirements and the discharge of hot effluents into adjacent water bodies result in higher levels of fossil fuel depletion and marine eutrophication. Furthermore, the deposition of inorganic chemicals, particularly sulfates, nitrates, and phosphates, in the atmosphere alters soil characteristics, leading to terrestrial acidification. The Doha case study indicates a terrestrial acidification impact of 68.8 kg $SO_2$-eq.

## 11.4 Conclusion

In conclusion, this study provides valuable insights into the environmental performance of water use systems in Qatar, utilizing a life cycle assessment (LCA) approach. By examining the various stages of the water use life cycle and considering associated environmental impacts, opportunities for sustainable improvements have been identified. The LCA analysis conducted serves as a fundamental reference point for establishing specific water policies and guidelines for buildings, using the quantitative scores obtained.

The findings underscore the significance of research strategies focused on enhancing desalination efficiency at a national level. Improving desalination processes can substantially reduce the water use impacts associated with buildings. In future studies, a comprehensive water use impact assessment for buildings should incorporate both embodied water and operational water. Incorporating both construction and operational stages is crucial for a more holistic impact assessment (Mannan and Al-Ghamdi 2020).

Furthermore, while this study concentrated on residential units, it is essential to conduct thorough assessments of water use impacts for various building types. This research paves the way for further investigations, including the analysis of embodied water, assessment of different building types, and exploration of diverse locations. Such comprehensive evaluations throughout a building's life cycle are necessary to obtain a holistic understanding of the overall impact of water use.

In summary, this study provides valuable insights into the environmental implications of water use systems, offering a foundation for developing sustainable water management strategies and promoting environmentally responsible practices in buildings.

## References

Al-Ghamdi SG, Bilec MM (2017) Green building rating systems and whole-building life cycle assessment: comparative study of the existing assessment tools. J Archit Eng 23:04016015. https://doi.org/10.1061/(asce)ae.1943-5568.0000222

Arpke A, Hutzler N (2006) Domestic water use in the United States: a life-cycle approach. J Ind Ecol 10:169–184. https://doi.org/10.1162/108819806775545312

Devkota J, Schlachter H, Apul D (2015) Life cycle based evaluation of harvested rainwater use in toilets and for irrigation. J Clean Prod 95:311–321. https://doi.org/10.1016/j.jclepro.2015.02.021

Forecasts IBMIS (2011) Qatar water report. Qatar water

ISO (2006) ISO 14040:2006. Environmental management—life cycle assessment—principles and framework. In: ISO 14040. https://www.iso.org/standard/37456.html. Accessed 6 June 2021

Lawler J, Mazzoni A, Shannak S (2023) The domestic water sector in Qatar 193–209. https://doi.org/10.1007/978-981-19-7398-7_11

Machado AP, Urbano L, Brito AG et al (2007) Life cycle assessment of wastewater treatment options for small and decentralized communities. Water Sci Technol 56:15–22. https://doi.org/10.2166/wst.2007.497

Mannan M, Al-Ghamdi SG (2020) Environmental impact of water-use in buildings: latest developments from a life-cycle assessment perspective. J Environ Manage 261. https://doi.org/10.1016/j.jenvman.2020.110198

Mannan M, Al-Ghamdi SG (2022) Water consumption and environmental impact of multifamily residential buildings: a life cycle assessment study. Buildings 12. https://doi.org/10.3390/buildings12010048

Mannan M, Alhaj M, Mabrouk AN, Al-Ghamdi SG (2019) Examining the life-cycle environmental impacts of desalination: a case study in the state of Qatar. Desalination 452:238–246. https://doi.org/10.1016/j.desal.2018.11.017

Ministry of Development Planning and Statistics (2017) Water statistics in the state of Qatar

Raluy RG, Serra L, Uche J, Valero A (2004) Life-cycle assessment of desalination technologies integrated with energy production systems. Desalination 167:445–458. https://doi.org/10.1016/j.desal.2004.06.160

Yoonus H, Al-Ghamdi SG (2020) Environmental performance of building integrated grey water reuse systems based on life-cycle assessment: a systematic and bibliographic analysis. Sci Total Environ 712. https://doi.org/10.1016/j.scitotenv.2020.136535

# Chapter 12
# Assessment of the Climate Change and Metal Depletion Impacts of a Cobalt Fischer–Tropsch Catalyst with Prospective Life Cycle Assessment

**A. E. M. van den Oever, Daniele Costa, and Maarten Messagie**

**Abstract** Fischer–Tropsch (FT) catalyst production, use, and end-of-life (EoL) phases are often omitted in Life Cycle Assessments (LCA) studies as the required data are not publicly available. Consequently, the environmental effects of these catalysts are unknown. This study presents the prospective LCA of a novel cobalt-based FT catalyst. The objectives were to evaluate future production pathways and identify a best-case and a worst-case foreground scenario. The foreground was modelled with upscaled data from lab-scale experiments and patents. The effects of different prospective background scenarios were also investigated. A superstructure database was constructed with the Python library premise for 2030 and 2050. The climate change impact ranged from 0.088 to 8.77 kg $CO_2$ eq/tonne syngas, and the metal depletion impact from 0.012 to 1.26 kg Fe eq/tonne syngas. The environmental impacts of the catalyst depended mainly on the catalyst loading and the EoL. The best-case scenario showed a high catalyst loading, regeneration at the EoL and autothermal reforming (ATR) of biomass for hydrogen production consumed in all processes. In the worst-case scenario, the catalyst is recycled, while the hydrogen is produced via ATR of natural gas. The background scenarios were less influential than the foreground scenarios.

**Keywords** Catalyst · Climate change · Metal depletion · Prospective life cycle assessment (LCA) · Upscaling

---

A. E. M. van den Oever (✉) · M. Messagie
Department of Electric Engineering and Energy Technology, Mobility, Logistics and Automotive Research Centre, Vrije Universiteit Brussel, Pleinlaan 2, 1050 Brussels, Belgium
e-mail: anne.van.den.oever@vub.be

D. Costa
VITO/EnergyVille, Boeretang 200, 2400 Mol, Belgium



N. S. Caetano and M. C. Felgueiras (eds.), *The 9th International Conference on Energy and Environment Research*, Environmental Science and Engineering,
https://doi.org/10.1007/978-3-031-43559-1_12

## 12.1 Introduction

Multiple Life Cycle Assessments (LCA) studies on synthetic liquid fuels produced via the Fischer–Tropsch (FT) process have been recently published. However, the FT catalyst production life cycle is often omitted or simplified since the data required to model these stages are generally proprietary and, therefore, not publicly available. Consequently, the environmental effects of these catalysts are unknown. This study presents the prospective LCA of a novel cobalt-based FT catalyst (Jeske et al. 2021) that enables the combined production of synthetic fuels and liquid platform chemicals (e.g., α-olefins, linear alcohols, among others), and the work was part of the project named 'Robust and Efficient processes and technologies for Drop In renewable FUELs for road transport' (REDIFUEL). The main objectives were to evaluate the climate change and metal depletion impacts of different future production pathways, identify a best-case and a worst-case foreground scenario, and provide eco-design recommendations for the catalyst.

## 12.2 Materials and Methods

The LCA methodology (ISO 2006) was used for the environmental assessment. The functional unit of the study is the conversion of one tonne of hydrogen-rich syngas in a biomass-to-liquid (BtL) plant with a 50 years lifetime. The assessment considered the years 2030 and 2050, having Europe as the geographical scope. The system boundaries were cradle-to-grave (Fig. 12.1). The effects of three different aspects on the environmental impacts of the catalyst were considered to identify the worst-case and best-case foreground scenario: the end-of-life (EoL) scenario, the hydrogen source, and the weight hourly space velocity (WHSV). These three different aspects affect the reference flows of the scenarios.

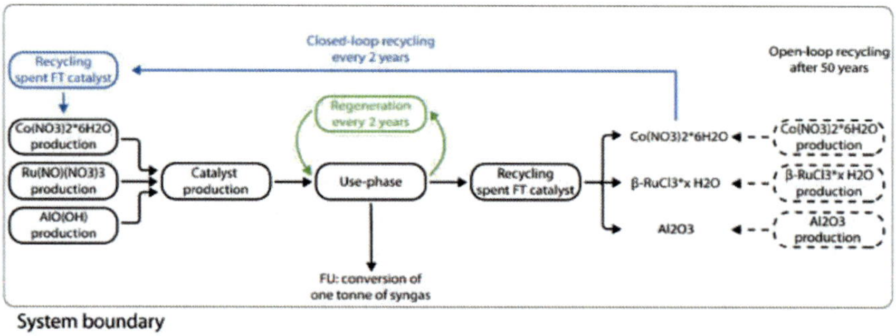


**Fig. 12.1** System boundaries. Two different end-of-life (EoL) options are modelled: **a** Periodic regeneration of the catalyst (green) and **b** recycling after two years with closed-loop recycling of Co (blue). All metals are recycled at the EoL of the biomass-to-liquids plant (open-loop)

The EoL scenarios considered were recycling and regeneration (Fig. 12.1). In the first scenario the catalyst has a lifetime of two years. It is assumed that the recycling occurs at the catalyst production factory and that the recovered cobalt is used for new catalyst production. The recovered ruthenium and aluminum are recycled in an open loop. Fresh catalyst is added to make up for the catalyst losses. In the second scenario, the catalyst can be reused directly in the BtL plant after two-yearly regeneration, and the catalyst's lifetime is extended to 50 years. The catalyst is recycled at the BtL plant's EoL, and all metals are recovered in an open loop. All life cycle stages required significant amounts of hydrogen. Therefore, different hydrogen production sources were considered: natural gas autothermal reforming (ATR), biogas ATR, biomass gasification, and electrolysis (photovoltaic). The WHSV expresses the weight of feed syngas per unit weight of the catalyst per hour and is inversely related to the catalyst loading in the FT-reactor. A high and low estimate for the WHSV were derived from lab-scale experiments.

In addition to assessing different foreground scenarios, the selected background model, Ecoinvent 3.7 cutoff (ISO 2006), was modified to represent two future scenarios. In the reference scenario, the implementation of climate change policies fails, and the world evolves following current trends (Wernet et al. 2016). In the 1.5 °C climate target scenario, mitigation policies are effectively implemented (Fricko et al. 2017). In addition, a base case was analyzed where all background data was derived from the original Ecoinvent 3.7 cutoff database.

As the Ecoinvent 3.7 cutoff database applies economic allocation, economic allocation was used to model the production of precursor materials, since their production is value-driven (Rogelj et al. 2018). However, the substitution approach was used for solving the multifunctionality of the recycled metals, accounting for the effects of a growing spent FT catalyst market. Data used to compile the Life Cycle Inventory (LCI) is disclosed in Sect. 12.3. Finally, the climate change impact was assessed with the IPCC 2013 GWP100 method (Ijassi et al. 2021) and the metal depletion impact with the Recipe 2008 method (IPCC 2013). The assessment was conducted with the Activity Browser software (Goedkoop and Huijbregts 2008).

## 12.3 Life Cycle Inventory

To implement the reference and 1.5 °C climate target scenarios in the background model, a superstructure prospective background database (Steubing et al. 2020) was constructed with the Python library premise (Sacchi et al. 2022) based on Ecoinvent 3.7 (ISO 2006) and output from the REMIND model (PIK, PSI 2020). The foreground model was derived from experiments during the REDIFUEL project and complemented with literature data. The full LCI and the script for generating the background database are available online (Luderer et al. 2015).

### 12.3.1 Precursor Production

The precursor materials for catalyst production are pseudo-boehmite (AlO(OH)), cobalt (II) nitrate hexahydrate ($Co(NO_3)_2 * 6H_2O$), and ruthenium (III) nitrosyl nitrate ($Ru(NO)(NO_3)_3$). The pseudo-boehmite is a co-product of the Ziegler process for fatty acid alcohol production (Oever 2022). Background data for the Ziegler process was taken from the superstructure database. Stoichiometrically, one molecule of C12 fatty acid alcohol and one molecule of aluminum oxide are co-produced. Economic allocation factors were based on Diblitz et al. (1998), ChemAnalyst (2021).

Background data regarding the production of the other reactants ($Co(NO_3)_2 * 6H_2O$ and $Ru(NO)(NO_3)_3$) was not available in the superstructure database. It was assumed that the $Co(NO_3)_2 * 6H_2O$ is supplied by recycling spent FT catalyst with a recovery rate of 97.75% (ISE 2021). Additional $Co(NO_3)_2 * 6H_2O$ is supplied from the market to compensate for the non-recovered cobalt. The overall reaction equation (Liu et al. 2014) used for the mass balance is shown in Eq. (12.1):

$$3Co + 8HNO_3 + 14H_2O \rightarrow 3Co(NO_3)_2 * 6H_2O + 2NO \tag{12.1}$$

The datasets for the reactants in Eq. (12.1) were taken from the superstructure database. The energy requirements for production were neglected. The mass balance of the $Ru(NO)(NO_3)_3$ solution production was derived from Yildiz (2017) and is given in Eq. (12.2):

$$Ru + 4HNO_3 \rightarrow Ru(NO)(NO_3)_3 + 2H_2O \tag{12.2}$$

Ruthenium is a high-value platinum group metal (PGM) and primary applications for ruthenium are electronics (44%) and catalysts (40%) (Fletcher et al. 1959). It is assumed that both applications represent half of the recycled ruthenium market mix. Hence, the dataset for PGM mining and concentration operations in the superstructure database was used to model Ruthenium production. An economic allocation factor for ruthenium was derived (Loferski et al. 2018). Of the total PGM supply, 23% consists of recycled materials (Fletcher et al. 1959) and the same share is assumed for ruthenium. The recycling processes are approximated with the recovery processes of electronic scrap and spent automobile catalysts.

### 12.3.2 Catalyst Production

To compile the LCI for the catalyst production, lab-scale data for catalyst production was obtained from the REDIFUEL project. This data was upscaled following (Nuss and Eckelman 2014). The manufacturing steps are pre-calcination of pseudo-boehmite, impregnation, calcination, and activation (Fig. 12.2).

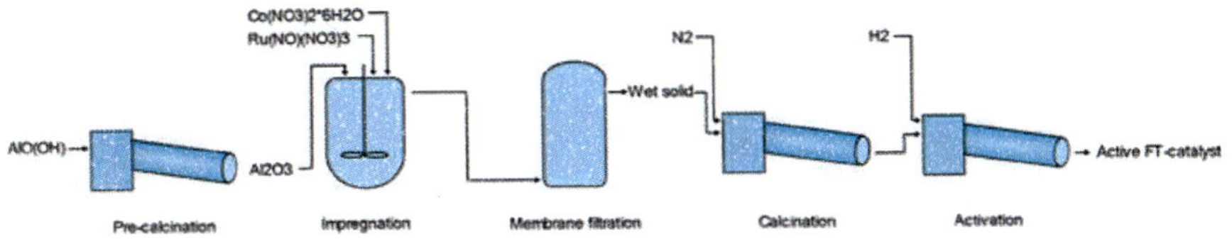


**Fig. 12.2** Flow chart of the industrial Fischer–Tropsch (FT)-catalyst production process

At commercial-scale production, the pre-calcination, calcination and activation steps occur in a rotary kiln, the impregnation in a 1000 L rotor–stator type homogenizer, followed by membrane filtration (Nuss and Eckelman 2014). The fuel for the rotary kiln was assumed to be natural gas (Piccinno et al. 2016). The heat demand of the lab-scale oven was taken as a conservative estimation of the industrial-scale rotary kiln. The stirring electricity required for the impregnation step in the rotor–stator type and the electricity demand for membrane filtration were calculated following (Nuss and Eckelman 2014). The background data for nitrogen, hydrogen, and the catalyst's production plant's infrastructure were derived from the superstructure database. It was assumed that a DeNOx unit with a NOx removal efficiency of 90% (Hofius et al. 1999) would be used at the production plant.

### 12.3.3 *Use-Phase*

The catalyst loading in the FT-reactor was calculated by dividing the syngas feed to the reactor by the WHSV. The lowest WHSV (5 $h^{-1}$) and the highest possible WHSV (33 $h^{-1}$) were derived from Jeske et al. (2021). The BtL plant was assumed to operate for 8000 h per year.

### 12.3.4 *End-of-Life Options*

The ex-situ regeneration is assumed every two years for the regeneration case, with a 100% activity recovery (Rytter and Holmen 2015). The regeneration steps include wax removal at high temperatures under a nitrogen flow, followed by hydrogenation and calcination/oxidation (VITO 2020). No quantitative data were available for the material and energy requirements. Hence, it was assumed that the regeneration step consumes the same amount of nitrogen, hydrogen and energy as the activation and calcination step of the primary catalyst production. The recycling process was modelled according to the patent PCT/CN2013/072119 (ISE 2021). The energy requirements were calculated according to Nuss and Eckelman (2014).

## 12.4 Results

The catalyst's climate change impact ranges from 0.088 to 8.77 kg $CO_2$ eq/tonne syngas (Fig. 12.3). The most critical parameter affecting the catalyst's climate change impact is the WHSV, followed by the EoL scenario. Low WHSVs (i.e., high catalyst loadings) and recycling lead to higher impacts than high WHSVs and regeneration. The hydrogen source also impacts climate change, but to a lesser extent. All renewable hydrogen sources lead to a decrease compared to hydrogen from the ATR of natural gas. The different prospective background databases affect the results, especially at low WHSV and for scenarios with recycling. As recycling is an energy-intensive process, the effects of a more renewable energy system are more pronounced in these scenarios.

The metal depletion impacts vary from 0.012 to 1.26 kg Fe eq/tonne syngas (Fig. 12.4). Low WHSV values and scenarios with recycling lead again to the highest impacts. Electrolysis leads to the highest metal depletion impacts from all hydrogen sources, whereas the ATR of natural gas and biomass gasification have the lowest impacts. Comparing Figs. 12.3 and 12.4 shows that the worst-case foreground scenario for climate change (WHSV = 5 $h^{-1}$, recycling, ATR natural gas) is also among the scenarios with the highest metal depletion impacts. In contrast, the scenario with the lowest metal depletion impacts is the best-case foreground scenario for climate change (WHSV = 33 $h^{-1}$, regeneration, ATR biogas). The different prospective background scenarios increase the metal depletion impacts compared to the original Ecoinvent database, due to the higher metal demand for renewable energy systems.

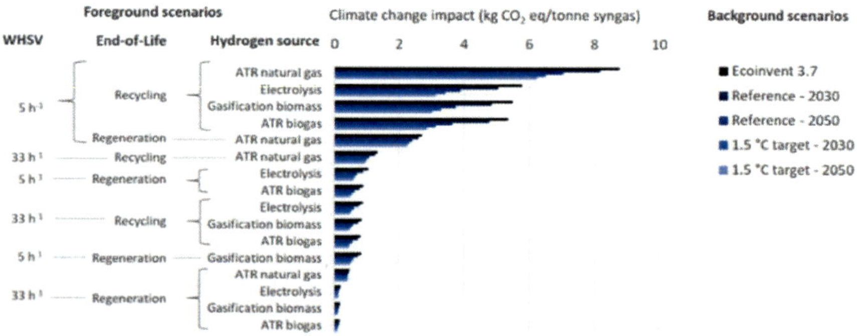


**Fig. 12.3** The climate change impact of converting one tonne of syngas with the analyzed Co-catalyst. Legend: *ATR* autothermal reforming; *WHSV* weight hourly space velocity

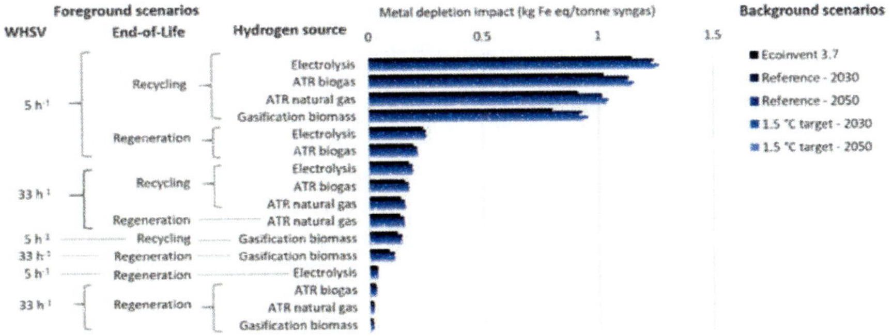


**Fig. 12.4** The metal depletion impact of converting one tonne of syngas with the analyzed Co-catalyst. Legend: *ATR* autothermal reforming; *WHSV* weight hourly space velocity

## 12.5 Conclusions

This work presented the LCA of a novel FT catalyst and the results showed that the climate change impact ranges from 0.088 to 8.77 kg $CO_2$ eq/tonne syngas, and the metal depletion impact from 0.016 to 1.14 kg Fe eq/tonne syngas. The best-case for the catalyst production is the foreground scenario with a WHSV of 33 $h^{-1}$, regeneration as EoL, and ATR of biogas as the hydrogen source. In the worst-case, the catalyst has a WHSV of 5 $h^{-1}$, recycling as EoL option, and ATR of natural gas as the hydrogen source. Although the prospective background database also affected the results, the different catalyst loadings and EoL options were more influential, making these parameters the primary concern of catalyst design and future prospective LCAs of Co-based catalysts.

**Acknowledgements** This work was supported by REDIFUEL, which has received funding from the European Union's Horizon 2020 research and innovation programme under the Grant Agreement no. 817612. The authors thank all REDIFUEL partners for the productive collaboration and also express their gratitude to Romain Sacchi for his help with premise.

## References

ChemAnalyst (2021) Fatty alcohol price trend and forecast. https://www.chemanalyst.com/Pricing-data/fatty-alcohol-1084. Accessed 4 Jan 2022

Diblitz K, Feldbaum T, Ludemann T (1998) Manufacturing of raw materials for the catalyst industry. Stud Surf Sci Catal 113:599–611. https://doi.org/10.1016/s0167-2991(98)80336-4

Fletcher JM, Brown PGM, Gardner ER, Hardy CJ, Wain AG, Woodhead JL (1959) Nitrosylruthenium nitrato complexes in aqueous nitric acid. J Inorg Nucl Chem 12:154–173. https://doi.org/10.1016/0022-1902(59)80106-8

Fricko O, Havlik P, Rogelj J, Klimont Z, Gusti M, Johnson N et al (2017) The marker quantification of the Shared Socioeconomic Pathway 2: a middle-of-the-road scenario for the 21st century. Glob Environ Chang 42:251–267. https://doi.org/10.1016/j.gloenvcha.2016.06.004

Goedkoop M, Huijbregts M (2013) ReCiPe 2008 A life cycle impact assessment method which comprises harmonised category indicators at the midpoint and the endpoint level. Report I: Characterisation

Hofius H, Karasch O, Georgiev L (1999) Calcination of hydrated alumina in a rotary kiln—has kiln fired by stoichiometric mixt. of natural gas and oxygen@burner, provides alumina for electrical insulators. DE4124581A1

Ijassi W, Ben Rejeb H, Zwolinski P (2021) Environmental impact evaluation of co-products: decision-aid tool for allocation in LCA. Int J Life Cycle Assess 26:2199–2214. https://doi.org/10.1007/s11367-021-01984-0

IPCC (2013) Anthropogenic and natural radiative forcing, chap 8. In: Climate change 2013. Physical science basis working group I contribution to the fifth assessment report of the intergovernmental panel on climate change. Cambridge University Press, pp 659–740. https://doi.org/10.1017/CBO9781107415324.018

ISE (2021) Average prices over the last few months for base metals. https://ise-metal-quotes.com/. Accessed 4 Jan 2022

ISO. ISO 14040:2006 (2006) Environmental management—life cycle assessment—principles and framework. International Organization for Standardization, Switzerland

ISO. ISO 14044:2006 (2006) Environmental management—life cycle assessment—requirements and guidelines. International Organization for Standardization, Switzerland

Jeske K, Kizilkaya AC, López-Luque I, Pfänder N, Bartsch M, Concepción P et al (2021) Design of cobalt Fischer-Tropsch catalysts for the combined production of liquid fuels and olefin chemicals from hydrogen-rich syngas. ACS Catal 11:4784–4798. https://doi.org/10.1021/acscatal.0c05027

Liu Q, Han Y, Song D, Xu L, Lai B (2014) Process for recovery of cobalt, ruthenium and aluminum from spent catalyst. Patent No. US 2014/0377151 A1

Loferski PJ, Ghalayini ZT, Singerling SA (2018) Minerals yearbook. Vol I—Platinum-group metals. USGS, USA

Luderer G, Leimbach M, Bauer N, Kriegler E, Baumstark L, Bertram C et al (2015) Description of the REMIND model (version 1.6). https://doi.org/10.2139/ssrn.2697070.

Nuss P, Eckelman MJ (2014) Life cycle assessment of metals: a scientific synthesis 9:1–12. https://doi.org/10.1371/journal.pone.0101298

Piccinno F, Hischier R, Seeger S, Som C (2016) From laboratory to industrial scale: a scale-up framework for chemical processes in life cycle assessment studies. J Clean Prod 135:1085–1097. https://doi.org/10.1016/j.jclepro.2016.06.164

PIK, PSI (2020) Premise. https://github.com/romainsacchi/premise

REDIFUEL (2020) Project. https://redifuel.eu/project/. Accessed 26 Nov 2020

Rogelj J, Popp A, Calvin K V., Luderer G, Emmerling J, Gernaat D et al (2018) Scenarios towards limiting global mean temperature increase below 1.5 °C. Nat Clim Chang 8:325–232. https://doi.org/10.1038/s41558-018-0091-3

Rytter E, Holmen A (2015) Deactivation and regeneration of commercial type Fischer-Tropsch co-catalysts—a mini-review. Catalysts 5:478–499. https://doi.org/10.3390/catal5020478

Sacchi R, Terlouw T, Siala K, Dirnaichner A, Bauer C, Cox B et al (2022) Prospective environmental impact assessment (premise): a streamlined approach to producing databases for prospective life cycle assessment using integrated assessment models. Renew Sustain Energy Rev 160:112311. https://doi.org/10.1016/j.rser.2022.112311

Steubing B, de Koning D, Haas A, Mutel CL (2020) The activity browser; an open source LCA software building on top of the brightway framework. Softw Impacts 3. https://doi.org/10.1016/j.simpa.2019.100012

van den Oever AEM, Costa D, Messagie M (2022) Fischer-Tropsch catalyst LCI. https://github.com/EVERGi-Brightway/Fischer-Tropsch-catalyst. Accessed 18 July 2022

VITO (2020) Selectieve katalytische reductie. https://emis.vito.be/nl/node/19361. Accessed 8 Sept 2021

Wernet G, Bauer C, Steubing B, Reinhard J, Moreno-Ruiz E, Weidema B (2016) The ecoinvent database version 3 (part I): overview and methodology. Int J Life Cycle Assess 21:1218–1230. https://doi.org/10.1007/s11367-016-1087-8

Yildiz Y (2017) General aspects of the cobalt chemistry. In: Maaz K (ed) Cobalt. IntechOpen, London. https://doi.org/10.5772/intechopen.71089

# Chapter 13
# Cooling Demand Under Climate Change and Associated Environmental Impacts

**Ammar M. Khourchid and Sami G. Al-Ghamdi**

**Abstract** The building sector is a major contributor to global greenhouse gas emissions, significantly impacting climate change. As a consequence of climate change, the demand for cooling in buildings is expected to rise, especially in countries with hot and arid climates where a substantial portion of energy is dedicated to cooling. This study aims to quantify the impact of climate change on the cooling demand of restaurant buildings in Qatar and assess the associated environmental implications. Future climate conditions were obtained from a fine-resolution regional climate model, specifically using the extreme Representative Concentration Pathway (RCP) 8.5 scenario. Energy analyses were conducted using OpenStudio software to compare reference and future climates. The simulation results were then used to perform a life cycle assessment. The findings indicate that by 2100, under the RCP 8.5 scenario, the annual cooling requirements of restaurant buildings in Qatar are projected to increase by 20%. Consequently, there will be a corresponding rise in power usage within these buildings, leading to increased environmental implications. Specifically, $CO_2$ emissions are expected to increase by 14.6 metric tons by 2100. The study underscores the necessity of incorporating renewable energy resources in power production and integrating climate change considerations into building designs. These measures are essential for reducing carbon emissions, ensuring energy security, and mitigating the adverse impacts of climate change.

**Keywords** Climate change · Cooling demand · Life cycle assessment

A. M. Khourchid · S. G. Al-Ghamdi (✉)
Division of Sustainable Development, College of Science and Engineering, Hamad Bin Khalifa University, Qatar Foundation, P.O. Box 34110, Doha, Qatar
e-mail: sami.alghamdi@kaust.edu.sa

S. G. Al-Ghamdi
Environmental Science and Engineering Program, Biological and Environmental Science and Engineering Division, King Abdullah University of Science and Technology (KAUST), Thuwal 23955-6900, Saudi Arabia

KAUST Climate and Livability Initiative, King Abdullah University of Science and Technology (KAUST), Thuwal 23955-6900, Saudi Arabia



N. S. Caetano and M. C. Felgueiras (eds.), *The 9th International Conference on Energy and Environment Research*, Environmental Science and Engineering,
https://doi.org/10.1007/978-3-031-43559-1_13

## 13.1 Introduction

The continuous increase in the average temperatures in the atmosphere has caused global warming, which causes a series of changes in the global climate and weather conditions. These changes occur as human activities continue to generate greenhouse gases (GHG) which trap the heat in the atmosphere (Alhorr et al. 2014). Among the anthropogenic GHG emissions, carbon dioxide ($CO_2$) is crucial because of its profusion and that it can trap for thousands of years in the atmosphere (Ali et al. 2020). Due to potential threats and consequences of climate change, several nations have taken action and put decarbonization strategies to reduce their carbon footprint. In this regard, the building sector has taken much attention as it contributes to more than 30% of global GHG emissions (Yang et al. 2021).

There is a strong linkage between energy consumption and $CO_2$ emissions. It is expected that the $CO_2$ level would increase significantly in response to an intense increase in energy consumption (Cloy and Smith 2015). During the life span of the buildings, the energy consumption in the operation phase is approximated to reach 80–90%. On the other hand, 10–20% of the total energy is consumed during raw material extraction, manufacturing, and construction phases (UNEP 2009). It was reported that in 2004 the GHG emissions associated with buildings had reached 8.6 billion metric tons $CO_2$ equivalent (eq), and it is anticipated to increase to 15.6 billion t $CO_2$ eq (by 26%) in 2030 under high emissions scenario (IPCC 2007).

Climate change and building cooling demand share a dual effect relationship, where the GHG emissions associated with building cooling contribute to climate change. On the other hand, cooling demand in buildings could increase due to climate change as a result of the expected increase in outdoor temperatures. Because of changes in ambient temperatures, climate change has a considerable impact on energy use for cooling in buildings (Radhi 2009). Building cooling energy consumption is positively linked with outside air temperature (Chakraborty et al. 2021). Compared to 2010, the global building cooling demand in 2100 buildings is expected to rise by 22–57%, dominating the building end-use energy (Levesque et al. 2018).

In hot climate countries such as the Middle East countries, the majority of the building energy is used for cooling services due to the significant difference between outdoor and indoor temperatures (Khourchid et al. 2023). For instance, it was estimated that up to 80% of the energy produced in the Middle East is attributed to cooling systems (Andric and Al-Ghamdi 2020). Dabaieh et al. (2015) stated that mechanical cooling consumes 70–80% of overall energy usage in hot climate countries like Egypt. Cooling systems in buildings in Qatar are responsible for more than half of the total energy usage (Kharseh and Al-Khawaja 2016). In Saudi Arabia, it is estimated that air conditioning systems consume about 65% of the total energy demand in buildings (Hasnain et al. 1999).

Climate change can substantially affect cooling demand in hot and dry regions such as the Gulf Cooperation Council (GCC) countries (Khourchid et al. 2022). Radhi (2009) assessed the effect of different climate change scenarios on residential buildings in the UAE. The results showed that the increase in cooling requirement of the building could reach 23.5% by 2100. In Qatar, Andric and Al-Ghamdi (2020) adopted the A2 emissions scenario to simulate the energy performance of a two-story villa building. They indicated that Qatar would have prolonged, frequent, and intense heat waves in the future, which could raise the consumption of building energy by 30%.

This study aims to quantify the increase in cooling demand due to climate change for full-service restaurant buildings and assess the associated environmental impacts, taking Doha, Qatar as a case study.

## 13.2 Methods

Figure 13.1 shows the methodology followed in this study in order to evaluate the climate change effect on restaurant buildings in Qatar. The framework of this methodology starts with modeling regional climate change for Doha, Qatar. The second step is energy simulation for the intended building. The last step is environmental impact assessment.

### *13.2.1 Reference Weather Data*

The reference weather file used in this study contains typical climate data for Qatar. This weather file included hourly climate data for 12 months collected from several years over a time period based on the consistency and the typicality of data to construct a typical meteorological year (TMY) (Hall et al. 1978). TMY is one of the most popular forms of meteorological data used to estimate building energy demand (Berardi and Jafarpur 2019). The Doha TMY used to evaluate the reference cooling demand of restaurant buildings includes hourly climatic data from 1985 to 2004.

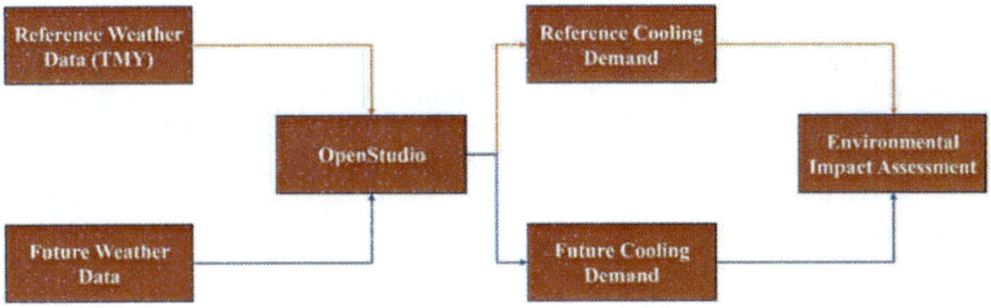


**Fig. 13.1** Study framework

### 13.2.2 Future Climate Data

This study examined the influence of climate change on restaurant building cooling demand using a variety of climatic variables. Relative humidity, dew point temperature, dry bulb temperature, wind speed, and atmospheric pressure were among the inputs. The year 2100 was analyzed under the high greenhouse gas emission scenario, i.e., Representative Concentration Pathways 8.5 (RCP-8.5). This pathway considers 8.5 W per $m^2$ of radiative forcing in 2100 compared to the pre-industrial period (1850–1900).

The future climate inputs used in this study were obtained from the Regional Climate Model (RCM) of Massachusetts Institute of Technology (MIT). The MIT RCM uses three Global Climate Models (GCMs) to determine its atmospheric boundary conditions, which were selected based on objective criteria for their depiction of water bodies in the region. These GCMs are Max-Planck Institute Earth System Model (MPI-ESM), Community Climate System Model V4 (CCSM4), and the Norwegian community Earth System Model (NorESM). Over southwest Asia, the original MIT RCM has a spatial resolution of 25 km. Nevertheless, the MIT RCM for Qatar has been improved to 12.5 km. The reader can refer to Pal and Eltahir (2016) for further information about the MIT RCM.

### 13.2.3 Building Energy Simulation

Energy simulation was conducted using Openstudio (v.3.3.0) software and integrated EnergyPlus (v.9.6.0). The prototype restaurant building is designed by the U.S. Department of Energy (DOE) according to ASHRAE standard 90.1 version 2016 (U.S. Department of Energy 2022).

### 13.2.4 Environmental Impact Assessment

The result of energy simulation for reference and future climate are used as inputs to conduct a life cycle assessment (LCA) following the ISO standard 14040 (ISO 2006) and 14044 criteria (ISO 2006). GaBi, a popular LCA program for sustainability study (Sphera 2022), was utilized. Using Gabi and based on the ReCiPe methodology, we evaluated the environmental impacts (per kWh) associated with the electricity consumed for building cooling.

## 13.3 Results and Discussion

The energy performance of full-service restaurant prototype buildings was assessed using OpenStudio and EnergyPlus integrated simulation engine to show how climate change can influence the building cooling requirements and what are the consequential environmental impacts.

### *13.3.1 Impact on Cooling Demand*

The energy simulation of a restaurant building in Doha, Qatar, was performed under reference and future climate conditions considering the emission scenario RCP-8.5. Figure 13.2 compares between the annual energy consumption for cooling in reference climate (TMY) and in 2100 under climate change scenario RCP 8.5. The results indicated that the annual cooling demand in TMY is 115,723 kWh/year. Nevertheless, under RCP-8.5, the annual cooling demand would rise to 139,295 kWh/year (20%) by 2100. Because Qatar relies on natural gas to generate power, an increase in building cooling demand might harm the local economy by reducing natural gas exports to meet local energy use. In this context, considering the climate change impact on the building sector in the design phase is essential to mitigate the expected increase in cooling demand.

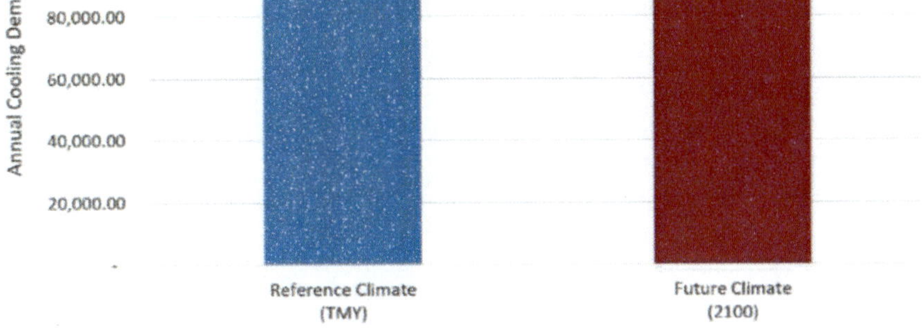


**Fig. 13.2** Impact of climate change on cooling demand

### 13.3.2 *Environmental Impacts*

The result obtained by the energy simulation was used to conduct LCA and evaluate the environmental impact related to the increase in the annual cooling energy consumption. The results shown in Table 13.1 confirmed the dual effect relationship discussed earlier. The increase in the annual cooling demand would contribute to climate change due to the increase in the amount of GHG emissions, which is expressed in kg of $CO_2$ eq. Between 1985–2004 and 2100, the $CO_2$ eq increased by 14.591 metric tons. This increase in the $CO_2$ eq only is equal to the amount of $CO_2$ produced by a typical passenger vehicle driven for three years (EPA 2021). Another vital impact category to a heavily reliant economy like Qatar's is fossil depletion. The fossil depletion impact is expected to rise from 2.94E+04 kg oil eq in TMY to 3.54E+04 kg oil eq by 2100. It is important to mention that the results obtained in this study are for one building restaurant only, which means the results would be much higher in larger-scale studies. These figures inspire Qatar and other GCC nations to shift their energy production to renewable energy sources in order to lessen their contribution to climate change and ensure future energy security.

**Table 13.1** Environmental impacts associated with the increase in cooling demand

| Impact category | Unit | Reference (1985–2004) | RCP8.5 (2100) |
|---|---|---|---|
| Climate change | [kg $CO_2$ eq] | 7.16E+04 | 8.62E+04 |
| Fossil depletion | [kg oil eq] | 2.94E+04 | 3.54E+04 |
| Terrestrial acidification | [kg $SO_2$ eq] | 1.10E+02 | 1.33E+02 |
| Freshwater eutrophication | [kg P eq] | 3.55E−04 | 4.28E−04 |
| Ozone depletion | [kg CFC-11 eq] | 8.69E−11 | 1.05E−10 |
| Freshwater ecotoxicity | [kg 1,4 DB eq] | 5.68E−02 | 6.84E−02 |
| Human toxicity | [kg 1,4-DB eq] | 9.29E+01 | 1.12E+02 |
| Lonising radiation | [kg U235 eq] | 1.03E + 00 | 1.24E + 00 |
| Marine ecotoxicity | [kg 1,4-DB eq] | 2.19E−01 | 2.63E−01 |
| Marine eutrophication | [kg N eq] | 7.17E+00 | 8.64E+00 |
| Metal depletion | [kg Fe eq] | 1.33E+01 | 1.60E+01 |
| Particulate matter formation | [kg PM10 eq] | 4.40E+01 | 5.29E+01 |
| Photochemical oxidant formation | [kg NMVOC eq] | 1.92E+02 | 2.31E+02 |
| Terrestrial ecotoxicity | [kg 1,4-DB eq] | 6.74E−02 | 8.11E−02 |
| Water depletion | [$m^3$] | 2.99E+01 | 3.59E+01 |

## 13.4 Conclusion

In conclusion, this study examined the future impact of climate change on the cooling energy consumption of a restaurant building located in Doha, Qatar. The findings revealed that climate change would have adverse effects on the cooling requirements of the building. Specifically, under the RCP-8.5 scenario, it is projected that the cooling demand will increase by 20% by the year 2100 compared to the reference climate.

This escalation in cooling demand would consequently lead to a rise in power usage within the building, resulting in various environmental consequences. Notably, the study estimates a substantial increase in $CO_2$ emissions, amounting to 14.591 metric tons between the typical meteorological year (1985–2004) and 2100. Additionally, the depletion of fossil fuels is expected to increase from 2.94E+04 to 3.54E+04 kg oil equivalent.

The study underscores the urgent need to incorporate renewable energy resources in electricity production and to incorporate climate change scenarios into building designs. These measures are crucial for effectively reducing carbon emissions, ensuring long-term energy security, and mitigating the potential impacts of climate change. By embracing sustainable practices and considering future climate projections, we can strive towards a greener and more resilient built environment.

**Acknowledgements** This publication was made possible by the National Priorities Research Program (NPRP) grant (NPRP12S-0212-190073) from the Qatar National Research Fund (QNRF), a member of Qatar Foundation (QF). Also, the research was supported by a scholarship from Hamad Bin Khalifa University (HBKU), a member of Qatar Foundation (QF). Any opinions, findings, and conclusions or recommendations expressed in these materials are those of the authors and do not necessarily reflect the views of QNRF, HBKU or QF.

## References

Alhorr Y, Eliskandarani E, Elsarrag E (2014) Approaches to reducing carbon dioxide emissions in the built environment: low carbon cities. Int J Sustain Built Environ 3(2):167–178. https://doi.org/10.1016/j.ijsbe.2014.11.003

Ali KA, Ahmad MI, Yusup Y (2020) Issues, impacts, and mitigations of carbon dioxide emissions in the building sector. Sustainability 12(18). https://doi.org/10.3390/SU12187427

Andric I, Al-Ghamdi SG (2020) Climate change implications for environmental performance of residential building energy use: the case of Qatar. Energy Rep 6:587–592. https://doi.org/10.1016/j.egyr.2019.09.030

Berardi U, Jafarpur P (2020) Assessing the impact of climate change on building heating and cooling energy demand in Canada. Renew Sustain Energy Rev 121(December 2019):109681. https://doi.org/10.1016/j.rser.2019.109681

Chakraborty D, Alam A, Chaudhuri S, Başağaoğlu H, Sulbaran T, Langar S (2021) Scenario-based prediction of climate change impacts on building cooling energy consumption with explainable artificial intelligence. Appl Energy 291(February):116807. https://doi.org/10.1016/j.apenergy.2021.116807

Cloy JM, Smith KA (2015) Greenhouse gas emissions. Ref Modul Earth Syst Environ Sci. https://doi.org/10.1016/B978-0-12-409548-9.05178-2

Dabaieh M, Wanas O, Hegazy MA, Johansson E (2015) Reducing cooling demands in a hot dry climate: a simulation study for non-insulated passive cool roof thermal performance in residential buildings. Energy Build 89:142–152. https://doi.org/10.1016/J.ENBUILD.2014.12.034

EPA (2021) Greenhouse gas emissions from a typical passenger vehicle. https://www.epa.gov/greenvehicles/greenhouse-gas-emissions-typical-passenger-vehicle. Accessed 15 Apr 2022

Hall IJ, Prairie RR, Anderson HE, Boes EC (1978) Generation of a typical meteorological year. In: Conference on analysis for solar heating and cooling, San Diego, CA, USA, 27 June 1978 [Online]. Available: https://www.osti.gov/biblio/7013202

Hasnain SM, Alawaji SH, Al-Ibrahim A, Smiai MS (1999) Applications of thermal energy storage in Saudi Arabia. Int J Energy Res 23(2):117–124. https://doi.org/10.1002/(SICI)1099-114X(199902)23:2%3c117

IPCC (2007) Fourth assessment report: climate change (AR4). https://www.ipcc.ch/report/ar4/syr/

ISO 14040 (2006) Environmental management e life cycle assessment principles and framework. https://www.iso.org/standard/37456.html

ISO 14044 (2006) Environmental management—life cycle assessment—requirements and guidelines. https://www.iso.org/standard/38498.html

Kharseh M, Al-Khawaja M (2016) Retrofitting measures for reducing buildings cooling requirements in cooling-dominated environment: residential house. Appl Therm Eng 98:352–356. https://doi.org/10.1016/j.applthermaleng.2015.12.063

Khourchid AM, Ajjur SB, Al-Ghamdi SG (2022) Building cooling requirements under climate change scenarios: impact, mitigation strategies, and future directions. Buildings 12(10):1519. https://doi.org/10.3390/buildings12101519

Khourchid AM, Al-ansari TA, Al-Ghamdi SG (2023) Cooling energy and climate change nexus in arid climate and role of energy transition. Buildings 13(4):836. https://doi.org/10.3390/buildings13040836

Levesque A, Pietzcker RC, Baumstark L, De Stercke S, Grübler A, Luderer G (2018) How much energy will buildings consume in 2100? A global perspective within a scenario framework. Energy 148:514–527. https://doi.org/10.1016/j.energy.2018.01.139

Pal JS, Eltahir EAB (2016) Future temperature in southwest Asia projected to exceed a threshold for human adaptability. Nat Clim Chang 6(2):197–200. https://doi.org/10.1038/nclimate2833

Radhi H (2009) Evaluating the potential impact of global warming on the UAE residential buildings—a contribution to reduce the $CO_2$ emissions. Build Environ 44(12):2451–2462. https://doi.org/10.1016/j.buildenv.2009.04.006

Sphera (2022) What is GaBi software?. https://gabi.sphera.com/international/overview/what-is-gabi-software/. Accessed 05 Apr 2022

U.S. Department of Energy (2022) Commercial building prototype models. https://www.energycodes.gov/prototype-building-models. Accessed 22 Mar 2022

UNEP (2009) Buildings and climate change: summary for decision makers [Online]. Available: https://wedocs.unep.org/20.500.11822/32152

Yang Y, Javanroodi K, Nik VM (2021) Climate change and energy performance of European residential building stocks—a comprehensive impact assessment using climate big data from the coordinated regional climate downscaling experiment. Appl Energy 298:117246. https://doi.org/10.1016/j.apenergy.2021.117246

# Chapter 14
# Environmental Feasibility of Second-Life Battery Applications in Belgium

**Maeva Lavigne Philippot, Dominik Huber, Daniele Costa, Jelle Smekens, and Maarten Messagie**

**Abstract** The growing electric vehicle and stationary storage markets raise the potential of second-life batteries (SLB) and question their environmental feasibility. Based on the life cycle assessment methodology, this study evaluates the impacts on climate change of three use cases (residential, industrial, and utility) for SLBs in Belgium. The assessed battery is a Renault Zoe40 pack, with lithium nickel manganese cobalt (NMC111) as the cathode active material. The residential use case is a domestic 4 kWp photovoltaic installation. The industrial use case is a 1,500 kWh installation providing behind-the-meter services. The utility use case contains SLBs with a total capacity of 20,000 kWh for in-front-of-the-meter services. The manufacturing and recycling impacts are allocated between the first and second life based on the delivered energy. As a result, the impact on climate change of the residential use case, the industrial use case, and the utility use case is 130.8 $gCO_2$ eq/kWh, 146.6 $gCO_2$ eq/kWh, and 168.3 $gCO_2$ eq/kWh, respectively. The residential use case performs better, as it delivers more energy than the two other use cases. Both the residential and the industrial use stages are close to the impacts of first-life batteries used as a benchmark in this study, while the utility use case impacts are 40% higher than the benchmark battery.

**Keywords** Life cycle assessment (LCA) · Li-ion battery · Repurpose · Stationary energy storage

M. Lavigne Philippot (✉) · D. Huber · D. Costa · J. Smekens · M. Messagie
Electric Vehicle and Energy Research Group (EVERGI), Department of Electrical Engineering and Energy Technology, Mobility Logistics and Automotive Technology Research Centre (MOBI), Vrije Universiteit Brussel, Pleinlaan 2, 1050 Brussels, Belgium
e-mail: maeva.philippot@vub.be



N. S. Caetano and M. C. Felgueiras (eds.), *The 9th International Conference on Energy and Environment Research*, Environmental Science and Engineering,
https://doi.org/10.1007/978-3-031-43559-1_14

## 14.1 Introduction

The increase in electric vehicles (EV) sold in the last decade will result in an increased amount of batteries reaching their end of life (EoL) (Melin 2018). Meanwhile, renewable energy sources are being incorporated into the electricity mixes, but their intermittent nature calls for increasing energy storage capacities (IEA 2021). Second-life batteries (SLB) can provide several services to the grid, whether physically located in front of the meter or behind it (Liu et al. 2020). Behind-the-meter batteries are installed on the customer side. They can be used for peak shaving, uninterrupted power supply, bill reduction, increased photovoltaic (PV) energy self-consumption, and backup power. In-front-of-the-meter batteries have bigger capacities and can be used for energy arbitrage, frequency regulation, and energy reserve, among other services (Liu et al. 2020). In Belgium, current policies encourage battery storage for PV panels, and future grid stability requires large energy storage systems (Gillabel et al. 2021). At the end of their automotive life, batteries still have storage capacity, raising the possibility of using them in a second life as stationary storage systems (Hossain et al. 2019).

The growing SLB market raises the question of their environmental impacts. Several cases have already been studied in the literature through Life Cycle Assessment (LCA) on a domestic scale (Faria et al. 2014; Kamath et al. 2020; Ioakimidis et al. 2019; Bobba et al. 2018; Philippot et al. 2022) and utility scales (Kamath et al. 2020; Canals Casals et al. 2017; Ahmadi et al. 2017; Schulz-Mönninghoff et al. 2021). This study assesses SLB in Belgium, considering three second-life use cases: residential, industrial, and utility.

## 14.2 Material and Methods

### *14.2.1 Goal and Scope*

LCA is a standardized tool to assess the environmental impacts of products or services over their life cycle (ISO 14044:2006 2006; ISO 14040:2006 2006). In this study, the functional unit (FU) is 1 kWh delivered by the battery. The system boundaries are set from cradle-to-grave (Fig. 14.1). The system boundaries include stages from the first life linked to the second life: repurposing, stationary use, maintenance, and collection. In the second use phase, the electricity to charge the batteries, in the domestic and industrial cases, is generated by PV installations (hatched in Fig. 14.1, which are partly considered in the evaluation based on the delivered energy by the PV panels).

As the study focuses only on SLB, the first use stage in the EV is excluded (in green in Fig. 14.1), while other processes and components (in blue in Fig. 14.1) are fully allocated to the second life. However, battery manufacturing, first collection, dismantling, and recycling are partially allocated to the system (shaded blue and

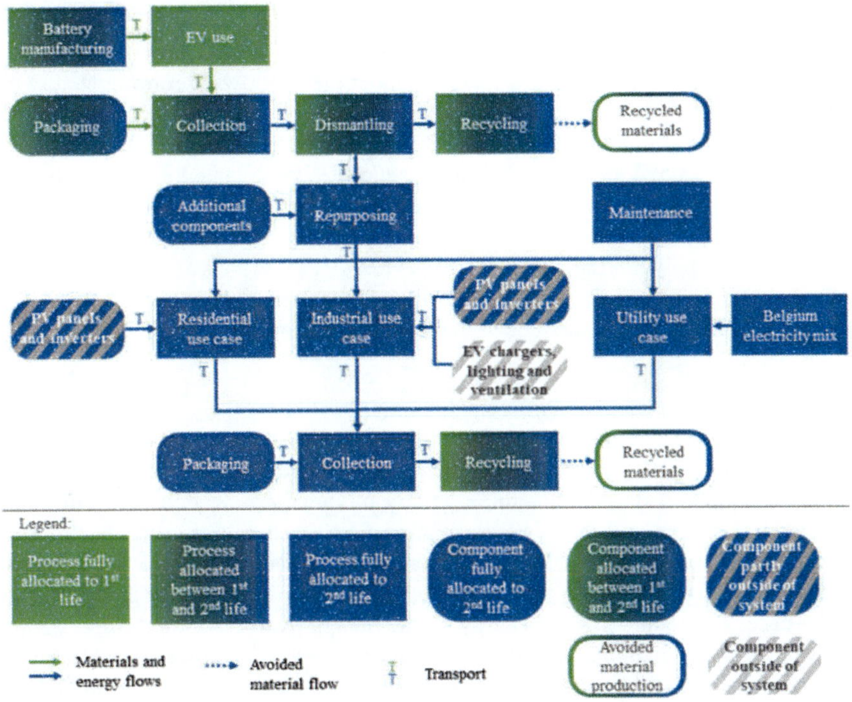


**Fig. 14.1** System boundaries. Legend: *EV* electric vehicle, *PV* photovoltaic

green in Fig. 14.1). Recycling includes the avoided primary material production and the burdens of the recycling process. The allocation rule is based on an allocation factor representing the ratio between the energy delivered during the second life and the total energy delivered.

## 14.2.2 *Life Cycle Inventory*

The collection stage is based on primary data. Background data is collected from the Ecoinvent database v3.7.1 (Wernet et al. 2016). The Renault Zoe is the reference vehicle, representing the 2019 Belgian fleet (EAFO 2020). The 41 kWh pack is assembled in France. It weighs 305 kg and contains 12 modules, with cells manufactured in South Korea. The battery components are modeled according to Ellingsen et al. (2014) and the use stages are based on the cell datasheet (Lima 2022). The cathode active material is lithium nickel manganese cobalt oxide ($LiNi_{1/3}Co_{1/3}Mn_{1/3}O_2$, NMC111), the main cathode active material in the global market (Van Mierlo et al. 2021) and is modeled with primary data (Philippot et al.

2018). As a matter of comparison, a benchmark battery is included. The chemistry is the same as the SLB. Other parameters are detailed in Table 14.1.

The vehicle consumption is 175 Wh/km under the Worldwide Harmonized Light Vehicles Test Procedure (Database 2022). Based on Belgian national statistics, the total mileage is 133,668 km (FEBIAC 2020), leading to a total delivered energy over the first life of 23,392 kWh. The battery's cycle life is calculated thanks to the cell datasheet (Lima 2022), and the capacity of the entire battery pack at the end of the first life is estimated as 36.5 kWh. The collection packaging is modeled as an impact-resistant plastic polyethylene box (KAISER+KRAFT 2022), useable 15 times, filled with a plastic bag to protect the container, and vermiculite as buffer material. The repurpose duration is 12 years due to the technical performance and potential safety issues.

The use cases for the SLBs are as follows:

Residential use case: an average domestic 4 kWp PV installation with an energy: power ratio of 2:1. The produced energy is assumed to be self-consumed at its maximum. As a result, only three battery modules are needed, and only one-quarter of the energy delivered during the first life is allocated to the first life (5,848 kWh). The yearly delivered energy by the battery is 1873 kWh. A new battery management system and an electric cabinet are needed for the residential use case. The total delivered energy by the battery over the second life is 22,478 kWh and the remaining capacity of the battery is 59% (Fig. 14.2). The Ecoinvent dataset for the PV installation is adjusted to match the 12-year lifetime of this study (Wernet et al. 2016; Frischknecht et al. 2020).

Industrial use case: the SLBs provide behind-the-meter services, such as peak shaving and an uninterrupted power supply. This use case is based on an industrial facility operator (Schulz-Mönninghoff et al. 2021). A container is needed for the batteries and another for the power electronics. The SLBs are grouped in a freight container in racks with a cooling system and another container is included with the power electronics. As in the residential use case, the manufacturing impacts of the PV panels and other system components were customized to the installation lifetime. The total delivered energy of the battery installation over the second life is 3,800 MWh, leading to a final capacity of 74%.

Utility-scale use case: the SLB provides any in-front-of-the-meter service, e.g., interacting with the secondary reserve market. The packs are mounted in

**Table 14.1** Benchmark battery parameters

| Parameter | Unit | Residential benchmark battery | Industrial benchmark battery | Utility benchmark battery |
|---|---|---|---|---|
| Capacity | kWh | 9.12 | 1,500 | 20,000 |
| Lifetime | Years | 15 | 15 | 15 |
| Weight | kg | 58.8 | 8,947 | 154,482 |
| Data source | | VARTA (2022) | da Silva Lima et al. (2021) | da Silva Lima et al. (2021) |

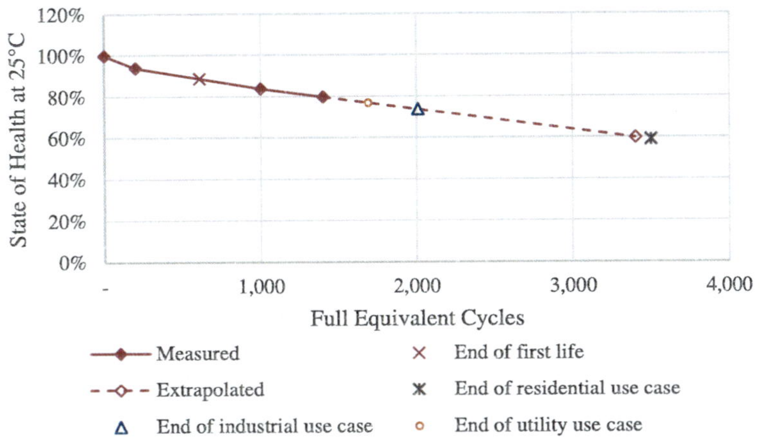


**Fig. 14.2** Battery remaining capacity at 25 °C. Based on Lima (2022)

containers with racks and a cooling system (Schulz-Mönninghoff et al. 2021). The power electronics are similar to the industrial use case and their inclusion in the system boundaries follows the same rule. Only electricity losses for charging and discharging are included. This installation allows a total amount of annually discharged electricity of 1,750 MWh per year. The EoL capacity is the highest among all the use cases: 77%.

The average distances of 55 km and 70 km are considered for the battery collection before the first dismantling and after the second uses, respectively. The dismantling to the module level includes electricity for manual dismantling. The repurposing before the second use includes a test and qualification protocol. This step requires 1 kWh of electricity for 1 kWh of battery capacity (Philippot et al. 2022). Next, the battery tray and retention are removed, and the energy for assembly is the same as for a new pack assembly.

A pyrometallurgical process combined with a hydrometallurgical process was considered for cell recycling (Dunn et al. 2012). Material recovery targets are 95% for cobalt, nickel, and copper, and 70% for lithium, based on the 2030 targets of the European draft regulation on waste batteries (European Parliament 2020).

### *14.2.3 Life Cycle Impact Assessment*

As the assessment focus on climate change (CC) impacts, the IPCC 2013 (100 years) is chosen as the life cycle impact assessment method.

## 14.3 Results and Discussion

The impacts on CC for the three use cases are shown in Fig. 14.3. The results of each case study are discussed in Sects. 3.1–3.3.

### 14.3.1 Residential Use Case

The SLB system impact on the CC is 130.8 $gCO_2$ eq/kWh, mostly originating from the PV system manufacturing (Fig. 14.3). The collection, dismantling, and repurposing represented less than 1% of the greenhouse gas emissions. Recycling the cells and avoiding producing raw materials decreases the total impact by 17%. The benchmark battery performs better because of the larger amount of energy delivered by the benchmark battery over 15 years, instead of 12 years for the SLB.

Comparing results with other studies is not straightforward, even when the application is similar (Faria et al. 2014; Kamath et al. 2020; Ioakimidis et al. 2019; Bobba et al. 2018; Philippot et al. 2022). Some studies have different FUs (Faria et al. 2014; Bobba et al. 2018), include the first life (Ioakimidis et al. 2019; Philippot et al. 2022), or exclude the EoL (Kamath et al. 2020). In addition, other studies use other life cycle impact assessment methods (Ioakimidis et al. 2019). For example, Kamath et al. (2020) do not give the absolute values for their residential use case. However, those previous studies conclude that repurposing a battery in a household with PV installation reduces the impact on CC (Faria et al. 2014; Kamath et al. 2020; Ioakimidis et al. 2019; Bobba et al. 2018; Philippot et al. 2022). Due to the allocation factor adopted in this study, almost 80% of the manufacturing environmental impacts of our residential use case are carried by the SLB, which is a high allocation factor

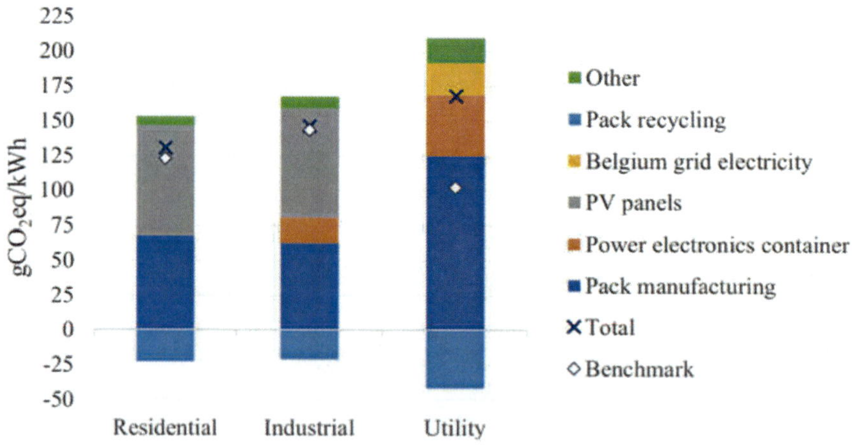


**Fig. 14.3** Impact on climate change of the three use cases. Legend: PV photovoltaic

compared to the scientific literature (Faria et al. 2014; Kamath et al. 2020; Ioakimidis et al. 2019; Bobba et al. 2018; Philippot et al. 2022).

### 14.3.2 Industrial Use Case

In the industrial use case, the SLB system impact on the CC is 146.6 $gCO_2$ eq/kWh. From this, the PV panels manufacturing is the main contributor (53%), followed by battery manufacturing (43%) and power electronics manufacturing (12%). The impact from the power electronics container is noteworthy (11%), driven mainly by the converter and its printed wiring boards. The collection, dismantling, repurposing, installation, and maintenance are not significant steps as they represent 0.7% or less of the total impact. Recycling lowers the total impact by 14%. The SLB impact on CC is higher than the benchmark battery by only 3 $gCO_2$ eq/kWh.

As for the residential use case, the comparison with studies with similar applications (Kamath et al. 2020; Canals Casals et al. 2017; Schulz-Mönninghoff et al. 2021) is not entirely fair, as first life may be included (Canals Casals et al. 2017; Schulz-Mönninghoff et al. 2021). Results from these studies range from 42 $gCO_2$ eq/kWh (Kamath et al. 2020) to more than 600 $gCO_2$ eq/kWh (Canals Casals et al. 2017). The lack of transparency in the inventory (Kamath et al. 2020) impedes explaining the reasons for the high variability.

### 14.3.3 Utility Use Case

The utility SLB emits 168.3 $gCO_2$ eq/kWh, primarily due to the battery pack manufacturing (75%). As for the other use cases, collection, dismantling, repurposing, and maintenance are not shown to be major contributors. The power electronics container is a bigger contributor than the Belgian electricity mix. The benchmark battery performs better than the SLB because it is lighter and provides more energy over its full life than the heavier SLB over its second life. As the repurpose duration is 12 years, this SLB is under-used and the remaining capacity at the end of the second life is high (77%). Also, there are more grid electricity losses during the SLB use than for the benchmark. The impacts of those losses are dependent on the electricity mix.

The obtained results are lower than studies that include the first life of the battery (Kamath et al. 2020; Ahmadi et al. 2017). For example, Kamath et al. (2020) conclude that peak shaving SLB has an impact on CC between 540 and 1370 $gCO_2$ eq/kWh and Ahmadi et al. (2017) found 250 $gCO_2$ eq/kWh when the SLB is charged with electricity from Ontario (Canada) grid. However, comparing to different applications may not be fair as the function of the product under study is different.

## 14.4 Conclusion

This study assessed the impact on the CC of three use cases for SLBs in Belgium: residential, industrial, and utility. A cradle-to-grave LCA was performed, including the assessment of new batteries as a benchmark. For all the cases, the FU is 1 kWh of delivered energy.

The residential use case has the lowest impact on the CC (130.8 $gCO_2$ eq/kWh) as the SLB delivers more energy over the life cycle, reaching 59% capacity at the end of its second life. The utility use case has the highest impact on CC, corresponding to 168.3 $gCO_2$ eq/kWh. It performs worse than the benchmark battery due to the lower technical performance of the SLB. In this use case, the SLB is under-used due to safety issues, and the final capacity is 77%.

Due to the allocation rule based on delivered energy, the pack manufacturing drives up the impact on the CC of the use cases, representing at least 43% of the greenhouse gas emissions. The PV panels of residential and industrial use cases also carry a large share of those impacts. The impact on the CC of the use cases depends on the battery lifetimes and charging electricity. Improving the dimensioning of the SLB would help reduce the impact on the CC: a SLB delivering more energy has lower impacts. In this study, the battery in the residential use case is used until its technical limits, while the other batteries are recycled for safety reasons.

The obtained results should be carefully analyzed as they depend on the analyzed battery chemistry. For example, other chemistries might lead to different battery performances, leading to different environmental performances. Furthermore, this assessment only considers CC impacts. Including other environmental impact categories would reveal further opportunities to improve their environmental performance.

**Acknowledgements** This work was supported by the Agency for Innovation and Entrepreneurship (VLAVIO) [grant number HBC.2019.0125].

## References

Ahmadi L, Young SB, Fowler M, Fraser RA, Achachlouei MA (2017) A cascaded life cycle: reuse of electric vehicle lithium-ion battery packs in energy storage systems. Int J Life Cycle Assess 22(1):111–124

Bobba S, Mathieux F, Ardente F, Blengini GA, Cusenza MA, Podias A, Pfrang A (2018) Life cycle assessment of repurposed electric vehicle batteries: an adapted method based on modelling energy flows. J Energy Storage 19:213–225

Canals Casals L, García BA, Aguesse F, Iturrondobeitia A (2017) Second life of electric vehicle batteries: relation between materials degradation and environmental impact. Int J Life Cycle Assess 22(1):82–93

da Silva Lima L, Quartier M, Buchmayr A, Sanjuan-Delmás D, Laget H, Corbisier D, Mertens J, Dewulf J (2021) Life cycle assessment of lithium-ion batteries and vanadium redox flow batteries-based renewable energy storage systems. Sustain Energy Technol Assess 46(101286)

EV Database. https://ev-database.org/car/1236/Renault-Zoe-ZE40-R110. Accessed 25 Jan 2022
Dunn JB, Gaines L, Barnes M, Sullivan J, Wang M (2012) Material and energy flows in the materials production, assembly, and end-of-life stages of the automotive lithium-ion battery life cycle. Argonne National Laboratory
EAFO. https://alternative-fuels-observatory.ec.europa.eu/transport-mode/road/belgium/vehicles-and-fleet. Accessed 2 July 2020
Ellingsen LAW, Majeau-Bettez G, Singh B, Srivastava AK, Valøen LO, Strømman AH (2014) Life cycle assessment of a lithium-ion battery vehicle pack. J Ind Ecol 18(1):113–124
European Parliament (2020) Proposal for a regulation of the European Parliament and of the council concerning batteries and waste batteries, repealing Directive 2006/66/EC and amending Regulation (EU) No 2019/1020
Faria R, Marques P, Garcia R, Moura P, Freire F, Delgado J, De Almeida AT (2014) Primary and secondary use of electric mobility batteries from a life cycle perspective. J Power Sour 262:169–177
FEBIAC. https://www.febiac.be/public/statistics.aspx?FID=23&lang=FR. Accessed 21 Oct 2020
Frischknecht R, Stolz P, Krebs L, de Wild-Scholten M, Sinha P (2020) Life cycle inventories and life cycle assessments of photovoltaic systems 2020 task 12 PV sustainability. IEA Report T12-19:2020
Gillabel J, Dams Y, Vanderreydt I (2021) Circular economy and the energy transition—potential of a Flemish circularity hub for EV Li-ion batteries. CE Center Publication N° 17
Hossain E, Murtaugh D, Mody J, Faruque HMR, Sunny MSH, Mohammad N (2019) A comprehensive review on second-life batteries: current state, manufacturing considerations, applications, impacts, barriers potential solutions, business strategies, and policies. IEEE Access 7:73215–73252
IEA (2021) Global energy review 2021. IEA Publications
Ioakimidis CS, Murillo-Marrodán A, Bagheri A, Thomas D, Genikomsakis KN (2019) Life cycle assessment of a lithium iron phosphate (LFP) electric vehicle battery in second life application scenarios. Sustainability (Switzerland) 11(9):2527
ISO. ISO 14040:2006 (2006) Environmental management—life cycle assessment—principles and framework. International Organization for Standardization
ISO. ISO 14044:2006 (2006) Environmental management—life cycle assessment—requirements and guidelines. International Organization for Standardization
KAISER+KRAFT International. https://www.export.kaiserkraft.com/drums-and-tanks/hazardous-goods-containers/pe-storage-and-transport-container-for-rechargeable-batteries/model-l-filling-weight-400-kg/p/M7957686/. Accessed 14 Feb 2022
Kamath D, Shukla S, Arsenault R, Kim HC, Anctil A (2020) Evaluating the cost and carbon footprint of second-life electric vehicle batteries in residential and utility-level applications. Waste Manage 113:497–507
Lima P. https://pushevs.com/2019/02/10/renault-zoe-ze-40-full-battery-specs/. Accessed 25 Jan 2022
Liu J, Hu C, Kimber A, Wang Z (2020) Uses, cost-benefit analysis, and markets of energy storage systems for electric grid applications. J Energy Storage 32:101731
Melin HE (2018) The lithium-ion battery end-of-life market—a baseline study. Global Battery Alliance
Philippot M, Smekens J, Van Mierlo J, Messagie M (2018) Life cycle assessment of silicon alloy based lithium-ion battery for electric vehicles. WIT Trans Built Environ 182
Philippot M, Costa D, Hosen MS, Senécat A, Brouwers E, Nanini-Maury E, Van Mierlo J, Messagie M (2022) Environmental impacts of the second life of an automotive battery: reuse and repurpose based on ageing tests. J Clean Prod 366:132872
Schulz-Mönninghoff M, Bey N, Nørregaard PU, Niero M (2021) Integration of energy flow modelling in life cycle assessment of electric vehicle battery repurposing: evaluation of multi-use cases and comparison of circular business models. Resour Conserv Recycl 174:105773

Van Mierlo J, Berecibar M, El Baghdadi M, De Cauwer C, Messagie M, Coosemans T, Hegazy O (2021) Beyond the state of the art of electric vehicles: a fact-based paper of the current and prospective electric vehicle technologies. World Electr Veh J 12(20)

VARTA. https://www.varta-ag.com/en/consumer/product-categories/energy-storage-systems/varta-element. Accessed 7 Jan 2022

Wernet G, Bauer C, Steubing B, Reinhard J, Moreno-Ruiz E, Weidema B (2016) The ecoinvent database version 3 (part I): overview and methodology. Int J Life Cycle Assess 21(9):1218–1230

# Chapter 15
# The Carbon Footprint of a Furniture Industry Facility: Evaluation of the Impact Progress Over 2013–2019

**Carolina Vicente, Dânia S. Ascenção, João R. Silva, and Luís M. Castro**

**Abstract** The increasing social and political pressure on the development of more sustainable industries has led to an increase in the importance given by organizations to their products' impact on the ecosystems. Carbon Footprint is a valuable tool in the assessment of those impacts. This study used Scope 1, 2 and 3 emissions to estimate the carbon emissions of a furniture industry plant and evaluate the evolution of those emissions in the 2013–2019 period. It was found that the main contributor to the studied plant carbon emissions is electricity consumption. In the scope of this work, it is highlighted the positive progress in the greenhouse gas emissions over the set period, mainly throughout 2017–2019, as a result of an important investment in solar energy.

**Keywords** Carbon footprint · Direct and indirect emissions · Furniture industry

C. Vicente · J. R. Silva · L. M. Castro (✉)
Instituto Politécnico de Coimbra, Instituto Superior de Engenharia de Coimbra, Rua Pedro Nunes, Quinta da Nora, 3030-199 Coimbra, Portugal
e-mail: mcastro@isec.pt

D. S. Ascenção
IKEA Industry Portugal, SA, Avenida Capital do Móvel, Nº 157, 4595-282 Penamaior, Portugal

L. M. Castro
Department of Chemical Engineering, Faculty of Sciences and Technology, CIEPQPF—Chemical Engineering Processes and Forest Products Research Center, University of Coimbra, Coimbra, Portugal

Laboratório SiSus, Instituto Politécnico de Coimbra, Instituto de Investigação Aplicada, Rua Pedro Nunes, Quinta da Nora, 3030-199 Coimbra, Portugal



N. S. Caetano and M. C. Felgueiras (eds.), *The 9th International Conference on Energy and Environment Research*, Environmental Science and Engineering,
https://doi.org/10.1007/978-3-031-43559-1_15

## 15.1 Introduction

Over the last decades, greenhouse gas (GHG) emissions from anthropogenic sources have been appointed as the primary cause of climate change. This way, accurate estimation of GHG emissions is essential to provide valid data to policymaking procedures (Notte et al. 2017). The carbon footprint (CF) is a vital tool to assess the anthropogenic activity's environmental impact (Čuček et al. 2012). CF presents the total amount of GHG, usually in carbon dioxide equivalent ($CO_2$ eq), emitted during a specific period or over the product life cycle (Biron 2020). Quantification of CF requires a detailed analysis of the process inputs and outputs and demands the specification of organizational and operational boundaries to determine the considered activities (WBCSD and WRI 2012).

This study pretends to estimate the carbon footprint resulting from a furniture industry plant activity over the 2013–2019 period and to evaluate the evolution of the environmental impacts of the studied facility. The case study is an industrial plant from a multinational company located in the northern part of Portugal. There are limited Portuguese perspectives on the carbon footprint of furniture factories. Thus, this paper addresses a gap in the literature and adds to informed practice in this sector. Therefore, the principal contribution of the presented case study is to address how sustainable solutions can reduce the carbon footprint generated inside the proposed boundaries of the production facility.

## 15.2 Methodology

According to the Intergovernmental Panel on Climate Change (IPCC), $CO_2$ emissions might be estimated by "combining the information on the extent to which a human activity takes place (called activity data or AD) with coefficients which quantify the emissions or removals per unit activity", as it is translated by Eq. 15.1.

$$Emissions(\mathrm{CO_2\,eq}) = AD \times EF \tag{15.1}$$

AD can relate, for example, to electricity consumption, traveled distance or the amount of produced waste. The emission coefficients or factors (EF) can be provided by different sources, namely the Department for Environment, Food and Rural Affairs (DEFRA), IPCC and GHG Protocol. DEFRA provides a complete database to estimate $CO_2$ emissions, referring to the UK reality. Consequently, in this study, it is assumed that the British reality is identical to the Portuguese one. CF can also be quantified using Global Warming Potential (GWP) values that allow converting a quantity of a GHG into $CO_2$eq (Al-Mansour and Jejcic 2017). GHG Protocol Corporate Standard states that GHG emissions can be categorized into three scopes. Scope 1 accounts for GHG emissions from sources owned or controlled by the company (direct emissions). Scope 2 relates to electricity indirect GHG emissions, in order words, the emissions from the generation of purchased electricity. Scope 3 is an optional reporting category as it accounts for emissions that are a consequence of the company's activities but whose sources are not owned or controlled by the organization.

In this study, the carbon footprint estimation is divided according to those scopes. CF quantification was performed applying Eq. 15.1. However, some of the considered activities have specificities that require an adjustment of Eq. 15.1 to allow the CF estimation. In this section, the setting of the studied activities for each category and the methodology used to determine each activity's impact are described in detail.

### 15.2.1 Scope 1: Direct GHG Emissions

For Scope 1 emissions, the GHG emissions from fixed sources, fugitive emissions, and company-owned vehicles, including forklift trucks, have been considered. The CF related to the emissions of fixed sources and the fugitive emissions is determined by Eq. 15.2. The considered fixed sources emissions are measured at the air extractors present at many stages of the production lines.

$$CF_{\text{Fixed Sources}}(CO_2\,\text{eq/year}) = \text{Emissions}_x \times \text{GWP}_x \tag{15.2}$$

where $\text{Emissions}_x$ represents the emissions of the pollutant x (kg/year) and $\text{GWP}_x$ represents the GWP of that pollutant (kg $CO_2$ eq $kg^{-1}$ GHG).

GWP values of the considered pollutants are presented in Table 15.1, with fugitive emissions corresponding to leaks on air conditioning, refrigeration and heating equipment.

The fixed sources include a chimney from a boiler for sawdust combustion. Biomass combustion resultant emissions are not considered on the CF determination (NCASI 2005). However, sawdust is also composed of non-organic substances, like the glues used on the particleboard production process. Therefore, the GHG emissions from this element are due to the combustion of the sawdust resin content and can be estimated by applying Eq. 15.3.

$$CF_{sawdust\ combustion}(CO_2\,\text{eq/year}) = RC \times m \times EF_{resin} \tag{15.3}$$

**Table 15.1** GWP of greenhouse and refrigerant gases

| GHG | CO | NOx | COV | $CO_2$ | R-407C | R-134a | R-410A | HFC-227ea | $SF_6$ | R-404A |
|---|---|---|---|---|---|---|---|---|---|---|
| GWP | 3[a] | 8[a] | 11[a] | 1[a] | 1774[b] | 1300[c] | 2088[b] | 3350[c] | 23,500[c] | 3922[b] |

[a]Johnke (1996), [b]Air Conditioning and Refrigeration Industry Board (ACRIB) (2015), [c]Protocol (2015)

**Table 15.2** Fuel emission factors related to net calorific value and NCV per mass of fuel (Union 2007)

| Fuel | NCV (TJ/t fuel) | EF (t $CO_2$ $TJ^{-1}$) |
|---|---|---|
| Diesel | 0.0430 | 74.0 |
| LPG | 0.0473 | 63.0 |
| Gasoline | 0.0443 | 69.2 |

**Table 15.3** Electricity consumption emission factors (kg $CO_2$ eq/kWh) (European Environment Agency 2022)

| 2013 | 2014 | 2015 | 2016 | 2017 | 2018 | 2019 |
|---|---|---|---|---|---|---|
| 0.318 | 0.299 | 0.364 | 0.296 | 0.353 | 0.310 | 0.244 |

where $RC$ is the resin content, expressed as a mass fraction, m represents the mass of sawdust consumed during a year (kg/year), and $EF_{resin}$ is the resin emission factor (kg $CO_2$ eq $kg^{-1}$ resin). According to the particleboard supplier, the resin content is 8–10%. In this study, it was assumed RC = 0.1 and $EF_{resin}$ = 1.1 kg $CO_2$ eq $kg^{-1}$ resin (National Council for Air and Stream Improvement Inc. (NCASI) 2005).

The forklift trucks of the studied organization worked with diesel, gasoline, or liquefied petroleum gas (LPG) until 2019, when all the forklifts, except one, were replaced with electric. As the organization fuel consumption data were expressed in mass terms, it was necessary to employ the net calorific value and the corresponding EF to estimate the contribution of the resultant emissions. Net Calorific Value (NCV) and EF values for the three fuels are presented in Table 15.2.

### 15.2.2 *Scope 2: Electricity Indirect GHG Emissions*

Scope 2 only accounts for the emissions generated from purchased public electricity, with the respective EF, presented in Table 15.3, calculated as the ratio of $CO_2$ eq emissions from public electricity production and gross Portuguese electricity production.

### 15.2.3 *Scope 3: Other Indirect Emissions*

For Scope 3, emissions from materials and waste transport, waste disposal, and employees commuting to and from work are considered. Although GHG Protocol Corporate Standard Emissions does not refer to emissions related to water consumption and wastewater treatment, in this study, both are considered. The assessment of the contribution of raw and discarded materials transport presupposes an estimation of the travelled distance (TD), Eq. 15.4.

$$TD_{materials\ and\ waste\ transport} = 2 \times NL \times d \quad (15.4)$$

where *NL* represents the number of loads and *d* the distance between the supplier or the waste management organization (km). For both of the cases, NL is estimated based on the 2019 available data. Likewise, the estimation of employees commuting impact to and from work also presumes an estimation of TD by Eq. 15.5.

$$TD_{employees\ commuting} = 2 \times d \times (1 - SR) \times NE \times WD \quad (15.5)$$

where *SR* is the share rate (determined by Eq. 15.6), *NE* is the number of employees, and *WD* is the annual number of working days.

$$SR = \frac{NE}{VPD} = \frac{NE}{\frac{NV}{WD}} \quad (15.6)$$

where *VPD* is the number of vehicles that park in the facility's parking lot each day and *NV* is the annual number of vehicles that entered the facility's parking lot.

Table 15.4 represents the emission factors provided by the DEFRA database for Scope 3 activities, except waste disposal.

The waste produced by the organization activity was categorized according to the DEFRA guidelines. Table 15.5 presents the emissions factors related to the different end-of-life (EoL) scenarios for the different discharged materials categories. Even though the EF values are presented in kg $CO_2$ eq/t, according to the DEFRA database guidelines, they refer to the emissions on the transportation to the waste management facilities. Consequently, in this context, EF cannot be used to compare different EoL scenarios since the energy gains obtained from valorization scenarios are not included. Whereas for landfills, the EF comprises transportation and landfill emissions.

**Table 15.4** Scope 3 activities, excluding waste disposal, emission factors (Department for Environment Food & Rural Affairs)

| Category | Emission factor | | | | | | | Units |
|---|---|---|---|---|---|---|---|---|
| | 2013 | 2014 | 2015 | 2016 | 2017 | 2018 | 2019 | |
| Water consumption | 0.344 | 0.344 | 0.344 | 0.344 | 0.344 | 0.344 | 0.344 | (kg $CO_2$ eq $m^{-3}$) |
| Wastewater treatment | 0.709 | 0.709 | 0.708 | 0.708 | 0.708 | 0.708 | 0.708 | (kg $CO_2$ eq $m^{-3}$) |
| Raw material and waste transport | 0.915 | 0.915 | 0.915 | 0.915 | 0.870 | 0.873 | 0.867 | (kg $CO_2$ eq $km^{-1}$) |
| Employees commuting | 0.190 | 0.189 | 0.186 | 0.187 | 0.182 | 0.181 | 0.177 | (kg $CO_2$ eq $km^{-1}$) |

**Table 15.5** Emission factors related to the different waste typologies and different end-of-life scenarios (Department for Environment Food & Rural Affairs)

| Equivalent waste category on DEFRA database | 2013 | 2014 | 2015 | 2016 | 2017 | 2018 | 2019 |
|---|---|---|---|---|---|---|---|
| Recycled residues | 21.00 | 21.00 | 21.00 | 21.00 | 21.80 | 21.38 | 21.35 |
| Commercial and industrial waste—landfill | 199.0 | 199.0 | 93.00 | 199.0 | 100.1 | 99.77 | 99.76 |
| Electrical items—batteries | 65.00 | 65.00 | 65.00 | 65.00 | 64.60 | 64.64 | 64.37 |
| Average construction—open loop | 1.40 | 1.40 | 1.40 | 1.40 | 1.40 | 1.37 | 1.37 |
| Asphalt—open loop | 1.40 | 1.40 | 1.40 | 1.40 | 1.40 | 1.37 | 1.37 |
| Non-recycled glass | 25.78 | 25.78 | 25.80 | 25.80 | 26.00 | 9.00 | 8.99 |
| Municipal waste—landfill | 289.8 | 289.8 | 459.0 | 421.0 | 588.9 | 586.5 | 586.5 |
| Electrical items—open-loop | 21.00 | 21.00 | 21.00 | 21.00 | 21.80 | 21.38 | 21.35 |
| Commercial and industrial waste—valorized | 21.00 | 21.00 | 21.00 | 21.00 | 21.80 | 21.38 | 21.35 |
| Construction—plasterboard | 21.00 | 21.00 | 21.00 | 21.00 | 21.80 | 21.38 | 21.35 |
| Construction—metals | 1.00 | 1.00 | 1.00 | 1.00 | 1.10 | 1.02 | 1.01 |

## 15.3 Results and Discussion

The CF estimations of Scope 1, 2, 3 activities are presented in Fig. 15.1. For Scope 1 emissions, fixed sources are the main contributor, and even though emissions fluctuate during the studied period, Scope 1 emissions decreased from 2013 to 2019. The fugitive emissions have a low impact on the CF since only the GHG quantities used to recharge equipment are considered. In this type of emissions, it is worth highlighting those that occurred in 2013 associated with the leak to the atmosphere of a significant amount of fluorinated greenhouse gas from a larger equipment. The contribution of company-owned vehicles presents a significant reduction in 2019 due to the transition to electric forklift trucks, which are now accounted for in Scope 2. Emissions related to the production of the consumed electricity are compared to the total electricity consumption in better detail in Fig. 15.2.

From Fig. 15.1, it is possible to notice that Scope 3 emissions increased between 2013 and 2014 and decreased in 2015. This decline can be explained by a production reduction that occurred in 2015 and resulted in a reduction of supplied raw materials and utility necessities, leading to a decrease in waste production. Additionally, waste transportation decreased significantly in 2019, seemingly resulting in a significant reduction of total emissions. The main contributor to Scope 3 is the raw material transport, whilst water consumption and treatment are minor contributors to the CF due to the reduced significance that this environmental descriptor assumes in this industry. Green procurement policies can significantly affect Scope 3 emissions, promoting the products' recyclability, durability, and packaging reduction, resulting in smaller quantities of waste and toxic disposable substances.

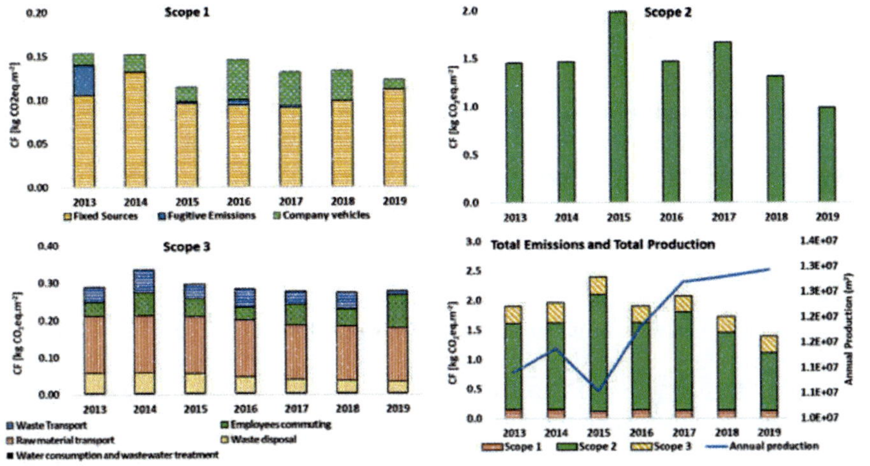


Fig. 15.1 Evolution of the Scope 1, 2 and 3 emissions effect on the CF between 2013 and 2019. Evolution of total CF per manufactured m$^2$ (total emissions) and evolution of annual production between 2013 and 2019

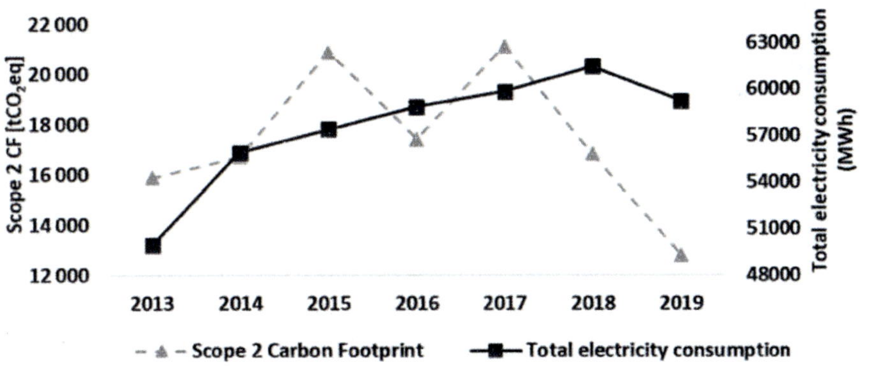


Fig. 15.2 Scope 2 emissions and total electricity consumption evolution from 2013 to 2019

The total CF, expressed per manufactured m$^2$ and the total annual production for each of the considering years are also represented in Fig. 15.1. A decrease in the total CF resulting from the plant activity is noticeable since 2017, with most being attributed to the reduction of emissions related to electricity consumption, which is the main contributor to the environmental impact of the studied plant. The change of electricity-related emissions imposes the evolution of the total carbon footprint. The importance of Scope 3 emissions can be worrying because this emissions typology refers to the activities that are less controlled by the company thus, being the most difficult to be reduced by company controlled decisions. Scope 1 emissions have diminished impact on total emissions CF, reducing the impact of the organization's decisions on the CF evolution.

Analyzing the data from Fig. 15.2, it is possible to notice a discrepancy between electricity consumption evolution and the CF related to Scope 2 emissions. Despite a consistent increase in electricity consumption between 2013 and 2018, the corresponding CF oscillates within the studied period. Scope 2 emissions are highly dependent on the country's approach for energy production. Therefore, EF changes over time, directly impacting the emissions of the factory. Additionally, in 2018, the electricity consumed in the facilities began to be partly self-produced, resorting to solar panels. Total electricity consumption reached a peak in 2018. However, this fact did not translate into a higher total emission per square of production (Fig. 15.1). In fact, production has been increasing since 2015 but CF decreases in about 25% from 2017 to 2018, and 32% from 2018 to 2019. In this regard, installing a photovoltaic plant with a total installed power of 5.8 MW, consisting of more than 18,000 pannels, had a clear impact on reducing electricity consumption and associated emissions. The self-produced electricity from renewable energies presents an EF of 0.0 kg $CO_2$ eq/kWh (Despacho (extrato) nº15293-D/2013 2013), contributing to the noticed divergence. The reduction of electric consumption in 2019 is greatly attributed to replacing the illumination system with LED lights, originating a possible decrease of 90% in the electric consumption related to illumination. Nevertheless, it was not possible to access the electricity consumption by utilization category. For this reason, it is not possible to conclude if the change performed on the illumination system was sufficient to compensate for the transition to electric forklift trucks impact on the factory energy consumption.

## 15.4 Conclusion

The carbon footprint evaluation allows identifying the main contributors to carbon emissions that result from industrial activity, providing valuable data for a conscious decision-making process. Studying CF evolution permits the ascertainment of the impact of the measures on GHG emissions, becoming a powerful tool to evaluate the efficiency of the decisions taken by organizations to increase the environmental sustainability of their activity.

From this study, it is possible to conclude that the carbon emissions related to the considered furniture industrial plant activity had a positive evolution over the last three years of the studied period, revealing the success of the various measures introduced by the company to reduce the carbon footprint of its activity. With the presented data, it is clear that the introduction of solar panels and the replacement of conventional illumination with LED contributed to a significant reduction in the specific energy consumption of the plant. The former, coupled with the introduction of electric forklifts and reduced waste transportation, greatly diminished the CF by the end of 2019. These results reflect the efficiency of the company's measures following the environmental impact reduction.

Lastly, the main contributor to GHG emissions is electric energy consumption, which has experienced a decrease in the last two years of the studied period due to the energy rationalization measures introduced and the company's commitment to renewable energies.

## References


2007/589/CE. European Union. https://doi.org/10.1111/j.1477-8947.2004.00084.x

Air Conditioning and Refrigeration Industry Board (ACRIB) (2015) 2014 F-gas regulation and GWP values

Al-Mansour F, Jejcic V (2017) A model calculation of the carbon footprint of agricultural products: the case of Slovenia. Energy 136:7–15. https://doi.org/10.1016/j.energy.2016.10.099

Biron M (2020) Recycling plastics: advantages and limitations of use. In: A practical guide to plastics sustainability. Elsevier, Amsterdam, pp 411–467. https://doi.org/10.1016/B978-0-12-821539-5.00009-4

Čuček L, Klemeš JJ, Kravanja Z (2012) A review of footprint analysis tools for monitoring impacts on sustainability. J Clean Prod 34:9–20. https://doi.org/10.1016/j.jclepro.2012.02.036

Department for Environment Food & Rural Affairs. Rules, guidance and support

Despacho (extrato) nº15293-D/2013 (2013) Ministério do Ambiente, Ordenamento do Território e Energia. Published in Diário da República, II Série-Nº234, 3 Dec 2013 (in portuguese)

European Environment Agency (2022) Greenhouse gas emission intensity of electricity generation. https://www.eea.europa.eu/data-and-maps/daviz/co2-emission-intensity-10/#tab-chart_2

GHG Protocol (2015) Global warming potential values

Johnke B (1996) Emissions from waste incineration. In: Intergovernmental panel on climate change, pp 455–468

La Notte A, Tonin S, Lucaroni G (2018) Assessing direct and indirect emissions of greenhouse gases in road transportation, taking into account the role of uncertainty in the emissions inventory. Environ Impact Assess Rev 69(July 2017):82–93. https://doi.org/10.1016/j.eiar.2017.11.008

National Council for Air and Stream Improvement Inc. (NCASI) (2005) Calculation tools for estimating greenhouse gas emissions from wood product facilities

WBCSD and WRI (2012) A corporate accounting and reporting standard. Greenhouse Gas Protocol, p 116

# Chapter 16
# Environmental Performance Comparison of Active Living Wall and Commercial Air Purifier: Life Cycle Assessment Study

**Mehzabeen Mannan and Sami G. Al-Ghamdi**

**Abstract** Indoor air pollution in buildings poses a significant risk to human health and well-being, as it comprises a wide range of particulate matter, gaseous contaminants, mold, and pollen. Various strategies have been employed to address indoor air quality concerns, such as minimizing pollution sources, dilution, utilizing air cleaning devices, and implementing vertical greening systems. However, integrating sustainability into these air cleaning methods remains a crucial challenge. In light of this challenge, this study conducts a life cycle assessment to compare the environmental performance of two indoor air cleaning methods: commercial air purifiers (CAP) and active living walls (ALW). The results reveal the potential environmental benefits of employing ALW in indoor spaces, compared to commercially available air purifiers. Although the production impacts of CAP are relatively lower than those of ALW in this study, the comprehensive evaluation of the entire product life cycle demonstrates the significantly higher overall impacts of CAP. This preliminary investigation aims to assist building professionals by providing a framework for estimating the environmental cost associated with two distinct indoor air purification methods. By highlighting the advantages of ALW over CAP, this study contributes valuable insights into sustainable approaches for enhancing indoor air quality.

**Keywords** Life cycle assessment · Active living wall · Commercial air purifier · Indoor air quality

---

M. Mannan · S. G. Al-Ghamdi (✉)
Division of Sustainable Development, College of Science and Engineering, Hamad Bin Khalifa University, Qatar Foundation, P.O. Box 34110, Doha, Qatar
e-mail: sami.alghamdi@kaust.edu.sa

S. G. Al-Ghamdi
Environmental Science and Engineering Program, Biological and Environmental Science and Engineering Division, King Abdullah University of Science and Technology (KAUST), Thuwal 23955-6900, Saudi Arabia

KAUST Climate and Livability Initiative, King Abdullah University of Science and Technology (KAUST), Thuwal 23955-6900, Saudi Arabia



N. S. Caetano and M. C. Felgueiras (eds.), *The 9th International Conference on Energy and Environment Research*, Environmental Science and Engineering,
https://doi.org/10.1007/978-3-031-43559-1_16

## 16.1 Introduction

In any building environment, indoor air quality (IAQ) is considered as one of the critical factors among indoor environmental qualities which ensures the health and wellbeing of the occupants as people tend to spend over 90% of their lifetime in different indoor spaces (Cincinelli and Martellini 2017). Therefore, it has become a particular challenge for the building professionals to reduce the indoor air pollutants exposure. To address the indoor air quality issues, several ways have been adopted so fur such as source minimization, dilution, use of air cleaning devices as well as vertical greening systems (Mannan and Al-Ghamdi 2021).

Compared to the other parts of the world, in Middle Eastern countries people tend to spend their time longer than average in indoor spaces as a result of harsh weather such as higher ambient temperature and humidity (Ali et al. 2016). In such condition, coupled with the ambient air pollution level, IAQ has become a great concern in this region. Studies on several indoor spaces in these counties revealed the dominating indoor air pollutants including $PM_{10}$, $PM_{2.5}$, $CO_2$, VOCs, $NO_2$, $SO_2$, and heavy metals (Amoatey et al. 2018). It has also been highlighted that most investigated buildings could not meet the ASHRAE (American Society of Heating, Refrigeration, and Air-conditioning Engineers) threshold values for indoor air velocity and ventilation level (Amoatey et al. 2020). These factors are highly influencing the market demands for air purifiers in Middle Eastern countries for past few years. Along with the indoor portable air purifiers, building professionals are showing high interest for different indoor vertical greening systems, such as active living wall (ALW), for combating indoor air pollutants which also adds aesthetic benefits for the occupants (Radić et al. 2019; Pettit et al. 2018). However, still there is very limited research results about the life cycle performance of these different indoor air purification methods from environmental perspective.

Hence, this study aims to assess the environmental impacts of a specific ALW system and a commercially available indoor air purifier through life cycle assessment (LCA) and finally compares the impacts of the two systems. LCA is a standard approach to evaluate the potential environmental loads of any product or process throughout its life cycle (Mannan and Al-Ghamdi 2019). The comparative results of this LCA study are particularly beneficial for Middle Eastern countries as the necessity for indoor air purification is much higher here.

## 16.2 Methodology

The LCA study is based on a ALW and CAP system which are located in a small office room (Fig. 16.1) inside an educational institute (University located in Doha, Qatar).The next section presents the details of the specific ALW and CAP systems that have been investigated in this study.

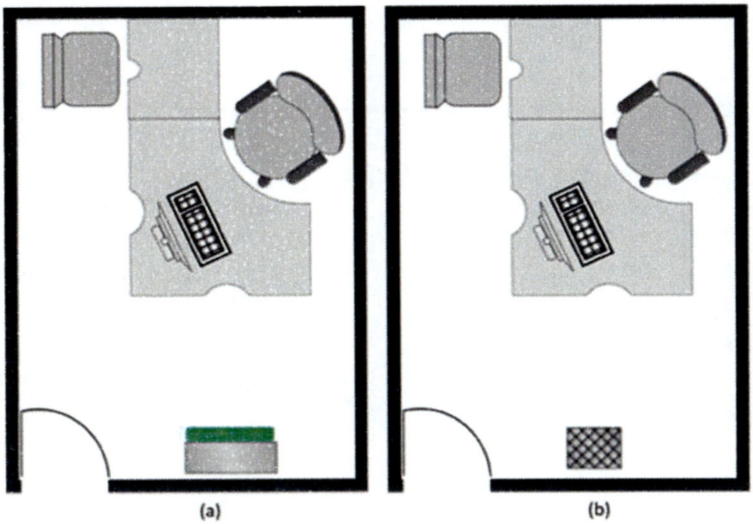


**Fig. 16.1** Layout of the office room with **a** ALW system and **b** CAP unit (not to scale)

### *16.2.1 Description of ALW and CAP*

ALW system is an innovative improved form of vertical greening systems which allows a forced air flow through the plants rooting zone and substrates to improve the removal capacity of indoor air pollutants (Pérez-Urrestarazu et al. 2016). This advanced greening technology has the potential to reduce both gaseous pollutants and particulate matters through adsorption and absorption process as well as has the ability to reduce indoor temperature and increase in humidity (Pérez-Urrestarazu et al. 2016; Llewellyn and Dixon 2019). As a part of sustainability goals, most Middle Eastern countries have taken the opportunity to implement indoor greening systems, hence indoor living walls specially the ALW is getting increasing attention. The ALW system used for this study is a felt based system where several felt layers have been used as substrate, supported by a PVC foam plate. Pre-grown plants have been selected and inserted into the pockets of felt layers. Automated irrigation system helps to supply the required water and nutrients while the axial fan on the top of the system ensures the sufficient airflow through the plants rhizosphere zone.

On the other hand, according to the air purifier manufactures and distributors, mechanical portable air purifiers that included HEPA (high efficiency particulate air filter) technology are the most popular among the Middle Eastern countries. HEPA filter is considered as one of the most effective filters for particulate matters which is one of the main concerns in this region. The unique glass threads in this filter have a diameter less than 1 μ which makes it capable to capture particulate matters as low as 0.3 μ in diameter (Vaughn and Ramachandran, 2002). However, HEPA filter alone is not efficient in removing any gaseous pollutants such as volatile organic compounds

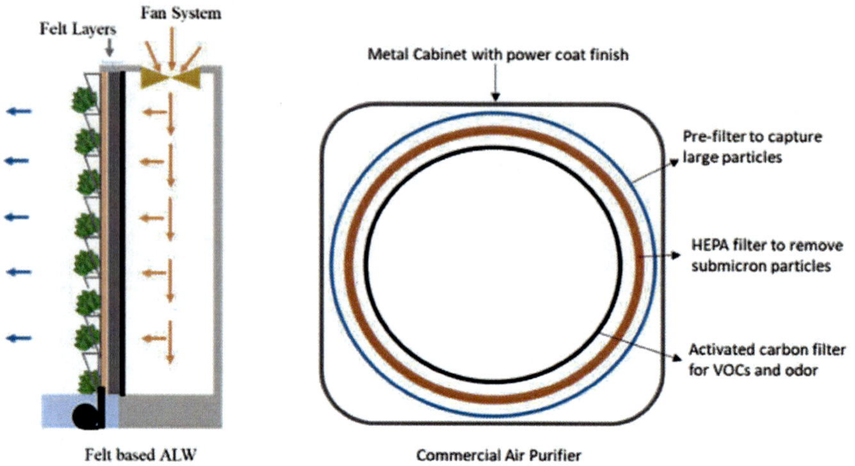


**Fig. 16.2** Schematic of ALW system (side view) and commercial indoor air purifier (top view)

(VOCs), therefore, indoor purifiers mostly are equipped with activated carbon based filters as a pre-filter or post-filter. In this study, a commercial HEPA filter has been used with a three stage filtering process including pre-filter, HEPA filter and finally carbon filter for VOC/odor removal (Fig. 16.2).

## *16.2.2 Basic Approach to LCA*

In this study, LCA has been used to evaluate the environmental burdens of ALW and CAP systems. LCA tool GaBi has been used to assess the environmental impacts. This section describes the basic approach of LCA. Generally, any LCA study consists of four steps: (1) goal and scope definition; (2) life cycle inventory analysis; (3) life cycle impact assessment; and (4) interpretation.

**Goal and Scope Definition**

Functional Unit: Since there is no direct comparison method for the efficiency of ALW systems and CAP systems, this study compares the two systems based on the capability to reduce the pollutants, specially the VOCs, particulate matters, from a predefined space. To fulfil the aim of the study, these two systems have been set in a small office room having a floor area of 8 $m^2$ (~86 $ft^2$). Hence, the functional unit was set to "purification of indoor air of the specified office room over a period of 12 months". Parameters related to the functional unit was chosen based on the CAP and ALW manufacturers data as well as the data from peer-reviewed articles (Pettit et al. 2019; Irga et al. 2017).

System Boundary: This comparative LCA study comprises the following life cycle phases of both ALW and CAP system: manufacturing, transportation and finally the use phase. As the study covers the life time of 12 months for each system, therefore, maintenance phase and disposal phase were not included in the system boundary. Moreover, the life cycle for vegetation system in ALW system was out of the boundary as pre-grown plants have been used for this study and nutrient solutions were not taken into account as having minimal impact (Ottelé et al. 2013).

**Life Cycle Inventory Analysis**
Data for ALW system has been collected from the manufacturer as well as from the inventory of Ottele et al. (2013) for some of the construction materials as they are similar in living wall system described in their study and from the previous work of the authors (Ottelé et al. 2013; Mannan and Al-Ghamdi 2020). All the construction materials and plants have been transported from manufacturer's site to the installation site by road inside Doha where average transportation distance is considered as 10 km. The installation of the construction materials has been done in the installation site by using simple mechanical tools, however, the installation process has been kept out of the system boundary as it has minor impacts (Feng and Hewage 2014). Irrigation system has been included in this LCA study where water consumption for plants is 2.5 l/day. For the CAP system, inventory data has been collected from the manufacturer's booklet. The production of the CAP system is mainly bench assembly where the assembly of different parts (pre-filer, HEPA filter, activated carbon filter, casing, fan, and switch) of CAP system is done by a person through hand assembly. The transportation distance has also considered as 10 km from manufacturer's office to installation site inside Doha. The inventory data have been listed in Tables 16.1, 16.2, and 16.3.

**Life Cycle Impact Assessment**
Life cycle impact assessment (LCIA) method ReciPe (midpoint method) was chosen in this LCA study. This LCIA step helps to translate the results from the life cycle

**Table 16.1** Inventory data for ALW construction materials

| ALW parts | Construction materials | Weight (kg/m$^2$ of ALW) |
|---|---|---|
| Structural bolt | Stainless steel | 0.13 |
| Structural-spacer brackets | Stainless steel | 0.19 |
| Supporting U section | Stainless steel | 4.62 |
| Foam plate | PVC | 7 |
| White fleece | Polypropylene | 0.3 |
| Wool fleece | Polyamide | 0.6 |
| PE fleece | Polyethylene | 0.045 |
| Black fleece | Polypropylene | 0.27 |
| Irrigation system | PE pipes and flexible tubes | 0.09 |
| Axial fan | Constructed from mild steel | 6 |
| Submersible pump | Stainless steel | 3 |
| Irrigation support tank | Stainless steel | 20.6 |

**Table 16.2** Inventory data for CAP construction materials

| CAP parts | Construction materials | Weight (kg)/CAP unit |
|---|---|---|
| Powder coated outside cabinet | Aluminum | 9.5 |
| Pre-filter | Impregnated carbon filter | 0.17 |
| HEPA filter | Glass fiber | 0.38 |
| Activated carbon filter | Activated carbon | 6.8 |
| Motorized impeller/blower | Stainless steel | 2.6 |

**Table 16.3** Operational data (for use phase) for ALW and CAP system

| Operation | Consumption data | Unit |
|---|---|---|
| Axial fan (ALW) | 735.8 | kWh/year |
| Water pump (ALW) | 10.95 | kWh/year |
| Lighting (ALW) | 367.9 | kWh/year |
| Water (ALW) | 1.28 | $m^3$/year |
| Operational energy (CAP) | 2190 | kWh/year |

inventory step into meaningful environmental impact scores by means of characterization factors. The following impact categories have been assessed for this study:

- Climate change (kg $CO_2$-eq.)
- Human toxicity (kg 1,4-DB eq.)
- Terrestrial acidification (kg $SO_2$-eq.)
- Fossil depletion (kg oil-eq.)
- Marine Eutrophication (kg N-equiv.)

## 16.3 Results and Discussions

This section presents the comprehensive LCA findings for both ALW and CAP systems and finally draws a comparative analysis. Figure 16.3 presents the comparative results between the two systems.

Results of this study showed potential reduction of environmental impact in case of using ALW for indoor spaces compared to commercially available air purifier. Although the impacts of the CAP production are comparatively lower to the ALW system in this study, however, the overall impacts are much significant when looking at the entire product life cycle. The total $CO_2$ emissions for the one year life cycle of ALW and CAP systems were found 820 and 1440 kg, respectively.

The highest impacts in each category are due to the use phase, except for human toxicity category. In this impact category, the production of steel based products are found mainly responsible for higher impact in the case of ALW while for CAP, the production of outside aluminum (Al) cabinet is the major factor for higher impact.

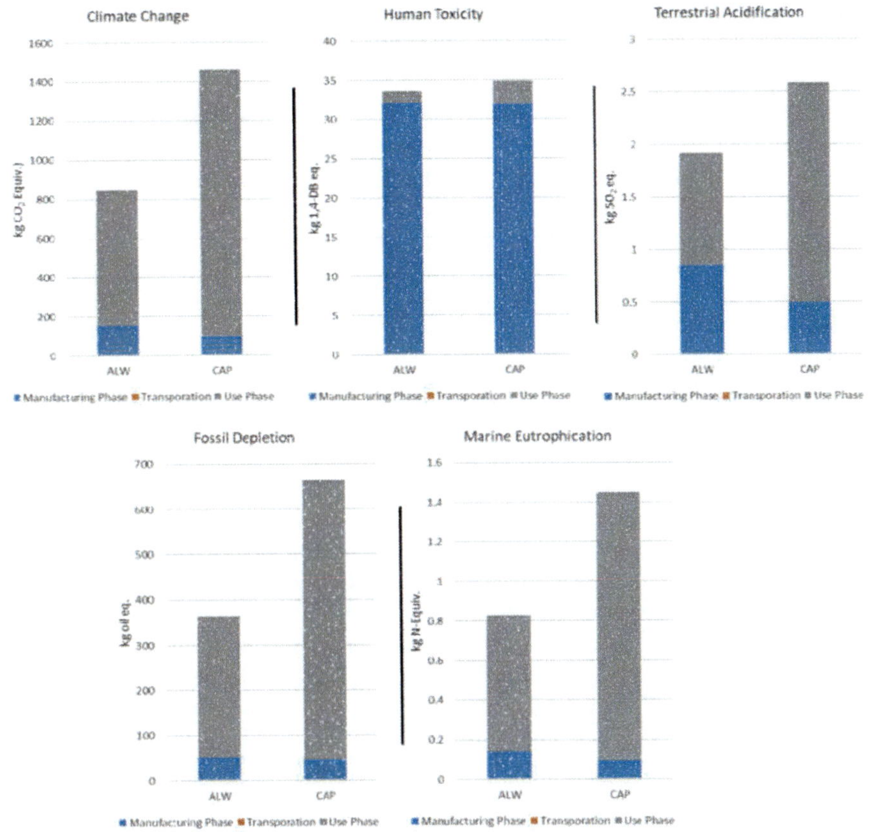


**Fig. 16.3** Selected life cycle impact category results comparison between indoor ALW and CAP system

For ALW system, production of irrigation supporting tank, axial fan, and irrigation pump resulted 63%, 19%, and 6% impacts, respectively, where for CAP system, Al cabinet alone was found to have 90% impact in human toxicity category.

For the rest of the categories, the highest percentage of impact lies in use phase which is due to the use of electricity. In the case of ALW system, the impacts from electricity use are 81%, 55%, 85%, and 82%, respectively, for the climate change category, terrestrial acidification, fossil depletion, and marine eutrophication category. The transportation phase created very minimal impact compared to production and use phases.

In each impact category, ALW system was found to have less environmental impact. For climate change, ALW system resulted 848 kg $CO_2$ equiv. where the number was 1491 kg $CO_2$-eq. for CAP system. Terrestrial acidification causes due to the deposition of inorganic elements (i.e. sulphates) in the soil which ultimately changes the acidity of the deposition area. In this study, the impacts resulted for ALW and CAP systems are found 1.92 and 2.58, respectively, for terrestrial acidification category. Fossil fuel depletion category estimates the damage to natural resources due to fossil fuel extraction. In this comparative study, ALW resulted 45% lower than the CAP system in fossil depletion category where ALW resulted nearly 43% lower than the CAP system in marine eutrophication category. However, for human toxicity category both ALW and CAP systems resulted in similar range, 33.5 and 34.9 kg 1,4-DB eq., respectively.

## 16.4 Conclusion

In conclusion, this preliminary study provides valuable insights into the environmental performance of two distinct indoor air purification systems, benefiting building professionals and air purifier designers. The life cycle assessment (LCA) of active living walls (ALW) serves as a starting point to identify key factors for reducing environmental impacts and optimizing future sustainable ALW production.

An important finding of this LCA study is the significant role played by the use phase in the life cycle of both air purification methods, with electricity consumption being the primary contributor to environmental impacts. To mitigate these impacts, the study recommends incorporating renewable energy sources alongside the conventional grid electricity supply. Additionally, for ALW systems, alternative materials such as hard wood, recycled plastic, or coated steel can replace stainless steel, which is identified as a dominant factor in environmental burden. Similarly, for commercial air purifiers (CAP), recycled plastic can serve as a sustainable alternative to aluminum-based outside cabinets.

While this study focused on comparing the systems' indoor air pollutant removal efficiency in a predefined space (specifically a small office room), future research should investigate their performance in different spaces, such as hall rooms and bedrooms, considering variations in size and characteristics. Furthermore, conducting long-term studies spanning, for example, 10 years, would enable the identification of potential impacts arising from the maintenance phase. Future comprehensive studies should also encompass the dismantling and disposal phases of these systems.

By considering these recommendations and conducting further research, the field of indoor air purification can progress towards more sustainable and environmentally friendly solutions, ultimately promoting healthier and safer indoor environments.

**Acknowledgements** This research was supported by a scholarship (210004673) from Hamad Bin Khalifa University (HBKU), a member of Qatar Foundation (QF). Any opinions, findings, and conclusions or recommendations expressed in this material are those of the author(s) and do not necessarily reflect the views of the HBKU or QF.

## References


Ali N, Ismail IMI, Khoder M et al (2016) Polycyclic aromatic hydrocarbons (PAHs) in indoor dust samples from cities of Jeddah and Kuwait: levels, sources and non-dietary human exposure. Sci Total Environ 573:1607–1614. https://doi.org/10.1016/j.scitotenv.2016.09.134

Amoatey P, Omidvarborna H, Baawain MS, Al-Mamun A (2018) Indoor air pollution and exposure assessment of the gulf cooperation council countries: a critical review. Environ Int 121:491–506. https://doi.org/10.1016/j.envint.2018.09.043

Amoatey P, Omidvarborna H, Baawain MS et al (2020) Association between human health and indoor air pollution in the Gulf Cooperation Council (GCC) countries: a review. Rev Environ Health 35:157–171. https://doi.org/10.1515/reveh-2019-0065

Cincinelli A, Martellini T (2017) Indoor air quality and health. Int J Environ Res Public Health 14. https://doi.org/10.3390/ijerph14111286

Feng H, Hewage K (2014) Lifecycle assessment of living walls: air purification and energy performance. J Clean Prod 69:91–99. https://doi.org/10.1016/j.jclepro.2014.01.041

Irga PJ, Abdo P, Zavattaro M, Torpy FR (2017) An assessment of the potential fungal bioaerosol production from an active living wall. Build Environ 111:140–146. https://doi.org/10.1016/j.buildenv.2016.11.004

Llewellyn D, Dixon M (2019) Can plants really improve indoor air quality? Compr Biotechnol 343–350. https://doi.org/10.1016/B978-0-444-64046-8.00228-7

Mannan M, Al-Ghamdi SG (2019) Life-cycle assessment of thermal desalination: environmental perspective on a vital option for some countries. In: Groundwater, sustainability hydro-climate/climate change, and environmental engineering—selected papers from world environmental and water resources congress 2019, pp 449–460. https://doi.org/10.1061/9780784482346.046

Mannan M, Al-Ghamdi SG (2020) Life cycle embodied energy analysis of indoor active living wall system. Energy Rep 6:391–395. https://doi.org/10.1016/j.egyr.2020.11.180

Mannan M, Al-Ghamdi SG (2021) Indoor air quality in buildings: a comprehensive review on the factors influencing air pollution in residential and commercial structure. Int J Environ Res Public Health 18:1–24. https://doi.org/10.3390/ijerph18063276

Ottelé M, Perini K, Haas EM (2013) Life cycle assessment (LCA) of green façades and living wall systems

Pérez-Urrestarazu L, Fernández-Cañero R, Franco A, Egea G (2016) Influence of an active living wall on indoor temperature and humidity conditions. Ecol Eng 90:120–124. https://doi.org/10.1016/j.ecoleng.2016.01.050

Pettit T, Irga PJ, Torpy FR (2018) Towards practical indoor air phytoremediation: a review. Chemosphere 208:960–974. https://doi.org/10.1016/j.chemosphere.2018.06.048

Pettit T, Irga PJ, Torpy FR (2019) The in situ pilot-scale phytoremediation of airborne VOCs and particulate matter with an active green wall. Air Qual Atmos Heal 12:33–44. https://doi.org/10.1007/s11869-018-0628-7

Radić M, Dodig MB, Auer T (2019) Green facades and living walls—a review establishing the classification of construction types and mapping the benefits. Sustainability 11. https://doi.org/10.3390/su11174579

# Chapter 17
# Investigating the Embodied Energy of Wall Assembly with Various Material Service Life Scenarios

**Abdul Rauf, Daniel Efurosibina Attoye, and Robert Crawford**

**Abstract** Studies have advocated that there is much less research on the impact of embodied energy. Researchers have asserted that a building's embodied energy can be as high as 60% of the life cycle energy. However, there is insufficient research and understanding of embodied energy impacts and its relationship with material specification and service life. This research aims to fill this gap by investigating the life cycle embodied energy of a villa in the United Arab Emirates with particular emphasis on the wall assembly. The findings show that the embodied energy impact of the wall structure and wall finishes was found to be 19.7% and 11.7% of the villa's life cycle embodied energy (LCEE), respectively. Alternative material service life (MSL) scenarios for the wall assembly shows that using minimum material service life (MSL) values results in a 54% increase in LCEE of the wall, and 74% increase in the LCEE of the villa. For maximum MSL, the findings show a 27% and 31% decrease in LCEE of walls and villa, respectively. Alternative wall finishes show that wallpaper as a replacement of water-based paint will increase the LCEE of the villa by 28%.

**Keywords** Life cycle embodied energy (LCEE) · Initial embodied energy (IEE) · Recurrent embodied energy (REE) · Wall assembly · Building materials · Input output based hybrid analysis (IOBHA)

---

A. Rauf (✉) · D. E. Attoye
United Arab Emirates University, Al Ain, UAE
e-mail: a.rauf@uaeu.ac.ae

R. Crawford
The University of Melbourne, Melbourne, Australia



N. S. Caetano and M. C. Felgueiras (eds.), *The 9th International Conference on Energy and Environment Research*, Environmental Science and Engineering,
https://doi.org/10.1007/978-3-031-43559-1_17

## 17.1 Introduction

In literature, several studies have investigated strategies that help to reduce the operational energy associated with heating, cooling, and other occupant-related activities during the building life (Birgisdottir et al. 2017; Mirabella et al. 2018). The benefit of this emphasis is that established protocols now exist which provide a clear pathway to reduce the negative impact and energy drain of buildings caused by operational phase. As a result, some studies have reported that 70% reduction in operational energy can be reached (Danish Government Strategy for Energy Renovation of Buildings 2014), while some other studies have proved the practicality of NetZero building designs (Crawley et al. 2009; Jiang et al. 2022; Kolokotsa et al. 2011).

In reality, the energy consumed by buildings may either be due to heating, cooling, lighting etc. (operational energy) or the extraction and processing of materials, as well as the construction of the building, and its life cycle maintenance. This second component is called the embodied energy. Although it has received less consideration, researchers in this field have reported that it plays a significant role in life cycle energy (Crawford 2014; Rauf 2016) and the environmental performance of buildings (Hu 1905).

In previous studies we assessed the impact of embodied energy caused by material replacement and maintenance-related activities which are carried out over the course of a building's service life in the Australian context (Crawford 2014; Crawford et al. 2010; Rauf and Crawford 2015). Also, we investigated the initial, recurrent and life cycle embodied energy of a villa in the United Arab Emirates context (Rauf et al. 2021). The preliminary literature review revealed that regions such as the UAE do not have sufficient data to provide a clear picture of embodied energy impacts (Rauf et al. 2021). Consequently, conducting studies on this topic, in such regions, are complicated and often overlooked (Tabet Aoul et al. 2018; Mawed et al. 2017). This study was thus, designed to fill this gap in literature and builds up on the UAE villa study (Rauf et al. 2021). To provide an in-depth assessment, wall assembly was selected as the focus of this investigation, with the aim of exploring possible ways to reduce its embodied energy. During this process, effects of variations in service life of the original and alternative materials on life cycle embodied energy were analysed.

## 17.2 Background

One primary challenge in building construction and practice is a mix of 'objective' professional specifications and 'subjective' client preferences. In some cases, major decisions made during the building's early design stage are simply based on the client's choice. One critical aspect of building design and construction is the design, selection, and specification of the various building assemblies such as walls, roofs and floors used in the construction. Nassar et al. (2003) conducted a study on some of these building assemblies, extrapolating them as design variables in line with multiple performance metrices. However, studies affirm that the critical contribution of building assemblies has not been well investigated from the aspect of embodied energy (Rauf 2016; Rauf and Crawford 2015; Nassar et al. 2003).

In previous studies, the service life of both materials and buildings have been reported has having a definitive impact on the embodied energy of buildings (Rauf 2016; Rauf and Crawford 2015; Rauf et al. 2021; Janjua 2021). However, building assemblies are often made of building materials which have different lifespans or service life in years. For example, a wall, has a structural component such as concrete, and its finishing such as paint. The service life of the concrete structure may be up to 50 years, or as long as the building's service life (Rauf 2016). On the other hand, the paint may only have a service life of 10 years or less. Consequently, the wall assembly's embodied energy impact can be significantly impacted by a different selection of materials; in this case stone tiles may last longer than paint and have lower embodied energy. Other considerations such as the need for proper construction detailing, serviceability, and replacement of some components are also important. As another example, poor workmanship or a defective window hinge may cause it to fall and break. Thus, there is also the need for periodic maintenance and visual inspection of building assemblies.

From a comparative point of view, some authors have reported that it is possible to reduce the embodied energy of a representative case study by up to 20% (Huberman and Pearlmutter 2008) and up to 50% (Reddy and Jagadish 2003) when then building materials used are changed for alternative options. Thus, an understanding of materials with lower embodied energy is definitively advantageous in building design, specification, and decision-making processes.

It is also important to note that several studies have explored the subtle but definite impact played by the service life of building materials in association with their embodied energy, and by extension, the embodied energy of the building assemblies. In previous studies, we have reported that the recurrent embodied energy of a building directly dependent on service life of a material (Rauf 2016; Rauf and Crawford 2015). The current study builds on the literature by bridging the gap: providing emphasis on the unexplored study context in relation to the subject matter, while connecting embodied energy research with building assemblies and finishes.

## 17.3 Method

The use of case study buildings in evaluating embodied energy has recently been used in multiple studies due to the strategic benefit of providing practical data based on practice-driven contributions (Dascalaki et al. 2021; García-Sanz-Calcedo et al. 2021; Tavares et al. 2019). In this paper, this approach was adopted to investigate the embodied energy of a UAE residential villa, giving particular emphasis to wall assemblies (both external and internal). In the study context, villas are the common building typology for citizens of UAE. The primary source of data was the villa's bill of quantities acquired from a local architectural design firm. This was used to provide

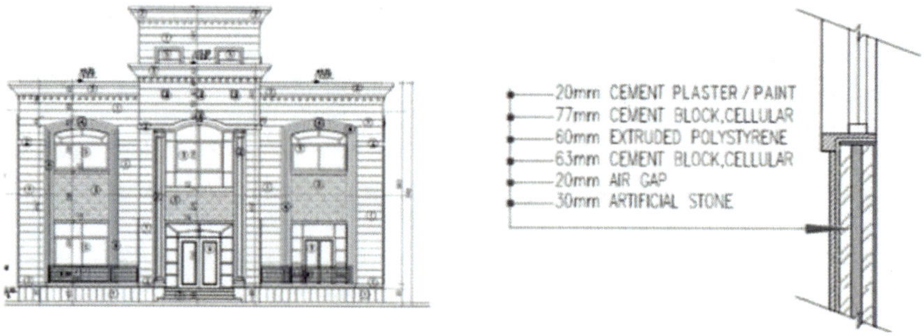


**Fig. 17.1** Front elevation and typical wall section of the case study villa

the exact description of material quantities used in order to guide the quantification of the life cycle embodied energy of the case study villa.

Selected case study villa is located Al Ain, UAE. It is a two-storey house with a total floor area of 532 $m^2$. The structural system is comprised of concrete columns and beams, with in-fill walls, as shown in Fig. 17.1.

### *17.3.1 Embodied Energy Calculation Procedure*

In literature, the calculation approaches are the *process analysis, input–output analysis,* and *hybrid analysis* which have been reviewed in previous studies (Rauf 2016; Crawley 2011). In this study, the *Input–Output Based Hybrid Analysis* (IOBHA) was selected since the hybrid approach combines process and input–output analysis approaches, simultaneously maximizes their strengths and minimizes their limitations, and improves the completeness, accuracy and reliability of the analysis (Crawley 2011; Dixit and Singh 2018; Venkatraj and Dixit 2021; Omar 2018). In summary, the assessment procedure starts with the process-based hybrid embodied energy based on the material quantities described in the bill of quantities. A detailed definition of the approach and the calculation steps for both the initial and recurrent embodied energy is provided in previous studies (Rauf 2016; Rauf et al. 2021).

In the current study, the focus is on the embodied energy of the wall assembly, which is made up of both the structural part and the finishes. Table 17.1 shows extraction from the bill of quantities, showing the materials used in wall structure and finishes. The Environmental Performance in Construction (EPiC) database (Crawford et al. 2019) was used to define energies intensities and environmental flows for all referenced building materials while the mathematical computation of the IOBHA was carried out using Microsoft Excel.

Table 17.1 Summary of materials used in the wall assembly

| Wall structure | Wall finishes |
|---|---|
| 20 cm solid block-60 $m^2$ @425 kg/$m^2$ | Plaster 1530 $m^2$ |
| 20 cm hollow block with insulation—440 $m^2$ | Wall finishing (W2) (washable emulsion paint) 1530 $m^2$ |
| 20 cm hollow block with insulation—440 $m^2$ | Wall finishing (W3) (ceramic tiles) 3 m height 278 $m^2$ |
| 20 cm hollow block-620 $m^2$ @275 g/$m^2$ | Wall finishing (W4) (ceramic tiles) 3 m height 58 $m^2$ |
| 10 cm hollow block-150 $m^2$ @275 kg/$m^2$ | Wall finishing (W5) (ceramic tiles) 3 m height 32 $m^2$ |

### *17.3.2 Additional Embodied Energy Assessment Scenarios*

To accommodate variations in design and material specifications, two additional assessment scenarios were created to facilitate a further investigation of the potential magnitude of the embodied energy associated with the wall assembly. In the first scenario, alternative material service life values were used to assess possible variations in the IEE, REE and LCEE of the wall assembly. This was based on a literature search to find out the lowest, average and highest material service life (MSL) values of the building materials used in the wall construction (Rauf 2016; Rauf et al. 2021). Thus, the average MSL was first used to calculate the IEE, REE and LCEE of the base case using the IOBHA, then two other MSL alternatives were used in a recalculation of the embodied energy aspects.

The second scenario was developed to assess the potential impact due to variations in the type of finishes used for the treatment of the walls of the villa. This variation provides insight to facilitate the reduction of the building's embodied energy based on changing the materials specified during the design phase. Combined, the two additional assessments provide a systematic protocol to evaluate the embodied energy impacts of the wall assembly.

## 17.4 Results and Discussion

The first level of analysis carried out was to investigate the embodied energy of the villa from a life cycle consideration. The results show that the IEE for the villa was 7390.5 GJ (56%) and the REE was 5690.01 GJ (43%). These values depict the base case of the study using average material service life values, calculated over a building lifespan of 50 years.

### *17.4.1 Embodied Energy of the Base Case*

With specific reference to the focus of this paper, Fig. 17.2 shows the IEE and REE of the wall assembly, distinguishing between the wall (structure) and the wall finishes as calculated for the base case described above. Figure 17.2a shows that for the IEE of the villa, the associated embodied energy of the wall structure was 1456 GJ (19.7%) while the associated embodied energy of the wall finishes was 866.4 GJ (11.7%). For the REE, the embodied energy values were 0 GJ (0%) for the wall structure and 942.5 GJ (13.8%) for the wall finishes with average material service life values. Recurrent embodied energy for the wall structure was based on assumption that its service life is equal to the service life of the case study building; thus, no material replacements happened during the building life cycle. Figure 2b combines these results to show the proportion of both the wall structure and finishes for the life cycle embodied energy demand over 50 years. The figure shows that although the wall structure was responsible for only 11% of the LCEE of the case study house, the wall finishes were responsible for 14% of its LCEE. This shows the importance of careful selection for wall structure as well as for its finishes since both consume a significant amount of energy throughout the building lifespan.

This result indicates three key issues for future consideration; firstly, the embodied energy associated with the finishes of a building assembly can be significantly greater than its structural part. Secondly, although the structure may be larger in terms of thickness, mass or volume, the finishes may still constitute a greater amount of embodied energy. Thirdly, in comparative terms, embodied energy is not directly proportional to the size or volume of a building component but may be influenced by other factors.

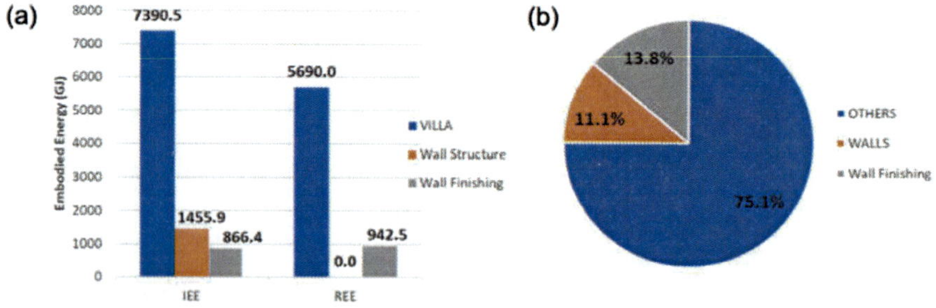


**Fig. 17.2** Embodied energy of wall assembly and villa for the base case with average material service life over 50 years **a** IEE and REE **b** LCEE

### 17.4.2 Scenario 1: Embodied Energy Variations Due to Alternative MSLs

Figure 17.3, shows the embodied energy associated with the lowest reported material service life values (MSL-MIN) for the building materials used. The figure shows that the IEE for both the wall structure and finishes were the same as the MSL-AVE values: 1456 GJ (19.7%) and 866.4 GJ (11.7%) respectively. However, for the REE, the respective embodied energy values were 0 GJ (0%) for the walls and 2700 GJ (17.5%).

Similarly, Fig. 17.4, shows the recalculated embodied energy associated with the highest reported material service life (MSL-MAX) for the building materials used in the construction of the wall assembly. For the IEE, the MSL-MAX values have no impact; thus, the results are the same with MSL-MIN and MSL-AVE. However, the structure and finishes have a different REE when compared. For the MSL-MAX, the REE were 0 GJ (0%) for the walls and 76 GJ (17.5%).

A first take of the recalculated embodied energy values for the MSL-MIN and MSL-MAX reflects three clear impacts and variations in the results due to the increase in the MSL. Firstly, there was a decrease in the LCEE of the wall assembly. Secondly, there was an increase in the proportion of the "structure" component's associated

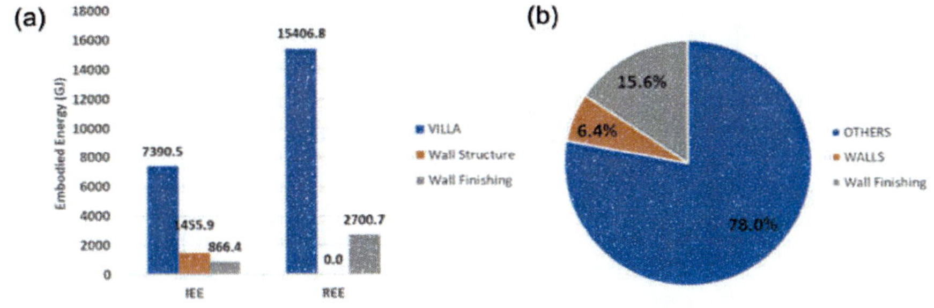


**Fig. 17.3** Embodied energy of wall assembly due to MSL-MIN **a** IEE and REE **b** LCEE

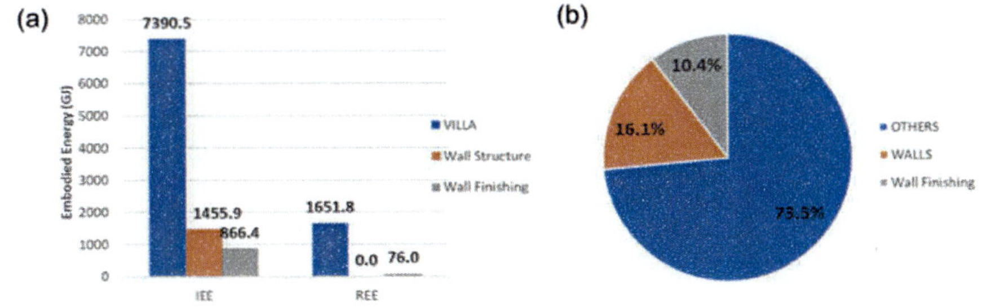


**Fig. 17.4** Embodied energy of wall assembly due to MSL-MAX **a** IEE and REE **b** LCEE

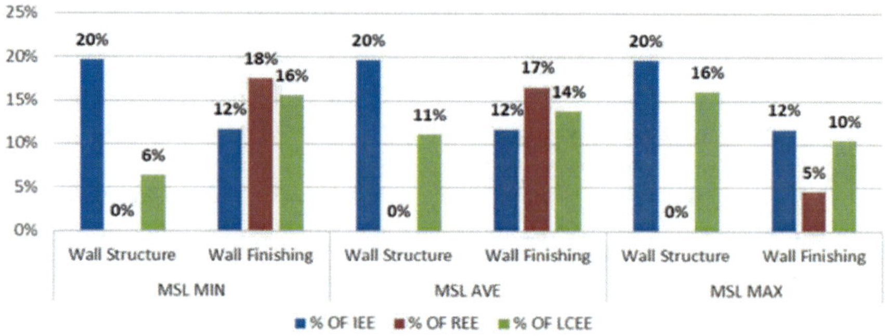


**Fig. 17.5** Comparative assessment of embodied energy by percentage for the MSL alternatives

embodied relative to the LCEE of the wall assembly. Thirdly, there was a decrease in the proportion of the "finishes" component's associated embodied relative to the LCEE of the wall assembly. To further elucidate these findings, the next section is devoted to a comparative assessment of the embodied energy magnitudes.

**Comparative Assessment of Embodied Energy Magnitude**
Figure 17.5 shows the comparative embodied energy figures and percentages due to the different material service life values. For each material service life variation, the figure shows the proportion of both the wall structure and wall finish relative to the initial, recurrent and life cycle embodied energies of the case study villa. It has already been stated that the assumption made relative to the IEE for this study, is that the material service life is does not have not any impact on it. Thus, the comparison in this section is therefore of value in considering the REE and LCEE.

For MSL-MIN, the figure shows that while the structure has no contribution to the REE, yet 18% of the villa's REE is associated with the wall finishes. Similarly, for the MSL-AVE and MSL-MAX, there is no embodied energy associated with the wall structure but respectively, 17 and 5% of the villa's REE is associated with the wall finishes. The figure shows that for all the variations of the MSL, and for all the comparative contributions to the recurrent and life cycle embodied energy of the villa, the wall finishes are associated with more embodied energy than the wall structure.

### *17.4.3 Scenario 2: Assessment of Alternative Wall Finishes*

The results of the IOBHA, MSL variations and the comparative assessments presented thus far, indicate clearly that the wall finishes play a significant role in the life cycle embodied energy of both the wall as an assembly and of the villa from the whole building point of view. Two alternative wall finishes were selected and the embodied energy of the wall assembly, relative to the villa was recalculated and compared with the original materials of the base case. For alternative #1, solvent-based paint was used in place of water-based paint in the base case while for alternative #2, wallpaper was the used. Other wall materials such as ceramic tiles for wet areas were retained. For these alternative materials, embodied energy variations were expected in the results due to differences in mass, delivered quantity, wastage factor and energy coefficient. To facilitate a focused IOBHA assessment similar to the base case, the average material service life option (MSL-AVE) was selected to review the alternative wall finish scenarios. Other parameters such as the building's service life and energy coefficients were kept constant.

**Embodied Energy Assessment of Alternative Finishes**

*Alternative 1 (ALT#1)*

The results of the IOBHA showed that for this alternative, the IEE of the villa rose by only 0.23% to become 7408 GJ and the REE rose by 1% to become 8406.5 GJ (see Fig. 17.6). For the IEE, the associated embodied energy of the wall structure was the same since no material was changed. However, the IEE of the wall finishes rose by 2% to 883.9 GJ. For the REE, the embodied energy value of the wall structure was unchanged but the REE of the wall finishes rose by 7% to 1012.4 GJ. These variations are due to the change from water-based to solvent-based paint which has a higher density, delivered quantity in mass, and a greater energy coefficient.

*Alternative 2 (ALT#2)*

Analysis results showed that the IEE of the villa rose by 12% to become 8304 GJ and the REE rose by 48% to become 8406.5 GJ. Figure 17.6 also shows that for the IEE, the associated embodied energy of the wall structure was also the same since no material was changed. However, the IEE of the wall finishes rose by 105% to

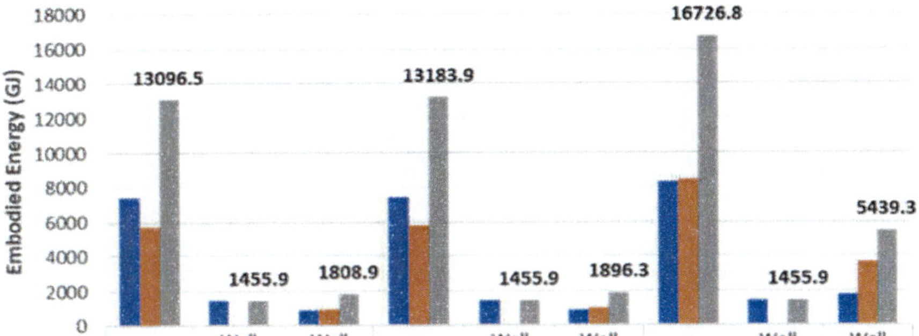


| | BASE CASE | | | ALT #1: Solvent-based paint | | | ALT #2: Wallpaer | | |
|---|---|---|---|---|---|---|---|---|---|
| | VILLA | Wall Structure | Wall Finishing | VILLA | Wall Structure | Wall Finishing | VILLA | Wall Structure | Wall Finishing |
| IEE | 7390.5 | 1455.9 | 866.4 | 7408.0 | 1455.9 | 883.9 | 8304.4 | 1455.9 | 1780.4 |
| REE | 5690.0 | 0.0 | 942.5 | 5759.9 | 0.0 | 1012.4 | 8406.5 | 0.0 | 3658.9 |
| LCEE | 13096.5 | 1455.9 | 1808.9 | 13183.9 | 1455.9 | 1896.3 | 16726.8 | 1455.9 | 5439.3 |

**Fig. 17.6** Embodied energy of Base case versus Alternative #1 and Alternative #2

become 1780.4 GJ. For the REE, the embodied energy value of the wall structure was unchanged. However, the REE of the wall finishes rose by 288% to 3659 GJ. These variations are due to change from water-based paint to wallpaper which has a greater energy coefficient (45.5 MJ/m$^2$ for wallpaper as compared to 8.7 MJ/m$^2$ for water-based paint), but a longer material service life.

Comparing the two alternatives, the impact of alternative #1 was a 28% increase in the villa's LCEE from 13,096.5 to 16,726.8 GJ while the impact of alternative #2 was a 1% increase in the villa's LCEE from 13,096.5 to 13,184 GJ. The results confirm that changes in the material specification of the wall finishes plays a large role in the associated embodied energy of the villa as a whole.

## 17.5 Conclusion

The objective of this study was to investigate the embodied energy of the entire wall assembly of a villa to provide an in-depth assessment of its initial, recurrent and life cycle embodied energy. The findings of the study show that for the base case, the embodied energy associated with the initial construction of the wall structure and wall finishes were found to be 1455.9 GJ/m$^2$ (19.7%) and 866.4 GJ/m$^2$ (11.7%) of the villa's IEE respectively. Also, the wall finishes are associated with 16.6% (942.5 GJ/m$^2$) of the recurrent embodied energy of the villa. The wall finishes consumed more embodied energy than the wall structure as seen when the total life cycle embodied energy was reviewed: finishes (1808.9 GJ/m$^2$ or 13.8% of the LCEE) as compared to the structure (1455.9 GJ/m$^2$ or 11.1% of the LCEE).

This study showed that increasing the material service life of the wall assembly leads to a drop in the LCEE of villa, indicating the advantages of prolonging the service life of materials. The study also showed that varying the type of wall finishes impacted the embodied energy of the finishes significantly. For the IEE, the variations were up to 105%, for the REE the variations were up to 288% and for the LCEE, the variations were up to 200%. This study also shows the significance of recurrent embodied energy and how it can be impacted by the selection of materials for the wall finishes, highlighting the importance of formulating and applying the strategies to increase their material service life.

**Funding** This work was supported by the United Arab Emirates University Start-Up Grant [G00000034].

## References

Birgisdottir H, Moncaster A, Wiberg AH, Chae C, Yokoyama K, Balouktsi M, Seo S, Oka T, Lützkendorf T, Malmqvist T (2017) IEA EBC annex 57 'evaluation of embodied energy and $CO_2$eq for building construction. Energy Build 154:72–80

Crawford RH (2011) Life cycle assessment in the built environment. Routledge
Crawford RH (2014) Post-occupancy life cycle energy assessment of a residential building in Australia. Archit Sci Rev 57:114–124
Crawford RH, Czerniakowski I, Fuller RJ (2010) A comprehensive framework for assessing the life-cycle energy of building construction assemblies. Archit Sci Rev 53:288–296
Crawford R, Stephan A, Prideaux F (2019) Environmental performance in construction (EPiC) database. University of Melbourne, Melbourne
Crawley D, Pless S, Torcellini P (2009) Getting to net zero. National Renewable Energy Lab (NREL), Golden, CO (United States)
Danish Government Strategy for Energy Renovation of Buildings (2014) The route to energy-efficient buildings in tomorrow's Denmark
Dascalaki EG, Argiropoulou P, Balaras CA, Droutsa KG, Kontoyiannidis S (2021) Analysis of the embodied energy of construction materials in the life cycle assessment of hellenic residential buildings. Energy Build 232:110651
Dixit MK, Singh S (2018) Embodied energy analysis of higher education buildings using an input-output-based hybrid method. Energy Build 161:41–54
García-Sanz-Calcedo J, de Sousa Neves N, Fernandes JPA (2021) Measurement of embodied carbon and energy of HVAC facilities in healthcare centers. J Cleaner Prod 289:125151
Hu M (2020) A building life-cycle embodied performance index—the relationship between embodied energy, embodied carbon and environmental impact. Energies 13:1905
Huberman N, Pearlmutter D (2008) A life-cycle energy analysis of building materials in the Negev desert. Energy Build 40:837–848
Janjua S (2021) Sustainability implication of residential building materials considering service life variability. PhD thesis, Curtin University
Jiang W, Ju Z, Tian H, Liu Y, Arıcı M, Tang X, Li Q, Li D, Qi H (2022) Net-zero energy retrofit of rural house in severe cold region based on passive insulation and BAPV technology. J Clean Prod 360:132198. https://doi.org/10.1016/j.jclepro.2022.132198
Kolokotsa D, Rovas D, Kosmatopoulos E, Kalaitzakis K (2011) A roadmap towards intelligent net zero- and positive-energy buildings. Sol Energy 85:3067–3084. https://doi.org/10.1016/j.solener.2010.09.001
Mawed M, Al Bairam I, Al-Hajj A (2017) Linking between sustainable development and facilities management strategies: an integrated approach for evaluating the sustainability of existing building in the UAE. ICSF 2017 Kingdom of Bahrain, p 33
Mirabella N, Roeck M, Ruschi Mendes Saade M, Spirinckx C, Bosmans M, Allacker K, Passer A (2018) Strategies to improve the energy performance of buildings: a review of their life cycle impact. Buildings 8:105
Nassar K, Thabet W, Beliveau Y (2003) A procedure for multi-criteria selection of building assemblies. Autom Constr 12:543–560. https://doi.org/10.1016/S0926-5805(03)00007-4
Omar WMSW (2018) A hybrid life cycle assessment of embodied energy and carbon emissions from conventional and industrialised building systems in Malaysia. Energy Build 167:253–268
Rauf A (2016) The effect of building and material service life on building life cycle embodied energy. PhD thesis
Rauf A, Crawford RH (2015) Building service life and its effect on the life cycle embodied energy of buildings. Energy 79:140–148
Rauf A, Attoye DE, Crawford RH (2021) Life cycle energy analysis of a house in UAE. In: Proceedings of the ZEMCH 2021 international conference, United Arab Emirates, pp 13–23
Reddy BV, Jagadish KS (2003) Embodied energy of common and alternative building materials and technologies. Energy Build 35:129–137
Tabet Aoul KA, Hagi R, Abdelghani R, Akhozheya B, Karaouzene R, Syam M (2018) The existing residential building stock in UAE: energy efficiency and retrofitting opportunities. In: Proceedings of the 6th annual international conference on architecture and civil engineering. ACE

Tavares V, Lacerda N, Freire F (2019) Embodied energy and greenhouse gas emissions analysis of a prefabricated modular house: the "Moby" case study. J Clean Prod 212:1044–1053
Venkatraj V, Dixit MK (2021) Life cycle embodied energy analysis of higher education buildings: a comparison between different LCI methodologies. Renew Sustain Energy Rev 144:110957

# Chapter 18
# Primary Energy and Carbon Emissions of Different Concrete Sandwich Panels

**Bruna Moura, Tiago Ramos da Silva, Nelson Soares, and Helena Monteiro**

**Abstract** This study evaluates the embodied non-renewable primary energy (NRe) and the global warming potential (GWP) of six different concrete sandwich panels (CSP) with the same thermal transmittance. The functional unit (FU) of 1 $m^2$ of CSP was considered for the *cradle-to-gate* life-cycle assessment (LCA). A high-performance concrete (HPC) layer was assumed for the outer and inner layers, and six different insulation materials were sandwiched in between: lightweight concrete (LWC), expanded-cork-panel (cork), glass wool, expanded polystyrene (EPS) 100% from virgin resources, EPS with 55% virgin material, and 100% recycled EPS. It was concluded that the scenario with the lowest GWP-value is the CSP with cork (30 kg of $CO_2$-eq/$m^2$), due to the carbon captured during the raw material growth. The CSP with virgin EPS is the scenario with the highest carbon footprint (68 kg of $CO_2$-eq/$m^2$), which could be significantly reduced if recycled EPS is used instead. The scenario with the lowest NRe-value is the CSP with recycled EPS (375 MJ/$m^2$), followed by the scenarios with LWC (411 MJ/$m^2$) and glass wool (424 MJ/$m^2$), respectively.

**Keywords** Concrete sandwich panel · Global warming potential · Insulation material · LCA · Non-renewable primary energy

## 18.1 Introduction

The construction sector is responsible for 40% of the worldwide energy consumption, of which 80% is used for heating and cooling (Yılmaz et al. 2022), releasing about 40% of greenhouse gases (GHG) in Europe (Torres-rivas et al. 2021). Significant

B. Moura · T. Ramos da Silva · H. Monteiro (✉)
Low Carbon & Resource Efficiency, R&Di, Instituto de Soldadura e Qualidade, R. do Mirante 258, 4415-491 Grijó, Portugal
e-mail: himonteiro@isq.pt

N. Soares
ADAI, Department of Mechanical Engineering, University of Coimbra, Coimbra, Portugal



N. S. Caetano and M. C. Felgueiras (eds.), *The 9th International Conference on Energy and Environment Research*, Environmental Science and Engineering,
https://doi.org/10.1007/978-3-031-43559-1_18

heat can be lost throughout the building's envelope if the walls and roofs are not well insulated. Thus, attention should be devoted to the choice of thermal insulation materials incorporated into the building's envelope to improve the thermal performance and energy efficiency of buildings. The production of materials also incorporates significant primary energy that should be considered. The careful selection of walls with superior thermal insulation properties should allow controlling the embodied energy and environmental impacts of buildings (Monteiro and Soares 2022). In this context, LCA can be employed to evaluate the potential environmental burdens (or trade-offs) of diverse construction components and/or envelope solutions.

CSP typically consist of two layers of concrete, an outer and inner, with a thermal insulation sandwiched in between. Although CSP are not a new technology, they are increasing in popularity due to their good thermal and acoustic insulation, fire resistance, durability, cost-effectiveness, suitability for modular and prefabricated construction, ease of installation, and low maintenance. However, since CSP can incorporate different concrete-based and thermal insulation layers, further research is required to assess the best configurations that will reduce the embodied energy and the environmental impacts of the panels (Resende et al. 2022). Some LCA studies have analyzed the environmental performance of different external walls, such as lightweight steel framing (LSF) (Tavares et al. 2021), wooden, double and single brick masonry, and concrete block (Monteiro and Freire 2012; Monteiro et al. 2021); but few have evaluated the environmental performance of CSP. On the other hand, some works only focused on sandwich panels for floors (Demertzi et al. 2020a), or roofs, that do not have the same function as CSP for walls.

LCA studies devoted to the assessment of the environmental burdens of CSP are still scarce in the literature. To address this gap, this work used a *cradle-to-gate* LCA methodology to evaluate the NRe and the GWP of six CSP scenarios with similar thermal transmittance but incorporating different insulation materials. The CSP under evaluation make use of HPC as outer and inner layers and vary on the insulation layer in between: LWC, cork, glass wool, EPS 100% from virgin resources, EPS with 55% virgin material, and 100% recycled EPS.

## 18.2 Materials and Methods

### *18.2.1 Goal and Scope Definition*

This study aims to compare the embodied NRe and the GWP of six CSP using a LCA methodology following ISO 14040 (ISO 2020a) and ISO 14044 (ISO 2020b). To assess the environmental performance of the six scenarios, the FU of 1 $m^2$ of CSP is considered within a *cradle-to-gate* approach. It means that all the environmental impacts from the extraction and acquisition of the raw materials until the end of the on-site production of the CSP will be considered. Figure 18.1 depicts the physical characteristics of the CSP and its production flowchart for each scenario. All the

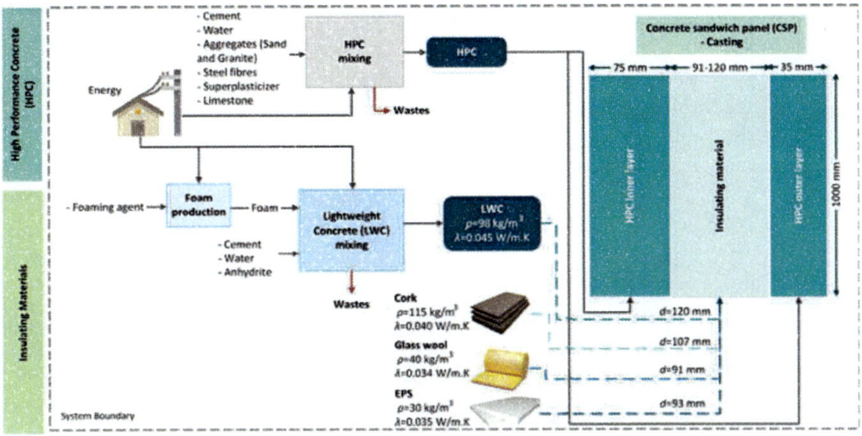


**Fig. 18.1** Flowchart and main properties of the production of the CSP under study. Note that $\rho$ stands for density; $\lambda$ for thermal conductivity; and $d$ for thickness

panels have similar thermal transmittance ($U = 0.33$ W/m$^2$ K), which is achieved by adjusting the thickness of the insulation layers so as to reflect the different values of the thermal conductivity of each material. The outer and inner layers are made of steel fiber reinforced HPC (75 and 35 mm, each). Six alternative insulation materials were sandwiched in between the HPC layers: LWC, cork, glass wool, EPS 100% from virgin resources, EPS with 55% virgin material, and 100% recycled EPS. The CSP production was assumed to be on-site, and it starts with the casting of the inner HPC layer to guarantee the load-bearing properties based on a recipe tested in a previous research work (Lameiras et al. 2013a, 2013b). Then, the insulation material is placed. Finally, the outer concrete layer is casted. The LWC layer is also considered to be produced on-site, where the foaming agent is mixed with cement, water, and anhydrite. The remaining materials are produced upstream and transported to the construction site.

### *18.2.2 Life Cycle Inventory (LCI)*

The LCI was gathered based on data collected in literature (Lameiras et al. 2013a, 2013b; Thomas et al. 2018) (Table 18.1). Afterwards, this information was completed with secondary data (e.g., transport, and Portuguese energy mix) from literature and the EcoInvent database (v3.8) using SimaPro software (v9.3). Moreover, when background environmental data was not available on EcoInvent database, information associated with embodied impacts of some materials was gathered from available environmental product declarations (EPD) (DAPHabitat 2016; Environmental Product Declaration (EPD) 2021). Regarding the transport stage, it was assumed

**Table 18.1** LCI for the six CSP under evaluation

| Layer | Scenario | Material | Quantity (kg/m$^2$) |
|---|---|---|---|
| Inner layer | 1–6 | HPC | 180.0 |
| Insulation layer | 1 | LWC | 30.0 |
| | 2 | Cork | 12.3 |
| | 3 | Glass wool | 1.4 |
| | 4 | EPS, virgin | 2.8 |
| | 5 | EPS, 55% virgin | 2.8 |
| | 6 | EPS, recycled | 2.8 |
| Outer layer | 1–6 | HPC | 84.0 |
| | | Glass fibers | 1.2 |

FU: 1 m$^2$ of CSP

that all materials were transported by a lorry for 25 km according to the Portuguese context.

### 18.2.3 *Life Cycle Impact Assessment (LCIA)*

To assess the potential environmental impacts of the six CSP configurations, two different LCIA methods were used (PRé Sustainability B. V. 2020): the Intergovernmental Panel on Climate Change (IPCC20201, v1.0) to evaluate the GWP of the life cycle of each scenario (in kg of $CO_2$-eq); the Cumulative Energy Demand (CED, v1.11) method to measure the NRe used (in MJ), based on the high heating values of fuels used along the life cycle.

## 18.3 Results and Discussion

Figure 18.2a shows the GWP for each CSP configuration. The cork-scenario is the one with the lowest environmental impacts (30 kg of $CO_2$-eq/m$^2$) considering the carbon uptake of this renewable plant-based material. Since cork oak plants absorb $CO_2$ along their life span, the carbon uptake embodied in cork may offset about half of carbon emissions of the cork-CSP. On the other hand, the CSP with virgin EPS is responsible for the highest carbon emissions (68 kg of $CO_2$-eq/m$^2$). Nevertheless, instead of using virgin raw material, a more environmentally friendly alternative can be selected, such as EPS with 55 or 100% of recycled material. These alternatives reduce the GWP of about 11 and 23% when compared with the EPS, virgin, respectively. In this context, the scenario with recycled EPS has the second lowest carbon

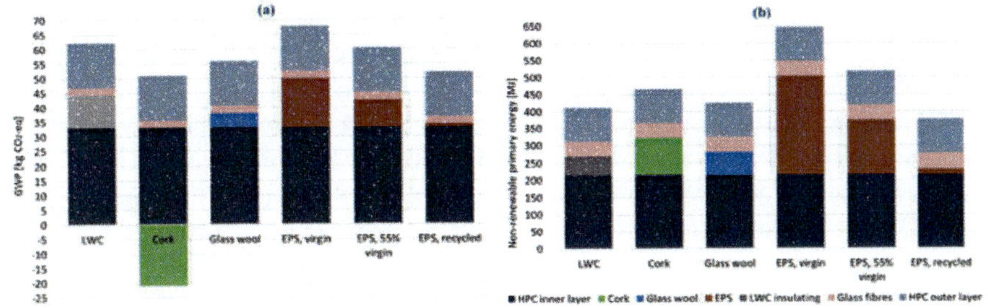


**Fig. 18.2** a GWP and b NRe of each CSP scenario; FU: 1 m$^2$ of CSP

emissions, being followed by glass wool (56 kg of $CO_2$-eq/m$^2$) and LWC (62 kg of $CO_2$-eq/m$^2$) scenarios.

Figure 18.2b depicts NRe-values used to produce each sandwich panel. The CSP configuration with the virgin EPS presents the worst energy performance having the highest NRe-value (646 MJ/m$^2$). However, if recycled EPS is considered, a reduction of 42% of embodied primary energy can be achieved, making it the best scenario, followed by the panel incorporating LWC (411 MJ/m$^2$). Compared to carbon emission results, the cork-panel does not present good primary energy results. This scenario uses 465 MJ/m$^2$ of NRe due to the raw material transformation and the manufacturing of expanded cork panels which is a high energy demand process. Moreover, results show that the best CSP depends on the impact category evaluated. In fact, as pointed out by Santos et al. (2021) and Demertzi et al. (2020a, b), different impact categories should be evaluated to assess different environmental impacts and potential benefits. Thus, to fully understand all the environmental impacts of each scenario, further impact categories should be investigated. As expected, in every scenario, the HPC layers are the main hot spot in both impact categories, representing from 49 to 76% of the environmental burdens. Therefore, reducing their impacts is very important. During the HPC mix local production, the electricity accounted for less than 1% of the impacts in both GWP and NRe categories. Nevertheless, the materials and their transport, are responsible for the majority of the impacts.

## 18.4 Conclusion

This work evaluated the primary energy and the carbon emissions of six different CSP with the same *U*-value. A *cradle-to-gate* LCA was performed considering a FU of 1 m$^2$ of CSP and two different impact categories: NRe and GWP. The CSP scenarios have outer and inner layers of HPC, and six different thermal insulation materials in between: LWC, cork, glass wool, EPS 100% from virgin resources, EPS with 55% virgin material, and 100% recycled EPS. It was concluded that the scenario with the lowest carbon emissions (lower GWP-value) is the CSP with cork (30 kg of $CO_2$-eq/

$m^2$), due to the carbon sequestration during the raw material growing. Despite the CSP with virgin EPS being the scenario with the highest carbon footprint (68 kg of $CO_2$-eq/$m^2$), EPS impacts could be significantly reduced if virgin raw materials were replaced by recycled materials. Regarding the NRe impact category, it was concluded that the CSP with recycled EPS is the scenario with the best environmental performance (375 MJ/$m^2$), followed by the scenarios with LWC (411 MJ/$m^2$) and glass wool (424 MJ/$m^2$). Further research is suggested to assess more impact categories.

**Acknowledgements** This work was financially supported by the European Community's H2020 Programme, under the grant agreement Nr. 814632 being aligned with the research project LIGHTCOCE.

## References

DAPHabitat (2016) Aglomerado de Cortiça Expandida (ICB) (in Portuguese)

Demertzi M, Silvestre J, Garrido M et al (2020a) Life cycle assessment of alternative building floor rehabilitation systems. Structures 26:237–246. https://doi.org/10.1016/j.istruc.2020.03.060

Demertzi M, Silvestre JD, Durão V (2020b) Life cycle assessment of the production of composite sandwich panels for structural floor's rehabilitation. Eng Struct 221:111060. https://doi.org/10.1016/j.engstruct.2020.111060

Environmental Product Declaration (EPD) (2021) Recycled aggregate products

ISO (2020a) ISO 14040:2006/Amd 1:2020a Environmental management—life cycle assessment—principles and framework. Amendment 1

ISO (2020b) ISO 14044:2006/Amd 2:2020b Environmental management—life cycle assessment—requirements and guidelines. Amendment 2

Lameiras R, Barros J, Azenha M, Valente IB (2013a) Development of sandwich panels combining fibre reinforced concrete layers and fibre reinforced polymer connectors. Part II : Evaluation of mechanical behaviour. Compos Struct 105:460–470. https://doi.org/10.1016/j.compstruct.2013.06.015

Lameiras R, Barros J, Valente IB, Azenha M (2013b) Development of sandwich panels combining fibre reinforced concrete layers and fibre reinforced polymer connectors. Part I: Conception and pull-out tests. Compos Struct 105:446–459. https://doi.org/10.1016/j.compstruct.2013.06.022

Monteiro H, Freire F (2012) Life-cycle assessment of a house with alternative exterior walls: comparison of three impact assessment methods. Energy Build 47:572–583. https://doi.org/10.1016/j.enbuild.2011.12.032

Monteiro H, Soares N (2022) Integrated life cycle assessment of a southern European house addressing different design, construction solutions, operational patterns, and heating systems. Energy Rep 8:526–532. https://doi.org/10.1016/J.EGYR.2022.02.101

Monteiro H, Freire F, Soares N (2021) Life cycle assessment of a south European house addressing building design options for orientation, window sizing and building shape. J Build Eng 39:102276. https://doi.org/10.1016/j.jobe.2021.102276

PRé Sustainability B. V. (2020) Simapro database manual—methods

Resende P, May M, Hallak T, Hiermaier S (2022) Bio-based/green sandwich structures: a review. Thin-Walled Struct 177:109426. https://doi.org/10.1016/j.tws.2022.109426

Santos P, Correia JR, Godinho L et al (2021) Life cycle analysis of cross-insulated timber panels. Structures 31:1311–1324. https://doi.org/10.1016/j.istruc.2020.12.008

Tavares V, Soares N, Raposo N et al (2021) Prefabricated versus conventional construction: comparing life-cycle impacts of alternative structural materials. J Build Eng 41. https://doi.org/10.1016/j.jobe.2021.102705

Thomas A, Anya V, Johannes H et al (2018) Ultra-lightweight foamed concrete for an automated facade application. Mag Concr Res 71:424–436. https://doi.org/10.1680/jmacr.18.00272

Torres-rivas A, Pozo C, Palumbo M et al (2021) Systematic combination of insulation biomaterials to enhance energy and environmental efficiency in buildings. Constr Build Mater 267:120973. https://doi.org/10.1016/j.conbuildmat.2020.120973

Yılmaz E, Aykanat B, Çomak B (2022) Environmental life cycle assessment of rockwool filled aluminum sandwich facade panels in Turkey. J Build Eng 50:104234. https://doi.org/10.1016/j.jobe.2022.104234

# Chapter 19
# Influence of Culture Medium on Carbon Footprint and Energy Requirement of Microalgae Lipid Production

**Roberto Novais, Teresa M. Mata, Leandro Madureira, Filipe Maciel, António A. Vicente, and António A. Martins**

**Abstract** Microalgae are considered an important alternative source for the production of lipids that are essential to the human organism. Nowadays, companies seek to reach an optimal solution—the balance between profit and sustainability, the latter coming from environmental performance. Therefore, this work compares the carbon footprint resulting from the use of different concentrations (M1, M2 and M4) of an organic fertilizer medium for obtaining lipids from microalgae. This LCA study considers a cradle-to-gate system, from cultivation and production to lipid extraction. A functional unit of 1 g of lipids was considered. The ReCiPe 2016 methodology

R. Novais · L. Madureira · F. Maciel · A. A. Vicente
CEB, Centre of Biological Engineering, University of Minho Campus de Gualtar, 4710-057 Braga, Portugal
e-mail: ae6345@uminho.pt

L. Madureira
e-mail: leandromadureira@ceb.uminho.pt

F. Maciel
e-mail: jfilipemaciel@ceb.uminho.pt

A. A. Vicente
e-mail: avicente@deb.uminho.pt

T. M. Mata (✉)
LAETA-INEGI, Associated Laboratory for Energy and Aeronautics, Institute of Science and Innovation in Mechanical and Industrial Engineering, R. Dr. Roberto Frias 400, 4200-465 Porto, Portugal
e-mail: tmata@inegi.up.pt

A. A. Martins
LEPABE, Faculty of Engineering, University of Porto (FEUP), R. Dr. Roberto Frias, S/N, 4200-465 Porto, Portugal

ALiCE, Faculty of Engineering, University of Porto, Rua Dr. Roberto Frias, 4200-465 Porto, Portugal

A. A. Martins
e-mail: aamartins@fe.up.pt

 
N. S. Caetano and M. C. Felgueiras (eds.), *The 9th International Conference on Energy and Environment Research*, Environmental Science and Engineering,
https://doi.org/10.1007/978-3-031-43559-1_19

was carried out to quantify the environmental impacts, with midpoint factors and an egalitarian perspective. According to the results M4 medium presents less carbon emissions than M1 and M2 media. Energy contributes to 98% of the GWP impacts. The lyophilizer represents 64.9% of the electricity consumption. In order to reduce the carbon footprint of this microalgae lipid production system, it is recommended to use a combination of less energy consuming lipid extraction methods and renewable energy sources, e.g. solar or wind energy.

**Keywords** Life cycle assessment · Lipidic production · Microalgae

## 19.1 Introduction

Microalgae are microscopic organisms that can photosynthesize, converting sunlight, carbon dioxide, and nutrients into organic compounds, including lipids (Corrêa et al. 2021). Microalgae are known to produce a wide range of lipids, including triacylglycerols, which are the main storage form of energy in cells. These triacylglycerols are rich in long-chain polyunsaturated fatty acids (LC-PUFAs), such as omega-6 or linoleic acid (LA), and omega-3 or α-linoleic acid (ALA), which are highly beneficial for human health (Saini and Keum 2018). These essential fatty acids have properties that protect against inflammatory and cardiovascular diseases (Minihane et al. 2016) and are beneficial for other physiological processes, including immune function, brain development and cognitive function. The usual problem is the unbalanced ingestion between n-6 and n-3 PUFA's (western culture ratio, 15.1 to 16.7:1), which ultimately leads to the production of eicosanoids that carry out inflammatory functions and vasoconstriction (Harnack et al. 2009; Simopoulos 2006). Omega-3 fatty acids are not efficiently synthesized by the human body and must be obtained from the diet (Oliver et al. 2020). Traditional sources, such as fish oil, are becoming increasingly unsustainable due to overfishing and concerns about environmental pollution. Fish and shellfish are the largest source of omega-3, but expose humans, especially infants, to the neurotoxic effects of methylmercury (Cholewski et al. 2018). Furthermore, governments have established severely restricted quotas in recent decades to protect and preserve these ecosystems (Oliver et al. 2020).

Marine microalgae are an alternative to fishing and offer a sustainable and potentially scalable solution for the production of omega-3 fatty acids. Also, they are rich sources of carbohydrates, proteins and lipids. Microalgae can be cultivated in controlled environments, such as photobioreactors or open ponds, using sunlight, carbon dioxide, and nutrient-rich water (Mata et al. 2010). By optimizing cultivation conditions and strain selection, microalgae can be engineered to produce higher levels of desirable lipids, including omega-3 fatty acids (Benavente-Valdés et al. 2016). Ongoing research and development efforts are focused on optimizing microalgae cultivation and lipid production processes to make them more environmentally friendly and economically viable and scalable for commercial production (Branco-Vieira et al. 2020).

Microalgae, such as *Pavlova gyrans*, consume atmospheric carbon dioxide, have low cultivation requirements (especially nutrients) and, from a biochemical point of view, are capable of contributing to a range of high-value products. Because of these advantages the lipids production from this microalga was analyzed in this work from the point of view of energy consumption and carbon footprint reduction (Morais Junior et al. 2020). In order to fully assess the carbon footprint reduction, it is important to consider the entire life cycle of the microalgae lipids production process, including factors such as the energy required for cultivation, harvesting, processing and transportation, following a Life Cycle Thinking approach. Life cycle assessment (LCA) is a systematic methodology for assessing the environmental impacts of a product, process or service over its entire life cycle, from raw material extraction to disposal (Huijbregts et al. 2017). Thus, performing an LCA allows for the identification of additional areas for carbon footprint reduction and guiding further improvements. It is also important to assess how the nutrients concentration of the culture medium can influence the carbon footprint of microalgae lipid production (Mata et al. 2022). Implementing solar energy and using energy efficient extraction method are also good alternatives for analysis.

## 19.2 Materials and Methods

This study follows the LCA methodology as described in the ISO 14040 (2006a) and 14044 (2006b) standards, which involves four interactive phases: (1) definition of the objective and scope, (2) life cycle inventory (LCI), (3) life cycle impact assessment (LCIA) and (4) Interpretation.

### *19.2.1 Goal and Scope Definition*

**Study objective.** The main objective of this work is to analyze the carbon footprint and energy demand of lipids production from microalgae *P. gyrans*.

**Target audience.** The target audience include but are not limited to companies and experts in the area of microalgae cultivation and lipidic production.

**Functional unit.** The functional unit (FU) of this attributive LCA study is 1 g of lipids extracted.

**System boundary.** This cradle-to-gate system encompasses the microalgae cultivation and biomass production stages, harvesting, drying and extraction of the lipids from microalgae biomass, as shown in Fig. 19.1.

Information and data required for this work, and the process and technologies implemented, were based on laboratory research, conducted in Portugal, described by Madureira (2019). Therefore, the cultivation and production of *P. gyrans* took

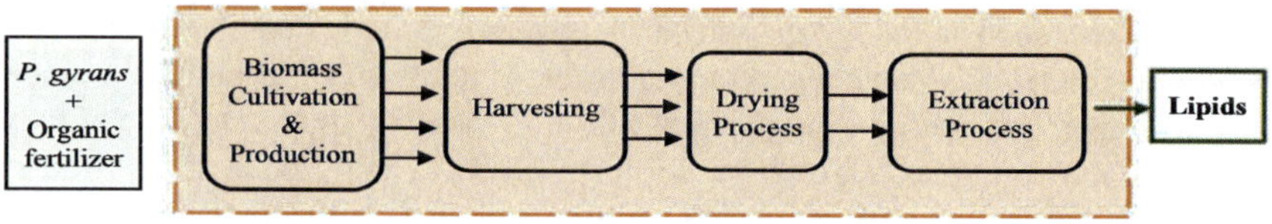


**Fig. 19.1** System boundary for the life cycle study of microalgae cultivation for lipids production

place in 250 mL Erlenmeyer flasks, but the volume of the culture medium used was 100 mL, for 14 days in autotrophic conditions, at room temperature and with a pH of 8.2. The harvesting and drying stages were performed respectively in a centrifuge and a lyophilizer. The lipids extraction from microalgae was performed according to the Bligh and Dyer method, with some modifications, as described in Madureira (2019).

### *19.2.2 Life Cycle Inventory (LCI)—Analysis and Data Sources*

Information and data for the life cycle inventory (e.g., raw materials, auxiliary materials and water) were obtained from experimental analysis conducted by Madureira (2019), and complemented with information from the literature and approximations and calculations made by the authors of this paper for some of the components used in the study. The organic fertilizer used as culture medium was approximated to the Conway's medium, since the owing company preferred to keep the medium composition confidential. However, to calculate the amounts of micronutrients and macronutrients the percentages (v/v) described by Madureira (2019) were taken into consideration. The auxiliary components and water in the system were also considered.

The energy consumed in this system consists mainly of the electricity consumption of the equipment used in the laboratory. Regarding the electricity, it was assumed the Portuguese electricity (PE) mix, as defined in the EcoInvent database. To obtain the energy values, *E*, in general, the Eq. 19.1 was used.

$$E = Poxt \tag{19.1}$$

The power, *Po*, was taken from the equipment itself (no energy efficiency functions was taken into consideration) and the equipment utilization time, *t*, was obtained from Madureira (2019).

Table 19.1 synthetizes the inventory data for the production of 1 g of lipids from microalgae *P. gyrans*. The medium M1, M2 and M4 differ in the micronutrients % (v/v) (shown with gray filling in the table).

To account for the transportation of components in 32-ton trucks, an average distance, $d_m$, of 200 km, was considered as it depends on the total mass of components

**Table 19.1** Inventory data for the production of 1 g of lipids from microalga *P. gyrans*

| Process | Input | Amount M1 | M2 | M4 | Unit |
|---|---|---|---|---|---|
| Cultivation and production | Energy | $5.44\times 10^3$ | $6.04\times 10^3$ | $5.35\times 10^3$ | [kWh/g] |
| | Water | $1.30\times 10^4$ | $1.45\times 10^4$ | $1.28\times 10^4$ | [mL/g] |
| | Potassium nitrate | $2.07\times 10^1$ | $2.30\times 10^1$ | $2.04\times 10^1$ | [mg/g] * |
| | Trisodium phosphate | $4.14\times 10^0$ | $4.60\times 10^0$ | $4.07\times 10^2$ | |
| | Ammonium molybdate | $3.72\times 10^{-1}$ | $4.14\times 10^{-1}$ | $3.66\times 10^{-1}$ | |
| | Copper (II) sulphate | $1.03\times 10^{-2}$ | $1.15\times 10^{-1}$ | $2.04\times 10^{-1}$ | |
| | Zinc chloride | $1.09\times 10^{-2}$ | $1.21\times 10^{-1}$ | $2.14\times 10^{-1}$ | |
| | Manganese (II) chloride | $9.31\times 10^{-4}$ | $1.03\times 10^{-2}$ | $1.83\times 10^{-2}$ | |
| | Iron (III) chloride | $3.36\times 10^{-3}$ | $3.74\times 10^{-2}$ | $6.62\times 10^{-2}$ | |
| | Cobalt (II) chloride | $1.03\times 10^{-2}$ | $1.15\times 10^{-1}$ | $2.04\times 10^{-1}$ | |
| | Boric acid | $8.64\times 10^{-2}$ | $9.60\times 10^{-1}$ | $1.70\times 10^0$ | |
| | Disodium EDTA salt | $1.16\times 10^{-1}$ | $1.29\times 10^0$ | $2.29\times 10^0$ | |
| Harvesting | Energy | $5.17\times 10^1$ | $5.75\times 10^1$ | $5.09\times 10^1$ | [kWh/g] |
| Drying | Energy | $1.04\times 10^4$ | $1.16\times 10^4$ | $1.03\times 10^4$ | |
| Extraction | Energy | $1.48\times 10^2$ | $1.65\times 10^2$ | $1.46\times 10^2$ | |
| | Water | $1.30\times 10^4$ | $1.45\times 10^4$ | $1.28\times 10^4$ | [mL/g] |
| | Chloroform | $9.47\times 10^2$ | $1.05\times 10^3$ | $9.32\times 10^2$ | |
| | Methanol | $4.74\times 10^2$ | $5.26\times 10^2$ | $4.66\times 10^2$ | |
| | Nitrogen gas | $1.71\times 10^3$ | $1.90\times 10^3$ | $1.68\times 10^3$ | [L/g] |

*Although these are dimensionless units, it's presented to the reader recognize the mass's component is diving by the FU

and distance travelled. The end-of-life cycle treatment was also considered and it only depends on the volume (expressed in $m^3$) of the culture media used.

### 19.2.3 *Life Cycle Impact Assessment (LCIA)*

To choose the appropriate methodology and calculate the values of carbon footprint that best suit the system considered a review of 11 articles was performed. It was decided to use the ReCiPe 2016 methodology following the Equalitarian perspective. All this was carried out with the aid of the software SimaPro (V 8.5.2) and its inventory databases, namely EcoInvent (V 3.5).

Due to lack of information in the database it was necessary to perform the modelling of some components. This was done according to the life cycle tree model, considering the reactants necessary for the production of these compounds (Cuéllar-Franca et al. 2016; Geisler et al. 2004). Using the stoichiometric reactions, the products masses were determined. With small adjustments of the mass of the product of interest, to the reference value of 1 kg, the mass values of the reagents were calculated.

## 19.3 Results and Discussion

### 19.3.1 Carbon Footprint Analysis

Three different concentrations of organic fertilizer medium were evaluated. Figure 19.2 shows the carbon footprint or the GWP impact, on each concentration (M1, M2 and M4).

As shown in the graph of Fig. 19.2 the M4 medium presents less $CO_2$ emissions, which corresponds to a smaller carbon footprint. This is explained mainly due to the higher lipid content obtained with the M4 medium (15.72 ± 2.53 wt%) as compared to the other two mediums (M1 with 15.47 ± 1.47 wt % and M2 with 13.92 ± 1.12 wt % of lipid content). Additionally, the relative values of the inventory items were analyzed, as shown in Fig. 3a. The "consumables" include all the chemicals used for lipids' extraction from microalgae biomass and the "components" include all the micro and macronutrients used in the cultivation and production of microalgae. The graph of Fig. 3b shows the electricity consumption per equipment unit. The results of Fig. 3a shows that energy is the dominant contributor to the GWP, representing 98% of the carbon emissions. Figure 3b shows that the lyophilizer is responsible for 64.9% of electricity consumption, followed by the orbital mixer (18.9%) and then by the artificial illumination (13.3%).

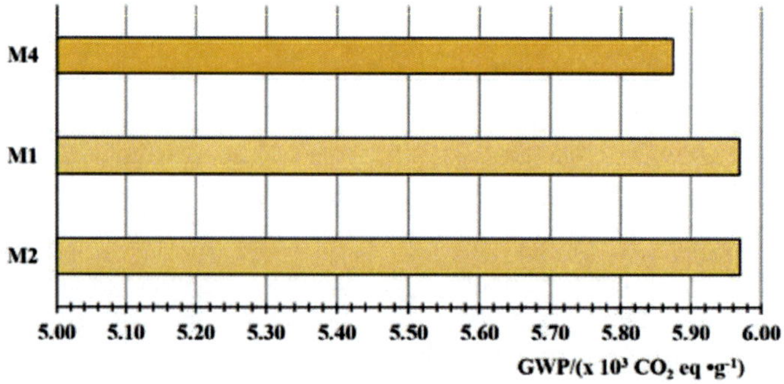


**Fig. 19.2** Carbon emissions of microalgae lipids production, for each culture medium concentration (M1, M2 and M4) of organic fertilizer

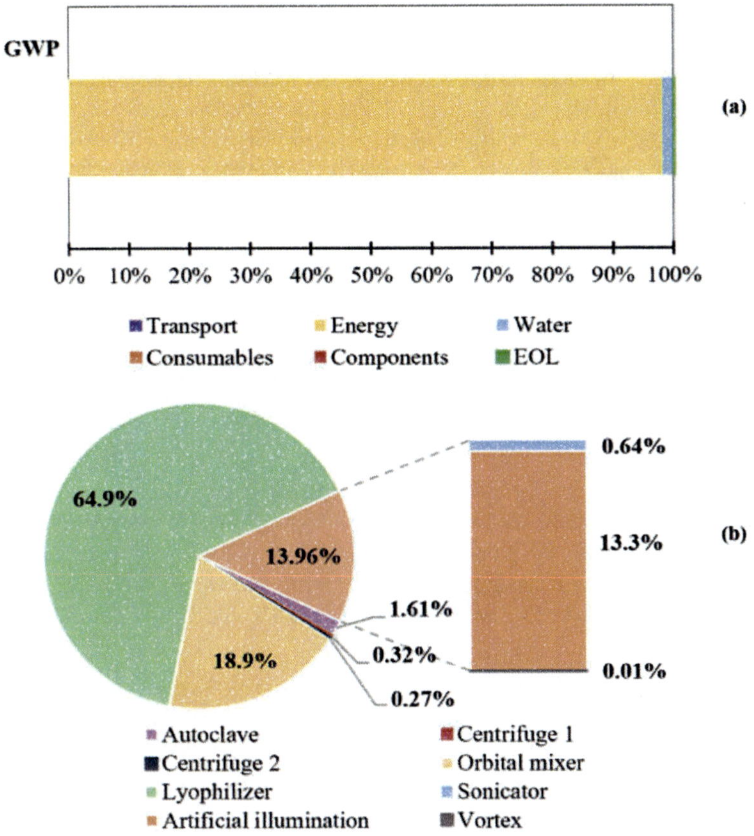


**Fig. 19.3** **a** Contribution of the inventory items to carbon footprint (GWP) and **b** electricity consumption per equipment unit

### *19.3.2 Improvement Proposal*

Because of the large contribution of energy consumption to the carbon footprint, it is necessary to analyze the origin of energy in order to find alternatives with lower carbon footprint. Thus, it was evaluated in this work the replacement of the Portuguese electricity (PE) mix by electricity exclusively produced by silicon photovoltaic solar panels (Solar Energy), which are the type of panels most used for smaller scale applications.

Figure 19.4 compares the contribution to carbon footprint (or GWP) of photovoltaic energy compared to the Portuguese electricity mix.

Figure 19.4 shows that the contribution to the carbon footprint of photovoltaic electricity is 6.2 times lower than that of the Portuguese electricity mix. Solar energy systems have improved in efficiency over the years, making them a cleaner and more

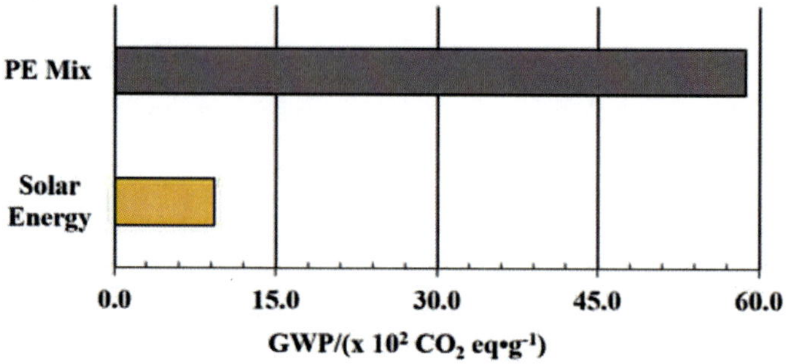


**Fig. 19.4** Carbon footprint of photovoltaic electricity (Solar energy) versus the Portuguese electricity (PE) mix

sustainable energy source. Photovoltaic energy is a good renewable energy alternative, instead of other renewable energy sources, such as wind energy, because of its ease of implementation. Solar energy systems may have simpler grid integration requirements, especially for smaller-scale installations. Also, solar panels are generally easier to install, with straightforward mounting and wiring processes than wind turbines and also allows varying and adjusting the installed power according to the production capacity.

## 19.4 Conclusion

In this paper, the carbon footprint and energy consumption of microalgae lipids production was evaluated. Three concentrations of an organic fertilizer were evaluated for microalgae cultivation. Results shows that the medium which contributes to a higher lipid production (M4) has a lower carbon footprint that the other mediums concentrations (M1 and M2), mainly due to the higher lipid yield. Energy consumption contributes to 98% of GWP. The equipment unit with the highest energy consumption is the lyophilizer, which is responsible for 64.9% of $CO_2$ eq emissions. The use of renewable energy sources such as solar energy, from photovoltaic panels, results in a reduction of the carbon footprint by 6.2 times in comparison with the Portuguese electricity mix. By switching to a less energy-consuming method, one can reduce the demand for fossil fuels and associated carbon emissions. Thus, alternatively by using solar energy for the lipid extraction process, one can take advantage of the high efficiency of solar panels, which directly convert sunlight into electricity without combustion or emissions.

**Acknowledgements** This work was financially supported by base Funding of the following projects: LA/P/0045/2020 (ALiCE), UIDB/00511/2020 (LEPABE) and UIDB/50022/2020 (LAETA), funded by national funds through FCT/MCTES (PIDDAC). António Martins gratefully acknowledges the Portuguese national funding agency for science, research and technology (FCT) for funding through program DL 57/2016—Norma transitória. Teresa Mata gratefully acknowledges the funding of Project NORTE-06-3559-FSE-000107, cofinanced by Programa Operacional Regional do Norte (NORTE2020), through Fundo Social Europeu (FSE). Filipe Maciel and Leandro Madureira gratefully acknowledge the project ALGAVALOR—Microalgas: produção integrada e Valorização da biomassa e das suas diversas aplicações (POCI-01-0247-FEDER-035234), cofinanced by Fundo Europeu de Desenvolvimento Regional (FEDER), Portugal 2020, through Programa Operacional Competitividade e Internacionalização (COMPETE2020), Programa Operacional Regional do Norte (Norte 2020), Programa Operacional da Região Centro (Centro 2020), Programa Operacional Regional de Lisboa (Lisboa 2020), Programa Operacional Regional do Alentejo (Alentejo 2020) and Programa Operacional do Algarve (CRESC ALGARVE 2020).

## References

Benavente-Valdés JR, Aguilar C, Contreras-Esquivel JC, Méndez-Zavala A, Montañez J (2016) Strategies to enhance the production of photosynthetic pigments and lipids in chlorophycae species. Biotechnol Rep 10:117–125. https://doi.org/10.1016/j.btre.2016.04.001

Branco-Vieira M et al (2020) Biotechnological potential of Phaeodactylum tricornutum for biorefinery processes. Fuel 268. https://doi.org/10.1016/j.fuel.2020.117357

Cholewski M, Tomczykowa M, Tomczyk M (2018) A comprehensive review of chemistry, sources and bioavailability of omega-3 fatty acids. Nutrients 10(11). MDPI AG, Nov. 2018. https://doi.org/10.3390/nu10111662

Corrêa PS, Júnior WGM, Martins AA, Caetano NS, Mata TM (2021) Microalgae biomolecules: extraction, separation and purification methods. Processes 1–43. https://doi.org/10.3390/pr9010010

Cuéllar-Franca RM, García-Gutiérrez P, Taylor SFR, Hardacre C, Azapagic A (2016) A novel methodology for assessing the environmental sustainability of ionic liquids used for $CO_2$ capture. Faraday Discuss 192:283–301. Royal Society of Chemistry. https://doi.org/10.1039/c6fd00054a

Geisler G, Hofstetter TB, Hungerbühler K (2004) Production of fine and speciality chemicals: procedure for the estimation of LCIs. Int J Life Cycle Assess 9(2):101–113. https://doi.org/10.1007/BF02978569

Harnack K, Andersen G, Somoza V (2009) Quantitation of alpha-linolenic acid elongation to eicosapentaenoic and docosahexaenoic acid as affected by the ratio of n6/n3 fatty acids. Nutr Metab 6. https://doi.org/10.1186/1743-7075-6-8

Huijbregts MAJ et al (2017) ReCiPe2016: a harmonised life cycle impact assessment method at midpoint and endpoint level. Int J Life Cycle Assess 22(2):138–147. https://doi.org/10.1007/s11367-016-1246-y

ISO 14044 (2006b) Environmental management—life cycle assessment—requirements and guidelines. International Organization for Standardization

ISO 14040 (2006a) Environmental management—life cycle assessment—principles and framework. International Organization for Standardization

Madureira LFF (2019) Use of agro-industrial by-products for *Pavlova* sp. culture and heterotrophic growth of *Nannochloropsis* sp. as relevant production strategies for oleaginous microalgae. Master thesis, Universidade do Minho Escola de Engenharia, Braga, Portugal

Mata TM, Martins AA, Caetano NS (2010) Microalgae for biodiesel production and other applications: a review. Renew Sustain Energ Rev. https://doi.org/10.1016/j.rser.2009.07.020

Mata TM, Rodrigues S, Caetano NS, Martins AA (2022) Life cycle assessment of bioethanol from corn stover from soil phytoremediation. Energ Rep 8:468–474. https://doi.org/10.1016/j.egyr.2022.01.059

Minihane AM et al (2016) Consumption of fish oil providing amounts of eicosapentaenoic acid and docosahexaenoic acid that can be obtained from the diet reduces blood pressure in adults with systolic hypertension: a retrospective analysis. J Nutr 146(3):516–523. https://doi.org/10.3945/jn.115.220475

Morais Junior WG, Gorgich M, Corrêa PS, Martins AA, Mata TM, Caetano NS (2020) Microalgae for biotechnological applications: cultivation, harvesting and biomass processing. Aquaculture 528:735562. https://doi.org/10.1016/j.aquaculture.2020.735562

Oliver L, Dietrich T, Marañón I, Villarán MC, Barrio RJ (2020) Producing omega-3 polyunsaturated fatty acids: a review of sustainable sources and future trends for the EPA and DHA market. Resources 9(12):1–15. MDPI AG, Dec. 2020. https://doi.org/10.3390/resources9120148

Saini RK, Keum YS (2018) Omega-3 and omega-6 polyunsaturated fatty acids: dietary sources, metabolism, and significance—a review. Life Sci 203:255–267. Elsevier Inc, https://doi.org/10.1016/j.lfs.2018.04.049

Simopoulos AP (2006) Evolutionary aspects of diet, the omega-6/omega-3 ratio and genetic variation: nutritional implications for chronic diseases. Biomed Pharmacother 60(9):502–507. https://doi.org/10.1016/j.biopha.2006.07.080

# Chapter 20
# Life Cycle Assessment and Evaluation of External Costs of the Italian Electricity Mix

**Benedetta Marmiroli, Maria Leonor Carvalho, Giulio Mela, Andrea Molocchi, and Pierpaolo Girardi**

**Abstract** This study combines the Life Cycle Assessment of the Italian electricity mix and external cost evaluation to develop an indicator useful for policy-making and evaluation. Even though external costs of air emissions are site specific, traditional LCA studies do not characterize environmental flows at geographical level. Therefore, we adapt the LCA model of the electricity consumption mix to include site specific information. The geographical characterization of LCA air emissions and benefit transfer techniques are used to calculate the indicator, which can be computed for different countries and economic aggregates and reference years, also into the future. In this work we apply the proposed methodology to two energy scenarios for Italy: one referring to 2019 and 2030. The first is based on the last available data, while the second on assumptions on the implementation of the European Union's Green New Deal. The average external cost of the electricity consumption mix is 0.057 €2019/kWh in the 2019 scenario and as low as 0.019 €2019/kWh for the 2030 scenario. We also find that, even though emissions linked to electricity production take place in many countries around the world, most externalities occur within national boundaries and are linked to direct emissions from thermoelectric power plants.

**Keywords** Italian electricity mix · Externalities · Life cycle assessment

## 20.1 Introduction

The European and Italian energy sectors are facing the challenge of a dramatic reduction of their greenhouse gas emissions. The environmental effectiveness of these efforts can be properly evaluated only considering the entire life cycle of energy assets

B. Marmiroli (✉) · M. L. Carvalho · G. Mela · A. Molocchi · P. Girardi
RSE Ricerca sul Sistema Energetico, via Rubattino 54, 20133 Milano, Italy
e-mail: benedetta.marmiroli@rse-web.it



N. S. Caetano and M. C. Felgueiras (eds.), *The 9th International Conference on Energy and Environment Research*, Environmental Science and Engineering,
https://doi.org/10.1007/978-3-031-43559-1_20

and a wide range of environmental impacts besides climate change, using a Life Cycle Assessment approach (LCA) (Carvalho et al. 2022). However, the wide set of indicators typically used in LCA, despite providing plenty of information, can generate confusion among decision-makers. An aggregated indicator measuring external costs linked to given process (i.e. energy generation) and expressed in monetary units can contribute in mitigating this problem (Girardi et al. 2020).

External costs are the monetary value of the damages generated by an economic or social activity impacting on third parties (e.g.: health damages or damages to agriculture) or on the environment (e.g.: on biodiversity or ecosystems), which are not already paid by the same activity (through insurance premiums, environmental taxes, compensation fees, etc.) (European Commission 2020; Molocchi 2020).

The aim of this paper is to develop a methodology to develop an indicator of the "pre-tax external costs" of a given process in an LCA framework and to apply it to two energy consumption mixes for Italy as case studies.

An LCA external cost indicator summarizes different types of externalities (air pollution and greenhouse gases) and different types of impact pathways (effects on health, agriculture, buildings' surfaces and biodiversity) related to different processes and geographical regions. Such an indicator can be applied in several areas of interest such as: energy transition policy, fiscal policy (environmental fiscal reform), cost–Benefit Analysis (CBA) (a proper calculation of external costs is the basis for a complete CBA), sustainable finance (as in Esposito et al. 1633).

In the case of electric energy, pre-tax external costs are the costs of damages due to electricity without subtracting any cost component already paid by electricity producers or consumers to prevent or compensate such damages (Parry et al. 2014). They are calculated through the impact-pathways approach, which links emission factors to the involved environmental media and then to the final damage receptors. The impact-pathway modelling approach was developed by several EU research projects (ExternE 1995, 1999; Bickel and Friedrich 2005; Preiss et al. 2008; Müller et al. 2010). The approach allows to quantify the risk of incurring in the negative effects of air emissions to human health, ecosystems, economic activities or assets, and estimate the willingness to pay to avoid or lower such risk. The proposed indicator is calculated for the 2019 Italian electricity consumption mix and for the forecasted 2030 mix, if decarbonisation targets set by the Green New Deal are met (RSE 2030 scenario, see Carvalho et al. 2022).

This paper is structured as follows. Section 20.2 outlines the methodology developed to develop the LCA external cost indicator and illustrate how it can be applied to evaluate consumption energy mixes. Section 20.3 illustrates and discusses the results. Section 20.4 concludes.

## 20.2 Materials and Methods

### *20.2.1 Life Cycle Assessment of the Electricity Mix*

The first step to evaluate the externalities along the electricity generation life cycle, is to model the life cycle of the annual electricity consumption mix in Italy: the current mix (2019 data) and the future mix (2030 mix, as defined by the Green Deal scenario developed by RSE (Carvalho et al. 2022). The mixes are modelled in the software SimaPro 9.1.1, using the Ecoinvent database v. 3.3. The functional unit is 1 kWh of electricity Gross National Consumption (GNC), which includes the total gross national electricity generation from all sources (excepted pumped hydro generation), plus electricity imports, minus exports. The data used to model the current mix are from Eurostat (2021) and from the Italian transmission system operator (Terna S.p.A). The most updated data refer to the year 2019. Fuel consumption is derived from Eurostat, while the technologies taken into consideration are those reported in the Terna annual statistical reports. Thermoelectric plants are modelled based on their technology (Terna 2019), and their type of fuel (Eurostat 2021). Their efficiency is calculated using fuel consumption and electricity production available at Eurostat (2021). Air emissions are obtained from EMAS environmental declarations. Wind and Photovoltaics plants are modelled calculating the load hours dividing production by the installed capacity (Eurostat 2021). For the electricity import, the average European production is considered. A more detailed description of the model can be found in Carvalho et al. (2022) and Gargiulo et al. (2020). In this study, an updated version of the above-mentioned models is developed, using data from 2019 and the 2030 Green Deal scenario.

### *20.2.2 Externality Evaluation*

Ideally, externality evaluation should be carried out through ad hoc studies based on stated (or revealed) preference methods. Such methods attempt to infer people's willingness to pay for a good or service not traded in markets (such as an environmental good). Performing such studies is however very expensive and time-consuming. It is therefore a common practice to exploit the information already available in the literature by adapting it to the study context, to take into account the socio-economic differences between the place where the primary study was performed and the place where the literature estimate has to be used.

In this study, we use the external cost estimates from the European Commission's Handbook on the external costs of transport (European Commission 2020). Given that many means of transportation are electrified, this handbook includes a section on the external costs of "well-to-tank emissions" (emissions related to electricity production and to other industrial processes needed to produce transport fuels). The handbook provides recommended input values for greenhouse gas emissions (in terms of $CO_2$ equivalent) and several air pollutants (NOx, NMVOC—Non-Methane Volatile Organic Compounds, $SO_2$, ammonia, PM2.5). The central value recommended by the European Commission Handbook for $CO_2$ emissions up to year 2030 is €2016 100/t. This is a global average value, which does not change with the place of emissions, but increases over time, depending on the year of emission (after 2030). The unit air pollution external costs provided by the Handbook represent the aggregate value of health effects, crop losses, material and building damages and biodiversity loss related to acidification and to eutrophication of ecosystems. Unit external costs refer to 2016 and vary according to the country of emission, age class of the exposed population and stack height.

In this study value transfer techniques are used to adjust the handbook values to the countries (or geographical aggregates) and years considered. Such techniques allow to adjust literature values (estimated for a given study country and year) so that they can be used also in other spatial and temporal contexts (in a so-called policy Country and/or year). Given that also externalities referring to a future scenario (2030) are evaluated, country-specific social discount rates are estimated and used to assign a present value (at the base year 2019) to the estimated external costs of the electricity mix in 2030.

The value transfer technique used is the unit value transfer with income adjustment. A relatively simple method that assumes that the willingness to pay for a given good is positively (but not always proportionally) correlated with per capita income:

$$V_{pt_1} = V_{st_0}\left(\frac{Y_{pt_1}}{Y_{st_0}}\right)^{\epsilon} \tag{20.1}$$

where $V_{pt_1}$ is the value transferred to the policy site $p$ in year $t_1$ (2019), $V_{st_0}$ is the primary estimate referring to the study site $s$ in year $t_0$, $Y_{pt_1}$ and $Y_{st_0}$ are the per capita GDP (based on purchasing power parity) of policy and study sites respectively and $\epsilon$ is the income elasticity of the willingness to pay for the good. $\epsilon$ depends on income levels, as suggested by Molocchi et al. (2021) who estimate it for all World Bank's income groups building following Navrud (2009). According to them, $\epsilon$ ranges from a minimum of 0.2 (high income Countries) to a maximum of 1.2 (low income countries. $V_{pt_1}$ is converted to 2019 price levels using the implicit GDP deflator of the study country (the Euro Area).

The value transfer is performed to both 2019 and 2030. For this reason, forecasts on per capita GDP and $\epsilon$ are made. Per capita GDP levels in 2030 are estimated starting from the last available data (2019) using the average annual per capita GDP growth rate of the previous 10 years. Once values are transferred to 2030, they need to be discounted back to 2019 using a social discount rate. In finance, discount rates reflect the opportunity cost of capital (European Commission 2014). In economic analysis, which evaluates the effects of a given project on societal welfare, financial discount rates should not be used since it reflects the opportunity cost of a single investor and not that of the society as a whole. For this reason, a social discount rate is required: the rate at which the society is willing to perform an inter-temporal trade off in terms of consumption (Groom and Maddison Pr 2019).

**Table 20.1** Cost factors [€$_{2019}$] breakdown by geographical area, 2019

| Factor costs [€$_{2019}$/kg] | NMVOC | $CO_2$eq | NOx | $SO_2$ | PM2.5 | Ammonia |
|---|---|---|---|---|---|---|
| Germany | 1.92 | 0.108 | 21.50 | 17.57 | 74.56 | 29.91 |
| European Union | 1.27 | 0.108 | 11.54 | 11.54 | 41.24 | 18.53 |
| Italy | 1.14 | 0.108 | 14.61 | 13.16 | 45.19 | 22.39 |
| Others | 0.73 | 0.108 | 6.65 | 6.65 | 23.74 | 10.67 |
| Russia | 1.13 | 0.108 | 4.09 | 8.33 | 45.47 | 12.84 |
| Libya | 0.56 | 0.108 | 1.37 | 2.76 | 6.69 | 2.08 |
| Algeria | 0.70 | 0.108 | 3.28 | 9.19 | 24.12 | 10.76 |
| Netherlands | 3.02 | 0.108 | 15.55 | 21.82 | 80.72 | 32.40 |
| North Sea | 2.42 | 0.108 | 11.26 | 11.05 | 36.19 | – |
| Mediterranean Sea | 0.53 | 0.108 | 3.16 | 9.69 | 25.92 | – |
| Indonesia | 0.67 | 0.108 | 6.07 | 6.07 | 21.69 | 9.75 |
| Malaysia | 1.04 | 0.108 | 9.41 | 9.41 | 33.60 | 15.10 |
| Sweden | 0.68 | 0.108 | 6.66 | 5.31 | 13.82 | 10.22 |
| France | 1.56 | 0.108 | 17.97 | 14.44 | 42.60 | 16.00 |
| Hungary | 0.90 | 0.108 | 17.23 | 11.15 | 36.70 | 21.28 |
| Spain | 0.74 | 0.108 | 5.15 | 7.15 | 20.92 | 6.73 |
| China | 0.78 | 0.108 | 7.08 | 7.08 | 25.31 | 11.37 |

The social discount rate is estimated through the social rate of time preference (SRTP), in turn calculated with the Ramsey's rule (Ramsey 1928):

$$SRTP = p + \eta g \tag{20.2}$$

where $p$ is the pure rate of time preference, $\eta$ is the marginal elasticity of consumption and $g$ the growth rate of per capita consumption.

External costs referring to countries and aggregates involved in the life cycle of a kWh consumed in Italy are shown in Tables 20.1 and 20.2.

### *20.2.3 Life Cycle Assessment of the Electricity Mix*

External cost factors depend on the site where air pollutant take place. Thus, emissions evaluated in the life cycle assessment need to be defined at the spatial level (national or regional scale), something that it is not generally done in conventional LCA studies. It is thus required to identify the country of origin of fuels, raw materials and components entering the value chain of Italian electricity. In the case of fossil and renewable fuels, data from SNAM (2019), the Ministry of economic development

**Table 20.2** Cost factors [€$_{2019}$] breakdown by geographical area, 2030

| Factor costs [€$_{2019}$/kg] | NMVOC | $CO_2$eq | NOx | SO2 | PM2.5 | Ammonia |
|---|---|---|---|---|---|---|
| Germany | 1.37 | 0.75 | 15.43 | 12.60 | 53.49 | 21.46 |
| European Union | 0.93 | 0.75 | 8.45 | 8.45 | 30.19 | 13.57 |
| Italy | 1.02 | 0.75 | 13.03 | 11.74 | 40.31 | 19.97 |
| Others | 0.57 | 0.75 | 5.21 | 5.21 | 18.62 | 8.37 |
| Russia | 0.65 | 0.75 | 2.37 | 4.82 | 26.32 | 7.43 |
| Libya | 0.31 | 0.75 | 0.75 | 1.51 | 3.67 | 1.14 |
| Algeria | 0.58 | 0.75 | 2.70 | 7.57 | 19.86 | 8.86 |
| Netherlands | 2.33 | 0.75 | 11.98 | 16.81 | 62.17 | 24.96 |
| North Sea | 1.84 | 0.75 | 8.58 | 8.42 | 27.57 | – |
| Mediterranean Sea | 0.43 | 0.75 | 2.58 | 7.90 | 21.12 | – |
| Indonesia | 0.41 | 0.75 | 3.69 | 3.69 | 13.20 | 5.93 |
| Malaysia | 0.66 | 0.75 | 6.02 | 6.02 | 21.49 | 9.66 |
| Sweden | 0.53 | 0.75 | 5.23 | 4.17 | 10.85 | 8.03 |
| France | 1.22 | 0.75 | 14.11 | 11.33 | 33.44 | 12.56 |
| Hungary | 0.56 | 0.75 | 10.68 | 6.91 | 22.75 | 13.20 |
| Spain | 0.57 | 0.75 | 3.99 | 5.54 | 16.20 | 5.21 |
| China | 0.38 | 0.75 | 3.46 | 3.46 | 12.35 | 5.55 |

(Ministero dello Sviluppo Economico 2019) and from the Energy Service Operator GSE (2018) are used define the origin of:

- Natural gas (Russia, Algeria, The Netherlands and Libya);
- Petroleum products (Russia, Libya, Algeria, The Netherlands, North Sea, Mediterranean Sea, Rest of the world);
- Coal (Rest of the world);
- Vegetable oil (Italy, France, Spain, Germany, Hungary, EU-28, Indonesia, Malaysia, rest of the world).

Fuels like wastes, biogas and derived gas are produced in Italy. Other components entering the value chain assumed to be produced abroad are photovoltaic panels and wind turbines. For these components the main producer/importer has been taken as a proxy for the origin of the component (China for PVs and EU-28 for wind turbines). Auxiliary products entering more than a single production chain have been assigned to regions (Europe, rest of the world, etc.) where their use is predominant. In Simapro the related emissions of these processes are assigned to the corresponding geographical group. To do so, datasets are transformed into “System processes” and assigned to the geographical groups using the “analyse group” tool. The emissions divided per geographical area are then multiplied for the corresponding factor costs.

## 20.3 Results and Discussion

The application of external cost factors to the life cycle emissions spatially characterized lead to the following external costs for the electricity consumption mix: 0.057 $€_{2019}$/kWh for the 2019 mix and 0.019 $€_{2019}$/kWh for the 2030 mix (Green Deal scenario). These results are obtained multiplying the emissions, differentiated by geographical area, by the corresponding national/regional external cost factors.

Figures 20.1 and 20.2 show the external costs per kWh obtained for the two mixes. The graph on the left shows the share of each pollutant to the overall external cost. The one on the right shows the contribution of the geographical area of emission. External costs are expressed in $€_{2019}$, since 2019 is the reference year chosen for the economic valuation of the study (Fig. 20.1 is referred to the 2019 electricity mix; Fig. 20.2 refers to the 2030 mix).

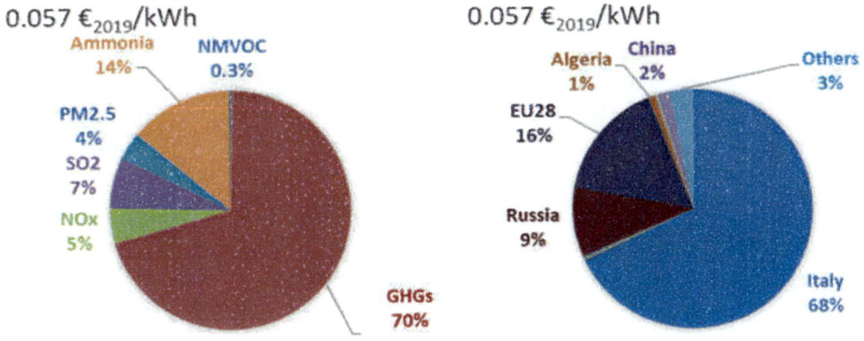


**Fig. 20.1** External costs of the 2019 electricity mix ($€_{2019}$/kWh). Breakdown by: pollutant (left); region of emission (right)

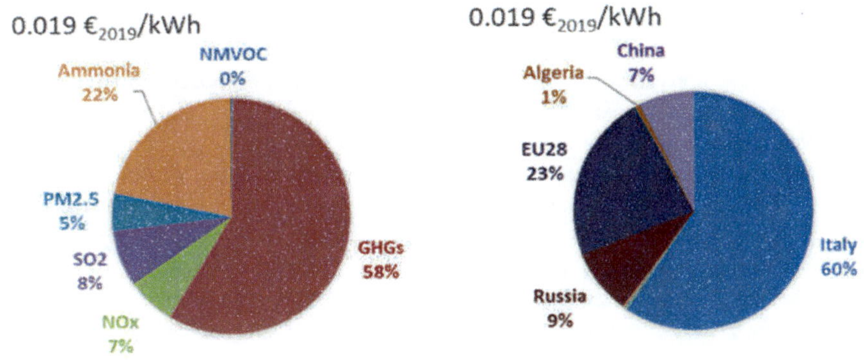


**Fig. 20.2** External costs of the 2030 electricity mix (Green Deal scenario) ($€_{2019}$/kWh). Breakdown by: pollutant (left); region of emission (right)

$CO_2$ is the major contributor to the external cost of electricity. It contributes from 60 to 70% to the overall cost. This contribution decreases in the 2030 scenario as it assumes that decarbonisation targets are met. According to the LCA analysis, most of the $CO_2$ emissions are from direct emissions of thermal power plants. Not surprisingly, in the regional breakdown (Fig. 20.1 right), Italy has the major share of LCA external costs. Interestingly, the second most important pollutant in term of contribution to LCA external costs is ammonia and not pollutants such as $SO_2$, PM2.5 and NOx, which are nonetheless more regulated. Figure 20.3 shows the analysis of the 2019 mix more in detail, highlighting the place of emission of each pollutant.

Greenhouse gas emissions occur mainly in Italy (70%). Italy is also the only country contributing to ammonia emissions (99%). According to LCA results, 99% of ammonia emissions are due to biogas production. Italy is also the first country in terms of LCA nitrogen oxides emissions. That is for the prevalence of direct emission in the use phase of thermal power plants (0.115 g NOx per 1 kWh in a CHP plant using natural gas). Other LCA emission sources are more evenly distributed across the globe: non-methane volatile organic compounds (NMVOC), $SO_2$ and PM2.5 take place in Italy, Russia and European Union. NMVOC emissions are mainly related to the production of natural gas, of which Russia is the main exporter to Italy. PM2.5 emissions are mainly due to the production of electricity from lignite although lignite is not used in Italy for electricity generation. These emissions are linked to the Italian electricity system through the energy consumed abroad to produce those fuels and components needed in the electricity supply chain. In particular, a relevant role is played by the energy required to compress Russian natural gas in pipelines.

## 20.4 Conclusion

In this work we integrate the LCA approach and EU harmonised external costs estimates from the best available literature (EC Handbook for external costs calculation) to develop an LCA based external cost indicator for the Italian electricity consumption mix. The methodology is based on the geographical characterisation of LCA process-related air emissions and on benefit transfer techniques. The proposed indicator can be calculated for all countries (or economic aggregates) in the world and different reference years, also in the future (depending on the scenario's time horizon). The LCA external cost indicator aggregates different types of externalities (air pollution and green-house gases) and different types of impact pathways (effects on health, agriculture, buildings' surfaces and biodiversity) related to different electricity generation processes and geographical regions.

In this work, we calculate the suggested indicator for two different electric energy mixes: one based on the last available data (2019) and one developed by RSE in a previous and recent study, which assumes that the 2030 decarbonisation targets identified by the Green New Deal for Italy are met. The average external costs of the electricity consumption mix in Italy is 0.057 and 0.019 $€_{2019}$/kWh in 2019 and 2030 respectively. In the 2030 Green New Deal scenario the average external costs

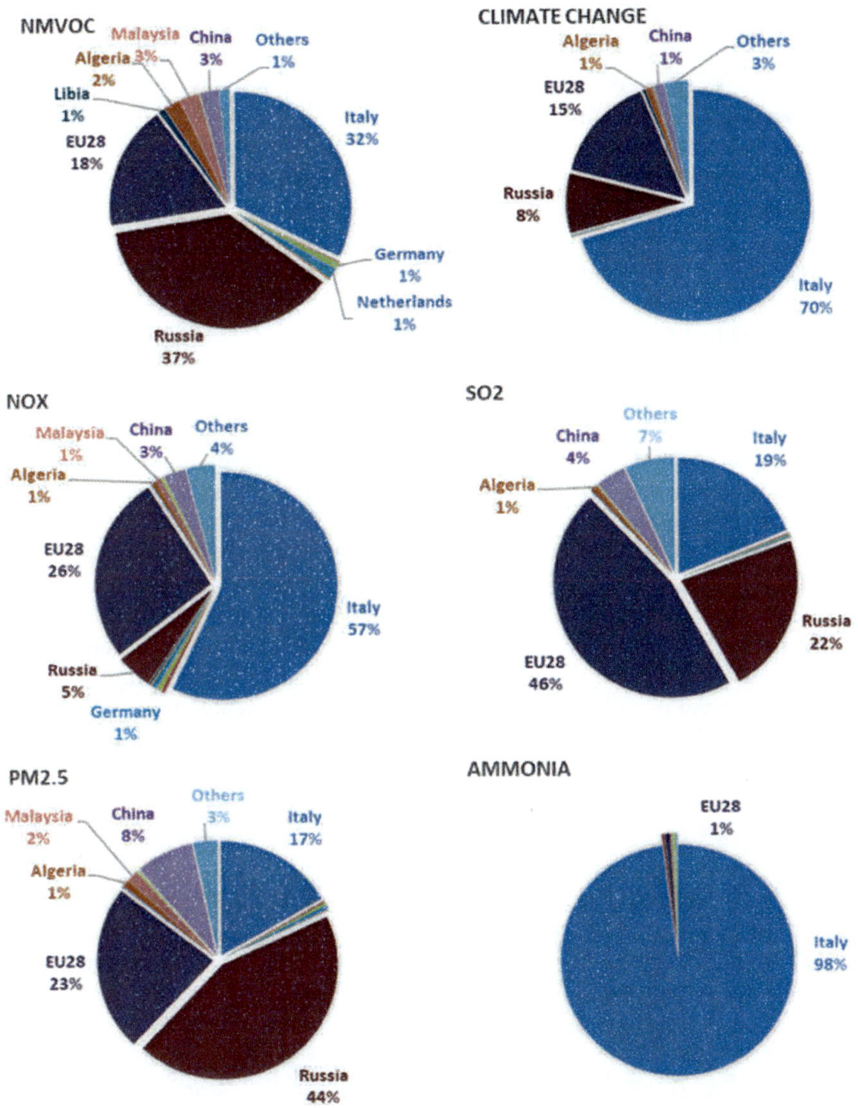


**Fig. 20.3.** 2019 Mix. Region contribution to environmental externalities

of LCA emissions linked to the Italian electricity consumption mix are expected to decrease considerably, notwithstanding world-wide spread processes related to the different supply chains of fuel sources and generation plant components. The proposed indicator seems very promising for many potential policy and economic applications such as:

- Energy transition policy. A long term zero external cost target for the electricity-related life-cycle emissions is consistent with the net zero carbon emissions target before 2050, adopted by the EU Green New Deal, while allowing to optimize also co-benefits related to air pollution mitigation.
- Fiscal policy. The indicator allows a "polluter pays comparison", where pre-tax external costs are compared with current environmental taxes or charges implicitly paid by the final consumers through the price formation chain (such as for emission permits under the Emission Trading Schemes). This can be useful to identify the needs for an environmental fiscal reform as suggested by International Monetary Fund.
- Cost Benefit Analysis (CBA) of public investments in energy transition projects, where electricity is one of the main drivers (electric vehicles' recharging stations, electrified rail transport, cold ironing in ports, etc.). A standardized calculation of the external costs related to electricity taken from the grid, used in energy transition projects, can make the implementation of CBA easier.
- Sustainable finance. The EU Taxonomy on environmentally sustainable activities requires a life-cycle carbon footprint indicator in the electricity sector for demonstrating its contribution to climate change mitigation (Commission Delegated Regulation (EU) 2021/2139 of 4 June 2021 supplementing Regulation (EU) 2020/85). Since the EU taxonomy has been designed to target air pollution mitigation as well, an LCA indicator integrating air pollution and climate change targets in the electricity sector would ease taxonomy implementation, while
- supporting the diffusion of sustainable finance instruments. An example of an application of the suggested indicator in sustainable finance is the paper by Esposito et al. on green mortgages to reduce energy consumptions in the building sector.

## References


Bickel P, Friedrich R (2005) ExternE—externalities of energy, methodology update. European Commission, Bruxelles

Carvalho ML, Marmiroli B, Girardi P (2022) Life cycle assessment of Italian electricity production and comparison with the European context. Energ Rep 8(3):561–568. https://doi.org/10.1016/j.egyr.2022.02.252

European Commission (2014) Guide to cost-benefit analysis of investment projects: economic appraisal tool for cohesion policy 2014–2020. European Commission, Brussels, Belgium

Esposito L, Mastromatteo G, Molocchi A, Brambilla PC, Carvalho ML, Girardi P, Marmiroli B, Mela G (2022) Green mortgages, EU taxonomy and environment risk weighted assets: a key link for the transition. Sustainability 14(1633). https://doi.org/10.3390/su14031633

European Commission (2020) Handbook on external costs of transport. https://doi.org/10.2832/27212

Eurostat (2021) Energy balances June 2021 edition

ExternE (1995) Externalities of energy, vol 1–6. European Commission DGXII, Bruxelles

ExternE (1999) Externalities of energy, vol 7–10. European Commission DGXII, Bruxelles

Gargiulo A, Carvalho ML, Pierpaolo G (2020) Life cycle assessment of Italian energy scenarios to 2030. Energies 15(3):3852. https://doi.org/10.3390/en13153852

Girardi P, Brambilla C, Mela G (2020) Life cycle air emissions external costs assessment for comparing electric and traditional passenger cars. Integr Environ Assess Manag 16. https://doi.org/10.1002/ieam.4211

Groom B, Maddison Pr D (2019) New estimates of the elasticity of marginal utility for the UK. Environ Resour Econ 72:1155–1182. https://doi.org/10.1007/s10640-018-0242-z

GSE (2019) Rapporto statistico 2018—fonti rinnovabili

Ministero dello Sviluppo Economico (2019) Produzione nazionale di idrocarburi—anno 2019

Molocchi A (2020) From production to consumption: an inter-sectoral analysis of air emissions external costs in Italy. Economics and Policy of Energy and the Environment (EPEE) (2):155–180. https://doi.org/10.3280/EFE2020-002007

Molocchi A, Mela G, Brambilla PC, Girardi P (2021) Fattori Di Costo per La Valutazione Delle Esternalità Dei Trasporti Nel Contesto Italiano. Rapporto Aggiuntivo Di Ricerca Di Sistema, Ricerca Sistema Energetico (RSE), Milan

Müller W, Preiss P, Klotz V, Friedrich R (2010) External cost values for EE SUT framework. Final report providing external cost values to be applied in an EE SUT framework. IER University Stuttgart, Stuttgart

Navrud S (2009) Value transfer techniques and expected uncertainties. Project Deliverable, SWECO

Parry IW, Heine MD, Lis E, Li S (2014) Getting energy prices right. From principle to practice. IMF, Washington DC, USA

Preiss P, Friedrich R, Klotz V (2008) Report on the procedure and data to generate averaged/aggregated data. Deliverable n. 1.1—RS 3a, NEEDS project, 6th FRP

Ramsey FP (1928) A mathematical theory of saving. Technical report

SNAM (2020) Report di Sostenibilità 2019

Terna (2019) Dati statistici sull'energia elettrica in Italia

# Chapter 21
# Life Cycle Energy and Climate Change Impacts of a Chicken Slaughtering Process

**Teresa M. Mata, José N. F. G. Rodrigues, Joaquim C. G. Esteves da Silva, and António A. Martins**

**Abstract** The industrial production of poultry meat intended for human consumption has experienced continuous growth worldwide. In order to better understand the environmental performance of this industry and to formulate proposals for environmental improvement, it is of utmost importance the development of Life Cycle Assessment (LCA) studies. Hence, this work aims to evaluate a process of slaughter and preparation of chicken meat for human consumption, in a Portuguese company, following a "gate-to-gate" approach, and focusing on the energy consumption and climate change impacts. The functional unit selected for the study is 1 kg of chicken live weight at the company gate. The inventory data was gathered from real industrial practice, considering 2020 as the reference year. For the potential environmental impacts assessment, information on the characterization factors of the various impact categories was obtained from the EcoInvent V3.5 database, using the SimaPro V8.5.2 software. The results show the variation in energy consumption and carbon footprint throughout the year, depending on the quantity of chickens processed. In particular, it shows that the process is more energy efficient the greater the quantity/weight of

T. M. Mata (✉)
LAETA-INEGI, Associated Laboratory for Energy and Aeronautics, Institute of Science and Innovation in Mechanical and Industrial Engineering, R. Dr. Roberto Frias 400, 4200-465 Porto, Portugal
e-mail: tmata@inegi.up.pt

J. N. F. G. Rodrigues · J. C. G. E. da Silva
Chemistry Research Unit (CIQUP), Institute of Molecular Sciences (IMS), DGAOT, Faculty of Sciences of University of Porto (FCUP), Rua do Campo Alegre s/n, 4169-007 Porto, Portugal
e-mail: jcsilva@fc.up.pt

A. A. Martins
LEPABE, Faculty of Engineering, University of Porto (FEUP), R. Dr. Roberto Frias, S/N, 4200-465 Porto, Portugal

ALiCE, Faculty of Engineering, University of Porto, Rua Dr. Roberto Frias, 4200-465 Porto, Portugal

A. A. Martins
e-mail: aamartins@fe.up.pt

 
N. S. Caetano and M. C. Felgueiras (eds.), *The 9th International Conference on Energy and Environment Research*, Environmental Science and Engineering,
https://doi.org/10.1007/978-3-031-43559-1_21

slaughtered chickens. On average, the total energy consumption per kg of live weight of chicken is 1.18 MJ, to which contributes 67% of electricity, 31% biomass and 2% of propane gas. Therefore, the electricity consumption is what contributes most to the climate change impacts. Thus, future improvements to be proposed should focus on using renewable energy as an alternative or increasing its percentage in the process.

**Keywords** Chicken slaughtering · Energy intensity · Global warming potential

## 21.1 Introduction

There is currently recognition by the scientific community that human activities have a direct impact on climate change. For this reason, the food industry has taken more proactive actions to analyze the potential environmental impacts of its activities. The production of poultry products is an economic sector that has experienced continuous growth worldwide. In 2017, these represented about 122 Mt produced worldwide, accounting for 37% of global meat production, with an expected increase of 40 Mt in 2028 (Costantini et al. 2021). By 2050, the production of poultry meat and eggs is expected to have the highest growth rate among the various livestock sectors. Accompanying this increase, there will be a high volume of by-products and associated waste streams, which needs to be properly managed and valorized (Kanani et al. 2020). Therefore, more studies are needed in this economic sector, to deepen its knowledge and understand its environmental impacts from a broader perspective of cause-effect and enabling the implementation of improvements.

Poultry meat, mainly represented by chicken (89%), plays an important role in human nutrition, especially in developing countries, due to various factors, such as the lower price in comparison to other meat types, it is widely available, unaffected by religious restrictions, and has a high nutritional value (Costantini et al. 2021). The production of chicken meat is heavily reliant on agricultural processes, in which the main contributors to the GHG emissions are feed production (37–53%) and poultry farming (19–39%), followed by transport of components (18.1%) and chicken meat processing (7.2%) (Tan et al. 2014). In addition, LCA studies have recognized poultry products as one of the most environmentally friendly, particularly with regard to energy consumption and global warming potential. For example, the amount of energy used to produce 1 kg of chicken (15–29 MJ) is lower than that used for producing 1 kg of pork (18–45 MJ) and beef (34–52 MJ). Also, when expressing energy use in terms of protein, milk production requires 37–144 MJ/kg, pork 95–236, chicken 80–152, eggs 87–107 and beef 177–273 MJ/kg. In which respects climate change impacts, the production of 1 kg of pork contributes 3.9–10, chicken 3.7–6.9 and beef 14–32 kg $CO_2$-eq, which are directly related with the combustion of fossil energy, $CH_4$ emission and amount of feed needed per kilogram of meat, which are higher for ruminants than for monogastric animals (Vries and Boer 2010). However, LCA studies in this sector are still scarce and are more focused on the agriculture life cycle stage (Prudêncio da Silva et al. 2014; Aan Den Toorn et al. 2017). More studies are needed in this sector, following what is mandated by current strategies to improve the sustainability of food production systems, as for example the European Union "Farm to Fork" (Farm2Fork) strategy (European Commission 2021), whose main objective is to make food systems healthier, fairer, more environmentally friendly (Martins et al. 2021; Monteiro et al. 2020; Moura et al. 2022), with innovative and more sustainable solutions, such as for example integrating renewable energy in the production system (Martins et al. 2020), closed circuit water utilization (Gorgich et al. 2019), soil phytoremediation (Mata et al. 2022), and adding value to the generated by-products (Mata et al. 2017, 2018; Martins et al. 2017). Hence, this work aims to study the life cycle energy and climate change impacts of slaughter and preparation of chicken meat for human consumption in a Portuguese company. In this study, the climate change impacts, also known as Global Warming Potential (GWP) or carbon footprint (Carneiro et al. 2022), are measured in units of carbon dioxide ($CO_2$) equivalents (or $CO_2$-eq), following the LCA methodology (Gautam et al. 2020).

## 21.2 Methods

### *21.2.1 Goal and Scope Definition*

**Study Goal**. This work aims to evaluate the energy consumption and climate change impacts (also named global warming potential or carbon footprint) associated with the process of slaughter and preparation of chicken meat for human consumption, in a Portuguese company, on a "gate-to-gate" approach, following the LCA methodology, according to ISO 14040 (2006) and ISO 14044 (2006).

**Functional Unit**. The functional unit (FU) chosen for this study is 1 kg of chicken live weight at the company gate. It was selected considering the FU commonly used in the published LCA studies in this area. This LCA study is attributive, as the environmental impacts are directly attributed to the functional unit.

**Scope and Assumptions**. The system boundary for this study is shown in Fig. 21.1, including the life cycle stages considered. The first process step is the reception by the company of the live chickens. They arrive in drawers that are arranged for the animal to be hung on a slaughter line. The line follows almost the entire slaughter process, starting with electrocution, continuing with the behead, bleeding, scalding, plucking, evisceration, refrigeration and evaluation of the animal meat. Then, it can be packaged whole or be dismantled and later packaged.

### *21.2.2 Inventory Analysis and Data Sources*

The inventory data was mainly obtained from real industrial practice, in a Portuguese company, with chicken slaughterhouse and meat preparation process for human

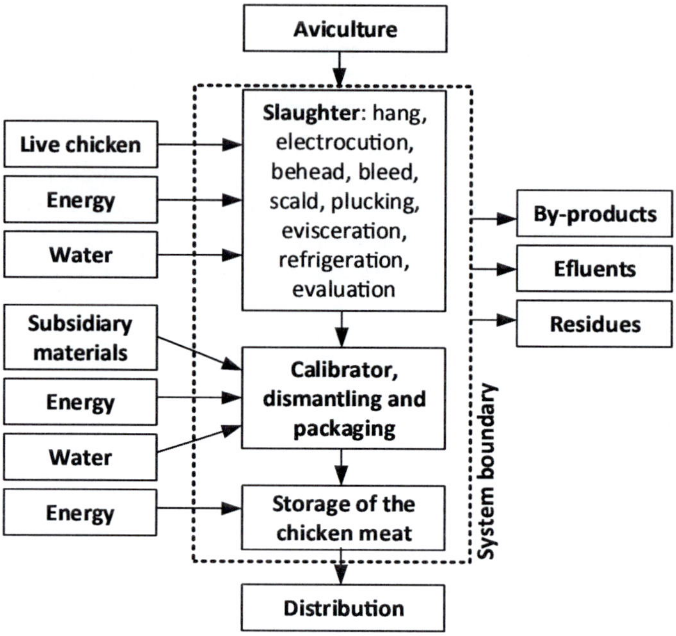


**Fig. 21.1** System boundary definition, including the life cycle stages considered for the LCA study

consumption. In particular, for the reference year 2020, the monthly values of the total live weight of chickens processed, as well as total energy consumption, materials used and waste generated were accounted for. The live weight of the chicken is measured at the entrance to the processing line. For the base case scenario, which is the current situation, the electricity consumption is assumed to be a combination of electricity coming from the national electricity grid and electricity locally generated using photovoltaic (PV) panels. The total PV electricity generated was estimated using geographic information of the company location, and using the webtool PVGIS (https://re.jrc.ec.europa.eu/pvg_tools/en/). Whenever possible, Portuguese conditions were considered, in particular for the energy supply and the production of raw and construction materials, and whenever it was not possible, European regions with similar economic and/or technological development were considered.

This LCA study follows a "gate-to-gate" analysis, from the reception of the live chicken in the slaughterhouse to the storage of the chicken meat. Thus, the life cycle steps considered include the chicken's slaughter, calibration, dismantling, packaging and storage. The life cycle steps of aviculture and distribution is out of this study scope. For the defined life cycle steps, all material, energy and water inputs and emissions to air, water and soil were considered, as they contribute to the energy consumption and greenhouse gas emissions. The generation of energy (electricity or fuel), and the production of auxiliary materials used in the process were also accounted for.

### 21.2.3 *Climate Change Impacts Evaluation*

The climate change impacts, carbon emissions or global warming potential (kg $CO_2$-eq/ FU) was evaluated following the ReCiPe 2016 v1.02 midpoint method (Huijbregts et al. 2016) and the methodologies defined in Garcia et al. (2014) and Martins et al. (2018) were taken into account. The emission factors were obtained from the Simapro V8.5.2 software, using the life cycle inventory database EcoInvent V 3.5.

## 21.3 Results and Discussion

Figure 21.2 presents for the year 2020, the monthly values of the energy consumption per functional unit (kg of chicken live weight), corresponding to the bars and the right-side vertical scale, and the live weight of chicken processed, corresponding to the solid line and the left-side vertical scale.

Figure 21.2 shows that there are strong variations in values throughout the year. Comparing the behaviour throughout the year for both parameters, it can be concluded that when the chicken live weight increases, the total energy consumed per kg of chicken live weight decreases, equivalent to an energy efficiency. Thus, it is possible to conclude that the process is more energy efficient the greater the quantity/weight of slaughtered chickens. A possible explanation for this is a more efficient use of the process equipment in the slaughtering and meat processing lines, in particular closer to their maximum capacity, as the energy consumption of the processing equipment seems to not depend significantly on the quantity (either mass or number) of chickens processed. On average, the total energy consumption per kg

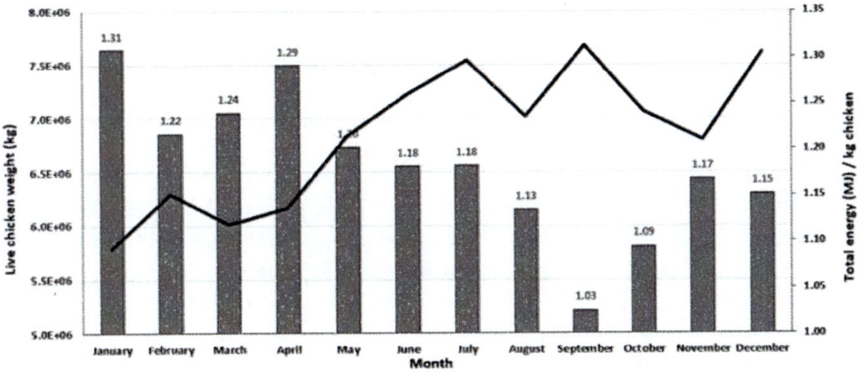


**Fig. 21.2** Monthly energy consumption per kilogram of chicken live weight (bars and right-side vertical scale) and kilograms of live weight of chickens at the company reception (solid line and left-side vertical scale), for year 2020

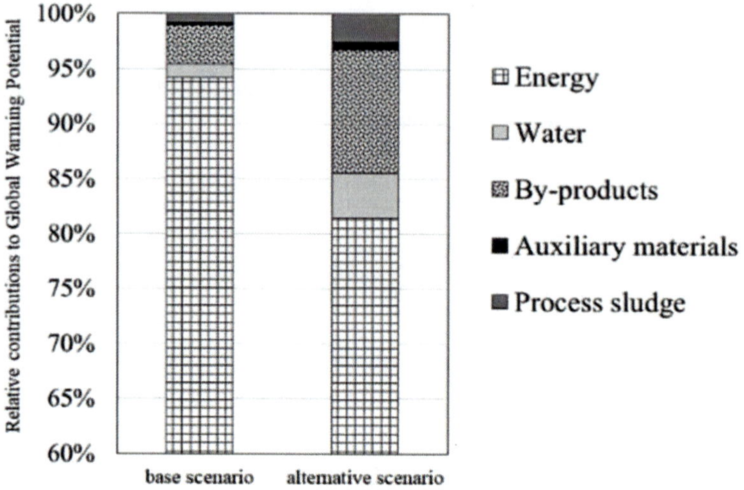


**Fig. 21.3** Relative contribution of the process' inputs (energy, water, animal by-products, auxiliary materials, sludge) to global warming potential, comparing two energy source scenarios: current electricity mix (base scenario), and all electricity from PV system (alternative scenario)

of chicken live weight is 1.18 MJ. This energy includes about 67% of electricity, 31% biomass and 2% of propane gas.

Since the electricity used in the process is the most significant contributor to the GWP, in order to reduce this impact, an alternative scenario was defined, considering that all electricity used in the process was obtained using PV panels. Thus, Fig. 21.3 shows the relative contribution of the process' inputs (energy, water, animal by-products, auxiliary materials, sludge) to the climate change or global warming potential (GWP), comparing the two energy source scenarios:

(1) the current electricity mix obtained from the national electricity grid (base scenario), and
(2) assuming that all the electricity is generated locally in the photovoltaic system (alternative scenario).

From Fig. 21.3 it can be concluded that, as expected, the energy/electricity consumption is the most relevant factor in the climate change impact, followed by processing of by-products generated in the process. For the alternative scenario, which considers only photovoltaic electricity, there is a reduction in the relative importance of energy, however it is still the dominant factor. While PV electricity is often considered carbon-free, when considering the energy and carbon emissions linked to the production of the raw materials used in the manufacture of PV panels, from a life-cycle perspective, the energy generated does indeed have some environmental impacts. Therefore, there is the need to use renewable energy with a lower life cycle environmental impact.

## 21.4 Conclusion

This work evaluates the life cycle energy consumption and climate change impacts associated with chicken slaughter and meat processing for human consumption in a Portuguese company. The study is based on real industrial data supplied by the company. It can be concluded that electricity consumption is the main factor (67%) controlling the life cycle energy and carbon emissions of the process. The calculations revealed a positive correlation between of the process energy efficiency and the total quantity/weight of chickens processed. The change from the actual electricity mix to completely PV generated electricity reduced the climate change impacts, yet energy remains the dominant factor contributing to the process' global warming potential. To further reduce the process carbon emissions there is a need to use other renewable electricity sources with lower life cycle carbon emissions when compared with the PV electricity.

**Acknowledgements** This work was financially supported by base Funding of the following projects: UIDB/00081/2020 (CIQUP), LA/P/0056/2020 (IMS-Institute of Molecular Sciences), LA/P/0045/2020 (ALiCE), UIDB/00511/2020 (LEPABE) and UIDB/50022/2020 (LAETA), funded by national funds through FCT/MCTES (PIDDAC). António Martins gratefully acknowledge the Portuguese national funding agency for science, research and technology (FCT—Fundação para a Ciência e a Tecnologia) for the financial support through program DL 57/2016—Norma transitória. Teresa Mata gratefully acknowledge the funding of Project NORTE-06-3559-FSE-000107, cofinanced by Programa Operacional Regional do Norte (NORTE2020), through Fundo Social Europeu (FSE).

## References

Aan Den Toorn SI, Van Den Broek MA, Worrell E (2017) Decarbonising meat: exploring greenhouse gas emissions in the meat sector. Energ Procedia 123:353–360. https://doi.org/10.1016/j.egypro.2017.07.268

Carneiro AL, Martins AA, Duarte VCM, Mata TM, Andrade L (2022) Energy consumption and carbon footprint of perovskite solar cells. Energ Rep 8:475–481. https://doi.org/10.1016/j.egyr.2022.01.045

Costantini M, Ferrante V, Guarino M, Bacenetti J (2021) Environmental sustainability assessment of poultry productions through life cycle approaches: a critical review. Trends Food Sci Technol 110:201–212. https://doi.org/10.1016/j.tifs.2021.01.086

de Vries M, de Boer IJM (2010) Comparing environmental impacts for livestock products: a review of life cycle assessments. Livest Sci 128:1–11. https://doi.org/10.1016/j.livsci.2009.11.007

EU Farm to Fork Strategy—for a fair, healthy and environmentally-friendly food system. European Commission. Available online https://ec.europa.eu/food/farm2fork_en. Accessed on 2 Aug 2021

Garcia R, Marques P, Freire F (2014) Life-cycle assessment of electricity in Portugal. Appl Energ 134:563–572. https://doi.org/10.1016/j.apenergy.2014.08.067

Gautam A, Mata TM, Martins AA, Caetano NS (2020) Evaluation of areca palm renewable options to replace disposable plastic containers using life cycle assessment methodology. Energ Rep 6:80–86. https://doi.org/10.1016/j.egyr.2019.08.023

Gorgich M, Mata TM, Martins A, Caetano NS, Formigo N (2019) Application of domestic greywater for irrigating agricultural products: a brief study. Energ Rep. https://doi.org/10.1016/j.egyr.2019.11.007

Huijbregts M, Steinmann ZJN, Elshout PMFM, Stam G, Verones F, Vieira MDM, Zijp M, van Zelm R (2016) ReCiPe 2016—a harmonized life cycle impact assessment method at midpoint and endpoint level. Report I: Characterization. Available online https://www.rivm.nl/bibliotheek/rapporten/2016-0104.pdf. Accessed on 12 May 2022

ISO 14040 (2006) Environmental management—life cycle assessment—principles and framework. International Organization for Standardization

ISO 14044 (2006) Environmental management—life cycle assessment—requirements and guidelines. International Organization for Standardization

Kanani F, Heidari MD, Gilroyed BH, Pelletier N (2020) Waste valorization technology options for the egg and broiler industries: a review and recommendations. J Clean Prod 262:121129. https://doi.org/10.1016/j.jclepro.2020.121129

Martins AA, Pinto F, Caetano NS, Mata TM (2017) Acidity reduction in animal fats by enzymatic esterification: economic and environmental analysis. Energ Procedia 136:308–315. https://doi.org/10.1016/j.egypro.2017.10.258

Martins AA, Simaria M, Barbosa J, Barbosa R, Silva DT, Rocha CS, Mata TM, Caetano NS (2018) Life cycle assessment tool of electricity generation in Portugal. Environ Dev Sustain 20:129–143. https://doi.org/10.1007/s10668-018-0179-y

Martins AA, Mota MG, Caetano NS, Mata TM (2020) Decentralized electricity storage evaluation in the Portuguese context. Electr J 33:106822. https://doi.org/10.1016/j.tej.2020.106822

Martins AA, Andrade S, Correia D, Matos E, Caetano NS, Mata TM (2021) Valorization of agro-industrial residues: bioprocessing of animal fats to reduce their acidity. Sustain 13:1–18. https://doi.org/10.3390/su131910837

Mata TM, Correia D, Pinto A, Andrade S, Trovisco I, Matos E, Martins AA, Caetano NS (2017) Fish oil acidity reduction by enzymatic esterification. Energ Procedia 136:474–480. https://doi.org/10.1016/j.egypro.2017.10.306

Mata TM, Rodrigues S, Caetano NS, Martins AA (2022) Life cycle assessment of bioethanol from corn stover from soil phytoremediation. Energ Rep 8:468–474. https://doi.org/10.1016/j.egyr.2022.01.059

Mata TM, Pinto F, Caetano N, Martins AA (2018) Economic and environmental analysis of animal fats acidity reduction by enzymatic esterification. J Clean Prod 184. https://doi.org/10.1016/j.jclepro.2018.02.253

Monteiro H, Moura B, Iten M, Mata TM, Martins AA (2020) Life cycle energy and carbon emissions of ergosterol from mushroom residues. Energ Rep 6:333–339. https://doi.org/10.1016/j.egyr.2020.11.157

Moura B, Monteiro H, Mata TM, Martins AA (2022) Life cycle energy and carbon emissions of essential oil extraction from Rosemary. Energ Rep 8:291–297. https://doi.org/10.1016/j.egyr.2022.01.063

Prudêncio da Silva V, van der Werf HMG, Soares SR, Corson MS (2014) Environmental impacts of French and Brazilian broiler chicken production scenarios: an LCA approach. J Environ Manage 133:222–231. https://doi.org/10.1016/j.jenvman.2013.12.011

Tan MQB, Tan RBH, Khoo HH (2014) Prospects of carbon labelling—a life cycle point of view. J Clean Prod 72:76–88. https://doi.org/10.1016/j.jclepro.2012.09.035

# Chapter 22
# Life Cycle Energy and Carbon Footprint of Native Agar Extraction from *Gelidium sesquipedale* Using Alternative Technologies

**Sara G. Pereira, Teresa M. Mata, Ricardo N. Pereira, José A. Teixeira, Cristina M. R. Rocha, and António A. Martins**

**Abstract** The search for sustainable extraction processes led to the development of innovative technologies with less environmental impact. Subcritical water extraction (SWE) and ohmic heating (OH) emerged as novel extraction technologies with higher selectivity and efficiency. The red algae *Gelidium sesquipedale* is mainly known for agar, its main hydrocolloid, widely used as a gelling agent in the food industry.

S. G. Pereira · R. N. Pereira · J. A. Teixeira · C. M. R. Rocha
CEB-Centre of Biological Engineering, University of Minho, Campus Gualtar, 4710-057 Braga, Portugal

LABBELS-Associated Laboratory, Braga, Guimarães, Portugal

S. G. Pereira
e-mail: sarapereira@ceb.uminho.pt

R. N. Pereira
e-mail: rpereira@deb.uminho.pt

J. A. Teixeira
e-mail: jateixeira@deb.uminho.pt

C. M. R. Rocha
e-mail: cmrocha@ceb.uminho.pt

T. M. Mata (✉)
LAETA-INEGI, Associated Laboratory for Energy and Aeronautics, Institute of Science and Innovation in Mechanical and Industrial Engineering, R. Dr. Roberto Frias 400, 4200-465 Porto, Portugal
e-mail: tmata@inegi.up.pt

A. A. Martins
LEPABE, Faculty of Engineering, University of Porto (FEUP), R. Dr. Roberto Frias, S/N, 4200-465 Porto, Portugal
e-mail: aamartins@fe.up.pt

ALiCE, Faculty of Engineering, University of Porto, Rua Dr. Roberto Frias, 4200-465 Porto, Portugal

 
N. S. Caetano and M. C. Felgueiras (eds.), *The 9th International Conference on Energy and Environment Research*, Environmental Science and Engineering,
https://doi.org/10.1007/978-3-031-43559-1_22

Although SWE and OH are considered to be more sustainable than existing extraction technologies, an objective and quantitative assessment of their environmental performance is needed. Hence, this work aims to evaluate the carbon footprint of SWE and OH for native agar extraction, following the life cycle assessment methodology. The study focuses on native agar extraction, on a "gate-to-gate" approach, comparing the two extraction technologies. Concerning the process operating conditions, the OH extraction technology operates for 180 min at 95 °C, while SWE operates for 1 s at 140 °C. The functional unit is 1 kg of agar extracted, considering the experimental variability. Results show that OH has a lower carbon footprint than SWE (about 30% less, even considering the experimental variability). Considering two different electricity scenarios, the OH carbon footprint is 45 and 6 kg $CO_2$ eq/kg agar for, respectively, the process powered by the Portuguese electricity mix and by Photovoltaic electricity.

**Keywords** Agar extraction · Carbon footprint · Sustainable technologies

## Nomenclature

| | |
|---|---|
| OH | Ohmic heating |
| MEF | Moderate electric fields |
| SWE | Subcritical water extraction |
| LCA | Life cycle assessment |
| LCI | Life cycle inventory |
| LCIA | Life cycle impact assessment |

## 22.1 Introduction

Seaweed is a natural, renewable resource with interesting nutrients and many health benefits. Among the great variability of species, agarophytes (red algae) are mainly recognized for their agar content, an hydrocolloid present in the cell wall, with good thickening and gelling properties, used mainly in the food industry and in biotechnological applications (Shannon and Abu-Ghannam 2019; Matos et al. 2021). Ohmic heating (OH), involving the application of moderate electric fields (MEF), and subcritical water extraction (SWE), are alternative extraction technologies, with several advantages such as: low cost, high selectivity, short extraction times, high energy transfer and extraction efficiency.

SWE is a hydrothermal treatment, based on water's properties at subcritical conditions, involving high temperatures (100–374 °C) and pressures (1–22 MPa). SWE can be considered an environmentally friendly technology in several aspects. It utilizes water as solvent, eliminating the need for organic solvents that may be harmful to human health and the environment, minimizing the generation of chemical waste and reducing the potential for environmental contamination. The selectivity of SWE allows for targeted extraction of desired compounds while minimizing the extraction of unwanted components, reducing waste and optimizing resource utilization. Its versatility in extracting valuable compounds from various raw materials promotes a sustainable use of resources. It's important to note that the environmental friendliness of subcritical water extraction also depends on specific operational parameters, such as temperature, pressure and extraction time, being always necessary to evaluate its environmental impact (Matos et al. 2021).

OH is a technology that, by applying moderate electric fields, improves the permeabilization of algae cell membranes and allows for rapid and uniform heating of a sample, which can result in reduced processing times and energy requirements (Álvarez-Viñas et al. 2021; Kusnadi and Sastry 2012; Pereira et al. 2016; Pereira and Vicente 2010). OH can be considered a green technology in certain contexts, being also essential to evaluate its environmental impact. It is important to note that the overall environmental impact of OH depends on various factors such as the energy source used, the specific application, and the life cycle analysis of the process equipment used. In order to maximize the green credentials of OH the electricity used should ideally come from renewable sources. It is important to assess the sustainability of these new green technologies in the early stages of the process development and assess whether it is feasible to scale it up to industrial scale. Therefore, careful consideration of factors such as energy source, equipment life cycle, and waste management is necessary to fully assess its environmental impact. For this purpose, Life cycle assessment (LCA) is a data-intensive methodology to evaluate the potential environmental impacts of a product or process, and to better understanding their life cycle stages. LCA depends on four main steps (Baumann and Tillman 2004): (i) goal and scope definition, (ii) life cycle inventory analysis (LCI), (iii) life cycle impact assessment (LCIA) and (iv) interpretation of results.

This work aims to assess and compare, on a life cycle thinking perspective, the carbon footprint of two innovative technologies, SWE and OH, for the extraction of native agar from seaweed. Results allow the identification of the process environmental hot-spots and possibilities for its optimization and improvement in order to further develop these technologies.

## 22.2 Materials and Methods

### *22.2.1 Raw Material Preparation*

The red seaweed *Gelidium sesquipedale* was supplied by Iberagar—Sociedade Luso-Espanhola de Colóides Marinhos S.A. (Coina, Portugal). Upon receipt, it was manually cleaned of all foreign and coarse materials (e.g., sand, debris, salt) and washed three times with distilled water to remove excess salt and smaller debris. It was then dried in the oven at 40 °C and stored in vacuum sealed bags at room temperature in a dry, dark place until use.

### 22.2.2 *Agar Extraction Processes*

The native agar extraction experiments were conducted using two reactor types:

(1) A 1.9 L, 4520 Stirred Pressurized Bench Top Reactor (Parr Instruments Company, Moline, Illinois, USA) was used for the SWE, without applying electric field, and
(2) A 400 mL pressurized PTFE (Polytetrafluoroethylene) reactor with two stainless steel electrodes was used for the OH extraction.

In both reactor types, the experiments were performed using *Gelidium sesquipedale* biomass, keeping the same parameters as follows:

- 100% distilled water as solvent,
- 1:30 of solid/solvent ratio,
- 400 mL final volume,
- 3.5 mS/cm conductivity.

A 20 kHz frequency and 1–12 V/cm electric field were used only for the OH extraction.

The extraction operating conditions selected for SWE were 140 °C for 1 s, while for OH 95 °C for 180 min were selected.

A conventional extraction (CE) using hot water was performed as a control (at 95 °C, for 180 min) (Öğretmen and Duyar 2018).

After each extraction protocol, the hot mixture obtained was filtered through a cotton filter cloth and frozen at −20 °C overnight. The native agar was separated from the solution upon freezing and thawing (syneresis water), washed, dehydrated with ethanol (96%), dried overnight at 60 °C and milled in a coffee grinder (150 W). The extraction procedures were carried out in triplicate.

### 22.2.3 *Goal and Scope Definition for the LCA Study*

**Study Goal**. This study aims to evaluate the carbon footprint of two innovative technologies, SWE and OH, for native agar extraction from *Gelidium sesquipedale*, following the LCA methodology, according to ISO 14040 (2006a) and ISO 14044 (2006b).

**Functional Unit**. The functional unit (FU) chosen for this study is 1 kg of dry agar powder. It was selected considering the FU commonly used in published LCA studies

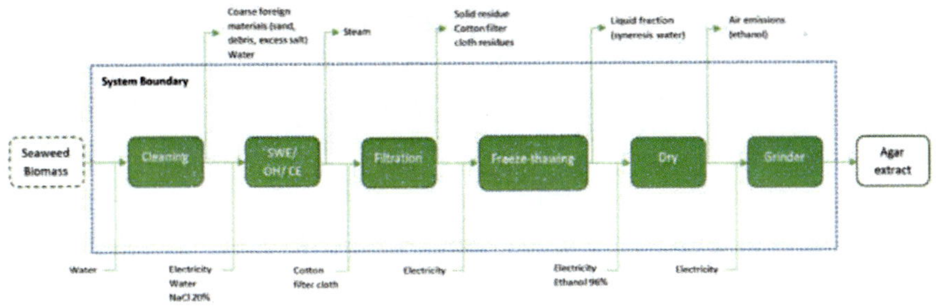


**Fig. 22.1** System boundary for the LCA study, showing the process steps for agar extraction and the fluxes of materials and energy across it

in this area. The study is attributive, as the carbon footprint will be allocated to the functional unit.

**Scope and Assumptions.** This study focuses on the seaweed preparation and agar extraction technologies, following a "gate-to-gate" approach. The remaining life cycle steps of seaweed cultivation, harvesting and extracts utilization, were not considered in the system boundary for the study, because its main goal is to compare the carbon footprint of the two agar extraction technologies, SWE and OH, and the remaining life cycle stages would have similar contribution.

Thus, the system boundary for the study is represented by the process diagram of Fig. 22.1.

This study considers the energy generation (only electricity is used in the process), the production of solvents and other materials used in the extraction process, and their transportation to the extraction site. Also, the carbon emissions associated with equipment's manufacture and the carbon emissions associated with the treatment of the liquid fraction were accounted for. Whenever possible, Portuguese conditions were considered, if not European conditions were used instead.

### *22.2.4 Life Cycle Inventory Analysis and Data Sources*

The inventory analysis was developed mostly based on primary data from laboratorial experiments conducted by the authors of this work, and using information from the equipment used, complemented whenever necessary with secondary data from the literature.

Inventory data for the electricity generation, production of solvents and other materials used in the agar extraction process, and also for their transportation, were obtained from the LCI database EcoInvent V3.5 (Wernet et al. 2016).

### 22.2.5 Environmental Impact Evaluation

The carbon footprint was estimated using the ReCiPe 2016 Midpoint V1.02 (Huijbregts et al. 2016) methodology using the equalitarian perspective. The emission factors for the carbon footprint evaluation were obtained from the Simapro V8.5.2 software, using the life cycle inventory database of EcoInvent V3.5.

As three replicas were done for each experimental set of conditions, it was possible to have a measure of the variability/uncertainty of the carbon footprint values of each extraction process. Thus, in the results section the average values of the variability are graphically presented.

## 22.3 Results and Discussion

In this work two different scenarios were analyzed for the electricity used in the agar extraction processes:

(1) Low voltage electricity, supplied by the Portuguese electrical distribution grid (Electricity Mix) and
(2) Electricity locally generated, using standard silicon photovoltaic solar panels (Electricity PV Renewable).

Even though the Portuguese electricity mix contains a very significant percentage of renewable electricity, around 50–60% of the annual consumption (APREN 2021), there is still a significant part generated using fossil fuels (in particular natural gas). Thus, it is expected that the substitution of the Portuguese electricity mix by 100% photovoltaic energy will lower the carbon footprint.

Table 22.1 shows the quantitative inventory of the energy required for the extraction technologies regarding a functional unit of 1 kg of dry agar powder extracted from seaweed biomass.

As shown in Table 22.1, the CE technology requires the most energy (3078 MJ) to produce 1 kg of agar, which is reflected in higher carbon emission values.

**Table 22.1** Inventory analysis of energy for extraction technologies considering 1 kg of dry agar powder as functional unit

| Extraction process | Energy inventory | | | |
|---|---|---|---|---|
| | Process (MJ) | Oven drier (MJ) | Freezer (MJ) | Grinder (MJ) |
| CE, 95 °C | 3078 ± 107 | 25.57 | 1.64 | 1.0 ± 0.04 |
| SWE, 140 °C | 1116 ± 93 | | | 0.8 ± 0.07 |
| OH, 95 °C | 370 ± 100 | | | 1.1 ± 0.21 |

CE = conventional extraction with hot water performed in this study as control; SWE = subcritical water extraction; OH = ohmic heating. Results are expressed in Megajoule (MJ). Data presented as mean ± standard deviation from three replicates

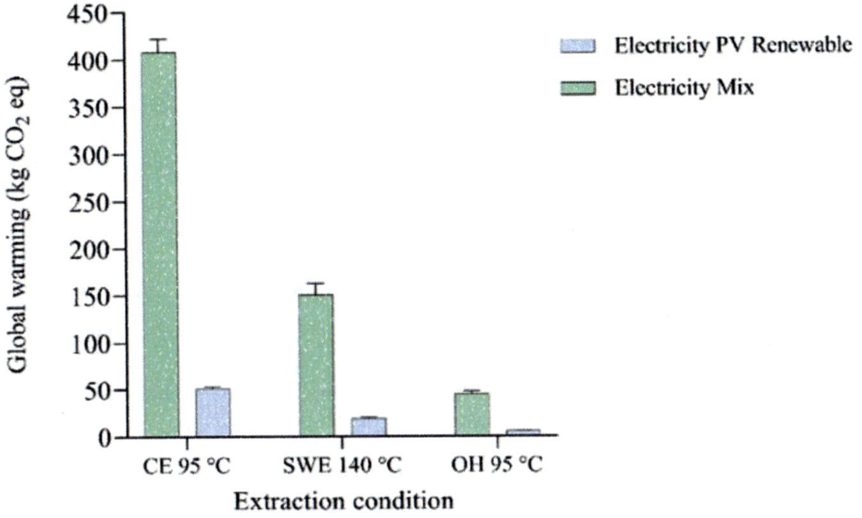


**Fig. 22.2** Comparison of the carbon footprint per functional unit of the three agar extraction technologies, conventional extraction (CE), subcritical water extraction (SWE) and ohmic heating (OH), including the variability values

Figure 22.2 presents the average values and the variability of the carbon footprint values for the extraction technologies considered in this work. As expected, high carbon footprint values per functional unit were obtained, since they report to experiments carried out in the laboratory. Significant reductions are expected when the process is scaled up to industrial production.

The graph of Fig. 22.2 shows that, comparing the three extraction technologies (CE, SWE and OH), and the two electricity scenarios analyzed (Portuguese electricity mix vs PV electricity), the extraction process by OH at 95 °C is the one that contributes the least to carbon footprint, followed by SWE at 140 °C and finally by CE at 95 °C. Even considering the experimental variability into the calculations, it is possible to conclude that OH is the best process with the lowest carbon footprint.

Regarding the change of the electricity source, the results show a significant reduction (about 88% lower) in the carbon footprint when using photovoltaic energy, even considering the experimental variability, confirming the hypothesis formulated earlier that this alternative energy source would decrease the process carbon footprint. This carbon footprint reduction (88%) is the same for the three extraction technologies (OH, SWE and CE), as it was only driven by a change in the type of electricity source used. Since the production of photovoltaic silicon solar cells panels is a very energy demanding process, and the life cycle carbon emissions of photovoltaic electricity generation were considered, the reduction achieved in this study can be even more significant if other types of renewable electricity sources can be used, with lower life cycle carbon emissions, such as wind energy.

Therefore, the OH carbon footprint is 45 and 6 kg $CO_2$ eq/kg agar for, respectively, the process powered by the Portuguese electricity mix and by PV electricity. The SWE carbon footprint is 150 and 19 kg $CO_2$ eq/kg agar for, respectively, the process powered by the Portuguese electricity mix and by PV electricity. And the CE carbon footprint is 408 and 51 kg $CO_2$ eq/kg agar for, respectively, the process powered by the Portuguese electricity mix and by PV electricity.

Although the SWE temperature (140 °C) is higher than for the CE process (95 °C), naturally requiring more energy to heat the system and thus, generating more carbon emissions, the SWE process is faster (1 s of processing time, non-isothermal) and the increased extraction efficiency compensates, leading to a lower carbon footprint. Comparing SWE with OH, it is important to note that SWE will have to operate in a pressurized system, with the corresponding higher equipment costs. Additionally, operating at higher temperature, increases the process energy intensity and thus, the contribution of SWE to the carbon footprint is greater than that of OH.

Similarly, Martínez-Sanz et al. (2020) compared the environmental performance (Global Warming Potential) between different extraction methods—conventional, ultrasound and microwave—using *Gelidium sesquipedale* biomass to produce 1 kg of agar, showing that the microwave technology consumed high amounts of electricity, the semi-refined processes (without alkaline pretreatment) had lower emissions of $CO_2$-eq than the refined process and the electricity was the major contributor to global warming. Vijay Anand et al. (2018) described that the processing of agar production from *Gracilaria edulis* is the step that contributed the most to the environmental impacts, ranging between 65 and 99% of the total impacts (with 54.9 kg $CO_2$-eq) and the electricity is the responsible for the main proportion of impacts. Other work described by Cassiani-Cassiani et al. (2018) performed an environmental analysis of agar production from *Gracilaria* sp., showing that the environmental impacts were higher when the electricity used is included in the assessment and the highest value obtained is 107 PEI (potential environmental impacts)/hr of atmospheric impacts.

## 22.4 Conclusion

This work evaluated the carbon footprint of agar extraction from the seaweed *Gelidium sesquipedale*, comparing alternative extraction technologies (subcritical water extraction and ohmic heating) with the conventional extraction. A "gate-to-gate" study was developed based on experimental data, complemented with information from the literature, analyzing the influence of the electricity source. Results showed that the OH extraction at 95 °C has the lowest contribution to carbon footprint, followed by SWE at 140 °C and then by the conventional extraction at 95 °C. The use of photovoltaic (solar) energy instead of the Portuguese electricity mix to power the extraction processes significantly reduces (by 88%) the carbon footprint of the three extraction technologies analyzed.

**Acknowledgements** This work was supported by the Portuguese Foundation for Science and Technology (FCT) under the scope of the strategic funding of UID/BIO/04469/2019 (CEB) unit and by FCT, the European Fund for Regional Development (FEDER) and COMPETE 2020—Competitiveness and Internationalization Operational Program, under the scope of the project OH2O (POCI-01-0145-FEDER-029145, PTDC/EQU-EQU/029145/2017). This work was financially supported by base Funding of the following projects: LA/P/0045/2020 (ALiCE), UIDB/00511/2020 (LEPABE) and UIDB/50022/2020 (LAETA), funded by national funds through FCT/MCTES (PIDDAC). Sara G. Pereira acknowledges FCT for the scholarship 2021.07623.BD. António Martins gratefully acknowledges FCT for the financial support through program DL 57/2016—Norma transitória. Teresa Mata gratefully acknowledges the funding of Project NORTE-06-3559-FSE-000107, cofinanced by Programa Operacional Regional do Norte (NORTE2020), through Fundo Social Europeu (FSE). Ricardo N. Pereira acknowledges FCT for its Assistant Research program under the scope of Scientific Stimulus Employment with reference CEECIND/02903/2017.

## References

Álvarez-Viñas M et al (2021) Subcritical water for the extraction and hydrolysis of protein and other fractions in biorefineries from agro-food wastes and algae: a review. Food Bioprocess Technol 14(3):373–387

APREN (2021) Associação de Energias Renováveis. Evolução da Potência Instalada e da Produção Elétrica em Portugal. [Online]. Available https://www.apren.pt/pt/energias-renovaveis/destaques. Accessed: 16-Jun-2022

Baumann H, Tillman A-M (2004) The Hitch Hiker's guide to LCA—an orientation in life cycle assessment methodology and application

Cassiani-Cassiani D, Meza-González DA, González-Delgado ÁD (2018) Environmental evaluation of agar production from macroalgae *Gracilaria* sp. Chem Eng Trans 70:2005–2010

Huijbregts M et al (2016) ReCiPe 2016. Natl Inst Public Heal Environ 194

ISO 14040 (2006a) Environmental management—life cycle assessment—principles and framework. International Organization for Standardization

ISO 14044 (2006b) Environmental management—life cycle assessment—requirements and guidelines. International Organization for Standardization

Kusnadi C, Sastry SK (2012) Effect of moderate electric fields on salt diffusion into vegetable tissue. J Food Eng 110(3):329–336

Martínez-Sanz M et al (2020) Alternative protocols for the production of more sustainable agar-based extracts from *Gelidium sesquipedale*. Algal Res 55

Matos GS, Pereira SG, Genisheva ZA, Gomes AM, Teixeira JA, Rocha CMR (2021) Advances in extraction methods to recover added-value compounds from seaweeds: sustainability and functionality. Foods 10:516

Öğretmen ÖY, Duyar HA (2018) The effect of different extraction methods and pre-treatments on agar yield and physico-chemical properties of Gelidium latifolium (Gelidiaceae, Rhodophyta) from Sinop Peninsula Coast of Black Sea, Turkey. J Appl Phycol 30(2):1355–1360

Pereira RN et al (2016) Effects of ohmic heating on extraction of food-grade phytochemicals from colored potato. LWT Food Sci Technol 74

Pereira RN, Vicente AA (2010) Environmental impact of novel thermal and non-thermal technologies in food processing. Food Res Int 43(7):1936–1943

Shannon E, Abu-Ghannam N (2019) Seaweeds as nutraceuticals for health and nutrition. Phycologia 58(5):563–577

Vijay Anand KG, Eswaran K, Ghosh A (2018) Life cycle impact assessment of a seaweed product obtained from *Gracilaria edulis*—a potent plant biostimulant. J Clean Prod 170:1621–1627

Wernet G, Bauer C, Steubing B, Reinhard J, Moreno-Ruiz E, Weidema B (2016) The ecoinvent database version 3 (part I): overview and methodology. Int J Life Cycle Assess 21(9):1218–1230

# Chapter 23
# Life Cycle Assessment of Nanotechnology: Carbon Footprint and Energy Analysis

**S. Alves, M. Gonçalves, Helena Monteiro, Bruna Moura, R. Godina, and J. Almeida**

**Abstract** Nanomaterials are widely applied to improve the performance of technological products and services due to their optical, magnetic, and electrical properties, and cost-efficiency. Although an extensive literature has focused on validating the technical feasibility of nanoparticle applications, the sustainability of these is an equally important topic, not deeply addressed yet. Therefore, this study aims to present a critical review of life cycle assessment studies concerning the use of nanomaterials in technologies. The gaps and barriers of these approaches are discussed, as well as the environmental performance of nanoparticles uses. Herein, the environmental hotspots of the processes involved are highlighted. The results demonstrated that energy-consuming stages are the main contributors to global warming and cumulative energy demand categories. In particular, the production of raw materials and the embodied energy in solvents and precursors contribute to 62% and 80%, respectively, of the cumulative energy demand impacts. To improve the environmental performance of nanomaterials applications, the adoption of the best available techniques is a key factor to mitigate the burdens observed. Furthermore, the upscaling of bench and pilot approaches and the reuse and recycling of the resources, in a circular economy perspective, could mitigate the environmental impacts in 20% and 73%, respectively.

S. Alves (✉)
Department of Mechanical and Industrial Engineering, School of Sciences and Technology, Universidade NOVA de Lisboa, Caparica Campus, 2829-516 Caparica, Portugal
e-mail: sdi.alves@campus.fct.unl.pt

S. Alves · M. Gonçalves · H. Monteiro · B. Moura · J. Almeida
Low Carbon and Resource Efficiency, R&Di, Instituto de Soldadura e Qualidade, 4415-491 Grijó, Portugal

R. Godina
UNIDEMI—Research and Development Unit for Mechanical and Industrial Engineering, Department of Mechanical and Industrial Engineering, NOVA School of Science and Technology, Universidade NOVA de Lisboa, 2829-516 Caparica, Portugal

Laboratório Associado de Sistemas Inteligentes, LASI, 4800-058 Guimarães, Portugal

 
N. S. Caetano and M. C. Felgueiras (eds.), *The 9th International Conference on Energy and Environment Research*, Environmental Science and Engineering,
https://doi.org/10.1007/978-3-031-43559-1_23

**Keywords** Energy · Life cycle assessment · Climate change · Cumulative energy demand · Nanomaterials · Sustainability

## 23.1 Introduction

Nanotechnology is a scientific research field that is booming, and it is expected to present significant cross-sectorial technological progress and breakthrough applications. In particular, nanomaterials (NM) have been employed to develop new coatings and surface treatments with advanced functionalities, lightweighting of components, increasing material strength, new thinner, flexible electronics, renewable energy generation films, and selective environmental remediation techniques to capture pollutants from air and wastewater streams, among others (Prakash Sharma et al. 2018).

Being an all-pervasive technology, with a vast number of potential applications that surpass sectoral boundaries, the market interest in such developments is steadily growing (Nizam et al. 2021). This burgeoning area will play a dominant role in global manufacturing in bioremediation, energy, medicine, construction, automotive, agriculture, and entertainment fields, which raises doubts regarding the potential environmental impacts associated to the new engineered NM and their future implications towards sustainability.

Nanomaterials can be applied during the removal and recovery of metals from wastewater. These elements, namely heavy metals, are related to harmful effects on public health and the environment, their eradication being crucial to promote sustainable interactions between human and water resources (Pesce et al. 2021). In recent years, efforts are being carried out to develop magnetic nanoparticles that can effectively capture metals from industrial wastewater. One example is the study developed by Pesce et al. (2021), where functionalized magnetic nanoparticles were tested to capture copper and zinc ions from wastewater. Furthermore, the magnetic aggregates regeneration through an electrochemical process, to recover the pollutants, was complied.

Nanomaterials are also used in renewable energy technologies to power electricity grids, where their integration can contribute to greater energy storage and conversion efficiency (Garvey et al. 2019). Other cases include high-energy nanoengineered batteries for next-generation electronics and implantable medical equipment (Dhingra et al. 2010), as well as energy production and storage-related applications, due to electrical and mechanical properties of nanotechnology (Wender and Seager 2011). This represents a step forward to stakeholders, researchers, and regulators, who are seeking sustainable alternative sources of energy (Garvey et al. 2019).

Monteiro et al. (2022) showed that the application of NM in concrete recipes can result in concrete blends having lower cement content, which is an energy- and carbon-intensive material. Additional reported benefits are a higher compressive strength, increased incorporation of secondary materials and/or additional functionalities as self-compacting, self-cleaning, self-sensing, and self-healing that are expected to have potential advantages during the use stage. Nano silica, known by its pozzolanic activity, is one of the most studied NM to be incorporated in concrete. Some studies present LCA of concrete with nano-silica (Monteiro et al. 2022).

The developments in NM are trying to promote more sustainable products, but on the other hand the energy and environmental implications of NM production and use are not fully understood yet. To analyze this, previous works have suggested the use of life cycle assessment (LCA) methodology to estimate the environmental burdens of NM throughout different stages of their life cycle: production, material use, potential release during lifetime till the end-of-life. However, these studies are still scarce. To address this gap, the present study is aligned to ongoing Horizon 2020 research projects PureNano—develop of fast and low-cost method for purification of spent plating baths and reuse of plating solutions and metal ions (PureNano-H2020 project 2022), Biomac—technologies and solutions using nano-enabled bio-based materials upscale and preparation for market applications (BIOMAC 2022) and Lightcoce—building an ecosystem for the up-scaling of lightweight multi-functional concrete and ceramic materials and structures (Lightcoce 2022), and reviews available of LCA studies in fields as environmental remediation, construction, and energy generation to identify the environmental benefits and critical hotspots of NM. The existing challenges of using LCA to analyze NM are also highlighted as areas for further research.

## 23.2 Methodology

A search on NM studies concerning an LCA perspective was performed (June 2022) with the online version of Scopus. With the search words "life cycle assessment", "nanomaterials" and "energy", a total of 64 scientific papers were obtained from 2006 to 2022. The dominant research areas were environmental sciences (20.3%), engineering (14.5%) and energy (13.4%), being the United States the country with the most scientific publications.

Different literature studies covering the application on different sectors were found, namely considering NM production, concrete incorporation NM, NM for heavy metal removal from wastewater generation, among others. However, the present work converged the analysis to topics that are being developed in ongoing research projects. The main drivers and barriers of NM uses are highlighted, namely concerning the environmental impacts and energy consumption.

During the past decade, to evaluate the environmental impacts of products and/or services, the LCA has been a methodology widely applied to understand the cumulative burdens that may emerge from raw materials acquisition to end-of-life alternatives, for instance. This systematic framework is defined by ISO 14040:2006 (ISO 2006) and ISO 1044:2006 (ISO 2006a), and it is organized in four main phases: (1) goal and scope definition; (2) life cycle inventory analysis (LCI); (3) environmental impact assessment (LCIA) and (4) interpretation. Even at an early stage of development, LCA allows identifying critical processes or hotspots to support decision-making towards the overall environmental performance improvement of the system (Almeida et al. 2021). LCA studies on NM application in the areas named before were gathered and analyzed. Despite the diverse nature of the applications, production methods, and LCA modeling assumptions, the results of the main LCA studies were compiled jointly with goal, case study, LCIA method used, and assumptions (boundaries, functional unit and production scale) to help to identify trends and areas of further research.

## 23.3 Results and Discussion

### *23.3.1 Life Cycle Assessment Approaches Applied to Nanotechnology*

The life cycle assessment studies found in the literature emphasize pioneering NM-based technologies with a focus on energy. Thus, a deeper analysis of the methodologies applied to evaluate the cumulative environmental impacts, as well as the main burdens that result from the NM inclusion in products and processes are summarized in Table 23.1.

The majority of the studies presented in Table 23.1 aimed to evaluate the environmental impact of emergent nanotechnologies, coupling a comparison with conventional technologies. Garvey et al. (2019), Tsang et al. (2018), Dhingra et al. (2010) and Wender et al. (2011) highlighted the energy trade-off associated with nanotechnology by the study of cumulative energy demand (CED), which describes the net life-cycle energy used to produce a material or product (Garvey et al. 2019).

A cradle-to-gate approach represents 86% of the system boundaries of the LCA studies analyzed. This means that all the impacts involved from raw materials acquisition to the final facility transportation were considered. The case study from Dhingra et al. (2010) shows the foreseen environmental benefits of nano-products in the use stage could not be representative for all the life cycle impacts, when, e.g., production and waste management steps, are not considered.

Regarding the production scale, the results were mainly determined at an industrial perspective, representing a more realistic scenario for an eventual technology implementation.

The environmental hotspots identified are often energy-consuming steps, namely due to the high-energy required during NM synthesis process (Dhingra et al. 2010). The energy consumed can range between 32 and 40 GJ per ton of titania produced by the sulfate process, and approximately 19 GJ for the chloride process (Osterwalder et al. 2006). Pini et al. (2017) highlighted the use of non-renewable energy as the second major contribution to the total environmental damage. Additionally, Bartolozzi et al. (2020) referred that energy-consuming processes are also the main

**Table 23.1** Case studies of nanomaterials uses and life cycle assessment methodologies

| Case study | LCA methodology | Goal | Boundaries | Functional unit | Scale of production | CC/CED per FU | Hotspots | Reference |
|---|---|---|---|---|---|---|---|---|
| Production of CNS | ILCD 2011 Midpoint+ | Optimize the production of CNS for water remediation | Cradle-to-gate | 1 kg of CNS produced | Three production systems, from the lab-level to a modeled scale-up system | Lab-scale: CC = 3343.13 kg $CO_2$ eq. Optimized lab-scale: CC = 2.93 × $10^3$ kg CO2 eq. Scale-up: CC = 27.4 kg $CO_2$ eq. | Energy-consuming processes; Final washing step procedures | Bartolozzi et al. (2020) |

(continued)

**Table 23.1** (continued)

| Case study | LCA methodology | Goal | Boundaries | Functional unit | Scale of production | CC/CED per FU | Hotspots | Reference |
|---|---|---|---|---|---|---|---|---|
| CFHS of $TiO_2$, $ZrO_2$, ZnO and of $LiFePO_4$ nanoparticles | CED and GWP | Assess sustainability of CFHS of nanoparticles and compare it with existing production technologies | Cradle-to-gate | 1 kg of nanoparticles dispersed in water | Industrial | $TiO_2$: CED = 149 (+90.7 for drying) MJ; GWP = 11.42 (+5.1 for drying) $CO_2$ kg.eq. $ZrO_2$: CED = 325 (+61.3 for drying) MJ; GWP = 16.16 (+3.41 for drying) $CO_2$ kg.eq. ZnO: CED = 347 (+224 for drying) MJ; GWP = 33.45 (+12.5 for drying) $CO_2$ kg.eq. $LiFePO_4$: CED = 240 (+211 for drying) MJ; GWP = 16.77 (+11.8 for drying) $CO_2$ kg.eq. | Raw materials (precursors and solvents); Heating and cooling | Stieberova et al. (2019) |

(continued)

**Table 23.1** (continued)

| Case study | LCA methodology | Goal | Boundaries | Functional unit | Scale of production | CC/CED per FU | Hotspots | Reference |
|---|---|---|---|---|---|---|---|---|
| Production and release of nano-Ag, nano-$TiO_2$, SWCNTs produced using CVD or Arc, and C60 produced using Pyrolysis or Arc | USEtox and CED | Characterize uncertainty in the life cycle ecotoxicity impacts of nanomaterial production and release | Cradle-to-gate | 1 kg of pristine nanomaterial (not modified or functionalized) | Industrial | Ag: CED = 8 $\times 10^3$ MJ; $TiO_2$: CED = 1 $\times 10^2$ MJ; SWCNTs CVD: CED = 2 $\times 10^6$ MJ; SWCNTs Arc: CED = 1 $\times 10^6$ MJ; C60 Pyrolysis: CED = 3 $\times 10^4$ MJ; C60 Arc: CED = 4,5 $\times 10^4$ MJ | Energy consumption | Garvey et al. (2019) |
| SFFS of nano-$TiO_2$ | ReCiPe v1.0.5; USEtox v2.0 and CED | Compare the SFFS of $TiO_2$ to that of a conventional precipitation method from an environmental and human health perspective | Cradle-to-gate | 1 kg of dry nano-$TiO_2$ powder | Laboratory | CED = 78 MJ; CC = 3 kg $CO_2$ eq. | Titanium isopropoxide (high embodied resource and energy demand); Transportation (high transport demands of ethanol) | Tsang et al. (2018) |
| Ecodesign of nano-$TiO_2$ functionalized porcelain stoneware tile production | IMPACT 2002+ and USEtox™ | Assess the environmental sustainability of nano-$TiO_2$ functionalized porcelain stoneware tile production | Cradle-to-grave | 1 $m^2$ of a nano-$TiO_2$ functionalized porcelain stoneware tile | Industrial | CC = 54.392 kg $CO_2$; Non-renewable energy = 782.569 MJ primary | GHG emissions; Non-renewable energy (natural gas, crude oil and hard coal emissions) | Pini et al. (2017) |

(continued)

**Table 23.1** (continued)

| Case study | LCA methodology | Goal | Boundaries | Functional unit | Scale of production | CC/CED per FU | Hotspots | Reference |
|---|---|---|---|---|---|---|---|---|
| Production of organic and silicon solar cells (alternatives: FTOinkjet—using inkjet printing in place of sputtering, PCBMdcb—using ortho-dichlorobenzene instead of toluene) | ReCiPe v1.0.5 and CED | LCA for a prospective organic solar photovoltaics technology and comparison with traditional inorganic modules (i.e., silicon solar cells) | Cradle-to-gate | 1 W-peak of electricity produced | Mixture of laboratory and pilot-scale | FTOinkjet: CC = $3.45 \times 10^{-2}$ kg $CO_2$; CED = $8.37 \times 10^{-1}$ MJ; PCBMdcb: CC = $3.83 \times 10^{-2}$ kg $CO_2$; CED = 2.30 MJ; Default: CC = $5.22 \times 10^{-2}$ kg $CO_2$; CED = 2.60 MJ | FTO film production (energy required during sputtering) | Tsang et al. (2016) |
| SWCNTs for lithium-ion batteries | Energy demand (manufacturing) | Prospective LCA for the scale-up of SWCNT-enabled Li-ion batteries | Cradle-to-gate | kWh battery storage capacity | Laboratory | 45 MWh | Energy requirements | Wender and Seager (2011) |
| Concrete brick units using Nano-silica | GWP | Reduction emission of $CO_2$ from bricks industry | Cradle-to-gate | 1 $m^3$ of concrete | Laboratory | 187–209 $CO_2$ emission kg | Energy requirements | Dawood and Mahmood (2021) |

(continued)

**Table 23.1** (continued)

| Case study | LCA methodology | Goal | Boundaries | Functional unit | Scale of production | CC/CED per FU | Hotspots | Reference |
|---|---|---|---|---|---|---|---|---|
| Optimization of self-compacting based hybrid fiber reinforced concrete | CML | Compare the environmental impacts of hybrid fiber reinforced self-compacting concrete added with fly ash and colloidal Nano silica and normal self-compacting concrete | Cradle-to-gate | 1 $m^3$ of concrete | Laboratory | GWP = 200–900 kg $CO_2$ eq. (maximum effect is imparted by the absence of nano silica, which is primarily due to the highest cement content) | Energy requirements during cement production | Mahapatra et al. (2021) |

CNS—cellulose nanosponges; CC—climate change; CFHS—continuous-flow hydrothermal synthesis; CED—cumulative energy demand; GWP—global warming potential; SWCNTs—single-wall carbon nanotubes; CVD—chemical vapor deposition; Arc—arc ablation; C60—60-carbon spherical fullerenes; SFFS—supercritical fluid flow synthesis; GHG—greenhouse gas; FTO—fluorine-doped tin oxide, PCBM—phenyl-C61-butyric acid methyl ester contributors to the environmental impacts reported. In this sense, to optimize the process, a scale-up model was developed. The environmental impacts were re-assessed and a mitigation of two orders of magnitude was observed, in comparison to the laboratory level. Furthermore, Miseljic and Olsen (2014) corroborated these results by analysing 29 LCA studies on metal, carbon, and composite engineered NM. Generally, NM products have demonstrated high energy demand and, therefore, accentuated environmental impacts in a cradle-to-gate approach, compared to conventional products (without NM). The life cycle energy requirements could be from 6 to 60 times higher (Khanna et al. 2007).

Studies on ecotoxicity were also carried out. Due to NM synthesis processes, Garvey et al. (2019) suggested that the energy consumption across the supply chain can result in greater ecotoxicity than the direct NM release to the ecosystems. These results may have limitations in nanotechnology fields. In particular, Wender et al. (2011) concluded that the current energy demand of nanomanufacturing makes commercial-scale application of nano-enabled batteries unfeasible. Nowadays, challenges on emerging technologies upscaling and maturity level increase (technology readiness level—TRL) need to be tackled, to turn laboratory and pilot developments suitable to enter in the global market (Rivera and Sutherland 2015).

However, there are solutions to overcome these barriers, coupling the use of nanotechnology with cleaner sources of energy. Continuous-flow hydrothermal synthesis (CFHS) was ranked among low energy consumption methods by Stieberova et al. (2019), compared with existing traditional production technologies. Supercritical fluid flow synthesis (SFFS) of $TiO_2$ presented a 30% decrease in CED category and a 55% reduction in climate change potential, in comparison to the conventional precipitation method. Tsang et al. (2016) estimated that organic solar photovoltaics require a 25-year minimum lifetime to reach parity, while amorphous silicon solar cell range from 1 to 9 years, representing a more sustainable energy source.

The use of nanostructured materials is being researched in the environmental nanoremediation field, by using building blocks for the design of nano-porous micro-dimensional systems. Thus, the eco- and health-toxicology impacts generally associated with the use of nanotechnologies are overcome (Bartolozzi et al. 2020).

The implementation of a safe(r) by design (SbD) approach considers human and environmental burdens and costs at the early stages of innovation processes. This allows the minimization of uncertainties along the innovation process instead of only at the final stage, promoting the continuous improvement of the processes (Jiménez 2020).

The use of NM in concrete was also addressed, in cradle-to-gate approaches. Dawood and Mahammod (2021) addressed the sustainability of concrete mixtures in bricks with NM, steel slag (SS) and glass powder (GP). When NM are added, even at low ratios (around 3%), flexural and compressive strengths could be improved. Additionally, the carbon emissions could be reduced up to 29% (cement replacement by SS) and 30% (cement replacement by GP), where GW potential impacts are due to the cement mix content (Dawood and Mahmood 2021). On the other hand, Mahapatra et al. (2021) produced a self-compacting concrete reinforced by hybrid fibre using NM, fly ash, crimped steel fibre and polypropylene fibre. Herein, the results demonstrated that conventional concrete have the highest environmental impact due to the high cement content (905 kg $CO_2$ eq/m$^3$). The lower the cement content, the lower the impact, where the sample having the lowest amount of cement demonstrated a GW = 358 kg $CO_2$ eq/m$^3$ (Mahapatra et al. 2021).

Regarding the LCA methodology in NM, some important premises should be considered to fully address the environmental impacts. In the goal and scope definition step, all nanorelated functionalities should be considered and covered by choosing a suitable functional unit and system boundary. In the inventory analysis, it is imperative to show generic and transparent data of the most relevant production routes of the NM applications and account for nanoparticles release in inventory modeling. Regarding the impact assessment, it is important to systematically establish missing LCIA characterization factors for NM release into the surrounded ecosystems (Hischier 2021).

## *23.3.2 Energy Analysis*

Bearing in mind the relevance of the energy paradigm, a detailed analysis focused on energy results is further presented. Bartolozzi et al. (2020), analyzed the energy impacts from cellulose nanosponges (CNS) production processes, at laboratory and industrial scales, as shown in Fig. 23.1. This example demonstrates the relevance of the energy component in the production of nanotechnology.

At the laboratory scale, energy-consuming processes, such as freeze-drying and heating, contribute to the most climate change impacts. Thus, to optimize these routes, scale-up models were developed for CNS production, leading to a 20% mitigation of the environmental impacts, in comparison to laboratory synthesis. Although

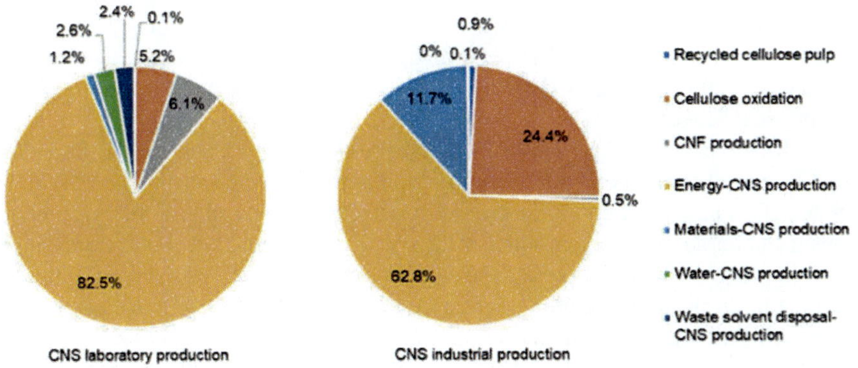


**Fig. 23.1** Comparison of the energy impacts distribution during laboratory and industrial production of CNS, in climate change impact category (CNS—cellulose nanosponges; CNF—cellulose nanofibers)

improvements were achieved through a scale-up approach, energy consumption still represents the largest share, contributing to 63% of the impact in the climate change category (Fig. 23.1).

Stieberova et al. (2019) studied the continuous-flow hydrothermal synthesis of NM. The CED of nanoparticles, with several functional applications, demonstrated dissimilar energy demands, and, consequently, different contributions from individual components and processes. The embodied energy in the precursor and heating and cooling procedures have the most significant impact on total CED (an average of 49% and 44%, respectively). Additionally, a higher output flow rate on the impact of heating and cooling stages was highlighted. Herein, similar amounts of energy were divided into larger volumes of production, meaning the output flow rate increase and, thus, an impact decrease (Stieberova et al. 2019).

Due to current concerns about potential environmental impacts of NM production and use outweigh their potential benefits, Garvey et al. (2019) compared the environmental impact of direct NM release with indirect impacts across the NM supply chain. According to an LCA realistic scenario, electricity production during the NM synthesis or purification steps is the primary contributor to ecotoxicity in single-wall carbon nanotubes (SWCNT) and 60-carbon spherical fullerenes (C60) produced by arc plasma method. On the other hand, embodied electricity, that is all other upstream energy inputs to material extraction and preparation, demonstrated a larger contribution for C60 produced by pyrolysis and for Nano-$TiO_2$, whereas the ecotoxicity impact for nano-Ag is linked to silver release during mining and precursor preparation (Garvey et al. 2019).

Tsang et al. (2018) studied the $TiO_2$ supercritical fluid flow synthesis. In the CED category, the most representative contribution was from solvents and precursors, totalizing 80% of the impacts. In fact, promising results were achieved, in comparison to similar technologies. The use of low embodied energy solvents, which could alleviate the CED impacts, and higher precursor concentrations (1.0 M instead of 0.5 M), that may minimize solvent needs and, consequently, associated energy and auxiliary material, are key factors for the sustainability optimization, whereas promoting the decrease of the surrounded ecosystems' impacts. Recycling and regeneration of alcohols were also assessed, resulting in a limited consumption of those resources. The solvent recycling alleviates CED by 73% in similar SFFS nano synthesis setups (Tsang et al. 2018).

In the life cycle of nano-$TiO_2$ functionalized porcelain stoneware tile, Pini et al. (2017) reported the second major contribution to total damage as the non-renewable energy. This impact category was primarily affected by natural gas (40%), crude oil (39%) and hard coal (9%) emissions. Regarding natural gas, the process with major environmental loads is the partial cycle (53%), which represents the core of the tile production cycle. The remaining emissions are caused by the use phase stage (32% from crude oil and 43% from hard coal emissions) (Pini et al. 2017).

Following the work developed by Tsang et al. (2016), the results show that default organic photovoltaic cell contribute to 62% of CED. This impact arose from the fluorine-doped tin oxide (FTO) film production (production/deposition of the electrode), due to the energy required during sputtering. The phenyl-C61-butyric acid methyl ester (PCBM) manufacture, the active layer anneal, and solar cell lamination also resulted in considerable impacts in CED category (Tsang et al. 2016).

Finally, the energy trade-off related to nano-enabled lithium-ion batteries was quantified. The results showed that the energy demand of SWNCT anode manufacturing by laser vaporization is two orders of magnitude larger than the sum of all other lithium-ion battery manufacturing processes. The energy requirements at a laboratory scale reflect the energy intensity of innovative nanomanufacturing processes. However, there is potential for improvement as nanomanufacturing processes are implemented at a commercial scale (Wender and Seager 2011).

## 23.4 Conclusions

Nanotechnology is currently a subject with technical, economic, social, and environmental interests, although there are challenges and limitations, namely from an LCA perspective:

There are controversial opinions regarding energy consumption impacts in nanotechnology: some authors explain that one of the benefits of nanotechnology is the low energy requirements, while other studies presented more promising results for conventional processes, without the use of NM. This highlights the need to deepen the understanding of this technology burdens.

The environmental impact of energy consumption in nanotechnologies is significant in the CED category. Therefore, further research should address the effectiveness of CED as an approximation for hotspot analysis.

The hotspots of nanotechnology usage are related to energy-consuming processes (that involve cooling and heating procedures, for instance) and materials. Therefore, more sustainable alternatives should be studied to improve current practices. Under the clean energy transition movement, these solutions could comply with the use of alternative sources of energy, biodegradable reagents, and the reuse of resources to promote a circular economy. The reduction or substitution of high-impact compounds and the adjustment of NM synthesis conditions are essential to minimize energy uses and, consequently, the related energy impacts.

Lower environmental impacts were observed when the production scale increase. In this sense, the study of scale-up approaches to alleviate energy consumption issues should be a top priority.

LCA studies on nanoparticles topic need to be developed, to tackle gaps and uncertainties in the processes and methodologies described. Through the knowledge increase on nanotechnologies hotspots, improvement strategies and new opportunities can be identified. Additionally, it is worth to note the importance of implementing the best available techniques to prevent and control industrial emissions to ensure a high level of environmental and human health protection. Even though the potential environmental impacts of NM have been a growing concern, minimal research on the LCA of NM and their environmental implications have been conducted. Thus, more life cycle assessment studies should be developed to support the sustainability of NM uses in different sectors.

**Acknowledgements** This project has received funding from European Union's Horizon 2020 research and innovation programme under grant agreement No 821431, within the PureNano project. The author would also like to acknowledge Biomac (Grant Agreement 952941) and Lightcoce (grant agreement 814632) European Union's Horizon 2020 Research and Innovation projects. Radu Godina acknowledges Fundação para a Ciência e a Tecnologia (FCT-MCTES) for its financial support via the project UIDP/00667/2020 and UIDB/00667/2020 (UNIDEMI).

## References

Almeida J, Magro C, Mateus EP, Ribeiro AB (2021) Life cycle assessment of electrodialytic technologies to recover raw materials from mine tailings. Sustainability 13:3915

Bartolozzi I, Daddi T, Punta C, Fiorati A, Iraldo F (2020) Life cycle assessment of emerging environmental technologies in the early stage of development: a case study on nanostructured materials. J Ind Ecol 24:101–115

BIOMAC. European sustainable BIO-based nanoMAterials Community. https://www.biomac-oitb.eu/en/normal/home. Last accessed 2022/07/13

Dawood ET, Mahmood MS (2021) Production of sustainable concrete brick units using nano-silica. Case Stud Constr Mater 14:e00498

Dhingra R, Naidu S, Upreti G, Sawhney R (2010) Sustainable nanotechnology: through green methods and life-cycle thinking. Sustainability 2:3323–3338

Garvey T, Moore EA, Babbitt CW, Gaustad G (2019) Comparing ecotoxicity risks for nanomaterial production and release under uncertainty. Clean Technol Environ Policy 21:229–242

Hischier R (2021) 17—Life cycle assessment of engineered nanomaterials. In: Njuguna J, Pielichowski K, Zhu H (eds) Health and environmental safety of nanomaterials, 2nd edn. Woodhead Publishing Series in Composites Science and Engineering. Woodhead Publishing, pp 443–458

Jiménez AS (2020) Safe(r) by design implementation in the nanotechnology industry. 11

Khanna V, Bakshi BR, Lee LJ (2007) Life cycle energy analysis and environmental life cycle assessment of carbon nanofibers production. In: Proceedings of the 2007 IEEE international symposium on electronics and the environment, pp 128–133

Lightcoce—European Research Program of Innovation. https://www.lightcoce-oitb.eu/en/normal/home. Last accessed 2022/07/13

Mahapatra CK, Pradhan S, Barai SV (2021) Influence of mechanical properties and $CO_2$ emissions on the optimization of self-compacting based hybrid fiber reinforced concrete. Procedia CIRP 98:145–150

Miseljic M, Olsen SI (2014) Life-cycle assessment of engineered nanomaterials: a literature review of assessment status. J Nanoparticle Res 16:2427

Monteiro H, Moura B, Soares N (2022) Advancements in nano-enabled cement and concrete: innovative properties and environmental implications. J Build Eng 56:104736

Nizam N, Hanafiah M, Woon KS (2021) A content review of life cycle assessment of nanomaterials: current practices, challenges, and future prospects. Nanomaterials

Osterwalder N, Capello C, Hungerbühler K, Stark WJ (2006) Energy consumption during nanoparticle production: how economic is dry synthesis? J Nanoparticle Res 8:1–9

Pesce R, Accogli A, Kostoula C, Ilare J, Panzeri G, Perecin CJ, Magagnin L (2021) Innovative magnetic aggregates for the removal of transition metals from industrial wastewater. Minerals 11:643

Pini M, Bondioli F, Montecchi R, Neri P, Ferrari AM (2017) Environmental and human health assessment of life cycle of nano$TiO_2$ functionalized porcelain stoneware tile. Sci Total Environ 577:113–121

Prakash Sharma V, Sharma U, Chattopadhyay M, Shukla VN (2018) Advance applications of nanomaterials: a review. Mater Today Proc 5:6376–6380

PureNano-H2020 project. https://www.purenano-h2020.eu/. Last accessed 2022/08/07

Rivera JL, Sutherland JW (2015) A design of experiments (DOE) approach to data uncertainty in LCA: application to nanotechnology evaluation. Clean Technol Environ Policy 17:1585–1595

Stieberova B, Zilka M, Ticha M, Freiberg F, Caramazana-González P, McKechnie J, Lester E (2019) Sustainability assessment of continuous-flow hydrothermal synthesis of nanomaterials in the context of other production technologies. J Clean Prod 241:118325

Tsang MP, Sonnemann GW, Bassani DM (2016) A comparative human health, ecotoxicity, and product environmental assessment on the production of organic and silicon solar cells: environmental assessment of organic and silicon solar cells. Prog Photovolt Res Appl 24:645–655

Tsang MP, Philippot G, Aymonier C, Sonnemann G (2018) Supercritical fluid flow synthesis to support sustainable production of engineered nanomaterials: case study of titanium dioxide. ACS Sustain Chem Eng 6:5142–5151

Wender BA, Seager TP (2011) Towards prospective life cycle assessment: single wall carbon nanotubes for lithium-ion batteries. In: Proceedings of the 2011 IEEE international symposium on sustainable systems and technology, pp 1–4

# Chapter 24
# Prospective Life Cycle Assessment of a Lignin Nanoparticle Biorefinery

**Luís Soares, Helena Monteiro, António A. Martins, Teresa M. Mata, and Joaquim C. G. Esteves da Silva**

**Abstract** The potential environmental impacts of producing lignin nanoparticles from wheat straw, on an industrial scale biorefinery, are evaluated following a prospective life cycle assessment (LCA) on a "cradle-to-gate" basis. This biorefinery includes emergent processes, such as organosolv, steam explosion and ultrasonication. The selected environmental impact categories include: climate change, cumulative energy demand, fine particulate matter formation, fossil resource scarcity, freshwater eutrophication and ecotoxicity, human toxicity, ionizing radiation, land use, mineral resource scarcity, ozone depletion and formation, terrestrial acidification and ecotoxicity and water use. The energy and chemical requirements highly influence the outcome of the proposed biorefinery's environmental performance. Process optimization is however key to unlocking the potential for sustainable production of lignin nanoparticles. Further studies must be carried out in order to provide a more comprehensive understanding of the potential environmental impacts of industrial scale lignin nanoparticle production.

**Keywords** Nanolignin · Prospective LCA · Wheat straw pretreatment

L. Soares · J. C. G. E. da Silva
Chemistry Research Unit (CIQUP), Institute of Molecular Sciences (IMS), DGAOT, Faculty of Sciences of University of Porto (FCUP), R. do Campo Alegre s/n, 4169-007 Porto, Portugal

H. Monteiro
Low Carbon and Resource Efficiency, R&Di, Instituto de Soldadura e Qualidade, 4415-491 17 Grijó, Portugal

A. A. Martins
LEPABE, Faculty of Engineering, University of Porto (FEUP), R. Dr. Roberto Frias, S/N, 4200-465 Porto, Portugal

ALiCE, Faculty of Engineering, University of Porto, R. Dr. Roberto Frias, 4200-465 Porto, Portugal

T. M. Mata (✉)
LAETA-INEGI, Associated Laboratory for Energy and Aeronautics, Institute of Science and Innovation in Mechanical and Industrial Engineering, R. Dr. Roberto Frias 400, 4200-465 Porto, Portugal
e-mail: tmata@inegi.up.pt



N. S. Caetano and M. C. Felgueiras (eds.), *The 9th International Conference on Energy and Environment Research*, Environmental Science and Engineering,
https://doi.org/10.1007/978-3-031-43559-1_24

## 24.1 Introduction

There is current consensus on the need to undergo a global transition from fossil resources with high greenhouse gas (GHG) emissions to renewable and more sustainable energy alternatives. At the forefront of this industrial transformation it is possible to find bio-based materials made from lignocellulosic biomass, the world's most abundant renewable resource. This is composed by main fractions of cellulose, hemicellulose and lignin, in varying quantities depending on the biomass type, forming a complex network of natural fibers in plant's cell walls. Each of these fractions has the potential to serve as fillers in a polymeric matrix, forming an emergent class of materials called nanocomposites (Sen MSE-M 2020).

Biologically, lignin has several functions, one of which is essentially to provide strength and integrity to the cell wall, chemically binding to other biomass fractions (Spiridon 2020). Other attractive qualities of lignin are the non-toxic nature, its biodegradability, thermal stability and antioxidant properties, which protect the plant from invading pathogens (Melro et al. 2018). All of these traits have made lignin a target for the potential manufacture of bio-based products that could replace materials made from fossil resources, for instance conventional plastics (Jawaid and Abdul Khalil 2011). Lignin is not found naturally unattached in the complex structure of biomass, thus needing to be extracted. For that purpose, several methods can be applied to fractionate lignocellulosic biomass, such as physical, chemical, physicochemical or biological treatments (Aftab et al. 2019). This paper focuses on a combination of two particular treatments: organosolv and steam explosion, as described by Matsakas et al. (2018). In the organosolv biomass pretreatment, an aqueous solution of an organic solvent (acid or alcohol) is used to treat the biomass, with or without addition of a reaction catalyst (Zijlstra et al. 2020). In the steam explosion pretreatment, biomass is subjected to severe conditions of temperature and pressure, followed by a quick decompression (Lora and Wayman 1978). Other existing biomass pretreatments require either the intensive use of chemicals and water, high energy intensity, or both, with associated environmental impacts.

This work aims to evaluate the potential environmental impacts of producing lignin nanoparticles from wheat straw, following the Life Cycle Assessment (LCA) methodology, as defined by the international standards ISO 14040 (2006a) and ISO 14044 (2006b). The LCA methodology comprises four steps: (i) Goal and scope definition; (ii) Life cycle inventory (LCI) analysis; (iii) Life cycle impact assessment (LCIA) and (iv) Interpretation. Thus, it is based on information such as mass and energy flows across a product or process's life cycle, in order to estimate its potential environmental impacts.

## 24.2 Materials and Methods

### *24.2.1 Goal and Scope Definition*

**Study Goal.** The main goal of this study is to estimate the potential environmental impacts of producing lignin nanoparticles from wheat straw, on a hypothetical industrial scale biorefinery, following a prospective LCA study.

**Functional Unit.** The functional unit (FU) selected for this LCA study is 1 kg of lignin nanoparticles produced from wheat straw.

**Study Scope and Assumptions.** A "cradle-to-gate" approach was followed, considering the life cycle steps from wheat straw production to lignin nanoparticles production, excluding its usage at consumer level. Thus, as shown in Fig. 24.1, first wheat straw is harvested by a tractor and kept at a warehouse, before being transported by diesel truck to the biorefinery. There, it is air dried and undergo size reduction by milling, in order to be prepared for the biomass pretreatment step. The biomass pretreatment is an organosolv-steam explosion hybrid method, described as follows: after loading the reactor with the biomass feedstock, treatment takes place at 180 °C, for 60 min, with a solvent mixture of ethanol (60%, v/v), applied together with sulfuric acid catalyst (1% wt). Then, the system is pressurized with steam at high temperatures for 4 min, after which, it is quickly depressurized to atmospheric levels. From this treatment, it results a slurry that is vacuum filtered to obtain two streams: a cellulose-rich solid fraction and a liquid fraction with dissolved lignin. Further processing of the solid stream is outside this study scope and should be investigated in future works.

From this point, there two distinct scenarios were evaluated:

(1) a baseline scenario, where the used solvent is discarded as waste, and
(2) similar to the previous scenario, but with solvent recovery.

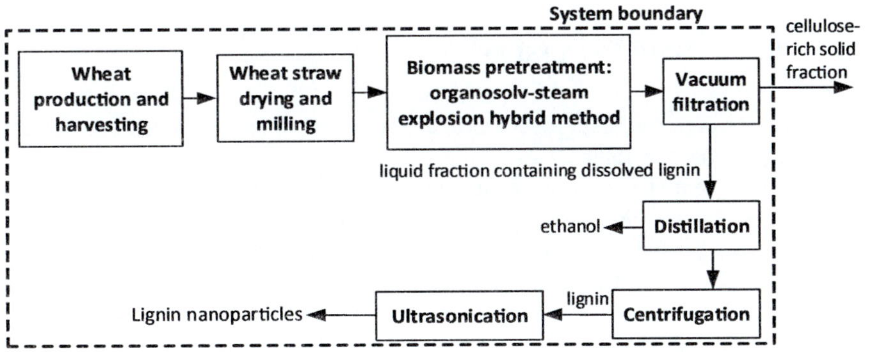


**Fig. 24.1** System boundary definition, including the life cycle stages considered for the LCA study

In the latter, the liquid fraction is distilled to recover ethanol for its reuse in the pretreatment stage and submitted to centrifugation in order to precipitate the dissolved lignin. Up to 98% of ethanol is recovered and reintegrated into the system, while the remaining is assumed to be lost by evaporation. The lignin yield is assumed to be 63% of the wheat straw's lignin content (18% wt). Finally, the lignin nanoparticles are produced by ultrasonication from the lignin fraction, with a yield corresponding to 57% of the wheat straw's lignin content.

### 24.2.2 *Life Cycle Inventory Analysis*

For the life cycle inventory analysis of the foreground processes, it was used data provided from a pilot plant of a BIOMAC project's partner in Sweden. For the background processes, it was used LCA databases, complemented with literature data (Koch et al. 2020; Borand and Karaosmanoğlu 2018; Lu et al. 2015; Yadav et al. 2021; Herry 2020; Garcia Gonzalez et al. 2017). In order to determine the environmental impacts attributed to the wheat straw production, a mass allocation method was considered (35%, w/w). In the Ecoinvent v3.8 database, the model selected was the Allocation at Point of Substitution (APOS), considering whenever possible, European conditions (RER). For energy, the low voltage Portuguese electricity mix was considered, supplied through the national energy grid.

### 24.2.3 *Life Cycle Impact Assessment*

The LCIA was conducted using the software SimaPro V9.3 and Microsoft Excel®. The potential environmental impacts' evaluation followed the ReCiPe method (Huijbregts et al. 2017), considering the impact categories presented in Table 24.1.

## 24.3 Results and Discussion

The life cycle inventory data for the production of lignin nanoparticles from wheat straw are presented in Table 24.2.

Then, the potential environmental impacts associated with the production of lignin nanoparticles from wheat straw, when solvent is recovered for further utilization, were evaluated per functional unit, as presented in Table 24.3.

Figure 24.2 shows the potential environmental impacts normalized in relation to the baseline scenario, for comparing the two scenarios, with and without solvent recovery, as described above.

Figure 24.3 shows the potential environmental impacts of the lignin nanoparticles value chain, considering the solvent recovery scenario, depicting the different life

**Table 24.1** Potential environmental impact categories of the ReCiPe method

| Acronym | Environmental impact category | Unit | Description |
|---|---|---|---|
| GWP | Climate change | kg $CO_2$-eq/FU | Accounts for the impact caused by the greenhouse gas emissions |
| PMFP | Fine particulate matter formation | kg $PM_{2.5}$-eq/FU | Considers the change in atmospheric concentration of particulate matter |
| FFP | Fossil resource scarcity | kg oil-eq/FU | Based on future cost increase of a fossil resource |
| FETP | Freshwater ecotoxicity | kg 1,4-DCB-eq/FU | Assesses the impact of a chemical emitted to freshwater |
| FEP | Freshwater eutrophication | kg P-eq/FU | Assess the damage caused by excessive use of nutrient-rich fertilizers (nitrogen, potassium and phosphorous) in the wheat straw cultivation |
| HTP | Human toxicity | kg 1,4-DCB-eq/FU | Considers the impact of toxic substances on human health |
| IRP | Ionizing radiation | kBq Co-60-eq/FU | Evaluates the increase in the absorbed dose of radiation |
| LOP | Land use | $m^2$yr crop-eq/FU | Refers to the relative species loss prompted by land use |
| ODP | Ozone depletion | kg CFC-11-eq/FU | Decrease in stratospheric ozone concentration |
| OFP | Ozone formation | kg NOx-eq/FU | Aggregation of ozone impacts on the health of ecosystems and humans |
| TAP | Terrestrial acidification | kg $SO_2$ | Measures the acidity in the soil |
| TETP | Terrestrial ecotoxicity | kg 1,4-DCB-eq/FU | Accounts for the potential impact of a chemical emitted to the soil |
| WCP | Water use | $m^3$/FU | Evaluates water consumed, subtracting the amount of water returned to the environment from the total amount of water brought into the system |
| SOP | Mineral resource scarcity | kg Cu-eq/FU | Quantifies the extra amount of ore produced |

FU = Functional unit; P = phosphorus; DCB-eq = dichlorobenzene equivalent; GWP = Global warming potential

cycle stages, from feedstock production to the lignin nanoparticles production via ultrasonication.

Figure 24.4 shows the relative contribution to the potential environmental impacts of wheat straw production (including the wheat straw cultivation, biomass harvesting by tractor and transportation by truck to the biorefinery site), considering the solvent recovery scenario.

**Table 24.2** Life cycle inventory for the production of lignin nanoparticles from wheat straw

| Life cycle stage | Unit | Input | Output |
|---|---|---|---|
| *Feedstock production* | | | |
| Wheat straw | kg | 7.89 | – |
| *Organosolv-Steam Explosion* (Koch et al. 2020) | | | |
| Energy | kWh | 13.8 | – |
| Ethanol | kg | 47.3 | – |
| Sulfuric acid | kg | 0.0789 | – |
| Water | kg | 31.6 | – |
| *Filtration* (Borand and Karaosmanoğlu 2018) | | | |
| Energy | kWh | 1.18 | – |
| Cellulose-rich fraction | kg | – | 3.08 |
| *Ethanol distillation* (Herry 2020) | | | |
| Energy | kWh | 5.86 | – |
| Ethanol | kg | – | 46.4 |
| *Centrifuge* (Matsakas et al. 2018) | | | |
| Energy | kWh | 0.0875 | – |
| Lignin | kg | – | 1.18 |
| *Ultrasonication* (Garcia Gonzalez et al. 2017) | | | |
| Energy | kWh | 4.40 | – |
| Lignin nanoparticles | kg | – | 1.00 |

Figure 24.5 shows the relative contribution to the potential environmental impacts of the pre-treatment stage (including electricity and water consumption, and production of EtOH and $H_2SO_4$), considering the solvent recovery scenario.

When comparing both scenarios, it is clear that solvent recovery is an essential mechanism to minimize the environmental impacts of producing lignin nanoparticles from wheat straw. Generally, solvent recovery contributed to decrease the contribution to the several environmental impact categories evaluated, except for ODP and LOP, where ethanol production is presumably not as important as other inputs. Overall, biomass pretreatment is the most important contributor to the environmental impact categories (except for ODP, LOP and HTP) associated with the lignin nanoparticles value chain, followed by feedstock production that is the most important contributor to ODP, LOP and HTP. In particular, WCP is mostly influenced by the pretreatment stage, showcasing that biomass pretreatment consumes much more water than wheat straw production.

In general, the remaining stages of the lignin nanoparticles value chain contributes less to the environmental impacts, but concerning WCP and IRP, distillation and ultrasonication outweigh feedstock production. When it comes to the wheat straw production stage, wheat straw cultivation is the main contributor to every impact category, due to fertilizers and pesticides usage, while harvesting by tractor and transportation to the biorefinery by truck present a lower contribution to the environmental impacts. Harvesting and transportation are the most important contributors to the TETP, IRP, SOP and WCP impact categories. At the pretreatment stage, it

**Table 24.3** Potential environmental impacts per 1 kg of lignin nanoparticles produced from wheat straw

| Environmental impact category | Unit | Feedstock production | Pretreatment | Others | Total |
|---|---|---|---|---|---|
| Climate change | kg $CO_2$ eq | 2.59 | 2.80 | 1.46 | 6.85 |
| Cumulative energy demand | kWh | – | 13.8 | 11.5 | 25.3 |
| Fine particulate matter formation | kg $PM_{2.5}$ eq | $2.39 \times 10^{-3}$ | $4.13 \times 10^{-3}$ | $2.62 \times 10^{-3}$ | $9.13 \times 10^{-3}$ |
| Fossil resource scarcity | kg oil eq | $2.79 \times 10^{-1}$ | 1.37 | $4.01 \times 10^{-1}$ | 2.05 |
| Freshwater ecotoxicity | kg 1,4-DCB eq | $1.17 \times 10^{-1}$ | $1.77 \times 10^{-1}$ | $1.29 \times 10^{-1}$ | $4.23 \times 10^{-1}$ |
| Freshwater eutrophication | kg P eq | $4.42 \times 10^{-4}$ | $1.11 \times 10^{-3}$ | $5.67 \times 10^{-4}$ | $2.12 \times 10^{-3}$ |
| Human toxicity | kg 1,4-DCB eq | 6.07 | 2.24 | 1.42 | 9.73 |
| Ionizing radiation | kBq Co-60 eq | $8.24 \times 10^{-3}$ | $1.21 \times 10^{-1}$ | $9.08 \times 10^{-2}$ | $2.20 \times 10^{-1}$ |
| Land use | $m^2$yr crop eq | 8.42 | $5.72 \times 10^{-2}$ | $3.76 \times 10^{-2}$ | 8.51 |
| Mineral resource scarcity | kg Cu eq | $1.63 \times 10^{-3}$ | $5.70 \times 10^{-3}$ | $2.56 \times 10^{-3}$ | $9.89 \times 10^{-3}$ |
| Ozone depletion | kg CFC-11 eq | $5.84 \times 10^{-5}$ | $7.70 \times 10^{-7}$ | $5.83 \times 10^{-7}$ | $5.97 \times 10^{-5}$ |
| Ozone formation | kg NOx eq | $9.95 \times 10^{-3}$ | $1.43 \times 10^{-2}$ | $7.79 \times 10^{-3}$ | $3.21 \times 10^{-2}$ |
| Terrestrial acidification | kg $SO_2$ | $1.45 \times 10^{-2}$ | $1.24 \times 10^{-2}$ | $7.97 \times 10^{-3}$ | $3.49 \times 10^{-2}$ |
| Terrestrial ecotoxicity | kg 1,4-DCB eq | 2.70 | 8.06 | 3.97 | 14.7 |
| Water use | $m^3$ | $1.16 \times 10^{-3}$ | $4.89 \times 10^{-2}$ | $1.34 \times 10^{-2}$ | $6.35 \times 10^{-2}$ |

is the electricity used to produce thermal energy, i.e. to heat the process units, that mostly contributes to the impact categories, followed by ethanol. The exceptions are WCP and FFP, for which the most important contributors are respectively, water and ethanol consumption. Nonetheless, ethanol production is also very significant to all impact categories, whereas water and sulfuric acid present a relatively low contribution to the environmental impacts.

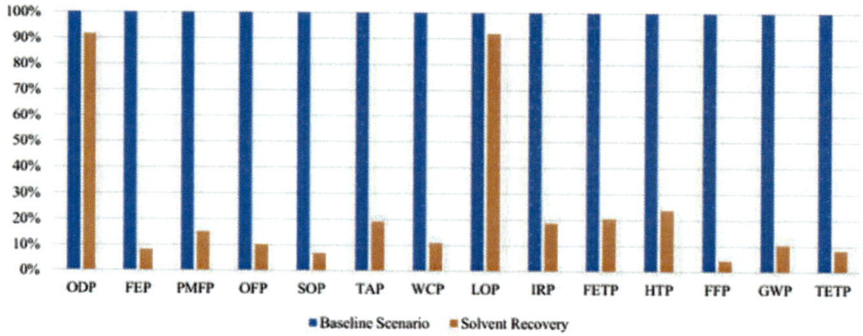


**Fig. 24.2** Potential environmental impacts normalized in relation to the baseline scenario, for comparing the two scenarios, with and without solvent recovery

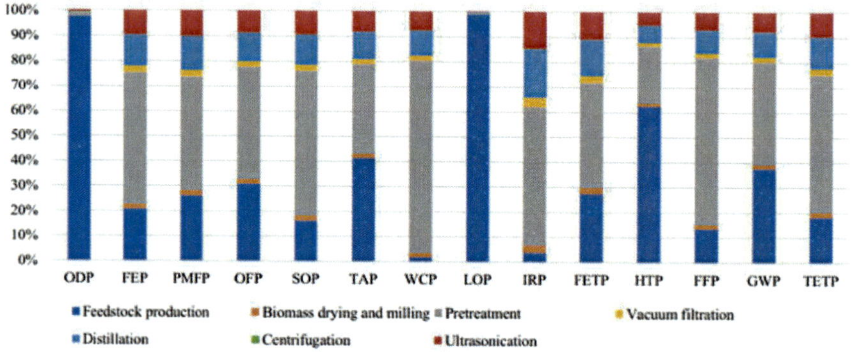


**Fig. 24.3** Potential environmental impacts of the lignin nanoparticles value chain, considering the solvent recovery scenario, depicting the different life cycle stages, from feedstock production to the lignin nanoparticles production via ultrasonication

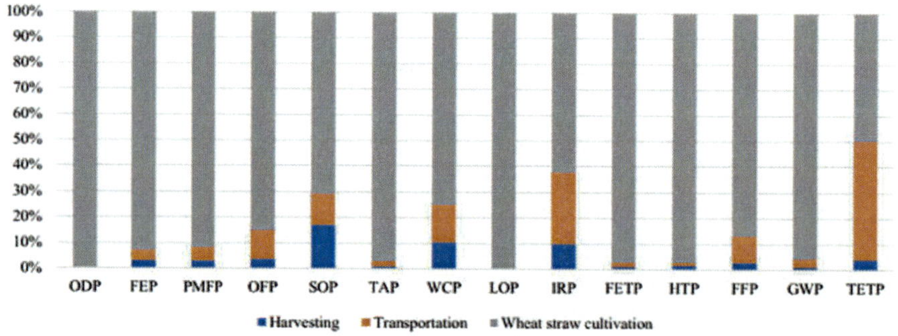


**Fig. 24.4** Relative contribution to the potential environmental impacts of wheat straw production (including wheat straw cultivation, biomass harvesting by tractor and transportation by truck to the biorefinery site), considering the solvent recovery scenario

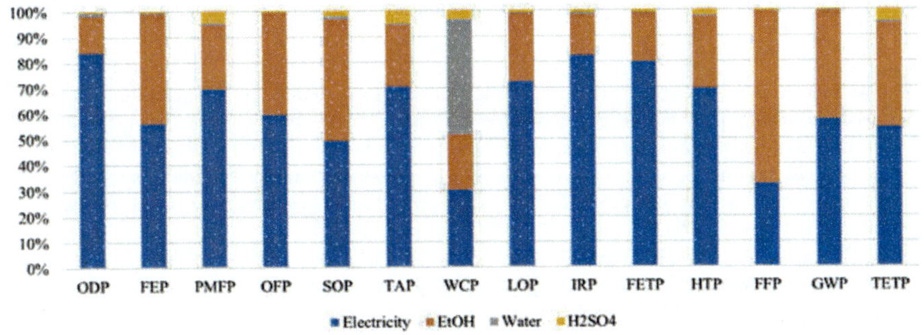


**Fig. 24.5** Relative contribution to the potential environmental impacts of the pre-treatment stage (including electricity and water consumption, and production of EtOH and $H_2SO_4$), considering the solvent recovery scenario

## 24.4 Conclusions

This LCA study addresses a biorefinery process for obtaining lignin nanoparticles from wheat straw. The environmental impact assessment highlights the importance of process optimization to minimize the environmental impacts. In particular, it shows that there is a tradeoff between ethanol recovery that has the potential to minimize the environmental impacts. Therefore, it is essential to recover ethanol, which improves the biorefinery's environmental performance. On the other hand, the source of energy used in the biorefinery is also important, as fossil energy have a higher contribution to the environmental impacts than renewable energy. Using bioethanol to pretreat biomass may reduce GWP even further, once electricity becomes the main driver of environmental impacts after solvent recovery is assured. Additionally, the location of the biorefinery has an influence on the impacts too, as for example in the case of WCP, as this process requires large amounts of water, and if the plant is operating in a place where water is abundant it can contribute to diminish water usage impacts. It is also desirable to make a more complete recovery of biomass, in addition to lignin, to make the biorefinery process more sustainable. For that purpose, more studies are needed to understand the hotspots of a biorefinery process more complex than the one here described, and to allow for new developments.

**Acknowledgements** This work was financially supported by: Project Horizon 2020—European Sustainable BIObased nanoMAterials Community (BIOMAC), Grant agreement ID: 952941. Base Funding of the following projects: UIDB/00081/2020 (CIQUP) and IMS (Institute of Molecular Sciences) LA/P/0056/2020, LA/P/0045/2020 (ALiCE), UIDB/00511/2020 (LEPABE) and UIDB/50022/2020 (LAETA), funded by national funds through the FCT/MCTES (PIDDAC). António Martins thanks FCT (Fundação para a Ciência e Tecnologia) for funding through program DL 57/2016—Norma transitória. Teresa Mata gratefully acknowledge the funding of Project NORTE-06-3559-FSE-000107, cofinanced by Programa Operacional Regional do Norte (NORTE2020), through Fundo Social Europeu (FSE).

## References

Aftab MN, Riaz F, Karadag A (2019) Different pretreatment methods of lignocellulosic biomass for use in biofuel production. In: Iqbal I (ed) Biomass for bioenergy. IntechOpen, Rijeka, Ch. 2. https://doi.org/10.5772/intechopen.84995

Borand MN, Karaosmanoğlu F (2018) Effects of organosolv pretreatment conditions for lignocellulosic biomass in biorefinery applications: a review. J Renew Sustain Energ 10:033104. https://doi.org/10.1063/1.5025876

Garcia Gonzalez MN, Levi M, Turri S, Griffini G (2017) Lignin nanoparticles by ultrasonication and their incorporation in waterborne polymer nanocomposites. J Appl Polym Sci 134:45318. https://doi.org/10.1002/app.45318.

Herry A (2020) Large scale lignin organosolv extraction. Rijksuniversiteit Groningen

Huijbregts MAJ, Steinmann ZJN, Elshout PMF, Stam G, Verones F, Vieira M et al (2017) ReCiPe2016: a harmonised life cycle impact assessment method at midpoint and endpoint level. Int J Life Cycle Assess 22:138–147. https://doi.org/10.1007/s11367-016-1246-y

ISO 14040 (2006a) Environmental management—life cycle assessment—principles and framework. International Organization for Standardization

ISO 14044 (2006b) Environmental management—life cycle assessment—requirements and guidelines. International Organization for Standardization

Jawaid M, Abdul Khalil HPS (2011) Cellulosic/synthetic fibre reinforced polymer hybrid composites: a review. Carbohydr Polym 86:1–18. https://doi.org/10.1016/j.carbpol.2011.04.043

Koch D, Paul M, Beisl S, Friedl A, Mihalyi B (2020) Life cycle assessment of a lignin nanoparticle biorefinery: decision support for its process development. J Clean Prod 245:118760. https://doi.org/10.1016/j.jclepro.2019.118760

Lora JH, Wayman M (1978) Delignification of hardwoods by autohydrolysis and extraction. Tappi [Technical Association of the Pulp and Paper Industry], v. 61

Lu D, Tabil GL, Wang D, Li X, Mupondwa KE (2015) Comparison of pretreatment methods for wheat straw densification by life cycle assessment study. Trans ASABE 58:453–64. https://doi.org/10.13031/trans.58.10510

Matsakas L, Nitsos C, Raghavendran V, Yakimenko O, Persson G, Olsson E et al (2018) A novel hybrid organosolv: steam explosion method for the efficient fractionation and pretreatment of birch biomass. Biotechnol Biofuels 11:160. https://doi.org/10.1186/s13068-018-1163-3

Melro E, Alves L, Antunes FE, Medronho B (2018) A brief overview on lignin dissolution. J Mol Liq 265:578–84. https://doi.org/10.1016/j.molliq.2018.06.021

Sen MSE-M (2020) Nanocomposite materials. Nanotechnology and the environment. IntechOpen. Rijeka, Ch. 6. https://doi.org/10.5772/intechopen.93047

Spiridon I (2020) Extraction of lignin and therapeutic applications of lignin-derived compounds. A review. Environ Chem Lett 18:771–785. https://doi.org/10.1007/s10311-020-00981-3

Yadav P, Athanassiadis D, Antonopoulou I, Rova U, Christakopoulos P, Tysklind M et al (2021) Environmental impact and cost assessment of a novel lignin production method. J Clean Prod 279:123515. https://doi.org/10.1016/j.jclepro.2020.123515

Zijlstra DS, Lahive CW, Analbers CA, Figueirêdo MB, Wang Z, Lancefield CS et al (2020) Mild organosolv lignin extraction with alcohols: the importance of Benzylic Alkoxylation. ACS Sustain Chem Eng 8:5119–5131. https://doi.org/10.1021/acssuschemeng.9b07222

# Part IV
# Modeling, Simulation, and Forecasting of Energy and Carbon Markets

# Chapter 25
# Demand Response Flexibility: Forecasts and Expectations for 2030 and 2050

**Débora de São José, Pedro Faria, and Zita Vale**

**Abstract** The change in the electric grid is a well-known and addressed topic. To achieve the very ambitious goals to prevent climate crises and to increase the participation of renewable generation without decreasing the reliability and security of the power system, demand side flexibility and demand response presents themselves as effective solutions to increase the needed flexibility. However, it is also necessary to make forecasts about the future of the system, which is more difficult for smaller loads in times of intense changes. This work addresses forecasts techniques and the predictions and expectations for achieving the ne zero emission 2050 scenario with focus on the European market. Even though the scenario presents itself as very challenging, the efforts are being made and the rollout plans, in many European countries are advanced. Changes in policies will also need to take a faster pace in the next few decades.

**Keywords** Demand response · Forecast · DR flexibility

## 25.1 Introduction

The electricity grid is changing and adapting to new restrictions imposed by environmental issues, and smart grids initiatives are also inducing the grid into a new scenario (Javed et al. 2012). Another observable change is the increase of integrated community energy systems, which brings the benefits of $CO_2$ reduction and increase in efficiency and self-sufficiency (São et al. 2021) beneficiating from smart grid technology and contributing with climate goals. However, to sustain this change, the grid is expected to get more resilient and cost efficient. For that, demand response (DR) programs are one solution to achieve the overall goal (Javed et al. 2012).

D. de São José · P. Faria · Z. Vale (✉)
Research Group on Intelligent Engineering and Computing for Advanced Innovation and Development (GECAD), Intelligent Systems Associated Laboratory (LASI), Polytechnic of Porto (P.PORTO), Rua Dr. António Bernardino de Almeida 431, 4200-072 Porto, Portugal
e-mail: zav@isep.ipp.pt

 
N. S. Caetano and M. C. Felgueiras (eds.), *The 9th International Conference on Energy and Environment Research*, Environmental Science and Engineering,
https://doi.org/10.1007/978-3-031-43559-1_25

Nevertheless, an accurate methodology must be stablished to have a successful and trustworthy DR (Panagiotidis et al. 2019). Furthermore, when aiming economic growth and environmental security in the future, electricity demand forecasting became imperative (Islam et al. 2020). Even though large loads' forecasting (e.g., city, whole grid) has been demonstrating high accuracy in its results, smaller loads (e.g., building, community, micro-grid, houses) has been presenting a very prominent volatility in its dynamics. Besides that, data deficiency in what concerns users' information was pointed out as one of the main reasons DR programs fail to achieve its goals (Javed et al. 2012). Thus, many techniques have been used for forecasting. Table 25.1 presents some of them.

Besides the evident need for forecasting, demand-side flexibility is becoming more relevant as countries take up more challenging climate goals. The power system transformation underway to support net zero goals is putting strain on both supply and demand, necessitating greater flexibility, which can increasingly be obtained through demand-side resources such as DR systems.

Lastly, this paper is organized in the following way. First, the introduction addresses demand response, demand forecasting and some issues and characteristics of the subject. The second topic addresses the methodology used in this research. The third topic covers DR forecasting considering different countries all over the globe. Followed by a section focused on the European scenario and, finally, the final section presents the conclusions of this work.

## 25.2 Methodology

For this analysis, two phases were conducted. The first one is a small introduction and review presenting the DR techniques available in the literature, its definition, and points of interest. The second one is a synthesis of the main demand response, demand response flexibility and smart meters information available. It was made to have a clear idea of what is expected and what is happening. This second part was divided into two segments, global and Europe, considering the characteristics of this research and the relevance and impact of European energy sector in the global scenario, besides its responsibility in addressing the climate crisis.

## 25.3 2050 Net Zero Emission and Global Expectations to Achieve It

The Net Zero Emission by 2050 scenario (NZE), already mentioned, NZE refers to reducing greenhouse gas emission to as close to zero as possible to prevent the worst impacts of climate crisis (United Nations 2022). Considering the fast pace required for the NZE, some assumptions/predictions were made based on what is necessary

**Table 25.1** Forecast approaches

| Names | Description |
|---|---|
| Time-series | Simplest model<br>Uses time series trend analysis for extrapolating the future energy requirement |
| Spline-based | Avoid over-parametrization<br>Rely on splines to describe the baseload |
| Multivariate state-space | Time-varying regression coefficients<br>Useful for analyzing DR |
| Semi-parametric | Predict the contribution of load from non-linear variables<br>Useful for DR volume evaluation |
| Regression | Used to forecast different energy sources requirement (include electricity) |
| Econometric | Correlate the energy demand with macroeconomic variables |
| Decomposition | Energy consumption approach<br>Energy intensity approach |
| ARIMA | Extensively used |
| Artificial systems | Benefit from artificial neural network<br>Able to capture unspecified non-linear relationships between external variables like weather<br>Criticized for leading to over-parametrized models<br>Not necessarily outperform linear regression models<br>Electricity load forecasting<br>Long term energy demand projections considering macroeconomic variables |
| Grey prediction | Simplicity<br>Ability to characterize unknown system<br>Need only a few data points |
| Input–output | Access how social and economic changes affect energy requirements and energy intensity |
| Fuzzy logic/Genetic algorithm | Fuzzy logic is used for short term electric load forecasting<br>Modern computational techniques using genetic algorithms are being adapted for load forecasting |
| Integrated | Some of the latest techniques (e.g., Bayesian vector auto regression) Support vector regression, ant colony, particle swarm optimization, ...<br>Used in energy demand analysis |
| Bottom-up | MARKAL is a dynamic technique which was originally developed as a least cost linear programming model<br>Long term energy demand and $CO_2$ emission for China forecast using TIMES G5 model<br>LEAP model is a bottom-up-type accounting framework which is used for forecasting |

Adapted from Antonopoulos et al. (2020), Austrian Energy Agency—AEA (2020), Dusparic et al. (2013), Garulli et al. (2015), Larsen et al. (2017), Ruiz et al. (2015), Suganthi and Samuel (2012), Waseem et al. (2019)

to achieve the ambitious goals needed to address the climate crisis. Based on that, Table 25.2 presents the expectations for 2030 and 2050.

Based on that, the increase in renewable generation expected to 2030 are, already imposing the need for more flexibility in the energy system. Besides, one issue that may affect its development and the achievement of such goals is the need to increase information and communication technology. This increase can bring benefits as more quality data, real-time visualization, among others, but also bring some concerns as the risks associated with failures of these communication channels, security of supply and power system resilience (IEA 2022; European Commission 2022a).

Finally, some steps are already being taken around the world in the direction of NZE 2050. Some examples are presented in Table 25.3.

Even with advances already started in several parts of the world, the challenges are so big, the goals so ambitious that it is very difficult to say that it would be achieved, and that it will reach/approximate to the NZE 2050 established scenario.

**Table 25.2** 2030 and 2050 expectations for NZE

| 2030 | 2050 |
|---|---|
| Have 500 GW of DR in the market | 15% of average annual demand can be shifted to some extent |
| 10 times increase (compared to 2020) of global inventory of flexible assets in all sectors (residential, commercial and industry) | Fast implementation of more ambitious policies |
| Further investments in smart grids are pointed as needed, however, no specific value is pointed out | Fast electrification of transport, heating, and others |
| Participation of electricity in the energy grid is also expected to increase to 26% (when compared to 20% in 2020) | Experience a change in the demand curve shape |
| Approximately 25% of total flexibility needed globally will come from battery storage and DR, especially in developed and developing markets | Approximately 50% of total flexibility needed will come from battery storage and DR, especially in developed and developing markets |
| Renewable participation on energy supply is expected to increase to over 60% (compared to 29% in 2020) | |
| Double the electricity system flexibility | |

Adapted from IEA (2022)

**Table 25.3** Actions per country

| Country | Action |
|---|---|
| Australia | Approved a wholesale DR mechanism, opening the DR market to consumers and aggregators |
| Belgium | Adopted a capacity remuneration mechanism that allows the participation of DR operators and is aimed at ensuring security of supply |
| Bosnia and Herzegovina | A project started with the installation of solar power plants on the roof of a primary school, with 25 kW capacity that 2 kW will be used for the needs of school and the rest will be directed to a power supply network that is enough for approximately 20 households |
| Chile | Launched a power system flexibility strategy focused on market design, regulatory frameworks, and system operation |
| China | Is in the process of issuing regulations on operations and ancillary services to encourage the use of storage, user-adjustable loads, load aggregators, virtual power plants and other resources in power system ancillary services |
| Colombia | Extended tax incentives to non-conventional sources of energy and energy efficiency projects, including smart metering and DR |
| Singapore | Is reviewing existing DR programs and schemes to improve remuneration methodologies, penalty, and compliance rules |
| Slovenia | The main existing action to influence consumers' behavior is the use of two different electricity prices, the high price 6 am to 10 pm, and the low price from 6 pm to 10 am (with adaptation during summertime) |
| United States | System operators in the six capacity and ancillary services markets were mandated to remove barriers to the participation of distributed energy resources of more than 100 kW, including DR, renewables, EVs and energy efficiency |

Adapted from Austrian Energy Agency—AEA (2020), IEA (2022)

## 25.4 Current and Expected European Smart Meters Rollout

After summarizing global expectations, this topic will focus on the European situation and future scenarios. Considering the already addressed need of information and communication technology to achieve the level of DR flexibility required, this section will focus on European smart meters rollout evolutions and plans to relate to demand side flexibility with prospects and future expectations. Figure 25.1 presents each country by its past, current, and expected rollout.

From Fig. 25.1, it is also important to highlight some particularities of few countries (European Commission 2022a, 2022b, 2014a, 2014b, 2014c; European Union Agency for the Cooperation of Energy Regulators and the Council of European Energy Regulators 2021; Smart Energy International 2022). Croatia and Slovakia, until now, did not present a positive rollout. Austria had a goal of achieving 80% by the end of 2020, however, only achieved 29% by the deadline. A similar situation happened with United Kingdom, the country had a 2019 as the deadline to achieve an 80% rollout rate, however, is only expected to achieve it by 2024. Finland, during the

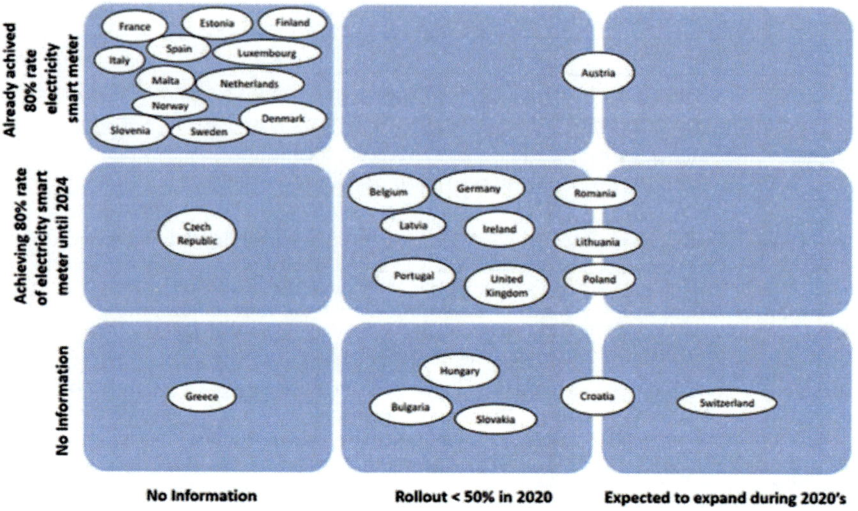


**Fig. 25.1** EU rollout by country. Adapted from European Commission (2022a, 2022b, 2014a, 2014b, 2014c), European Union Agency for the Cooperation of Energy Regulators and the Council of European Energy Regulators (2021), Smart Energy International (2022). Note 1: Approximately 34.8 million smart electric meters units were installed in the bloc in 202. However, installations decline by 10% in 2020 due to COVID-19. Note 2: Some countries were not considered due to its dimension or not being part or European Union. Note 3: Norway was considered, even though it is not part of European Union

last years, became one of the main responsible for growth of smart meters in Europe. The French market for demand-side flexibility has grown by, approximately, 0.6 GW in 2021. In Germany, it is expected for the market to slowly expand until 2032. Italy is already halfway through its second-wave rollout of smart electricity meters and Sweden is also performing a large scale second-wave rollout. Based on that, one reason that may happen is the small presence of aggregators observed in European Union. The presence of aggregators could encourage, for example, the participation and attendance of prosumers.

So far, aggregators exist in 19 out of 28 MSs, as in the following graphic. However, according to European Union Agency for the Cooperation of Energy Regulators and the Council of European Energy Regulators (2021), only eight European Union state members have aggregators able to operate autonomously from the supplier and only ten have residential consumers and aggregators are able to participate in the electricity market. Therefore, progress in policies would be crucial to improve this scenario, as pointed out in Table 25.3.

## 25.5 Conclusion

The reasons behind the need to adapt the energy sector as a whole and the electric sector make it a global issue and an urgent one. Very ambitious goals were created with a challenging deadline. For that, forecasting the outcome is necessary to make new policies and to know the possible challenges in the future. During forecasts, the method is very relevant, but also is the data. Smart meters are a big deal in collecting users' data and allowing them to collaborate increasing the systems flexibility and reliability. That create a dynamic system with active roles from the generation to the end user. From this review and analysis, one can observe the urgent need for new policies implementation. Also, that most countries in Europe are in a fast pace to achieving the 80% rate on smart meter until 2024. Some countries had a bad start in 2020 but are expected to catch up during the 20's decade, which is very important considering the milestones for 2030. Finally, global, and local efforts are being observed in the direction of NZE and, to achieve that, we will need more quality data to produce better forecasts, especially for local loads.

**Acknowledgements** This work has received funding from the EU Horizon 2020 research and innovation program under project TradeRES (grant agreement No 864276). The authors acknowledge the work facilities and equipment provided by GECAD research center (UIDB/00760/2020) to the project team. Pedro Faria received funding from FCT, CEECIND/01423/2021.

## References

Antonopoulos I, Robu V, Couraud B, Kirli D, Norbu S, Kiprakis A et al (2020) Artificial intelligence and machine learning approaches to energy demand-side response: a systematic review. Renew Sustain Energ Rev 130:10989

Austrian Energy Agency—AEA (2020) Survey on existing demand-response (DR) actions and collective actions in the heating and cooling sector and overview of legal and other requirements and challenges

de São José D, Faria P, Vale Z (2021) Smart energy community: a systematic review with metanalysis. Energ Strateg Rev 36

Dusparic I, Harris C, Marinescu A, Cahill V, Clarke S (2013) Multi-agent residential demand response based on load forecasting. In: 2013 1st IEEE conferences for sustainable development. IEEE, pp 90–6

European Commission (2014a) Cost-benefit analysis and state of play of smart metering deployment in the EU-27 Accompanying the document Report from the Commission Benchmarking smart metering deployment in the EU-27 with a focus on electricity

European Commission (2014b) Benchmarking smart metering deployment in the EU-27 with a focus on electricity

European Commission (2014c) Cost-benefit analyses and state of play of smart metering deployment in the EU-27

European Commission. Climate Action. 2050 Long-Term Strategy. https://ec.europa.eu/clima/eu-action/climate-strategies-targets/2050-long-term-strategy_en. Last accessed 2022/05/9

European Commission. Smart Electricity Systems and Interoperability. Smart Metering Deploy Eur Union. https://ses.jrc.ec.europa.eu/smart-metering-deployment-european-union. Last accessed

2022/05/9
European Union Agency for the Cooperation of Energy Regulators and the Council of European Energy Regulators (2021) Annual Report on the Results of Monitoring the Internal Electricity and Natural Gas Markets in 2020. Energy Retail Markets and Consumer Protection Volume
Garulli A, Paoletti S, Vicino A (2015) Models and techniques for electric load forecasting in the presence of demand response. IEEE Trans Control Syst Technol 23:1087–1097
IEA, Demand Response. https://www.iea.org/reports/demand-response. Last accessed 2022/05/12
Islam MA, Che HS, Hasanuzzaman M, Rahim NA (2020) Energy demand forecasting. Energ Sustain Dev 105–23. Elsevier
Javed F, Arshad N, Wallin F, Vassileva I, Dahlquist E (2012) Forecasting for demand response in smart grids: an analysis on use of anthropologic and structural data and short term multiple loads forecasting. Appl Energ 96:150–160
Larsen EM, Pinson P, Leimgruber F, Judex F (2017) Demand response evaluation and forecasting—methods and results from the EcoGrid EU experiment. Sustain Energ Grids Netw 10:75–83
Panagiotidis P, Effraimis A, Xydis GA (2019) An R-based forecasting approach for efficient demand response strategies in autonomous micro-grids. Energ Environ 30:63–80
Ruiz N, Claessens B, Jimeno J, López JA, Six D (2015) Residential load forecasting under a demand response program based on economic incentives. Int Trans Electr Energ Syst 25:1436–1451
Smart Energy International. Smart Energy International. https://www.smart-energy.com/. Last accessed 2022/05/11
Suganthi L, Samuel AA (2012) Energy models for demand forecasting—a review. Renew Sustain Energ Rev 16:1223–1240
United Nations. Net Zero Coalition, https://www.un.org/en/climatechange/net-zero-coalition. Last accessed 2022/05/12
Waseem M, Lin Z, Yang L (2019) Data-driven load forecasting of air conditioners for demand response using Levenberg–Marquardt algorithm-based ANN. Big Data Cogn Comput 3:36

# Chapter 26
# Multivariate Weather Derivatives for Wind Power Risk Management: Standardization Scheme and Trading Strategy

**Takuji Matsumoto and Yuji Yamada**

**Abstract** With the recent introduction of wind power generation of various scales due to its promise as a green energy resource, effectively managing the risk of fluctuations in wind power generation revenues has become an important issue. Against this background, this study introduces several weather derivatives based on wind speed and temperature as underlying assets and examines their effectiveness. In particular, we propose new standardized derivatives with higher-order monomial payoff functions, such as "wind speed cubic derivatives" and "wind speed and temperature cross derivatives." In contrast to the existing nonparametric derivatives, the minimum variance hedging problem to find the optimal contract amount of these standardized derivatives is reduced to estimating a linear regression. We also develop a market trading model to put the proposed standardized derivatives into practical use and clarify the real-world implications of standardizing weather derivatives. Furthermore, to make trading more efficient, we propose a "product selection" strategy utilizing the "variable selection" approach of LASSO regression. Empirical analysis confirms hedging effectiveness comparable to existing nonparametric derivatives and reveals the effectiveness of the proposed derivatives standardization scheme as well as their trading strategies.

**Keywords** Empirical simulations · Generalized additive model · LASSO regression · Optimal hedging · Weather derivatives · Wind power

---

T. Matsumoto (✉)
Faculty of Transdisciplinary Sciences for Innovation, Kanazawa University, Ishikawa 920-1192, Japan
e-mail: mtakuji@staff.kanazawa-u.ac.jp

Y. Yamada
Faculty of Business Sciences, University of Tsukuba, Tokyo 112-0012, Japan



N. S. Caetano and M. C. Felgueiras (eds.), *The 9th International Conference on Energy and Environment Research*, Environmental Science and Engineering,
https://doi.org/10.1007/978-3-031-43559-1_26

## 26.1 Introduction

Against the backdrop of the global movement toward achieving carbon neutrality, the introduction of wind power, a typical green energy source, has been steadily progressing. According to an IEA report, global wind power capacity is increasing at an accelerating rate, and the trend is expected to continue (IEA 2021). In addition, small wind turbines, which can be installed on home sites, have attracted attention in recent years, and its trading market is expected to grow steadily (Report Ocean 2022). In this way, the number of players is expected to increase and diversify worldwide, and thus, the need to manage the risk of fluctuations in power generation revenues is becoming increasingly apparent.

There are two main types of risks associated with fluctuating earnings in the wind power business: price and volume. For managing price risk, financial instruments, such as power futures and bilateral contracts, are often available as well as support measures including feed-in tariffs. However, when it comes to volume risk, there are few financial instruments available (despite the great need for them). For example, if a wind power generator has a contract to supply a certain amount of electricity generated at a fixed price, it will incur losses when it is unable to generate electricity due to poor wind conditions.

Among few instruments for treating such volume risks, the Wind Power Futures (WPF) have been traded on the European Energy Exchange (EEX) and Nasdaq (note that the EEX's WPF was delisted in 2020), which are contracts with a weighted average of area-wide wind power generation in Germany/Austria as the underlying asset. Since the underlying asset is the amount of electricity generated, the WPF may be considered effective for large-scale wind power generators in the targeted area. However, more specific and sophisticated products (derivatives or forwards/futures) may be required for individual power producers including small-scale wind farms or even households (prosumers) with small wind power generators operating in a specific area.

Instead of using general instruments such as the WPF, it would be more effective if we construct a flexible instrument tailored to the specific weather conditions of the generator's business area using only observed weather values as the underlying asset. Furthermore, such weather derivatives would ensure transparency as a financial instrument, since the generator's situation would not affect the underlying asset. Based on the above considerations, we propose an approach to use "weather" derivatives, emphasizing transparency and flexibility in product design.

There have been several studies on derivatives for wind power including not only the WPF (Benth and Pircalabu 2018; Gersema and Wozabal 2017), which actually exists, but also European put-type quanto option (Benth et al. 2018), the wind put barrier option (Rodríguez et al. 2021) and wind call option (Kanamura et al. 2021); these have been studied in the context of pricing rather than its effectiveness for reducing the risk (i.e., hedging) based on derivative instruments. A unique derivative for wind power risk was proposed by Yamada (2008), where the effectiveness of derivatives based on nonparametric regression using the Generalized Additive Model (GAM) (Hastie and Tibshirani 1990) was demonstrated. Subsequently, nonparametric regression hedging techniques have been applied to retailers, solar power companies, and others, and have been demonstrated to have high hedging effect (Matsumoto and Yamada 2018, 2021a, 2021b), respectively. These studies take the approach of calculating optimal derivative payoffs by introducing derivatives on weather indexes, where the hedger optimizes the payoffs (for nonparametric derivatives) or the contract volume (for standard derivatives). In the present study, we construct a new hedging strategy which not only combines these findings, but also multivariates the method of Yamada (2008) (that dealt with univariate derivatives only with wind speed prediction errors as the underlying asset). Moreover, we propose a market model for the new standardized derivatives.

For constructing an elaborate hedging model for wind power business, it would also be useful to review previous research on wind power forecasting. Broadly speaking, the wind power forecasting methods can be classified into time series models and exogenous variable models (Foley et al. 2012). In recent years, the latter has emerged due to advances in computing technology and artificial intelligence, and several models have been proposed using machine learning (ML) methods such as ANN and deep learning (Hanifi et al. 2020). Wind power generation is strongly correlated with wind speed, where intensity nonlinearities are known to exist. In general, the wind power curve is often described in the form of the cube of wind speed (Ko et al. 2015), and in Marčiukaitis et al. (2017), various models of nonlinear regression with wind speed as an explanatory variable have been compared and validated. More recently, it has been reported that temperature is the next most commonly used explanatory variable after wind speed (Hanifi et al. 2020). Furthermore, some studies have demonstrated the effect of using temperature to improve accuracy using ML methods (Aksoy and Selbaş 2021; Bilal et al. 2018; Giorgi et al. 2011), but in these studies, the structural effect of temperature and wind speed (including their interaction) on wind power generation has not been clarified.

Based on the above reviews, this study proposes weather derivatives on wind speed and temperature for wind power businesses. In particular, in order to capture the complex interaction between wind speed and temperature, we apply two-dimensional tensor product spline functions (Wood 2017) and estimate nonparametric payoffs for derivatives with high hedging effect. In addition, we introduce a variety of standard derivatives (with payoffs defined by monomial functions) and show that a combination of standard derivatives can approximately replicate the payoffs of highly nonlinear derivatives. Although "higher-order" standard-type derivatives have been proposed in previous studies, such as the squared error derivative of temperature (Matsumoto and Yamada 2021a, 2021b), in this study, we introduce new composite standard derivatives such as the "cubic derivatives of wind speed" and the "cross-derivatives of temperature and wind speed." In the empirical analysis, the effectiveness of the proposed method is demonstrated by dealing with data from Denmark, where wind power generation is particularly prevalent among European countries with abundant wind power capacity, in anticipation of its potential for practical use. Also, in order to clarify the practical merits of standardizing derivatives and the concept for their social implementation, a concrete market transaction model is developed and discussed. Furthermore, new perspectives for hedgers' decision making, such as product selection strategies using LASSO regression, are provided. Comprehensively, standardized derivatives with highly objective weather indices as underlying assets allow various market players to trade flexibly as needed, and thus, the proposed methodology offers implications for promising hedging measures in an increasingly decentralized electricity market.

The structure of this study is as follows. After constructing the market and hedging models in Sect. 26.2, the estimation results of the models and hedging effects are verified using real data in Sect. 26.3. Finally, in Sect. 26.4, we present our conclusions.

## 26.2 Methods

### *26.2.1 Market Trading Model*

In this section, we first develop a market model and discuss the possible trading schemes for the proposed weather derivatives. This paper addresses the two types of derivatives: nonparametric (made-to-order) derivatives and standard derivatives. Figures 26.1 and 26.2 provide a conceptual comparison of both derivative schemes, focusing on the market trading models and the payoff functions (with the problem to be solved), respectively.

Non-parametric derivatives take a bilateral contractual form in which an individual optimal derivative payoff is sought for each hedger, and both the seller and the buyer agree to it. On the other hand, in a standard derivative trading scheme, the seller (i.e., an insurance company) may offer derivatives with predetermined payoff functions (i.e., monomials of the weather indexes as the payoff functions) for any buyers, whereas the buyers may agree to purchase them and include in their portfolio. In this situation, the buyer is supposed to choose the kinds and the trading amount

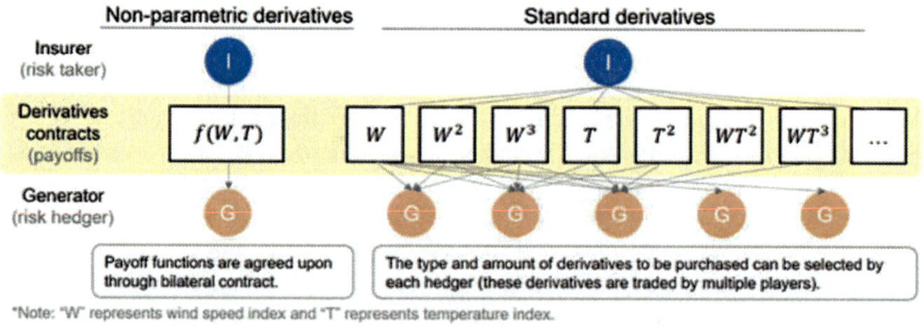


**Fig. 26.1** Market trading model for wind derivatives

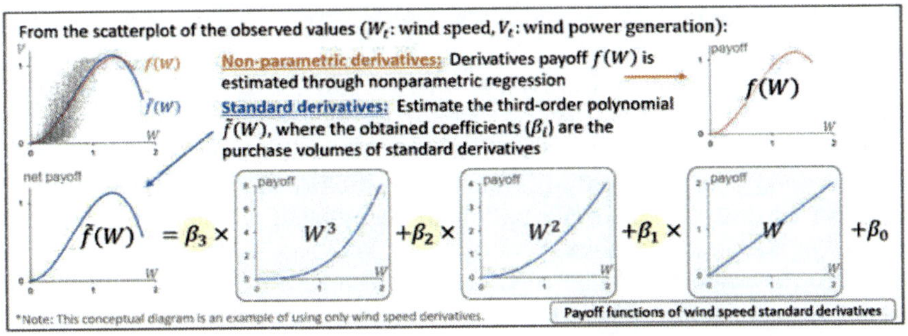


**Fig. 26.2** Conceptual diagram of methods for determining standard (non-parametric) derivatives' trading volume (payoff functions)

of derivatives based on their own generation patterns and the correlation with the used weather indexes. Also, such standardization would allow hedgers (buyers) to trade only those instruments that are necessary according to their own risk tolerance. In addition, unlike non-parametric derivatives, where payoffs are determined on a made-to-order basis, standard derivatives allow each product to be traded by multiple players. Similarly, if standardized derivatives are sold by any sellers, the market principle would bring the price closer to a fair price and at the same time, many players would participate, increasing the liquidity of the market.

Note that instead of setting the insurance company as a potential seller, a more efficient trading scheme might be established if baseload power generators (who sell baseload power generations in the wholesale market) are assumed to be the sellers of derivatives (risk underwriters). This is because when the derivatives' payout is higher (i.e., when wind generation is lower), baseload generator's revenue from the wholesale power price also tends to be higher (due to reduced power supply), conveniently offsetting cash flows. In this transaction scheme, an insurance company could be involved as an intermediary rather than a simple seller. Such a scheme has been considered in previous studies (Matsumoto et al. 2021; Yamada and Matsumoto 2021) for electricity derivatives, temperature derivatives, or solar radiation derivatives, but it could also be applied to wind derivatives. This study, which focuses on standardization schemes, does not conduct empirical analysis of such schemes, but this is an issue for future study.

### 26.2.2 Minimum Variance Hedging Problem

Hedgers aim to minimize the variance of the net cash flows of the electricity sales revenue (wind power generation volume) $V_t$ and the derivative payoff, $payoff(W_t, T_t)$, at each point in time, regardless of which scheme is chosen. In other words, the minimum variance hedging problem to be solved is as follows. (For a detailed exposition, see Yamada and Matsumoto 2021)

$$\text{minimize}: \text{Var}[V_t - payoff(W_t, T_t)] \tag{26.1}$$

where the term $payoff(W_t, T_t)$ refers to the payoff function itself for nonparametric derivatives and to the net payoff of a replicating portfolio of derivatives for a standard derivative scheme. The latter corresponds to the problem of finding the "contract volume" of standard derivatives. Note that Eq. (26.1) has degrees of freedom with respect to the payoff's location (first order moment), so it is possible to impose constraints on the mean value. For example, if a zero-mean constraint is imposed on the payoff itself (i.e., $\overline{payoff(W_t, T_t)} = 0$), it can be designed as a derivative with zero expected payoff that may not require premium payment at contract time. However, in this study, for convenience, we impose a zero-mean constraint on the hedging error $\varepsilon_t := V_t - payoff(W_t, T_t)$ (i.e., $\overline{\varepsilon_t} = 0$). In this case, at the time of the derivative contract, the buyer must pay the seller a premium equal to the expected payoff (the seller may require an additional risk premium), which must be calculated separately for each individual derivative. However, the value of the premium equivalent to the expected payoff could be obtained in a simplified manner, for example, by substituting historical weather data for the same period into the estimated payoff function.

### *26.2.3 Non-parametric Derivatives*

In the following, we construct concrete hedging models. The minimum variance hedging problem (26.1) for non-parametric derivatives, i.e., the problem of optimizing the derivatives' "payoff function" $payoff(W_t, T_t) := f(W_t, T_t)$, corresponds to estimating the following GAM:

$$V_t = f(W_t, T_t) + \varepsilon_t \tag{26.2}$$

where $f$ is the two-dimensional tensor product spline function estimated by GAM (Wood 2017) and $\varepsilon_t$ is the residual with mean 0. Since GAM estimation minimizes the sum of squares of the residual under the smoothing condition of $f$, it is synonymous with (26.1), the variance minimization of $\varepsilon_t$.

Note that it is possible to design "prediction error derivative" (i.e., a derivative with weather prediction errors as the underlying asset) by removing the time-dependent trend from the underlying weather index (see, e.g., Matsumoto and Yamada 2021a). However, in this study, we prioritized the intuitive and easy-to-understand estimation of cross-trend estimation of wind speed and temperature (e.g., payoff function shown in Sect. 26.3), and used the weather index as the underlying asset as is.

### 26.2.4 Standard Derivatives

Next, as to the standard derivatives' hedging model, a problem is to optimize "purchase volume" of the derivatives. Based on the idea for approximating the payoff functions of nonparametric derivatives, $f(W_t, T_t)$, using lower-order polynomials denoted by $\tilde{f}(W_t, T_t)$, the following model is formulated.

$$V_t = \tilde{f}(W_t, T_t) := \sum_{0 \leq i \leq I, 0 \leq j \leq J} \beta_{ij} W_t^i T_t^j + \varepsilon_t. \tag{26.3}$$

where $W_t^i T_t^j (i = 0, \ldots I; j = 0, \ldots, J)$ are the payoff functions for different standard derivatives. By estimating Eq. (26.3) with OLS, we can estimate the optimal (in the context of minimum variance hedging) contract volume $\beta_{ij}$ for each standard derivative. Note that the term when $i, j = 0$ is a constant, which can be regarded as a discount bond (with purchase unit $\beta_{00}$). Note also that wind power output is generally modeled by the third-order formula for wind speed (Ko et al. 2015). In addition, it is known that wind turbines have the tendency to reduce their power output when the temperature becomes too hot or too cold, due to a decrease in air pressure or freezing of components, suggesting an inverse U-shaped relationship with respect to temperature (see e.g., https://www.windturbinestar.com/wind-turbine-annual-output.html). Based on the above consideration, this study adopts $I = 3$, $J = 2$ in the hedging model (26.3).

These standardized derivatives can replicate payoffs that express nonlinearities the more they are incorporated, but increasing them too much may reduce the out-of-sample period fit, i.e., the hedging effect, due to so-called over fitting. At this point, a strategy of estimating Eq. (26.3) with LASSO regression (Tibshirani 1996) may be even more effective: in LASSO regression, the residual sum of squares with the L1 norm constraint on the regression coefficient vector is minimized. This allows variables with low relationships to be automatically removed from the model (i.e., setting the regression coefficient to 0) (Tibshirani 1996). In other words, "variable selection" using LASSO regression corresponds to a rational "product selection" strategy in the context of standardized derivatives trading, which allows hedgers to efficiently select only the necessary (effective) derivatives. The empirical analysis will also examine the effectiveness of this strategy.

## 26.3 Results

In this section, we present the results of estimating the payoff function and calculating the hedging effect. Here, we use the following observed values of wind power generation and weather in Denmark.

- Wind power generation [MWh]: actual power generation in Eastern Denmark (DK2) (Downloaded from https://www.nordpoolgroup.com/en/Market-data1/).
- Wind speed [m/s] and temperature [°C]: Observed values at Copenhagen Airport (Downloaded from http://rp5.ru/metar.php?metar=EKCH).

The estimation period (in-sample) and validation period (out-of-sample) are 2019 and 2020, respectively, using hourly data for all 24-h days from January 1 to December 31. The estimated trend shown in Sect. 26.3.1 is obtained from the in-sample period data only, while the hedging effect shown in Sect. 3.2 is calculated using the out-of-sample period data. While taking into account that sufficient historical data is often not available in practical situations, we estimate the payoffs from only one year of in-sample data and calculate the hedging effect for the following year, also for the purpose of confirming the high robustness of the proposed method. The hedging effect used in this study is defined as 1 − VRR, using the variance reduction ratio (VRR) defined as follows.

$$\text{VRR} := \text{Var}[V_t - payoff(W_t, T_t)]/\text{Var}[V_t] \tag{26.4}$$

### 26.3.1 Estimated Trend (Non-parametric Derivatives)

The trend of the estimated nonparametric derivative ($f(W_t, T_t)$ in Eq. (26.2)) is shown in Fig. 26.3, where, as mentioned in Sect. 2.3, a cubic form in the wind speed direction and a near quadratic form in the temperature direction are observed. For example, the sensitivity of wind power generation to wind speed is low when wind speed is low, but increases as wind speed increases, and then decreases when wind speed becomes too high. On the other hand, the effect of temperature peaks at around 10 °C, and the power generation decreases at both higher and lower temperatures.

### 26.3.2 Measurement of Hedging Effects

In this section, we measure hedging effect over the out-of-sample period; Fig. 26.4 plots cumulative hedging effect, and the hedging instruments included in each hedging model are shown in Table 26.1 (note that the operators "+" and "*" are not given strict definitions, but are used for convenience to correspond intuitively to the long regression equations). Both the nonparametric and standard types show that the hedging effect is enhanced with each combination of derivatives. The hedging effect increases from 62.8% when only "wind speed futures" are used ("W1") to 65.1% when "wind speed cubic derivatives" are included (to 65.4% when wind speed nonparametric derivatives are used), and "temperature squared derivatives" increases to 66.5% (to 67.2% using nonparametric derivatives of temperature), and to 67.2%

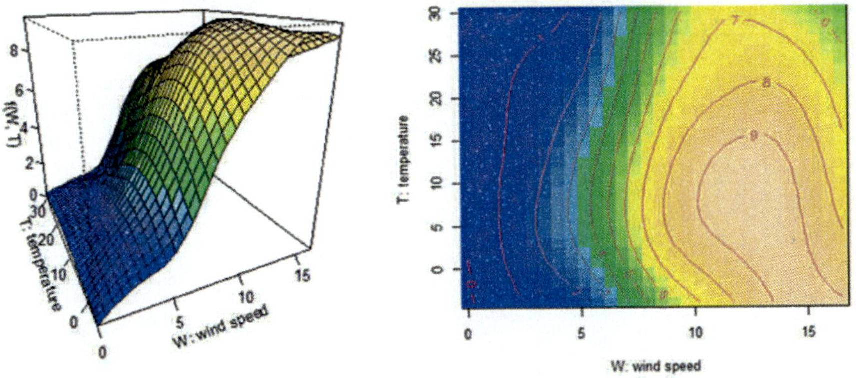


**Fig. 26.3** Estimated trends of nonparametric derivatives (left: perspective, right: contour)

when all "cross-derivatives of wind speed and temperature" are included (to 67.8% using two-dimensional nonparametric derivatives). Among the "cross-derivatives of wind speed and temperature" (six types in all), the improvement by the derivatives with the order of wind speed being one (i.e., $WT + WT^2$) is relatively large.

The above results are similar by month, as shown in Table 26.2. Among them, the effectiveness of the wind speed cubic derivative should be noteworthy, as it improved the hedging effect in all months, and in some months by more than 10%, compared to the case using only wind speed futures. Note that August in the validation year was the month in which nonlinearities and interactions were stronger, with an 81.0% improvement in hedging effect when including cross derivatives with temperature $WT + WT^2$ (first order for wind speed).

Finally, we test the effectiveness of the "product selection" strategy by LASSO regression described in Sect. 2.3. Since this strategy is expected to be particularly effective when trying to trade multiple standard derivatives, we use the "W3*T2"

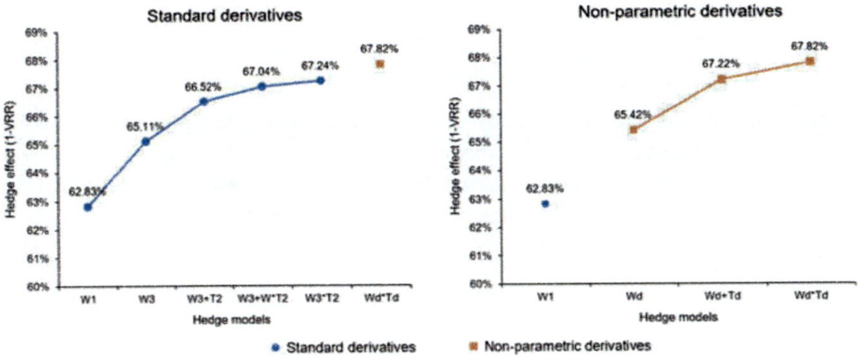


**Fig. 26.4** Hedging effects of standard derivatives and non-parametric derivatives

**Table 26.1** Components of hedging instruments by hedging model

| Hedge products | | Standard derivatives model | | | | | Non-parametric derivatives model | | |
|---|---|---|---|---|---|---|---|---|---|
| | | "W1" | "W3" | "W3 + T2" | "W3 + W * T2" | "W3 * T2" | "Wd" | "Wd + Td" | "Wd + Td" |
| Standard derivatives | $W$ | ✓ | ✓ | ✓ | ✓ | ✓ | | | |
| | $W^2 + W^3$ | | ✓ | ✓ | ✓ | ✓ | | | |
| | $T + T^2$ | | | ✓ | ✓ | ✓ | | | |
| | $WT + WT^2$ | | | | ✓ | ✓ | | | |
| | $W^2T + W^2T^2 + W^3T + W^3T^2$ | | | | | ✓ | | | |
| Non-parametric derivatives | $f_W(W)$ | | | | | | ✓ | ✓ | |
| | $f_T(T)$ | | | | | | | ✓ | |
| | $f(W, T)$ | | | | | | | | ✓ |

**Table 26.2** Monthly hedging effect and improvement rate against the "W1" model

| | | Jan (%) | Feb (%) | Mar (%) | Apr (%) | May (%) | Jun (%) | Jul (%) | Aug (%) | Sep (%) | Oct (%) | Nov (%) | Dec (%) |
|---|---|---|---|---|---|---|---|---|---|---|---|---|---|
| Hedge effect (1 − VRR) | "W1" | 63.9 | 73.2 | 62.1 | 56.8 | 55.3 | 38.6 | 62.8 | 18.2 | 60.6 | 61.7 | 67.0 | 60.3 |
| | "W3" | 64.2 | 75.9 | 68.3 | 58.4 | 56.5 | 41.5 | 65.7 | 26.5 | **64.3** | 62.2 | 69.9 | **60.6** |
| | "W3 + W * T2" | **64.6** | **76.8** | **68.7** | **58.5** | **58.2** | **44.0** | **67.6** | **32.9** | 62.7 | **62.7** | **70.5** | 59.5 |
| Improvement rate (to "W1") | "W3" | 0.4 | 3.7 | 10.0 | 2.8 | 2.1 | 7.5 | 4.5 | 45.4 | **6.1** | 0.8 | 4.4 | **0.5** |
| | "W3 + W * T2" | **1.1** | **4.8** | **10.7** | **2.8** | **5.3** | **14.0** | **7.5** | **81.0** | 3.5 | **1.5** | **5.2** | −1.3 |

model (with the original hedging effect of 67.24%) as an example and estimate it with LASSO regression instead of normal OLS. As a result, the coefficients of $WT$ and $W^2T^2$ are estimated as zero (i.e., the decision not to contract these standard type derivatives can be made), and the hedging effect is calculated to be 67.25%, which is equivalent (slightly higher) than the original model.

The LASSO model makes sense particularly in terms of the ability to reduce the number of hedging instruments purchased (while ensuring the same level of hedging effect). In actual exchange trades, it often happens that a smaller number of trading instruments would be much more desirable in terms of transaction costs. Hence, hedgers may be able to save transaction fees by utilizing a LASSO regression-based product selection strategy provided in this study.

## 26.4 Conclusion

This study demonstrated the effectiveness of a pre-existing nonparametric derivative portfolio hedging strategy applied to wind power generation business. The main contributions of this study are summarized as follows:

- Proposing new high-order standardized derivatives such as wind speed cubic derivatives and wind speed/temperature cross-derivatives, and demonstrating their effectiveness.
- Developing a specific market trading model, we discussed in detail the practical benefits of standardizing weather derivatives.
- Proposed a "product selection" strategy utilizing LASSO regression to streamline hedger's trading and demonstrated its effectiveness.
- In addition, we visualized mixed trend in wind power generation relative to temperature and wind speed that have been largely unexplored, and provided implications in the context of wind power forecasting models.

Derivatives with fine time granularity, such as those addressed in this study, may become more in demand once P2P power trading platforms are in place. For example, on a platform using blockchain technology, where settlements (payments) are executed immediately, highly granular derivatives would be very useful for small players who wish to minimize fluctuations in their digital wallets. Design methods and standardization schemes for sophisticated weather derivatives, such as those proposed in this study, are expected to have certain implications in the electricity market, where the transaction granularity will become increasingly finer in space and time.

**Acknowledgements** This work was funded by a Grant-in-Aid for Scientific Research (A) 20H00285, Grant-in-Aid for Challenging Research (Exploratory) 19K22024, and Grant-in-Aid for Young Scientists 21K14374 from the Japan Society for the Promotion of Science (JSPS).

## References

Aksoy B, Selbaş R (2021) Estimation of wind turbine energy production value by using machine learning algorithms and development of implementation program. Energ Sour Part A Recovery Utilization Environ Effects 43(6):692–704

Benth FE, Pircalabu A (2018) A non-Gaussian Ornstein-Uhlenbeck model for pricing wind power futures. Appl Math Finan 25(1):36–65

Benth FE, Di Persio L, Lavagnini S (2018) Stochastic modeling of wind derivatives in energy markets. Risks 6(2):1–21

Bilal B, Ndongo M, Adjallah KH, Sava A, Kébé CM, Ndiaye PA, Sambou V (2018) Wind turbine power output prediction model design based on artificial neural networks and climatic spatiotemporal data. In: 2018 IEEE international conference on industrial technology, pp 1085–1092

De Giorgi MG, Ficarella A, Tarantino M (2011) Assessment of the benefits of numerical weather predictions in wind power forecasting based on statistical methods. Energy 36(7):3968–3978

Foley AM, Leahy PG, Marvuglia A, McKeogh EJ (2012) Current methods and advances in forecasting of wind power generation. Renew Energ 37(1):1–8

Gersema G, Wozabal D (2017) An equilibrium pricing model for wind power futures. Energ Econ 65:64–74

Hanifi S, Liu X, Lin Z, Lotfian S (2020) A critical review of wind power forecasting methods—past, present and future. Energies 13(15):3764

Hastie T, Tibshirani R (1990) Generalized additive models. Chapman & Hall, Boca Raton, FL, USA

IEA (2021) Renewables. Analysis and forecast to 2026. https://www.iea.org/reports/renewables-2021. Accessed 15 June 2022

Kanamura T, Homann L, Prokopczuk M (2021) Pricing analysis of wind power derivatives for renewable energy risk management. Appl Energ 304:117827

Ko W, Hur D, Park JK (2015) Correction of wind power forecasting by considering wind speed forecast error. J Int Council Electr Eng 5(1):47–50

Marčiukaitis M, Žutautaitė I, Martišauskas L, Jokšas B, Gecevičius G, Sfetsos A (2017) Non-linear regression model for wind turbine power curve. Renew Energ 113:732–741

Matsumoto T, Yamada Y (2018) Cross hedging using prediction error weather derivatives for loss of solar output prediction errors in electricity market. Asia Pac Financ Mark 26:211–227

Matsumoto T, Yamada Y (2021a) Simultaneous hedging strategy for price and volume risks in electricity businesses using energy and weather derivatives. Energ Econ 95:105101

Matsumoto T, Yamada Y (2021b) Customized yet standardized temperature derivatives: a non-parametric approach with suitable basis selection for ensuring robustness. Energies 14(11):3351

Matsumoto T, Bunn DW, Yamada Y (2021) Pricing electricity day-ahead cap futures with multifactor skew-t densities. Quant Finan 22(5):835–860

Report Ocean (2022) Small wind power market size, share and trends analysis—global opportunity analysis and industry forecast 2030

Rodríguez YE, Pérez-Uribe MA, Contreras J (2021) Wind put barrier options pricing based on the Nordix index. Energies 14(4):1177

Tibshirani R (1996) Regression shrinkage and selection via the lasso. J Roy Stat Soc Ser B (Methodol) 58(1):267–288

Wood SN (2017) Generalized additive models: an introduction with R. Chapman and Hall, New York

Yamada Y (2008) Optimal hedging of prediction errors using prediction errors. Asia Pac Financ Mark 15:67–95

Yamada Y, Matsumoto T (2021) Going for derivatives or forwards? Minimizing cashflow fluctuations of electricity transactions on power markets. Energies 14(21):7311

# Chapter 27
# Assessment of Potential Tidal Power Sites in the Seto Inland Sea, Japan Using Multi-criteria Evaluation

**Morhaf Aljber, Ginga Sakanoue, Jae-Soon Jeong, Jonathan Salar Cabrera, and Han Soo Lee**

**Abstract** This paper is an analytical study of the optimal locations for tidal power generation in the Seto Inland Sea (SIS) in Japan. Multiple criteria were considered and evaluated to determine the optimal locations in the SIS to deploy tidal turbines. The tidal current speed was considered in a position of 30 m depth below the sea surface, given that most of the current commercial turbines are functioning between 20 and 50 m underwater. Also, the distance from ports and the coastlines was included for the installation and maintenance of tidal devices. The distance from existing power generation plants is also considered for the outcome energy to be easily inserted into the grid. Further, renewable energy resources from wind and surface waves are also considered in the analysis for the potentiality of a combined power generation. The national parks and restricted areas were eliminated from the selection process. The assessment process was conducted using the analytic hierarchy process (AHP) in combination with GIS-based spatial analysis. The results illustrated that Kurushima Strait, Bisan Strait, Naruto Strait, and Akashi Strait are suitable locations for tidal power generation. Moreover, Bungo Channel and Kii Channel are highlighted as potential sites for wind-wave power generation sites.

**Keywords** Analytic hierarchy process (AHP) · Multi-criteria analysis · Renewable energy · Tidal power generation · Wind-wave power generation

M. Aljber · G. Sakanoue · J.-S. Jeong · J. S. Cabrera · H. S. Lee (✉)
Transdisciplinary Science and Engineering Program, Graduate School of Advanced Science and Engineering, 1-5-1 Kagamiyama, Higashi-Hiroshima, Hiroshima 739-8529, Japan
e-mail: leehs@hiroshima-u.ac.jp

H. S. Lee
Center for the Planetary Health and Innovation Science (PHIS), The IDEC Institute, Hiroshima University, Hiroshima, Japan



N. S. Caetano and M. C. Felgueiras (eds.), *The 9th International Conference on Energy and Environment Research*, Environmental Science and Engineering,
https://doi.org/10.1007/978-3-031-43559-1_27

## 27.1 Introduction

Tidal currents hold the potential to contribute generously to the power sector in several countries around the world, where the coastal terrains offer unique positions in terms of current circulation, speed, and bathymetry properties (Roberts et al. 2016). Recently, the popularity of tidal power generation has increased rapidly and for this technology to be economically effective, optimum distribution of the tidal farm should be deployed (Divett et al. 2016). Nevertheless, selecting an optimum site for the tidal farm is the first step.

The reciprocal interaction between water current and the tidal turbine has always been drawing the attention of science and research to reach an outstanding tidal turbine design that harnesses the largest amount of energy. McAdam et al. (2013), experimentally proved that by applying certain conditions to perceive the merits of varied blade pitch, and blockage proportion, the extracted energy could disproportionately exceed Lanchester-Betz limits for kinetic efficiency.

Since the Fukushima Daiichi nuclear power plant accident in 2011, most nuclear power plants in Japan were terminated for inspection (Bricker et al. 2017). Simultaneously, an innovative approach has been released to assess and investigate potential substituted power to compensate for the electrical system of the country. Concerning the rising environmental issues, a large body of research and assessment has been continuously developing to estimate several sites for tidal power in the Japanese national water. The New Energy Development Organization (NEDO), universities, and research institutions in Japan conducted a general estimation of potential sites for tidal power generation based on tidal current velocity and density and a postulated amount of energy that could be extracted was also provided. Following such results, Coles et al. (2018) conducted a study to estimate the potential power that could be harnessed from the Naru strait offshore Nagasaki prefecture in the western part of Kyushu Island using the Finite Volume Community Ocean Model (FVCOM). Another general estimation for Goto Archipelago, which includes Naru Strait, was held by Waldman et al. (2017) as a primer step for further study. However, the previously mentioned research did not consider other factors that could affect the optimal position of tidal power generation, like distance from the existing power plants and ports, excluding preserved and restricted areas and other factors.

The Seto Inland Sea (SIS) in Japan is considered one of the most identical spots for tidal power generation due to the unique characteristics of waves, tides, and tidal circulations Coles et al. (2018). The SIS is the largest channel-shaped enclosed coastal sea in the western part of Japan, with approximately 23,000 km$^2$, a length of approximately 500 km, and an average depth of roughly 38 m. The SIS is connected to the outer Pacific Ocean via Kii Channel and Bungo Channel and to the Korea Strait via the Kanmon Strait. It includes approximately 700 Islands and a number of waterways (Lee et al. 2015). Thus, the tidal currents are high and strong, which provides the most favourable place for tidal current power generation.

Taking into account the spatial distribution of the islands in the SIS and the fact that most of these islands are inhabited, there is daily usage of the water around. It is not possible to select randomly one of the sites for this technology depending just on the tidal current's properties. Although some studies have suggested the potential deployment of a tidal device in a particular location in the SIS (Nakamura et al. 2013), there was almost no holistic approach that includes all possible locations, without mentioning a combined power generation in which resources from winds, waves and tides are concurrently evaluated. Moreover, one should consider the restricted areas for environmental protection and the distance from ports and harbours to construct and install tidal devices and the maintaining processes later. With respect to the installation, water depth also should be considered because most commercial tidal power devices/turbines are designed to operate between 20 and 50 m deep underwater. Lastly, the distance from the existing power station or substation should be considered to connect directly to the existing grid and avoid supplementary constructions.

To achieve the aforementioned, the GIS-based Multi-Criteria Evaluation (MCE) methodology was applied in this study. The spatial analysis of potential power generation sites is crucial as it allows for the visual presentation of site suitability along with quantitative site assessments (Marsh et al. 2021). Thus, the analytic hierarchy process (AHP), one of the most commonly used MCE methods, was applied in combination with GIS-based spatial analysis to identify the optimal locations that satisfies all the considered criteria for the potential deployment of tidal power generation technology. In addition, the potential sites for combined wind-wave power generation in the SIS were evaluated.

## 27.2 Methodology

To identify the optimal location in the SIS, two major steps were considered as depicted in Fig. 27.1. The first step is conducting an online survey to calculate the relative weight for each criterion. The second step is GIS-based spatial analysis to determine the optimal location by considering the relative weight for each criterion.

### *27.2.1 Analytical Hierarchy Process (AHP)*

The Analytical Hierarchy Process (AHP) is a commonly used method for decision-making analysis and evaluation, especially in the field of social science and economics. However, a combined with GIS estimation to assess the potentiality of tidal turbines is rarely used (Marsh et al. 2021). AHP is a keystone for a better understanding of the importance of each one of the considered criteria, and it offers the possibility to evaluate each one of the criteria by comparing it with others of the same level. Following the concept of AHP, a paired comparison designed survey was prepared and conducted online to reach and obtain the opinions of experts in the field of energy and ocean engineering, along with academic and governmental officers. The survey was built in a hierarchical structure, in which the criteria were separated

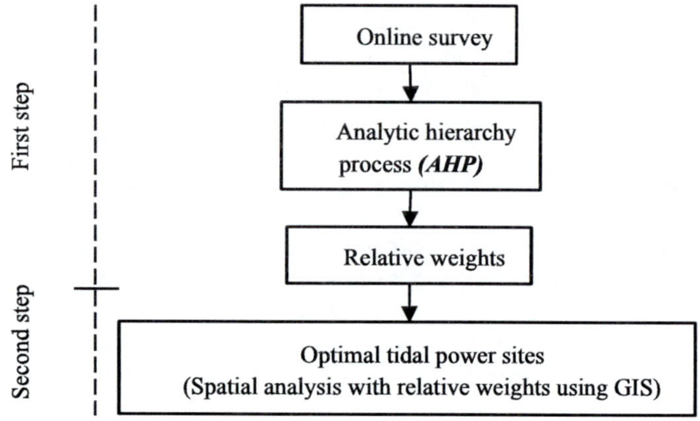


**Fig. 27.1** Flow chart of the study

**Table 27.1** The relative weights obtained from AHP method

| Criteria | | Relative weight |
|---|---|---|
| Main criteria | Wind | 0.151 |
| | Wave | 0.166 |
| | Tidal currents | 0.264 |
| Sub-criteria | Distance from ports | 0.035 |
| | Distance from coastlines | 0.052 |

into levels of importance. Hence, the main criteria were the wind, wave, and tidal currents. Sub-criteria were the distance from ports, electrical power generation, and consumption areas. The water depth and other factors were considered as well in sub-criteria. Thus, the participants were requested to give a rate to each criterion within the same level of importance compared with other criteria. After collecting the results of the survey, the AHP method was applied to calculate the relative weight of each criterion. Table 27.1 presents the relative weights for the main criteria and sub-criteria resulting from the AHP.

### 27.2.2 *Analytical GIS-Based Spatial Evaluation*

The second step aims to analytically identify the optimal locations in the SIS for wind-wave combined power generation, and to identify the optimal locations for only tidal current power generation. Therefore, the suitability index for optimal sites was calculated. The relative weights obtained using the AHP method were multiplied by their representative layers in the GIS. The suitability index (SI) is calculated as in Eq. (27.1):

$$SI = \sum_{i=1}^{n} Rw_i * F_i \tag{27.1}$$

where $Rw_i$ are the relative weights for the layer, and $F_i$ is the representative layer.

The optimal sites estimation was obtained using the raster calculator from the spatial analyst tools in ArcMap. This approach produced a new layer with a unique batch of colours classifying the possible locations from the highest to the lowest so that a more coherent conception could be perceived out of this process for the best sites in the SIS.

In the GIS-based spatial analysis, the shapefile format data were imported into the GIS map. Then, the main criteria layers, such as wind, wave, and tide, were converted into raster files. Hence a perceptible distribution of the values within each layer was obtained. Then, the implicit accumulated distance tool was utilised on the sub-criteria layers to divide them into several parts of importance, from the closest to the furthest in case of ports and vice versa in case of restricted areas. Lastly, the depth was extracted out of the raster file.

The third step is the reclassification of the values within each raster layer by which a unique number from 1 to 5 was assigned to each range of values. Thus, 5 was given to the largest value of wind speed and wave height and 1 for the smallest. Also, 5 was given to the closest distance from ports and electrical stations. On the other hand, 1 was assigned for the closest points to the restricted areas and 5 for the furthest.

The fourth step was to multiply those final obtained layers with their relative weights using the Raster Calculator tool. Thus, the resulting layer classified the optimal locations for wind-wave-tide combined power potential sites. By following the same manner, a separate estimation was conducted for tidal current power generation and wind-wave power generation.

## 27.3 Results and Discussion

The results of the estimation process identified several promised locations in the SIS for a combined potential, wind-wave-tide (Fig. 27.2). The Bungo Channel, the western entrance of the SIS, was identified as a promising position. Due to the water volume exchange with the outer Pacific Ocean and the high wind, Bungo Channel experiences a relatively higher wave compared to the inner side of the SIS. Therefore, there is a possibility of a wind-wave combined power generation. Given that the existence of the ports in the southwestern part of Shikoku Island such as, Uwajima port and Sukumowan port, and in the northeastern part of Kyushu Island such as, Saiki port and Tsukumi port, provides an advantage for construction and maintenance of a floating wind power generation. Also, the neighbouring towns provide a possibility to connect directly to the electrical grid. A separate estimation was conducted considering only wind and wave as alternatives for the main criteria, to investigate more the potentiality of wind-wave combined power generation. The

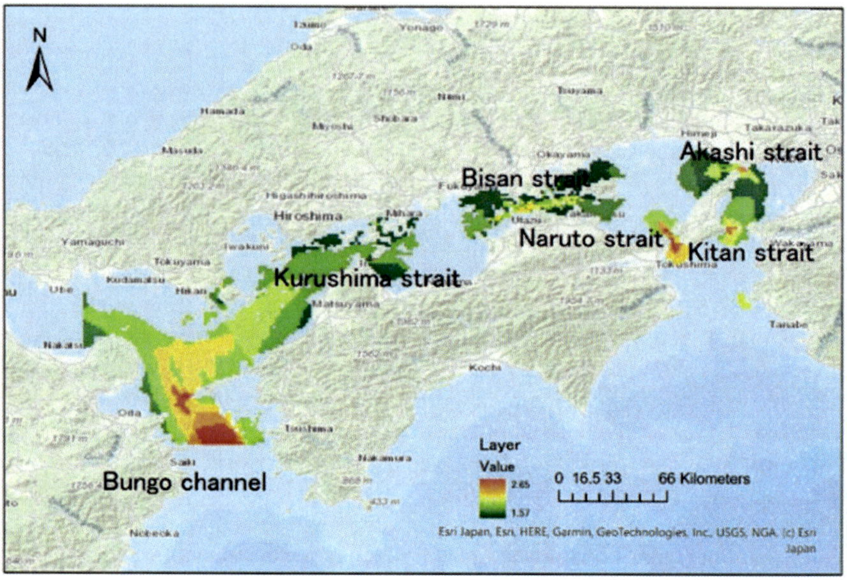


**Fig. 27.2** The optimal locations for wind-wave-tide combined power generation starting from the western part of SIS respectively Bungo Channel, Kurushima Strait, Bisan Strait, Naruto Strait, Akashi Strait, Kitan Strait

obtained results identified the Kii Channel, the north-eastern entrance of SIS, along with the Bungo Channel (Fig. 27.3), as positions with high potential for wind-wave power generation. On the other hand, those positions do not hold a high potential for a tidal current power generation, due to the wide-opened volume of water and properties of the entering waves (Lee et al. 2015).

Starting from the middle part of SIS, Kurushima Strait and Bisan Strait have been identified as potential locations (Fig. 27.4a). Considering the characteristics of the tidal currents in those positions and the distance from other criteria, it is accepted to consider the two straits as potential locations for tidal current power generation. In addition, the resulted map identified the three straits around Awaji Island in the eastern part of the SIS, Akashi Strait, Naruto Strait, and Kitan Strait, as potential locations for tidal power generation (Fig. 27.4b). The three straits are well-known for high-velocity tidal currents. Moreover, the neighbouring towns provide the possibility to connect to the electrical grid.

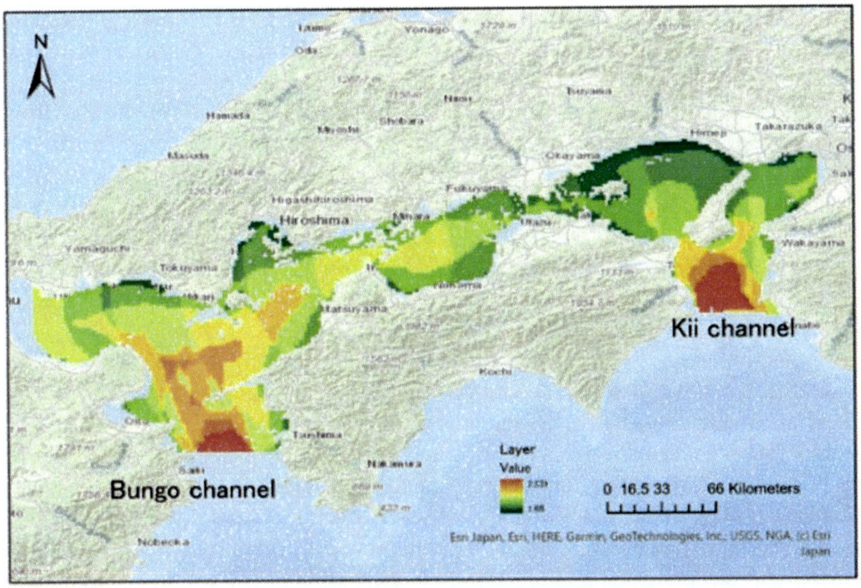


**Fig. 27.3** The optimal locations for wind-wave case, Bungo Channel and Kii Channel

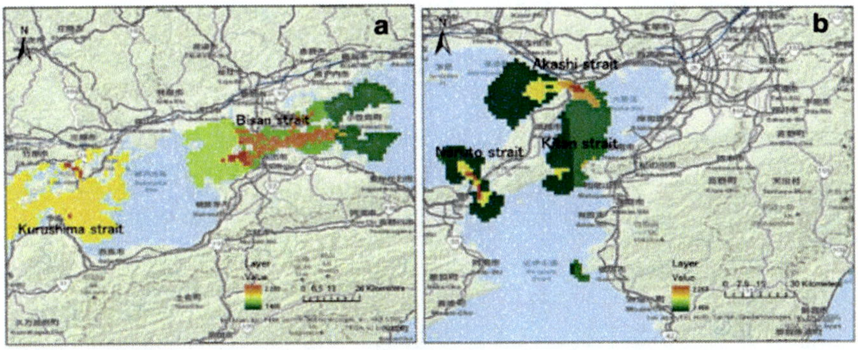


**Fig. 27.4** The optimal locations for tidal power generation only case. **a** Kurushima strait and Bisan strait in the middle part of the SIS, **b** Naruto strait, Akashi strait, and Kitan strait in the eastern part of the SIS

## 27.4 Conclusion and Future Work

This study aimed to identify the optimal locations for a wind-wave-tide combined power generation using multi-criteria evaluation. The criteria were divided into two levels of importance. The first level contains the main criteria for marine renewable energy resources, offshore winds, waves, and tidal currents. The second level contains the sub-criteria, such as the distance from ports, electrical power generation, consumption areas, the water depth, and restricted areas. Three cases of estimation were considered. In the first estimation, a combined wind-wave-tide power generation was conducted. The results presented Bungo Channel, Kurushima Strait, Bisan Strait, Akashi Strait, Naruto Strait, and Kitan Strait as possible locations. To improve the results and identify more specifically, the second case considered wind-wave only for the main criteria with reserving the sub criteria. The results identified the Bungo Channel and Kii Channel as potential locations. In the third case, an estimation for tidal current alone as main criteria with reserving sub criteria was conducted. It resulted in Kurushima Strait, Bisan Strait, Akashi Strait, Naruto Strait, and Kitan Strait illustrating high potentials for tidal current power generation.

Although this study presented a robust approach to identify the possible locations in the SIS for a wind-wave-tide combined power generation, the main goal is to assess the potentiality of tidal power generation. More criteria should be considered to achieve better results, such as navigational routs, ports' size, population, and so on. Furthermore, the study could be expanded to the entire coastal area of Japan to estimate the optimal locations for the wind, wave, and tide potential power generation.

Finally, a comprehensive numerical simulation with Adaptive Mesh Refinement (AMR) will be conducted and used to study the optimal tidal turbine configuration and the optimal tidal array distribution in the designated location to have robust calculation of the potential power contribution to the electrical system in Japan.

## References

Bricker JD, Esteban M, Takagi H, Roeber V (2017) Economic feasibility of tidal stream and wave power in post-Fukushima Japan. Renew Energ 114:32–45. https://doi.org/10.1016/j.renene.2016.06.049

Coles D, Walsh T, Kyozuka Y, Oda Y (2018) Tidal turbine array design and energy yield assessment for Naru Strait, Japan

Divett T, Vennell R, Stevens C (2016) Channel-scale optimisation and tuning of large tidal turbine arrays using LES with adaptive mesh. Renew Energ 86:1394–1405. https://doi.org/10.1016/j.renene.2015.09.048

Lee HS, Shimoyama T, Popinet S (2015) Impacts of tides on tsunami propagation due to potential Nankai Trough earthquakes in the Seto Inland Sea, Japan. J Geophys Res Oceans 120(10):6865–6883. https://doi.org/10.1002/2015JC010995

Marsh P, Penesis I, Nader JR, Cossu R (2021) Multi-criteria evaluation of potential Australian tidal energy sites. Renew Energ 175:453–469. https://doi.org/10.1016/j.renene.2021.04.093

McAdam RA, Houlsby GT, Oldfield MLG (2013) Experimental measurements of the hydrodynamic performance and structural loading of the Transverse Horizontal Axis Water Turbine: Part 1. Renew Energ 59:105–114. https://doi.org/10.1016/j.renene.2013.03.016

Nakamura T, Abe H, Husain F (2013) Energy conversion efficiency of a vertical water turbine under combined actions of wave and current

Roberts A, Thomas B, Sewell P, Khan Z, Balmain S, Gillman J (2016) Current tidal power technologies and their suitability for applications in coastal and marine areas. J Ocean Eng Mar Energ 2(2):227–245. https://doi.org/10.1007/s40722-016-0044-8

Waldman S, Yamaguchi S, O'Hara Murray R, Woolf D (2017) Tidal resource and interactions between multiple channels in the Goto Islands, Japan. Int J Marine Energ 19:332–344. https://doi.org/10.1016/j.ijome.2017.09.002

# Chapter 28
# Numerical Analysis of Cooling Characteristics of Battery Pack Through an Integrated Liquid Spray and Air Cooling System

**Patcharin Saechan and Isares Dhuchakallaya**

**Abstract** Due to the requirement of a long driving range of electric vehicles, the batteries must be packed ever more densely to increase the energy density. Thus, enormous heat generated is required to dissipate appropriately. This highlights the design challenge for thermally managing the densely packed batteries. In this study, the air cooling system integrated with non-electrically conductive liquid spray was used to enhance the cooling performance. The transient thermal response of a 40-cell lithium-ion battery pack was simulated to investigate the effect of the injection rate on the heat removal performance. From this study, it shows that the integrated liquid spray and air cooling system is highly capable of reducing the maximum temperature as well as the temperature uniformity of the battery pack compared to the dry air cooling scheme. When the HFE flow rate increased, the ability to reduce temperature became weaker. The optimum HFE flow rate should be determined. Thoroughly, spray cooling can be used securely for thermally managing the high heat flux of the densely-packed batteries.

**Keywords** Electric vehicles · Liquid spray cooling · Lithium-ion · Simulation · Thermal management

P. Saechan
Faculty of Engineering, King Mongkut's University of Technology North Bangkok, Bangsue, Bangkok 10800, Thailand

I. Dhuchakallaya (✉)
Center of Excellence in Computational Mechanics and Medical Engineering, Faculty of Engineering, Thammasat School of Engineering, Thammasat University, Klong-Luang 12120, Pathumthani, Thailand
e-mail: disares@engr.tu.ac.th



N. S. Caetano and M. C. Felgueiras (eds.), *The 9th International Conference on Energy and Environment Research*, Environmental Science and Engineering,
https://doi.org/10.1007/978-3-031-43559-1_28

## 28.1 Introduction

So far, the demand for Lithium-ion batteries (LIBs) to power electric vehicles (EVs) has been exponential growth because of its remarkable characteristic, i.e., high specific energy, long life span, and low self-discharge rate. For safe high power density and fast charging concerns, a battery thermal management system (BTMS) should be carefully designed to maximize the power capacity and lifespan of the battery. For optimized battery performance, the operating temperature of the LIBs is necessary to be managed within 25–40 °C, and the non-uniformity of temperature across the cells should not exceed 5 °C (Pesaran 2002). These thermal impacts are then the stringent issues of the BTMS to ensure that the batteries operate efficiently and last safely.

Currently, there are primarily two working media including air and liquid used successfully in the BTMS. An air-based cooling system can be classified into natural and forced convection. The main advantages of this approach are low-cost manufacturing, simple design, and low environmental impact. Nevertheless, this system still suffers from some weaknesses such as low thermal capacity and non-uniformity of temperature during the heavy load of cooling. Recent studies have reported that the battery pack operating at a low discharge rate can be dissipated successfully with natural convection, but at a high discharge rate of 2C the air cooling was not economical for thermal management (Saechan and Dhuchakallaya 2022).

The liquid-based system is generally able to buffer and dissipate a larger amount of heat than the air-based system, due to their superior heat transfer coefficient. The cooling liquid generally used in the BTMS can be classified into two groups: non-conductive and conductive liquid coolants.

Non-electrically conductive liquid coolants are in direct contact with the batteries, where the batteries are immersed in the liquid (Li et al. 2022). Thus, safety issues related to a short circuit is necessary be concerned. Conversely, for indirect cooling, the electrically conductive liquids are not brought into direct contact with the batteries. This method is simple to implement owing to the battery pack placed in the fluid jacket. However, the cooling performance of this technique is controlled by the convective boundary. The thermal resistances of the material layers cause the temperature gradients. Noticeably, the thermal properties of the non-electrically conductive liquids are relatively inferior to those of common coolants. Furthermore, the BTMS with liquid-based are still restricted owing to several drawbacks, i.e., the safety, leakage, more weight, additional components.

Although the aforementioned cooling approaches are successfully used, but still with inevitable drawbacks, leading to the research gap for improving the cooling performance. Owing to the inferior thermal properties of non-electrically conductive liquid, the viability of a direct liquid cooling scheme strongly depends on the convective heat transfer coefficient. This can be accomplished by injecting a small quantity of liquid into the main airflow. The liquid droplets then absorb heat from the air and hot surface of the batteries. Due to a massive surface-area-to-volume ratio coupled with the wide droplet dispersion across the battery surface, a high heat transfer coefficient is certainly achieved.

The inlet air pre-cooling with water spray affected the cooling performance was numerically investigated (Alkhedhair et al. 2013; Montazeri et al. 2015). It showed that a higher air speed decreased the evaporation rate of droplets. Larger droplets were potentially less cooling efficient, but they had a wider dispersion ability, resulting in a more effective heat and mass exchange. Thus, the cooling performance can improve by decreasing the droplet size and broadening the droplet size distribution.

Recently, the water spray was applied to pre-cool the inlet air for the BTMS (Yang et al. 2019; Saw et al. 2018). It reported that the water spray cooling system can control the battery temperature within the desired operating temperature range for 3C-rate of charging. A higher water flow rate can reduce the maximum temperature, but caused a large temperature difference in the batteries. Besides, the feasibility of humid air cooling applications for the battery pack was experimentally proved (Zhao et al. 2020; Youssef et al. 2021). The relative humidity of the ambient air had a significant effect on the cooling performance. The ambient air with lower relative humidity provided a better cooling performance.

Although the cooling concept of liquid spray on the hot surface has been explored for years, the applications of liquid spray cooling to the BTMS are rather limited. Most research did not mention the safety concern for the long-term use because the water used as the coolant is electrically conductive fluid. Thus, the primary objective is to investigate the viability of non-electrically conductive liquid spray cooling of the densely-packed batteries to meet the temperature requirements. The flow rate of liquid was also examined to optimize the usage of liquid spray.

## 28.2 Battery Pack and Cooling System

Battery pack is comprised of many series and parallel connected batteries to achieve a desired voltage and capacity. In this study, a rectangular (5 × 8 cells) pack of cylindrical batteries NCR18650B with a capacity of 3400-mAh was cooled by a forced-air coupled with liquid spray cooling system as shown in Fig. 28.1. An inline layout of batteries with a center-to-center spacing of 21 mm was used. The inlet air flow was of uniform velocity at 2 m/s and the liquid droplets were injected with 4-nozzle under a constant pressure. The nozzles were mounted at the top of fluid domain above the battery pack. The two-phase flow made of liquid droplets suspended in air was then simulated.

In this study, the cylindrical orthotropic thermal conductivities of the LIB in radial, tangential and longitudinal directions are $k_r = 0.951$ W/m °C, $k_\theta = k_z = 37.106$ W/m °C, respectively. The density and specific heat capacity of the LIB are $\rho = 3602.12$ kg/m$^3$ and $c_p = 776.59$ J/kg °C (Saechan and Dhuchakallaya 2022). According to the coolant assessment of Mudawar et al. (2008), Novec fluid HFE-7100 was evaluated as the suitable cooling liquid for the direct-liquid-cooling system in high-flux applications. This evaluation was based on the performance rating with

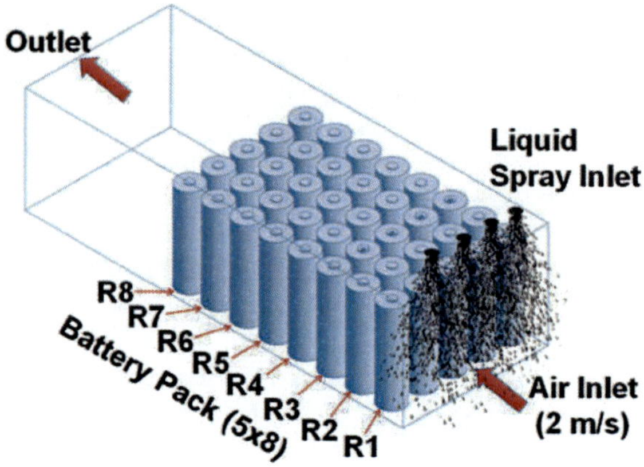


**Fig. 28.1** Schematic diagram of the battery pack and spray cooling system

regard to the thermal, environmental, and safety aspects. Hence, HFE-7100 is utilized for spray cooling to dissipate high-heat-flux from the pack of batteries. The properties of the test liquid 3M™ Novec™ Engineered Fluid HFE-7100 are $k = 0.069$ W/m °C, $\mu = 1.168 \times 10^{-3}$ Pa s, $\rho = 1510$ kg/m$^3$ and $c_p = 1183$ J/kg °C (3M™ Thermal Management Fluids for Military and Aerospace Apps).

## 28.3 Heat Generation Rate and Simulation Method

Throughout the charging or discharging processes, the battery generates heat, and its temperature consequently increases. Heat generation in the battery can be attributed to two main sources. One is the reversible heat which is dominated by a change in entropy during the electrochemical reactions inside the battery. Thus, it might be called "entropic heat". This can be either endothermic or exothermic, depending on the state being opposite for charge and discharge. The entropic heat is given as:

$$\dot{Q}_{rev} = -I\left(T\frac{\mathrm{d}U_{OC}}{\mathrm{d}T}\right) \tag{28.1}$$

where $I$ is the current, which sign is defined as positive when discharged. $T$ is the cell temperature, $\mathrm{d}U_{OC}/\mathrm{d}T$ is the entropy thermal coefficient, and $U_{OC}$ is the open-circuit voltage.

The other source is irreversible heat. This is caused by the voltage drop due to electron and ion transport and electrochemical reactions. It can be estimated from Joule heating of the internal resistance. Therefore, this might be called "ohmic heat" and is always exothermic. The ohmic heat can be expressed as:

$$\dot{Q}_{irr} = I(U_{OC} - U) = I^2R \tag{28.2}$$

where $U$ is the cell voltage, and $R$ is the internal resistance of the cell. Thus, the total heat generation can be rewritten as:

$$\dot{Q}_{gen} = \dot{Q}_{rev} + \dot{Q}_{irr} = I^2 R - IT\frac{dU_{OC}}{dT} \tag{28.3}$$

At low current process, the contribution of entropic heat is of the similar order to ohmic heat. Conversely, at higher currents, the ohmic heat dominates, and the heat generation rates during charge and discharge are quite similar. Due to the internal resistance ($R$) and entropy thermal coefficient ($dU_{OC}/dT$) being function of the cell temperature and state of charge ($SOC$), the experimental results of Xie et al. (2018) were used to formulate the correlations between them.

A three-dimensional model for transient simulation of the battery pack is executed using ANSYS Fluent. For liquid spray cooling, the two-phase flow behavior is characterized by the Eulerian-Eulerian approach. Both air and liquid droplets are solved in the Eulerian framework. Since there are two fluids present in the continuum, the transport of mass or momentum has to be weighted by the volume fraction of the dispersed phase. The coupling between phases is accomplished through the pressure and the interphase exchange terms. The set of conservation equations is solved separately for each phase in the Eulerian framework. The continuity, momentum, and energy equations for phase $k$ are expressed as (Vujanović et al. 2015):

$$\frac{\partial(\alpha_k \rho_k)}{\partial t} + \nabla \cdot (\alpha_k \rho_k \underset{k}{\vec{v}}) = \sum_{l=1, l\neq k}^{n} \Gamma_{kl} \tag{28.4}$$

$$\frac{\partial(\alpha_k \rho_k \vec{v}_k)}{\partial t} + \nabla \cdot (\alpha_k \rho_k \vec{v}_k \vec{v}_k) = -\alpha_k \nabla p + \nabla \cdot \alpha_k (\tau_k + \tau_k^t) + \alpha_k \rho_k \vec{g} + \sum_{l=1, l\neq k}^{n} M_{kl} + \vec{v}_k \sum_{l=1, l\neq k}^{n} \Gamma_{kl} \tag{28.5}$$

$$\frac{\partial(\alpha_k \rho_k h_k)}{\partial t} + \nabla \cdot (\alpha_k \rho_k \vec{v}_k h_k) = \nabla \cdot \alpha_k (q_k + q_k^t) + \alpha_k \rho_k \vec{g} \cdot \vec{v}_k + \alpha_k \rho_k \theta_k + \alpha_k \tau_k : \nabla \vec{v}_k + \alpha_k \frac{\partial p}{\partial t} + \sum_{l=1, l\neq k}^{n} H_{kl} + h_k \sum_{l=1, l\neq k}^{n} \Gamma_{kl} \tag{28.6}$$

As a requirement of the conservative condition, the sum of all phasic volume fractions must equal 1. The terms $\Gamma_{kl}$, $M_{kl}$, and $H_{kl}$ are the mass, momentum, and energy exchange coefficients between phases $k$ and $l$. A uniform droplet size distribution of 10 μm was assumed as the initial spray characteristic. The spray is assumed to be diluted, so the collisions of droplets may be neglected. The drag function of Schiller and Naumann and the lift model of Saffmen-Mei were used for the momentum exchange coefficients, and the Ranz-Marshall correlation was used for the heat transfer coefficient of droplets. More details of these source terms are described in Vujanović et al. (2015), ANSYS (2020).

The heat generation rate of the battery pack can be found in Eq. (28.3), and the governing equation for heat conduction in the battery cell can be written as:

$$\rho_b c_{p,b} \frac{\partial T_b}{\partial t} = \frac{k_{b,r}}{r} \frac{\partial}{\partial r}\left(r \frac{\partial T_b}{\partial r}\right) + \frac{k_{b,\theta}}{r^2} \frac{\partial^2 T_b}{\partial \theta^2} + k_{b,z} \frac{\partial^2 T_b}{\partial z^2} + \dot{Q}_{gen} \quad (28.7)$$

Those governing equations were solved based on the Phase Coupled SIMPLE (Semi-Implicit Method for Pressure Linked Equations) algorithm for pressure–velocity coupling. The Realizable $k - \varepsilon$ turbulence model was selected to capture the flow field of the mixture. Walls were modeled as a non-slip boundary condition, and the enhanced wall treatment was used for the near-wall boundaries. The second-order upwind was used for the spatial discretization of the pressure, momentum, energy, and turbulent terms. The heat source of the battery was created by the User Define Function. The uniform velocity of 2 m/s and temperature of 27 °C were set as the inlet air condition. The liquid spray of monodisperse HFE-7100 droplets was injected at the temperature of 25 °C. The pressure-outlet boundary condition was assigned to the outlet.

## 28.4 Results and Discussion

The fully charged battery pack was discharged at 2C-rate. The ambient air was introduced at a velocity of 2 m/s, and the liquid HFE was injected from the nozzles. The cell temperature variation in each row of the battery pack during discharge is presented in Fig. 28.2. The cell temperature increased linearly in the first stage of discharge, and then the temperature growth rate dropped significantly in the middle stage. In the final stage of discharge, the temperature of the battery increased rapidly. This is because the high entropy thermal coefficient and internal resistance of the battery in the low SOC caused a higher heat generation rate.

Furthermore, the cell temperatures in each row increased almost linearly along the flow path, except in the first row. This might be because the turbulence caused by the fluid flowing through the first row promoted the heat transfer in the subsequent rows as explained in Saechan and Dhuchakallaya (2022). Moreover, HFE droplets contacted with the cell surface of the frontmost row were less than with the later rows.

Figure 28.3 shows the behavior of HFE droplets colored by velocity streamline and cell temperature distribution along the flow channel. It can be seen that the HFE droplets with a higher flow rate can move farther into the pack of batteries. The temperature of the battery was clearly reduced as the HFE droplets passed. This is due to the high heat transfer coefficient of liquid droplets. Noticeably, the injection of HFE with low flow rate in Fig. 3a was of slight importance in reducing the temperature of the battery pack. This might be that most droplets injected were carried away by

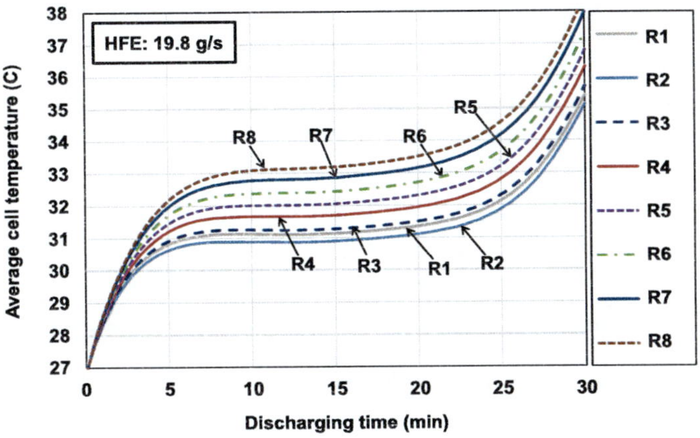


**Fig. 28.2** Time history of cell temperature during discharge at 2C-rate

the air stream, before contacting the battery surface. Hence, the temperatures in each row increased along the flow path.

The effect of the HFE mass flow rate on the cooling performance of the battery pack was investigated and the simulated results are presented in Fig. 28.4. At the end of discharge, the cell temperatures decreased with increasing the mass flow rate of HFE. This also diminished the non-uniformity of temperature distribution across the battery pack. As seen, the mass flow rate of HFE should be greater than 30 g/s in total to guarantee that all cells in the dense pack can operate under the optimal working temperature range, where the maximum temperature and the maximum temperature difference among battery cells are restricted at about 40 and 5 °C, respectively. Although the maximum cell temperature and temperature difference generally decreased with increasing the HFE flow rate, it is not economical for the excessive HFE injection in the aspect of thermal management. Compared with a dry

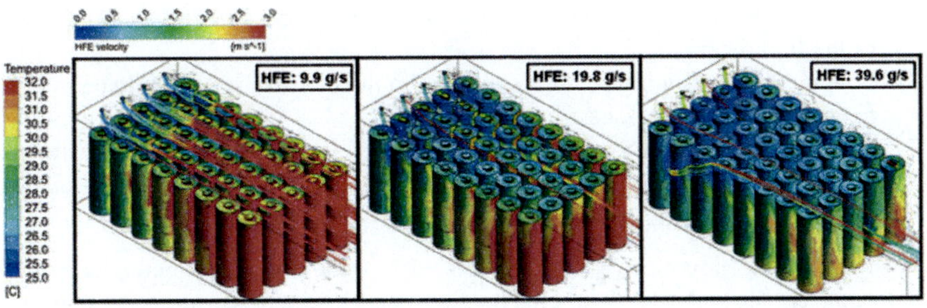


**Fig. 28.3** Temperature contours and HFE streamlines for different HFE injection rates at the air inlet of 2 m/s and discharge rate of 2C

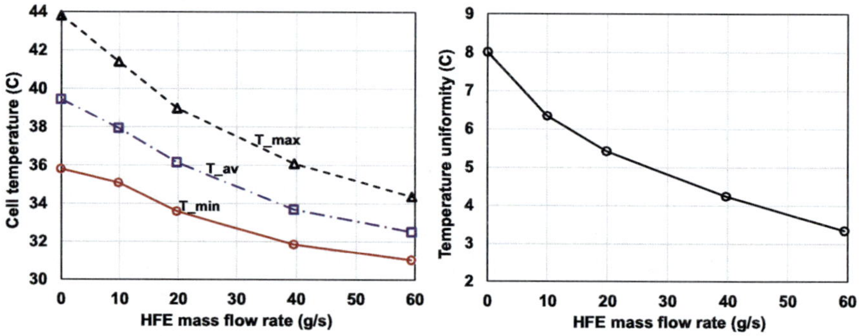


**Fig. 28.4** Effect of HFE mass flow rate on the temperature distribution of the battery pack

air-cooling case, where there was no HFE injection, both the maximum temperature and uniformity of temperature were higher than the thermal requirements in the design of BTMS. Therefore, the HFE spray cooling is clearly capable of improving the performance of the air-cooled BTMS.

## 28.5 Conclusion

In this work, the cooling characteristic of the densely-packed Li-ion batteries under the integrated liquid spray and air cooling system was numerically investigated. Uniform size distribution of non-electrically conductive liquid was injected to enhance the heat transfer performance. According to the simulation results, the following conclusions could be drawn:

(1) Non-electrically conductive spray played a significant role in reducing both the maximum temperature and temperature difference of the densely-packed batteries.
(2) The battery thermal impacts can be mitigated by increasing the mass flow rate of the liquid spray.
(3) To improve the cooling performance, the liquid jet movement should be steered to contact the cell surface.
(4) The higher mass flow rate of the spray had less effect on the temperature reduction. This is not economical for thermal management. The liquid spray quantity, therefore, need to be optimized.

To summarize, non-electrically conductive spray cooling has the potential to thermally manage the high heat flux of the densely-packed batteries at an acceptable cost. To ensure practical implementation, this hybrid cooling system will experiment later on.

**Acknowledgements** This research was funded by National Science, Research and Innovation Fund (NSRF) and King Mongkut's University of Technology North Bangkok with Contract no. KMUTNB-FF-65-52, and Thammasat University Research Fund and Thailand Science Research and Innovation Fundamental Fund with Contract no. TUFT 37/2565.

## References

3M™ Thermal Management Fluids for Military and Aerospace Apps. https://multimedia.3m.com/mws/media/569860O/3mtm-thermal-management-fluids-for-military-aerospace-apps.pdf. Last accessed 2022/05/10

Alkhedhair A, Gurgenci H, Jahn I, Guan Z, He S (2013) Numerical simulation of water spray for pre-cooling of inlet air in natural draft dry cooling towers. Appl Therm Eng 61(2):416–424

ANSYS (2020) Fluent Theory Guide. Canonsburg, PA

Li Y, Zhou Z, Hu L, Bai M, Gao L, Li Y, Liu X, Li Y, Song Y (2022) Experimental studies of liquid immersion cooling for 18650 lithium-ion battery under different discharging conditions. Case Stud Therm Eng 34:102034

Montazeri H, Blocken B, Hensen JLM (2015) CFD analysis of the impact of physical parameters on evaporative cooling by a mist spray system. Appl Therm Eng 75:608–622

Mudawar I, Bharathan D, Kelly K, Narumanchi S (2008) Two-phase spray cooling of hybrid vehicle electronics. In: 11th intersociety conference on thermal and thermomechanical phenomena in electronic systems, pp 1210–1221

Pesaran AA (2002) Battery thermal models for hybrid vehicle simulations. J Power Sour 110(2):377–382

Saechan P, Dhuchakallaya I (2022) Numerical investigation of air cooling system for a densely packed battery to enhance the cooling performance through cell arrangement strategy. Int J Energ Res 46(14):20670–20684

Saw LH, Poon HM, Thiam HS, Cai Z, Chong WT, Pambudi NA, King YJ (2018) Novel thermal management system using mist cooling for lithium-ion battery packs. Appl Energ 223:146–158

Vujanović M, Petranović Z, Edelbauer W, Baleta J, Duić N (2015) Numerical modelling of diesel spray using the Eulerian multiphase approach. Energ Convers Manage 104:160–169

Xie Y, Li W, Yang Y, Feng F (2018) A novel resistance-based thermal model for lithium-ion batteries. Int J Energ Res 42(14):4481–4498

Yang Y, Yang L, Du X, Yang Y (2019) Pre-cooling of air by water spray evaporation to improve thermal performance of lithium battery pack. Appl Therm Eng 163:114401

Youssef R, Hosen MS, He J, Jaguemont J, De Sutter L, Van Mierlo J, Berecibar M (2021) Effect analysis on performance enhancement of a novel and environmental evaporative cooling system for lithium-ion battery applications. J Energ Storage 37:102475

Zhao R, Liu J, Gu J, Zhai L, Ma F (2020) Experimental study of a direct evaporative cooling approach for Li-ion battery thermal management. Int J Energ Res 44(8):6660–6673

# Chapter 29
# Thermodynamic Equilibrium Modelling of Glycerol Gasification

**Ana Almeida, Elisa Ramalho, Albina Ribeiro, Carlos Pinho, and Rosa Pilão**

**Abstract** The modeling of the gasification process using the thermodynamic chemical equilibrium of the process is an important tool when it is intended to obtain preliminary results or to scale-up an experimental installation. In this work, the gasification process of crude glycerol using steam as the gasification agent was modeled using stoichiometric and non-stoichiometric chemical equilibrium models. The effect of the gasification temperature on the equilibrium composition of the producer gas was evaluated. The simulation results were compared with the experimental results obtained in a downdraft fixed bed reactor. The results obtained showed that the two models predict the equilibrium composition in a similar way. They also showed that the gasification reactor is operating under conditions deviating from chemical equilibrium.

**Keywords** Gasification · Glycerol · Thermodynamic equilibrium

## 29.1 Introduction

The methodologies for modeling gasification processes undoubtedly comprise important preliminary tools, especially when one intends to scale-up this type of process. The development and application of mathematical models, when validated with experimental data, allows understanding and predicting optimal conditions of performance and viability of a given production process, avoiding unnecessary costs and time losses (Basu 2018). Simpler mathematical models, such as the one that considers the thermodynamic equilibrium of the process, are a useful

A. Almeida · E. Ramalho · A. Ribeiro · R. Pilão (✉)
CIETI, ISEP, Instituto Politécnico do Porto, Rua Dr. António Bernardino de Almeida 431, 4200-072 Porto, Portugal
e-mail: rmp@isep.ipp.pt

C. Pinho
CEFT/FEUP, Faculdade de Engenharia da Universidade do Porto, Rua Dr. Roberto Frias, 4200-465 Porto, Portugal



N. S. Caetano and M. C. Felgueiras (eds.), *The 9th International Conference on Energy and Environment Research*, Environmental Science and Engineering,
https://doi.org/10.1007/978-3-031-43559-1_29

tool when only preliminary results are intended, and if their inherent limitations are recognized. Under chemical equilibrium conditions, the reaction system has maximum entropy and minimum Gibbs free energy. The thermodynamic equilibrium model can be implemented using two different approaches: stoichiometric and non-stoichiometric. Several researchers have formulated equilibrium models to simulate biomass gasification.

Ferreira et al. (2019) performed a review focused on thermodynamic equilibrium models. The authors concluded that almost all equilibrium models are modified by the inclusion of empirical correction factors that improve the model's predictability, but that compromises the generality of application.

A thermodynamic equilibrium model to simulate a fluidized bed biomass gasifier considering the conversion of tar and charcoal was developed by Rupesh et al. (2015). The simulation results were compared with the experimental results and the predictability of the model was improved by multiplying the equilibrium constants with suitable coefficients. A modified stoichiometric equilibrium model for a downdraft gasifier to simulate the gasification process was developed by Htut et al. (2014). The developed model considers the mass and energy balances of the gasifier, coupled with two equilibrium reactions. The model validation was done by comparison with experimental published result. Ayub et al. (2020, 2017) carried out studies where they compared several different thermodynamic stoichiometric models to determine which ones are the most appropriate. In these studies, factors were used to correct the equilibrium constants of the methane and water gas shift reactions. Sharma and Agarwal (2019) presented a non-stoichiometric equilibrium model to analyze the effect of process parameters on the gasification of Indian coal with high ash content.

Gambarotta et al. (2018) presented a non-stoichiometric equilibrium model to simulate the biomass gasification process in downdraft gasifiers. The model was validated by comparing its results with odder simulation models and with experimental data found in the literature. The thermodynamic chemical equilibrium of glycerol steam reforming based on the Gibbs free energy minimization method was analyzed by Ismaila et al. (2021). The steam/glycerol feed ratios and the reaction temperatures were shown to have a considerable influence on the equilibrium composition of the products and on solid carbon formation.

The present work intends to model the gasification process of crude glycerol taking into account the simplified thermodynamic equilibrium. The effect of temperature on the producer gas composition and gasification parameters was studied and the results predicted by the model were compared with the experimental results. The focus of this simulation is to try to understand if the experimental installation under analysis is, at some point or condition, operating in the proximity of thermodynamic and chemical equilibrium.

## 29.2 Methodology

Given the complexity of the gasification process, the models presented were developed based on the following considerations and simplifications:

- The gasifier is under steady state;
- The residence time is enough to reach the chemical and thermal equilibrium (Ferreira et al. 2019; Villetta et al. 2017);
- The operating pressure is atmospheric pressure (1 atm);
- Solid carbon (C) formation is negligible (Htut et al. 2014; Barman et al. 2012), since for all experimentally tested conditions the solid phase yield was negligible;
- The formation of light hydrocarbons ($C_nH_m$) in the producer gas was not considered;
- The producer gas, behaves as an ideal gas [1, 2, 10];
- Tars are generically presented as benzene ($C_6H_6$) (Rupesh et al. 2015; Aydin et al. 2017; Abuadala et al. 2010).

### *29.2.1 Stoichiometric Model*

Taking into account the elementary composition of the glycerol sample (on dry basis), its empirical formula was determined and the global gasification reaction of the sample, using steam as a gasification agent, is given by:

$$\begin{aligned} C_1H_{2.48}O_{0.99} + 0.89\, H_2O \rightarrow n_1H_2 + n_2CO \\ + n_3CO_2 + n_4CH_4 + n_5H_2O + n_6\text{Tars} \end{aligned} \quad (29.1)$$

To determine the stoichiometric coefficients *ni*, six equations are needed, three of which are obtained by the carbon, hydrogen and oxygen balances:

$$\textbf{Carbon balance}: 1 = n_2 + n_3 + n_4 + 6n_6 \quad (29.2)$$

$$\textbf{Hydrogen balance}: 2.48 + 2 \times 0.89 = 2n_1 + 4n_4 + 2n_5 + 6n_6 \quad (29.3)$$

$$\textbf{Oxygen balance}: 0.99 + 0.89 = n_2 + 2n_3 + n_5 \quad (29.4)$$

The quantity of tars obtained was estimated by applying an empirical correlation (Abuadala et al. 2010):

$$\eta_{tars} = 35.98\, \exp^{-0,00298 \times (T + 273.15)} \quad (29.5)$$

where, $\eta_{tars}$ is the tar yield (% w/w), and T is the temperature (°C).

The following gasification reactions and respective equilibrium constants were selected:

(i) Water gas shift reaction (WGS): $CO + H_2O \rightleftharpoons CO_2 + H_2$

$$K_{e-WGS} = \frac{n_3 \times n_1}{n_2 \times n_5} \tag{29.6}$$

(ii) Methane steam reforming1 (MSR1): $CH_4 + H_2O \rightleftharpoons CO + 3H_2$

$$K_{e-MSR1} = \frac{n_2 \times n_1^3}{n_t^2 \times n_4 \times n_5} \tag{29.7}$$

(iii) Methane steam reforming2 (MSR2): $CH_4 + 2H_2O \rightleftharpoons CO_2 + 4H_2$

$$K_{e-MSR2} = \frac{n_3 \times n_1^4}{n_t^2 \times n_4 \times n_5^2} \tag{29.8}$$

(iv) Methane dry reforming (MDR): $CH_4 + CO_2 \rightleftharpoons 2CO + 2H_2$

$$K_{e-MDR} = \frac{n_2^2 \times n_1^2}{n_t^2 \times n_4 \times n_3} \tag{29.9}$$

where:

$$n_t = \sum n_i \tag{29.10}$$

The values of the respective equilibrium constants, for the temperature range between 650 and 950 °C, were calculated but they are also available in the literature (Software and Databases 2010–2016).

Six combinations of these reactions were tested (Table 29.1), which together with the previous equations (Eqs. 29.2–29.5) allow to predict the equilibrium composition of the producer gas, for the tested temperatures. The system was solved using the Solver—Microsoft Excel for Windows.

**Table 29.1** Combinations of reactions used in the stoichiometric model

| Comb. 1 | Comb. 2 | Comb. 3 | Comb. 4 | Comb. 5 | Comb. 6 |
|---|---|---|---|---|---|
| MSR1 | WGS | MSR2 | WGS | MSR1 | MSR1 |
| WGS | MDR | MDR | MSR2 | MSR2 | MDR |

### 29.2.2 Non-stoichiometric Model

The global Gibbs free energy for the gasification product comprising N species, considering constant temperature and pressure, can be calculated by Basu (2018):

$$G_{global}(n_i, T) = \sum_{i=1}^{N} n_i \Delta G^{\circ}_{f,i} + \sum_{i=1}^{N} n_i RT \ln\left(\frac{n_i}{\sum n_i}\right) \tag{29.11}$$

where $\Delta G^{\circ}_{f,i}$ is the Gibbs free energy of formation of species $i$ at standard pressure of 1 bar, $R$ is the universal constant of perfect gases and $T$ is the temperature.

The minimization of $G_{global}(n_i, T)$ together with the mass balances of each element involved, as presented in (Eqs. 29.2–29.4), allows to estimate the equilibrium producer gas composition, for the tested temperatures. The function was minimized using the *Solver—Microsoft Excel for Windows.*

For the $i$ species under study, the standard Gibbs free energies of formation were calculated but they can be obtained from the literature (Coelho and Costa 2008).

## 29.3 Results

### 29.3.1 Producer Gas Composition

The effect of temperature on the equilibrium composition of the glycerol gasification process, using steam as the gasification agent, was simulated using the stoichiometric and non-stoichiometric models. For the stoichiometric model, six combinations of two equilibrium reactions were tested, as specified in Table 29.1. The results obtained showed that the equilibrium composition of the producer gas is independent of the chosen pair of reactions. Therefore, the results obtained using combination 1 in will be used as a reference for the analysis to be carried out. The results of this simulation are shown in Fig. 29.1, where the results obtained with the non-stoichiometric simulation model are also represented. In this figure, the experimental producer gas compositions for 800, 850, 900 and 950 °C were also represented (Almeida et al. 2019).

The results obtained showed that the two models used, predict in a similar way the producer gas equilibrium composition of the gasification process under study.

For the range of tested temperatures there was no significant variation in the concentration of $H_2$ in the producer gas, with its value stabilizing at 60% for temperatures above 700 °C. Regarding the concentration of CO, it was observed an increase in its value from about 20 to 30% with the increase in temperature. The concentration of $CO_2$ had an opposite behavior, decreasing with the increase in temperature from about 17–18 to 10%. The presence of $CH_4$ in the equilibrium composition of the producer gas is null, for gasification temperatures above 700 °C.

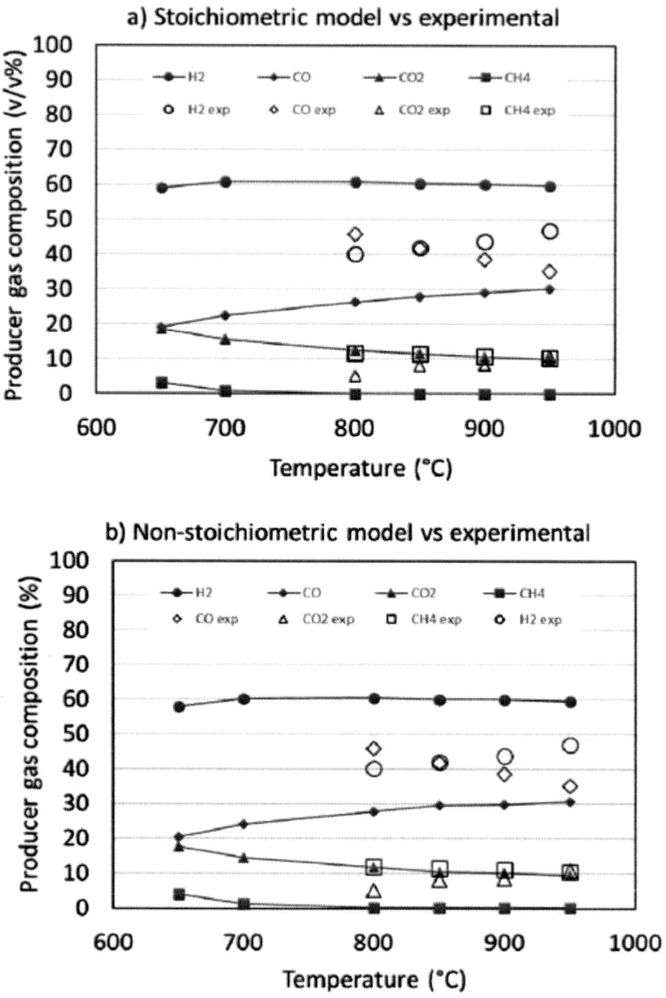


**Fig. 29.1** Effect of temperature on producer gas composition (dry base)

The first obvious observation is that the gasification reactor is not operating close to chemical equilibrium, since there is a significant difference between the values predicted by the models and the experimental values for all the compounds of the producer gas. However, the deviation between the predicted compound concentrations for equilibrium and the experimentally measured concentrations tends to decrease with the temperature rise.

For the tests carried out at 950 °C, it is verified that the experimentally measured concentrations for CO and $CO_2$ are close to the equilibrium concentrations predicted by the models. The concentrations of $H_2$ and $CH_4$ are still far from the values predicted by the models. The models overestimate the $H_2$ concentration and underestimate the $CH_4$ concentration in the producer gas. This behavior may be related to the fact that gasification reactions are influenced by kinetics and mass transfer and therefore some components never reach equilibrium.

The model accuracy was verified using the statistical parameter root mean square error, *RMSE*. The high values obtained for the RMSE (15.29–8.43) confirm that there is a large error between the model and measured results and that this error decreases with increasing temperature.

### 29.3.2 Gasification Parameters

The evaluation of a gasification process is normally carried out using performance parameters. In this study, five gasification parameters defined in a previous study (Almeida et al. 2019) were used: dry gas yield, higher heating value (HHV) of producer gas, cold gas efficiency, carbon and hydrogen conversion efficiencies. The equilibrium results of the non-stoichiometric model were used to predict the values of the gasification parameters and compare them with the results obtained experimentally, in the temperature range under study (Almeida et al. 2019).

Regarding to the higher heating value of producer gas (Fig. 29.2), the model predicts equilibrium values lower than those obtained experimentally. This result is related to the difference in the composition of the producer gas predicted by the model and obtained in the experimental tests. In Fig. 29.2 are also presented the results for the dry gas yield predicted by the model and the ones obtained experimentally. The values of this parameter predicted by the model are much higher than the values obtained in the experimental tests, for the range of tested temperatures. This marked difference is because the model predicts, for the equilibrium, a much higher gas phase production yield (78–76%) than that obtained in the experimental tests (50–55%).

The results for the odder 3 gasification parameter are represented in Fig. 29.3. The cold gas efficiency is a measure of the chemical energy in the raw material that is transferred for the producer gas. The model predicts equilibrium values greater than 100% while experimentally, only for the temperature of 950 °C values above 100% were obtained.

This results from the contribution of water present in the feeding to gas phase production. The water, fed to the reactor as an oxidizing agent in the gasification process, can act together with the raw material as a source to produce hydrogen.

The values estimated by the model for the hydrogen conversion efficiency, are greater than 120%, which confirms this possibility. In practice, the highest values were around 85% for the temperature of 950 °C.

Finally, values of 100% are obtained for carbon conversion efficiency using the model, while experimentally the highest value obtained was 84%. This behavior results from the simplification assumed in the model that the production of solid carbon is negligible. The modelling does not predict the formation of tars at equilibrium, which also contributes to this result.

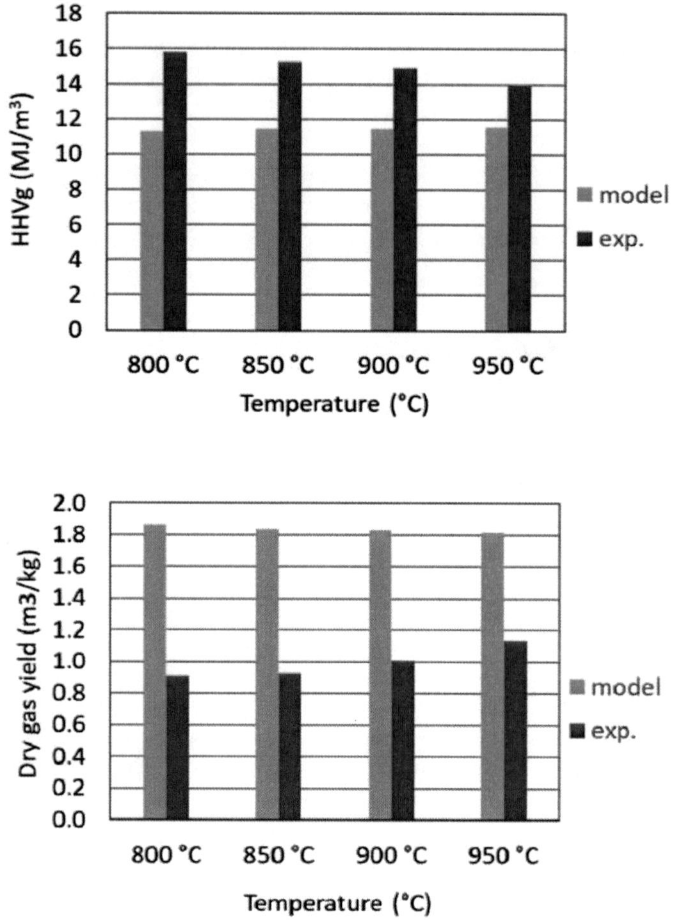


**Fig. 29.2** Effect of temperature on $HHV_g$ and dry gas yield

## 29.4 Conclusions

This work aimed to thermodynamically simulate the gasification process of crude glycerol after salt removal, in a continuous fixed bed reactor using steam as oxidizing agent. This modeling, bearing in mind all the inherent limitations, allows to position the experimental results in relation to the situation of existence of chemical equilibrium.

The equilibrium composition of the producer gas was estimated using stoichiometric and non-stoichiometric thermodynamic models. The results obtained showed that the calculated composition of the produced gas does not depend on the homogeneous gasification reactions chosen for the application of the stoichiometric model.

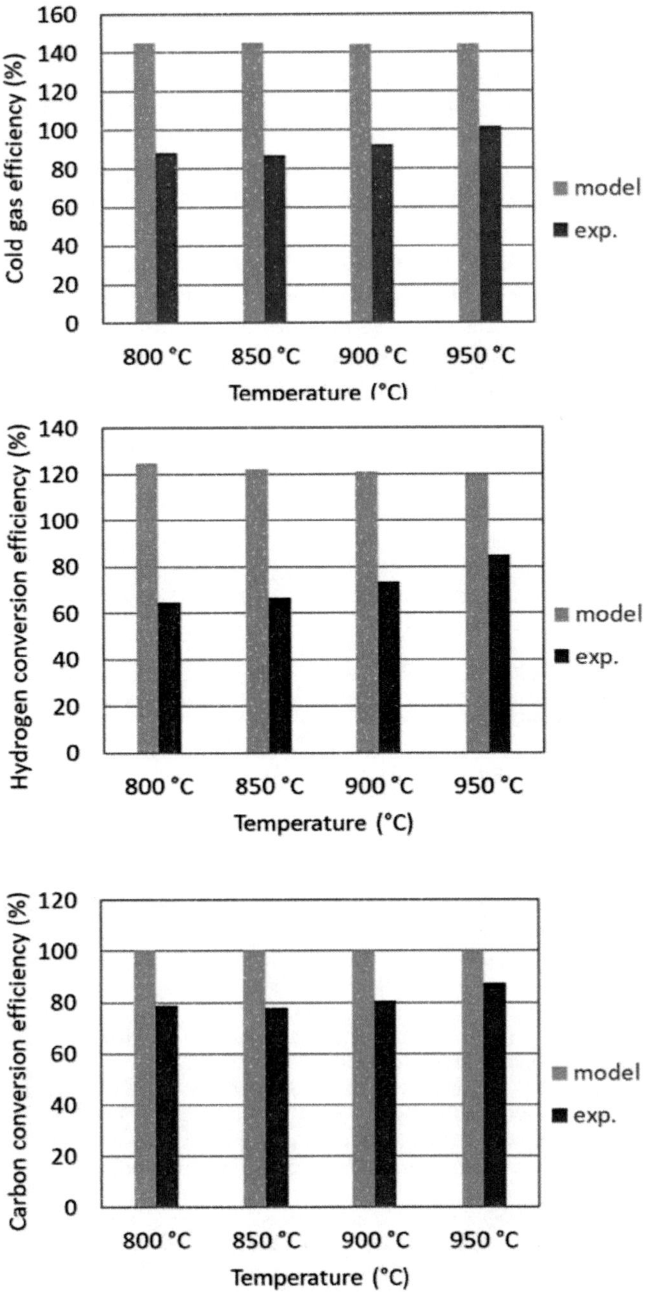


**Fig. 29.3** Effect of temperature on the hydrogen conversion efficiency

The results also showed that the predictions made with a model considering the existence of chemical equilibrium, cannot be compared with experimental results obtained in the real reactor working under current conditions. Reaching chemical equilibrium in the reactor would probably require much longer residence times.

The limitations inherent to the pure thermodynamic model, frequently mentioned in the literature, were also reflected in this study: the overestimated calculation of the $H_2$ and CO concentrations in gas phase and the underestimated $CH_4$ and $CO_2$ concentrations, are quite clear.

As a result, it was found that the model predicts the production of a gas with lower HHV than the gas experimentally obtained. Regarding the other gasification parameters under study, it is observed that they are overestimated when compared to those obtained experimentally, registering, however, a decrease in their deviation with the increase in the gasification temperature.

**Acknowledgements** This work was supported by Portugal 2020—POCI-01-0145-FEDER-024067 and by Multiyear funding of FCT—Fundação para a Ciência e Tecnologia (grant UIDB/04730/2020).

## References

Abuadala A, Dince I, Naterer GF (2010) Exergy analysis of hydrogen production from biomass gasification. Int J Hydrogen Energy 35(10):4981–4990

Almeida A, Pilão R, Ribeiro A, Ramalho E, Pinho C (2019) Gasification of crude glycerol after salt removal. Energy Fuels 33(10):9942–9948

Aydin ES, Yucel O, Sadikoglu H (2017) Development of a semi-empirical equilibrium model for downdraft gasification systems. Energy 130:86–98

Ayub HM, Park SJ, Binns M (2020) Biomass to syngas: modified non-stoichiometric thermodynamic models for the downdraft biomass gasification. Energies 13(21):5668

Barman NS, Ghosh S, De S (2012) Gasification of biomass in a fixed bed downdraft gasifier—a realistic model including tar. Bioresour Technol 107:505–511

Basu P (2018) Biomass gasification, pyrolysis and Torrefaction: practical design and theory, 3rd edn. Elsevier

Coelho P, Costa M (2008) Combustão. 2nd ed., Editora Orion

Ferreira S, Monteiro E, Brito P, Vilarinho C (2019) A holistic review on biomass gasification modified equilibrium models. Energies 12(1):1–31

Gambarotta A, Morini M, Zubani A (2018) A non-stoichiometric equilibrium model for the simulation of the biomass gasification process. Appl Energy 227:119–127

Htut YM, Win MM, Khine MM (2014) Modification and application of a stoichiometric equilibrium model for downdraft gasification system. In: 36th international proceedings on conference on software engineering

Ismaila A, Chen X, Gao X, Fan X (2021) Thermodynamic analysis of steam reforming of glycerol for hydrogen production at atmospheric pressure. Front Chem Sci Eng 15(1):60–71

Rupesh S, Muraleedharan C, Arun P (2015) A comparative study on gaseous fuel generation capability of biomass materials by thermo-chemical gasification using stoichiometric quasi-steady-state model. Int J Energy Environ Eng 6(4):375–384

Sharma V, Agarwal VK (2019) Equilibrium modeling and optimization for gasification of high-ash Indian coals by the Gibbs free energy minimization method. Process Integr Optim Sustain 3:487–504

FactStage Thermochemical Software and Databases 2010–2016. Calphad, 54. www.factstage.com. Last accessed 21 Nov 2020

La Villetta M, Costa M, Massarotti N (2017) Modelling approaches to biomass gasification: a review with emphasis on the stoichiometric method. Renew Sustain Energy Rev 74:71–88

# Chapter 30
# Comprehensive Modeling and Evaluation of the Feasibility of the EU Energy Transition Concerning the Development of the Installed Capacity of Different Energy Sources Until 2050

**Adam Kubín and Lukáš Janota**

**Abstract** This paper presents the methodology and findings of a comprehensive study that in the selected key time milestones (2030, 2040, 2050) assess the ongoing energy transformation and its impacts within the interconnected European electricity system and overall energy security. The study models the expected development of the installed capacities of individual types of energy sources, respecting the steps leading to meeting the EU's ambitious binding climate and energy goals, as well as individual countries' national plans like development of RES or phase out of coal and nuclear power plants. The majority output of the complex methodology is the evaluation of the state of shortages or surpluses of electrical energy in the European electricity system in the selected key time milestones. The model outputs show that increasing integration of RES may cause higher requirements for maintaining the EU's energy security. The current most significant barriers to the fulfillment of climate and energy goals for maintaining the EU's energy security include not the lack of generation resources but the inflexibility of the electricity system and the lack of technologically and economically efficient seasonal accumulation. It will be essential to involve all types of technologies providing flexibility and all energy market participants, including households.

**Keywords** Energy transformation · Energy security · EU climate and energy policy · Coal phase-out · Interconnected EU power grid · EU energy mix

A. Kubín (✉) · L. Janota
Faculty of Electrical Engineering, Czech Technical University in Prague, Technická 2, 166 27 Prague, Czech Republic
e-mail: kubinada@fel.cvut.cz

 
N. S. Caetano and M. C. Felgueiras (eds.), *The 9th International Conference on Energy and Environment Research*, Environmental Science and Engineering,
https://doi.org/10.1007/978-3-031-43559-1_30

## 30.1 Introduction

The European Union is at a critical energy-climate crossroads, where it must respond more dynamically than ever to the unpredictable changes and challenges of today's world, such as the adverse effects of the COVID-19 pandemic or the ongoing Russian aggression in Ukraine (European Union 2022). These unexpected and uncontrollable situations directly affect the available instruments of action, change existing long-term strategies, and initiate more ambitious future steps to meet the EU's binding climate and energy policy objectives beyond the Paris Agreement. That is why we need to develop new technologies and innovative solutions to maintain the direction of sustainable energy, energy security, protection and prevention of climate change, and, finally, to increase the resilience of the European Union and its people (Calvo-Gallardo et al. 2021).

In 2021, the European Union imported more than 40% of its total gas consumption, 27% of fuel oil, and 46% of coal from Russia (The European Commission 2022). According to the European Commission, about 60% of energy comes from Russia, and the annual cost of this energy is about 99 billion euros. In March 2022, the European Commission issued the REPowerEU plan, which sets out steps and measures to drastically reduce the EU's dependence on fossil fuel and gas imports from Russia (European Commission 2022a). This ambitious plan is based on three main pillars: diversification of supply, reduction of fossil fuel and gas consumption and increasing the use and pace of RES development. Complete independence from energy imports from Russia should be achieved by the latest decade's end.

The EU is demonstrating its position in the fight against climate change with a legal commitment of 2021 to achieve complete carbon neutrality by 2050 and raise interim climate and energy targets by 2030 (European Commission 2021). By 2030, greenhouse gas emissions should be reduced by 55%, increasing the share of RES in final energy consumption from 32 to 40% (European Parliament 2018). In the light of the gas crisis and the Russian-Ukrainian war dispute, it is currently being discussed to increase the target for RES to a very ambitious 45%. Most Member States have already set a binding coal phase-out date to meet these targets, with most having until 2033.

The ongoing decarbonization is strongly affecting other sectors than the energy and heavy industry sectors. In the coming decades, a significant increase in energy consumption is expected, caused mainly by the growing number of electric cars and heat pumps, which are expected to increase by up to 10 million units in the next five years. The expected development of annual electricity consumption within the EU countries, including Norway, the United Kingdom, and Switzerland, based on prognosis of European Commission is shown in Fig. 30.1 (European Commission 2022b). The graph in Fig. 30.1 shows that consumption in 2050 could increase by up to 66% compared to 2021. Furthermore, it is necessary to emphasize that the volume of electricity consumed will increase and the highest load on the network, which requires standby resources, storage technologies, and effective management over time. The crucial question is whether we can fully replace this lost production

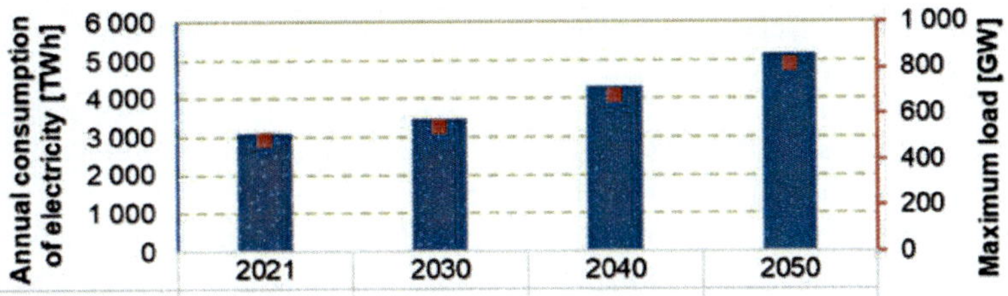


| | 2021 | 2030 | 2040 | 2050 |
|---|---|---|---|---|
| Annual consumption of electricity [TWh] | 3 108 | 3 459 | 4 312 | 5 163 |
| Maximum load [GW] | 489 | 545 | 679 | 813 |

**Fig. 30.1** Expected development of electricity consumption and load in the EU

capacity of conventional sources with renewable energy sources and suitable modern technologies to ensure energy security and self-sufficiency at sustainable final energy prices.

Many authors have already addressed this issue by examining the necessary measures, support schemes or strategic scenarios to achieve the EU's climate and energy goals while maintaining energy security (Hainsch 2022; Tagliapietra et al. 2019; Rodríguez-Fernández et al. 2022).

However, the current situation lacks a comprehensive deterministic mathematical model based on national plans for future energy mixes to assess whether energy transformation under ambitious conditions is realistically feasible while maintaining the security of supply. In this paper, we demonstrate the results of the proposed deterministic model, which examines and evaluates the EU's self-sufficiency in ensuring sufficient energy and output power over time, assuming the development of intermittent RES integration. Within our model, we work with simulated aggregate production and consumption across the interconnected EU energy system related to the identified strategic time milestones 2030, 2040, and 2050. As part of our assessment of the EU's energy resource adequacy, we also include the impact of performance in the model in Norway, Switzerland, and the United Kingdom.

## 30.2 Materials and Methods

The research method and analysis of this paper consist of two separate parts. The first phase deals with identifying and analyzing the current state of the installed capacities of the EU resource mix (including NOR, CH, and GB) and the course of aggregated hourly energy production over time. The current energy status of the interconnected European system was evaluated according to data from the ENTSO-E Transparency portal (Network and of Transmission System Operators for Electricity (ENTSO-E) 2022). To analyze the current state, data of the installed capacity of each category of resources in each country as of early 2022 and time series of hourly electricity generation in each country for the year 2021 are used. Missing data in the time series are extrapolated, clearly erroneous data are estimated to minimize the error compared to the actual missing values.

Forecasts of the future EU energy mix within the selected significant time milestones (the years 2030, 2040, and 2050) are based on an in-depth and comprehensive analysis of the large number of available official resources, in particular Member States' national climate and energy plans, Member States' strategy papers, national action plans, and sustainability and RES development plans or even official government statements (European Commission 2022a). To achieve the highest possible relevance of future predictions of the development of the installed capacity of individual energy sources, we performed an additive analysis covering the areas of individual types of energy sources. We used available information and data from the Global Energy Monitor (Global Energy Monitor 2022) and Wind Europe portals (Wind Europe 2022) and the conclusions from the S&P Global Commodity insight analysis (Edwardes-Evans 2022).

In the case of missing information on the development of installed capacity, especially in the case of photovoltaic and wind power plants, we estimated the future installed capacity of these types of sources based on the coefficient of compensation RES and conventional sources $k_{com}$. The $k_{com}$ coefficient gives us the necessary installed capacity of renewable energy sources to cover electricity production and maintain supply security to replace 1 MW of conventional capacity. The average value of the coefficient $k_{com}$ is around 4.9, which means that to cover the loss of 1 MW of a conventional source, we must connect to the system at least 4.9 MW of the installed capacity of intermittent renewable sources.

In the second part of the research, a deterministic model is developed based on identified and prepared data sets, which evaluate the future energy balance in the interconnected European energy system. The determination of energy balance is calculated based on a deterministic comparison of aggregated hourly energy production during the year with the expected aggregated electricity consumption. Hourly production greater than consumption indicates a surplus of power (production) in the power system, hourly production less than consumption indicates a shortage of production power in the system.

To determine the deterministic time series of hourly energy production of each resource category in the monitored time milestones (2030, 2040 and 2050), the baseline production time series from 2021 are used. These baseline time series are adjusted proportionally in the ratio of the installed capacity of resources each time milestone (2030, 2040 or 2050) to the installed capacity at present. The time series of hourly electricity consumption are determined in the same way for each year. The values in the baseline time series of 2021 are proportionally increased in the ratio of the annual consumption of the years 2030, 2040 or 2050 to the annual consumption in 2021.

The proposed deterministic model respects the basic key parameters of the power system and its elements. Model respects the different time of utilization of the maximum installed power of energy sources, the variable nature of production, and its controllability. In the case of intermittent RES, the unavailability of installed capacity is almost exclusively driven by the unavailability of primary energy. For conventional sources, the unavailability of power at specific hours to cover consumption respects the provision of balancing services, repair, and maintenance, breakdowns, shutdowns for fuel replacement, thermal energy supply, or greening. The model does not have a stochastic character and therefore does not consider changes in climatic conditions, the randomness of phenomena in the system, or changes in energy and fuel prices in the markets.

The model's output identifies the number of hours with non-supply of electricity and insufficient available standby power in the interconnected system per year. In its calculations, the model does not consider the more widespread use of accumulation technologies, and its output can be understood as the identification of the need for the development of accumulation in the future concerning the growing share of RES in the EU energy source mix.

## 30.3 Results and Discussion

### *30.3.1 The Current State of the EU Energy Mix*

The identified current state of the EU energy mix and the volume of annual energy produced by type of energy source is represented by Fig. 30.2. Around 55% of the EU's energy mix is renewable. In 2021, they accounted for 36.6% of the total annual volume of electricity produced. However, Europe's interconnected energy system remains operationally dependent on conventional fossil fuel sources, which account for 32% of installed generation capacity. In 2021, these sources produced a total of 33.5%. The current EU energy mix has an emission intensity of 237 g $CO_2$eq/kWh. The most used type of source in the evaluation according to the installed capacity is gas sources (224 GW; 19.8%), followed by wind power plants (221 GW; 19.6%), solar power plants (178 GW; 15.7%), hydroelectric power plants (154 GW; 13.6%), coal sources (115 GW; 10.1%) and nuclear power plants (112 GW; 9.9%). From a country perspective, most gas sources are installed in Italy (44 GW), wind and solar power plants in Germany (63.8 and 58.5 GW), hydropower plants in Norway (29.3 GW), coal power plants in Germany (37.9 GW) and nuclear power plants in France (61.4 GW).

In evaluating sources according to the volume of annual electricity, the situation is significantly different compared to the installed capacity. Most electricity was produced from nuclear sources (807 TWh; 25.6%), followed by gas sources (540 TWh; 17.2%), hydroelectric power plants (505 TWh; 16.1%), coal sources (447 TWh; 14.2%), wind farms (443 TWh; 14.1%) and solar power plants (158 TWh; 5.0%). The largest exporters in absolute numbers in 2021 were France (44.4 TWh) and Sweden (25.4 TWh), the largest importers were Italy (42.4 TWh) and the UK (25.1 TWh). According to the identified situation, it can be stated that although RES have a majority share of installed capacity in the total energy mix, their contribution to the total annual energy produced is not proportional. This is due to their

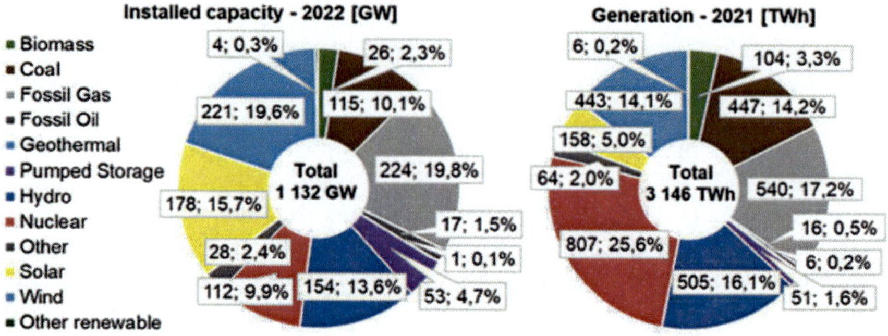


**Fig. 30.2** EU energy mix in 2022 and annual electricity production in 2021 (own processing based on ENTSO-E transparency data)

intermittent nature and strong dependence on climatic and geographical conditions. Nuclear power plants have a standard annual utilization time of installed capacity of 7500 h, compared to photovoltaic power plants around 1100 h and on-shore wind farms 2200 and off-shore 4400. A graphical representation comparing the course of electricity generation within the time of individual sources is shown in Fig. 30.3. Despite the maximum production of PV power plants, whose installed capacity is about 1.5 times greater than nuclear sources, the contribution to the energy produced in the monitored week was only 6.15 TWh, ie about half of what nuclear sources (12.37 TWh).

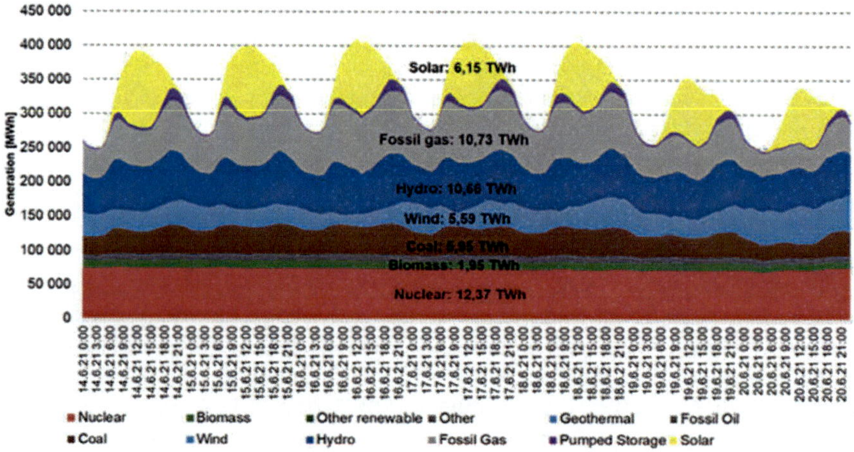


**Fig. 30.3** Aggregated generation in EU 27 + (UK, NOR, CH) within 14.6.2021–20.6.2021 (own processing based on ENTSO-E transparency data)

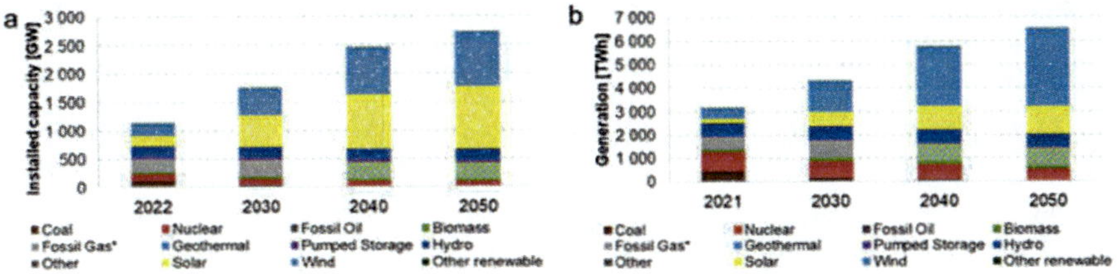


**Fig. 30.4** **a** Installed capacity outlook (GW); **b** total generation outlook (TWh)

### *30.3.2 Prediction of the Future EU Energy Mix*

The prediction of the future composition of installed generation capacity and the overall share of electricity generation in the EU is shown in Fig. 30.4. From the predictions and modeling of the future composition of the EU energy mix and the total share of annual energy production, the trend of dynamic growth of RES is evident. The highest increase in photovoltaic and wind power plants is expected in the future, the maximum technical potential of which is far from being fulfilled. The increase in production from wind farms over the years is due to a higher proportion of off-shore power plants. The identified course of EU development and transformation of the energy mix is in line with the identified EU climate and energy policy until 2050. By 2030, the development of RES will be most affected by Germany's energy strategy and transformation. In the case of gas energy sources, in connection with the current situation and a high degree of uncertainty, a constant installed capacity is maintained until 2030.

The development of gas sources can be expected only in the next two decades after year 2030. The current gas-fired power plants will not be shut down and the newly built ones will already use hydrogen and biomethane as fuel or will have Carbon Capture and Storage technology. No significant increase can be expected in terms of their contribution to energy production since they should be the primary sources involved in balancing the power balance in the interconnected system. The development of nuclear resources is mainly influenced by the shift away from nuclear energy in Germany and the expected increase in installed capacity in countries such as the United Kingdom, Finland, Slovakia and the Czech Republic.

### *30.3.3 Evaluation of the Proposed Model*

According to the proposed model and performed simulations for selected years, the results are shown in Fig. 30.5, and it is pretty clear that in the future, the certainty of securing the required volume of electricity in the European interconnected energy system will decrease. In 2030, according to the simulations, a total of 458 h per year is a lack of electricity in the electricity network. The situation is deteriorating due

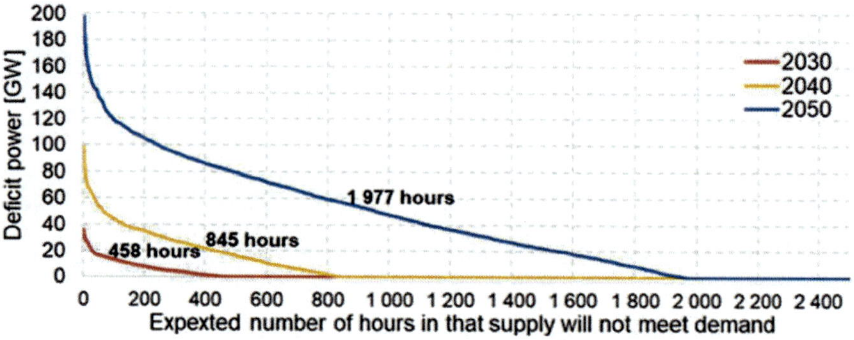


**Fig. 30.5** Simulated duration curve of insufficient standby power in the EU energy system

to the predicted change in the EU energy mix, as evidenced by the values for 2040, where energy shortages in the system occur for 845 h and in 2050 the total time of inadequate energy supply will reach 1977 h.

From the graphical output of the power balance simulation in the interconnected EU system in Fig. 30.6, it is evident that the number of hours of lack of sufficient output power in the system will increase. The most significant deficit occurs in September when up to 200 GW of aggregate power will be missing in the energy system. An important conclusion of this simulation is that the lack of performance is not seasonal but occurs throughout the year. The volume of undelivered electricity also has an increasing tendency, as evidenced by the increase in the volume of missing energy from 2030 3.8 TWh, to 19.8 TWh in 2040 to a serious 106.8 TWh in 2050.

The future problem that the EU energy system must face and to which we must adapt will not only be the lack of available energy in the grid, but especially the surplus energy from renewable sources. This excess energy in the network is graphically presented in Fig. 30.7. The energy will remain in order of magnitude more than it

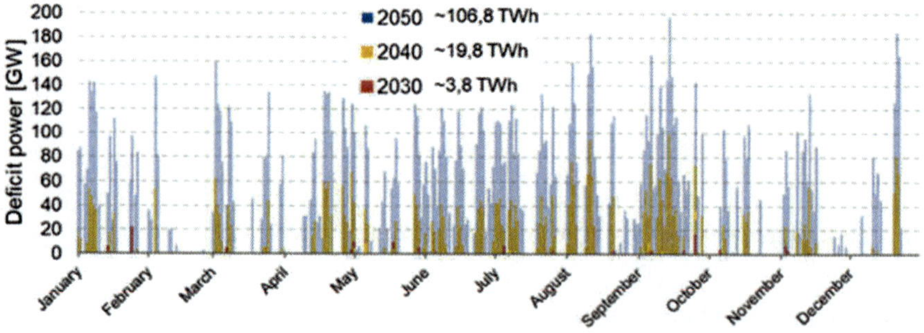


**Fig. 30.6** Evaluation and identification of hourly steps of lack of electrical energy and its volume to meet 100% of demand within EU energy system

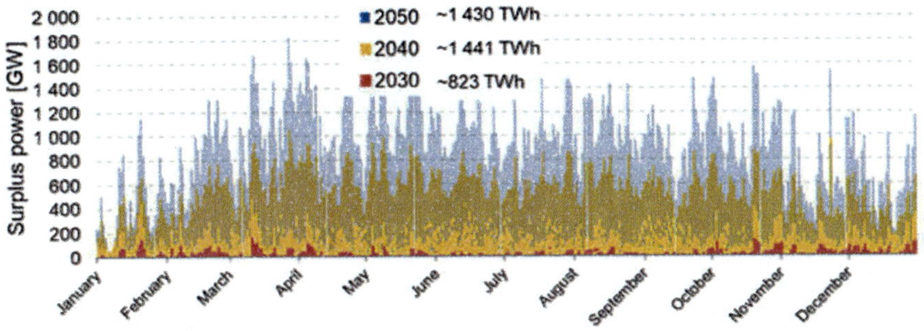


**Fig. 30.7** Simulation and identification of hourly steps with excess power in an interconnected European power grid

**Table 30.1** The need for additional installed capacity of variant sources to cover demand

| Technology | 2030 | 2040 | 2050 |
|---|---|---|---|
| Wind [GW] | + 205 | + 500 | + 1 000 |
| Nuclear [GW] | + 33 | + 84 | + 187 |
| Fossil gas [GW] | + 110 | + 297 | + 600 |
| Energy storage [GW] | + 28 | + 74 | + 165 |

will lack, confirmed by the volume of excess energy in 2040 and 2050, more than 1400 TWh.

The following Table 30.1 contains the results of a simulation that evaluates how much additive installed capacity of individual types of energy sources in the observed years would be needed to ensure sufficient energy for 99.9% of the year. The simulation considers the proportional allocation of resources on the territory of the Member States according to the volume of missing energy over time.

As already analyzed, the main problem in the resulting annual volume is not the lack of electricity, but the lack of usable flexibility and seasonal accumulation in the EU electricity system, which would allow us to use energy arbitrage and peak shaving over time effectively.

## 30.4 Conclusion

This work aimed to identify composition of the EU energy resource mix and evaluate future energy conditions within the interconnected EU electricity network. To fulfill these goals, we designed a deterministic mathematical model that simulates and evaluates the resulting energy balance, the number of hours when demand does not meet energy supply, and the annual volume of surplus or lack of electricity for the years 2030, 2040 and 2050. The model considered the basic operating parameters of various sources, the nature of production over time, and their controllability. Increasing integration of installed RES capacity may cause higher requirements for maintaining the power balance and the stability of the operation of the EU energy system. For the year 2050, around 1970 h, endangered network operation was identified, which can be critical for securing energy supply and maintaining system stability. This situation may escalate into a large blackout, which in the current situation is preceded by conventional combustion and gas power plants.

The operation of the energy system and the maintenance of the power balance in the coming decades will require an enormous increase the number of technologies providing flexibility in different time horizons and at all voltage levels of the electricity system. It is necessary to actively involve all market participants in the management of the system, including households, which have a vast potential to regulate consumption at the lowest voltage levels or locally consume energy and thus minimize unwanted energy overflows over time. According to simulations after 2040, all storage technologies should have such storage capacity that they can store and perform energy arbitrage of up to 1400 TWh per year. The current most significant barriers to the safe and timely fulfillment of climate and energy goals for maintaining the EU's energy security include not the lack of funds or energy sources but the inflexibility of the electricity system and the lack of technologically and economically efficient seasonal accumulation.

**Acknowledgements** Authors would like to thank prof. CSc Jaroslav Knápek and Ph.D. Tomáš Králík for their supervision and constant valuable feedback.

**Funding** This work was supported by the Grant Agency of the Czech Technical University in Prague, grant No. SGS21/118/OHK5/2T/13.

## References

Calvo-Gallardo E, Arranz N, Fernández de Arroyabe JC (2021) Analysis of the European energy innovation system: contribution of the framework programmes to the EU policy objectives. J Clean Prod 298. https://doi.org/10.1016/j.jclepro.2021.126690

European Union (2022) Russia's war on Ukraine: implications for EU energy supply, Mar 2022 [online]. Available http://www.europarl.europa.eu/thinktank

European Commission (2021) European climate law. Off J Eur Union 2021:17 [online]. Available https://eur-lex.europa.eu/legal-content/EN/TXT/?uri=CELEX:32021R1119

European Commission (2022a) National energy and climate plans: EU countries' 10-year national energy and climate plans for 2021–2030. https://ec.europa.eu/info/energy-climate-change-environment/implementation-eu-countries/energy-and-climate-governance-and-reporting/national-energy-and-climate-plans_en#final-necps

European Commission (2022b) Communication from the commission to the European Parliament, The European Council, The Council, The European Economic and Social Committee and The Committee of the Regions: REPowerEU Plan, vol COM(2022)

European Network of Transmission System Operators for Electricity (ENTSO-E) (2022) Installed capacity per production type. https://transparency.entsoe.eu/generation/r2/installedGenerationCapacityAggregation/show

European Parliament (2018) Directive (EU) 2018/2001 of the European Parliament and of the Council on the promotion of the use of energy from renewable sources. Off J Eur Union 2018(L328):82–209

Edwardes-Evans H (2022) Europe's gas bridge to energy transition is crumbling: S&P global report [online]. Available https://www.spglobal.com/commodityinsights/en/market-insights/latest-news/electric-power/040522-europes-gas-bridge-to-energy-transition-is-crumbling-s-p-global-report

Global Energy Monitor (2022) Building an open guide to the world's energy system. https://globalenergymonitor.org/

Hainsch K et al (2022) Energy transition scenarios: what policies, societal attitudes, and technology developments will realize the EU green deal? Energy 239. https://doi.org/10.1016/j.energy.2021.122067

Rodríguez-Fernández L, Carvajal ABF, de Tejada VF (2022) Improving the concept of energy security in an energy transition environment: application to the gas sector in the European Union. Extr Ind Soc 9. https://doi.org/10.1016/j.exis.2022.101045

Tagliapietra S, Zachmann G, Edenhofer O, Glachant JM, Linares P, Loeschel A (2019) The European union energy transition: key priorities for the next five years. Energy Policy 132:950–954. https://doi.org/10.1016/j.enpol.2019.06.060

The European Commission (2022) In focus: reducing the EU's dependence on imported fossil fuels. Energy. https://ec.europa.eu/info/news/focus-reducing-eus-dependence-imported-fossil-fuels-2022-abr-20_en#:~:text=In2021%2C43.5%25 of the, and the US (6.6%25)

Wind Europe (2022) National energy & climate plans [online]. Available https://windeurope.org/2030plans/

# Chapter 31
# Towards Multiscale Modeling to Predict Diatom Metabolites Production for Biofuels and High-Value Compounds

**Monique Branco-Vieira, Nídia S. Caetano, Alex Ranieri J. Lima, and Nadine Töpfer**

**Abstract** The utilization of diatom biomass as a renewable resource for the production of high-value compounds has gained significant attention in the pharmaceutical, nutraceutical, and energy industries. To promote the feasibility and rapid expansion of this industrial sector, it is essential to develop accurate and multiscale models for predicting the production of microalgae biomass components. In this study, we present the initial steps towards the development of a multiscale model capable of forecasting microalgae production in pilot-scale photobioreactors. The framework integrates a genome-scale metabolic network model of the diatom *Phaeodactylum tricornutum* with pilot-scale bubble column photobioreactor parameters. By integrating experimental data and taking into account site-specific climate conditions over a one-year operation, the model predicts biomass production under varying light

M. Branco-Vieira (✉)
Julius Kühn-Institute, Federal Research Centre for Cultivated Plants, Institute for Strategies and Technology Assessment, Stahnsdorfer Damm 81, 14532 Kleinmachnow, Germany
e-mail: monique.branco@julius-kuehn.de

M. Branco-Vieira · N. S. Caetano
Faculty of Engineering, LEPABE-Laboratory for Process Engineering, Environment, Biotechnology and Energy, University of Porto (FEUP), R. Dr. Roberto Frias S/N, 4200-465 Porto, Portugal

Faculty of Engineering, ALiCE-Associate Laboratory in Chemical Engineering, University of Porto, R. Dr. Roberto Frias S/N, 4200-465 Porto, Portugal

N. S. Caetano
CIETI, School of Engineering (ISEP), Polytechnic of Porto (P. Porto), R. Dr. Antonio Bernardino de Almeida 431, 4249-015 Porto, Portugal

A. R. J. Lima
Center of Scientific Development, Butantan Institute, Av. Vital Brasil, 1500 - Butantã, São Paulo 05503-900, Brazil

N. Töpfer
Institute for Plant Sciences, University of Cologne, COPT Center Luxemburger Str. 90, 50939 Köln, Germany



N. S. Caetano and M. C. Felgueiras (eds.), *The 9th International Conference on Energy and Environment Research*, Environmental Science and Engineering,
https://doi.org/10.1007/978-3-031-43559-1_31

intensities, temperatures, and day lengths. This comprehensive framework provides valuable insights into the metabolic behavior and cellular dynamics of microalgae, enabling spatiotemporal analysis for optimizing industrial processes. This work can support further advancements in the field and provide valuable insights that can pave the way for the accelerated growth and sustainability of the microalgae-based industry.

**Keywords** Genome-scale metabolic model · Flux balance analysis · *Phaeodactylum tricornutum* · Pilot-scale · Microalgae

## 31.1 Introduction

The energy transition from a fossil-based economy to a circular and renewable bioeconomy is considered one of the ultimate requirements for global sustainability. Algae and terrestrial plants play a crucial role in the global carbon cycle by absorbing $CO_2$ from the atmosphere through photosynthesis. The percentage of $CO_2$ uptake can vary depending on various factors such as species, environmental conditions, and physiological characteristics. In general, it is estimated that land-based ecosystem absorbed approximately 30% of the $CO_2$ emitted by human activities in the last decade (Pathak et al. 2022). Concerning algae, it has been reported that they serve as a significant biological platform for $CO_2$ capture and storage. Specifically, certain species of microalgae have demonstrated the ability to assimilate approximately 10–30% of $CO_2$ derived from industrial flue gases (Branco-Vieira et al. 2022; Deprá et al. 2020). This estimation is based on a combination of direct measurements, experimental studies, and modeling approaches.

Diatoms are a group of unicellular microalgae that inhabit a wide range of environments, including both fresh and saline waters, and wet soils (Branco-Vieira et al. 2018a). Due to their interesting biochemical composition and high productivity, they have been employed for biotechnological purposes across multiple applications and have proven to be suitable feedstock for several industries, including pharmaceutical, nutraceutical, feed, food, and energy.

The marine diatom *Phaeodactylum tricornutum* is highly productive, exhibits high photosynthetic efficiency, rapid growth rates, and can grow in a wide range of environments. These characteristics make these organisms a promising platform for producing biofuels and high-value compounds (Dhanker et al. 2022).

Despite the commercial exploitation having arisen decades ago, there is a lack of predictive models for the *in-silico* simulation of microalgal growth at a large-scale production scheme. This is primarily due to the multiscale nature of parameters that regulate the behavior of cells within a photobioreactor (PBR).

Kinetic models have been largely applied for modeling microalgal growth and they share several features relevant for the integration into multiscale models of phototrophic organisms (Ova Ozcan and Ovez 2020). Typically, these models rely on the Monod and Droop models, which are commonly employed to investigate the dynamic microalgae growth as a function of $CO_2$ and nutrient constraints (Branco-Vieira et al. 2022). Nevertheless, the limited availability of enzyme kinetic data imposes a hurdle for the development of comprehensive kinetic models (Westermark and Steuer 2016), preventing the elucidation of the metabolic phenomena governing growth when various environmental constraints are imposed on the model.

Large-scale metabolic models, derived from annotated genomes, are progressively used to describe the metabolic network behavior of different organisms. These so-called genome-scale metabolic (GSM) models describe the organism´s metabolism based on a stoichiometric representation of the metabolic network and can be analyzed using constraint-based optimization methods, such as flux balance analysis (FBA), its extension parsimonious FBA (pFBA), and flux variability analysis (FVA). FBA relies on an optimality assumption, typically the maximization of growth, to predict metabolic fluxes through the metabolic network (Orth et al. 2010). On the other hand, the pFBA aims to identify a flux distribution that maximizes the growth rate while minimizing the overall sum of flux. FVA is used to determine the potential range of values for each metabolic flux at optimum.

The dynamic growth of microalgae in a PBR environment presents a myriad of challenges and is subject to various parameters that significantly impact cell behavior and phenotype. However, the literature lacks a comprehensive exploration of integrating reactor-scale dynamics with metabolic models to accurately estimate microalgae growth.

The development of a mathematical framework that combines metabolic models with cellular behavior within a PBR is a promising route for predicting the growth and accumulation of metabolites in microalgae. This approach can enhance the industrial applications of diatoms under a biorefinery concept to produce biofuels and high-value compounds. Furthermore, in silico prediction of phenotypic changes in microalgae under different constraints offers significant advantages in terms of time and cost-effectiveness compared to traditional experimental methods. This approach can provide insights into the underlying mechanisms driving phenotypic changes in microalgae, thus supporting further strain engineering and methods for stress-induced metabolites overexpression.

In this work, we outline the initial steps in developing a robust multiscale framework to simulate the year-long growth of *P. tricornutum* within a PBR operation.

## 31.2 Framework Development

To develop the framework, we used experimental data obtained from the cultivation of *P. tricornutum* under natural environmental conditions to constrain the previous published GSM model of *P. tricornutum* iLB1034 (Broddrick et al. 2019a). The model also considers the weather data and the cellular photon uptake under different solar irradiance.

The biomass production output of the GSM model, indicative of metabolic changes, was upscaled to the PBR level by constraining it with various PBR-scale parameters including culture volume, dimensions, surface area, and cultivation time (Fig. 31.1).

### 31.2.1 Culture Conditions and Biochemical Characterization

The diatom *P. tricornutum* was cultivated in an outdoor bubble column PBR as described in Branco-Vieira et al. 2018b. The four PBRs of 200 L each were filled with culture medium containing replete modified Guillard's f/2 formulation (Guillard and Ryther 1962) prepared in natural seawater. The microalgae were exposed to direct sunlight and supplemented with atmospheric $CO_2$. Solar irradiance, quantified as photosynthetically active daily averaged irradiance (PAR) in μmol photons $m^{-2}$ $s^{-1}$, was monitored using the sensor QSPL-2100 (Biospherical Instruments Inc.). After 14 days of cultivation, the microalgae were harvested, and the total macromolecule components including proteins, carbohydrates, lipids, and pigments were quantified using the methodology described in Branco-Vieira et al. (2020).

### 31.2.2 Modeling Photophysiological Constraints and Experimental Data Integration

We employed the *P. tricornutum* GSM model to simulate the metabolic processes underlying the growth of autotrophic microalgae. Using FBA, we predicted the maximal growth rates and steady-state metabolic fluxes through the reaction network. The optimization problem representing this analysis is as follows:

$$\begin{aligned}&\text{Maximize } \nu_{\text{biomass}}\\&\text{Subject to:}\\&S \cdot \nu = 0\\&lb_i \leq \nu_i \leq ub_i.\end{aligned}$$

The stoichiometric matrix (S) represents the reaction network; $\nu$ is the vector of reaction fluxes which maximizes the objective function $\nu_{biomass}$ within specified lower and upper flux boundaries (lb, ub).

The experimental data on biomass components ratios (pigments, lipids, proteins, and carbohydrates) was integrated with the GSM model to establish a context-specific biomass objective function. These experimental results correspond to the biomass produced over the fourteen-day experiment. The values for DNA, RNA, and membrane lipids remained unchanged from the original GSM model publication (Broddrick et al. 2019a).

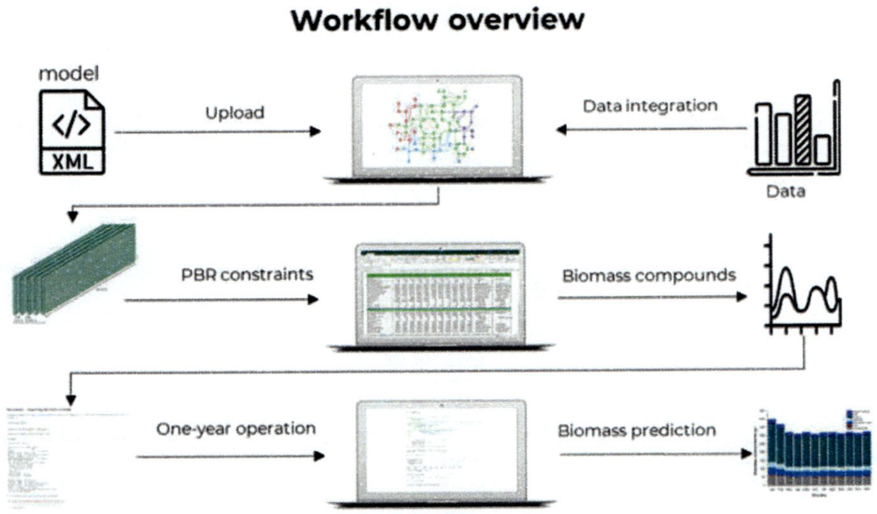


**Fig. 31.1** Representation diagram of workflow applied in this study. The initial step involves integrating the experimental dataset into the GSM model. Additionally, the constraints specific to the PBR are incorporated into the model to predict biomass production. To predict biomass production over the course of one year, the model is further constrained using monthly solar irradiance and temperature data throughout the year

We modeled the photon absorption flux in the PAR spectral range (400–700 nm) using the method described in Broddrick et al. 2019a. In summary, the photon uptake was modeled in terms of the biomass-normalized photon absorption rate (BPA) in μmol photons dry weight $(DW)^{-1}$ $h^{-1}$. This modeling approach took into account the experimentally measured concentration of chlorophyll-*a* ($mgChla$), the biomass dry weight ($gDW$), the solar irradiance ($I_0$) in μmol photons $m^{-2}$ $s^{-1}$ and the chlorophyll-a normalized optical absorption cross-section (*Chla**) in $cm^2$ mg chlorophyll-$a^{-1}$:

$$BPA = \frac{mgChla}{[gDW]} \times I_0 Chla^* \tag{31.1}$$

*Chla** was derived by using the *Chla* specific spectral absorption coefficient of 2.303 (Wagner et al. 2006), the average of specific absorption spectra in the PAR range ($A\gamma$) of *P. tricornutum* cells acclimated to 876 μmol photons $m^{-2}$ $s^{-1}$ (Jallet et al. 2016) and the experimentally measured concentration of Chla (Broddrick et al. 2019b):

$$Chla^* = 2.303 \frac{A\gamma}{[mgChla]} \tag{31.2}$$

We calculated the incident light by considering the light intensity within the PBR at the midpoint of the simulated time interval. To constrain the PSII reaction, we employed the Platt equation for photosynthesis prediction (PP), utilizing oxygen evolution values obtained from Broddrick et al. 2019a.

$$PP = P_s \times \left(1 - e^{-\alpha I/P_s}\right), \tag{31.3}$$

where $P_s$ is the maximum photosynthetic rate. Since the Platt equation describes a hyperbolic relationship between photosynthesis and irradiance, the parameter α represents the initial slope of the hyperbolic curve of photosynthesis at low light intensity.

Constraints on cyclic electron flow, chloroplast water-water reactions, energetic coupling between the chloroplast and mitochondria, and carbon accumulation during the light period were set as previously reported by Broddrick et al. 2019a. In this model, the storage carbon was accounted by using a reaction consists of triglyceride (TAG) and β-1,4-glucan molecules with a mass ratio of 3:1. This was added to the biomass objective function using a biomass metabolite (carbon_storage_c). These molecular level constraints were then combined with the PBR parameters to model the photoautotrophic growth of diatom in the cultivation level.

### *31.2.3 Modeling Microalgae Growth in PBR and Upscaling to One-Year Cultivation*

The photoautotrophic growth was simulated for averaged 200 L culture of *P. tricornutum*, using the same physicochemical conditions as described in Sect. 2.1 and photophysiological constraints modeled in Sect. 2.2. The photon delivery rate at PBR scale was determined considering total volume of culture medium in the PBR and the total surface area of the PBR (2.75 $m^2$). The integrated experimental biomass content (gDW) was used to run the growth simulation. The model was then used to predict the biomass production in terms of gDW per 14 days. The simulations were performed using COBRApy (Ebrahim et al. 2013) with the FBA (Lewis et al. 2010) and the GLPK Optimizer version v4.65.

The optimal growth temperature for *P. tricornutum* was set to 20 °C. A temperature factor was incorporated to account for suboptimal growth caused by exponential limitations due to suboptimal temperature (Branco-Vieira et al. 2020):

$$T_f = e^{-K(T - T\,opt)^2}, \tag{31.4}$$

where $T$ is the actual temperature (°C); $T_{opt}$ is the species-specific optimal temperature (°C) and K is an empirical constant which was set to 0.004 as reported by James and Boriah (2010).

We conducted the upscaling process from cellular metabolic changes to the cultivation level in a PBR over the course of a one-year operation. This process involved parameterizing the model with various light regimes, temperatures, and hours of daylight based on the monthly average parameters specific to the city of Concepción, Chile. The specific parameters used for this parameterization are presented in the Table 31.1.

In order to simulate one-month biomass production, we assumed a steady state and applied FBA to determine the biomass production during a 14-day cultivation period. To calculate the monthly biomass quantity, we considered the average daily cellular growth rate, taking into account the number of days in each month and the corresponding temperature factor. For each time point in the simulation, following a batch mode cultivation approach, we performed the following steps: (1) updated the model with the parameters specified in Table 31.1; (2) calculated the growth rate using FBA; and (3) determined the biomass production over the course of the months. The cumulative monthly biomass production represents the projected annual production.

**Table 31.1** Average climate parameters to model microalgae growth over one year

| Month | Days of the month | Temperature (°C)[a] | Global radiation (μmol photons $m^{-2}$ $s^{-1}$)[a] | Sunshine (h)[a] | Temperature factor[b, c] |
|---|---|---|---|---|---|
| January | 31 | 18.4 | 1381 | 14.4 | 0.98 |
| February | 28 | 17.1 | 1341 | 13.5 | 0.96 |
| March | 31 | 15.2 | 1244 | 12.3 | 0.90 |
| April | 30 | 12.3 | 959 | 11.1 | 0.77 |
| May | 31 | 12.9 | 638 | 10.2 | 0.82 |
| June | 30 | 8.6 | 559 | 9.7 | 0.62 |
| July | 31 | 9.3 | 543 | 9.9 | 0.62 |
| August | 31 | 10.4 | 678 | 10.8 | 0.67 |
| September | 30 | 12.1 | 979 | 11.9 | 0.77 |
| October | 31 | 12.6 | 1128 | 13.1 | 0.82 |
| November | 30 | 15.1 | 1298 | 14.1 | 0.90 |
| December | 31 | 16.3 | 1352 | 14.6 | 0.94 |

[a] Chile (2011)
[b] Branco-Vieira et al. (2020)
[c] James and Boriah (2010)

## 31.3 Results and Discussion

In this study, we examine the production of diatom biomass during the one-year operation of a PBR using a multiscale approach that incorporates different spatiotemporal scales (Fig. 31.2). The multiscale framework encompasses the metabolic changes occurring at the molecular level within the cells (gDW $h^{-1}$ $cell^{-1}$), the dynamic behavior of the cultivation phase at the lab scale (gDW $day^{-1}$ $L^{-1}$) and the pilot-scale PBR's microalgae biomass production over monthly (gDW $month^{-1}$ $L^{-1}$) and yearly (gDW $year^{-1}$ $m^{-3}$).

The molecular level simulation used the *P. tricornutum* GSM model iLB1034 (Broddrick et al. 2019a) which was constrained based on light uptake (Fig. 3a) and experimental data for biomass ratios (Fig. 3b) in order to predict the diatom growth. The biomass composition was used to update the model's biomass objective function, with lipids accounting for 9.08%, carbohydrates 7.85%, proteins 38.4%, and pigments 5.5%. The remaining biomass percentage, which includes ashes, was not integrated into the model. Moreover, values for DNA, RNA, and membrane lipids were adopted from the original GSM model publication (Broddrick et al. 2019a). The molecular-level output provided the predicted growth rate in gDW $day^{-1}$ (Fig. 3c).

To simulate the cultivation at the PBR level, we mathematically integrated PBR parameters and climate data into the model. The total volume and surface area of the PBR determined the area exposed to solar radiation, enabling the calculation of the available surface for photon uptake during the dynamic diatom growth.

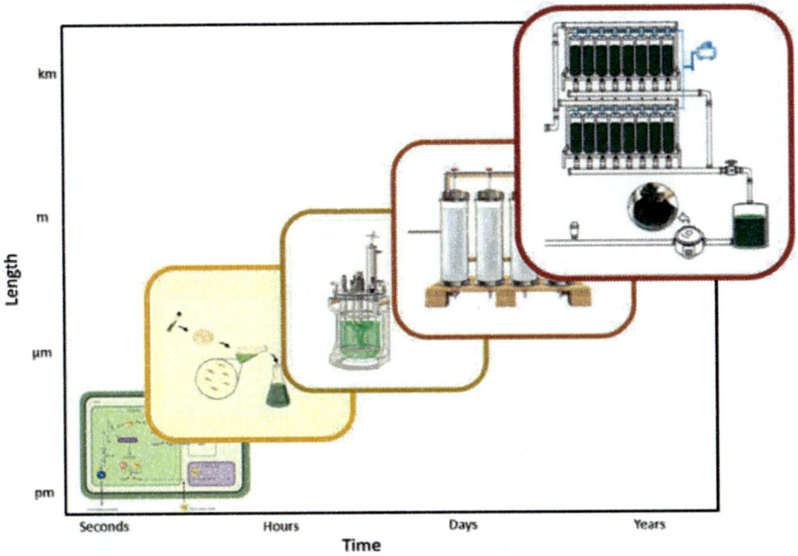


**Fig. 31.2** The multiscale modeling approach is illustrated in the representation diagram, which account various time and spatial scales. This multiscale framework encompasses the metabolic changes occurring at the molecular level within the cell (gDW $h^{-1}$ $cell^{-1}$), the dynamic behavior of the cultivation phase at the lab scale (gDW $day^{-1}$ $L^{-1}$) and the biomass production of microalgae at the pilot-scale PBR, both on a monthly (gDW $month^{-1}$ $L^{-1}$) and yearly basis (gDW $year^{-1}$ $m^{-3}$)

Monthly biomass production was estimated using climate parameters specific to the City of Concepción, Chile. Biomass accumulation over the 14-day cultivation period served as both model validation and a means for refining and adjusting the predictive results (Fig. 2d). The differences between the experimental biomass curve and the modeled curve were mainly observed during the stationary phase, which occurred between days 12 and 14. This discrepancy was primarily attributed to the model's constraints not accurately representing this phase of cultivation (Fig. 3d).

We used the biomass production rate obtained on the cultivation-level simulations to scale up biomass production over the course of months and one year. Each time point simulation represented one month of the year, updated with averaged monthly climate parameters. The average daily growth rate was then used to calculate the monthly biomass production. The model predicted a maximum production of 62.52 gDW carbohydrates, 231.15 gDW proteins, 42.62 gDW membrane lipids, 2.40 gDW carbon storage (TAG and β-1,4-glucan), and 33.17 gDW pigments (Fig. 3e), when considering a one-month simulation.

Subsequently, the process was modeled to simulate the one-year operation of the PBR, with constraints applied to account for monthly light insensitivities, temperature variations, and the number of days (Fig. 3f). The highest biomass value was observed during the South American summer season, specifically in January and

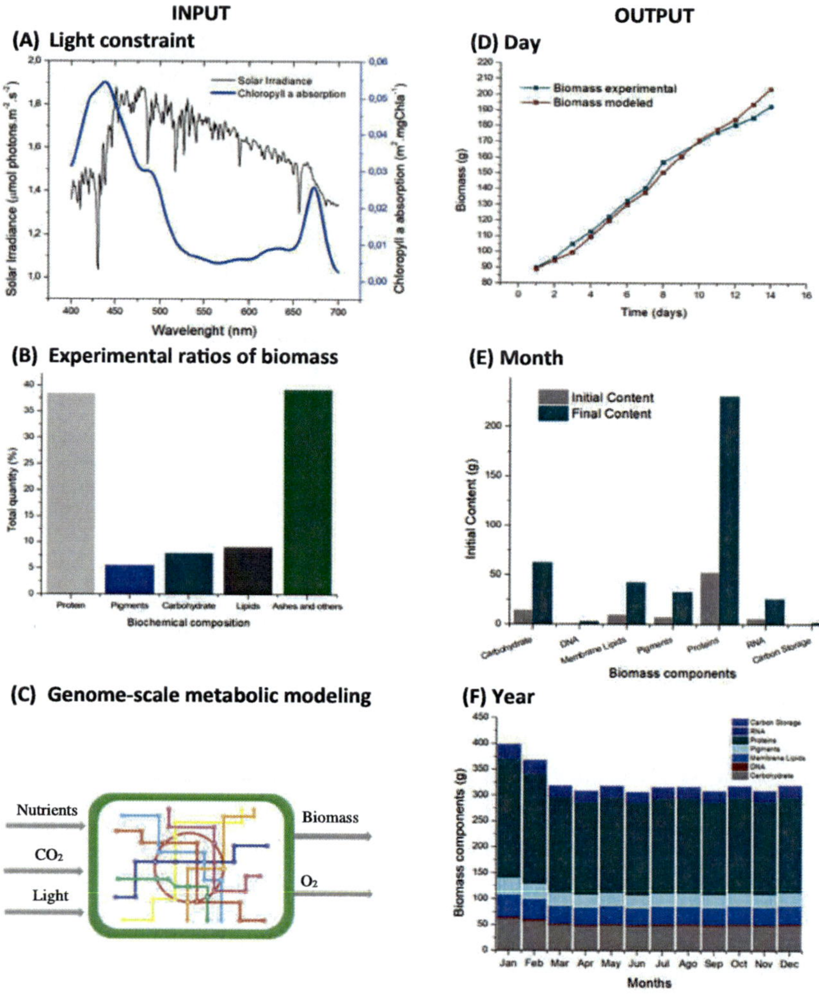


**Fig. 31.3** Input and output of the developed framework to predict biomass production and components ratios over different time-spatial scales. **a** Light constraints; **b** microalgae biomass experimental data integrated into the GSM; **c** schematic representation of GSM, highlighting the input and output of simulations; **d** model validation with experimental data; **e** simulation of biomass production and components over one month; **f** simulation of biomass production over one-year PBR operation

February. This observation indicates that higher light intensities and temperatures play crucial roles in driving microalgae growth.

Many studies have employed kinetic models to dynamically simulate the growth of *P. tricornutum* (Ova Ozcan and Ovez 2020). In their analysis, Ova and Ovez (2020) investigated the predictive effects of Ratkowsky's square root model and Monod-type model, assessing how temperature and light intensity influence the growth of *P. tricornutum*. However, these models fall short in incorporating reactor-scale dynamics on microalgae metabolism at the molecular level. Consequently, their ability to predict the impact of environmental fluctuations on the microalgae phenotype is limited.

Modeling microalgae-based commodity production offers significant potential for enhancing productivity and system sustainability. While there have been efforts to create modeling frameworks for improved predictability, a thorough evaluation that spans from the molecular level to pilot-scale operations is essential. This comprehensive assessment is required to inform the design of microalgae products, taking into account their influence on both biomass productivity and ecosystem sustainability.

Predicting the productivity of growth cultures in PBRs presents significant challenges, mainly due to the need to integrate a wide range of experimental data, which includes physiological information and climate parameters. To address this challenge effectively, there is a need for integrative and multiscale models. Multiscale modeling of phototrophic growth typically involves the application of various mathematical and computational approaches. However, the complexity of this task often makes it challenging to address within a single computational framework. Thus, the development of integrative and multiscale models is crucial for accurately predicting the productivity of microalgae cultures in PBRs and advancing our comprehension of the underlying processes (Westermark and Steuer 2016).

## 31.4 Conclusions and Future Prospects

Accurate microalgae growth models play a crucial role in the design and optimization of industrial production processes for valuable compounds. In this study, a comprehensive framework was developed to predict biomass production over the course of one year of facility operation. The framework combines a metabolic model with dynamic growth simulations of diatoms at the pilot reactor scale. The integration of parameters at the PBR scale with metabolic modeling improved the model's accuracy, providing a deeper insight into how the PBR affects the microalgae's phenotype and biomass productivity under varying light intensities and temperature conditions.

By having the capacity to analyze metabolic behavior and cellular dynamics across different scales, ranging from molecular level to pilot-scale operation, this study provides a robust foundation for future advancements and innovations in the field. By leveraging this knowledge, researchers and industry can make well-informed decisions to enhance the growth and sustainability of the microalgae-based industry, ultimately leading to the production of high-value compounds in a more efficient and environmentally friendly manner.

Future improvements to the model will encompass the integration of mass transfer and reactor gas exchange constraints, consideration of diverse nutrient availability and cellular uptake, and accounting for different phases of the microalgae growth. Furthermore, it is essential to include varying pigment ratios throughout the year to better assess the impact of light intensity on microalgae growth. Additionally, the model could benefit from refinements that encompass cellular self-shading and photoinhibition during batch mode growth simulations. These enhancements will establish a robust framework for more extensive exploration of microalgae biomass potential in the production of biofuels and high-value compounds through a cascade of optimized processes.

**Acknowledgements** This work was supported by: Base Funding—UIDB/04730/2020 of the Center for Innovation in Engineering and Industrial Technology—CIETI; LA/P/0045/2020 (ALiCE) and UIDB/00511/2020—UIDP/00511/2020 (LEPABE) funded by national funds through FCT/MCTES (PIDDAC) and by the Deutsche Forschungsgemeinschaft (DFG) under Germany's Excellence Strategy EXC 2048/1, Project number 390686111.

## References

Branco-Vieira M, San Martin S, Agurto C et al (2018a) Biochemical characterization of *Phaeodactylum tricornutum* for microalgae-based biorefinery. Energy Procedia 153:466–470. https://doi.org/10.1016/J.EGYPRO.2018.10.079

Branco-Vieira M, San Martin S, Agurto C et al (2018b) Potential of *Phaeodactylum tricornutum* for biodiesel production under natural conditions in Chile. Energies 11:54. https://doi.org/10.3390/en11010054

Branco-Vieira M, Lopes MPC, Caetano N (2022) Algae-based bioenergy production aligns with the Paris agreement goals as a carbon mitigation technology. Energy Rep 8:482–488. https://doi.org/10.1016/J.EGYR.2022.01.081

Branco-Vieira M, San Martin S, Agurto C et al (2020) Biotechnological potential of *Phaeodactylum tricornutum* for biorefinery processes. Fuel 268. https://doi.org/10.1016/j.fuel.2020.117357

Broddrick JT, Du N, Smith SR et al (2019a) Cross-compartment metabolic coupling enables flexible photoprotective mechanisms in the diatom *Phaeodactylum tricornutum*. New Phytol 222:1364–1379. https://doi.org/10.1111/NPH.15685

Broddrick JT, Welkie DG, Jallet D et al (2019b) Predicting the metabolic capabilities of *Synechococcus elongatus* PCC 7942 adapted to different light regimes. Metab Eng 52:42–56. https://doi.org/10.1016/J.YMBEN.2018.11.001

Chile UD (2011) Explorador solar DGF. In: Ministerio de Energia. http://walker.dgf.uchile.cl/Explorador/Solar2/. Accessed 25 Jan 2018

Deprá MC, Dias RR, Severo IA et al (2020) Carbon dioxide capture and use in photobioreactors: the role of the carbon dioxide loads in the carbon footprint. Bioresour Technol 314:123745. https://doi.org/10.1016/J.BIORTECH.2020.123745

Dhanker R, Kumar R, Tiwari A, Kumar V (2022) Diatoms as a biotechnological resource for the sustainable biofuel production: a state-of-the-art review. 38:111–131. https://doi.org/10.1080/02648725.2022.2053319

Ebrahim A, Lerman JA, Palsson BO, Hyduke DR (2013) COBRApy: constraints-based reconstruction and analysis for python. BMC Syst Biol 7:1–6. https://doi.org/10.1186/1752-0509-7-74/FIGURES/2

Guillard RRL, Ryther JH (1962) Studies of marine planktonic diatoms. I. *Cyclotella nana* Hustedt, and *Detonula confervacea* (cleve) Gran. Can J Microbiol 8:229–239. https://doi.org/10.1139/m62-029

Jallet D, Caballero MA, Gallina AA et al (2016) Photosynthetic physiology and biomass partitioning in the model diatom *Phaeodactylum tricornutum* grown in a sinusoidal light regime. Algal Res 18:51–60. https://doi.org/10.1016/J.ALGAL.2016.05.014

James SC, Boriah V (2010) Modeling algae growth in an open-channel raceway. J Comput Biol 17:895–906. https://doi.org/10.1089/cmb.2009.0078

Lewis NE, Hixson KK, Conrad TM et al (2010) Omic data from evolved *E. coli* are consistent with computed optimal growth from genome-scale models. Mol Syst Biol 6:390. https://doi.org/10.1038/MSB.2010.47

Orth JD, Thiele I, Palsson BO (2010) What is flux balance analysis? Nat Biotechnol 28(3):245–248. https://doi.org/10.1038/nbt.1614

Ova Ozcan D, Ovez B (2020) Evaluation of the interaction of temperature and light intensity on the growth of *Phaeodactylum tricornutum*: Kinetic modeling and optimization. Biochem Eng J 154:107456. https://doi.org/10.1016/J.BEJ.2019.107456

Pathak M, Slade R, Shukla PR et al (2022) Technical summary. In: Shukla PR, Hasija A, Lisboa G, et al (eds) Climate change 2022: mitigation of climate change. contribution of working group III to the sixth assessment report of the intergovernmental panel on climate change. Cambridge University Press, Cambridge, UK and New York, NY, USA

Wagner H, Jakob T, Wilhelm C (2006) Balancing the energy flow from captured light to biomass under fluctuating light conditions. New Phytol 169:95–108. https://doi.org/10.1111/J.1469-8137.2005.01550.X

Westermark S, Steuer R (2016) Toward multiscale models of cyanobacterial growth: a modular approach. Front Bioeng Biotechnol 4:95. https://doi.org/10.3389/FBIOE.2016.00095/BIBTEX

# Part V
# Energy and Environment

# Chapter 32
# Recyclability of Wind Turbines: Overview of Current Situation and Challenges

**Nacef Tazi and Youcef Bouzidi**

**Abstract** The European energy system is increasingly incorporating renewable energies as a result of the wind power market's exponential rise. However, from a life cycle standpoint, there still exists certain issues with wind turbines, especially in end-of-life management. In this study, an overview of the current state of a wind turbine's end-of-life management is provided. Turbines' end-of-life management practices are a difficult endeavor for completing materials loops. The current state of play demonstrates that some materials used in wind turbines are still diverted towards energy recovery or landfills rather than more virtuous management such as recycling, remanufacturing or reuse, which hinder the recyclability and thus circularity of wind turbines. Blades, rare earth elements used in permanent magnets, and critical raw materials used in alloyed components are preventing wind turbines from achieving high recyclability. The remaining difficulties to improve recycling and close the material loop for wind turbines are then highlighted.

**Keywords** Blades · Circularity · End-of-life management · Rare earth elements · Critical raw materials · Recyclability · Wind turbine

## 32.1 Introduction

The main cumulative renewable energy sources in Europe (EU) are provided by wind and water (Eurostat 2022), yet both the shift to clean energy technologies and the EU energy carbon neutrality depend heavily on wind energy. With a cumulative

N. Tazi (✉)
Circular Services, 12 Rue Thomas Dupuy, 31300 Toulouse, France
e-mail: nacef.tazi.2018@utt.fr; naceftazi@gmail.com

N. Tazi · Y. Bouzidi
Interdisciplinary Research on Society-Technology-Environment Interactions, University of Technology of Troyes, Troyes, France
e-mail: youcef.bouzidi@utt.fr



N. S. Caetano and M. C. Felgueiras (eds.), *The 9th International Conference on Energy and Environment Research*, Environmental Science and Engineering,
https://doi.org/10.1007/978-3-031-43559-1_32

installed capacity of more than 200 GW, the EU is paving the way for the use of renewable energy sources and the transition away from fossil fuels (Euro Commission 2022). Onshore and offshore locations in Europe have installed wind turbines (WT), both of which help the EU achieving its energy roadmap targets (Euro Commission 2012). WT is a heterogeneous system that couples numerous parts and components comprised of various materials. These components aid in the creation of clean and renewable energy. Nevertheless, a thorough life-cycle analysis reveals that some materials have the potential to harm the environment and require intensive energy to source and manufacture (Majewski et al. 2022). Additionally, their End-of-Life (EoL) management does not allow the deployment of virtuous R-strategies that support higher circularity goals, such as reusing or remanufacturing (EMF 2013; Morseletto 2020).

Specifically, this is the case for blades made of composite materials (Liu and Barlow 2050), or Rare Earth Elements (REEs) used in permanent magnets. Still, flows for WT's EoL management routes diverge to energy recovery, incineration, and landfill. Thus, WT recyclability is hindered. The ability to be decommissioned, gathered, separated, and recycled is all that recycling of WT materials entails. Hence, it is crucial to evaluate WT's recyclability.

Afterwards, the initial goal of this paper is to give a general review of the current state of wind turbine and WT material EoL management. The main obstacles to improve WT's recyclability are then listed.

## 32.2 Wind Turbine Bill of Materials

For both onshore and offshore sites, there exists two primary technical designs of wind turbines on the market: gearbox technology (geared) and direct drive (gearless) one. Although geared WT still hold the majority of the market, direct-drive WT share is rapidly rising. Less maintenance is needed for the latter, which also results in a smaller nacelle and less weight for the turbine. Although hybrid models do exist, their market adoption is quite modest. Permanent magnets and rare earth materials are mainly used in direct drive WT. In onshore and offshore applications, synchronous generators (DDSG) are one of the most used direct drive WTs. Figure 32.1 displays the typical material bill for such an onshore WT model. More than 76% of the weight of WT in foundations is made of concrete, making it the most common material used in WT foundations. Steel (and its alloys) is the second most frequently utilized material. WT nacelle and rotor also use various materials such alloys, iron, copper, and aluminum. The blades of wind generators are generally made of Glass Fiber Reinforced Plastics (GFRP). WT permanent magnets and tower magnets employ rare earth elements (REEs) and boron (B).

Despite supply constraints in the EU and the rising criticality of these materials designated as strategic or critical raw materials (CRM), their use in the wind energy sector rose significantly over the last decade (WTO 2015; Euro Commission 2021). Figure 32.2 lists the most common REEs utilized in WT permanent magnets. In

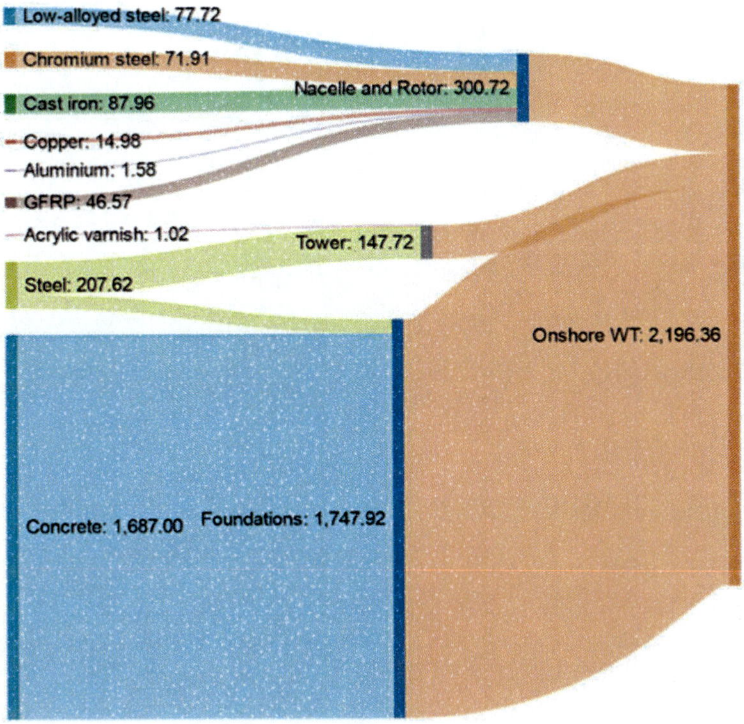


**Fig. 32.1** Bill of materials used for DDSG WT. Average estimations from Schreiber et al. (2019). Units in tonnes. Figure made with SankeyMatic

principle, 150–200 kg are used on average for every MW of capacity. And while Dysprosium (Dy) and Terbium (Tb) are the most frequently used heavy REEs in WT magnets, Neodymium (Nd) and Praseodymium (Pr) are the most commonly employed light REEs (light REEs are lanthanides with the lowest atomic numbers, and heavy REEs are classified considering their higher atomic weights relative to light REEs). Gadolinium is also another heavy rare earth used in WT. Even so, current and upcoming WT generations would gradually decrease their use of light REEs and phase out heavy REEs as a step towards less REEs use in wind energy technologies (Vestas 2021).

Still, current stocks and flows of REEs in the EU wind sector can be toughly estimated from the 210 GW installed capacity. Other components, such as lubricants and sulfur hexafluoride (SF6), also are utilized in WT, although at lower rates than those shown in Fig. 32.1. While more than 80% of the materials used in wind turbines can be recycled, some are lost in the recycling process or diverted to landfill and energy recovery. The management of WT materials' current EoL is then described in general.

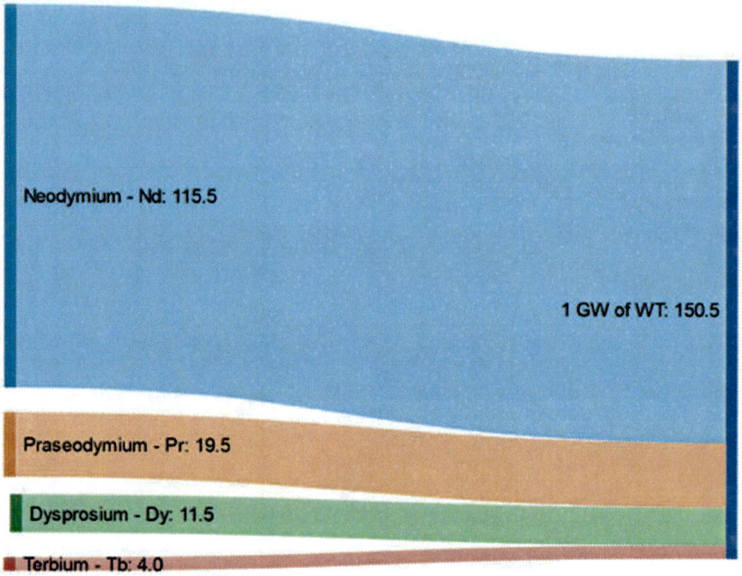


**Fig. 32.2** Bill of REEEs used for onshore and offshore WT. Average estimations for 1 GW of installed WT. Gadolinium was not captured in this sankey. Data from Alves Dias et al. (2020). Units in t/GW. Figure made with SankeyMatic

## 32.3 Current Situation of EoL Management: Reuse, Recycle, Recover and Landfill

Wind turbine decommissioning is not regulated in Europe. However, its management is guaranteed to either rehabilitate the wind farm site or to repower the wind farm and increase its. Resources that can be reintroduced into the economy are gathered by WT. All materials are not, however, recycled (Tazi 2018; Jani et al. 2020).

In wind farms, concrete is the most used material. The potential for recycling this material is very significant. However, it is frequently not completely removed from the soil (Tazi 2018), and it is neither cost effective nor ecologically sustainable to carry it from remote wind farm locations to recycling facilities. Its employment in repowered wind farms is particularly problematic because it requires on-site recycling, and its ability to be included into new concretes as a substitute of natural aggregates and sands is restricted subject to regulatory, standards or performance restrictions. Concrete foundations are usually either left in place or landfilled after being excavated.

In general, metals are recovered and treated according to the waste framework directive.[1] They may be reused and recycled readily, and they can be reinjected into secondary markets. A separate waste flow for recycling is possible due to the

[1] https://environment.ec.europa.eu/topics/waste-and-recycling/waste-framework-directive_en.

massive size of the WT system and its components. Even yet, certain partial metal flows (such those that were combined with concrete) may end up in landfills. The materials used in alloyed components, such as manganese (Mn), molybdenum (Mo), chromium (Cr), cobalt (Co), and nickel (Ni), are still included in these flows but are typically recycled together with ferrous and non-ferrous metal flows and therefore are not usually targeted for recycling.

Regarding REEs, they are typically not retrieved from magnets. The procedures currently in use are to disassemble and separate these materials, which require significant time and energy. Some approaches, such as disintegration (degradation and demagnetization carried out in a hydrogen atmosphere), permit recovering NdFeB magnet waste or powder for use in new magnets (Chowdhury et al. 2021; Coelho et al. 2021). However, such approaches are currently targeting only small magnets.

The majority of blades are made of composites, GFRP, and carbon fiber reinforced polymers, coupled with adhesives and balsa wood. The current deployed recycling methods do not recover blade materials (Jani et al. 2020). Typically, they are either delivered to landfills or cement co-processing facilities for energy recovery.

The WEEE directive[2] generally manages other Electronics and Electrics components from the wind turbine. In terms of WT's overall weight, electronics and electrics components make up about 1%. Fluids, lubricants, and SF6 have distinctive EoL management approaches as well and are often treated prior to the decommissioning of WT. Recyclability potential of wind turbines can be reduced if these waste flows are inadequately managed due to their comparatively high risk of contamination.

The market for second-hand replacement parts in the wind energy industry is still under development. On a global basis, initiatives are being made to offer used goods for the industry.[3]

Finally, repurposed solutions are prompt and mostly focus on tower and blade fragments that are converted into bike shelters, urban furniture, or included in bridge structures (Stone 2022).

As a result, not all material loops in WT could be sustained and closed.

## 32.4 Challenges in Recyclability of Wind Turbines

To help achieve a closed material loop and supply flows to secondary markets, a significant portion of wind turbine materials can be adequately recyclable. To assure greater WT recycling, there are still challenges in the wind energy value chain. The main difficulties are shown in Fig. 32.3 and are explained below.

WT are generally located in remote areas, which render recycling and proper EoL management of some WT materials not environmental nor economically effective. This is true for concrete, which has a range of ecological profitability distance of less

---

[2] https://environment.ec.europa.eu/topics/waste-and-recycling/waste-electrical-and-electronic-equipment-weee_en.

[3] See example of: https://en.wind-turbine.com/.

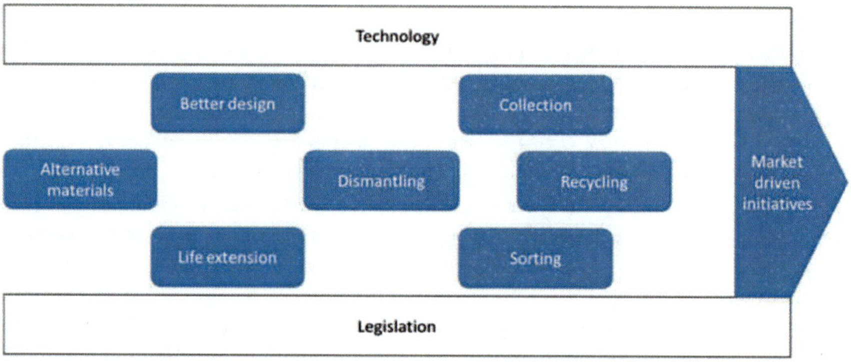


**Fig. 32.3** Main challenges for better recyclability of WT

than 100 km (Ben Fraj and Idir 2017). On-site excavation, recycling, and utilization activities are all fairly scarce on-site. Besides, no specific provision or measure exist in order to legally bind the wind farm to completely recycle WT foundations.

When it comes to recycling, the collection phase may seem routine, but this activity ensures proper management of CRM and REEs in WT. Such a concern was associated to industry incentives for high-quality recycling (Reck and Graedel 2012). Design is also recognized as a major step in enhancing recyclability and move towards design for circularity. When the service lifespan is reached, the majority of raw materials, including metals, diverge to waste management and follow the life cycle of a single use product. Such a situation necessitates sectoral and policy measures in order to close material loops on the one hand and transition to higher R-strategy on the other (Morseletto 2020). The recyclability of WT blades and less recyclable materials such components containing REEs would both be improved by design provisions like design for recycling or design for circularity.

WT recyclability currently faces significant challenges related to recycling methods and costs. One way to look at it is that the recycling technologies used for WT material recycling are somewhat constrained and overtaken by all the new design and manufacturing procedures. On the other hand, to close material loops, the cost of recycling is still a frontier that needs incentives and legislative support. The absence of eco-design and design for circularity principles further emphasizes this phase shift. To achieve improved circularity of products, it is vital to distribute the burdens along the value chain in order to shut material loops. Additionally, there should be incentives for using cutting-edge technologies for recycling and sorting. If EoL initiatives are successful, contamination of WT wastes can also be controlled.

The wind energy sector has already highlighted the first measures to reduce the usage of REEs from the design phase. Such procedures ought to be made standard, and they ought to be supported by enforceable legal measures that address this problem. CRM management need to be covered by a cross-sectoral EU regulation to guarantee proper handling and recycling of these materials. The recyclability of WT can also be improved by using alternative materials. Blades, alloyed parts, and

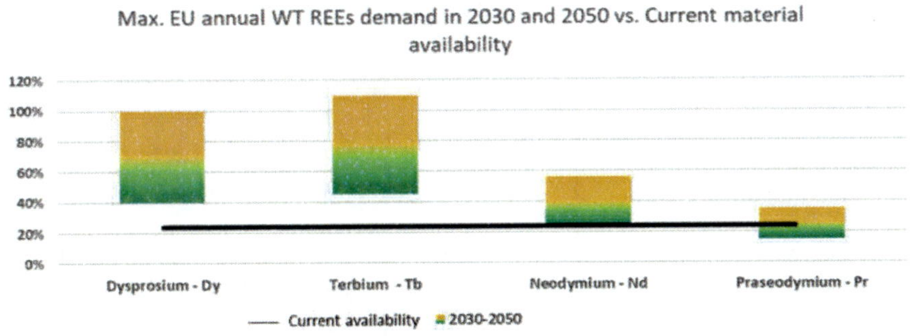


**Fig. 32.4** Maximum EU annual WT REEs demand in 2030 and 2050; compared with current material availability. Gadolinium was not captured in this figure. Figure adapted from Carrara et al. (2020)

REEs are examples of them. The wind energy industry needs stronger and more appropriate regulatory measures from the perspective of execution in order to fully realize WT's potential for recycling and/or circularity. These actions can be taken anywhere throughout the value chain and can be made stronger by standards.

To assure the use of recyclable materials in WT, policy measures can be used at the design phase. They can also be used as tools to extend the lifespan of wind turbines.

In order to monitor and guarantee effective EoL management (including WT recycling), policy tools can also be used. Legal tools can also promote secondary markets that improve recycling operations. These measures ought to lower supply-related risks and due diligence.

According to Fig. 32.4, when compared to the existing supply, Dy and Tb will continue to be in high demand in 2030 and 2050. The rising demand for such minerals for clean technology would also increase the supply risk. However, it is anticipated that with current practices, the yearly EU REEs requirement for wind turbine use may not be met. Overcoming these obstacles might reduce this risk.

## 32.5 Conclusion

In summary, this paper illustrates the current state of play around a wind turbine's end-of-life management and highlights the difficulties and main gaps that need to be addressed to improve recyclability of wind turbines and circularity performances in the wind energy industry. This paper discusses various end-of-life activities, such as inter-alia decommissioning, recycling, reusing, and repurposing. The use of non-recyclable materials for wind turbine blades, such as composites and glass fibers, is the main factor impeding their capacity to be recycled.

Additionally, wind turbines do not fully recover rare earth elements, essential raw materials, and other strategic and critical materials for alloyed components.

These materials continue to present considerable challenges for a wind turbine to achieve high levels of recyclability. The lack of sustainable design principles and the advancement of collection, sorting, and recycling technologies are the remaining difficulties.

Finally, appropriate policy instruments that encourage investments and actions towards higher recyclability of wind turbines have to be woven into the plan to attain higher recyclability of wind turbines.

Everything is a work in progress to tackle these issues from a circular economy perspective.

**Acknowledgements** The paper was markedly improved by the comments of the anonymous reviewers.

**Funding** This research did not receive any specific grant from funding agencies in the public, commercial, or not-for-profit sectors.

## References

Alves Dias P, Bobba S, Carrara S, Plazzotta B (2020) The role of rare earth elements in wind energy and electric mobility. https://doi.org/10.2760/303258

Ben Fraj A, Idir R (2017) Concrete based on recycled aggregates—recycling and environmental analysis: a case study of Paris' region, vol 157. https://doi.org/10.1016/j.conbuildmat.2017.09.059

Carrara S, Alves Dias P, Plazzotta B, Pavel C (2020) Raw materials demand for wind and solar PV technologies in the transition towards a decarbonised energy system. JRC. https://doi.org/10.2760/160859

Chowdhury AN, Deng S, Jin H, Prodius D, Sutherland JW, Nlebedim IC (2021) Sustainable recycling of rare-earth elements from NdFeB magnet Swarf: techno-economic and environmental perspectives, vol 9, no 47

Coelho F, Abrahami S, Yang Y, Sprecher B, Li Z, Menad NE, Bru K, Marcon T, Rado C, Saje B, Sablayrolles ML, Decottignies V (2021) Upscaling of permanent magnet dismantling and recycling through VALOMAG project, vol 5, no 1

European Commission (2021) European Commission, directorate-general for internal market, industry, entrepreneurship and SMEs, 3rd raw materials scoreboard: European innovation partnership on raw materials. Publications Office

European Commission (2012) Energy roadmap 2050. Publications Office of the European Union

European Commission (2022) Onshore and offshore wind [online]. Available https://energy.ec.europa.eu/topics/renewable-energy/onshore-and-offshore-wind_en. Accessed 15 May 2022

EMF (2013) Towards the circular economy vol. 1: an economic and business rationale for an accelerated transition

Eurostat (2022) Renewable energy statistics [online]. Available https://ec.europa.eu/eurostat/statistics-explained/index.php?title=Renewable_energy_statistics#Wind_and_water_provide_most_renewable_electricity.3B_solar_is_the_fastest-growing_energy_source. Accessed 15 July 2022

Jani HK, Kachhwaha SS, Nagababu G, Das A (2020) A brief review on recycling and reuse of wind turbine blade materials, vol 62, no 7124–7130

Liu P, Barlow CY (2017) Wind turbine blade waste in 2050, vol 62. https://doi.org/10.1016/j.wasman.2017.02.007

Majewski P, Florin F, Jit J, Stewart RA (2022) End-of-life policy considerations for wind turbine blades, vol 164, no 112538

Morseletto P (2020) Targets for a circular economy, vol 153, no 1. https://doi.org/10.1016/j.resconrec.2019.104553

Reck BK, Graedel TE (2012) Challenges in metal recycling, vol 337. https://doi.org/10.1126/science.1217501

Schreiber A, Marx J, Zapp P (2019) Comparative life cycle assessment of electricity generation by different wind turbine types, vol 233. https://doi.org/10.1016/j.jclepro.2019.06.058

Stone M (2022) Engineers are building bridges with recycled wind turbine blades [online]. Available https://www.theverge.com/2022/2/11/22929059/recycled-wind-turbine-blade-bridges-world-first. Accessed 15 May 2022

Tazi N (2018) Evaluation de la disponibilité et de l'Impact Environnemental des Parcs Eoliens (EDIPEO). Université de Technologie de Troyes

Vestas (2021) Sustainability report 2021, vol 1. https://www.vestas.com/en/sustainability/reports-and-ratings

WTO (2015) China—measures related to the exportation of rare earths, Tungsten and Molybdenum—DS432 [online]. Available https://www.wto.org/english/tratop_e/dispu_e/cases_e/ds432_e.htm. Accessed 15 July 2021

# Chapter 33
# Drawing Behavioural Insights from Members of Social Innovations in the Energy Sector Through Cluster Analysis: A Comparative Study in Portugal

**Sofía Mulero-Palencia and Alejandro Hernández Serrano**

**Abstract** Social innovations in the energy sector (SEI) play an important role in the energy transition. This paper contributes to a comprehensive understanding of how SEI members use electricity, examining daily electricity consumption combined with contextual information from questionnaires for 65 users of a cooperative and a crowdfunding platform in Portugal. The research, based on an analysis of seasonal profiles of hourly data over a year using clustering techniques, allows the identification of five continuous and different behaviour patterns in each case study, four of them common. The results indicate that there are factors that determine each segment, especially the type of electrical service contracted, the heating/cooling systems available, the presence of storage systems and the characteristics of temperature control habits. All users had previously applied some measure of energy efficiency, suggesting a commitment to energy efficiency practices and environmental concerns.

**Keywords** Behavioural aspects · Cluster analysis · Electricity consumption · Residential · Social innovation · Questionnaire

## 33.1 Introduction

The global energy system is undergoing a major transformation, with the goal of achieving at least a 55% reduction in greenhouse gas emissions compared to 90s levels to achieve climate neutrality by 2050 (IRENA, IRENA 2019). and sustainable technologies, or Climate change is a fact (Tollefson 2022), and to confront it, the Paris agreement established the objective of keeping the increase in global average

S. Mulero-Palencia (✉) · A. H. Serrano
CARTIF Technology Centre, Parque Tecnológico de Boecillo, 47151 Valladolid, Spain
e-mail: sofmul@cartif.es

 
N. S. Caetano and M. C. Felgueiras (eds.), *The 9th International Conference on Energy and Environment Research*, Environmental Science and Engineering,
https://doi.org/10.1007/978-3-031-43559-1_33

temperature below 2 °C (European Commission 2030). Numerous efforts have been made to curb this problem, which come hand in hand with the use of clean the electrification of different sectors such as transport. This means that the expected growth in electricity demand is expected to go from 23 to 30–31% 1 by 2030 (ENTSO-E 2020). Specifically, the values around buildings are not negligible, accounting for about 40% of total energy consumption in Europe and 36% of greenhouse gas (GHG) (European Commission 2020).

In this context, current policies are formulated with the purpose of improving the way energy is used from an efficiency and sustainability perspective. A context favoured by the advancement of technology and in which new actors are appearing. Moreover, as a fundamental part of this transition towards clean energy, the concept of energy democracy is gaining ground and social innovation projects in the renewable energy sector or (SIE) come into play, such as Cooperatives or Crowdfunding Platforms, which allow the consumer play a more active role (Matschoss et al. 2021). Social innovations in the energy sector can be seen as a driver of change capable of addressing societal challenges, such as population aging, growing inequality, climate change, globalisation and digitalisation (Wittmayer et al. 2022).

This research aims to create a better understanding of the socioeconomic aspects related to energy, to better know how social innovations may favour more active consumption behaviours. The analysis focuses on the effects of factors at the household level referring to the characteristics of the occupants, from aspects of the dwelling to the equipment available, heating habits or attitudes towards more efficient and environmentally friendly consumption practices.

Some research has focused on finding the best technical approach in order to perform this user segmentation. Key relevant information about the data is usually lost, especially in case of averaging or aggregation processes as explained in McLoughlin et al. (2015), then data mining techniques allow data to be segmented before those processes are applied. Zhou et al. include the most widely used clustering methods for load classification for smart grid environment, based on K-means algorithm, Fuzzy C-Means (FCM), hierarchical clustering and Self-Organization Mapping (SOM) in Zhou et al. (2013). Other researchers highlight the need to connect those profiles to contextual features as in Gouveia and Seixas (2016), selecting physical details about dwellings, the HVAC equipment or the level of income of the occupants. Viegas et al. demonstrate that the use of survey information contributes to significantly increasing accuracy of the classification task (Viegas et al. 2016). Finally, other research focuses on the analysis of similarity metrics in order to determine which method suits better after performing time series clustering (Iglesias and Kastner 2013). Not all customers contribute in the same way to the peak system (Azaza and Wallin 2017), and it is necessary to identify different behaviours, especially those that have a greater responsibility in this regard, and how to find ways to improve and provide energy reduction recommendations. Finally, users committed to energy efficiency practices and with strong environmental concerns may be more likely to reduce their electricity demand. Then, a thorough analysis of surrounding factors, including home properties and occupant features, can help determine a set of common patterns and associated typical behavioural characteristics, also considering the different business models in which they are involved. This provides the opportunity to identify similarities and differences between regular and engaged users in the renewable energy sector.

## 33.2 Materials and Methods

A better understanding of consumer demand and use of electricity is needed to drive measures to improve energy efficiency and promote the reduction of electricity consumption. As described in the introduction section, members of the social innovations can be seen as drivers of the energy transition, so identifying behavioural characteristics in this group of consumers and comparing them with ordinary ones in the renewable sector can help establish a roadmap for progress in this field. Then, this research is built around four main premises: the first one indicates that the members of social innovations may use electricity in a different way than the common users; the second hypothesis suggests that SEI members may have a set of characteristics and habits in common. Moreover, SEI people should be more familiar with energy efficiency practices and environmental issues, so they would likely make better use of electricity in the home. Finally, differences related to the associated business model and geographical location could appear.

### *33.2.1 Case Study Participants*

A total of 112 users participated in the smart meter installation campaign to collect information on electricity consumption for a period of one year, starting in April 2020. Additionally, a total of 65 users answered a questionnaire. After combining both data sources and ruling out incomplete or erroneous responses, the final number of participants selected for this analysis was reduced to 22 for the Coopernico cooperative and 43 for the GoParity crowdfunding platform case study.

The questionnaire was composed following premises and results of the previous literature, and the consolidated version underwent some final changes in the implementation due to local adaptation. The type of questions included can be classified into three main sections: household equipment, heating habits and attitudes towards behavioural change. Additionally, the general population was characterised using data from relevant and reliable reports, institutions, and databases. These actions allowed creating some reference profiles and extracting expected average characteristics that may be behind the defined profile. The hourly electricity profile of the control group in Portugal was defined considering the information from the Entidade Reguladora dos Serviços Energéticos (ERSE) (ERSE 2019). European and national databases (European Commission 2019, 2016; Enerdata 2015) were consulted to complete the image according to the main characteristics incorporated in the designed questionnaire.

### 33.2.2 *Methodology*

This section describes the four-step methodology used for consumer segmentation. Smart metering data provided by case study participants was combined with a questionnaire designed with the aim of collecting significant characteristics of the contextual information and behaviour of the occupants. A depth-analysis is conducted based on clustering, to find out a set of profiles representative for a group of participants along the different periods of the year, considering seasonality. This information is later crossed with the data provided by the questionnaires, to find main determinants and characteristics of members of each of the patterns found. Each step is detailed below:

1. **Data acquisition and manipulation.** Smart meters are installed for all the participants for a year-monitoring period. A collaborative approach is followed to collect information on electricity consumption and household characteristics, ensuring the anonymization and security of information. In this stage, missing and erroneous values are detected and cleaned to guarantee the reliability and quality of the information for the different data sources.
2. **Data exploration.** The analysis strategy is established, and average user profiles per season are calculated considering the type of day. The research focuses on finding continuous weekday profiles along the year, and their evolution.
3. **Clustering analysis.** Participants are classified to obtain a set of representative curves per kind of user. The analyses is performed on seasonal average and normalised profiles per household, using classification algorithms (K-means and hierarchical clustering). Results are compared and differences in the obtained results in relation with the number of groups depending on the considered period under analysis are explored.
4. **Other features analysis.** The information from the questionnaires is combined with the corresponding curves obtained, in order to identify the most representative characteristics of the participants of the group.

## 33.3 Results and Discussion

Following the methodology already described, a small set of profiles was found to segment users according to the way in which they make use of electricity. The analysis of the consumption profiles throughout the four seasons of the year was done per case study, seeking the continuity of the profiles. The number of profiles could differ among the different periods, due to two main reasons: first, the number of participants may not always the same depending on the selected time slot, since due to technical problems some information could not be retrieved, causing that the amount of data could not be enough to be considered. In those cases, some users might be discarded for the cluster analysis for that season period. Secondly, users may behave differently along the year for different reasons (leaving the household for long periods, habits change, etc.). Then, the found profile may not be continuous. However, the number of patterns reached for the analysed period was always five. There is a correspondence in terms of similarity for four of the five patterns obtained in both case study. The following tables show examples of correspondence of these analogous profiles in quantitative terms (Table 33.1), and the related qualitative information recovered through the questionnaire (Tables 33.2, 33.3 and 33.4).

The distribution of electricity consumption throughout the day is analysed considering daily profiles of 24 values, and daily profiles of 4 values, including morning, noon, evening and night aggregated periods. Most of them maintain a continuous 24-h shape during the different seasons.

Some of the key aspects revealed by the analysis indicate that main differences between profiles are related to features such as the contracted electricity service (type of rate, amount of power contracted). Moreover, this kind of behaviour could

**Table 33.1** Example of similar electricity profiles behaviour found for summer and winter season periods for both case study

| Electricity consumption profiles | | |
|---|---|---|
| *Coopernico P5* | | |
| Spring | Winter | Consumption distribution along the day [%] |
| 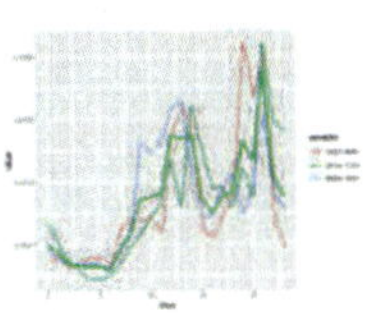 | 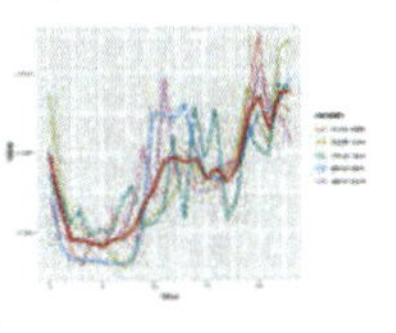 | 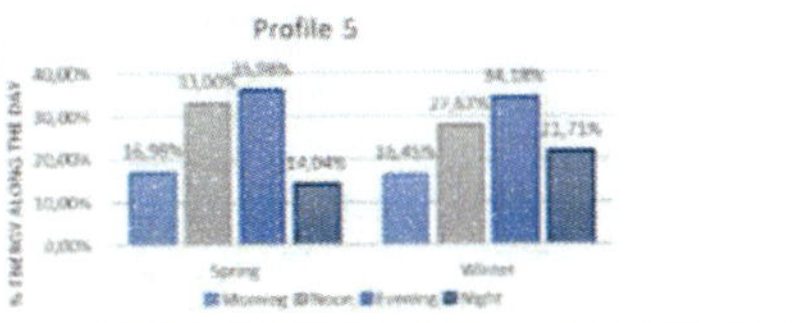 |
| *GoParity P5* | | |
| Spring | Winter | Consumption distribution along the day [%] |
| 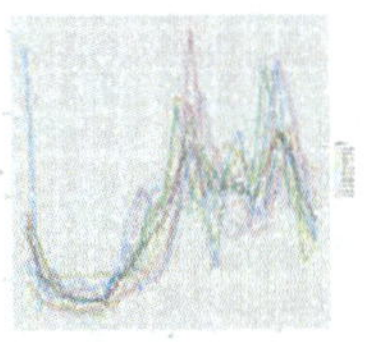 | 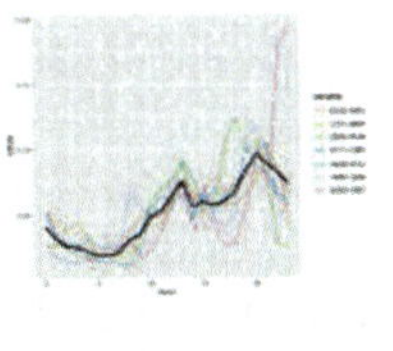 | 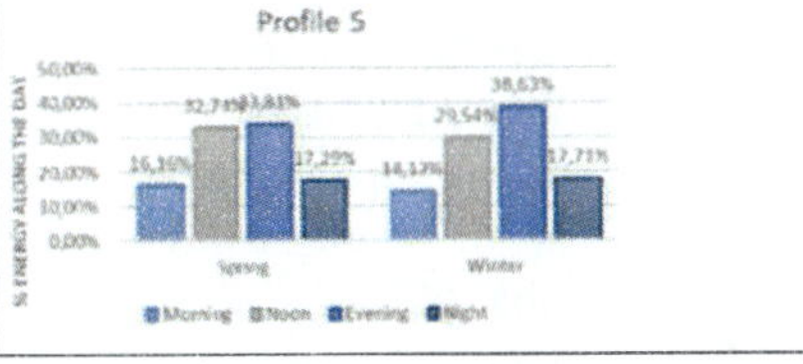 |

**Table 33.2** Example of variables characterising households, for two of the compared profiles

| | Type [%] | | Number inhabitants [%] | | | | |
|---|---|---|---|---|---|---|---|
| | Apartment | House | 1 | 2 | 3 | 4 | ≥ 5 |
| Coopernico P5 | 0.67 | 0.33 | 0.00 | 0.33 | 0.00 | 0.00 | 0.67 |
| GoParity P5 | 0.70 | 0.30 | 0.00 | 0.30 | 0.10 | 0.50 | 0.10 |
| Coopernico P3 | 0.67 | 0.33 | 0.17 | 0.25 | 0.25 | 0.25 | 0.00 |
| GoParity P4 | 0.55 | 0.45 | 0.05 | 0.32 | 0.32 | 0.27 | 0.05 |

**Table 33.3** Example of variables characterising behavioural aspects, for two of the compared profiles

| | Temperature control [%] | | Energy efficiency measures applied [%] | | | | |
|---|---|---|---|---|---|---|---|
| | Manual | Programmable thermostat | Building retrofitting | Change tariff | Change energy use | Lower power | HVAC system |
| Coopernico P5 | 0.67 | 0.00 | 0.00 | 1.00 | 0.00 | 0.33 | 0.00 |
| GoParity P5 | 0.60 | 0.10 | 0.30 | 0.50 | 0.20 | 0.30 | 0.10 |
| Coopernico P3 | 0.58 | 0.25 | 0.33 | 0.58 | 0.58 | 0.42 | 0.50 |
| GoParity P4 | 0.64 | 0.23 | 0.41 | 0.41 | 0.59 | 0.45 | 0.22 |

**Table 33.4** Example of variables characterising heating aspects, for two of the compared profiles

| | Energy systems [%] | | | | | | Storage systems [%] | |
|---|---|---|---|---|---|---|---|---|
| | Electric heating | Air conditioning | Gas boiler | Biomass boiler | Central heating | No energy system | Thermal | Electrical |
| Coopernico P5 | 0.67 | 0.00 | 0.67 | 0.00 | 0.00 | 0.00 | 0.00 | 0.33 |
| GoParity P5 | 0.30 | 0.20 | 0.20 | 0.10 | 0.00 | 0.20 | 0.20 | 0.00 |
| Coopernico P3 | 0.33 | 0.50 | 0.25 | 0.17 | 0.00 | 0.08 | 0.08 | 0.08 |
| GoParity P4 | 0.36 | 0.41 | 0.27 | 0.09 | 0.00 | 0.00 | 0.11 | 0.08 |

be conditioned by aspects related to the equipment, such as the type of heating/cooling system available at the household and the potential storage systems. Finally, habits in heating aspects can also be mentioned, namely the kind of temperature control (manual, by programmable thermostat). Something relevant according to the research is that all the users had applied some type of energy efficiency measure at home, mainly changes in electricity rate, contracted power or HVAC system, depending on the group analysed.

The second part of the research focused on discovering differences between the patterns from members of the social innovations in the energy sector and the general public. The profile representing the control group was included in the cluster analysis to determine which profile shows the most similar behaviour in electrical terms from each case study. This would allow characterising the people behind the expected profile for Portugal. Table 33.5 shows these results.

The main findings are the following:

**Table 33.5** Similar electricity profiles behaviour found for summer and winter season periods for both case study vs control group defined for Portugal

| Electricity consumption profiles | | |
|---|---|---|
| *Coopernico P3* | | |
| Spring | Winter | Consumption distribution along the day [%] |
| 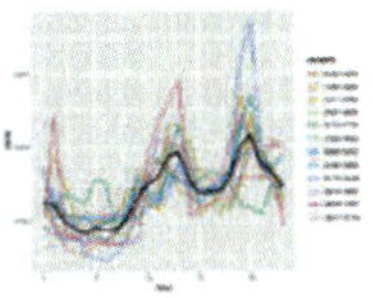 | 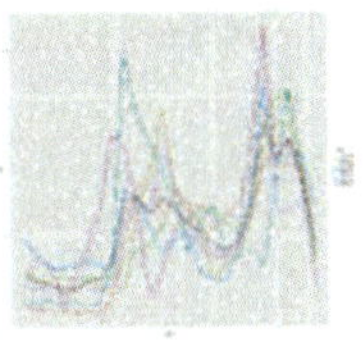 | 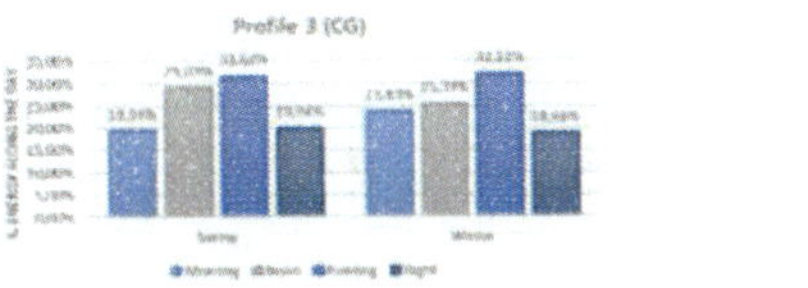 |
| *GoParity P4* | | |
| Spring | Winter | Consumption distribution along the day [%] |
| 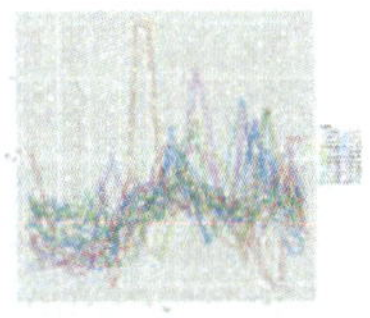 | 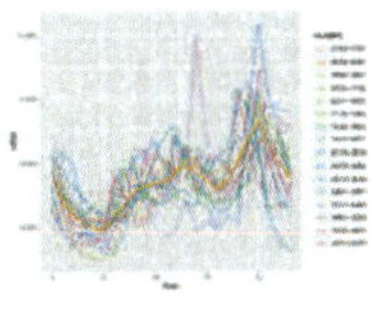 |  |
| *Control Group Portugal* | | |
| Spring | Winter | Consumption distribution along the day [%] |
| 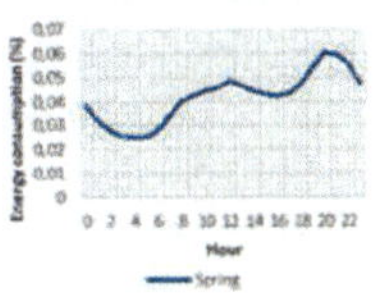 | 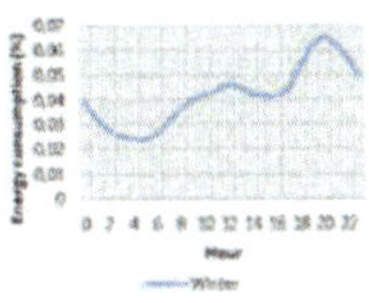 | 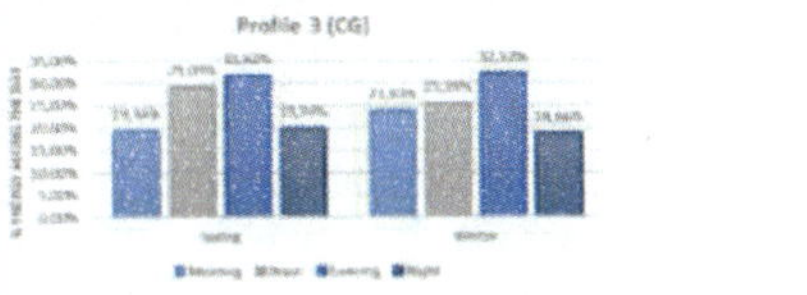 |

- In terms of dwelling features, the most represented typology is the apartment followed by the house. Household size seems very diverse, and these results are difficult to compare with the information from the control group, as is the year of construction.
- As in the control group, there is a low presence of clothes dryer at home for both cases, and the dishwasher is shown as the second least used appliance in the home. In the case of the control group defined for Portugal, the most frequent heating system is gas followed by other main systems, while biomass or district heating are in the last positions.
- Most users contracted a single rate, and there are some who have contracted a rate for two schedules. On the other hand, almost all users have applied some energy efficiency measure at home. The most used measures are the following; change of electricity rate, reduction of contracted power, building retrofitting and HVAC system change. However, the expected annual rate of major renovation in Portugal is around 0%.

There are many benefits that this type of analysis may offer, from help and improvement in energy planning strategies, to support for the implementation of effective energy policies. It is important to design energy efficiency programs appropriate to the type of user, according to their behaviour, and carry out campaigns to capture their attention. In addition, this knowledge makes it possible to offer electrical services and functionalities adapted to real needs, as well as the transmission of advice and improvement messages. The benefit can be double, and thanks to this, users could be more aware of their own actions, and reflect on how they should act to improve.

## 33.4 Conclusions

This document explains how quantitative information on electricity consumption and other contextual information can help characterise users behind the data. The retrieved information was carefully explored, continuing with a seasonal cluster analysis per case study, to segment users and find a set of profiles that could be representative. Afterwards, a socio-economic and behavioural explanation was added to the profiles obtained, based on three main categories: home equipment, heating habits and attitudes towards behaviour change. The research delivered five electricity consumption patterns for both case study, four of which have significant similarities in terms of energy and some differences according to other explanatory features. Moreover, a control group profile for general public behaviour was defined for Portugal and compared with the most similar pattern in both cases. The analysis depicted some relationships between the profiles and the contextual information, especially connected to the contracted electricity service (tariff type, amount of power), the available heating/cooling systems or the presence of storage systems. Other findings indicate that some control habits may be decisive, and that all users had previously applied some energy efficiency measure in the home. This connects with the context analysed, with participants belonging to two SEIs: the cooperative Coopernico and the crowdfunding platform GoParity, both in Portugal. The research examines how SEI members use electricity differently compared to the expected behaviour for the country. This can accelerate the offer of new functionalities adapted to users, while disseminating good practices from SEI members that could be adopted by the general population.

**Declaration of Competing Interest** The authors declare that they have no known competing financial interests or personal relationships that could have appeared to influence the work reported in this paper.

**Acknowledgements** The work presented is based on research activities conducted within the framework of the H2020 European Commission project SocialRES under contract No. 837758. All information related to the SocialRES project is available on the website https://socialres.eu (accessed on 6 June 2022). The authors wish to thank all the consortium partners, whose contribution, helpful remarks, and fruitful observations were invaluable for the development of this work.

The content of the paper is the sole responsibility of the authors and does not necessary reflect the views of the European Commission.

**Funding** This research was funded by the H2020 European Commission project SocialRES (Grant Agreement No. 837758).

## References

Azaza M, Wallin F (2017) Smart meter data clustering using consumption indicators: responsibility factor and consumption variability. Energy Procedia 142:2236–2242. https://doi.org/10.1016/j.egypro.2017.12.624

European Commission (2019) EUROSTAT. Data Browser. https://ec.europa.eu/eurostat/databrowser/

European Commission (2020) In focus. Energy efficiency in buildings. https://ec.europa.eu/info/news/focus-energy-efficiency-buildings-2020-lut-17_en

European Commission (2013) Climate action. 2030 climate & energy framework. https://ec.europa.eu/clima/eu-action/climate-strategies-targets/2030-climate-energy-framework_en

European Commission (2016) EU buildings factsheets. Building stock characteristics. https://ec.europa.eu/energy/eu-buildings-factsheets_en

Enerdata (2015) Zebra2020 data tool. Energy efficiency trends in buildings. https://zebra-monitoring.enerdata.net/

ENTSO-E (2020) TYNDP 2020—scenario report, p 48 [online]. Available https://www.entsos-tyndp2020-scenarios.eu/download-data/

ERSE (2019) ERSE website. Entidade Reguladora dos Serviços Energétics. https://www.erse.pt

Gouveia JP, Seixas J (2016) Unraveling electricity consumption profiles in households through clusters: combining smart meters and door-to-door surveys. Energy Build 116:666–676. https://doi.org/10.1016/j.enbuild.2016.01.043

Iglesias F, Kastner W (2013) Analysis of similarity measures in times series clustering for the discovery of building energy patterns. Energies 6(2):579–597. https://doi.org/10.3390/en6020579

IRENA, IRENA (2019) Global energy transformation: a roadmap to 2050

Le Zhou K, Yang SL, Shen C (2013) A review of electric load classification in smart grid environment. Renew Sustain Energy Rev 24:103–110. https://doi.org/10.1016/j.rser.2013.03.023

Matschoss K, Mikkonen I, Gynther L, Koukoufikis G, Uihlein A, Murauskaite-Bull I (2022) Drawing policy insights from social innovation cases in the energy field. Energy Policy 161:112728. https://doi.org/10.1016/j.enpol.2021.112728

McLoughlin F, Duffy A, Conlon M (2015) A clustering approach to domestic electricity load profile characterisation using smart metering data. Appl Energy 141:190–199. https://doi.org/10.1016/j.apenergy.2014.12.039

Tollefson (2022) Climate change is hitting the planet faster than scientists originally thought. Nature. https://doi.org/10.1038/d41586-022-00585-7

Viegas JL, Vieira SM, Melício R, Mendes VMF, Sousa JMC (2016) Classification of new electricity customers based on surveys and smart metering data. Energy 107(2016):804–817. https://doi.org/10.1016/j.energy.2016.04.065

Wittmayer JM, Hielscher S, Fraaije M, Avelino F, Rogge K (2022) A typology for unpacking the diversity of social innovation in energy transitions. Energy Res Soc Sci 88:102513. https://doi.org/10.1016/j.erss.2022.102513

**Sofía Mulero-Palencia** Conceptualisation, Methodology, Resources, Investigation, Data curation, Formal analysis, Software, Validation, Visualisation, Writing—original draft, Writing—review and editing, Supervision, Project administration and Funding acquisition.

**Alejandro Hernández-Serrano** Investigation, Software, Validation, Visualisation. All authors have read and agreed to the published version of the manuscript.

# Chapter 34
# Efficiency of Environmental Measures in Portuguese Healthcare Institutions Using Stochastic Frontier Analysis

**José Chen-Xu and Victor Moutinho**

**Abstract** Attention has in recent years shifted towards the environmental impact of human activities and the need to reach zero net carbon impact, including in the healthcare sector, being essential to review management processes to ensure their environmental sustainability. This study aims to evaluate the efficiency of environmental sustainability measures of institutions providing healthcare in Portugal's National Health Service associated with water, energy, and waste management. A cross-sectional study was implemented in 24 institutions, with application of a Stochastic Frontier Analysis. The significant model for electricity showed photovoltaic panels to be an efficient measure, whereas LED lamps, solar panels and $CO_2$ emissions quantification showed an association with inefficiency, with potential for investment. While the model for water consumption was not significant, water reuse and pre-treatment showed to be adequate. For total waste production, the model highlighted the relevance of green purchase and adequate waste management. When disaggregating for group IV waste, the model showed technical inefficiency of measures despite not being significant. While there are differences between primary and secondary care, there is margin for efficiency improvement, especially regarding energy and waste management, working with health managers and partners. Further research is needed to strengthen environmental policy changes.

**Keywords** Energy · Environment · Healthcare · Sustainability · Waste management · Water

---

J. Chen-Xu (✉)
Public Health Research Centre, National School of Public Health, NOVA University of Lisbon, Av. Padre Cruz, 1600-560 Lisbon, Portugal
e-mail: jc.xu@ensp.unl.pt

FEP—School of Economics and Management, University of Porto, Rua Dr. Roberto Frias, 4200-464 Porto, Portugal

V. Moutinho
Department of Management and Economics, NECE—Research Center for Business Sciences, University of Beira Interior, R. Marquês de Ávila e Bolama, 6201-001 Covilhã, Portugal

 
N. S. Caetano and M. C. Felgueiras (eds.), *The 9th International Conference on Energy and Environment Research*, Environmental Science and Engineering,
https://doi.org/10.1007/978-3-031-43559-1_34

## 34.1 Introduction

With the progression of climate change and the consequent adoption of a European Climate Law (European Commission 2021), organizations committed to lowering the carbon footprint. This means that all sectors, including healthcare institutions (HI), must adapt to climate change, but also have a role in the mitigation of its effects.

Buildings represent about 35% of global energy consumed, leading to almost 40% of energy-related carbon dioxide ($CO_2$) emissions, playing an important role in cost-effective reduction of carbon footprint and sustainable transformation (Li et al. 2022). Healthcare buildings have specific requirements that must guarantee continuous activity to attend medical needs of a population, which can increase energy consumption to twice of that of general buildings (Shen et al. 2019). Overall, the environmental impact can go up to 7% of total global impacts. Common practices of energy waste in this sector include inefficiency with ventilation and air conditioning, especially in unoccupied spaces, and failure to maintain equipment and check for air/water leaks. Further reported issues related with internal heat losses, inefficient lighting, difficulties in managing energy use in specific wards with higher consumption, and low staff knowledge (Friedericy et al. 2019). Institutions also consume a large amount of water throughout their life cycle, which can go up to 1200 L/bed/day. This may be influenced by the infrastructure size, accessibility to water, climate, and even cultural factors (Rothausen and Conway 2011). A proportion of wastewater in hospitals contains toxic and non-biodegradable components, and some infectious pollutants and antibiotics, negatively impacting human and aquatic life. The wastewater is often discharged without any treatment or directed to a specific water treatment plant if there is one (Kumari et al. 2020). In terms of waste, the use of material resources leads to the production of ~ 7000 tons of hospital waste daily and an annual cost of ten billion dollars in managing it (Senay and Landrigan 2018). Despite existing waste management regulations in hospitals, there is a lack of monitoring and quantification, incentives for recycling and reuse habits. Furthermore, around 85% of waste produced in a general hospital is non-hazardous (group I and II), often placed in the biological waste container, increasing costs and environmental impact through inadequate treatment. Mismanagement of hazardous waste, as chemotherapy drugs and needles, mainly from group IV, leads to a negative impact on environment and health. This includes direct discharge of chemicals and other toxic products to sewage or water treatment plants, often unprepared to process or eliminate such substances (Neveu and Matus 2007).

These factors contribute to a greater impact in the environment, aggravating climate change. Efforts have been made to invest in more resilient, sustainable, and environmentally friendly institutions, without compromising medical performance. Initiatives include water and energy saving and efficiency measures (McGain and Naylor 2014), such as automatic control systems, flow reducers or sensors for taps and lighting, and reduction of waste generation, by reprocessing single-use materials, decreasing material use in the source, and improving waste segregation (Kallio et al.

2020). Organizational commitment, through training, raising awareness, and implementing ecological practices among the health workforce is vital (Pisters et al. 2017). Further strategies involve health managers, as environmental measures can bring organizational and financial benefits (Kaplan et al. 2012). This includes improving building materials for lowering heat loss (Langstaff and Brzozowski 2017), and transitioning to renewable energy, either by direct purchase or recurring to photovoltaic panels (Burch et al. 2021). Institutions have also started adopting sustainability reports to monitor their progress. However, HI are now striving for zero carbon footprint (European Commission 2021). For this to happen, the impact of environmental measures needs to be evaluated and replaced for more cost-effective practices, which includes urban planning policies for energy efficiency and renewable energies, improving performance, adaptability and resilience of buildings (UN Environment and International Energy Agency 2017).

There is a need for a baseline assessment on a local level, namely in HI buildings since its conception to maintenance, in a life-cycle perspective. When designing a green hospital, one must consider energy efficiency, water management and sustainability planning (Wood et al. 2016). For existing buildings, the evaluation of environmental sustainability must include comprehensive methods, such as an analysis of environmental parameters by using a survey (Romero and Carnero 2019) or a system based on indicators measuring environmental performance, with a strategic focus on solutions (Pasqualini Blass et al. 2017). Despite these tools, specific approaches which reflect real-world data are still needed.

Thus, this study evaluates the efficiency of environmental measures in reducing electricity and water consumption, and waste production, in Portuguese HI, by applying a Stochastic Frontier Analysis. This will increase understanding of factors needing improvement or investment, useful for decision-making and environmental policy changes carried out by HI management bodies.

## 34.2 Materials and Methods

A cross-sectional ecological study was developed with one null hypothesis (H0): There are no technical inefficiency components within the environmental sustainability measures applied in heath institutions influencing electricity and water consumption, and total and group IV waste production.

The population comprised of HI that provide health care within Portuguese National Health Service (SNS), which include 49 hospitals and 54 primary healthcare clusters. From these those that did not provide the required information were excluded, resulting in a total of 24 HI.

Data from 2019 was extracted from Sustainability, Financial and Non-Financial Reports and then matched with data from SNS Transparência about HI characteristics. Information on energy, water and waste was obtained from governmental monitoring reports. Data on sustainability measures related to green purchase were retrieved from the public procurement platform. The remaining data was obtained through a questionnaire, implemented in HI between January and August 2020.

Ethical approval was obtained from the Ethics Committee for Health of the respective HI. According to agreed confidentiality requirements, HI were attributed a code: secondary (H) or primary care (A).

The models included the following outcomes: Model 1—electricity consumption (MWh/year); Model 2—water consumption ($m^3$/year); Model 3—total waste production (ton/year); and Model 4—production of group IV waste (ton/year). Environmental sustainability measures were utilized as independent variables, including quantification of $CO_2$ emissions; energy consumption reduction measures; photovoltaic and solar panels; type of lamps; water consumption reduction measures; water treatment or reuse; water pre-treatment system before discharge to treatment plants; reuse practices; groups I + II waste treatment; group III waste treatment; group IV waste treatment; use of a waste treatment guide; green purchase policy; waste management within the Institution.

A descriptive analysis characterized HI's current state and environmental practices. The Stochastic Frontier Analysis (SFA) was then applied, an economic modelling method that evaluates of technical efficiency of production. This regression method involves a logarithmic transformation of the technical efficiency related to each of the observed outcomes, $y_{it}$, to the production frontier, $f\ (X_{it};\ \beta)$, with a random error term $v_{it}$. In this study a Cobb–Douglas SFA is followed, given by either of the equations:

$$\ln \ln y_{it} = \beta_0 X_{it} + v_{it} - u_{it} \tag{34.1}$$

$$\varepsilon_{it} = \exp \exp(\beta_0 X_{it}) \times \exp \exp(v_{it}) \times \exp \exp(-u_{it}) \tag{34.2}$$

where $\exp \exp(\beta_0 X_{it})$ represents the deterministic component, $\exp \exp(v_{it})$ represents the noise and $\exp \exp(-u_{it})$ represents inefficiency. The composed error term $\varepsilon_{it}$ is the difference of a normally distributed disturbance, where $v_{it}$ represents measurement and specification error, and $u_{it}$ denotes inefficiency. Moreover, $v_{it}$ and $u_{it}$ are assumed to be independent of each other and across observations. The disturbance $v_{it}$ accounts for random unobserved factors and can be positive or negative. The majority of the SFA models is directed towards the prediction of inefficiency effects. Technical Efficiency ($TE_{it}$) is measured by $\exp \exp(-u_{it})$, varying from zero to one.

The different models considered different specifications for the technical inefficiency effects. In this study, the SF models were estimated according to the assumptions connected to the inefficiency behavior, applying cross-sectional models of the SFA (Belotti et al. 2013):

$$y_i = \alpha + x_i\beta + \varepsilon_i = 1, \ldots, N \tag{34.3}$$

$$\varepsilon_i = v_i - u_i \tag{34.4}$$

$$\upsilon_i \sim \mathrm{N}(0, \sigma^2 v) \quad (34.5)$$

$$u_i \sim \mathrm{F} \quad (34.6)$$

where $y_i$ represents the logarithm of the output of the $i$th productive unit, $x_i$ is a vector of inputs, and β is the vector of technology parameters. The composed error term $\varepsilon_i$ is the difference of a normally distributed disturbance, $\upsilon_i$, which represents the measurement and specification error, and a one-sided disturbance, $u_i$, representing inefficiency. These are assumed to be independent from each other and independent and identically distributed across observations. The assumption of a F distribution of the inefficiency term is needed to make the model estimable. The $TE_{it}$ was estimated assuming a half-normal distribution, $u_i \sim \mathrm{N} + 0, \sigma^2$ v, , and a Truncated Normal distribution. This is conducted by a two-step process that involves (1) the estimation of model parameters, obtained by maximizing the log-likelihood function; (2) estimation points of inefficiency, obtained through the mean of the conditional distribution. All statistical analyses were conducted in STATA.

## 34.3 Results and Discussion

The sample comprised of 24 HI, including 10 secondary care and 14 primary care institutions. The North region had most institutions (n = 9, 37.5%). Their size is proportional to the number of workers and outpatients, with an average of 1094 (± 869) workers and 814,221 (± 550,412) appointments.

Regarding the outcomes, the electricity consumption presented an average annual electricity consumption of 4100.02 (± 6657.03) MWh and water consumption of 44,360.8 (± 54,336,3) $m^3$. Waste originated a total amount of 282 (± 442) tons, of which 12.6 (± 24.7) tons were group IV waste. Analysis of environmental measures were also carried out. Concerning energy-related measures, only four HI reported either implementing measures aiming at reducing energy consumption, using photovoltaic or solar panels (16.7%). Ten reported using LED lamps (41.7%), while the majority still had fluorescent lamps (n = 13, 54.1%). Despite this, eight HI were able to perform the quantification of $CO_2$ emitted (33.3%). Regarding water consumption, specific measures such as tap sensors were reported by half (n = 12). Other measures included the reuse of water (n = 3, 12.5%) and water pre-treatment system before discharge (n = 4, 16.7%). As for waste management, most HI implemented reuse practices (n = 20, 83.3%). Regarding waste treatment, some still deposited group I + II waste in landfill (n = 7, 29.2%), while others reported concomitant recycling (n = 7, 29.2%). Most HI utilized autoclave for treatment of group III waste (n = 14, 58.3%), whereas group IV waste was treated by incineration in most cases (n = 21, 87.5%). Institutions were also questioned regarding organizational waste management measures, with more than half owning a waste treatment guide (n = 13, 54.1%), however most HI subcontracted waste management to external specialized companies (n = 23, 95.8%). Some had a green purchase policy or considered this aspect in purchases (n = 7, 29.2%).

Four Stochastic Frontier Cross-sectional models were then developed (Tables 34.1 and 34.2), considering distribution models for inefficiency as Half Normal and Truncated Normal. In the Truncated distribution, all measured variables in Models 1, 2 and 3 were significant, while the Half-Normal distribution showed statistical significance for all measures used for Model 1 and 3.

In the analysis of individual measures on electricity consumption (model 1), there were positive and significant effects in HI that quantify $CO_2$ emissions, when compared to those that do not. This is contrary to expectations. In turn, the effect of photovoltaic panels was negative and significant, whereas the solar panels showed a positive effect in electricity consumption. The type of lamps also had a positive effect, with most HI consuming less LED lamps and more of the other types. Overall, model 1 estimations using the Truncated distribution allowed the rejection of the null hypothesis. The use of photovoltaic panels showed to be effective and should be further implemented in other HI, as evidenced in other studies (Burch et al. 2021). LED lamps revealed to positively influence efficiency (Friedericy et al. 2019), contrary to the results of this study. This might be due to their partial adoption in the HI. Moreover, measures required workers' involvement, such as turning off the lights in unoccupied spaces, and regular monitoring of energy expenditure, not corresponding to their potential value as described in the literature (Kaplan et al. 2012). Other measures, as building infrastructure interventions and ventilation

**Table 34.1** Results for the Stochastic Frontier Cross-sectional Model 1 and Model 2

| Model 1: electricity | Truncated | Half normal | Model 2: water | Truncated | Half normal |
|---|---|---|---|---|---|
| Quantification of $CO_2$ | 2.44*** | 2.16*** | Reuse of water | − 0.86*** | − 0.44** |
| Energy reduction measures | 0.72*** | 1.03*** | Water reduction measures | 1.32*** | 1.46*** |
| Photovoltaic panels | − 2.53*** | − 2.96*** | Water pre-treatment | − 0.26*** | 0.34 |
| Type of lamps | 0.96*** | 0.68** | Solar panels | − 1.12*** | − 0.56* |
| Constant | 11.15*** | 11.99*** | Constant | 10.94*** | 10.45*** |
| *Sigma u* | 41.74 | 1.72*** | *Sigma u* | 2.47 | 1.40*** |
| Sigma v | 1.67e − 07 | 2.52e − 07*** | Sigma v | 4.47e − 08 | 0.46 |
| Lambda | 2.83e + 09*** | 6.81e + 06*** | Lambda | 5.54e + 07*** | 3.02*** |
| H0: no inefficiency component | Z = − 3.04*** | | H0: no inefficiency component | Z = 0.10 | |

*Note* All series of variables were transformed to natural logarithms
*, **, *** denotes 10%, 5%, and 1% levels of statistical significance, respectively

**Table 34.2** Results for the stochastic frontier cross-sectional model 3 and model 4

| Model 3: total waste | Truncated | Half normal | Model 4: group IV waste | Truncated | Half normal |
|---|---|---|---|---|---|
| Quantification of $CO_2$ | 3.93*** | 3.23*** | Quantification of $CO_2$ | 6.02*** | 2.84*** |
| Reuse practices | – 0.65*** | – 0.65*** | Reuse practices | – 1.47 | – 1.48* |
| Groups I + II waste treatment | – 0.27*** | – 0.50** | Group IV waste treatment | 0.72 | 0.60 |
| Group III waste treatment | 1.34*** | 0.99** | External waste management | 0.10 | 0.17 |
| Waste treatment guide | 2.03*** | 1.68*** | Waste treatment guide | 1.09* | 1.33** |
| Green purchase | – 1.79*** | – 0.74** | | | |
| Constant | 3.80*** | 4.50*** | Constant | 0.47 | 0.83 |
| Sigma u | 43.62 | 1.66*** | *Sigma u* | 53.49 | 2.37*** |
| Sigma v | 3.32e – 10 | 5.38e – 08*** | Sigma v | 1.16*** | 1.14*** |
| Lambda | 1.31e + 11*** | 2.85e + 07*** | Lambda | 46.19 | 2.08** |
| H0: no inefficiency component | Z = 0.13*** | | H0: no inefficiency component | Z = – 1.09 | |

*Note* All series of variables were transformed to natural logarithms
*, **, *** denotes 10%, 5%, and 1% levels of statistical significance, respectively

need to be further evaluated. When analyzing model 2, most measures significantly contributed to decrease water consumption. However, behavioral measures to reduce water consumption had a positive and significant effect, which reveals that these were not effective in reducing water consumption. Despite the model not being significant, this pointed to the importance of these measures, namely the internal circuit treatment or reuse of water, and a pre-treatment system for water, which is in line with the existing literature (Kumari et al. 2020).

The model 3 for total waste production showed a significant negative effect of green purchase, reuse practices, and group I + II waste treatment. The $CO_2$ emissions quantification, the use of a waste treatment guide, and group III waste treatment showed a positive effect, contrary to expectations. However, the Truncated model justified the H0 rejection, allowing to incorporate relevant measures to Hospital Management and to the sample. The group IV waste production (model 4) had a positive effect of $CO_2$ emissions quantification, revealing that most HI were not effective in performing it. The negative effect of reuse practices, as well as the positive effect of group IV waste treatment and use of a waste treatment guide, were not significant. There is a need to invest in improving the waste treatment guide, quantification of $CO_2$ emissions, and better define group III and IV waste treatment, in line with existing evidence (Ryan-Fogarty et al. 2016). However, internal metrics for waste management, group III waste treatment and recycling did not correspond to the literature (Neveu and Matus 2007). Additionally, the evaluation of quality of the reusable materials and repurposing practices were not evaluated. Another factor was the active role of management and healthcare professionals (Kallio et al. 2020), which was not taken into account in this study.

The efficiency scores were derived from the SFA estimations. Three hospitals stood out in terms of efficiency on electricity consumption. When ranking the sample, these were placed first with an average efficiency above 99.99% for the same estimates of inefficiency. Regarding water consumption, five HI showed the highest scores, ranking first place with an average efficiency above 99.99%. In turn, three institutions presented the highest scores of efficiency, however when ranking, these three had a scoring between 61.04 and 79.55% for inefficiency border estimations. The scoring related to the efficiency in total waste production, for Truncated distribution showed that six HI were ranked the highest with maximum efficiency (scoring = 100%). It must also be pointed out, according to the estimation with the Half Normal distribution, these HI were in the first place with an average score of about 99.99% of maximum efficiency for these same border estimations. When further analyzing this data, the efficiency scores in management of group IV waste, only two stood out with moderate levels in the scores, one with 66.86% and 69.96% and another with 60.33% and 60.93%, under the Truncated and Half Normal respectively.

The study's weaknesses were mainly due to its retrospective nature and selection bias. Another factor was the difference between primary and secondary care, which presented different levels of activity and environmental measures. Also noteworthy is the absence of some data on the variables requested, which hindered the sampling process. The pandemic might also have contributed to the low participation of health institutions, as efforts were directed to the COVID-19 pandemic mitigation. Despite this, the present study brought novelty by applying SFA to environmental measures in hospitals, which has been done in environmental aspects in other contexts or in hospitals with other metrics, but not combined. It would be important to carry this analysis with a higher number of HI to make the model more robust and more representative of Portuguese institutions. Furthermore, a cost-effectiveness analysis would be relevant to evaluate their true economic and environmental benefit, further strengthening the recommendations and replicate the measures in other institutions.

## 34.4 Conclusion

This article reinforced the importance of efficient environmental measures, such as pre-treatment of water, reuse of materials, and use of photovoltaic panels. Another aspect highlighted was the importance of local power decisions in enacting these changes, bringing the attention to the importance of working together with health administrators and workers in designing and implementing environmental practices and policies addressing climate change, especially on waste and energy management, with the common goal of reaching zero net carbon.

Future research might also focus on translating the efficiency into economic analyses, which would help decision-makers lead climate change mitigation within hospitals.

**Acknowledgements** This research was supported by Professor Cristina Chaves from the Faculty of Economics, University of Porto, who contributed by connecting the authors and with the brainstorming phase of this study.

## References

Belotti F, Daidone S, Ilardi G, Atella V (2013) Stochastic frontier analysis using stata. Stand Genomic Sci 13(4):719–758

Burch H, Anstey MH, McGain F (2021) Renewable energy use in Australian public hospitals. Med J Aust 215:160

European Commission (2021) Regulation (EU) 2021/1119 of the European Parliament and of the Council of 30 June 2021 establishing the framework for achieving climate neutrality and amending Regulations (EC) No 401/2009 and (EU) 2018/1999 ('European Climate Law')

Friedericy HJ, Weiland NHS, Van Der Eijk AC, Jansen FW (2019) Manieren om de $CO_2$-voetafdruk van de OK te verlagen. Ned Tijdschr Geneeskd 163:D4095

Kallio H, Pietilä A, Kangasniemi M (2020) Environmental responsibility in nursing in hospitals: a modified Delphi study of nurses' views. J Creat Behav 29(21–22):4045–4056

Kaplan S, Sadler B, Little K, Franz C, Orris P (2012) Can sustainable hospitals help bend the health care cost curve? Issue Brief Commonw Fund 29:1–14

Kumari A, Maurya NS, Tiwari B (2020) Hospital wastewater treatment scenario around the globe. Current developments in biotechnology and bioengineering, Elsevier, 549–70

Langstaff K, Brzozowski V (2017) Managing environmental sustainability in a healthcare setting. Healthc Manage Forum 30(2):84–88

Li K, Ma M, Xiang X, Feng W, Ma Z, Cai W et al (2022) Carbon reduction in commercial building operations: A provincial retrospection in China. Appl Energy 306:118098

McGain F, Naylor C (2014) Environmental sustainability in hospitals—a systematic review and research agenda. J Health Serv Res Policy 19(4):245–252

Neveu CA, Matus CP (2007) Management of hazardous waste in a hospital. Rev Med Chil 135(7):885–895

Pasqualini Blass A, da Costa SEG, de Lima EP, Borges LA (2017) Measuring environmental performance in hospitals: a practical approach. J Clean Prod 142(1):279–289

Pisters P, Bien B, Dankner S, Rubinstein E, Sheriff F (2017) Supporting hospital renewal through strategic environmental sustainability programs. Healthc Manage Forum 30(2):79–83

Romero I, Carnero MC (2019) Environmental assessment in health care organizations. Environ Sci Pollut Res 26(4):3196–3207

Rothausen SGSA, Conway D (2011) Greenhouse-gas emissions from energy use in the water sector. Nature Clim Change 1:210–219

Ryan-Fogarty Y, O'Regan B, Moles R (2016) Greening healthcare: systematic implementation of environmental programmes in a university teaching hospital. J Clean Prod 126:248–259

Senay E, Landrigan PJ (2018) Assessment of environmental sustainability and corporate social responsibility reporting by large health care organizations. JAMA Netw Open 1(4):e180975

Shen C, Zhao K, Ge J, Zhou Q (2019) Analysis of building energy consumption in a hospital in the hot summer and cold winter area. Energy Procedia 158:3735–3740

UN Environment and International Energy Agency (2017) Towards a zero-emission, efficient, and resilient buildings and construction sector. Global status report 2017
Wood LC, Wang C, Abdul-Rahman H, Jamal Abdul-Nasir NS (2016) Green hospital design: integrating quality function deployment and end-user demands. J Clean Prod 112:903–913

# Chapter 35
# Climate Change Mitigation and Adaptation in Military Organizations: The Case of the Portuguese Air Force

**Joana Pinto and Carlos Páscoa**

**Abstract** The Portuguese government, as part of the 22nd Conference of the Parties (COP22) of the United Nations Framework Convention on Climate Change (UNFCCC) in 2016, has committed to ensure net zero greenhouse gas (GHG) emissions by the end of 2050 as a way of complying with the Paris Agreement. For that purpose, the Portuguese government approved a set of essential documents that stand as a guide for national organizations. Complying to that, the Portuguese Air Force Chief-of-Staff identified sustainability as key in the strategy for 2022/24, giving way to the development of the Air Force roadmap for carbon neutrality in 2050 (RCN2050PrtAF), which is divided in three main phases—Mapping, Reduction and Neutrality. Each of these phases included several tasks like estimation of the de GHG emissions and the carbon sequestration capacity (mapping phase) and then definition of the reductions needed, the priorities of action, the possible scenarios of reduction, and the key indicators to monitor the evolution and achievement of the goal—net zero emissions—(reduction and neutrality phases).

**Keywords** Adaptation · Anthropogenic emissions · Climate change · Greenhouse effect · Mitigation · Portuguese air force

## 35.1 Introduction

The fifth assessment report (AR5) of the Intergovernmental Panel on Climate Change (IPCC) stated that human influence on the climate system is unequivocal and recent anthropogenic emissions of Greenhouse Gases (GHG) are the highest in history. The report provided scientific input into the Paris Agreement where, in 2015, under the UNFCCC, the Nations agreed to limit the increase of the global average temperature

J. Pinto (✉) · C. Páscoa
Portuguese Air Force, Av. da Força Aérea Portuguesa, 2614-506 Amadora, Portugal
e-mail: joana.guerreiro.pinto@gmail.com



N. S. Caetano and M. C. Felgueiras (eds.), *The 9th International Conference on Energy and Environment Research*, Environmental Science and Engineering,
https://doi.org/10.1007/978-3-031-43559-1_35

to 1.5 °C (United Nations (UN) 2022). This is an ambitious objective that requires societal transformations and joint efforts that allow the reduction of between 80 and 90% in global GHG emissions.

The European Union (EU) drew an ambitious plan to face climate change, as a way of complying with the Paris Agreement, that was the baseline for the various European countries. Portugal was one of the countries that developed a Roadmap for Carbon Neutrality in 2050 (RCN2050) (Presidency of the Council of Ministers (PCM) 2019), and the corresponding intermediate plan (Presidency of the Council of Ministers (PCM) 2020a). Because of the subsequent acts, the Armed Forces are compelled to incorporate in their strategic and operational planning the risks inherent in climate change and measures to reduce greenhouse gas emissions (Portuguese Parliament 2021), which reinforces the need for interaction between National Defence and environmental protection. The military organizations can be seen in two main ways in the context of climate change and decarbonization: the military organization conducting Defence, Security military operations and humanitarian aid and the military organization that has units, buildings, transports, and consumes energy and water and generates waste. Both can be affected by climate change. Thus, military organizations need to contribute and as far upstream as possible and, in the case of the Portuguese Air Force (PrtAF), it was developed a Roadmap for Carbon Neutrality in 2050 (RCN2050PrtAF) (Portuguese Air Force 2022). This paper summarizes the methodology of the RCN2050PrtAF and its development.

### 35.1.1 *Objective, Research Question and Methodology*

The objective of the RCN2050PrtAF is to lay the foundations for the development in military organizations of such plan and to show the importance that the environment has in the Portuguese Armed Forces, namely for the PrtAF. Giving the Portuguese government objective to achieve carbon neutrality by 2050, and considering that the PrtAF is part of the state and a contributor for GHG emissions, it was identified the research question (RQ): Can the PrtAF achieve carbon neutrality with current policies? To answer the RQ, the methodology adopted followed the IPCC guidelines (Intergovernmental Panel on Climate Change (IPCC) 2022), and the principles from Bowcott et al. (2021) and is detailed in paragraph 3.

### 35.1.2 *Structure*

The paper is structured as follows: Paragraph 2 presents the literature review focusing on academic knowledge, regulatory documents, and case studies; Paragraph 3 describes the RCN2050PrtAF model, outlines the environmental areas and Key Performance Indicators (KPI) used, and the way proposed to achieve carbon neutrality; Paragraph 4 concludes and presents future directions.

## 35.2 Literature Review

To act as a basis to methodology design, several open access articles were consulted and analysed. The main resource was taken from IPCC as well as the European and national publications, since they present a reference for the Portuguese objectives that must be attained by national organizations, both civil and military. The following sub-paragraphs give more detail about these documents.

### *35.2.1 IPCC Guidelines*

The IPCC guidelines were first published in 2006 (Intergovernmental Panel on Climate Change (IPCC) 2022) and refined in 2019. The objective of the guidelines is to infer emissions based on parameters associated with activities, for example the amount of fuel burnt for a determined unit and the GHG emissions related. The guidelines are supported by the IPCC inventory software, by the Emission Factor (EF) Database, by reports of expert meetings and by Frequently Asked Questions site.

### *35.2.2 The European Union Climate Change Acts and the Portuguese Acts*

The European Climate Law (Union 2021), translates the objectives laid in the European Green Deal (Union 2019), ensuring net zero emissions of GHG by 2050. The Portuguese government, in 2016, has committed to ensuring the neutrality of its greenhouse gas emissions by the end of 2050 as a way of complying with the Paris Agreement, UN (2015). In that context, Portugal developed a Roadmap for Carbon Neutrality in 2050 (RCN2050) whose principles and objectives gave origin to a set of doctrine publications and acts (for example National Energy and Climate Plan 2030—PNEC2030—and Efficiency Resources Program for the Public Sector—ECO.AP2030) (Presidency of the Council of Ministers (PCM) 2020b). The RCN2050 defines three scenarios to achieve carbon neutrality and fosters circular economy and the emphasis on smart cities. These documents define the national objectives for carbon neutrality and are summarized in Table 35.1.

**Table 35.1** National objectives for carbon neutrality

| References | Objective (base line is 2005) |
|---|---|
| RCN2050 and PNEC2030 | Reduce 55% of GHG emissions by 2030 |
| RCN2050 | Reduce 75% of GHG emissions by 2040; reduce 90% of GHG emissions by 2050 |
| ECO.AP2030 | Reduce 40% of primary energy (energy that can be used directly) consumption by 2030; by 2030, 10% of electricity consumption comes from self-production systems based on renewable energy sources; reduce 20% of water consumption by 2030; reduce 20% of material consumption (waste production and paper consumption) by 2030, achieve 5% energy and water renewal rate by 2030. Considers the number of buildings where energy and/or water interventions have taken place |

## 35.3 Results—Climate Change Mitigation and Adaptation Model Used for the PrtAF

The methodology used consisted of three main phases—Mapping, Reduction and Neutrality. At the mapping phase it was identified the GHG emission sources and estimated their values, as well as estimated the carbon sequestration capacity of the forest. After that, it was possible to answer the first research question: current policies are not carbon–neutral compatible. Therefore, some other questions emerged: How much GHG emissions do we need to reduce and at what pace? What are the priority sources to act on? How will we assure we're on the good way to achieve the carbon–neutral scenario? To answer these questions, there were established the second and third phases—reduction and neutrality, respectively. The reduction phase quantifies the decrease in GHG emissions needed, characterizes the actual energy systems, prioritizes the sources to act on and identifies three possible scenarios of reduction—off-track, peloton and yellow jersey. Neutrality phase identifies the KPI that allow to monitor the neutrality path. Table 35.2 summarizes the definition model described and the following paragraphs detail the actions taken in each of the three main phases.

### *35.3.1 Mapping*

The objective was to identify the current state of the carbon emissions and sequestration balance by estimating the GHG emissions and the carbon sequestration capacity of the forest. The calculations (both GHG emissions and carbon sequestration capacity) were made according to the IPCC guidelines for national greenhouse gas inventories, tiers 1 and 2. To estimate GHG emissions, it was necessary to define the calculation limits. This was done by following the GHG Protocol proposal, which classifies GHG emissions according to their scope: scope 1, direct emissions from sources owned or controlled by the organization; scope 2, indirect emissions that

**Table 35.2** RCN2050PrtAF definition model

| Phase | Action | Description | Indicator |
|---|---|---|---|
| Mapping | Identify GHG emission sources | Identify the GHG emission sources | No. of sources |
| | Estimate GHG emissions | Define the emissions scope to limit the calculation method (GHG protocol)<br>Use the more specific data (emission factors) possible | GHG emissions [t $CO_2$ e/year] |
| | Estimate carbon sequestration capacity | Define the calculation method based on the data available. In the PrtAF case it was used the "gains-losses" method, from the IPCC guidelines, once forest area and species were the only data available | C sequestration capacity [t C/year] |
| Reduction | Define the reduction goals.<br>Prioritize actions<br>Define project packages<br>Define scenarios | Determine how much needs to be reduced to achieve net zero emissions<br>Define what sources are priority to act on with a decision matrix<br>Deliver the project packages to the corresponding technical team<br>Define de possible scenarios of reduction to better understand, over time, in which path the organization is heading and what (if) need to be adjusted | Emissions of GHG of each source [t $CO_2$ e/year] |
| Neutrality | Monitor the KPI<br>Adjust (if needed) | Given the temporal distance, it is crucial to monitor the reduction phase and the sequestration capacity and adjust what is needed. This task helps to better understand, over time, in which path the organization is heading | Net emissions of GHG [t $CO_2$ e/year] |

result from the consumption of electricity or steam, or from heating or cooling the organization's facilities/buildings and scope 3, emissions from third party sources not controlled by the organization. In this study, the estimations were made under the emission scope 1 and 2.[1]

To estimate the carbon sequestration capacity of the PrtAF's forest it was selected the tier 1 calculation method defined by the IPCC and selected the "*Gains-Losses*" method. This method calculates $CO_2$ emissions and removals resulting from changes in the biomass carbon stock. Gains include biomass growth through the conversion of carbon to biomass and are marked with a plus sign (positive stock changes). Losses, mainly represented by logging, correspond to biomass carbon emissions to the atmosphere, and are marked with a minus sign (negative stock changes). The

[1] Scope 3 emissions were not included because the aim is to assess emissions that can be controlled by the PrtAF, as well as to avoid double counting of emissions (in the case where PrtAF suppliers also estimate their own GHG emissions).

biomass specific factors used were the same used by APA (Portuguese Environment Agency) in the National Inventory Report. In short, the essential steps were the following: (i) Identify standards. In the PrtAF case, the IPCC guidelines and the National GHG Emissions Inventory from APA were followed; (ii) Identify the calculation period. It was selected the period from 2005 to 2021 and called "history"; (iii) Define the calculation limits, where emissions accounting starts and ends (scopes); (iv) Analyse the data available and identify which tier to used. It was used the tiers 1 and 2; (v) Identify the EF and specific factors. The country specific factors, when available, were used; the IPCC default factors were used in the other cases, and; (vi) Do the estimations. In the present case, it was made under the emission scope 1 and 2.

The mapping phase results had the following emission distribution: aviation fuel—81%, electric energy—7%, Gas—6%, Diesel—5% and Fluorinated gases—1%. Regarding the calculations, it was estimated, for 2021, a sum of GHG emissions of 78.4 t $CO_2$e and a sequestration capacity of 58.2 t $CO_2$. This phase also produced two studies about the impacts that mobility, Calaixo et al. (2022), and forest, Correia et al. (2022), have in climate change.

### *35.3.2 Reduction*

The reduction phase quantifies the decrease in GHG emissions needed based on the goal of the net zero emissions by 2050, characterizes the actual energy systems and then prioritizes the reduction sources to act on. This prioritization was made with a decision matrix, that is presented in Table 35.3, and was constructed based on two critical factors: (i) The relationship with the operation—considered a critical factor because the operation cannot be compromised at any time and the reduction of GHG emissions in this component will require specific solutions and, eventually, coordination with external entities. Three levels of relationship were used, representing from no relation to direct relation with the operation (level I, II and III, respectively); The carbon impact—also considered a critical factor since the equivalent $CO_2$ emissions from the energy systems represent a major impact on the carbon neutrality objective. It was also used three levels of impact, based on the GHG emissions of each system, namely from low to high.

After defining the two critical factors, and considering that the decarbonization present threats but also opportunities, the decision matrix consists in three stages of priority as follows, and is presented in Table 35.3: High priority—actions that have, cumulatively, a relationship level I with the operation and high carbon impact on the following: (i) THREAT, systems whose operation is based on fossil fuels and whose emission factor (EF) is equal to or greater than 3 tons of $CO_2$ e/toe and, (ii) OPPORTUNITIES, replace fossil fuel systems with an EF equal to or greater than 3 tons of $CO_2$ e/toe by a solution based on renewable energy sources and/or EF less than 1.5 tons of $CO_2$ e/toe. In addition, energy efficiency measures were also considered high priority, and the production of electricity through renewable sources

**Table 35.3** Decision matrix, based on threats and opportunities

| | | | Relation to the operation | | |
|---|---|---|---|---|---|
| | Level | | Level I<br>Emissions not linked to Mission-Critical Capabilities (MCC) | Level II<br>Emissions linked to MCC but can be addressed without any impact to mission | Level III<br>Emissions related to MCC; a decrease in emissions would affect those capabilities |
| Carbon impact | Threat | High<br>Systems based on fossil fuels and EF $\geq 3$ [t $CO_2$ e/toe] | High<br>Diesel fuel [SHW + AC]<br>FG [equipment] | High<br>Diesel fuel [mobility] | Medium<br>JP8 [aircraft]<br>Diesel fuel [generators] |
| | | Medium<br>Systems based on fossil fuels and 3 [t $CO_2$ e/toe] $<$ EF $\geq 1.5$ [t $CO_2$ e/toe] | High<br>Gas [SHW + AC] | Medium | Low |
| | | Low<br>Systems operating includes renewable energy sources and EF $< 1.5$ [t $CO_2$ e/toe] | Medium<br>EE [illumination] | Low<br>EE [network consumption] | Low |
| | Opportunity | High<br>Change a fossil fuel system (EF $\geq 3$ [t $CO_2$ e/toe]) by a solution based on renewable energy sources and/or EF $< 1.5$ [t $CO_2$ e/toe] | High<br>EE [SHW + AC] | High<br>Electricity [mobility] | Medium |

(continued)

Table 35.3 (continued)

| | | Relation to the operation | | |
|---|---|---|---|---|
| Level | | Level I<br>Emissions not linked to Mission-Critical Capabilities (MCC) | Level II<br>Emissions linked to MCC but can be addressed without any impact to mission | Level III<br>Emissions related to MCC; a decrease in emissions would affect those capabilities |
| | Medium<br>Change fossil fuel system (3 [t $CO_2$ e/toe < EF ≥ 1.5 [t $CO_2$ e/toe]) by renewable energy sources and/or EF < 1.5 [t $CO_2$ e/toe] | High | Medium | Low |
| | Low<br>Change a system whose EF < 1.5 [t $CO_2$ e/toe] by renewable energy sources and/or EF also EF < 1.5 [t $CO_2$ e/toe] | Medium | Low<br>EE [production] | Low |

is also included as a measure of energy efficiency since the consumption of electricity from the network is reduced; Medium priority—are those actions that present an equivalent relationship between the two critical factors, that is, level I of relationship with operation and low carbon impact level, or level II of relationship with operation and medium carbon impact or level III of relationship with the operation and high carbon impact; Low priority—represent the other cases, not previously mentioned.

After identifying the needs of reduction and defining priorities, the actions that contribute to common objectives were grouped in project packages to expedite their subsequent implementation. The project packages were divided by typology—infrastructure, mobility, and air operations—and must be delivered to the correspondent technical team, considering that the interventions can take place at different levels—mode of operation, replacement of equipment and/or fuels and technology. The actions that must be studied to implement include: (i) For infrastructure—eliminate heating diesel; reduce diesel generators; reduce gas heating; produce green energy; increase energy efficiency; (ii) For mobility—replace vehicles; increase the use of

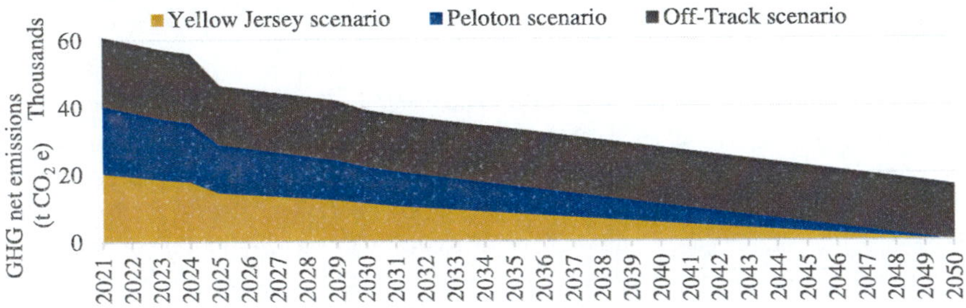


**Fig. 35.1** Evolution of greenhouse gases net emission for each scenario

embedded biofuels; ensure efficiency and sufficiency in driving, and (iii) For air operations—reduce GHG emissions associated with aviation fuel.

Subsequently, and following the methodology of the national roadmap (RCN2050), the national reduction scenarios were used to better understand, through time, in which path PrtAF is heading, as follows: (i) Scenario Off-Track, reflects the projection of energy consumption and GHG emissions that would be achieved by 2050 while maintaining current policies. It is not compatible with carbon neutrality nor national objectives; (ii) Scenario Peloton, represents reductions in energy consumption and GHG emissions primarily motivated by external impositions and/or incentives, not well planned, but still compatible with carbon neutrality; (iii) Scenario Yellow Jersey, represents structural and transversal changes implemented proactively. It is compatible with carbon neutrality and contributes to the achievement of national objectives.

Finally, based on the previous assumptions, the possible evolution of the GHG net emissions for each scenario were projected and are presented in Fig. 35.1. Here it is clear that maintaining the current policies will not allow the carbon–neutral goal, which is represented by the Off-Track scenario (in grey). By opposition, the Yellow Jersey scenario (in yellow), is carbon–neutral compatible and, according to the projections made, represents an energy dependence of 44% and a dependence on fossil sources of only 14% (not considering aircraft component).

### *35.3.3 Neutrality*

The Neutrality phase identified the KPI to monitor the path delineated for each scenario, namely environmental indicators related to economy and environment, water, soil and biodiversity, waste, and environmental risks. The validation of the current state will be made by the Regulation of the Air Force Environmental Management System that describes how the various Air Force entities relate to each other and what their environmental management responsibilities are.

## 35.4 Conclusion

The PrtAF developed the RCN2050PrtAF to materialize the necessary efforts to reduce its GHG emissions and offset the remaining ones. When the Mapping phase was completed, carbon neutrality was identified as a possible (but) difficult scenario. The PrtAF GHG emissions are a consequence of using aviation fuels, diesel, gas, electricity, fluorinated gases, and wastewater treatment. In 2021, the sum of emissions was 78.4 t $CO_2$e and the sequestration capacity of 58.2 t $CO_2$. Within the three possible emission reduction scenarios the Yellow Jersey is the one that best defines a sustainable and resilient organization. It is carbon–neutral compatible, with an energy dependence of 44% and a dependence on fossil sources of only 14%.

## References

Bowcott H, Gatto G, Hamilton A, Sulivan E (2021) Decarbonizing defense: imperative and opportunity. Aerospace & Defense Practice. McKinsey & Company. Available at https://www.mckinsey.com/industries/aerospace-and-defense/our-insights/decarbonizing-defense-imperative-and-opportunity. Last accessed 2021/07

Calaixo C, Páscoa C, Pinto J (2022) Measures to make the Portuguese Air Force a Carbon Neutral Organization. Revista de Ciências Militares, May X(1):233–260

Correia R, Páscoa C, Pinto J (2022) Optimising the Portuguese Air Force's Carbon Sequestration Potencial. Revista de Ciências Militares, May X(1):167–200

European Union (2019) COM(2019) 640. The European Green Deal, Brussels

European Union (2021) Regulation (EU) 2021/1119. European Climate Law. Brussels, Belgium

Intergovernmental Panel on Climate Change (IPCC) (2022) Guidelines. https://www.ipcc-nggip.iges.or.jp/support/support.html. Last accessed 2022/01

Portuguese Parliament (PP) (2021) Climate law bases, republic diary no. 253/2021, series I of 2021-12-31: 5–32, Lisboa, Portugal

Portuguese Air Force (2022) Roadmap for carbon neutrality 2050 (RCN2050PrtAF), Portuguese air force chief of staff, Alfragide, Portugal

Presidency of the Council of Ministers (PCM) (2020a) National plan for energy and climate 2030a (PNEC2030), PCM resolution no. 53/2020, republic diary no. 133/2020, series I of 2020-07-10: 2–158, Lisboa, Portugal

Presidency of the Council of Ministers (PCM) (2020b) Resource efficiency program in public administration for 2030, PCM resolution no. 104/2020, series I of 2020-11-24: 5–14, Lisboa, Portugal

Presidency of the Council of Ministers (PCM) (2019) Roadmap for carbon neutrality 2050 (RNC2050), PCM resolution no. 107/2019, republic diary no. 123/2019, series I of 2019-07-01: 3208–3299, Lisboa, Portugal

United Nations (UN) (2022) Paris agreement. Signed in 2015-12-12. https://unfccc.int/sites/default/files/english_paris_agreement.pdf. Last accessed 2022/01

# Chapter 36
# Development of Polyethersulphone Mixed Matrix Zeolite Membranes Functionalized with Ionic Liquids and Deep Eutectic Solvents for $CO_2$ Separation

**J. S. Cardoso, Z. Lin, P. Brito, and L. M. Gando-Ferreira**

**Abstract** Mixed matrix membranes (MMM) combine the flexibility of polymers and the strength and durability presented by inorganic solids. In an economically point of view, the advantages of membrane separation are low capital investment and space requirements, high process flexibility and lower energy consumption, helping for a more cost-effective separation process and providing a high separation degree. The molecular sieves based on nano-sized silicoaluminophosphates (SAPO) appear as one of the main materials in MMM for gas separation because the pore size of chabazite (CHA) (0.38 nm) is near the kinetic diameter of gases like $H_2$ (0.29 nm), $CO_2$ (0.33 nm), $N_2$ (0.36 nm), CO (0.37 nm), $CH_4$ (0.38 nm) and reduced crystal size improves the dispersion and decreases interfacial defects. Doping SAPO-34 are intended to increase the potential of these solids. The use of isomorphic substitution by transition metals (Fe, Ni, Co, Mn), results in materials with different acidity that differ from the original SAPO in interactions with other compounds. Besides, the addition of ionic liquids (IL) or Deep Eutectic Solvents (DES) with high affinity and selectivity to $CO_2$, onto the particle surface and then dispersing it in a polymer membrane can enhance the separation characteristics, resulting in better permeation and selectivity properties.

J. S. Cardoso (✉) · L. M. Gando-Ferreira
CIEPQPF, Department of Chemical Engineering, Faculty of Sciences and Technology, University of Coimbra, Pólo II, Rua Silvio Lima, 3030-790 Coimbra, Portugal
e-mail: jonathancardoso@ipb.pt

Z. Lin
Department of Chemistry, CICECO, University of Aveiro, 3810-193 Aveiro, Portugal

P. Brito
Centro de Investigação de Montanha (CIMO), Instituto Politécnico de Bragança, Campus de Santa Apolónia, 5300-253 Brgança, Portugal



N. S. Caetano and M. C. Felgueiras (eds.), *The 9th International Conference on Energy and Environment Research*, Environmental Science and Engineering,
https://doi.org/10.1007/978-3-031-43559-1_36

**Keywords** Mixed matrix membranes · $CO_2$ separation · Ionic liquids · Deep eutectic solvent · Permeability · Selectivity

## 36.1 Introduction

Reducing $CO_2$ emissions have become a key issue in environmental protection and sustainable development. To stop global warming and its consequences, thus, it is necessary to define strategies to reduce greenhouse gas emissions to the atmosphere. Currently, due to industrial development accompanied by a progressive consumption of energy, the society is highly dependent on fossil fuels, namely coal, oil, and natural gas, to produce electricity, in industry and transport (EIA IEO 2019). The use of these resources represents a significant impact on the evolution, both for the good (at a social, technological, and economic level) and profound consequences at an environmental level due to their burning. This practice is responsible for high $CO_2$ emissions to the atmosphere, representing about 80% of the total emissions of this gas. Data from 2014 show that, in Europe, the energy sector was responsible for around 55% of $CO_2$ emissions into the atmosphere, followed by the transport sector, with 23%, the agricultural sector with 10% and industrial sector, which represents about 9% of emissions. As a result, there has been a phenomenon of greenhouse effect, responsible for the increase of the average terrestrial temperature in the last years, putting at risk the environment and the human well-being in the long term. The consequences of this phenomenon are well known rising sea levels, floods, hurricanes, and droughts (Nabais 2016; Etxeberria-Benavides et al. 2018).

Matrix mixed membranes (MMM) containing zeolites have potential for separating $CO_2$ owing to their superior thermal, mechanical, and chemical stability, good erosion resistance, and stability at high $CO_2$ pressures (Gao et al. 2016). Zeolites have inorganic crystalline structures with uniform-sized pores of molecular dimensions. The molecular sieves based on silicoaluminophosphates (SAPO) appear as one of the main materials in MMM for gas separation because the pore size of SAPO-34 with chabazite structure (CHA) (0.38 nm) is near the kinetic diameter of gases like $H_2$ (0.29 nm), $CO_2$ (0.33 nm), $N_2$ (0.36 nm), CO (0.37 nm), $CH_4$ (0.38 nm). The silicoaluminophosphates are zeolite compounds formed by $TO_4$ (T = Si, Al, or P) and widely used in the catalysis and separation processes because of their pore diameter. The production of dopped SAPO-34 with transition metals (Fe, Ni, Co, Mn), named MeAPSO-34, leads to a zeolite with higher negative surface charge (Zhong et al. 2017; Sena et al. 2011).

Pressure Swing Adsorption (PSA), Temperature Swing Adsorption (TSA) and cryogenic distillation are mostly used industrially in the separation for $CO_2$, but their use implies high energy consumption. Mixed Matrix Membranes are hybrid membranes containing a solid with regular pores and the dispersive media that can be polymer, ceramic, carbon, and others (Peydayesh et al. 2013; Junaidi et al. 2014). The use of zeolites and other silicate derivatives as constituents of MMM in gas separation such as presented in Fig. 36.1, has lower energy consumption, is modular and can

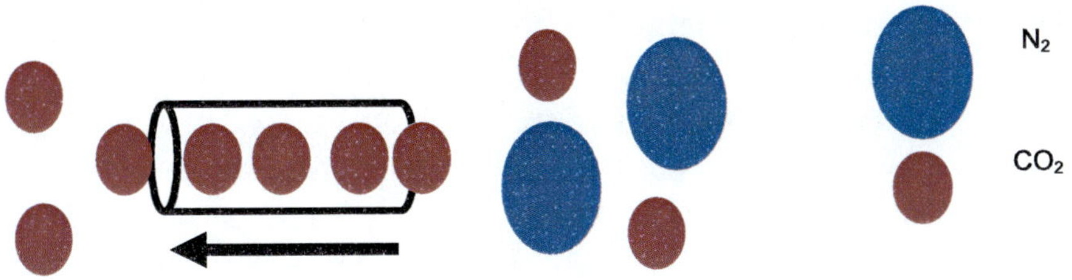


**Fig. 36.1** Molecular sieve permeation schematics through zeolite pore for $CO_2$ and $N_2$

be connected to traditional processes, besides showing high separation performance (Junaidi et al. 2015).

The use of Ionic Liquids (IL) and Deep Eutectic Solvents (DES) arises to improve the zeolite-polymer matrix interaction, as presented in Fig. 36.2, promoting a link between the zeolite surface (inorganic) and the polymer matrix (organic) due to its organic–inorganic dual nature. Those additives can be used via direct addition in the casting solution or through a post-impregnation process, and reduces interfacial voids, which increase the separation performance of MMM (Hudiono et al. 2011; Hu et al. 2018; Asghari et al. 2018; Huang et al. 2015).

**Fig. 36.2** Zeolite-ionic liquid interaction

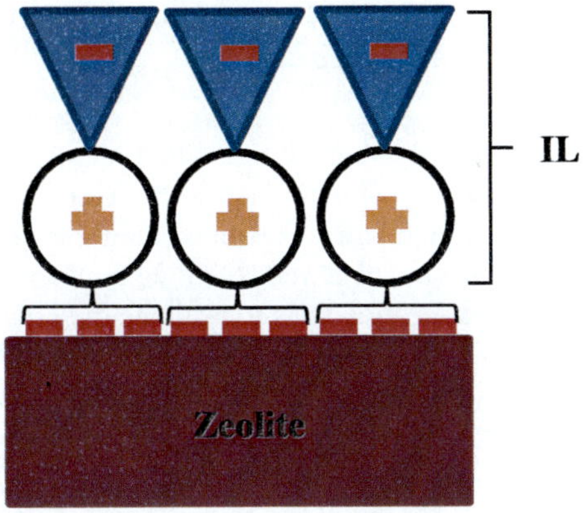

## 36.2 Materials and Methods

### *36.2.1 Synthesis and Characterization of SAPO-34 and MeAPSO-34*

The synthesis of the SAPO-34 samples, a gel having the molar composition 1 $Al_2O_3$; 1 $P_2O_5$; 0.6 $SiO_2$; 1.5 Morpholine; 0.5 TEAOH; 70 H2O was prepared. For MeAPSO-34 samples, the same gel composition is used, and Ni, Mn, Co, or Fe nitrate is added at ratio Me/Al of 0.0075. All gels were aged for 24 h under mixing and then dried at 90 °C for 24 h. The samples were prepared by adding water to dried gel in the proportion 1:1 and heated at 200 °C with different the reaction time. The solid obtained was centrifuged, washed, dried at 100 °C in an oven and calcined at 560 °C for 8 h.

The samples were characterized via powder X-ray diffraction for the analysis of the crystalline structure of the synthesized materials, pore size analysis to determine the pore distribution and surface area to verify the textual properties of the samples after the structural modifications.

### *36.2.2 Synthesis and Characterization of Mixed Matrix Membranes*

20%w/w Polyethersulfone (PES) membrane was prepared using 20%w/w zeolite, based on the polymer mass, in NMP and mixed for 24 h. The resulting solution was degassed for 60 min. The membrane was casted with a knife of 0.2 mm, then dried at 90 °C for 8 h and at 160 °C for 24 h under vacuum. The samples were characterized by gas permeation tests to evaluate the separation performance.

### *36.2.3 Performance of Membranes Through Gas Permeation Tests*

Gas permeation experiments were conducted using pure gases in order to determine the gas permeability and diffusivity coefficient. A schematic representation of the permeation system is shown in Fig. 36.3. The membranes were cut into 47 mm disks and mounted on a steel permeation cell and evaluated at feed pressure at room temperature.

The permeability results will be compared to each other while the selectivity results represent the ratio between the fluxes of two distinct pure gases in the same membrane to evaluate the performance of the membranes produced in relation to the permeability and selectivity.

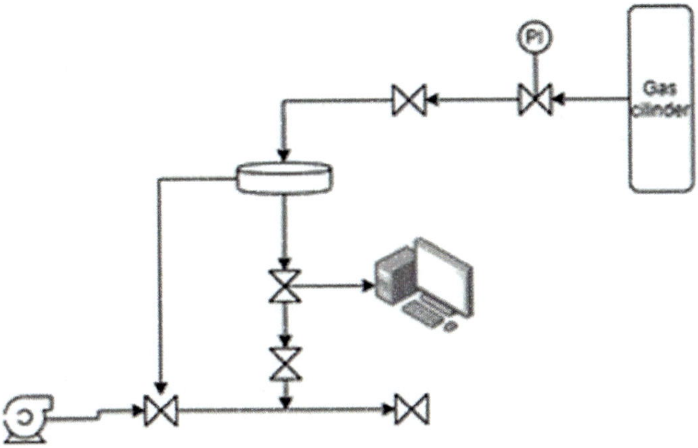


**Fig. 36.3** Schematics of the permeation system

### *36.2.4 Post-impregnation Procedure*

The membranes tested were placed in a kitasato, and vacuum was applied for a period of 1 h to remove the air from the pores. The additive was introduced with a syringe, through a septum and the membrane stays under vacuum for 1 h to allow the additive to fill the membrane pores. The membrane was removed, and dried using a tissue paper.

## 36.3 Results and Discussion

### *36.3.1 SAPO-34 and MeAPSO-34 Synthesis and Characterization*

SAPO-34 was synthetized using the dry-gel synthesis following the proportion 1:1 in mass, using different reaction times (12, 16, 24 and 48 h) presented in Fig. 36.4.

The results showed that in all reaction times the SAPO-34 was obtained, although the samples with reaction time of 24 h or more, presented a higher purity as presented in XRD when compared to a CHA standard x-ray dispersion analysis (see Fig. 36.4). The particles exhibited an average size of 0.857 μm in LDS analysis. The adsorption–desorption nitrogen isotherms are presented in Fig. 36.5 and the textural properties of the zeolites is in Table 36.1.

The adsorption–desorption nitrogen isotherms presented a well-defined microporous structure for all samples prepared.

The textural properties show that all materials presented a large amount of microporous with a high surface area compared to others SAPO-34 zeolites in literature. The

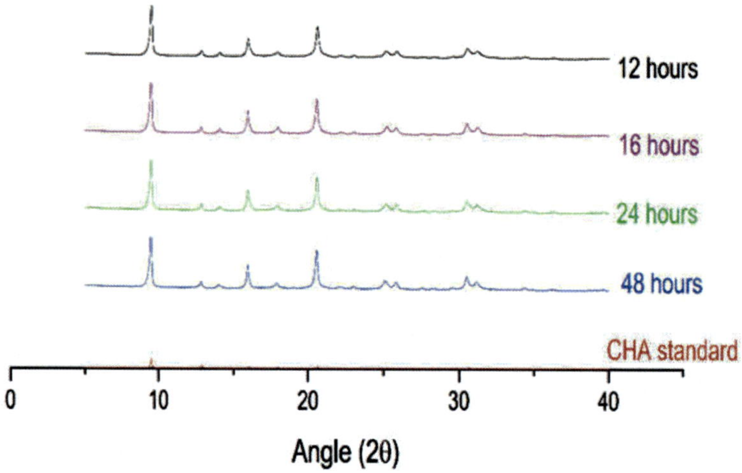


**Fig. 36.4** XRD for 12-, 16-, 24- and 48-h reaction time and CHA standard

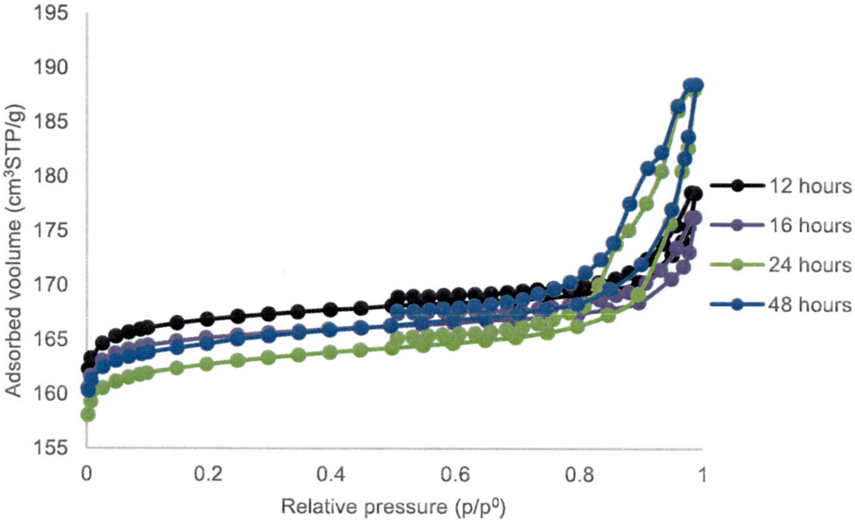


**Fig. 36.5** Adsorption–desorption nitrogen isotherms for 12, 16, 24 and 48 h of reaction time

sample with 24 h of reaction time was selected since it presented a higher purity even though with a bigger particle size. With that, another sample was prepared following the same procedure, named JS1, although changing the silicon source from LUDOX AS-40 to LUDOX HS-40. The XRD and LDS analysis are presented in Fig. 36.6.

**Table 36.1** Textural properties of the materials

| Sample (h) | $S_{BET}$ ($m^2\ g^{-1}$) | $S_{Langmuir}$ ($m^2\ g^{-1}$) | $S_{ext}$ ($m^2\ g^{-1}$) | $S_{mic}$ ($m^2\ g^{-1}$) | $V_{mic}$ ($mm^3\ g^{-1}$) | $V_{mic}/V_{Total}$ (%) |
|---|---|---|---|---|---|---|
| 12 | 477 | 725 | 6 | 471 | 0.257 | 94.4 |
| 16 | 470 | 718 | 5 | 465 | 0.254 | 94.7 |
| 24 | 464 | 707 | 9 | 455 | 0.249 | 87.8 |
| 48 | 470 | 715 | 10 | 460 | 0.251 | 88.3 |

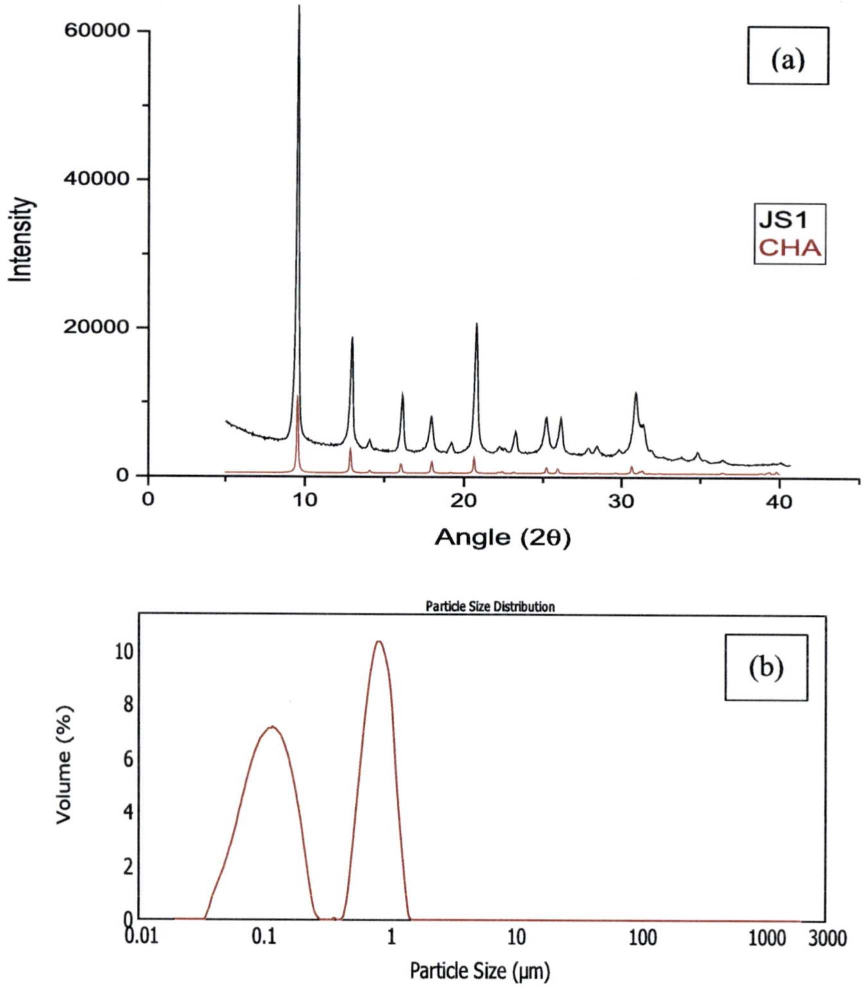


**Fig. 36.6** **a** XRD and **b** LDS for JS1 sample

Sample JS1 presented the same purity as the samples from LUDOX AS-40, however, the average particle size obtained was smaller, 0.198 μm, which is suitable to be used as the filler for the PES mixed matrix membranes.

### *36.3.2 Permeation Tests*

Initially, different neat PES membranes were prepared to evaluate the best synthesis procedure. The final procedure adopted was as described in the Materials and

Methods section, providing a homogeneous film without imperfections. The permeation curves are illustrated in Fig. 36.7.

The average value obtained for permeability though three replicates from the same membrane sheet were 17.40 GPU to N2 and 55 GPU to $CO_2$ with $CO_2/N_2$ selectivity of 3.20, which can be related to values in literature. Table 36.2 presents the permeability and ideal selectivity for $CO_2$ and $N_2$ pure gases.

The initial tests using the DES ChCl/Urea 1:2 as a post-impregnant showed an increase between 4 and 12 times of $CO_2/N_2$ selectivity, specially decreasing the $N_2$ permeability post-impregnation.

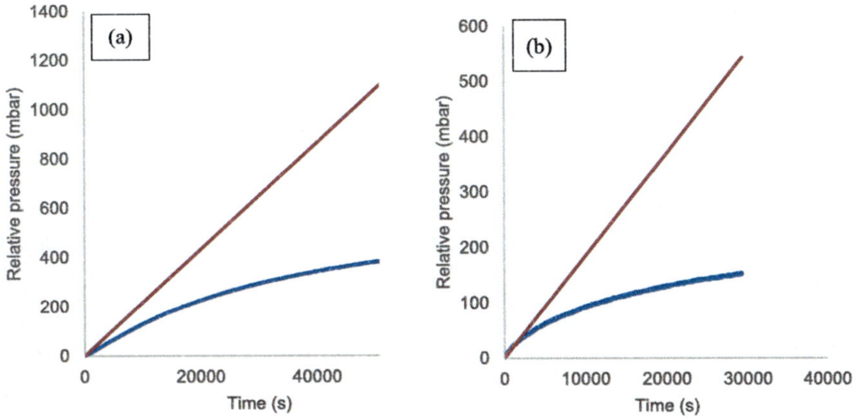


**Fig. 36.7** Permeability of **a** $CO_2$ and **b** $N_2$ through neat PES (20%w/w)

**Table 36.2** Permeability and ideal selectivity for $CO_2$ and $N_2$ pure gases

| Before post-impregnation | | | |
|---|---|---|---|
| Sample | $CO_2$ permeability (GPU) | $N_2$ permeability (GPU) | $CO_2/N_2$ selectivity |
| 1 | 50.17 | 17.07 | 2.94 |
| 2 | 68.08 | 16.33 | 4.17 |
| 3 | 46.76 | 18.80 | 2.49 |
| After post-impregnation | | | |
| Sample | $CO_2$ permeability (GPU) | $N_2$ permeability (GPU) | $CO_2/N_2$ selectivity |
| 1 | 46.58 | 13.70 | 3.40 |
| 2 | 43.30 | 2.52 | 17.18 |
| 3 | 38.78 | 0.90 | 43.32 |

## 36.4 Conclusion

It is stated for various authors that the use of mixed matrix membranes presents an exciting potential for further investigation and development in separation performance. The methodology for nano-sized SAPO-34 preparation was defined and the solid with the desired characteristics obtained to be used as filler in PES polymer membranes.

The use of post-sealing impregnation can be a healing procedure for poor compatibility after synthesis, those impregnations can be performed using ILs, with few studies showing that a post-impregnation in a MMM using [emim][Tf2N] can improve the selectivity, and no studies tried to use deep eutectic solvents (DES) to replace those ILs.

This work presents an innovation as preliminary results showing that 1:2 ChCl/Urea DES used as a post-sealing compound can reduce $N_2$ permeation and improve $CO_2/N_2$ separation performance by 4–12 times in neat PES membranes. Those results will be evaluated using PES/SAPO-34 membranes to verify if can also improve the separation performance in mixed matrix membranes.

**Funding** The authors gratefully acknowledge the fundings from the Strategic Project of CIEPQPF (UIDB/00102/2020), CICECO-Aveiro Institute of Materials (UIDB/50 011/2020, UIDP/50 011/2020 and LA/P/0 0 06/2020), CIMO (UIDB/00690/2020 and UIDP/00690/2020) and SusTEC (LA/P/0007/2021), financed by Fundação para a Ciência e Tecnologia (FCT) through national funds. J. S. Cardoso is also grateful for the financial support of the FCT through the PhD grant (SFRH/BD/148170/2019).

## References

Asghari M, Dashti A, Rezakazemi M, Raji M, Sodeifian G (2018) Polyurethane-SAPO-34 mixed matrix membrane for CO2/CH4 and CO2/N2 separation. Chin J Chem Eng 27:322–334. https://doi.org/10.1016/j.cjche.2018.03.012

Etxeberria-Benavides M, David O, Johnson T, Łozińska MM, Orsi A, Wright PA, Mastel S, Hillenbrand R, Kapteijn F, Gascon J (2018) High performance mixed matrix membranes (MMMs) composed of ZIF-94 filler and 6FDA-DAM polymer. J Memb Sci 550:198–207. https://doi.org/10.1016/j.memsci.2017.12.033

Gao B, Yang M, Qiao Y, Li J, Xiang X, Wu P, Wei Y, Xu S, Tian P, Liu Z (2016) A low-temperature approach to synthesize low-silica SAPO-34 nanocrystals and their application in the methanol-to-olefins (MTO) reaction. Catal Sci Technol 6:7569–7578. https://doi.org/10.1039/c6cy01461e

Hu L, Zhou J, Li Y, Cheng J, Cen K, Liu J (2018) CO2 absorption and diffusion in ionic liquid [P66614][Triz] modified molecular sieves SBA-15 with various pore lengths. Fuel Process Technol 172:216–224. https://doi.org/10.1016/j.fuproc.2017.12.022

Huang Y, Yu M, Carreon MA, Song Z, Zhou R, Li S, Feng X, Zhou SJ, Zong Z, Meyer HS (2015) SAPO-34 Membranes for N2/CH4 separation: Preparation, characterization, separation performance and economic evaluation. J Memb Sci 487:141–151. https://doi.org/10.1016/j.memsci.2015.03.078

Hudiono YC, Carlisle TK, LaFrate AL, Gin DL, Noble RD (2011) Novel mixed matrix membranes based on polymerizable room-temperature ionic liquids and SAPO-34 particles to improve CO2 separation. J Memb Sci 370:141–148. https://doi.org/10.1016/j.memsci.2011.01.012

EIA IEO (2019) International energy outlook 2019 with projections to 2050. Choice Reviews Online 2019:85. https://doi.org/10.5860/CHOICE.44-3624

Junaidi MUM, Khoo CP, Leo CP, Ahmad AL (2014) The effects of solvents on the modification of SAPO-34 zeolite using 3-aminopropyl trimethoxy silane for the preparation of asymmetric polysulfone mixed matrix membrane in the application of CO2 separation. Microporous Mesoporous Mater 192:52–59. https://doi.org/10.1016/j.micromeso.2013.10.006

Junaidi MUM, Leo CP, Ahmad AL, Ahmad NA (2015) Fluorocarbon functionalized SAPO-34 zeolite incorporated in asymmetric mixed matrix membranes for carbon dioxide separation in wet gases. Microporous Mesoporous Mater 206:23–33. https://doi.org/10.1016/j.micromeso.2014.12.013

Nabais AR (2016) Preparação e caracterização de membranas de matriz mista para separação de CO2

Peydayesh M, Asarehpour S, Mohammadi T, Bakhtiari O (2013) Preparation and characterization of SAPO-34 - Matrimid®5218 mixed matrix membranes for CO2/CH4separation. Chem Eng Res Des 91:1335–1342. https://doi.org/10.1016/j.cherd.2013.01.022

Sena FC, De Souza BF, De Almeida NC, Cardoso JS, Fernandes LD (2011) Influence of framework composition over SAPO-34 and MeAPSO-34 acidity. Appl Catal A Gen 406:59–62. https://doi.org/10.1016/j.apcata.2011.08.010

Zhong J, Han J, Wei Y, Tian P, Guo X, Song C, Liu Z (2017) Recent advances of the nano-hierarchical SAPO-34 in the methanol-to-olefin (MTO) reaction and other applications. Catal Sci Technol 7:4905–4923

# Chapter 37
# Tourism and Air Pollution in Italian Regions

**Sara Ciarlantini, Mara Madaleno, Margarita Robaina, Alexandra Monteiro, Carla Gama, Maria João Carneiro, and Celeste Eusébio**

**Abstract** This study intends to explore the impact of tourism on air pollution at a regional level in Italy, and also to investigate the evidence of a tourism-induced Environmental Kuznets Curve (EKC) for Italian regions, including variables as an economic indicator (GDP), the energy consumption, and the number of nights spent at tourist accommodation establishments from both residents and foreign tourists. Most of the studies found in the literature investigate this relationship on a national scale, while this research focuses on a regional basis. The analysis is conducted using a set of three air pollutants (NOx, PM10, and PM2.5—the most critical in terms of air quality), over two different periods for comparison purposes: 2000–2008 and 2009–2018. The Levin-Lin-Chu unit root test proves the variables to be stationary, while the Pedroni cointegration test shows that they are integrated. A common main econometric model is employed to check the relationship among the variables: the Pooled OLS Estimator; the Granger panel causality test is conducted to see the causality among them. The tourism-induced EKC hypothesis is not validated, even if the findings show a decreasing relationship between economic growth and environmental pollution. Results also show slightly few differences between the two analyzed periods.

**Keywords** Air pollutants · Emissions · Environmental kuznets curve · Italy · Region · Tourism

S. Ciarlantini (✉)
DIST—Interuniversity Department of Regional and Urban Studies and Planning, Polytechnic University of Turin and University of Turin, Turin, Italy
e-mail: sara.ciarlantini@polito.it

M. Madaleno · M. Robaina · M. J. Carneiro · C. Eusébio
GOVCOPP and Department of Economics, Management, Industrial Engineering and Tourism, University of Aveiro, Aveiro, Portugal

A. Monteiro · C. Gama
CESAM and Department of Environment and Planning, University of Aveiro, Aveiro, Portugal



N. S. Caetano and M. C. Felgueiras (eds.), *The 9th International Conference on Energy and Environment Research*, Environmental Science and Engineering,
https://doi.org/10.1007/978-3-031-43559-1_37

## 37.1 Introduction

The tourism industry is one of the main economic activities in the world. Tourism is considered an "engine of economic growth" (Vita et al. 2015): its development requires huge investments, especially in infrastructures (e.g. airports, roads), and other tourism services (such as resorts, restaurants, hotels, shops); its activity creates hundreds of million jobs (World Travel Tourism Council (WTTC) 2020) representing 10% of global employment and contributing for trillions of dollars to the world economy every year (Hsieh and Kung 2013).

Tourism is among the most important industries for Mediterranean countries, and it is slightly gaining importance over the years, as shown by the increasing portion of GDP that this sector represents. Italy is among the most visited European destinations (the third most visited country in Europe) according to the World Tourism Organization (World Tourism Organization–UNWTO 2018).

This sector is, like any other industry, a substantial contributor to environmental degradation (Ozturk et al. 2016), as it puts pressure on the quality of the environment by depleting its natural resources. Tourism can induce large pressure on the environment in the form of soil erosion, degradation of historic sites and monuments, deterioration and reduction of green fields, and loss of natural habitat, biodiversity, and landscape. The typical environmental problems for which local people suffer the most are air, marine, visual, and noise pollution, and large quantities of waste products (Ozturk et al. 2016; Lee et al. 2015; Vehbi and Doratli 2010). The environmental effects caused by the tourism industry are greater than other service sectors ones (except for hazardous industrial waste), particularly the impacts caused on the air quality (Hsieh and Kung 2013). This is because an increment in tourism activities comes with an increased demand for energy for numerous activities such as catering, accommodation, infrastructures construction, management of tourist attractions (Katircioğlu 2014a, b; Wu and Shi 2011), and especially air and road transportation since tourism activities are strictly connected to it (Hsieh and Kung 2013). All these high energy-consuming activities negatively affect the environment across countries in the form of different air pollutants (Shaheen et al. 2019), especially greenhouse gases emissions, above all $CO_2$ ones, which are an inevitable by-product of tourism activities (Bella 2018) and make the tourism sector one of the main causes of climate change (Shaheen et al. 2019). In fact, according to the World Tourism Organization (World Tourism Organization and International Transport Forum–UNWTO 2019), the tourism sector is responsible for 4.6% of global warming, and this is the reason why the tourism industry is referred to as the "industry without a chimney" (Hsieh and Kung 2013).

Besides the majority of the papers regarding the impact of tourism on atmospheric emissions are focused on $CO_2$ pollutants—a greenhouse gas with climate change effects (World Tourism Organization and International Transport Forum–UNWTO 2019)—when the purpose is to address air quality, this is not the proper pollutant to be considered, but the ones that affect human exposure and legislated in terms of human health protection. This paper aims to emphasize these air pollutants as proxy measures for environmental quality, namely the most critical ones in terms of urban areas: Nitrogen Oxide (NOx) and Particulate Matter (PM10 and PM2.5).

Even though (Lee et al. 2015) report that tourism has significant effects on environmental quality, while the quality of the environment has no significant impact on tourism, most of the literature contributes with opposite results. Multiple factors or attributes are influencing a destination's success, which may include the level of local prices, the safety at the destination, and most of the time the environmental conditions, which are considered relevant determinants in the selection of destinations by visitors (Fernandez 2020; Huybers and Bennett 2000). (Tang et al. 2019) conducted a research, which showed that tourism demand is sensitive to environmental pollution. Climatic variables and weather conditions are crucial to tourists' choices about which destination to select. In particular, air quality influences physical and mental comfort, hence it is one of the major criteria to assess the suitability of tourism activities and to choose the touristic destination. An example is provided by Wang et al. (2018), which refers that the severity of air quality (low air quality) in the place of origin negatively influences outbound tourism demand. Through empirical econometric models, this hypothesis has proven to be supported, showing that air pollution is expected to strongly influence outbound tourist flows. A paper regarding the city of Beijing found that air pollution harms the city's inbound tourist arrivals in the long run. This means that air pollution could decrease Beijing's inbound tourism in the long run, while the variations in air quality do not influence Beijing's tourist arrivals in the short run (Tang et al. 2019).

A particular concept not too much explored concerning the tourism-environment relationship is the Environmental Kuznets Curve (EKC). According to Stern (2004), "EKC is a hypothesized relationship between various indicators of environmental degradation and per capita income level that exhibits an inverted U-shape during the process of economic development of an economy", therefore the EKC hypothesis explains the relationship between the quality of the environment and economic growth (Gamage et al. 2017). Within this context, the focal point of the present study is to empirically investigate the relationship between NOx, PM10, PM2.5 emissions, economic growth, local and foreign tourism, and energy consumption through the construction of an EKC model for the Italian regions, over two sub-periods: 2000–2008 and 2009–2018. This research is based on previous studies, whose aim was to investigate the relationship among these variables and the existence of a tourism-induced EKC at a national level, generally focusing on $CO_2$ emissions, therefore neglecting the relevant air pollutants for air quality, which is also a consequence of tourism development. For this reason, this study incorporates emission data for the most critical pollutants regarding air quality (NOx and PM), contributing to the existing literature by providing analysis on a regional scale.

## 37.2 Data Description and Methodology

To find shreds of evidence for a tourism-induced EKC, different variables have been selected for determining the linkages between environmental pollution, economic development, tourism growth, and energy consumption. The data presented are annual time series covering the period 2000–2018, for Italian regions using NUTS 2, considering the current NUTS 2016 classification (https://ec.europa.eu/eurostat/web/gisco) for a total of 20 regions.

This study employs the per capita Gross Domestic Product in million euros at constant prices as a proxy for the per capita income. Regarding the tourism data, the number of nights spent at tourist accommodation establishments (i.e. hotels, holiday and other short-stay accommodations, camping grounds, recreational vehicle parks, and trailer parks) is used; the tourism data is divided into foreign and local tourists' visits, using per capita values. The data for GDP and tourism were obtained from the Eurostat Data Browser (2020). The energy use variable is proxied by energy consumption in kWh/capita and it is gathered from Terna (https://www.terna.it/). The dependent variable is environmental degradation, which is proxied by emissions in metric tons per capita. The pollutants selected are NOx, PM10, and PM2.5, which are all local air pollutants (Rasli et al. 2018). The source of the pollutants data is the EMEP/CEIP (Co-operative Program for Monitoring and Evaluation of long-range transmission of air pollutants in Europe) website (https://www.ceip.at/ceip-reports) and the gap-filled gridded emissions were obtained with a $0.1^\circ \times 0.1^\circ$ (longitude/latitude) resolution. The relationship between air pollutants, economic development, tourism growth, and energy consumption is explored by using the EKC model outlined in Eq. 37.1:

$$\begin{aligned} Pollutant_{it} = \beta_0 + \beta_1 GDP_{it} + \beta_2 GDP_{it}^2 + \beta_3 TOUFO_{it} \\ + \beta_4 TOURE_{it} + \beta_5 ENE_{it} + \varepsilon_{it} \end{aligned} \quad (37.1)$$

Pollutant denotes the NOx, PM10, and PM2.5 per capita emissions; GDP and GDP2 refer to per capita GDP and the squared term of per capita GDP, respectively. TOUFO represents the nights spent at tourist accommodation establishments by foreign people, whereas TOURE from resident tourists; ENE refers to energy consumption. The subscript i characterizes the region, while the subscript t denotes the years, and ε is the error term. The natural logarithm of all variables is used in the econometric analysis.

According to the EKC theory, we expect β1 to be greater than zero and β2 to be less than zero. We expect that the sign of β5 is positive since energy consumption tends to lead to an increase in emissions. We do not predict the sign of β3 and β4 since existing literature has documented mixed results. The econometric analysis begins with the panel unit root test, adopting the Levin-Lin-Chu one (Levin et al. 2002), to check the stationarity of the data to avoid spurious regressions. The cointegration test is analyzed afterward to check the long-run relationship between all the variables. For this purpose, the Pedroni cointegration test is selected (Pedroni

2004). Once confirmed the cointegration relationship, the Italian data is discussed through the estimations' analysis. The Pooled OLS Estimator model (Sayrs 1989) is selected for all the pollutants (NOx, PM10, and PM2.5). After these tests, the Granger causality test is conducted through the Dumitrescu–Hurlin panel Granger causality test (Dumitrescu and Hurlin 2012). This test suggests whether a short-run causal relationship exists among the variables, as the long-run relationships between them are already explored with the cointegration test.

## 37.3 Results

The panel time series should present the stationary property to have economically meaningful and reliable estimates of the explanatory variables. As some of these panel data contain unit roots at their levels, the test was run again on these variables' first differences. All the variables are stationary at the first level, with a significance of 1%, except for the NOx and PM2.5 pollutants, which become stationary at first differences. PM10 emissions at first differences, instead, are more stationary than level variables (even though the differenced version has a significance of over 20%).

The long-run relationship among the variables is tested with the Pedroni cointegration test, whose results show that the null hypothesis of no cointegration is rejected at the 1% significance level: all the variables employed in this analysis for the Italian regions are cointegrated for all three air pollutants. This confirms the long-run relationship between NOx, PM10, and PM2.5 emissions, economic growth, tourism, and energy consumption in the Italian regions.

Once the cointegration relationship is confirmed, the panel analysis continues with evaluating the presence of a tourism-induced EKC for the 2000–2008 period and possible differences with 2009–2018 one. The expectations are that the results are better for the second period, as an improvement of the environmental situation over the years is expected, coherent with the policies and plans implemented for the reduction of atmospheric emissions (National Emission Ceilings Directive 2001/81/CE (European Commission 2001); Renewable Energy Directive 2009/28/CE (European Commission 2009).

For both periods, no evidence of EKC for the Italian regions is found, even though a decreasing relationship between environmental pollution (for the three pollutants) and economic growth appears. The results are given in Table 37.1.

In the first sub-period, the goodness of fit is not high throughout the three pollutants, as the higher Adjusted R-squared is 53%. Even so, the results illustrate that all the coefficients of the explanatory variables are statistically significant, except for the economic variables, particularly concerning NOx emissions. Even though, the GDP's coefficient is negative and so is its square for all the pollutants, implying that there is no evidence of the Environmental Kuznets Curve between per capita GDP and the three per capita pollutants. Italy presents a decreasing relationship between economic growth and environmental pollution.

**Table 37.1** Pooled OLS Estimator outputs

| | | NOx | | PM10 | | PM2.5 | |
|---|---|---|---|---|---|---|---|
| | | Coef | *p*-val | Coef | *p*-val | Coef | *p*-val |
| 2000–2008 | GDP | −0.790 | 0.624 | −2.228 | 0.181 | −2.026 | 0.250 |
| | GDP2 | −0.174 | 0.476 | −0.445 | 0.078 | −0.417 | 0.118 |
| | TOUFO | −0.346 | 0.000 | −0.461 | 0.000 | −0.380 | 0.000 |
| | TOURE | 0.339 | 0.000 | 0.476 | 0.000 | 0.388 | 0.000 |
| | ENE | 0.790 | 0.000 | 0.758 | 0.000 | 0.708 | 0.000 |
| | | $R^2$ | 0.478 | $R^2$ | 0.547 | $R^2$ | 0.490 |
| | | Adj. $R^2$ | 0.460 | Adj. $R^2$ | 0.531 | Adj. $R^2$ | 0.472 |
| 2009–2018 | GDP | −4.369 | 0.000 | −3.836 | 0.000 | −3.621 | 0.001 |
| | GDP2 | −0.769 | 0.000 | −0.719 | 0.000 | −0.680 | 0.000 |
| | TOUFO | −0.357 | 0.000 | −0.417 | 0.000 | −0.360 | 0.000 |
| | TOURE | 0.301 | 0.000 | 0.410 | 0.000 | 0.382 | 0.000 |
| | ENE | 0.011 | 0.740 | 0.036 | 0.308 | 0.031 | 0.381 |
| | | $R^2$ | 0.413 | $R^2$ | 0.461 | $R^2$ | 0.434 |
| | | Adj. $R^2$ | 0.398 | Adj. $R^2$ | 0.448 | Adj. $R^2$ | 0.420 |

The tourism outputs demonstrate that an increase in foreign tourists leads to a decrease in NOx, PM10, and PM2.5 emissions in the Italian regions, respectively by 0.35%, 0.46%, and 0.38% for every 1% growth of non-Italian visitors, for the first sub-period. Instead, domestic tourists' increment leads to a rise in atmospheric pollution, by approximately the same amount they diminished in the foreign tourists' case for every 1% increase in Italian visitors. The PM emissions are the ones that decrease/increase the most when, respectively, the number of foreign and local tourists increases. Concerning the energy variable, a 1% increase in energy consumption inevitably leads to a rise in the level of emissions (NOx is the pollutant that gets incremented the most: by about 0.80%). In the Pooled OLS Estimator for the second sub-period, about 40–45% of the variation of the outputs is explained by the other variables in the model. Even though, the variables for the explanation of the pollutants are significant, except for the energy variable which has high p-values, as Table 37.1 shows. Even for this second sub-period, there is no evidence of an EKC for the Italian regions, although a negative relationship between economic growth and pollutant emissions is still found.

The causality test is employed to check the relationship among all the variables (pollutants, economic growth, tourism variables, and energy consumption). When applying the Granger causality test to the Italian regions for the 2000–2008 period, bidirectional causality is revealed between NOx and both the two tourism categories (see Table 37.2).

Concerning PM10 emissions, unidirectional causality is seen running from pollutants emissions to the (local and foreign) tourism variables. Finally, for the PM2.5 pollutant, bidirectional causality is presented between the pollutant and foreign

**Table 37.2** Dumitrescu–Hurlin panel Granger causality test

| | 2000–2008 | | | | 2009–2018 | | | |
|---|---|---|---|---|---|---|---|---|
| | W-stat | *p*-val | Result | Conclusion | W-stat | *p*-val | Result | Conclusion |
| TOUFO– > NOx | 0.392 | 0.055 | Yes | Bidirectional causality between NOx and Foreign tourism | 71.794 | 0.000 | Yes | Bidirectional causality between NOx and Foreign tourism |
| NOx– > TOUFO | 20.925 | 0.001 | Yes | | 35.925 | 0.000 | Yes | |
| TOURE– > NOx | 0.278 | 0.022 | Yes | Bidirectional causality between NOx and Local tourism | 18.656 | 0.006 | Yes | Bidirectional causality between NOx and Local tourism |
| NOx– > TOURE | 31.735 | 0.000 | Yes | | 23.423 | 0.000 | Yes | |
| TOUFO– > PM10 | 23.573 | 0.000 | Yes | Unidirectional causality from Foreign tourism to PM10 | 21.466 | 0.000 | Yes | Unidirectional causality from Foreign tourism to PM10 |
| PM10– > TOUFO | 13.272 | 0.301 | No | | 10.221 | 0.944 | No | |
| TOURE– > PM10 | 23.924 | 0.000 | Yes | Unidirectional causality from Local tourism to PM10 | 13.091 | 0.328 | No | No causality |
| PM10– > TOURE | 0.515 | 0.125 | No | | 0.794 | 0.515 | No | |
| TOUFO– > PM2.5 | 35.410 | 0.000 | Yes | Bidirectional causality between PM2.5 and Foreign tourism | 23.316 | 0.000 | Yes | Unidirectional causality from Foreign tourism to PM2.5 |
| PM2.5–> TOUFO | 15.372 | 0.089 | Yes | | 10.233 | 0.941 | No | |
| TOURE–> PM2.5 | 23.086 | 0.000 | Yes | Unidirectional causality from Local tourism to PM2.5 | 13.653 | 0.248 | No | No causality |
| PM2.5–> TOURE | 0.517 | 0.127 | No | | 0.781 | 0.488 | No | |

tourism. For this latter category, only weak evidence of a relationship is seen coming from the pollutant. Concerning domestic tourism, from the test, a one-way causality appears from the touristic variable to the pollutant. For the second sub-period, foreign tourism shows to strongly influence all three pollutants, as seen in Table 37.2. Bidirectional causality is found between NOx emissions and both tourism categories, foreign and local tourism. Instead, no significant unidirectional causalities are found for PM10 and PM2.5 pollutants, to both local and foreign tourism.

## 37.4 Conclusions

The findings fail to document any evidence supporting the EKC hypothesis for any of the air pollution variables at a regional level, therefore they do not reveal any inverted-U shape relationship between environmental degradation and output growth driven by tourism for any of the two sub-periods. Even so, the results point out that tourism development has mixed impacts (both positive and negative) on the different air pollutants.

Understanding the relationship between per capita emissions, economic growth, tourism, and energy consumption is significant for policymakers and countries' governments. It is imperative to recognize the vital importance of the tourism sector and its expansion to the regional economies, and also its big negative impact on the environment. It is fundamental to review the current tourism-related environmental policies, laws, and regulations, complementing and modifying them if necessary. Well-planned, coordinated strategies and clear guidelines for sustainable tourism development and emissions reduction should be implemented. Furthermore, a transformation from carbon-intensive tourism into green sustainable tourism should be pursued, through incentives and financial support.

The role of environmental policies should be that of pursuing the sustainability objective at all levels of development, including the adoption of best practices to invest in tourists' education, which will increase their sensibilization towards environmental problems and enable them to behave responsibly. The state of the environment affects the quality of everyone's life, its protection should be considered a priority in every society's agenda.

**Acknowledgements and Funding** Thanks are due to Professor Vito Frontuto (UniTo) for having read and contributed to the correction of the article. This research was funded by the FCT/MCTES through the ARTUR research project (POCI-01–0145-FEDER-029374). We also acknowledge the financial support of CESAM (UIDP/50017/2020 + UIDB/50017/2020), and that of GOVCOPP (UIDB/04058/2020) + (UIDP/04058/2020), through national funds.

## References

Bella G (2018) Estimating the tourism induced environmental Kuznets curve in France. J Sustain Tour 26(12):2043–2052

Centre on Emission Inventories and Projections (EMEP/CEIP). https://www.ceip.at/ceip-reports

De Vita G, Katircioğlu ST, Altinay L, Fethi S, Mercan M (2015) Revisiting the environmental Kuznets curve hypothesis in a tourism development context. Environ Sci Pollut Res 22:16652–16663

Dumitrescu EI, Hurlin C (2012) Testing for Granger non-causality in heterogeneous panels. Econ Model 29(4):1450–1460

European Commission (2001) European commission directive 2001/81/CE on national ceilings of certain atmospheric pollutants. https://eur-lex.europa.eu/legal-content/EN/TXT/?uri=CELEX%3A32001L0081

European Commission (2009) European commission directive 2009/28/CE on the promotion of energy use from renewable sources. https://eur-lex.europa.eu/legal-content/EN/TXT/PDF/?uri=CELEX:32009L0028

Eurostat Data Browser (2020). https://ec.europa.eu/eurostat/web/main/data/database. Last accessed Dec 2020

Fernandez JAS et al (2020) Determinants of tourism destination competitiveness in the countries most visited by international tourists: proposal of a synthetic index. Tourism Manag Perspect 33

Gamage SKN, Kuruppuge RH, Ul Haq I (2017) Energy consumption, tourism development, and environmental degradation in Sri Lanka. Energ Sources, Part B: Econ, Plann Policy 12(10):910–916

Geographic Information System of the Commission (GISCO). https://ec.europa.eu/eurostat/web/gisco

Hsieh HJ, Kung S (2013) The linkage analysis of environmental impact of tourism industry. Procedia Environ Sci 17:658–665

Huybers T, Bennett J (2000) Impact of the environment on holiday destination choices of prospective UK tourists: implications for tropical North Queensland. Tour Econ 6(1):21–46

Katircioğlu ST (2014a) International tourism, energy consumption, and environmental pollution: The case of Turkey. Renew Sustain Energy Rev 36:180–187

Katircioğlu ST (2014b) Testing the tourism-induced EKC hypothesis: the case of Singapore. Econ Model 41:383–391

Lee H, Verances JB, Song W (2015) The tourism-environment causality. Int J Tourism Sci 9(3):39–48

Levin A, Lin C, Chu CJ (2002) Unit root tests in panel data: asymptotic and finite-sample properties. J Econometrics 108(1):1–24

Ozturk I, Al-Mulali U, Saboori B (2016) Investigating the environmental Kuznets curve hypothesis: the role of tourism and ecological footprint. Environ Sci Pollut Res 23:1916–1928

Pedroni P (2004) Panel cointegration: asymptotic and finite sample properties of pooled time series tests with an application to the PPP hypothesis. Economet Theor 20(3):597–625

Rasli A, Qureshi MI, Isah-Chikaji A, Zaman K, Ahmad M (2018) New toxics, race to the bottom and revised environmental Kuznets curve: the case of local and global pollutants. Renew Sustain Energy Rev 81:3120–3130

Sayrs Lois W (1989) Pooled time series analysis, vol 70, Chapter 2. Sage Publications, California, pp 8–16

Shaheen K et al (2019) Dynamic linkages between tourism, energy, environment, and economic growth: evidence from top 10 tourism-induced countries. Environ Sci Pollut Res 26:31273–31283

Stern D (2004) The rise and fall of the environmental kuznets curve. World Dev 32(8):1419–1439

Tang J et al (2019) Does air pollution decrease inbound tourist arrivals? The case of Beijing, Asia Pacific. J Tourism Res 24(6):597–605

Terna—Rete Elettrica Nazionale. https://www.terna.it/

Vehbi BO, Doratli N (2010) Assessing the impact of tourism on the physical environment of a small coastal town: girne, Northern Cyprus. Eur Plann Stud 18(9):1485–1505

Wang L et al (2018) Effect of air quality in the place of origin on outbound tourism demand: disposable income as a moderator. Tour Manage 68:152–161

World Tourism Organization and International Transport Forum–UNWTO (2019) Transport-related $CO_2$ Emissions of the tourism sector—Modelling results. https://www.e-unwto.org/doi/epdf/10.18111/9789284416660

World Tourism Organization–UNWTO (2018) European Union tourism trends. https://www.e-unwto.org/doi/pdf/10.18111/9789284419470

World Travel & Tourism Council (WTTC) (2020) https://wttc.org/. Last accessed Nov 2020

Wu P, Shi P (2011) An estimation of energy consumption and $CO_2$ emissions in tourism sector of China. J Geog Sci 21(4):733–745

# Chapter 38
# Level of Awareness and Knowledge Regarding Climate Change Among the People of Dammam, Saudi Arabia

**Abdulaziz I. Almulhim and Khalid Mohammed Almatar**

**Abstract** Climate change (CC) is one of the most significant threats to human well-being—affecting all countries, including Saudi Arabia. Addressing CC requires efforts from both the public and government, and understanding people's CC awareness is critical for devising effective policies to mitigate CC risks. This study is aimed at determining people's CC knowledge and awareness. A cross-sectional survey was conducted in Dammam and involved 310 people who consented to participate. A pretested questionnaire was distributed to collect data about CC. Only a few respondents were aware of CC causes and effects (14%). The significant CC causes identified were deforestation (28.1%), population growth (24.2%), urbanization (18.7%), vehicles (17.7%), and greenhouse gases (11.3%). Most of the respondents were very concerned about CC issues (67%) but (78.4%) had poor knowledge of the Sustainable Development Goals related to CC. Overall, the respondents had moderate knowledge of CC, indicating the need for an effective targeted strategy with a comprehensive action plan for a larger population to reduce the impact of CC.

**Keywords** Climate change · Awareness · Knowledge · Sustainable development goals · Saudi Arabia

## 38.1 Introduction

With profound global and regional consequences, climate change (CC) is impacting almost every country, and in the coming decades, it poses a monumental threat to societies and economies (Kabir et al. 2016). CC is one of the most significant hazard to humankind, threatening the survival of civilization (Ebi et al. 2021; Bollettino et al.

A. I. Almulhim (✉) · K. M. Almatar
Department of Urban and Regional Planning, College of Architecture and Planning, Imam Abdulrahman Bin Faisal University, P.O. Box 1982, Dammam 31451, Saudi Arabia
e-mail: aialmulhim@iau.edu.sa



N. S. Caetano and M. C. Felgueiras (eds.), *The 9th International Conference on Energy and Environment Research*, Environmental Science and Engineering,
https://doi.org/10.1007/978-3-031-43559-1_38

2020). Since the start of the Industrial Revolution, greenhouse gas (GHG) emissions have been increasing. It is projected that by 2030, the average global surface temperature will increase by 3 °C; by 2050, the sea level will rise by 0.1–0.3 m—resulting in an increased occurrence of severe climate events, such as droughts and cyclones (Bollettino et al. 2020; Poortinga et al. 2019). According to the Paris Agreement, several countries worldwide have committed to limiting global warming rate to less than 2 °C (Wei et al. 2020). However, GHG emissions continue to increase significantly. CC substantially contributes to human mortality and morbidity and directly impacts individuals and public health (Xu et al. 2020). CC effects include air pollution, extreme weather conditions, food insecurity, and increasing temperatures. Scientific studies have confirmed that these changes impact health in a complex manner, resulting in allergies, infectious diseases, heat injuries, mental disorders, and malnutrition (Ebi et al. 2021; Xu et al. 2020; Eguiluz-Gracia et al. 2020).

The SDGs highlight CC as a priority (Streimikis and Baležentis 2020). SDG 13 mainly focuses on efforts to address CC and motivate all countries to undertake adequate measures to reduce GHG emissions (Streimikis and Baležentis 2020; Elder and Olsen 2019). Despite increased concern about this issue and its effect on health, only a handful of studies have explored public perceptions. It is essential for policymakers, researchers, and other stakeholders to actively involve the public. However, implementing CC policies is a significant challenge mainly due to the poor understanding of the various resultant cultural and socioeconomic effects (Wijk et al. 2020). Over the last few years, CC knowledge has improved significantly, and the concern level has also increased. Although CC knowledge developed in various developed countries, skepticism about the severity and existence of CC is low among the general public worldwide (Streimikis and Baležentis 2020; Elder and Olsen 2019; Wijk et al. 2020). Understanding the public perceptions of CC is imperative to encourage widespread participation and create successful educational strategies and communication to promote climate mindfulness.

Since the 1980s, studies on people's perceptions of CC have significantly increased (Capstick et al. 2015). In 2014, a cross-country study conducted in the US, China, and Canada demonstrated that people have good knowledge of CC (Berardi 2017). However, people's current CC understanding remains unclear in Saudi Arabia, as very few studies conducted in the country have assessed public awareness of CC. The general public is vulnerable to the long-term adverse effects of CC, and their involvement in any action related to CC is important. Therefore, this study is essential, as changing deep-rooted behaviors and habits detrimental to the earth needs a significant understanding of CC. Therefore, this study was conducted to determine the public's knowledge of CC in Dammam City. This understanding of CC should spur a positive attitudinal change in the public. The results of this survey will provide valuable material for researchers and policymakers to better respond to the CC challenge.

## 38.2 Materials and Methods

Quantitative methods were used to collect and analyze the study data. In particular, a questionnaire was used to collect information in a survey involving 310 individuals. This study's population was the general public in Dammam City, and respondents were recruited through mailing lists, websites, and social media accounts. In this study, only individuals who were willing to participate were included. Regarding validity, the questionnaire was pretested in a pilot study involving a small group of respondents. The questionnaire consists of the following sections: demographic information, respondents' CC knowledge, and strategies. All filled out questionnaires were double-checked to ensure no missing information, and any error was corrected immediately before being entered into the computer. Using a special coding system, the data were analyzed using the Statistical Package for Social Sciences version 21.

## 38.3 Results

In this study, questionnaires were distributed among 310 respondents to determine their CC knowledge, of which 190 (61.3%) were female. Most of the respondents, 93 (30%), were between 20 and 30 years, 131 (42.3%) each between 30 and 40 years and between 40 and 50 years, and 20 (6.5%) between 50 and 60 years. Only 12 (3.9%) were older than 60 years. Most of the respondents were single, 168 (54.2%), and had a college or university degree, 145 (47%). Only 20 (6%) had a postgraduate degree (Table 38.1).

### *38.3.1 Respondents' Knowledge of Climate Change*

There were four questions related to CC knowledge. As shown in Table 38.2, most of the respondents, 178 (57.4%), stated that they had heard of the term "climate change." Among the total respondents, 67 (21.6%) had some idea of the Sustainable Development Goals related to CC, whereas 243 (78.4%) were unaware. The significant causes of CC highlighted were deforestation, 87 (28.1%); population growth, 75 (24.2%); vehicles, 55 (17.7%); urbanization, 58 (18.7%), and GHGs, 35 (11.3%).

**Table 38.1** Respondents' demographic information

| Variable | Category | Frequency (Percentage) |
|---|---|---|
| Sex | Male | 120 (38.7) |
| | Female | 190 (61.3) |
| Age | 20–30 | 93 (30) |
| | 30–40 | 131 (42.3) |
| | 40–50 | 54 (17.4) |
| | 50–60 | 20 (6.5) |
| | > 60 | 12 (3.8) |
| Marital status | Single | 168 (54.2) |
| | Married | 97 (31.3) |
| | Divorced | 45 (14.5) |
| Education | Primary | 20 (7) |
| | Secondary | 85 (27) |
| | Intermediate | 40 (13) |
| | College/university | 145 (47) |
| | Postgraduate | 20 (6) |

**Table 38.2** Respondents' knowledge of CC

| Questions | Answers | Frequency (Percentage) |
|---|---|---|
| Have you ever heard of the term "CC"? | Yes | 178 (57.4) |
| | No | 132 (42.6) |
| Do you have any idea of the impact of CC on the environment? | Yes | 96 (31.0) |
| | No | 214 (69.0) |
| Do you know the Sustainable Development goals related to CC? | Yes | 67 (21.6) |
| | No | 243 (78.4) |
| In your opinion, what is the cause of CC? | Population growth | 75 (24.2) |
| | Greenhouse gases | 35 (11.3) |
| | Vehicles | 55 (17.7) |
| | Deforestation | 87 (28.1) |
| | Urbanization | 58 (18.7) |

### 38.3.2 Respondents' Levels of Concern Regarding Climate Change

Regarding CC concern level, most of the respondents, 208 (67%), were very worried. Seventy-one (23%) were less worried, 22 (7%) were not worried, and 9 (3%) stated that they did not care (Fig. 38.1).

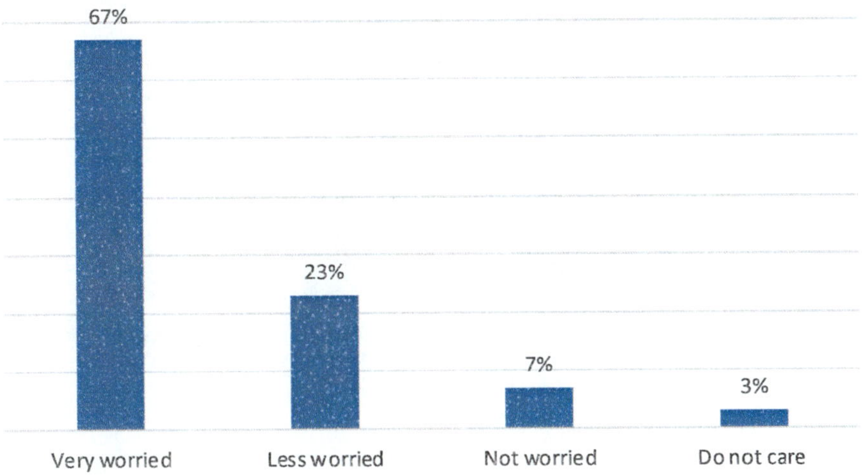


**Fig. 38.1** Respondents' levels of concern about CC

Each respondent's overall knowledge was determined. Only a small number, 44 (14%), was knowledgeable regarding CC causes and effects, whereas 102 (33%) were found to have no such knowledge. Among all respondents, 65 (21%) had knowledge of CC causes but not effects, as shown in Fig. 38.2.

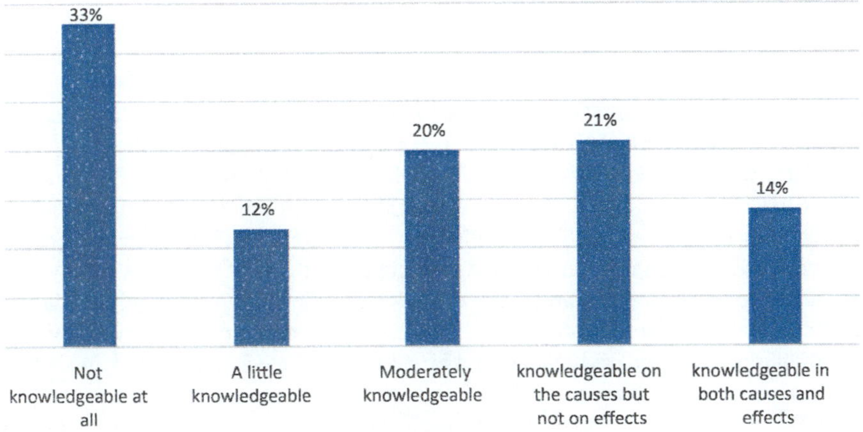


**Fig. 38.2** Knowledge of the effects and causes of CC

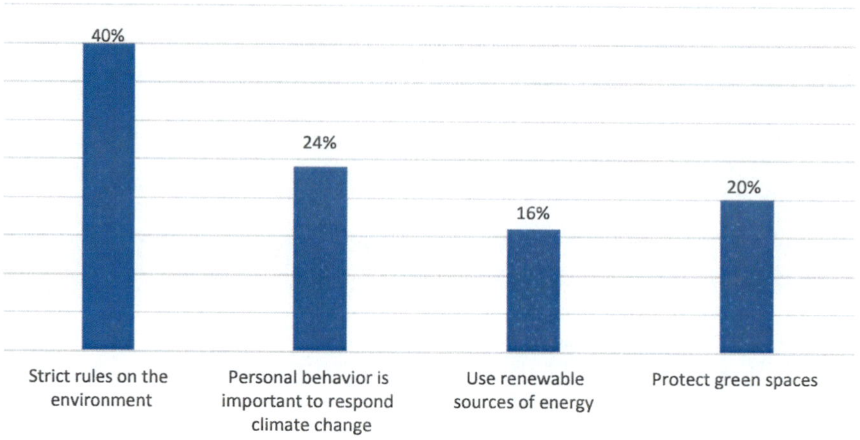


**Fig. 38.3** Strategies to respond the impact of CC

### *38.3.3 Strategies to Reduce the Effects of Climate Change*

Regarding strategies to reduce the impact of CC, 124 (40%) respondents stated that strict rules on the environment would be effective, while 75 (24%) highlighted the importance of personal behavior. Fifty (16%) respondents stated that renewable sources of energy could help mitigate the impact of CC (Fig. 38.3).

## 38.4 Discussion

Climate awareness efforts targeting the general public may prove to be an effective strategy for enhancing and instilling adaptive behaviors and addressing the challenges associated with CC (Elsharkawy et al. 2023; Ghanem 2022; Sarewitz 2011). Variables related to CC, such as knowledge and awareness, are multidimensional and complex. Determining the general public's knowledge gap in this context is important to ensure proper public involvement and mitigate the adverse effects of CC. This survey was conducted to determine people's CC knowledge and the coping strategies to address the impact of CC. There is a consensus among the scientific community that CC is mainly human-induced and related to a broad range of environmental changes that may adversely affect human health. Most of the respondents were quite concerned about the impact of CC. The results of the present study demonstrate public concern for CC, which calls for an effective targeted strategy with a comprehensive action plan for a larger population to encourage the willingness to act.

In this study, the respondents showed a moderate level of CC understanding, including its cause and effects. The findings are consistent with those of studies conducted in other countries (Ghanem 2022; Sarewitz 2011; Salem et al. 2022).

Only a few people in this study were aware of Sustainable Development Goal 13: Climate Action. This may be because most of the Saudi government's CC policies and initiatives target stakeholders and organizations, with limited involvement of the general public. This may also explain the respondents' poor knowledge of CC strategies.

Another finding of this survey was that only 11.3% of the respondents identified GHGs as the cause of CC, which is alarming. Although a larger percentage associated CC with vehicles, most respondents were unaware that increased GHG and $CO_2$ atmospheric concentrations result in CC. Hence, CC awareness campaigns should particularly promote knowledge of the association between CC and GHGs. The CC prevention strategies found in this study were strict environmental rules, personal behavior, and renewable energy sources. Other studies have produced similar results, indicating the adoption of renewable energy sources as the primary preventive strategy to reduce the impact of CC (Ghanem 2022). In our study, relevant education about CC was found to be the most significant factor that can help change people's behavior regarding CC. Further studies should be conducted to determine people's practices to address CC and the association between knowledge and practice. This would further help in formulating mitigation measures to reduce the impact of CC.

## 38.5 Conclusion

The collective findings demonstrated that the respondents had moderate knowledge of CC causes and effects. The respondents agreed that CC is a present and future problem, and that more CC information is needed. These results indicate the need for an effective CC communication system to increase people's knowledge, whether business or education, and therefore enhance their involvement in current CC mitigation measures worldwide. The survey results will help policymakers in making the decisions needed to reduce the negative impact of CC. Further research is needed at the national level, as this will provide a clear picture of the knowledge and practices of the whole population.

## References

Berardi U (2017) A cross-country comparison of the building energy consumptions and their trends. Resour Conserv Recycl 123:230–241

Bollettino V et al (2020) Public perception of climate change and disaster preparedness: Evidence from the Philippines. Clim Risk Manag 30:100250

Capstick S et al (2015) International trends in public perceptions of climate change over the past quarter century. Wiley Interdisc Rev Clim Change 6(1):35–61

Ebi KL et al (2021) Extreme weather and climate change: population health and health system implications. Annu Rev Public Health 42(1):293–315

Eguiluz-Gracia I et al (2020) The need for clean air: the way air pollution and climate change affect allergic rhinitis and asthma. Allergy 75(9):2170–2184
Elder M, Olsen SH (2019) The design of environmental priorities in the SDG s. Global Pol 10:70–82
Elsharkawy SA, Elsheikh AA, Refaat LAR (2023) Knowledge, perception, and practices regarding climate change among students of Al-Azhar University for Girls in Cairo, Egypt. J Publ Health 1–10
Ghanem A (2022) Assessment knowledge, perception, and behaviors towards climate change among universities youth in Egypt. Athens J Mediterr Stud 8:1–16
Kabir MI et al (2016) Knowledge and perception about climate change and human health: findings from a baseline survey among vulnerable communities in Bangladesh. BMC Public Health 16:1–10
Poortinga W et al (2019) Climate change perceptions and their individual-level determinants: a cross-European analysis. Glob Environ Chang 55:25–35
Salem MR et al (2022) Climate change-related knowledge and attitudes among a sample of the general population in Egypt. Front Publ Health 10:1047301
Sarewitz D (2011) Does climate change knowledge really matter? Wiley Interdisc Rev Clim Change 2(4):475–481
Streimikis J, Baležentis T (2020) Agricultural sustainability assessment framework integrating sustainable development goals and interlinked priorities of environmental, climate and agriculture policies. Sustain Dev 28(6):1702–1712
van Wijk M et al (2020) Perception and knowledge of the effect of climate change on infectious diseases within the general public: a multinational cross-sectional survey-based study. PLoS ONE 15(11):e0241579
Wei Y-M et al (2020) Self-preservation strategy for approaching global warming targets in the post-Paris Agreement era. Nat Commun 11(1):1624
Xu R et al (2020) Wildfires, global climate change, and human health. N Engl J Med 383(22):2173–2181

# Chapter 39
# Perceptions of Forest Experts on the Impact of Wildfires on Ecosystem Services in Portugal

**Renata Pacheco and João Claro**

**Abstract** In Mediterranean Europe, one of the expected consequences of climate change is the intensification of wildfire events. Given the importance of forests in helping regulate climate and the many ecosystem services they provide, it is crucial to identify how wildfires might impact them. In this context, the present work aims to identify the wildfire impacts caused to the ecosystem services in Portugal. This is done through a survey directed to Portuguese fire experts. Using The Economics of Ecosystems and Biodiversity (TEEB) definitions, experts were asked to share their perceptions on the fire impacts to forest ecosystem services in the short and long-term and indicate which services they feel require more policies to mitigate the impacts. The results showed that all ecosystem services are impacted to various degrees and different lengths of time. Regulating services were overall the most affected group and the most in need of specific policies. This study helped identify fire impacts, policy needs, and priorities in the perception of the experts in Portugal, which is valuable for guiding future research in various knowledge fields, especially related to raising awareness about behavioral adaptation to prevent and mitigate wildfire impacts in a changing climate.

**Keywords** Ecosystem services · Expert perception · Impact assessment · Portugal · Wildfire

R. Pacheco (✉) · J. Claro
INESC TEC and Faculdade de Engenharia, Universidade do Porto, Campus da FEUP, Rua Dr. Roberto Frias, 4200-465 Porto, Portugal
e-mail: renatapacheco@fe.up.pt



N. S. Caetano and M. C. Felgueiras (eds.), *The 9th International Conference on Energy and Environment Research*, Environmental Science and Engineering,
https://doi.org/10.1007/978-3-031-43559-1_39

## 39.1 Introduction

Climate change is considered one of the main challenges of the century, affecting many aspects of the environment and society. Among the most notorious consequences predicted for Mediterranean Europe is an increase in wildfire risk, an extension of the fire season, and an intensification of extreme events during the fire season, possibly leading to more intense and frequent fires (Raftoyannis et al. 2014; Ruffault et al. 2020). The Mediterranean region is considered a biodiversity hotspot, and its forests have provided numerous ecosystem services to human societies for millennia (Roces-Díaz et al. 2021). The European Commission acknowledges that the biodiversity crisis and the climate crisis are linked. Climate change accelerates the degradation of the natural world through more extreme climatic events, such as wildfires, while the unsustainable use and loss of nature are drivers of climate change (European Commission 2020). In this context, there is increasing consensus that forestry is an effective way to mitigate climate change. Forests can be valuable in reducing greenhouse gas emissions and increasing their absorption, storing carbon-containing chemicals for indefinite periods, acting as a carbon sink (Liu and Wu 2017). Given the importance of the ecosystem services forests provide, the increasing number of wildfires is turning post-fire forest management into an ever more crucial topic (Olsen and Shindler 2010), as over the past two decades, fires have progressively impacted human lives, assets, and ecosystem services, such as air quality and long-term carbon storage (Moritz et al. 2014). Within Mediterranean Europe, Portugal has the highest number of wildfires and the second with the larger burnt area (Parente et al. 2018).

In Portugal, the Institute for Nature Conservation and Forests (ICNF) divides the recovery of burnt areas into three phases: "emergency stabilization," "restoration and rehabilitation," and "long-term."

Emergency stabilization happens right after the fire or even during the fire-fighting phase, and it aims to control erosion and protect the hydrographic network, infrastructure, and the most sensitive stations and habitats. Restoration and rehabilitation take place in the two years following the fire. The damage and the response of ecosystems are assessed, salvage logging, biophysical recovery actions, and even reforestation of more sensitive areas might be performed. Finally, in the long-term phase, definitive recovery projects are planned and implemented; usually, three years after the fire has occurred.

The Economics of Ecosystems and Biodiversity (TEEB) define ecosystem services as "the direct and indirect contributions of ecosystems to human well-being." One of TEEB's goals is to provide a better understanding of the economic significance of the loss of these services and the consequences of policy inaction on halting biodiversity loss at various scales. This framework provides the scientific basis for the economics of ecosystems and biodiversity and results that are useful for policymakers, administrators, businesses, or consumers (Fisher and Christie 2010).

In fire-prone landscapes, recognizing the nuanced roles that fire plays in ecosystem services is essential to distinguish the trade-offs amongst people's needs and desires and the sustainability of complex socioecological systems (Vukomanovic and Steelman 2019). However, despite the growing literature on the subject, the roles that fire plays in ecosystem services have not yet been thoroughly identified (Vukomanovic and Steelman 2019).

In this context, the present work aims to identify the wildfire impacts caused to the ecosystem services in Portugal. This is done through an online survey directed to Portuguese fire experts. They were asked to share their perceptions on the fire impacts on forest ecosystem services in the short and long-term. They were also asked to indicate which services they believe more policies are needed to mitigate the fire impacts. The methods employed and results are presented in the following sections.

## 39.2 Materials and Methods

Experts are often required to share their knowledge in informal ways to raise awareness about wildfires and prevention measures. Despite its importance, this information is not always incorporated in scientific research.

In this sense, surveys are frequently used in studies that have the object of eliciting perceptions from expert groups, which allows their tacit knowledge to be brought to the academic environment (Raftoyannis et al. 2014; Molina et al. 2018), and thus is an adequate method for the goal of this study.

The survey was conducted via an electronic form sent to official bodies of wildfire experts. It had an introduction to the questionary and two sections. The first described the respondent in terms of: Institution (in which they work); and Position (they occupy currently).

In the second section, three questions were asked: From your perspective, which of the following ecosystem services are impacted by wildfires in Portugal in the short-term?; and From your perspective, which of the following ecosystem services are impacted by wildfires in Portugal, in the long-term (more than 3 years after the fire)?. All the 22 TEEB ecosystem services were enumerated following these questions. A Likert scale correlated with each service, from which the respondent had to decide from 0 (no impact) to 5 (severe impact).

Lastly, to gain perspective on Portugal's current post-fire policy state, we asked the following question: From your perspective, is there a lack of policies or guidelines to address the impacts to one of these services? Select as many alternatives as you see fit. Once again, all 22 TEEB ecosystem services were listed, but for this question "checkboxes" were used, and respondents were instructed to select as many services as they believed were relevant. In the next section, the results of the survey are presented.

## 39.3 Results and Discussion

The survey was sent to three different institutions with roles in wildfire management and policy in Portugal, and it was made available for two weeks. A total of 12 experts answered the survey. Table 39.1 shows the characterization of the respondents.

Figure 39.1 shows respondents' answers for the first two questions regarding the short and long-term impacts.

The survey was directed to ICNF; AGIF—Agency for Integrated Rural Fire Management, whose institutional mission is the "planning, strategic coordination and assessment of the integrated rural fire management system"; and ForestWISE, an R&D institution with emphasis on interdisciplinary knowledge transfer that focuses on rural fires and the valorization of forest (market and non-market) products and services. One respondent was also associated with another institution, the University of Trás-os-Montes and Alto Douro (UTAD). Most of the respondents were practitioners (6), followed by managers (3) and researchers (2). One respondent occupied a different position and answered as "Other." More than half of the respondents work at ICNF (7), which is the main responsible for dealing with post-fire issues.

Regarding the second section of the survey, Fig. 39.1 presents the answers to the two questions on the impacts of fire on forest ecosystem services, in the short and long (more than 3 years after the fire) terms, clustered around the categories of ecosystem services: Provisioning, Regulating, Habitat, and Cultural and Amenity. As previously mentioned, in the Likert-type scale used, "0" represents "No impact," and "5" represents "Severe impact." The percentages refer to the number of respondents to choose that level of impact.

Regarding the Provisioning services, the results indicate that the impacts tend to diminish somewhat over time but remain mostly on the side of severe, even three years after the fire. It is noticeable that the impact on the "Raw materials" service is one of the most significant and enduring, with half the respondents believing that the impact is still severe in the long-term.

The Regulating services were overall the most impacted in the short-term, and most experts thought that they improve their condition in the period following the wildfire. As for the Habitat services, most experts signaled that despite being quite severely impacted, they recover a bit in the period after the wildfire. The Cultural and Amenity services have more varied results. For some, the impacts are thought

**Table 39.1** Characterization of the respondents of the survey

| Institution | Position | | | |
|---|---|---|---|---|
| | Manager | Practitioner | Researcher | Other |
| AGIF | 1 | – | – | – |
| ForestWISE | – | 1 | 1 | 1 |
| ICNF | 2 | 5 | – | – |
| UTAD | – | – | 1 | – |

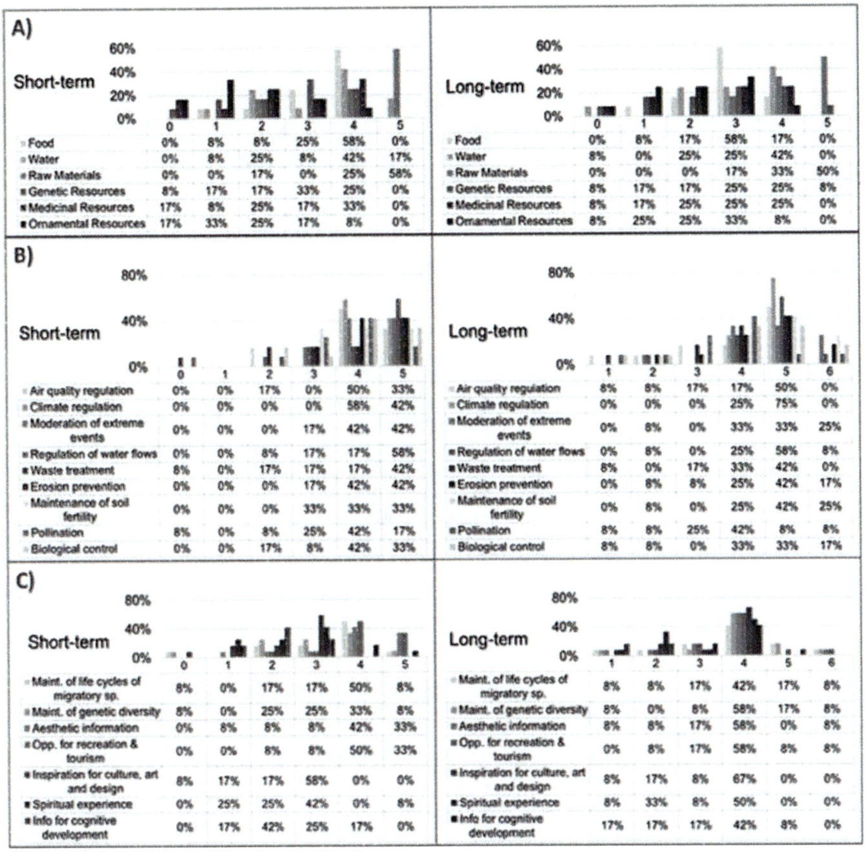


**Fig. 39.1** Short and long-term fire impacts. Zero means "No impact" and five "Severe impact." The letters on the top left refer to: **a** Provisioning Services; **b** Regulating Services; **c** Habitat, and Cultural and Amenity Services

to increase over time; for others, there is an improvement, while a few continue essentially unaltered. Regarding post-fire policies, Fig. 39.2 shows the answers to the third question of the survey.

Given the wide range of impacts and their persistence following wildfires, policies can play a vital role in restoring the environment. Figure 39.2 indicates the policy needs identified by the respondents. It is somewhat coherent that since all services were considered to be impacted by wildfires, all were deemed to need specific policies.

It can be highlighted that Regulating services seemed to be thought of as the group most in need of specific policies. This is coherent with the fact that this group was perceived as the most impacted by the wildfires, both in the short and long-term.

Despite the focus of this study being on fire impacts, these policies might not necessarily aim at recovering the impacted land. In this sense, European guidelines

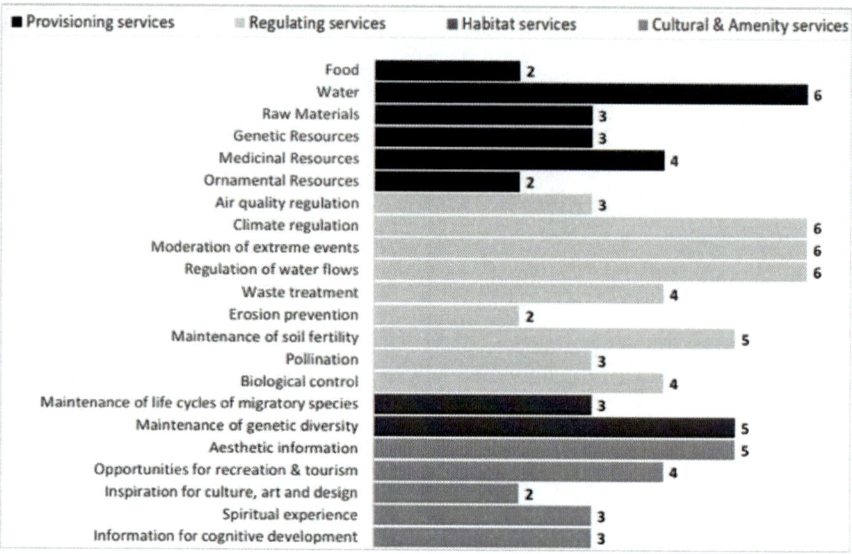


**Fig. 39.2** Impacted ecosystem services that need specific policies or guidelines, according to the respondents' perceptions

are shifting their focus from fire suppression to prevention and increasing the awareness and preparedness of populations at risk (Faivre et al. 2018). Therefore, it would make sense for environmental policy to follow the same trend as it is better to avoid fires and their impacts than mitigate them.

## 39.4 Conclusions

Climate change is considered one of the most pressing challenges for humanity, and it impacts the natural and built environment in many deleterious ways. In Portugal's case, this is most prominent in the form of increasing wildfire occurrences.

The present study used a survey to gather expert knowledge regarding the impacts caused by wildfires on the forest ecosystem services. The results indicate that all the 22 TEEB ecosystem services are impacted to various degrees and different lengths of time.

The Regulating services were overall the most affected group and presented the most lasting impacts. Not surprisingly, this was also the group of services that are considered to require more remediation policies. Namely, Climate regulation has an important role in this context, as forests help balance carbon cycles (Liu and Wu 2017), which in turn tend to decrease wildfire occurrences.

This study helped identify the most notorious wildfire impacts, policy needs, and priorities in the perception of Portuguese fire experts. This information will be further investigated in upcoming studies, and it is valuable for guiding future research in various knowledge fields, aiming to value and protect biodiversity, prevent wildfires, and avoid climate change. Furthermore, these results can serve as the basis for future work to address the issue of fire prevention over mitigation and provide knowledge to raise awareness in view of behavioral adaptation to prevent wildfires, among other pressing concerns in the context of a changing climate.

**Acknowledgements** We would like to acknowledge all twelve respondents of the survey for their kind and valuable share of knowledge that made this study possible.

**Funding** This work was financially supported by Operation NORTE-08–5369-FSE-000045 co-funded by the European Social Fund (FSE) through NORTE 2020—Programa Operacional Regional do NORTE. This work was also financed by National Funds through the Portuguese funding agency, FCT—Fundação para a Ciência e a Tecnologia within project: UIDB/50014/2020.

## References

European Commission (2020) EU Biodiversity Strategy for 2030: Bringing nature back into our lives. Brussels

Faivre N, Cardoso Castro Rego FM, Moreno Rodríguez JM et al (2018) Forest fires—Sparking firesmart policies in the EU

Fisher B, Christie M (2010) The economics of ecosystem and biodiversity: ecological and economic foundations

Liu J, Wu F (2017) Forest carbon sequestration subsidy and carbon tax as part of China's forestry policies. Forests 8:1–15. https://doi.org/10.3390/f8030058

Molina JR, Moreno R, Castillo M, Rodríguez y Silva F (2018) Economic susceptibility of fire-prone landscapes in natural protected areas of the southern Andean Range. Sci Total Environ 619–620:1557–1565. https://doi.org/10.1016/j.scitotenv.2017.11.233

Moritz MA, Batllori E, Bradstock RA et al (2014) Learning to coexist with wildfire. Nature 515:58–66. https://doi.org/10.1038/nature13946

Olsen CS, Shindler BA (2010) Trust, acceptance, and citizenagency interactions after large fires: influences on planning processes. Int J Wildl Fire 19:137–147. https://doi.org/10.1071/WF08168

Parente J, Pereira MG, Amraoui M, Tedim F (2018) Negligent and intentional fires in Portugal: spatial distribution characterization. Sci Total Environ 624:424–437. https://doi.org/10.1016/j.scitotenv.2017.12.013

Raftoyannis Y, Nocentini S, Marchi E et al (2014) Perceptions of forest experts on climate change and fire management in European Mediterranean forests. Iforest 7:33–41. https://doi.org/10.3832/ifor0817-006

Roces-Díaz JV, Vayreda J, De Cáceres M et al (2021) Temporal changes in Mediterranean forest ecosystem services are driven by stand development, rather than by climate-related disturbances. For Ecol Manage 480:118623. https://doi.org/10.1016/j.foreco.2020.118623

Ruffault J, Curt T, Moron V et al (2020) Increased likelihood of heat-induced large wildfires in the Mediterranean Basin. Sci Rep 10:1–9. https://doi.org/10.1038/s41598-020-70069-z

Vukomanovic J, Steelman T (2019) A systematic review of relationships between mountain wildfire and ecosystem services. Landsc Ecol 34:1179–1194. https://doi.org/10.1007/s10980-019-00832-9

# Chapter 40
# Coated Catalyst Plates for Effective Degradation of Industrial Effluents via Innovative Photocatalytic Reactor

**Mahmoud Samy, Mohamed Gar Alalm, Manabu Fujii, and Mona G. Ibrahim**

**Abstract** The evaluation of different catalysts performance for the degradation of effluents from pharmaceutical and agrochemical industries was performed. The response surface method was used to specify the optimum values of operating parameters (e.g., pH and flow rate). The optimum operating parameters were pH = 6.8 and flow rate = 108.6 mL/min for pharmaceutical wastewater, whereas the pH and flow rate values of 6.5 and 108.6 mL/min, respectively achieved the highest degradation efficiency for agrochemical wastewater using a composite of carbon nanotubes (CNTs) and lanthanum vanadate ($LaVO_4$). The mineralization rates of pharmaceutical wastewater were 89.5, 84, 74, 69.8 and 60.5% under five succeeding runs using CNTs/$LaVO_4$, whereas the total organic carbon (TOC) removal efficiencies of agrochemical wastewater were 94.3, 93.1, 85.8, 76.2 and 73.2% in the five following runs. The degradation rates were 0.0114, 0.0156 and 0.0084 $min^{-1}$ using a composite of

M. Samy (✉) · M. Fujii · M. G. Ibrahim
Environmental Engineering Department, Egypt-Japan University of Science and Technology, (E-Just), New Borg El Arab City, Alexandria 21934, Egypt
e-mail: mahmoud.samy@ejust.edu.eg

M. Fujii
e-mail: fujii.m.ah@m.titech.ac.jp

M. G. Ibrahim
e-mail: mona.gamal@ejust.edu.eg

M. Samy · M. G. Alalm
Public Works Engineering Department, Faculty of Engineering, Mansoura University, Mansoura 35516, Egypt
e-mail: m_gar_alalm@mans.edu.eg

M. Fujii
Department of Civil and Environmental Engineering, Tokyo Institute of Technology, Meguro-Ku, Tokyo 152-8552, Japan

M. G. Ibrahim
Environmental Health Department, High Institute of Public Health, Alexandria University, Alexandria 21544, Egypt



N. S. Caetano and M. C. Felgueiras (eds.), *The 9th International Conference on Energy and Environment Research*, Environmental Science and Engineering,
https://doi.org/10.1007/978-3-031-43559-1_40

Matériaux de l'Institut Lavoisier metal organic framework-53 (MIL-53(Al)) and zinc oxide (ZnO) (MIL-53(Al)/ZnO), CNTs/$LaVO_4$ and pristine titanium dioxide ($TiO_2$), respectively in the case of pharmaceutical wastewater, whereas the degradation rates were 0.017, 0.0196 and 0.0104 $min^{-1}$ using MIL-53(Al)/ZnO, CNTs/$LaVO_4$ and pure $TiO_2$, respectively in the case of agrochemical wastewater.

**Keywords** Operating conditions · Photodegradation · Photocatalysis · Real wastewaters

## 40.1 Introduction

The frequent need of pharmaceuticals for the medication of animals and humans as well as the wide use of pesticides for pests fight have attracted wide attention due to the environmental concerns of the release of pharmaceuticals and pesticides to water streams without sufficient treatment (Samy et al. 2023a; Mensah et al. 2022). The existence of pharmaceuticals in water streams leads to the growth of new bacteria that are able to resist antibiotics, whereas the spread of pesticides in surface water represents a great threat to humans and animals (Samy et al. 2023b). Due to the biopersistence of pharmaceuticals and pesticides, conventional biological treatment processes cannot be considered viable for the degradation of these pollutants (Pinna et al. 2021). Moreover, coagulation-sedimentation processes cannot effectively remove pharmaceuticals and pesticides and these processes generate large amounts of sludge requiring post treatment (Hussein et al. 2018). Moreover, adsorption and membrane filtration technologies have been employed for the degradation of pesticides and antibiotics, but adsorption and membrane technologies have some shortcomings such as the ceaseless regeneration of sorbents and their high cost as well as the nonstop need of energy, fouling and scaling of membranes (Salama et al. 2022). According to the drawbacks of the conventional treatment processes, researchers have focused on the development of new techniques that can efficiently transform bio-resistant toxins to safe compounds. The evaluation of the treatment of pharmaceutical and pesticide solutions has been conducted using advanced oxidation techniques (AOPs) (Wang et al. 2020).

Photocatalysis process has proved its superiority among other AOPs towards the aforementioned contaminants. Moreover, the degradation of contaminants can be achieved via photocatalysis using solar light. The mechanism of reactive oxygen species (ROS) generation has been discussed in detail in the literature (Samy et al. 2023c). The generated ROS are capable of the oxidation of contaminants to non-toxic intermediates. The commonly used semiconductors (e.g., titanium dioxide ($TiO_2$) and zinc oxide (ZnO)) suffer from the problems of wide bandgap and incessant concourse of charge carrier (Pinna et al. 2021). Different semiconductors were synthesized to overcome the drawbacks of conventional catalysts such as a composite of carbon nanotubes (CNTs) and lanthanum vanadate ($LaVO_4$) (CNTs/$LaVO_4$) and a composite of Matériaux de l'Institut Lavoisier metal organic framework-53 (MIL-53(Al)) and zinc oxide (ZnO) (MIL-53(Al)/ZnO). Vanadate materials such as $LaVO_4$ can be excited using visible light; however, these materials face the problem of fast concourse rate of holes and electrons (Samy et al. 2020a). Therefore, carbon nanotubes (CNTs) were composited with the pristine $LaVO_4$ to work as an electron acceptor guaranteeing the detachment of charge carrier.

The employment of metal–organic-frameworks (MOFs) as photocatalysts has been investigated for the degradation of pollutants due to their large surface area and the availability of pores (Li et al. 2020). The degradation of various pollutants was attained using MIL-53(Al) due to the unique previously mentioned characteristics of MOFs (Li et al. 2020). MIL-53(Al) was composited with the pristine ZnO to reduce the bandgap and the connection probability between charge carriers. The majority of studies have dealt with the suspended-mode of photocatalysts; however, this mode reported some limitations according to our previous work (Samy et al. 2020b). The synthesized catalysts were loaded on glass plates to face the problems of suspended-mode using polysiloxane due to its characteristics as well as the problems of other attachment methods (Samy et al. 2020b). Furthermore, previous studies only concentrated on the preparation of new photocatalysts and the evaluation of their performances for the degradation of various pollutants (Loeb et al. 2019). The literature did not pay attention to the design of photocatalytic reactors widening the gap between application and research (Loeb et al. 2019). Therefore, the construction of an innovative reactor was carried out in order to treat real wastewaters paving the way for the application of photocatalysis process.

In this study, various catalysts were employed to treat effluents from agrochemical and pharmaceutical industries using the designed photo-reactor. The specification of optimum values of operating parameters was conducted via response surface method (RSM). The assessment of the performance of the coated-plates under successive runs was conducted. Moreover, the degradation kinetics were studied.

## 40.2 Materials and Methods

### *40.2.1 Experimental Procedures*

Pure $TiO_2$ was purchased from Sigma-Aldrich. The chemicals used in the preparation of CNTs/$LaVO_4$ and MIL-53(Al)/ZnO were mentioned in our previous studies (Samy et al. 2020a, c). The preparation as well as the characterization of the aforementioned catalysts were discussed elsewhere (Samy et al. 2020a, c). All experiments were performed in an innovative photoreactor as shown in Fig. 40.1. The lengths, widths and heights of different components of the reactor, positions of coated plates and light source, were previously discussed in our previous study (Samy et al. 2020b). Moreover, the steps needed to prepare the coated plates were previously explained elsewhere (Samy et al. 2021). The volume of contaminated solution was 200 mL and the recirculation of flow continued for 2 h. The illumination time was normalized based on the equation mentioned in our previous work (Samy et al. 2020b). The pharmaceutical and agrochemical wastewaters were compiled from two companies located in Alexandria. The total organic carbon (TOC) values of the industrial effluents during treatment were estimated using total organic carbon analyzer (Shimadzu TOC-L, Japan). Before TOC analysis, all samples were diluted to 10, so the obtained TOC values were multiplied by 10 to have the actual values. The compounds in the real pharmaceutical and agrochemical wastewaters were recognized using liquid chromatography-tandem mass spectroscopy as mentioned in our previous study (Samy et al. 2021). The compounds (e.g., entecavir, desloratadine, gliclazide, etoricoxib, lansoprazole, tadalafil, sitagliptin phosphate, irbesartan, losartan potassium, glibenclamide, ferric saccharate, metformin, pregabalin, ciprofloxacin, gemifloxacin, acemetacin, fluticasone propionate, atorvastatin, moxifloxacin, dapoxetine, febuxostat, levofloxacin, montelukast sodium, azithromycin, fluconazole, clopidogrel, glimepiride, escitalopram, cefaclor) were detected in pharmaceutical wastewater, whereas the compounds (e.g., methomyl, fenitrothion, malathion, lufenuron, copper oxychloride, dimethoate, ethoprophos, chlorpyrifos-methyl, alpha-cypermethrin, mancozeb, penconazole, hexaflumuron, carbendazim, profenofos, abamectin) were found in agrochemical wastewater.

**Design of Experiments**. RSM was used to accomplish outstanding association between studied parameters that cannot be fulfilled using usual experiment design. A specific number of experiments was carried out to attain the required interaction between the studied parameters. Table 40.1 shows the coded and actual values of pH and flow rate. The linkage between the removal efficiency, pH and flow rate can

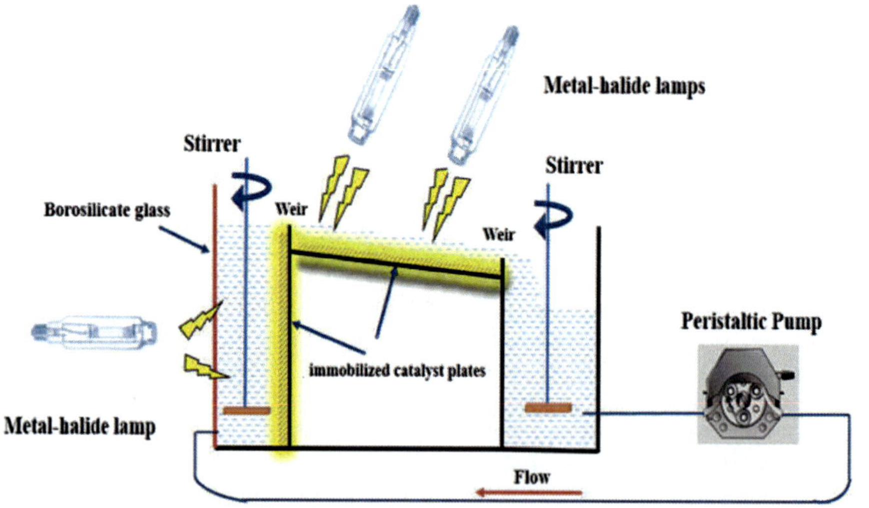


**Fig. 40.1** Details of the photoreactor

Table 40.1 Coded and actual values of pH and flow rate

| Operating conditions | Units | Levels | | |
|---|---|---|---|---|
| | | −1 | 0 | 1 |
| pH | – | 3 | 7 | 11 |
| Flow rate | mL/min | 50 | 100 | 150 |

be portrayed by a quadratic equation as explained elsewhere c. The strength and the regression study of the models were managed by analysis of variance (ANOVA) and Minitab © software, respectively. The allocation of the maximum removal efficiency as an aim in the software resulted in the specification of optimum values of studied parameters.

## 40.3 Results and Discussion

### 40.3.1 *Comparison of the Pristine $TiO_2$, CNTs/$LaVO_4$ and MIL-53(Al)/ZnO*

Figure 40.2 demonstrates the removal percentages of TOC of wastewaters from pharmaceutical and agrochemical industries using CNTs/$LaVO_4$, MIL-53(Al)/ZnO and pristine $TiO_2$. The initial obtained TOC values were 15.5 ± 0.5 mg/L for agrochemical wastewater and 17 ± 0.5 mg/L for pharmaceutical wastewater. The highest TOC removal efficiencies of 89.5% and 94.3% were attained using CNTs/$LaVO_4$ for pharmaceutical and agrochemical wastewaters, respectively. The superiority of the aforementioned catalyst was as a result of its aptitude to absorb visible light. Moreover, the compositing of CNTs with $LaVO_4$ participated in the decline of the joint between charge carrier (Samy et al. 2020a). In all following experiments, CNTs/$LaVO_4$ was employed to degrade real wastewaters.

**Optimization of Operating Parameters.** Equation (40.1, 40.2) shows the connection between the TOC removal ratio of pharmaceutical and agrochemical wastewaters and pH and flow rate:

$$R_1(\%) = 4.71 + 15.88\,M + 0.532\,Z - 1.180\,M^2 - 0.002494\,Z^2 + 0.00144\,MZ \tag{40.1}$$

$$R_2(\%) = 13.23 + 13.265\,M + 0.6915\,Z - 1.0144\,M^2 - 0.003142\,Z^2 - 0.00150\,MZ \tag{40.2}$$

where $R_1$ and $R_2$ are the TOC mineralization ratios of pharmaceutical and agrochemical wastewaters, respectively. M and Z refer to the pH of the solution and the flow rate (mL/min), respectively. The validity of the models for the application was

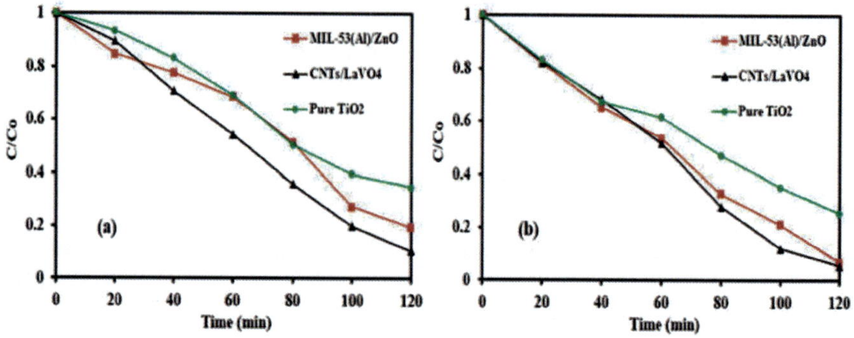


**Fig. 40.2** TOC removal efficiencies using MIL-53(Al)/ZnO, CNTs/LaVO$_4$ and pure TiO$_2$ (pH = 7; flow rate = 100 mL/min) for **a** pharmaceutical wastewater and **b** agrochemical wastewater

evaluated by estimating P-values and F-values using analysis of variance (ANOVA). The small P and high F values assured the appropriateness of the model.

The influence of pH and flow rate on the removal percentage of pharmaceutical and agrochemical wastewaters can be depicted by contour lines as presented in Fig. 40.3. The highest degradation efficiency was accomplished at pH of 6.8 and 6.5 in the case of pharmaceutical and agrochemical wastewaters, respectively. At low pH values, the removal efficiency may decline due to the decrease of hydroxyl ions causing the decline of ROS generation. In alkaline medium, the formed carbonates from the trap of carbon dioxide in the solution disabled the activity of hydroxyl radicals. So, neutral medium was favorable for the degradation of wastewaters (Chen et al. 1998).

The optimum flow rate was 108.6 mL/min for pharmaceutical and agrochemical wastewaters. Generally, the increase of flow rate enhanced the degradation percentage due to the achievement of considerable interaction between catalyst particles and flow layers. However, the rise of flow rates values above 108.6 mL/min for pharmaceutical and agrochemical wastewaters inhibited the degradation efficiency due to the

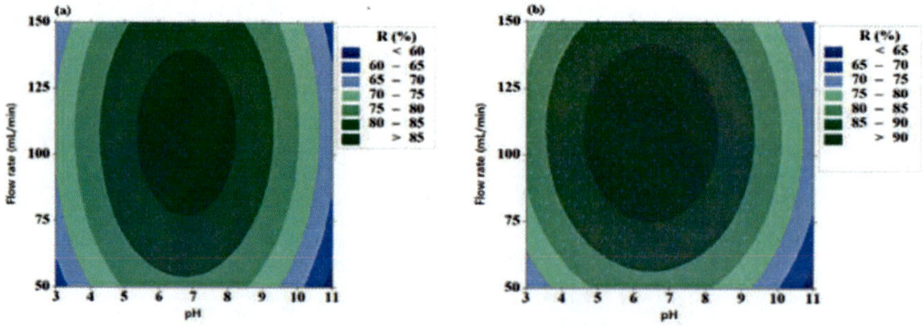


**Fig. 40.3** Effect of pH and flow rate on the degradation percentages of **a** pharmaceutical wastewater and **b** agrochemical wastewater

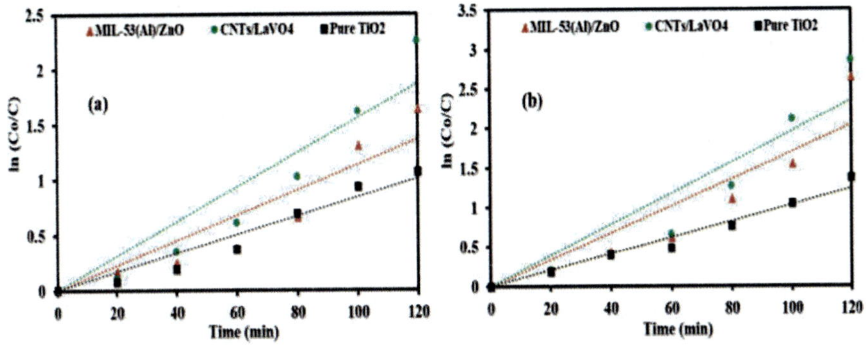


**Fig. 40.4** Degradation kinetics of MIL-53(Al)/ZnO, CNTs/$LaVO_4$ and pure $TiO_2$ (pH = 7; flow rate = 100 mL/min) for **a** pharmaceutical wastewater and **b** agrochemical wastewater

abrasion of coated catalyst particles. Moreover, at greater flow rate values, the time of contact between flow layers and catalyst particles decreased.

*Degradation kinetics of $TiO_2$, CNTs/$LaVO_4$ and MIL-53(Al)/ZnO and Stability of Coated Catalyst Plates.* Langmuir–Hinshelwood model was employed to determine the degradation rates using MIL-53(Al)/ZnO, CNTs/$LaVO_4$ and bare $TiO_2$ as shown in Fig. 40.4. Equation (40.3) describes the relation between degradation rate ($K_r$), initial TOC ($C_o$) and remaining TOC (C) at time (t):

$$\mathrm{Ln}\left(\frac{C_o}{C}\right) = k_r\, t \tag{40.3}$$

The measured degradation rates of pharmaceutical wastewater were 0.0114, 0.0156 and 0.0084 $min^{-1}$ using MIL-53(Al)/ZnO, CNTs/$LaVO_4$ and pure $TiO_2$, respectively, whereas the degradation rates of agrochemical wastewater were 0.017, 0.0196 and 0.0104 $min^{-1}$ using MIL-53(Al)/ZnO, CNTs/$LaVO_4$ and pure $TiO_2$, respectively. CNTs/$LaVO_4$ showed the highest degradation rate among other catalysts. These results are in congruent with the deductions from the previously discussed comparison among different catalysts.

The TOC removal ratios of pharmaceutical and agrochemical effluents in the five succeeding runs using coated CNTs/$LaVO_4$ plate were estimated as shown in Fig. 40.5. The removal efficiencies were 89.5, 84, 74, 69.8 and 60.5% in the case of pharmaceutical wastewater in the five following runs, whereas 94.3, 93.1, 85.8, 76.2 and 73.2% were the mineralization ratios of TOC for agrochemical wastewater in the five succeeding runs. The aforementioned results confirmed the stability of the prepared coated plates as well as the excellent bind between the catalyst particles and the plates. However, the reduction in the removal efficiency of TOC was due to the loss of active sites in the following runs.

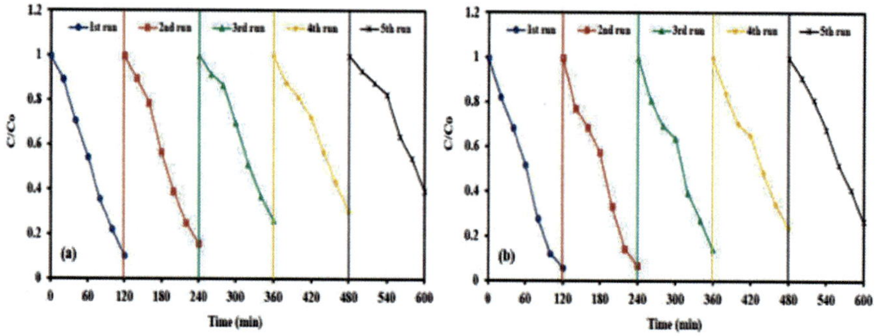


**Fig. 40.5** TOC removal efficiencies in five consecutive runs using CNTs/$LaVO_4$ (pH = 7; flow rate = 100 mL/min) for **a** pharmaceutical wastewater and **b** agrochemical wastewater

## 40.4 Conclusions

The highest TOC removal efficiency was attained using CNTs/$LaVO_4$ in the case of pharmaceutical and agrochemical wastewaters. The optimum values of pH (6.8 for pharmaceutical wastewater and 6.5 for agrochemical wastewater) and flow rate (108.6 mL/min for pharmaceutical and agrochemical wastewaters) were estimated using RSM. CNTs/$LaVO_4$ showed the highest degradation rates of 0.0156 and 0.0196 $min^{-1}$ among other catalysts in the case of pharmaceutical and agrochemical wastewaters, respectively. The removal efficiencies of TOC were 89.5, 84, 74, 69.8 and 60.5% in the five succeeding runs in the case of pharmaceutical wastewater, while the degradation ratios were 94.3, 93.1, 85.8, 76.2 and 73.2% in the five following runs for agrochemical wastewater.

**Acknowledgements** The first author is thankful to Egyptian Ministry of Higher Education (MOHE) for providing him a full scholarship.

**Funding** Egypt-Japan University of Science and Technology saved the required equipment and tools for this research.

## References

Chen D, Ray AK, Chen D, Ray AK (1998) Photodegradation kinetics of 4-nitrophenol in $TiO_2$ suspension. Water Res 32:3223–3234

Hussein M, Gar Alalm M, Eletriby H (2018) Removal of natural organic matter and turbidity in drinking water by modified flaxseed husk. Int J Environ Sci Nat Resour 11:1–4

Li Y, Wang Y, He L, Meng L, Lu H, Li X (2020) Preparation of poly(4-vinylpyridine)-functionalized magnetic Al-MOF for the removal of naproxen from aqueous solution. J Hazard Mater 383:121144

Loeb SK, Alvarez PJJ, Brame JA, Cates EL, Choi W, Crittenden J, Dionysiou DD, Li Q, Li-Puma G, Quan X, Sedlak DL, David Waite T, Westerhoff P, Kim JH (2019) The technology horizon for photocatalytic water treatment: sunrise or sunset? Environ Sci Technol 53:2937–2947

Mensah K, Samy M, Mahmoud H, Fujii M, Shokry H (2022) Rapid adsorption of sulfamethazine on mesoporous graphene produced from plastic waste: optimization, mechanism, isotherms, kinetics, and thermodynamics. Int J Environ Sci Technol. https://doi.org/10.1007/s13762-022-04646-2

Pinna M, Binda G, Altomare M, Marelli M, Dossi C, Monticelli D, Spanu D, Recchia S (2021) Biochar nanoparticles over $TiO_2$ nanotube arrays : A green co-catalyst to boost the photocatalytic degradation of organic pollutants. Catalysts 11(9):1048. https://doi.org/10.3390/catal11091048

Salama S, Samy S, Shokry H, El-Subruiti G, El-Sharkawy G, Hamad H, Elkady M (2022) The superior performance of silica gel supported nano zero-valent iron for simultaneous removal of Cr (VI). Sci Rep 12:1–19. https://doi.org/10.1038/s41598-022-26612-1

Samy M, Ibrahim MG, Gar Alalm M, Fujii M (2020a) Effective photocatalytic degradation of sulfamethazine by CNTs/$LaVO_4$ in suspension and dip coating modes. Sep Purif Technol 235:35516. https://doi.org/10.1016/j.seppur.2019.116138

Samy M, Ibrahim MG, Gar Alalm M, Fujii M, Diab KE, Elkady M (2020b) Innovative photocatalytic reactor for the degradation of chlorpyrifos using a coated composite of $ZrV_2O_7$ and graphene nano-platelets. Chem Eng J 395:124974. https://doi.org/10.1016/j.cej.2020.124974

Samy M, Ibrahim MG, Gar Alalm M, Fujii M (2020c) MIL-53 (Al)/ZnO coated plates with high photocatalytic activity for extended degradation of trimethoprim via novel photocatalytic reactor. Sep Purif Technol 249:117173. https://doi.org/10.1016/j.seppur.2020.117173

Samy M, Ibrahim MG, Fujii M, Diab KE, Elkady M, Gar Alalm M (2021) CNTs/MOF-808 painted plates for extended treatment of pharmaceutical and agrochemical wastewaters in a novel photocatalytic reactor. Chem Eng J 406:7. https://doi.org/10.1016/j.cej.2020.127152

Samy M, Mensah K, El-Fakharany EM, Elkady M, Shokry H (2023a) Green valorization of end-of-life toner powder to iron oxide-nanographene nanohybrid as a recyclable persulfate activator for degrading emerging micropollutants. Environ Res 223:115460. https://doi.org/10.1016/j.envres.2023.115460

Samy M, Kumi G, Salama E, Elkady M, Mensah K, Shokry H (2023b) Heterogeneous activation of persulfate by a novel nano-magnetite/ZnO/activated carbon nanohybrid for carbofuran degradation: toxicity assessment, water matrices, degradation mechanism and radical and non-radical pathways. Process Saf Environ Prot 169:337–351. https://doi.org/10.1016/j.psep.2022.11.038

Samy M, Gar Alalm M, Khalil MN, Ezeldean E, El-Dissouky A, Nasr M, Tawfik A (2023c) Treatment of hazardous landfill leachate containing 1,4 dioxane by biochar-based photocatalysts in a solar photo-oxidation reactor. J Environ Manage 332:117402. https://doi.org/10.1016/j.jenvman.2023.117402

Wang Z, Srivastava V, Wang S, Sun H, Thangaraj SK, Jänis J, Sillanpää M (2020) UVC-assisted photocatalytic degradation of carbamazepine by Nd-doped $Sb_2O_3$/$TiO_2$ photocatalyst. J Colloid Interface Sci 562:461–469

# Chapter 41
# Determination and Quantification of Nitrogen Species During the Different Stages of Dairy Wastewater Treatment in a Sequencing Batch Reactor

**João F. C. Silva, Carolina Vicente, João R. Silva, Anabela M. Moreira, and Luís M. Castro**

**Abstract** In this study, nitrogen removal analyses of ammonium, nitrite and nitrate, as well monitorization of dissolved oxygen (DO), at different phases of a sequencing batch process, were carried out to investigate the process of nitrification–denitrification on a synthetic dairy wastewater. Total nitrogen (TN) removal efficiency increased with each cycle with 84.2%, 89.1%, 90.9% and 92.3%, for day 1, 2, 3 and 4, respectively. Most of the nitrogen removed occurred due to assimilation during cell growth. Ammonium, nitrite, nitrate at the end of cycle 4 showed significant variation when compared with cycle 1, with 41.4%, 78.6% and 70.4% reduction, respectively. This work allows to conclude that SBR is an effective method of reducing the nitrogen content of dairy wastewater.

**Keywords** Dairy wastewater · Denitrification · Nitrification · Nitrogen · Sequencing batch reactor · Wastewater treatment

---

J. F. C. Silva · C. Vicente · J. R. Silva · A. M. Moreira · L. M. Castro (✉)
Instituto Politécnico de Coimbra, Instituto Superior de Engenharia de Coimbra, Rua Pedro Nunes, Quinta da Nora, 3030-199 Coimbra, Portugal
e-mail: mcastro@isec.pt

L. M. Castro
CIEPQPF—Chemical Engineering Processes and Forest Products Research Center, Department of Chemical Engineering, Faculty of Sciences and Technology, University of Coimbra, Coimbra, Portugal

Instituto Politécnico de Coimbra, Instituto de Investigação Aplicada, Laboratório SiSus, Rua Pedro Nunes, Quinta da Nora, 3030-199 Coimbra, Portugal



N. S. Caetano and M. C. Felgueiras (eds.), *The 9th International Conference on Energy and Environment Research*, Environmental Science and Engineering,
https://doi.org/10.1007/978-3-031-43559-1_41

## 41.1 Introduction

The dairy industry holds significant economic importance worldwide. However, as it generates between 4 and 11 million tons of wastewater worldwide annually, it is considered the largest source of industrial wastewater in the food sector, especially in Europe (Kolev Slavov 2017; Ahmad et al. 2019). The presence of nitrogen in dairy wastewaters is primarily due to the high concentration of milk proteins, and it can be present in its organic forms (proteins, urea, nucleic acids) or as ions (ammonium, nitrate, nitrite) (Demirel et al. 2005). However, the oxidized forms of nitrogen are rare in these effluents (Sant'Anna Jr 2011). Nitrogen accumulation can cause eutrophication in aquatic environments, such as rivers and lakes (Jia and Yuan 2016).

It is necessary to stem sustainable and efficient alternatives to treat those effluents. Although mechanical and physicochemical techniques can be applied, usually, it is given preference to biological technologies due to the high biodegradability of the dairy wastewater contaminants (Goli et al. 2019). Nitrification and denitrification are commonly employed in biological methods to mitigate the effects of discharging high concentrations of these compounds into the environment (McCarty 2018). Two distinct groups of bacteria drive nitrification: the ammonia-oxidizing bacteria (AOB) and the nitrite-oxidizing bacteria (NOB). The former can be significantly influenced by environmental conditions, such as temperature (T), pH and DO (Yan et al. 2019). Since nitrifying bacteria are strictly aerobic, low oxygen concentrations can inhibit partial or total activity in this process (Couvert et al. 2019). Nitrification occurs in two steps. On the first step (Eq. 41.1), the bacteria from the genus Nitrosomonas oxidize ammonia into nitrite. This reaction product is posteriorly converted into nitrate by bacteria from the genus Nitrobacter (Eq. 41.2) (Thakur and Medhi 2019).

$$NH_4^+ + \frac{3}{2}O_2 \rightarrow NO_2^- + 2H^+ + H_2O \tag{41.1}$$

$$NO_2^- + \frac{1}{2}O_2 \rightarrow NO_3^- \tag{41.2}$$

Nitrate resulting from the nitrification is converted via the denitrification process (Eq. 41.3) into nitrous oxide and atmospheric nitrogen released to the atmosphere (Thakur and Medhi 2019). This process is commonly carried out by facultative anaerobes, such as gram-negative classes of α, β and γ Proteobacteria and some gram-positive bacteria such as Bacillus licheniformis (Medhi et al. 2017). If denitrification is incomplete, intermediate compounds, such as NO and $N_2O$ can be released (Albrecht et al. 1997).

$$NO_3^- \rightarrow NO_2^- \rightarrow NO \rightarrow N_2O \rightarrow N_2 \tag{41.3}$$

Microorganisms involved on denitrification are heterotrophs and need an external carbon source to reduce $NO_2^-$ and $NO_3^-$. Nevertheless, an excess of carbon sources might induce an additional biomass growth that increases nitrogen assimilation. This phenomenon contributes to the system overall nitrogen removal (Li and Irvin 2007). Equation 41.4 represents the biomass formation process (Ray et al. 2019). Considering a general composition of the biomass as $C_5H_7NO_2$, a carbon mass/nitrogen mass ratio of, approximately, 60:14 is verified. This way, on the cell growth process, for each gram of nitrogen, 4.29 g of carbon are necessary.

$$NH_4^+ + 1.18\, C_6H_{12}O_6 + HCO_3^- + 2.06\, O_2 \rightarrow C_5H_7O_2N + 6.06\, H_2O + 3.07\, CO_2 \quad (41.4)$$

Wagner et al. (2015) reported that nitrogen assimilation by biomass for cell growth contributes to nitrogen removal. Mosquera-Corral et al. (2011) estimated, from a nitrogen mass balance, that 90% of the nitrogen removal was a consequence of biomass growth, for relatively low C:N ratios. Garrido et al. (2001) reported that for organic matter/nitrogen ratios lower than 5 the main nitrogen removal mechanism is biomass growth.

This work aimed to ascertain and quantify the nitrogen species present in the reactional content at the different stages of dairy industry wastewater treatment in a Sequential Batch Reactor (SBR). During this process, the reactor content is subjected to aerobic and anaerobic phases, as well as substrate abundance and scarce, whereby the reactions that occur vary along with time. In this way, the reactor conditions stipulate the nitrogen species concentrations. Depending on the surrounding environmental factors and treatment stage, nitrogen can be found inside the reactor in the form of ammonium ($NH_4^+$), nitrite ($NO_2^-$), nitrate ($NO_3^-$) or organic.

## 41.2 Material and Methods

The identification and quantification of the nitrogen species of synthetic dairy wastewater were performed on a sequencing batch bioreactor with 4 L of working volume. The experiment after a period of acclimatation lasted 4 cycles operating within the same conditions. A pump provided an airflow of 2.1 L/min, allowing the reactor aeration. Samplings of 50 mL of the reactional content were performed on strategic moments, before the start of a new cycle, during the aerated reaction time and after 2 h of sludge settling. The ElectroLab Fermac 320 computer program performed the control of the system and the automatic acquisition of DO and T data during each sequential batch. The feed was composed of low-fat milk with a dilution of 1:80, with a composition of 0.388 mg/L N-$NH_4^+$, 0.0280 mg/L N-$NO_2^-$, 0.406 mg/L N-$NO_3^-$ and 104 mg N/L, thus nitrogen is mainly in its organic form.

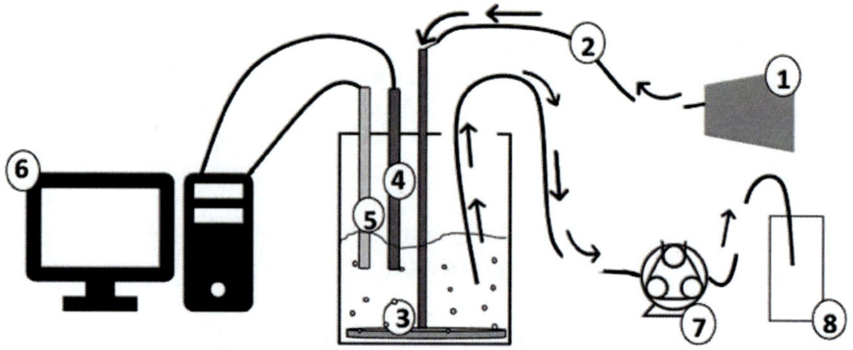


**Fig. 41.1** Schematic representation of the experimental apparatus. 1-Air pump; 2-Air pipe; 3-Air entrance on the bioreactor; 4-DO probe; 5-T probe; 6-Computer system; 7-Peristaltic pump; 8-Treated wastewater reservoir

### *41.2.1 Experimental Apparatus and Procedures*

One week before the beginning of this study, the aerated reactor was fed with 2 L of activated sludge and 2 L of synthetic effluent to allow the biomass to adapt to the new environmental conditions, under intermittent aerobic and anaerobic conditions with the last occurring for 15 min in each hour. This allowed for microbial adaptation and sludge rapid settling. Before the first cycle started, the aeration was turned off and sludge sedimentation was allowed for 2 h. Then, 1 L of the resultant supernatant was discharged and 1 L of diluted milk was added to initiate the first cycle.

Each cycle of 7.5 h was initiated by a feeding phase, in which 1 L of wastewater was manually added at once to the bioreactor. The subsequent "reaction" phase had a duration of 5 h and occurred under aerobic conditions. After this phase, the air pump was turned off, following 2 h of sedimentation taking place under anoxic conditions. The batch ended with the discharge of 1 L of treated wastewater using a peristaltic pump for 30 min. The sludge would then rest under anaerobic conditions for a period of 16.5 h until the beginning of the next cycle. Figure 41.1 represents the schematic of the experimental setup.

### *41.2.2 Analytical Methods*

The 50 mL samples were filtered and Total Nitrogen (TN) was measured by chemiluminescence with oxidative combustion, using Shimadzu Total N Measuring Unit TNM-1. Total Carbon (TC) was measured by nondispersive infrared sensor (NDIR) and chemiluminescence, using Shimadzu TOC-VCPH/CPN Total Organic Carbon Analyzer and TOC-Control V software. Ammonium, Nitrate and Nitrite were determined by molecular absorbance spectrometry, using a SKALAR segmented flow autoanalyzer. Nitrate and nitrite determination are based on Method 4500-NO3-F and Method 4500-NO2-B of the Standard Methods for the Examination of Water and Wastewater, respectively. Ammonium concentration is determined by the SKALAR method, based on the continuous flow method, resorting to an active chloride compound (sodium salicylate and sodium dichloroisocyanurate) and absorbance reading at 660 nm (Rodier et al. 2016).

## 41.3 Results and Discussion

From Fig. 41.2, it is possible to verify that the increase of DO values at the beginning of the batch was quicker on days 1 and 2. Similarly, the decrease of those values at the end of the "reaction" phase was also more abrupt. However, a sequence of increase, stabilization and reduction of DO values was verified every single day. The maximum oxygen concentration in the aeration phase was 6.30 mg $O_2$/L on day 3. Ammonium concentration decreases during the aerobic stage. Zhou et al. (2002) have shown that under anoxic conditions, biomass growth is nitrate mediated. Due to the low denitrification that was verified in that study, the authors concluded that the growth was dependent on nitrate metabolism to form ammonium. These results might explain the slight incrementation in ammonium concentration under anoxic conditions verified by data from Fig. 41.2.

As observed by other authors (Obaja et al. 2003; Li et al. 2019), nitrite experienced an increase of concentration at the beginning of each sequential batch, followed by a decrease. Nitrate concentration behavior is similar during the extent of the experiment, showing an increase in concentration during the aerobic stage, resulting from the nitrite oxidation and a decrease during the anaerobic phase, in which this species is converted into nitrogen gas forms or ammonium. Therefore, one can confirm that nitrification and denitrification occurred, and dissimilatory nitrate reduction to ammonium might have occurred. Comparing the nitrogen species concentration of the last sample analyzed on days 2 and 3 with the first one of days 3 and 4, respectively, it is possible to notice a decrease in nitrite and nitrate concentration and an increase in the ammonia concentration. The nitrate reduction to ammonium might explain this circumstance, which can occur under anoxic conditions in the presence of a carbon source (Burgin and Hamilton 2007).

It is interesting to notice that the nitrate concentration decreases each day. It ends the first cycle with 5.87 mg$NO_3^-$/L and completes the fourth cycle with 1.74 mg $NO_3^-$/L, a decrease of 70.4%. This corroborates the idea that nitrate is converted to nitrite and consequently is converted into a gaseous form ($N_2$) since ammonium and nitrate concentrations do not increase, and TN decreases with time. Likewise, the nitrite concentration lowers, from 2.56 mg$NO_2^-$/L at the end of day 1, to 0.548 mg $NO_2^-$/L at the end of day 4, a decrease of 78.6%. Regarding ammonium, the concentration remains relatively low, from 0.411 to 0.241 mg $NH_4^+$/L, decreasing 41.4%. These values indicate that the proposed SBR cycle effectively removes the different forms of nitrogen from dairy wastewater.

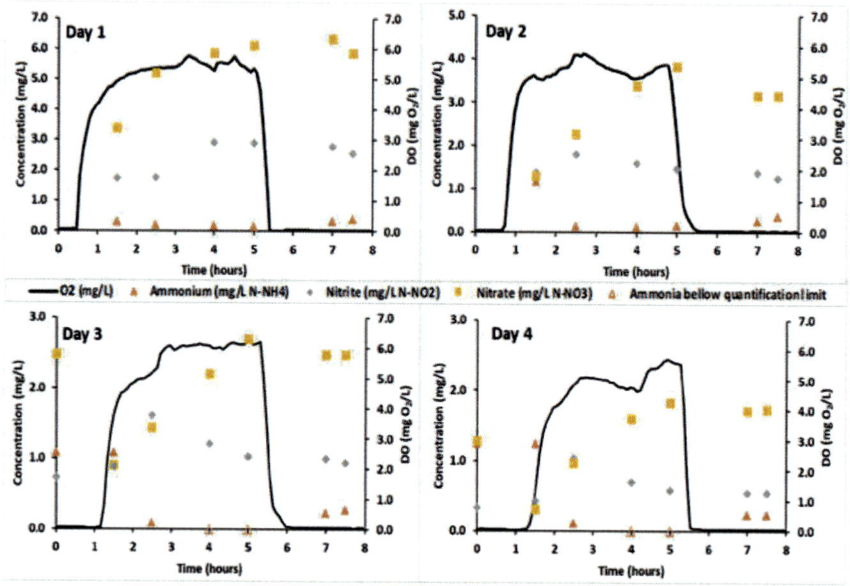


**Fig. 41.2** DO, $NH_4^+$, $NO_3^-$, and $NO_2^-$ concentrations (mg/L) evolution during each of the four cycles

From Fig. 41.3, it is possible to perceive that the TN values decrease during the batch time. It is also possible to acknowledge that TN content decreases from day 1 to day 4. The nitrogen removal efficiency obtained was 84.2%, 89.1%, 90.9% and 92.3%, for days 1, 2, 3 and 4, respectively. This evolution can be explained by biomass acclimation to environmental conditions. On days 1 and 2, a slight increase is verified between the samples corresponding to 4 h and 5 h of batch.

Although, as shown above, the nitrification/denitrification process is observed in the experiments carried out, the nitrogenous species are mainly in organic form since the total nitrogen is, in all samples, considerably higher than the sum of the

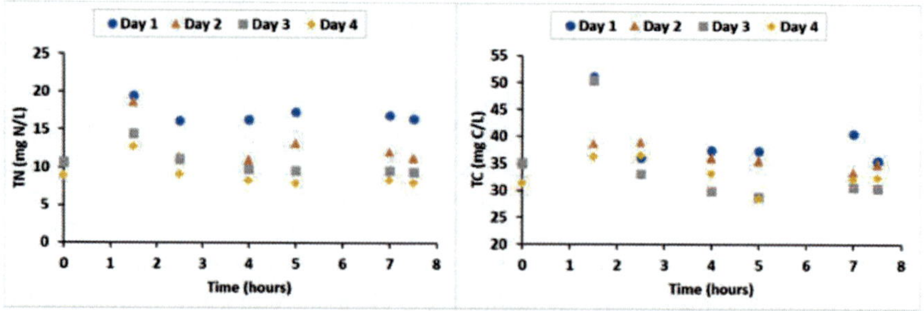


**Fig. 41.3** TN (mgN/L) and TC (mgC/L) evolution for the 4 batch cycles

nitrogen content present in $NO_3^-$, $NO_2^-$ and $NH_4$. Thus, the removal of total nitrogen observed between the influent placed in the reactor and the treated effluent removed, always greater than 80%, results essentially from the decrease in the concentration of organic nitrogen. This reduction, which cannot be explained by the denitrification processes mentioned above, is certainly due to the use of nitrogenous species in the construction of cellular material that forms the activated sludge. During this study C:N ratios in the affluent were lower than 5. Thus, according to Garrido et al. (2001), it is possible to infer that the main cause of nitrogen removal is assimilation due to heterotrophic cell growth, TC and TN evolution similarities (Fig. 41.3) might confirm that C and N assimilation occur as consequence of biomass growth. Once the C:N ratio of the reactional content were lower than 4.29, carbon constitutes the limiting nutrient, making denitrification a process dependent on the presence of this component. Thus, the absence of this nutrient limits the biomass decrease of TN while no significant production of nitrite or nitrate could be observed (Schryver and Verstraete 2009). On the other hand, heterotrophs have a growth rate that is 10 times higher than that of nitrifying bacteria (Hargreaves 2006). This way, short SBR cycles such as the ones verified in this study, promote the heterotrophs growth rather than that of nitrifiers. This fact can also contribute to explain the reduced observed nitrification rate.

## 41.4 Conclusion

The nitrification process was easily verifiable during each cycle, as the nitrate concentration increases during the aerobic stage, and ammonium concentration decreases during the same period. DO was directly responsible for removing ammonia, showing a notorious effect in the process of nitrification. After two and a half hours of aeration almost all ammonium has been converted to NO2- and NO3-. Due to the fact that denitrification is slower than nitrification (Dinçer and Kargı 2000), this phenomenon is difficult to observe considering only the sedimentation phase. Nevertheless, it is observable a slight reduction of $NO_3$ and $NO_2^-$, and an increase of $NH_4^+$ concentrations during the anoxic phase. Additionally, when considering the time-lapse between one cycle and another (anaerobic phase), denitrification and nitrate reduction to ammonium are observed. TN removal efficiency increased with each cycle, starting on 84.2% on day 1, 89.1% for day 2, 90.9% on day 3 and ending at 92.3% on day 4. The observed efficiency is mainly due to nitrogen assimilation during cell growth. Nitrification and denitrification only contribute partially to nitrogen removal. It is proven that SBR is an appropriate and reliable treatment system to perform the joint removal of organic and ammonia nitrogen in dairy industry wastewater.

## References

Ahmad T, Aadil RM, Ahmed H, Rahman U, Soares BCV, Souza SLQ, Pimentel TC, Scudino H, Guimarães JT, Esmerino EA et al. (2019) Treatment and utilization of dairy industrial waste: a review. Trends Food Sci Technol 88:361–372. https://doi.org/10.1016/j.tifs.2019.04.003

Albrecht A, Ottow JCG, Benckiser G, Sich I, Russow R (1997) Incomplete denitrification (NO and N 2 O) from nitrate by streptomyces violaceoruber and S. Nitrosporeus revealed by acetylene inhibition and 15 N gas chromatography-quadrupole mass spectrometry analyses. Naturwissenschaften 84:145–147. https://doi.org/10.1007/s001140050365

Burgin AJ, Hamilton SK (2007) Have we overemphasized the role of denitrification in aquatic ecosystems? A review of nitrate removal pathways. Front Ecol Environ 5:89–96. https://doi.org/10.1890/1540-9295(2007)5[89:HWOTRO]2.0.CO;2

Couvert O, Divanac'h M-L, Lochardet A, Thuault D, Huchet V (2019) Modelling the effect of oxygen concentration on bacterial growth rates. Food Microbiol 77:21–25. https://doi.org/10.1016/j.fm.2018.08.005

De Schryver P, Verstraete W (2009) Nitrogen removal from aquaculture pond water by heterotrophic nitrogen assimilation in lab-scale sequencing batch reactors. Bioresour Technol 100:1162–1167. https://doi.org/10.1016/j.biortech.2008.08.043

Demirel B, Yenigun O, Onay TT (2005) Anaerobic treatment of dairy wastewaters: a review. Process Biochem 40:2583–2595. https://doi.org/10.1016/j.procbio.2004.12.015

Dinçer AR, Kargı F (2000) Kinetics of sequential nitrification and denitrification processes. Enzym Microb Technol 27:37–42. https://doi.org/10.1016/S0141-0229(00)00145-9

Garrido JM, Omil F, Arrojo B, Méndez R, Lema JM (2001) Carbon and nitrogen removal from a wastewater of an industrial dairy laboratory with a coupled anaerobic filter-sequencing batch reactor system. Water Sci Technol 43:249–256. https://doi.org/10.2166/wst.2001.0144

Goli A, Shamiri A, Khosroyar S, Talaiekhozani A, Sanaye R, Azizi K (2019) A review on different aerobic and anaerobic treatment methods in dairy industry wastewater. J Environ Treat Tech 7:113–141

Hargreaves JA (2006) Photosynthetic suspended-growth systems in aquaculture. Aquac Eng 34:344–363. https://doi.org/10.1016/j.aquaeng.2005.08.009

Jia H, Yuan Q (2016) Removal of nitrogen from wastewater using microalgae and microalgae-bacteria consortia. Cogent Environ Sci 2:1275089. https://doi.org/10.1080/23311843.2016.1275089

Kolev Slavov A (2017) Dairy wastewaters—General characteristics and treatment possibilities—A review. Food Technol Biotechnol 55. https://doi.org/10.17113/ftb.55.01.17.4520

Li B, Irvin S (2007) The roles of nitrogen dissimilation and assimilation in biological nitrogen removal treating low, mid, and high strength wastewater. J Environ Eng Sci 6:483–490. https://doi.org/10.1139/S07-001

Li C, Liu S, Ma T, Zheng M, Ni J (2019) Simultaneous nitrification, denitrification and phosphorus removal in a sequencing batch reactor (SBR)under low temperature. Chemosphere 229:132–141. https://doi.org/10.1016/j.chemosphere.2019.04.185

McCarty PL (2018) What is the best biological process for nitrogen removal: when and why? Environ Sci Technol 52:3835–3841. https://doi.org/10.1021/acs.est.7b05832

Medhi K, Singhal A, Chauhan DK, Thakur IS (2017) Investigating the nitrification and denitrification kinetics under aerobic and anaerobic conditions by paracoccus denitrificans ISTOD1. Bioresour Technol 242:334–343. https://doi.org/10.1016/j.biortech.2017.03.084

Mosquera-Corral A, Arrojo B, Figueroa M, Campos JL, Méndez R (2011) Aerobic granulation in a mechanical stirred SBR: treatment of low organic loads. Water Sci Technol 64:155–161. https://doi.org/10.2166/wst.2011.483

Obaja D, Macé S, Costa J, Sans C, Mata-Alvarez J (2003) Nitrification, denitrification and biological phosphorus removal in piggery wastewater using a sequencing batch reactor. Bioresour Technol 87:103–111. https://doi.org/10.1016/S0960-8524(02)00229-8

Ray S, Scholz M, Haritash AK (2019) Kinetics of carbon and nitrogen assimilation by heterotrophic microorganisms during wastewater treatment. Environ Monit Assess 191:451. https://doi.org/10.1007/s10661-019-7599-5

Rodier J, Legube B, Eaux Naturelles MN, Résiduaires E, de Mer E (2016) L'analyse de l'eau; Dunod, Paris

Sant'Anna Jr GL (2011) Tratamento Biológico de Efluentes: Fundamentos e Aplicações. Eng Sanit e Ambient 16:IV–IV. https://doi.org/10.1590/S1413-41522011000200002

Thakur IS, Medhi K (2019) Nitrification and denitrification processes for mitigation of nitrous oxide from waste water treatment plants for biovalorization: challenges and opportunities. Bioresour Technol 282:502–513. https://doi.org/10.1016/j.biortech.2019.03.069

Wagner J, Guimarães LB, Akaboci TRV, Costa RHR (2015) Aerobic granular sludge technology and nitrogen removal for domestic wastewater treatment. Water Sci Technol 71:1040–1046. https://doi.org/10.2166/wst.2015.064

Yan L, Liu S, Liu Q, Zhang M, Liu Y, Wen Y, Chen Z, Zhang Y, Yang Q (2019) Improved performance of simultaneous nitrification and denitrification via nitrite in an oxygen-limited SBR by alternating the DO. Bioresour Technol 275:153–62. https://doi.org/10.1016/j.biortech.2018.12.054

Zhou Z, Takaya N, Nakamura A, Yamaguchi M, Takeo K, Shoun H (2002) Ammonia fermentation, a novel anoxic metabolism of nitrate by fungi. J Biol Chem 277:1892–1896. https://doi.org/10.1074/jbc.M109096200

# Chapter 42
# Effect of Temperature on the Thermolysis of Waste Polyethylene Terephthalate (PET) and Its Application in Methylene Blue Removal

**Kenneth Mensah, Hatem Mahmoud, Manabu Fujii, and Hassan Shokry**

**Abstract** Polyethylene terephthalate (PET), a major plastic waste, was recycled into solid carbon by thermal decomposition under autogenic pressure. PET was carbonized at 500–1000 °C and the effect of temperature on the properties and yield of the synthesized carbons or char was studied and discussed using SEM, TEM, EDX, XRD, FTIR, and BET. The synthesized carbons (SC) exhibit a range of microporous to mesoporous structures with a pore volume and surface area ranging from 0.2029 to 0.0573 $cm^3\ g^{-1}$ and 448.88 to 3.3 $m^2\ g^{-1}$ respectively. SC was employed in the removal of methylene blue (MB) as a model dye and the effect of contact time was investigated. SC produced at 500 °C demonstrated the highest adsorption capacity of 77 $mg\ g^{-1}$ due to its high surface area, porous structure, and surface functional groups.

**Keywords** Recycling · Plastic waste · Thermolysis · Graphene · Surface area · Dye adsorption

---

K. Mensah (✉) · H. Mahmoud · H. Shokry
Environmental Engineering Department, Egypt-Japan University for Science and Technology, New Borg El-Arab City, Alexandria 21934, Egypt
e-mail: kenneth.mensah@ejust.edu.eg

H. Mahmoud
Department of Architecture Engineering, Faculty of Engineering, Aswan University, Aswan 81542, Egypt

M. Fujii
Department of Civil and Environmental Engineering, Tokyo Institute of Technology, Meguro-Ku, Tokyo 152-8552, Japan

H. Shokry
Electronic Materials Research Department, Advanced Technology and New Materials Research Institute, City of Scientific Research and Technological Applications (SRTA-City), Alexandria, Egypt

 
N. S. Caetano and M. C. Felgueiras (eds.), *The 9th International Conference on Energy and Environment Research*, Environmental Science and Engineering,
https://doi.org/10.1007/978-3-031-43559-1_42

## 42.1 Introduction

Plastics have been widely used in both households and industries due to their numerous applications (Nimako et al. 2020). Plastics are characterized by many desirable properties such as low cost, lightness, formability, and durability. The production of plastics has therefore increased and is expected to triple by 2050 (Rhodes 2018). Enormous quantities of end-of-life plastics are consequently generated annually. Polyethylene terephthalate (PET) accounts for 18% of the world's plastic production and it is estimated that 1 million plastic bottles which are predominantly PET are discarded every minute and this is expected to double within 20 years (Magnier et al. 2019). These waste plastics are generally immune to natural degradation cycles and therefore pose severe threats to the environment due to a lack of proper management (Rhodes 2018).

Recycling is recommended as the best solution to treat waste plastics because recycling waste plastics reduces environmental pollution, conserves resources and energy, and produces significant socio-economic benefits (Rhodes 2018; Shokry 2019). PET contains about 75% total carbon and several studies have demonstrated the potential upcycling of waste PET (Rhodes 2018; Shokry et al. 2019; Elkady et al. 2020). Nevertheless, a small amount of PET waste is commercially recycled, and the rest is left without recycling (Chinchillas-Chinchillas et al. 2020). It is therefore expedient to investigate further potential uses of waste PET (Rhodes 2018).

However, dyes such as methylene blue (MB) are complex synthetic organic compounds used mainly in the textile industry (El-Kady et al. 2018). Dyes are highly stable, bio-recalcitrant, and toxic and thus are harmful to lives when their effluent is discharged into the environment without proper treatment (Elkady et al. 2020). Technologies used in the treatment of dyes include adsorption, membrane filtration, Fenton and Fenton-like processes, photocatalytic degradation, precipitation, and ion exchange (Samy et al. 2020). Adsorption is a simple and efficient dye removal technology but there is a need to investigate cheaper sources of adsorbent for the treatment of dye-containing effluents (Shokry et al. 2019). Recycling carbon-rich waste materials such as plastics into adsorbents must therefore attract much interest (Asante et al. 2020).

In our work, PET waste is upcycled into high-value carbon by thermal decomposition from 500 to 1000 °C under autogenic pressure. The effect of temperature on the crystallinity, surface functional groups, morphology, surface area, yield, and purity of the SC are investigated using various analytical techniques. The potential application of SC is demonstrated with MB removal and some adsorption parameters are studied.

## 42.2 Experimental Procedure

### 42.2.1 Synthesis of Carbon from Waste PET

Waste PET bottles were washed with distilled water, sun-dried, and then shredded. A 420 mL stainless steel 316 autoclave was filled with 10 g of shredded PET and sealed tightly to prevent the inflow and outflow of gases from the system. The autoclave is placed at the center of a muffle furnace for batch thermal decomposition at a heat ramping rate of 27 °C/min till 500, 600, 700, 800, 900, and 1000 °C under autogenic pressure. The target temperature was maintained for 2 h. The setup was afterward left to cool overnight. The synthesized char (SC) was crushed and stored in a desiccator. The SC was labeled by SC-thermolysis temperature. For instance, SC-500 refers to char synthesized at 500 °C.

### 42.2.2 MB Adsorption Test

A 100 mg $L^{-1}$ MB solution was contacted with 0.2 mg/mL of SC for 1 h under stirring. The effect of contact time was investigated with all SC adsorbents. After a specified contact time, the slurry was filtered and the concentration of MB was measured using a UV–vis spectrophotometer at a wavelength of 660 nm. All tests were conducted at room temperature and atmospheric pressure.

## 42.3 Results and Discussion

### 42.3.1 Characteristics of SC

The yield of SC was calculated using Eq. 42.1 and the result is illustrated in Fig. 42.1. SC-500 reported the lowest yield of about 15%. This is ascribed to the lack of the C–H and C–C bonds breakdown at this temperature (Pol et al. 2009). The yield however increases significantly and almost doubles (28%) for SC-600 which implies that increasing temperature contributed to the further breakdown of gaseous carbon radicals for solid C to remain as the product (Pol 2010). A yield of 30%, 32%, 32.4%, and 33% was obtained for SC-700, 800, 900, and 1000 respectively. Though the yield of SC increased with temperature as more C–H and C–C bonds break, the increment after 700 °C thermolysis was relatively steady which suggests that some gaseous hydrocarbons remained despite higher thermolysis temperatures. However, these gases serve as reducing agents keeping the carbon in solid form (Pol et al. 2009).

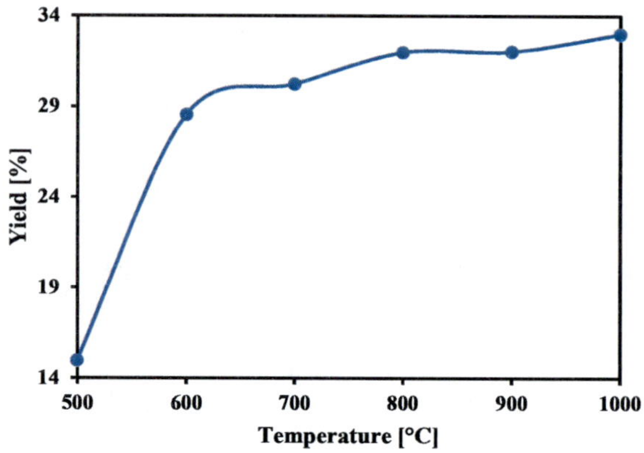


**Fig. 42.1** The yield of SC from the PET waste at various temperatures

$$Yield = \frac{mass\,of\;char\;product\;(\mathrm{g})}{mass\,of\;PET\;feed(10\,\mathrm{g})} \times 100\% \qquad (42.1)$$

Scanning electron microscopy (SEM) was conducted to assess the morphology of SC as shown in Fig. 42.2. SC-500 exhibits amorphous blocky morphology with some voids in between. SC-600 has similar amorphous morphology with a few fibrous structures. However, amorphous structures in SC-700 and SC-800 are blockier and clustered, and entangled. SC-900 and SC-1000 show similar blocks of irregular sizes and shapes depicting a possible improvement in graphitization and agglomeration of sheets to form the blocky structures with increasing temperature. However, it is observed that spaces between the materials become wider with temperature and this phenomenon is affirmed and expounded in the BET analysis.

Figure 42.3 shows transmission electron microscopy (TEM) images of SC. SC-500 has entangled and rippled sheets while SC-600 shows distorted sheets with few rippled structures. SC-700 has sheets that are rippled but dark portions suggest agglomeration of the graphene sheets thus the blocky nature of the SEM images. The conversion of amorphous morphologies to sheet-like structures suggests an improvement in graphitization with increasing temperature. This trend is confirmed by the XRD analysis. Furthermore, the increasing number of shadows in the SC-800, SC-900, and SC-1000 confirms the increasing agglomeration of graphene sheets with increasing temperature.

Elemental analysis of SC by EDX spectroscopy reveals 97.56, 97.09, 96.30, 94.71, 93.29, and 92.81% carbon in SC-500, 600, 700, 800, 900, and 1000 °C respectively with traces of oxygen (about 2%) and iron (Fe). The reduction in the purity of SC with temperature is attributed to the increase in Fe content as temperature increases which is primarily due to thermal corrosion of the stainless reactor by expansion and contraction during heating and cooling.

XRD patterns of SC are shown in Fig. 42.4. The broad flat peak at 2θ from 18.04 to 20.44° which corresponds to the (002) carbon plane affirms the production of

**Fig. 42.2** SEM images of SC

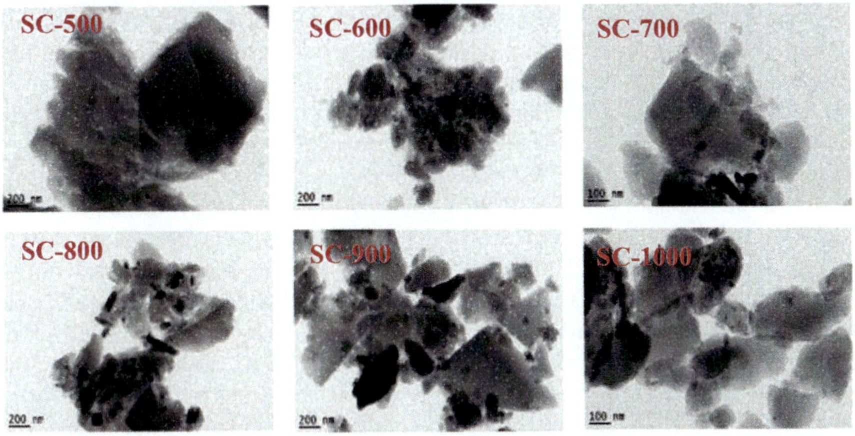


**Fig. 42.3** TEM images of SC

amorphous carbon and the lack of carbon growth in SC-500. Nevertheless, from SC-600 to SC-1000, this peak shifts to 2θ value of ~26° and becomes stronger indicating the progressive conversion of amorphous carbon to graphitic carbon with increasing temperature (Mensah et al. 2021). However, the broadness of the peaks suggests the production of amorphous graphite which is due to the termination of sp2 bonds of carbon atoms or cross-linking of the hexagonal planar units by oxygen that was present during the carbonization, thus disrupting any order in the graphitic plane. Such graphite typically has diffuse sets of inter-layer distances that are averagely larger than those in crystalline graphite (Shen and Lua 2013). The peaks at 2θ values

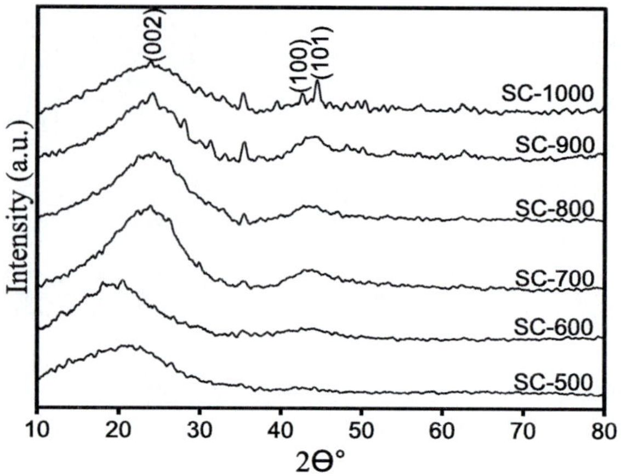


**Fig. 42.4** XRD pattern of SC

of ~ 42° and 44° corresponding to the (100) and (101) diffraction planes of graphite respectively are present from SC 700 to SC-1000 and strongest at SC-1000. This trend also confirms the temperature-graphitization effect. The peak at 2θ of approximately 44° depicts a stacking order of graphene sheets (Shen and Lua 2013). The presence of other peaks is suggested to have originated from the presence of a few impurities such as Fe in the SC as described by the EDX analysis.

The BET surface area, average pore diameter, and total pore volume of SC are summarized in Table 42.1. SC-500, 600, and 700 exhibit microporous morphology while SC-800, 900, and 1000 exhibit a mesoporous structure according to IUPAC classification. The surface area of SC decreased from 448.88 $m^2\ g^{-1}$ for SC-500 to 3.3066 $m^2\ g^{-1}$ for SC-1000 while the pore volume decreased from 0.2029 to 0.0573 $cm^3\ g^{-1}$ for SC-500 and SC-1000 respectively. The high pore volume observed in SC-500 and 600 can be attributed to the increase in the number of pores due to the arrangement of carbon layers resulting in the creation of micropores (Daud et al. 2001). Furthermore, activation of the surface and bulk of the char due to the presence of large quantities of hydrocarbon radicals and some hydrogen and water vapor at 500 and 600 °C can cause the amorphous nature and high pore volume and high surface area of their char products (Pol 2010; Essawy et al. 2017). However, the reduction in pore volume and surface area from SC-700 to 1000 is attributed to the formation and agglomeration of graphene sheets and the complete scission of hydrocarbon radicals (Shokry et al. 2019). Meanwhile, the increase in mean pore diameter from SC-700 to SC-1000 can be attributed to the dominant occurrence of the pore widening effect than the pore-opening effect, resulting in a decrease of the specific surface area and micropore volume (Shokry et al. 2019; Samy et al. 2019).

The FT-IR patterns of SC are shown in Fig. 42.5. The peak at 1600 $cm^{-1}$ is attributed to C = C stretching which is a fundamental property of sp2 graphite. The high intensity of this peak in SC-500 intensity is enhanced by the existence of oxygen atoms in the form of phenol or ether groups (Tseng et al. 2008). This characteristic peak in all SC affirms the synthesis of graphite by the PET thermolysis as depicted in

**Table 42.1** Surface area analysis of SC

| Sample | BET surface area ($m^2g^{-1}$) | Mean pore diameter (nm) | Total pore volume ($cm^3g^{-1}$) |
|---|---|---|---|
| SC-500 | 448.88 | 1.808 | 0.2029 |
| SC-600 | 361.03 | 1.810 | 0.2116 |
| SC-700 | 156.45 | 1.646 | 0.0644 |
| SC-800 | 128.01 | 2.94 | 0.0694 |
| SC-900 | 15.907 | 39.42 | 0.0577 |
| SC-1000 | 3.3066 | 69.835 | 0.0573 |

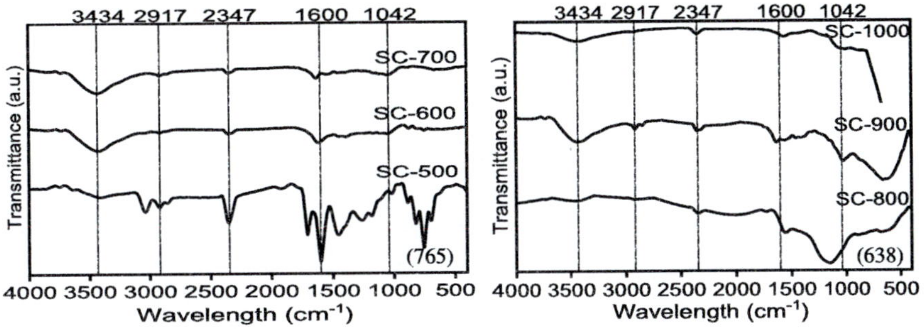


**Fig. 42.5** FT-IR spectra of SC

the XRD analysis. The peak at 3434 $cm^{-1}$ is ascribed to O–H stretching vibrations. The signal at 2917 $cm^{-1}$ may be assigned to C-H stretching vibrations indicating the removal of hydrogen from the PET during the thermolysis (Shokry et al. 2019). The higher temperature thermolysis of PET resulted in the decomposition of several C-H molecules thus the decrease in the intensity of the peak around 2900 $cm^{-1}$ with temperature. The peak at 2347 $cm^{-1}$ is ascribed to C = C bending while the peak at 1042 $cm^{-1}$ is attributed to C-O stretching. The intense peak at 765 $cm^{-1}$ for SC-500 is due to O–H stretching vibrations. SC, therefore, contains oxygen functional groups that are favorable for organic contaminants removal (Elkady et al. 2020). Moreover, such oxygen surface functional groups are frequent in graphene materials (Essawy et al. 2017). However, at higher thermolysis temperatures (≥800 °C), the surface oxygen groups tend to be removed increasing the surface basicity of the SC (Shokry et al. 2019).

### 42.3.2 Adsorption Studies

From Fig. 42.6, the adsorption ($q_t$) of MB was initially rapid for all SC and then steady (qe) after 50 min for SC-500 to 700, 40 min for SC-800 and 900, and 20 min for

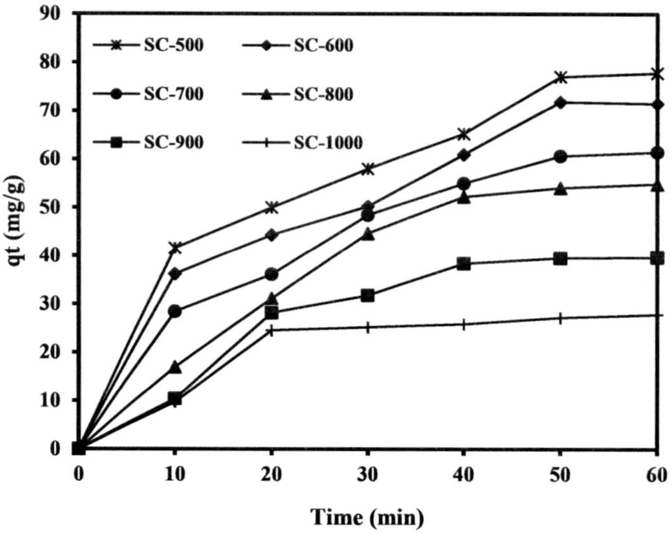


**Fig. 42.6** Effect of contact time on the adsorption of MB at 100 mg $L^{-1}$ MB solution, 0.2 mg $mL^{-1}$ of SC

SC-1000. This phenomenon is attributed to the free unoccupied sites available in the carbons at the initial stages of the adsorption which with time becomes occupied and makes it difficult for further occupation due to repulsive forces between the adsorbate and the remaining MB in solution. An equilibrium adsorption capacity ($q_e$) of 77, 72, 60, 52, 38, and 26 mg $g^{-1}$ was obtained for SC 500, 600, 700, 800, 900, and 1000 respectively. Though SC-900 and 1000 demonstrated lower adsorption capacities due to their lower pore volume and low surface area, their initial dye removal was rapid due to their larger mesopore diameters. However, these pores become filled quickly and adsorption becomes stagnated due to lack of free sites. Meanwhile, besides the high surface area, the presence of several polar functional groups in SC-500 can also contribute to its high performance.

## 42.4 Conclusion

The effect of temperature on the thermolysis of waste PET under autogenic pressure is analyzed and discussed using different analytical techniques. The process generally produced high-purity carbon (graphene) with an increase in yield as temperature increases. SEM, TEM, XRD, and FT-IR results indicate that higher temperature thermolysis of PET produces higher crystalline carbon allotropes (from amorphous to graphitic carbon as temperature increases). However, the graphitic sheets are agglomerated with reducing BET surface area and pore volume as temperature increases.

Adsorption test revealed that amorphous carbon synthesized below 600 °C is the most suitable for MB removal. It is concluded that PET converted into highly porous amorphous carbons is more suitable for organic pollutants adsorption than the crystalline product.

## References


Asante BNP, Nimako KO, Mensah K, Koshy P, Dankwah JR (2020) Calcination behaviour of Nsuta pyrolusite ore in the presence and absence of end-of-life polystyrene. In: Proceedings of 6th UMaT biennial international mining and mineral conference. pp 281–288

Chinchillas-Chinchillas MJ, Gaxiola A, Alvarado-Beltrán CG, Orozco-Carmona VM, Pellegrini-Cervantes MJ, Rodríguez-Rodríguez M, Castro-Beltrán A (2020) A new application of recycled-PET/PAN composite nanofibers to cement-based materials. J Clean Prod 252:1198272–1198281

Daud WMAW, Ali WSW, Sulaiman MZ (2001) Effect of carbonization temperature on the yield and porosity of char produced from palm shell. J Chem Technol Biotechnol 76:1281–1285

El Essawy NA, Ali SM, Farag HA, Konsowa AH, Elnouby M, Hamad HA (2017) Green synthesis of graphene from recycled PET bottle wastes for use in the adsorption of dyes in aqueous solution. Ecotoxicol Environ Saf 145:57–68

El-Kady MF, El-Aassar MR, El Batrawy OA, Ibrahim MS, Hassan HS, Fakhry H (2018) Equilibrium and kinetic behaviors of cationic dye decolorization using poly (AN-co-Py)/ZrO2 novel nanopolymeric composites. Adv Polym Technol 37:740–752

Elkady M, Shokry H, Hamad H (2020) New activated carbon from mine coal for adsorption of dye in simulated water or multiple heavy metals in real wastewater. Materials 13:1–18

Magnier L, Mugge R, Schoormans J (2019) Turning ocean garbage into products—Consumers' evaluations of products made of recycled ocean plastic. J Clean Prod 21:584–598

Mensah K, Mahmoud H, Fujii M, Shokry H (2021) Upcycling of polystyrene waste plastics to high value carbon by thermal decomposition. Key Eng Mater 897:103–108

Nimako KO, Dwumfour A, Mensah K, Koshy P, Dankwah JR (2020) Calcination behaviour of Nsuta rhodochrosite ore in the presence and absence of end-of-life high density polyethylene. Ghana Min J 20:22–35

Pol VG (2010) Upcycling: Converting waste plastics into paramagnetic, conducting, solid, pure carbon microspheres. Environ Sci Technol 44:4753–4759

Pol SV, Pol VG, Sherman D, Gedanken A (2009) A solvent free process for the generation of strong, conducting carbon spheres by the thermal degradation of waste polyethylene terephthalate. Green Chem 11:448–545

Rhodes CJ (2018) Plastic pollution and potential solutions. Sci Prog 101:207–260

Samy M, Mossad M, El-Etriby HK (2019) Synthesized nano titanium for methylene blue removal under various operational conditions. Desalin Water Treat 165:374–381

Samy M, Alalm MG, Mossad M (2020) Utilization of iron sludge resulted from electro-coagulation in heterogeneous photo-fenton process. Water Pract Technol 15:1228–1237

Shen Y, Lua AC (2013) A facile method for the large-scale continuous synthesis of graphene sheets using a novel catalyst. Sci Rep 3:1–6

Shokry H (2019) Role of preparation technique in the morphological structures of innovative nanocation exchange. J Market Res 8:2854–2864

Shokry H, Elkady M, Hamad H (2019) Nano activated carbon from industrial mine coal as adsorbents for removal of dye from simulated textile wastewater: operational parameters and mechanism study. J Market Res 8:4477–4488

Tseng RL, Tseng SK, Wu FC, Hu CC, Wang CC (2008) Effects of micropore development on the physicochemical properties of KOH-activated carbons. J Chin Inst Chem Eng 39:37–47

# Chapter 43
# The Interest of Dairy Farmer on Extension Activity Related to Adopt the Mobile Anaerobic Digester Technology at East Java, Indonesia

**Aris Winaya, Sutawi, Herwintono, Ali Mahmud, and Telys Kurlyana**

**Abstract** The farmer extension activity was allowed to increase livestock productivity, like in dairy farmers. In the backyard and semi-intensive system of dairy cattle raising, the socio-economic identification of farmers was important for the success of the new technology adopted. Hence, the purpose of this study was to know the socio-economic background of the dairy farmer, included age, education level, raising cattle experience, and the number of cattle occupied was affected to the farmer interested in the extension activities related to adopt a mobile anaerobic digestion (AD) technology. The survey method with stratified random sampling was applied by involved 93 dairy farmers of Batu City, East Java, Indonesia. The depth interview for data collection and followed by multiple regression analysis. The socio-economic variables were significantly ($p < 0.05$) correlated to the interested in an extension activity. While the partial *t*-test showed that the age of farmer and education level were significantly ($p < 0.05$) on interest but farming experience and the number of cattle were not significantly ($p > 0.05$) to the interest of extension. The education level was a major role in the interested extension. Hence, the educational level could be the main consideration in the intervention of extension related to adopting the mobile AD technology.

**Keywords** Anaerobic digester · Dairy farmer · Extension · Interest · Manure

A. Winaya (✉) · Sutawi · Herwintono · A. Mahmud · T. Kurlyana
Faculty of Agriculture and Animal Science, University of Muhammadiyah Malang, Malang 65144, Indonesia
e-mail: winaya@umm.ac.id



N. S. Caetano and M. C. Felgueiras (eds.), *The 9th International Conference on Energy and Environment Research*, Environmental Science and Engineering,
https://doi.org/10.1007/978-3-031-43559-1_43

## 43.1 Introduction

The dairy livestock product is the one kind of agricultural product that could support the family income of small holder farmer in Indonesia. The fresh milk production of Indonesia until 2020 was reached 947,685.36 tons (BPS 2021). But on the contrary, Indonesia is still imported fresh milk around 80% to cover the national supply (Budiani and Sudirman, 2020). The factors that caused in low productivity of dairy cattle due to the most rearing management system were in the backyard and semi-intensive. In this system commonly farmers hold a small number of cattle (two to three cattle) and household intervention was dominantly (Susanty et al. 2019) and the lack of capital is the main problem that farmer could not increase the number of cattle ownerships. Hence, dairy farmers need help to access the financial resources for cattle ownership possibilities. Furthermore, it could reduce the gap in milk production by increasing cattle ownerships.

On the other hand, the global dairy sector was the second largest contributor to greenhouse gas emissions in the world (Gerber et al. 2013). Thus, every country must be aware of its responsibility for the sustainable development of global dairy products. With the worldwide environmental concerns and growing awareness of a circular economy, the focus of government attention should be on environmental policies related to sustainable manure and sewage management (MSM). However, the great diversity of MSM practices causes more difficulties and complexity in the sustainability and further policy evaluation of MSM (Zhang et al. 2020). The population of dairy cattle in Indonesia until 2020 was around 568,265 heads and East Java province was the highest population around 295,141 heads (BPS 2021). If the dairy cow manure waste is around 30 kg per head (Tsai and Liu 2016), thus the potency of dairy manure waste in East Java in 2020 reached 17,047,950 kg or 17,047 tons per day. While, in Batu city the dairy manure waste until 2020 reached around 380,520 kg (BPS 2021). Hence, this was a huge number waste for biogas production, but factually there was a small number of the manure (16%) that was processed to convert biogas production (Puwantini 2014).

The development of rural biogas utilization at the household level has several benefits, like improving sanitation, improving rural ecology, more efficient and effective energy production, reducing greenhouse gas emissions, and increasing agricultural products. Thus, the installation and application of biodigesters on farms by using dairy cattle manure as raw material for generating biogas will become a cheaper energy source and a good waste management option (Alberdi et al. 2018). However, the expansion of biogas technology among agricultural households was slow diffusion rates in many developing countries (Putra et al. 2017). In accordance with Indonesia Presidential Instruction No. 1 of 2006, related to the policy of supplying bioenergy raw materials, the Ministry of Agriculture has implemented a policy to increase the utilization of agricultural waste biomass and the development of biogas from animal manure. In Indonesia around 6000 digesters have been installed up to 2009 and more than 5000 installations have been installed between 2009 and 2013

(Putra et al. 2019). With more than 14 million smallholder households, the proportion of digesters installed in farmer households was only 0.07% which is considered very small (BPS 2013). Compared to other Asian countries such as China and Nepal, the rate of diffusion in Indonesia was very slow (Mwirigi et al. 2014). Meanwhile, small-holder dairy farming was still less attention in manure waste management due to the lack of simple and cheap technology-related manure waste management. The aimed of this study was to know the dairy farmer's interest in extension intervention, especially on desires to adopt the mobile AD equipment that constructed by our faculty team. Hence, it was important to explore the diverse socio-economic profiles of dairy farmers if they using the mobile AD technology to utilize the manure waste for biogas sources.

## 43.2 Materials and Methods

### *43.2.1 Study Location and Sampling*

This study was conducted at Batu City, East Java province, Indonesia (112° 17′10.90″ –122° 57′11″ East and 7° 44′55.11″ – 8° 26′35.45 South Latitude) (Fig. 43.1). The purposive random samples were applied for chosen at eight villages of Batu city region. The villages consideration since most of the inhabitant were a dairy cattle farmer and the member of dairy milk cooperation institution, also were selected based on socio-economic cultural background.

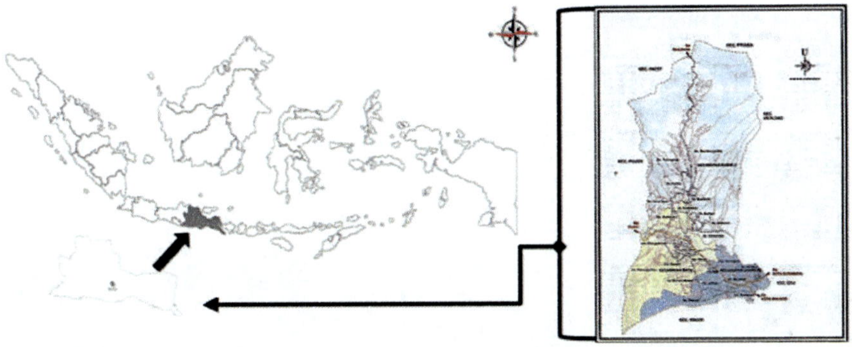

**Fig. 43.1** Sampling area of dairy cattle farmers at Batu city, East Java, Indonesia

### 43.2.2 Research Design and Statistics Analysis

An ex-post-facto cause to affect research design was used in this study. Ex-post-facto research is systematic empirical research that the researcher does not have direct control over the independent variable because the situation has already occurred or because it is inherently non-manipulable. In the ex-post-facto design, treatment is determined not by manipulation but by selection (Garai et al. 2017; Ryan 2013).

$$\mathrm{n} = \frac{N}{\left(1 + Ne^2\right)} \tag{43.1}$$

with, n = sample size; N = population size; $e$ = design margin of error (level of precision 10%) or standard value to 0.1.

By using the standard formula listed above, the required sample size becomes:

$$\mathrm{n} = \frac{1.251}{1 + 1.251(0, 1)^2} \text{ or n} = 92, 59 \text{ farmers (approx.93 farmers)}$$

The 93 small-scale dairy farmers were equally distributed in the eight villages of Batu city. Dairy farmer respondents were selected randomly from each village up to the total sample number were 93 respondents. The primary data were collected through interviews by selecting small-scale dairy farmers from eight villages. The instrument of the study to assess the interest of dairy farmers in extension activity was contains both closed and opened questions in an attempt to adequately generate research-related data. The questionnaire attempted to produce data on the age of the dairy farmer, the education level of the dairy farmer, the experience of raising dairy cattle, and the number of dairy cattle kept by the farmer. Figure 43.2 illustrated the situation of extension activity of dairy farmers based on focus group discussion (a, b), and an example of dairy cattle small-scale raising system (c, d).

The dependent variable in this study was the dairy farmer interested in extension activity related to a mobile AD technology. While the independent variables were included the age of the dairy farmer, the education level of the dairy farmer, the

(a) (b) (c) (d)

**Fig. 43.2** The extension activity of dairy farmer was based on Focus Group Discussion (**a, b**) and a dairy cattle small-scale raising system (**c, d**)

experience of the dairy cattle raised, and the number of dairy cattle ownerships. Table 43.1 shows the variables and their measurements of this study (Fig. 43.2).

The simple descriptive analysis of collected data was applied, which covered all response variables and provided the basic features for further analysis. The data was processed first before the analysis, it was necessary to edit, code, classify and tabulate which it could be analyzed. Data editing was necessary to ensure that the data was accurate, complete, compiled to facilitate coding and tabulation, and consistent with other facts collected. Coding was done to ensure that certain answers from respondents could be placed in one and only one cell in a certain set of categories (Kothari 2019).

To determine the factors that influence the dairy farmer's interest in the extension activity related to the mobile AD equipment in the study area, a multiple linear regression model using the ordinary least square (OLS) method was applied.

$$Ln\,Y = Ln\,a + b1\,ln\,X_1 + b2\,ln\,X_2 + b3\,ln\,X_3 + b4\,ln\,X_4 + e \tag{43.2}$$

*Ln Y* The interest of dairy farmer on extension activity.

**Table 43.1** Definition and measurement of variables

| Variable | Operational definition | Indicator | Measurements scale |
|---|---|---|---|
| *Dependent variable* | | | |
| The interest of dairy farmer in extension activity (Y) | The desire that encourages farmers to attend or participate in extension programs | 1. Extension subjects<br>2. Extension methods<br>3. Extension objectives<br>4. Facilities & infrastructures<br>5. Extension agencies ethics | used Likert scale; 1 to 5,<br>5 = strongly agree<br>4 = agree<br>3 = moderate<br>2 = less agree<br>1 = disagree |
| *Independents variables* | | | |
| Age of dairy farmer ($X_1$) | Life span as measured by years, calculated from birth | By year | Ratio scale |
| Education level of dairy farmer ($X_2$) | Farmer educational levels starting from kindergarten to undergraduate level | The length of farmer study time by year | Ratio scale |
| Experience of dairy cattle raising ($X_3$) | The expertise of dairy farmer in dairy cattle raising due to farmer has been done in this field for a long time | The length of dairy farmer has been raised dairy cattle by year | Ratio scale |
| Number of dairy cattle raised by farmer ($X_4$) | The population or the number of dairy cattle raised by farmers at the study | The dairy cattle raised by farmer in head | Ratio scale |

**Fig. 43.3** Anaerobic digester systems, **a** sub-marine digester system, commonly model applied by the dairy farmer with cattle ownership more than 7 heads; and **b** a mobile digester system which introduced in this study

| | |
|---|---|
| $a$ | Constant. |
| $b1,b2,b3,b4$ | $\ldots bn$ Regression coefficient. |
| $Ln\ X_1$ | Age of dairy farmer (year). |
| $Ln\ X_2$ | Education level of dairy farmer (year). |
| $Ln\ X_3$ | Experience of dairy cattle raising (year). |
| $Ln\ X_4$ | Number of dairy cattle raised by farmer (head). |
| $\varepsilon$ | Error term. |

The level of statistical significance of the variable was tested by using a *t-test* at a 5% level of significance. The multiple coefficients of determination ($R^2$) were used to measure the variation in the dependent variable defined by the independent variables. The value of $R^2$ is between 0 (zero) to 1 (one). A small value of $R^2$ means that the ability of the independent variable is very limited to explain the variation of the dependent variable. A value close to one means that the independent variable provides almost all the information to predict the variation of the dependent variable (Ghozali 2016). The F test was carried out to predict the effect of age, education level, experience of dairy cattle raising, and the number of dairy cattle raised by farmers to the interest in extension activities by simultaneously (jointly). The F test described all the independent variables were included in the model by a jointly affected dependent variable.

## 43.3 Results and Discussion

### *43.3.1 Characteristics of Dairy Cattle Farmer*

The age of the dairy farmer was a significant effect ($p < 0.05$) on the farmer interested in extension activities with a negative constant value (−0.141) (Table 43.2). It was illustrated that the older age of farmer, the lower interest in participating in extension and vice versa. This was by Baba et al. (2011), stated that the older farmer, the lower of participation in the extension activity. This condition was usually related to the physical ability decreases in older age thus this would also influence the farmer attended to the extension activities. The level of work productivity was in line with the age level and the productivity was stable around 45–60 years old in terms of a working team. However, it will continue until 65 years old on the individual worker (Börsch-Supan and Matthias 2016). Generally, the small-scale farming job was tended to the individual worker than the worker team. Hence, by the increasing age level, the farmers would feel confidence in raising cattle and no need for additional expertise. In Malaysia, farmers between 40 and 55 years old are generally more experienced in agricultural practices because they have learned on-farm from a young (Shafiai and Moi 2015).

Furthermore, the education level was significantly affected ($p < 0.05$) on the farmers interested in extension activities with a negative coefficient (−0.048). Hence the higher of education level, the lower the interest in extension activities. The low education level was not restraint the interest of farmers to get information through the extension activities. The level of education also would affect the level of farmers to adopt the extension intervention. Farmers who have higher education levels would feel more confident in their abilities, skills, and knowledge and more active in

**Table 43.2** Characteristic of dairy cattle farmer

| Variables | Coefficient | SE | *t* | Significancy |
|---|---|---|---|---|
| Age | −0.141 | 0.061 | −2.317 | 0.024[a] |
| Education level | −0.048 | 0.016 | −3.117 | 0.003[a] |
| Experience | 0.13 | 0.025 | 0.501 | 0.618[ns] |
| Number of cattle | −0.018 | 0.015 | −1.226 | 0.225[ns] |
| Adjusted $R^2$ | 0.126 | | | |
| Adjusted $R^2$ of age | 0.012 | | | |
| Adjusted $R^2$ of education level | 0.035 | | | |
| Adjusted $R^2$ of Experience | −0.012 | | | |
| Adjusted $R^2$ of number cattle | −0.008 | | | |
| F | 3.123 | | | |
| F sig | 0.022[a] | | | |

*Note* significance level at 5% [a]; ns: not significance

seeking information about their livestock business over various sources of information, including the internet. According to Yektiningsih et al. (2019) study that education level was significantly influenced by the adoption of technology on zero waste management of dairy cattle. As in Vietnam, farmers with higher education prefer to take the advantage of new technologies and find it easier to obtain the information (Chi et al. 2011). However, the continuance of extension activities will facilitate most farmers who were not get formal school or lower education to improve their knowledge by changed the mindset that knowledge was not only attained from formal education but also non-formal ways (Badar et al. 2015).

Generally, Indonesian dairy farmer has experience in dairy cattle raising more than 30 years. Similarly, the majority of Malaysian farmers have experienced above 20 years (Masso and Man 2016). But it was not guarantee that the farmer experienced would an affects interest in the participation of extension activities. This recent study showed that the experience of raising dairy cattle was not significantly affected ($p > 0.05$) to the farmers interested in extension activities. This result was according with previous study by Supriyanto et al. (2020) at Salaman district, Magelang Regency, Central Java, Indonesia that experience of cattle raising was no significant effect on the interested farmer to extension activities.

According to Taslim (2010), most dairy cattle raising in Indonesia was still in small number around 1 to 3 heads per household. Although there were farmers which have more than 7 cattle but it was a limited number. The diversity of dairy cattle ownership was influenced by differences in socio-economic conditions, such as the level of technology, capital capacity, availability of labor, and the area of land occupied. Unlike in Thailand, dairy cattle were previously dominated by small-farms with 5–10 heads, but recently the number of dairy cattle has increased to 20 cows per farmer (Knips 2021).

The size number of livestock ownership could help dairy farmers increasing income and meets the needs (Makatita 2013); Aji et al. (2019). Furthermore, it was possible farmers interested in extension could be assisted develop their dairy business. Nevertheless, in this study indicated that the number of livestock was not significantly affected ($p > 0.05$) to the farmer interested in extension activities. This situation occurred might that the dairy cattle raising in Batu City was not the primary business or secondary income. Hence, the number of dairy cattle ownerships has not endorsed the farmers to interest in extension activities regarding improving their rearing management, including a different type of AD technology. Rogers (2013) stated that the internal factors that might influence the adoption of innovations in term of the internal characteristic of farmers, included age, education level, number of family dependents, the intensity of extension activity, and the courage to take a risk.

### 43.3.2 *The Effect of Socio-Economic Variables to the Farmer Interested on Extension Activity*

The F test was carried to determine the independent variables, namely age, education level, experience in dairy cattle raising, and the number of livestock occupied which simultaneously (jointly) influenced the dependent variable or farmer's interest in an extension activity. This recent study showed that socio-economic background simultaneously significantly affected ($p < 0.05$) farmers interested in an extension activity. Thus, the regression model can be applied to predict the farmer interested in extension activities was affected by age, education level, experience in raising livestock, and the number of dairy cattle occupations by simultaneous (jointly).

Furthermore, based on the determination value (adjusted $R^2$) showed that education level was the highest value by 0.035. This was mean that education level should be a primary consideration of socio-economic background for the intervention of the extension activity on dairy cattle farmer. According to Rangkuti (2019) that the characteristic factors of innovation users were included education level, experience, business scale, and productivity since this was an important factor in the innovation adoption process in rural areas. Thus, if the dairy productivity is need to be increased by dairy farmers, they must be modernized in terms of knowledge, adoption and their personal, social and economic characteristics must be improved (Panchbhai et al. 2017).

## 43.4 Conclusion

Socio-economic variables which consisting of age, education level, the experience of dairy cattle raising, and the number of cattle occupied was a significant effect on the interest of dairy farmers to participate in the extension activities related to the adoption of a mobile AD technology. The educations level was the highest influence to the interest of dairy farmers in this study and the particular strategies to increase the adoption of innovation through the intervention of extension activities were included human capital resources both farmers and external sources such as livestock ownership, the environment, and the government policies. The extensive training is better accompanied by demonstrations on innovation intervention; strengthening farmer institutions; upgraded of extension workers quality, variation of media delivering, and methods of delivering information. Besides increasing the added value of dairy farmers, the use of biogas will encourage household independence in accessing fuel (biogas). If this is within a community of a region, it is possible will lead to a level of energy independence in the livestock-based rural areas.

**Acknowledgements** The authors gratefully thank the Rector of the University of Muhammadiyah Malang for the conducted the research successfully. Also, dairy cattle farmers at Batu city, East Java, Indonesia for supported as extension program participants.

**Funding** This work was supported by the University of Muhammadiyah Malang, Indonesia through The Institutional Research Application of Science and Technology Program of 2020/2021 by contract number: E.2.a/132/BAA-UMM/IV/2020.

## References

Aji GPW, Mastuti S, Hidayat NN (2019) Analisis kinerja ekonomi usaha sapi perah di Kecamatan Baturaden Kabupaten Banyumas [Economic performance analysis of dairy cattle business In Baturaden District, Banyumas Regency]. J Angon 1(1):38–47. http://jnp.fapet.unsoed.ac.id/index.php/angon/article/view/250/196

Alberdi HA, Sagala SAH, Wulandari Y, Srajar SL, Nugraha D (2018) Biogas implementation as waste management effort in Lembang Sub-district, West Bandung District. IOP Conf Ser Earth Environ Sci 158:012031. https://doi.org/10.1088/1755-1315/158/1/012031

Baba S, Isbandi T, Mardikanto Waridin (2011) Faktor-faktor yang mempengaruhi tingkat partisipasi peternak sapi perah dalam penyuluhan di Kabupaten Enrekang [Factors affecting participation level on extension of dairy farmer in Enrekang Regency]. J Ilmu dan Teknologi Peternakan (JITP). https://doi.org/10.20956/jitp.v1i3.680

Badar GA, Sri R, Sondi K (2015) Technical, social, and economic factors that influence the acceptance of grazing sheep business. UNPAD E-J Students 4(1):1–14. http://journal.unpad.ac.id/ejournal/article/view/5814/3096

Börsch-Supan A, Matthias W (2016) Productivity and age: Evidence from work teams at the assembly line. J Econ Ageing 7:30–42. S2212828X15000304. https://doi.org/10.1016/j.jeoa.2015.12.001

BPS [Badan Pusat Statistik] [Indonesian Central Statistics Bureau] (2013) Agricultural statistics based on census 2013. http://st2013.bps.go.id/dev2/index.php/site/index. [in Bahasa Indonesia]

BPS [Biro Pusat Statistik][Indonesian Central Bureau of Statistics] (2021) Produksi susu segar menurut provinsi (ton) 2018–2020] [Fresh milk production based on province (tons) 2018–2020]. Available at: https://www.bps.go.id/indicator/24/493/1/produksi-susu-segar-menurut-provinsi.html. (In Bahasa Indonesia). Last accessed 24 Mar 2021

Budiani NK, Sudirman IW (2020) Analysis of effect of consumption, production, and inflation levels on the import of milk in Indonesia. Int J Educ Soc Sci Res 3(01):143–149. https://doi.org/10.37500/IJESSR.2020.3012

Chi TTN, Tuyen TQ, Mai TTN, Khang NV, Tinh LV, Anh LTN (2011) Effect of community-based farmer groups with the same preference on adoption of technology. Omonrice 18:157–166. http://www.clrri.org/ver2/uploads/noidung/18-20.pdf

Garai S, Garai S, Maiti S, Meena BS, Ghosh MK, Bhakat C, Dutta TK (2017) Impact of extension interventions in improving livelihood of dairy farmers of Nadia district of West Bengal, India. Trop Anim Health Prod 49(3):641–648. https://doi.org/10.1007/s11250-017-1244-5

Gerber PJ, Steinfeld H, Henderson B, Mottet A, Opio C, Dijkman J, Falcucci A, Tempio G (2013) Tackling climate change through livestock: a global assessment of emissions and mitigation opportunities. Food and Agriculture Organization of the United Nations (FAO), Rome. ISBN: 9789251079201. http://www.fao.org/.../i3437e00.htm

Ghozali I (2016) Aplikasi Analisis Multivariate dengan Program IBM SPSS 23 [Aplication of multivariate analysis using IBM SPSS 23 program]. Badan Penerbit Universitas Diponegoro. Semarang. Cetakan Ke-8, 464 hlm

Knips V (2004) Review of the livestock sector in the Mekong countries. http://www.fao.org/ag/againfo/resources/en/publications/sector_reports/lsr_mekong.pdf. Last accessed 06 Mar 2021

Kothari CR (2019) Research methodology: methods and techniques, Fourth Edition. New Age International (P) Ltd., Publisher, New Delhi, India, p 480

Makatita J (2013) Hubungan antara karakteristik peternak dengan skala usaha pada usaha peternakan kambing di Kecamatan Leihitu Kabupaten Maluku Tengah [The relationship between farmer characteristics and business scale in goat farming business in Leihitu District, Central Maluku Regency. Agrinimal 3(2):78–83. https://ejournal.unpatti.ac.id/ppr_paperinfo_lnk.php?id=716

Masso WYA, Man N (2016) Maturity level of rural leaders in selected paddy farming technologies in Muda Agricultural Development Authority (MADA)—Malaysia. Asian Soc Sci 12(7):10–16. https://doi.org/10.5539/ass.v12n7p10

Mwirigi J, Balana BB, Mugisha J, Walekhwa P, Melamu R, Nakami S, Makenzi P (2014) Socio-economic hurdles to widespread adoption of small-scale biogas digesters in Sub-Saharan Africa: a review. Biomass Bioenergy 70:17–25. https://doi.org/10.1016/j.biombioe.2014.02.018

Panchbhai GJ, Siddiqui MF, Sawant MN, Verma AP, Parameswaranaik J (2017) Correlation analysis of socio-demographic profile of dairy farmers with knowledge and adoption of animal husbandry practices. Int J Curr Microbiol App Sci 6(3):1918–1925. https://doi.org/10.20546/ijcmas.2017.603.218

Putra RARS, Liu Z, Lund M (2017) The impact of biogas technology adoption for farm households—empirical evidence from mixed crop and livestock farming systems in Indonesia. Renew Sustain Energy Rev 74:1371–1378. https://doi.org/10.1016/j.rser.2016.11.164

Putra ARS, Pedersen SM, Liu Z (2019) Biogas diffusion among small scale farmers in Indonesia: an application of duration analysis. Land Use Policy 86:399–405. https://doi.org/10.1016/j.landusepol.2019.05.035

Puwantini TB (2015) Pemanfaatan limbah usaha ternak sapi perah untuk biogas mendukung kemandirian energi di perdesaan: kasus di Desa Bendosari, Malang, Jawa Timur [Utilization of Dairy Cattle Waste for Biogas in Supporting Energy Security in Rural Areas: A Case Study in Bendosari Village, Malang, East Java]. Prosiding Seminar Nasional Hari Pangan Sedunia Ke-34: Pertanian-Bioindustri Berbasis Pangan Lokal Potensial. [National seminar on world food day: Bioindustry-agriculture based on local food potential Makassar, 4 Nov 2014. https://pse.litbang.pertanian.go.id/ind/pdffiles/(PROS_2014_MP_10_SET_Tri%20Bastuti.pdf. [abstract in English]

Rangkuti PA (2019) Analisis peran jaringan komunikasi petani dalam adopsi inovasi traktor tangan di Kabupaten Cianjur, Jawa Barat [Analysis of the role of farmer communication networks in the adoption of hand tractor innovations in Cianjur Regency, West Java]. J Agroeconomics 27:45–60. https://doi.org/10.21082/jae.v27n1.2009.45-60

Rogers E (2013) Diffusion of innovations. Fifth Edition. Free Press, New York, London, Toronto, Sydney

Ryan TP (2013) [Wiley series in probability and statistics] Sample size determination and power (Ryan/Sample). Methods Determining Sample Sizes. pp 17–55. https://doi.org/10.1002/9781118439241.ch2

Shafiai MHM, Moi MR (2015) Financial problems among farmers in Malaysia: Islamic agricultural finance as a possible solution. Asian Soc Sci 11(4):1–16. https://doi.org/10.5539/ass.v11n4p1

Supriyanto, Haryadini AF, Nurdayati (2020) Analisis faktor yang mempengaruhi minat peternak dalam mengembangkan ternak kambing. [Analysis of factors affecting farmers' interest in developing goats raising]. J Pengembangan Penyuluhan Pertanian. https://doi.org/10.36626/jppp.v17i32.543

Susanty A, Bakhtiar A, Puspitasari NB, Susanto N, Handjoyo DKS (2019) The performance of dairy supply chain in Indonesia: a system dynamics approach. Int J Product Perform Manag 68(6):1141–1163. https://doi.org/10.1108/IJPPM-09-2018-0325

Taslim (2010) Pengaruh faktor produksi susu usaha ternak sapi perah melalui pendekatan analisis jalur di Jawa Barat [The impact of factor on dairy production small-holder with path analysis in West Java]. J Ilmu Ternak 10(1):52–56. https://doi.org/10.24198/jit.v10i1.461

Tsai WT, Liu SC (2016) Thermochemical characterization of cattle manure relevant to its energy conversion and environmental implications. Biomass Convers Biorefinery 6(1):71–77. https://doi.org/10.1007/s13399-015-0165-7

Yektiningsih E, Suryaminarsih P, Hidayat R (2019) Adoption of agricultural innovations in the context of zero waste: The case of dairy cattle biogas waste. Eurasian J Biosci 13(2):861–864. http://ejobios.org/download/adoption-of-agricultural-innovations-in-the-context-of-zero-waste-the-case-of-dairy-cattle-biogas-7182.pdf

Zhang J, Zhang L, Wang M, Brostaux Y, Yin C, Dogot T (2020) Identifying key pathways in manure and sewage management of dairy farming based on a quantitative typology: a case study in China. Sci Total Environ 143326. https://doi.org/10.1016/j.scitotenv.2020.143326

# Chapter 44
# Exploring the Environmental Significance of *Il-Magħluq Ta' Marsaskala*: A Study on Water Quality Within a Special Area of Conservation

**Dale Bartolo, Juan José Bonello, Francesca Spagnol Gravino, and Raymond Caruana**

**Abstract** *Il-Magħluq ta' Marsaskala* is a small saline marshland in the south of Malta. It is characterized with brackish water hosting peculiar salt-tolerant marsh communities and protected species that include the Mediterranean killifish, *Aphanius fasciatus*. This Natura 2000 Special Area of Conservation (SAC) consists of two different types of Annex I (Council *Directive* 92/43/EEC) habitats which are Habitat 1150* and Habitat 1410. This study focuses on water quality characterization and its implications within this SAC. Monitoring of water quality and abundance of the killifish was carried out over a period of one year (November 2017—November 2018). Temperature, pH, and dissolved oxygen were taken on-site while phosphates and nitrates were analyzed in the laboratory. The fluctuations observed in the water quality parameters can be attributed to seasonal variations, and possibly to anthropogenic activities surrounding the study area. The water quality seems also to be affecting the killifish population since a total of 22 individuals were recorded during this study. The density of killifish is excessively low when compared to data collected from similar studies within the same study area in previous years.

**Keywords** *Aphanius fasciatus* · Brackish · Malta · Mediterranean · Special area of conservation · Wetlands

---

D. Bartolo (✉) · J. J. Bonello · F. S. Gravino
Institute of Applied Sciences MCAST, Triq Kordin, Raħal Ġdid, Malta
e-mail: dale.bartolo.a101296@mcast.edu.mt

R. Caruana
Aquaculture Directorate, Forti San Luċjan, Triq il-Qajjenza, Marsaxlokk, Malta



N. S. Caetano and M. C. Felgueiras (eds.), *The 9th International Conference on Energy and Environment Research*, Environmental Science and Engineering,
https://doi.org/10.1007/978-3-031-43559-1_44

## 44.1 Introduction

*Il-Magħluq ta' Marsaskala* is a small saline marshland in the south-eastern region of Malta. 'Magħluq' means 'closed' in Maltese, in this case a body of water separate from the sea. It is made up of brackish water hosting peculiar salt-tolerant marsh communities. It is also the habitat of some protected species such as the Maltese national fish, *A. fasciatus*. *Il-Magħluq* site consisted of two interconnected fishponds lined with layered stones. The brackish water found here forms because of freshwater from the valley system, as well as water from the surrounding fields and rainwater entering the pool that combines with the seawater that enters through the connection under the road (MEPA 2012).

Even though the marshland is separated from the sea by a narrow road, there is still a connection to the marine environment via two pipes laid beneath the road. Therefore, sea water incursions are common during storms and bad weather. Whilst the saline marshland is surrounded by an urban area, the pool also has agricultural land on its western and southern sides. This research will focus on characterizing the water quality of this Natura 2000 Special Area of Conservation (SAC) and studying the anthropogenic activity that may affect it.

*Il-Magħluq* site is one of the two remaining marshes in the south of Malta, supporting a community of species based on various salt tolerant and wetland plants such as rushes (MEPA 2012). As previously mentioned, the SAC also hosts the Mediterranean killifish, *A. fasciatus*. This species is one of three *Aphanius* species found in the Mediterranean basin and is currently distributed in the saline coastal waters of the central and eastern Mediterranean, in salt flats and occasionally in inland fresh water (Wildekamp 1993). The population of the *A. fasciatus* is only found in a handful of sites, all of which are hydrologically isolated from each other (Deidun et al. 2002).

This study complements similar studies done on the same site since pressure on the water quality and the fauna present in this water body is always increasing; pressures include run-off from the nearby agricultural fields, alien species and illegal dumping (MEPA 2012). Such studies are essential because according to EUNIS, even though this species' threat status is of least concern, it still requires significant conservation and restoration measures to make it viable in the long-term, or to improve the quality and availability of its habitat. Subsequently, such studies will ensure adequate management of both the area and the species.

## 44.2 Materials and Methods

Water quality data was collected monthly from *Il-Magħluq* site between November 2017 and November 2018. The site was first divided into three different points (Fig. 44.1) to reflect the different ponds that characterize the study area and for

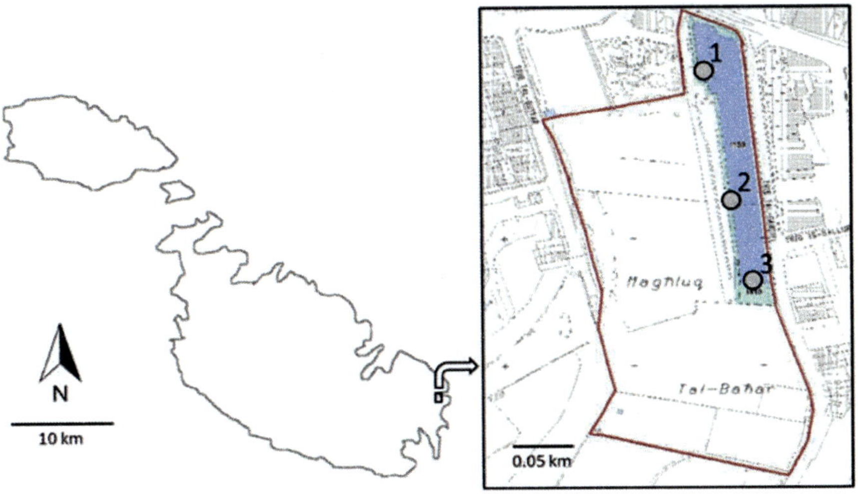


**Fig. 44.1** The study site and the selected sampling points. The border around the study site depicts the Natura 2000 site boundary

each point the same approach was used: temperature, dissolved oxygen and pH data were collected *in-situ*; whilst phosphates and nitrates were analyzed in the laboratory.

During the same sampling period and from the same sampling points, traps were temporarily installed so the *A. fasciatus* population could be recorded. A capture-release method was used. Meteorological information was recorded per session on the day before on the premise that weather and wind data could potentially affect both the water quality and the killifish itself. The presence of alien species such as ducks and mullet and other materials were also recorded.

After the raw data was compiled, a preliminary analysis was done to identify any patterns. For the statistical analysis, SPSS ver. 24 (IBM Corp.) was used.

## 44.3 Results and Discussion

During the one-year sampling period only 22 individuals of *A. fasciatus* were captured (Fig. 44.2a–e). The discrepancy is quite significant when compared to the 417 individuals recorded in the same study area in 2011 (Zammit-Mangion et al. 2011). Out of the 22 individual that were recorded, 15 were females while only 5 were male. The other 2 could not be identified since they were still juvenile. Individuals were captured in May and July. It is noted that in comparison to the study previously carried out in 2011, from the 417 fish captured in that study, 60% of them were females (Zammit-Mangion et al. 2011). According to the authors, reproduction and recruitment should proceed seamlessly due to high presence of females within the population. However, the data collected in this study suggests otherwise, with several features in fact reflecting a highly vulnerable population, such as a very low presence of juveniles and the overall small population.

Species from the *Aphanius* genus are known to be tolerant to a wide range of abiotic factors, such as wide temperature ranges, namely 18 to 37 °C (Frenkel and Goren 1997, 2000). However, one of the reasons that could be linked to such low numbers of fish captured in this study is that the species might struggle to maintain the reproductive system due to the impacts from anthropogenic activities in close proximity to the study area. Similar observations and attributions to anthropogenic

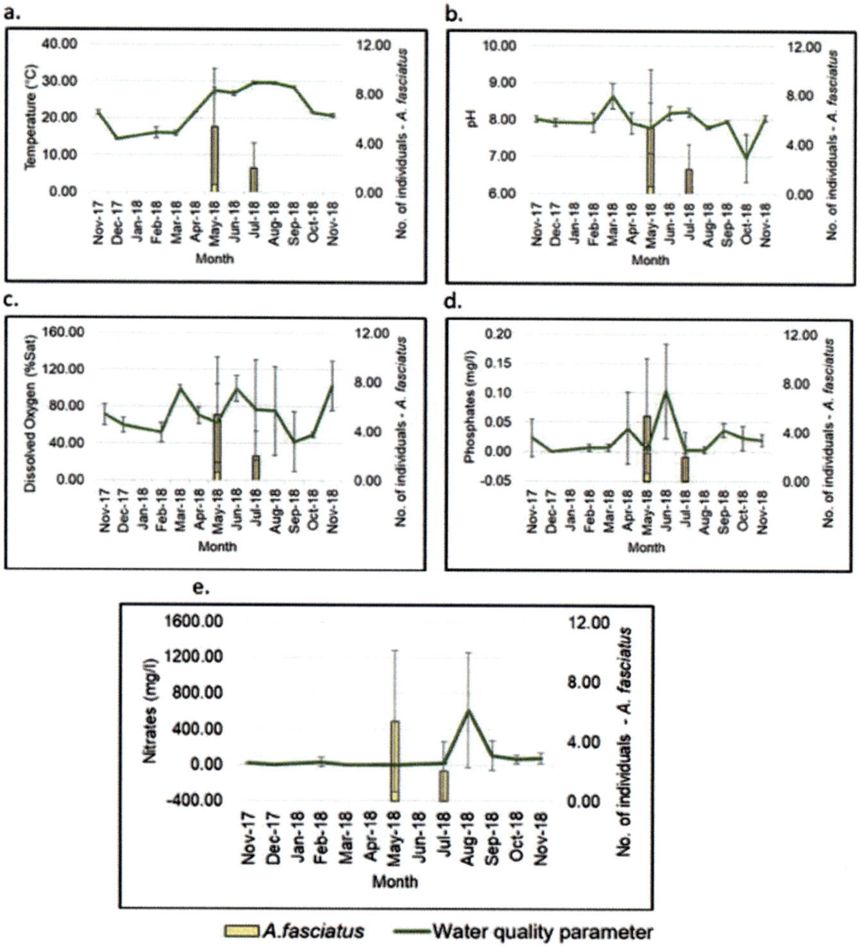


**Fig. 44.2** The mean monthly water quality parameters (based on the results obtained from the three sampling points), compared to the mean abundance of the *A. fasciatus*. The error bars represent the standard deviation from the mean. a. Temperature [°C]; b. pH; c. Dissolved Oxygen [%Sat]; d. Phosphates [mg/l]; e. Nitrates [mg/l]

activity were made in studies on different species from the *Aphanius* genus (Varó et al. 2008; Keivany 2013).

Studies have also suggested that predation could play a vital role in population and distribution (Clavero et al. 2007; Keskin 2007). Due to the connection to the sea, a number of predatory species such as the European seabass (*Dicentrarchus labrax*) together with omnivorous fish species like Gilt-head seabream (*Sparus aurata*) can also be found in the study area and subsequently affect the population of *A. fasciatus*. Such predatory fish can also impact the killifish recruitment due to feeding on the killifish eggs attached to the floating algal mats.

The water quality data was analyzed using Kruskal–Wallis H Test and whilst there was no statistically significant difference between the different sampling points, there was a statistical significance with regards to the months. The seasonal variation can also be seen in Fig. 44.2a–e. The annual mean temperature value during the one-year sampling period was that of 22.8 (±5.4 °C), with the highest temperature recorded in August at 29.7 °C. On the other hand, the study area recorded lower temperatures in the months of December, February, and March with 14.3, 16.1 and 16.0 °C, respectively.

Temperature levels influence both the biological and chemical features of surface water. The temperature also affects the dissolved oxygen level within the waterbody (Smith et al. 2010). Large inter-seasonal variations can be noted for different physical parameters. This is especially true for the water temperatures taken during the intense summer months of July and August where temperatures ranged between 29.2 and 29.9 °C and from 29.4 to 30.0 °C respectively. These were nearly double those recorded during the month of December, where temperatures ranged from 14.2 to 14.5 °C. The same pattern was recorded in the 2011 (Zammit-Mangion et al. 2011), where the same area was studied and whereby temperatures ranged from 13.9 to 32.1 °C.

Large inter-seasonal fluctuations in temperature can limit the population of any given species. Given that *A. fasciatus* is able to tolerate a wide range of temperature, indirect effects attributed to fluctuations of temperature can still have negative implications on species and their habitat. Such indirect effects can consist of temperature influence on toxic substances (Smith et al. 2010). In this site, toxic substances can find their way in the waterbody through the run-off from the nearby agricultural fields, as well as from the adjacent roads and from illegal dumping of waste material. An example of this would be ammonia, potentially originating from fertilizers from the surrounding fields; toxicity of ammonia (measured as total ammonia) tends to increase as temperature increases (EPA 1999).

In terms of dissolved oxygen during the sampling period there was a mean saturation value of 71.6% sat. (±30.5% sat.). September 2018 registered the lowest monthly mean value of 42.0% saturation dissolved oxygen while the highest monthly mean value was obtained in November 2018 with 102.5% saturation.

As previously mentioned, an implication associated with temperature variation is its direct effects on dissolved oxygen; high temperature results in low amount of dissolved oxygen while high amount of dissolved oxygen can be a consequence of low temperatures (Brown et al. 2000). A high amount of dissolved oxygen in a waterbody can be harmful to aquatic life. In such scenarios, although it is not that common, fish can suffer from “gas bubble disease”. Such bubbles can block the flow of blood through blood vessels causing death. External bubbles can also arise and can be seen on skin, fins, and other tissues (Lenntech 2021).

During the one-year sampling period the mean pH value was that of 7.9 ($\pm$0.5); the lowest being recorded in October 2018 with a pH value of 7.0, and the highest was recorded in March 2018 with a pH value of 8.6. From April 2018 onwards, the Malta Cleansing and Maintenance Department started cleaning the site. This was done as part of a project to maintain and manage the area. Cleaning consisted of the removal of oils, plastic, other materials, and residues that could have been found in the waterbody (Falzon 2019). The frequent cleaning could have also caused the pH of the water to change monthly, an observation that was highlighted by a study in similar conditions (Gomes et al. 2016).

Apart from dissolved oxygen, temperature also influences the pH level in a waterbody. Low pH can encourage the solubility of heavy metals. As the concentration of heavy metals increases, their toxicity also increases and as a result, this can limit the growth and reproduction while increasing the mortality rate of such endangered species. The influence of pH is also seen on ammonia; at low pH levels ammonia combines with $H^+$ to produce an ammonium salts ($NH_4^+$ salts). However, when the pH level rises the reverse reaction occurs producing the more toxic $NH_3$ and subsequently the mortality rate will rise due to an increase in ammonia concentration (Eddy 2005).

For phosphates, an annual mean of 0.02 mg/l ($\pm$ 0.04 mg/l) was observed. The highest value was recorded in June 2018 with a concentration of 0.10 mg/l. The lowest point was recorded in July 2018 with 0.0004 mg/l. The annual mean value of nitrates was 84.2 mg/l ($\pm$172.2 mg/l). It is noted that March and May of 2018 registered a mean value of less than 3.3 mg/l ($\pm$5.8 mg/l), while August (2018) registered the highest average with 620.0 g/l ($\pm$640.9 mg/l).

Phosphorus pollution from point sources has gradually become less significant and the total phosphorus concentration in lakes has declined to around 0.02 mg P/l in 2018 (EEA 2021). However, diffuse runoff from agricultural land is still considered to be a major source of phosphorus in many European lakes. High amounts of phosphates can result in eutrophic conditions which will eventually impact and decrease the amount of dissolved oxygen in the waterbody (Aloe et al. 2014). Similarly, high concentrations of nitrates, which could be the outcome of increase in run-off from the nearby agricultural fields, can directly yield eutrophic conditions. Nitrates can also indirectly induce strong phosphate eutrophication in wetlands during which oxygen levels in the water will eventually decrease thus impacting the mortality rate of the living organisms as well (Smolders et al. 2009). According to the Water Framework Directive (2000/60/EC) and the Nitrates Directive (91/676/EEC) surface freshwater bodies with a concentration of more than 50 mg/l of nitrates are considered to be polluted or at risk of pollution. This value was exceeded between August and November 2018. During this period it was observed that the sampling point 3 had the highest nitrate concentration from the three sampling points. Given that this sampling point is in closer proximity to adjacent agricultural fields on its western and southern sides, the implications of rainwater run-off might be more predominant at this point.

The water quality characteristics highlighted in this study could imply that the overall small size of the population of the killifish is due to stress-induced mortality, arising from the large seasonal fluctuations in abiotic factors attributed to anthropogenic activity. Such fluctuations are the consequence of several factors; such factors also consist of natural attributes, such as freshwater runoff at the mouth of the valley after heavy rainfall, as well as anthropogenic ones, such as engineering works at the wetland-sea interface, fertilizer and pesticide-contaminated runoff from adjacent fields, and illegal dumping of substances into the wetland (MEPA 2012). Besides abiotic conditions, biotic factors such as predation can also influence the population of the killifish (Clavero et al. 2007; Keskin 2007).

## 44.4 Conclusion

The results demonstrated that seasonal variation and possible anthropogenic activity play a vital role in the water quality of the study area. During the one-year sampling period the presence of the killifish was also recorded. The waterbody in this study underwent substantial seasonal abiotic fluctuations within the habitat, as well as anthropogenic disturbance in the form of engineering works, chemical run-off from the surrounding fields and illegal dumping of waste material. A total of 22 *A. fasciatus* individuals were recorded. When comparing the number of individuals recorded in this study with the 417 individuals recorded in a similar study in 2011 it is clear that the *A. fasciatus* population at the *Il-Magħluq* marshland site is highly vulnerable, even though according to EUNIS this species' threat status is of least concern. Such an alarming scenario clearly shows the need for further studies in trying to understand the biotic and abiotic factors that could contribute to the population decline of the killifish. Subsequently, these studies would promote the preservation of both the Special Area of Conservation and the population of the killifish. The result of the study gives an insight of what is going on with respect to the existing management plan and it can potentially be useful to improve on current legislation and policies to protect both the wetland and the inhabiting organisms, such as the killifish.

**Acknowledgements** The authors are grateful to the Environment and Resources Authority (ERA) for granting them the necessary permits to handle the legally protected *A. fasciatus* and to allow water sampling to take place in a Natura 2000 site.

## References

Aloe AK, Bouraoui F, Grizzetti B, Bidoglio G, Pistocchi A, Petrescu AM (2014) Managing nitrogen and phosphorus loads to water bodies: characterization and solutions towards macro-regional integrated nutrient management. Publications Office of the European Union, Luxembourg

Brown E, Peterson A, Kline-Robach R, Smith K, Wolfson L (2000) Developing a watershed management plan for water quality: An introductory guide. Millbrook Printing, Michigan

Clavero M, Blanco-Garrido F, Prenda J (2007) Population and microhabitat effects of interspecific interactions on the endangered Andalusian toothcarp (*Aphanius baeticus*). Environ Biol Fishes 78:173–182

Deidun A, Arcidiacono I, Tigano C, Schembri PJ (2002) Present distribution of the threatened killifish *Aphanius fasciatus* (Actinopterygii, Cyprinodontidae) in the Maltese Islands. Central Mediterr Nat 3:177–180

Eddy F (2005) Ammonia in estuaries and effects on fish. J Fish Biol 67:1495–1513

EEA (2021), Indicator Assessment—Nutrients in freshwater in Europe. https://www.eea.europa.eu/data-and-maps/indicators/nutrients-in-freshwater/nutrients-in-freshwater-assessment-published-10 (Accessed 13 July 2021)

EPA (1999) Update of ambient water quality criteria for ammonia. U.S. Environmental Protection Agency, Washington DC. EPA-822-R-99-014

Falzon G (2019) Upgrading of Marsascala's "Il-Magħluq" inlet to commence after summer. https://www.tvm.com.mt/en/news/se-jitnaddaf-u-jissebbah-il-upgrading-of-marsascalas-il-maghluq-inlet-to-commence-after-summermaghluq-ta-wied-il-ghajn-wara-s-sajf/ (Accessed: 24 Feb 2019)

Frenkel V, Goren M (1997) Some environmental factors affecting the reproduction of *Aphanius dispar* (Rüppell, 1828). Hydrobiologia 347:197–207

Frenkel V, Goren M (2000) Factors affecting growth of killifish, *Aphanius dispar*, a potential biological control of mosquitoes. Aquaculture 184:255–265

Gomes HI, Mayes WM, Rogerson M, Stewart DI, Burked IT (2016) Alkaline residues and the environment: a review of impacts, management practices and opportunities. J Clean Prod 112:3571–3582

Keivany Y (2013) Threatened fishes of the world: *Aphanius isfahanensis* Hrbek, Keivany & Coad, 2006 (Cyprinodontidae). Int J Ichthyol 19:67–70

Keskin E (2007) Molecular evidence for the predation of critically endangered endemic *Aphanius transgrediens* from the stomach contents of world wide invasive *Gambusia affinis*. Mitochondrial DNA Part A 27:1210–1215

Lenntech (2021) Why oxygen dissolved in water is important. https://www.lenntech.com/why_the_oxygen_dissolved_is_important.htm (Accessed 30 May 2021)

MEPA (2012) Natura 2000 management plan (SAC)—Il-Magħluq tal-Baħar ta' Marsaskala. In: Rural development programme for 2007–2013

Smith AJ, Delorme LD (2010) Ostracoda. In: Thorp JH, Covich AP (eds) Ecology and classification of North American freshwater invertebrates. Academic Press, Cambridge, pp 725–771

Smolders A, Lucassen E, Bobbink R, Roelofs J, Lamers L (2009) How nitrate leaching from agricultural lands provokes phosphate eutrophication in groundwater fed wetlands: the sulphur bridge. Biogeochemistry 98:1–7

Varó I, Amat F, Navarro J (2008) Acute toxicity of dichlorvos to *Aphanius iberus* (Cuvier & Valenciennes, 1846) and its anti-cholinesterase effects on this species. Aquat Toxicol 88:53–61

Wildekamp RH (1993) The genus *Aphanius Nardo*. In: Watters BR (ed) A world of killies, atlas of the oviparous cyprinodontiform, fishes of the world, Vol I. American Killifish Association, Hishawaka Indiana, pp 19–67

Zammit-Mangion M, Deidun A, Vassallo-Agius R, Magri M (2011) Management of threatened *Aphanius fasciatus* at Il-Maghluq, Malta. In: Proceedings of the 10th international conference on the mediterranean coastal environment, MEDCOAST

# Part VI
# Energy Efficiency

# Chapter 45
# Increasing Energy Efficiency of Electro-Hydraulic Oil Systems to Reduce Industrial Carbon Emissions

**Adriano A. Santos, António Ferreira da Silva, Carlos Felgueiras, and Filipe Pereira**

**Abstract** Oil-hydraulic system drives are conventionally performed by constant-speed electric engines, and typically operate with constant displacement pumps as their central transmission element, and oversized pressures are limited by relief valves. On the other hand, in applications where different speeds of approach and work are needed, different flows can be obtained with regenerative circuits, flow control valves, servo pumps or by different pump operating speeds. Consequently, a higher energy consumption is imposed by the relief flow valve. These and other approaches have been a growing focus of study in industry and academia, due to their potential for substantial increases in energy efficiency and reduction of electrical energy consumption and $CO_2$ emissions. This paper presents a study of the potential economic gains associated with the electric drive of mobile or industrial (static) hydraulic systems characterized by variable drive units. With this drive topology, the electrical consumption associated with the engine drive is compensated, increasing energy efficiency over conventional systems, while reducing industrial carbon emissions and pressures on the dead phases of the operating cycle.

**Keywords** Carbon emissions reduction · Energy efficiency · Energy efficiency of hydraulic system · Energy savings · Hydraulic pumps · Variable-speed drive

---

A. A. Santos (✉) · A. F. da Silva
CIDEM, School of Engineering (ISEP), Polytechnic of Porto (P.Porto), Rua Dr. António Bernardino de Almeida 431, Porto, Portugal
e-mail: ads@isep.ipp.pt

INEGI—Institute of Science and Innovation in Mechanical Engineering and Industrial Engineering, R. Dr. Roberto Frias S/N, Porto, Portugal

C. Felgueiras · F. Pereira
CIETI—School of Engineering (ISEP), Polytechnic of Porto (P.Porto), Rua Dr. António Bernardino de Almeida 431, Porto, Portugal



N. S. Caetano and M. C. Felgueiras (eds.), *The 9th International Conference on Energy and Environment Research*, Environmental Science and Engineering,
https://doi.org/10.1007/978-3-031-43559-1_45

## 45.1 Introduction

As a result of declining fossil energy resources and constantly rising fuel prices, as well as of increasingly stringent emission standards, industrial equipment manufacturers are constantly faced with the search for energy saving alternatives. In hydraulic oil systems, either mobile or industrial (static), the hydraulic pumps used to operate actuators (linear or rotary) are normally driven by electric engines. However, although this is a generalized approach, pump drive systems are responsible for a large part of the energy consumed worldwide by electric engines, reaching 70% of the total electric energy consumed in certain industrial facilities (Doe 2004). Therefore, and taking these indicators into account, it must be assumed that there are numerous opportunities to reduce energy consumption in systems with hydraulic pumps, through the adoption of more sustainable practices. These practices would be aimed not only at consumption efficiency but also towards creating adaptability towards various needs, retrofitting, and operating practices. On the other hand, it must be also considered that pump flow control is normally carried out through bypass systems, flow control valves or by varying the pump flow. If the use of a by-pass or a flow valve results in a large loss of energy transmitted to the fluid, the speed variation and consequently the flow supplied to the installation becomes the most efficient solution. The flow variation can be performed by variable flow pumps (high cost) or by a variable speed drive (VSD) using a variable frequency drive (VFD) (Doe 2004), by a set of constant displacement "digital" pumps, of different sizes, working in parallel (Heitzig et al. 2012) or by retrofitting, improving engine efficiency.

There are several systems in which the use of speed controllers for electric drives can bring considerable energy savings. This includes applications that operate with fluids such as pumps, compressors, fans with varying flow requirements, and conveyors, machine tools, etc. (Almeida et al. 2005) so, this is an area where equipment retrofitting can improve the energy efficiency of electric drive motors.

This paper presents several approaches to the problem of energy saving in industrial installations. Some approaches to the hydraulic methodologies used for manufacturing, in each of the activity sectors, and the efficiency of motor drives and hydraulic scenarios are presented in Sect. 45.2. In Sect. 45.3, an analysis of possible gains in consumption reduction is presented for some scenarios of manufacturing and industrial facilities. In Sect. 45.4, the possible energy efficiency improvements achieved, and the gains obtained will be summarized.

## 45.2 Efficient Electro-Hydraulic Systems

Ever since Pascal's physical principle, the basic law of pressure transfer for fluids at rest, was known, its development and applicability has spanned nearly four centuries (Xu et al. 2020). Based on this principle, it was possible to perform linear and rotary movements, by hydraulic transmission, which consequently led to the most diverse uses of hydraulics in mobile and industrial applications. To complement this trend, technological advances in manufacturing, electronics and control systems have implemented more precise, faster, and more efficient controls and therefore more energy is available. This will be the basis for the search and improvement of energy efficiency in the activation of hydraulic systems, an object of study at both industrial and academic levels in the last decade (Schmidt and Hansen 2022). This can be confirmed by the proposals presented by different manufacturers of hydraulic equipment such as digital displacement technology by Danfoss Power Solutions (2022), digital flow control of multi chamber cylinders by Volvo Construction Equipment (2020), development of variable-speed pump by Bosch Rexroth (2022), efficient floating cup piston pumps by Bucher Hydraulics (2022), energy recovery (Li et al. 2021), among others.

Despite all these efforts to reduce electricity or fossil energy consumption, we must keep in mind that the energy used in such drives is usually generated by a centralized source: an internal combustion engine, burning fuel, or a high-powered electric engine (Quan et al. 2014), consuming electrical energy. So, any measures, however small, that aim to improve the efficiency of hydraulic systems, will have a significant impact on the energy efficiency of industrial equipment. This small increment will be part of a whole that will contribute to closing the gaps in the implementation of environmental aspects in the 2020s, which generally involve the use of renewable energy, electric vehicles, building and efficient industrial motors retrofits (IEA 2021a). Figure 45.1 presents the next objectives for the Stated Policies Scenario (STEPS), the Announced Pledges Scenario (APS) and Net Zero Emissions Scenario (NZE) for the years 2020s.

In this sense, the contribution of this work to the reduction of industrial $CO_2$ emissions involves the retrofit of the drive systems, either by replacing them with more efficient engines or using a VFD. These interventions, though small, will become a

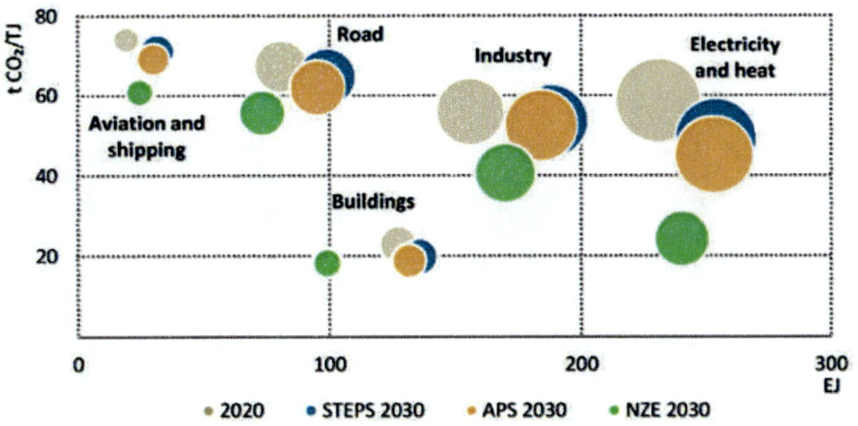


**Fig. 45.1** Energy use, carbon intensity and $CO_2$ emissions by sector and scenario, 2020–2030 12, (t $CO_2$/TJ = tons of carbon dioxide per terajoule; EJ = exajoule)

major contribution because the global market for hydraulic equipment is expected to reach US$ 51.65,310 million in 2027. In other words, a growth in the compound annual growth rate (CAGR) of 3.8% from 2020 to 2027 is expected (AMR 2027). So, any change of the equipment and any "*change in the speed or the displacement of the pump, pressure and volumetric flow will be matched with the need of loads*" (Quan et al. 2014), that is, suitability of consumption at productive needs.

### *45.2.1 Electric Engine Driven Systems*

Electric engines are equipment's used to convert the electrical energy received from the grid into mechanical drive energy. Of all types of motors, electric engines are the ones with the most advantages of use, as they combine the advantages of using clean energy with the simplicity of control, simple construction, and great adaptability to different loads (Fernandes et al. 2016). However, whenever possible, variable speed drives should be installed to adapt the engine's operation to the specific needs of the process. On the other hand, and given their versatility, electric engines play a fundamental role in the industry (85 billion electric engines in the European Union—EU), which is reflected in the high energy costs associated with their operation; they are responsible for 65–70% of industrial electricity consumption (WEG 2021). In Portugal, the consumption of electric engines distributed by the principal application (in 2013) is shown in Fig. 45.2a. Applications in which fluid movement occurs represent 62% of the total electrical consumption of industrial engines.

Recognizing the fact that electric engines have a high impact on global electricity consumption, most industrialized countries have adopted a set of Minimum Energy Performance Standards (MEPS) to reduce consumption in industrial equipment (engines, compressors, pumps, transformers, etc.) (IEA 2021b). On the other hand, as can be inferred from the (estimated) global market shares of industrial electric engines by energy efficiency class in 2020 (see Fig. 45.2b), the gains that can

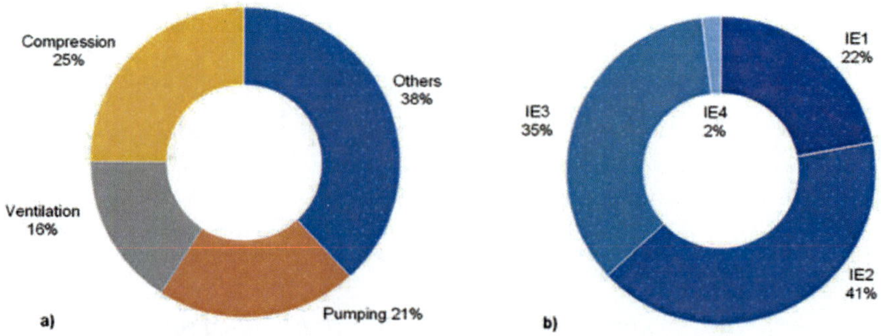


**Fig. 45.2** **a** Electricity consumption by sector of application in Portugal (ADENE 2013); **b** Engine distribution by energy efficiency class

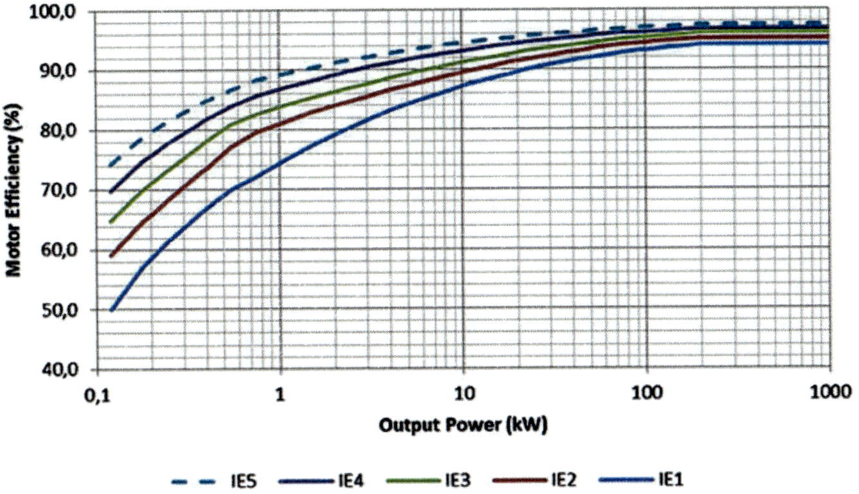


Fig. 45.3 Efficiency levels, classification standard and curves for 50 Hz, 4 poled motors (De Almeida et al. 2019)

come from simple measures can be significant, since classes IE1 and IE2 represent the largest share, with 63%. Regarding this issue, the European Union, through Regulation (EU) 2019/1781, imposes, since July 2021, ecological concession requirements applicable to electric engines and variable speed drives that, depending on the power and usage profile, a more efficient engine will generate economic gains that can vary between a few and several tens of thousands of euros (European Union 2019).

Currently, the scope for further energy savings involving motors is high, since the IE4 class represents a small share of the global market so far, at 2% (Fig. 45.2b). On the other hand, it should also be considered that the engine efficiency classes (Direct-on-Line, DOL) are defined as IE code in IEC 60034-30-1 (2014) which standardizes the efficiency classes. This standard defines four classes, IE1 to IE4, (see Fig. 45.3) where IE1 is the least efficient class and IE4 the most efficient class. The IE5 class, not yet defined in detail, shows gains of about 20% in relation to the IE4 class (De Almeida et al. 2019). It is also important to mention that the new Regulation (EU) 2019/1781 (European Union 2019) imposes minimum levels of IE4 efficiency, for motors with certain polarities and powers, starting from 1 July 2023 thus, the energy gains will be significant and the reduction in emissions will accompany this decline.

### *45.2.2 Hydraulic Oil Systems*

A conventional hydraulic system (Fig. 45.4) consists, in its most simplistic design, of a hydraulic unit (with a variable—flow regulator with mechanical flow adjustment—or constant pump), a control element (directional valve) and actuators (cylinders and/

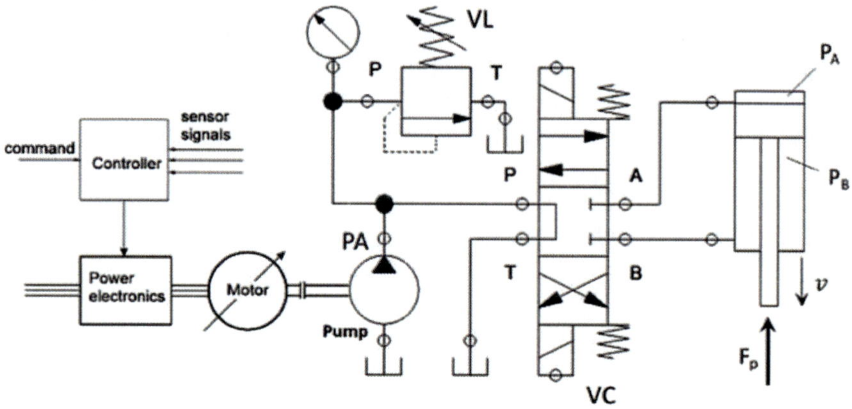


**Fig. 45.4** Basic architecture of an electrohydraulic system with a flow control unit

or hydraulic motors). Thus, the energy consumption of a hydraulic system includes not only the electrical losses associated with the drive, but mainly the losses due to throttling (Lyu et al. 2019) and overflow (Mahato and Ghoshal 2021). Of these, the overflow loss is caused by the relief valve (VL), used to limit the maximum working pressure of the system, and keeping a constant pressure (Li et al. 2021), high flow and hydraulic power is returned directly to the tank, leading to a costly wasting of potential energy. On the other hand, the use of hydraulic drives with variable speed control (VSD) and constant positive displacement pumps becomes a successful and widely used concept in the industry (Zagar et al. 2020), and continues to grow due to the development of high-speed control systems, fast-response engines, and software capable of responding to various needs as well as reducing energy consumption (Scroggins 2018).

Traditionally, hydraulic systems are driven by constant speed induction motors with hydraulic power controlled by regulating valves or variable displacement pumps. However, while this is a typical approach, the motor will draw up to 50% of full load current even when the system is not under load (Scroggins 2018). In addition to these constructive characteristics, hydraulic units are generally undersized to meet the maximum needs of the work cycle, which is conducive to wasting. It should also be considered that in several industrial applications require fast speeds in the approach phase and slow speeds with high loads in the work phase (Zagar et al. 2020), and different flow rates are obtained by a regenerative circuit, servo pumps (Song et al. 2022) or by different pump operating speeds. Thus, based on the variable speed control associated with a high-performance motor, it will be possible to reduce energy consumption, with different pressures and flows (at low speeds), and recovery energy by the reducing loss in the relief valve (Li et al. 2021).

The development of industrial equipment or the retrofit of the currently machines in operation can lead to an improvement in Energy and Resource Efficiency in companies, and the energy savings with a variable-speed pump dives can reach up to 80%

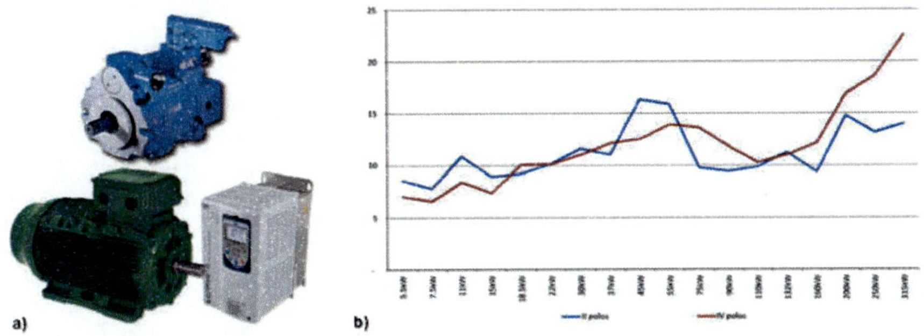

**Fig. 45.5** **a** Hydraulic speed control elements; **b** Driver IE4 payback in months (Costa , 2013)

(Scroggins 2018). In Portugal and according to Een (2018), "*more than 90% of companies are SMEs, making this potential even greater, since there have been few relevant improvements in this sector*". Thus, replacing current drives with IE4 minimum efficiency motors, mandatory from 1 July 2023, combined with variable-speed drives (Fig. 45.5a) translates into high energy savings, reduced $CO_2$ emissions and a quick payback (Fig. 45.5b).

## 45.3 Economic Analysis

A quantitative analysis of the volumes that may be involved in the acquisition (2.5%), maintenance (1.5%), and operation (energy, 96%) of the IE4 class electric engine, significant energy savings can easily be identified. So, considering the national reality and the data provided by PORDATA (2022), the consumption of electricity for the manufacturing industry in 2020 was 15, 867, 025, 045 kWh, which will translate into an average annual value of € 2,925,879,417—weighted average prices of electricity in industry, in Portugal (0.1844 Euros/kWh) considering an annual consumption between 20 and 500 MWh. On the other hand, it should be noted that only a small part of the national industry has already adopted high-performance drives, the margins for energy consumption become effectively very concrete. Based on Fig. 45.2b), we ascertained that IE3 class electric engines represent a 32% share of utilization that can be retrofitted to IE4. This conversion, with a payback of about 8 months for a 4 kW engine (Fig. 45.5b), will represent a gain of about 2.5% (see Fig. 45.3). Therefore, the annual gain obtained would be around € 21,066,331.81, 114,242,580.3 kWh less (45.1):

$$
\begin{aligned}
Gain &= Power[\text{kWh}] \times IE3\% \times SME\% \times \Delta\eta\% \\
Gain &= 15867025045 \times 0.32 \times 0.9 \times 0.025 = 114242580.3\,\text{kWh}
\end{aligned}
\tag{45.1}
$$

which would be equivalent to a reduction of around 29 245.11 *ton* of $CO_2$ per year.

## 45.4 Conclusions

Hydraulic oil as a work source can be found in all industrial sectors, whether it be in manufacturing, services, agriculture, or construction. On the other hand, with the prospect of growth in the global market for hydraulic equipment, many of the measures presented and developed in recent years can be transformed into effective energy gains and the reduction of industrial $CO_2$ emissions.

The retrofit of the machinery park, even if partial, will lead to a reduction in energy consumption, through the use of more efficient drives, also contributing to commercial gains with reduced financial returns. If we also consider the conceptual changes in the equipment, circuits in the flow source, reduced pressures in the dead points of operation, these will be a substantial contribution to the reduction of electricity consumption and to the increase of the equipment's work lifespan.

The combination of all these methodologies with variable speed drive (VSD), with pumps of constant or variable flow, will lead to energy gains and consequent reductions in industrial carbon emissions. Anticipating regulation, at least at national and European level, is a significant step towards meeting the targets imposed by the European community for the next years.

**Acknowledgements** We acknowledge the financial support of FCT—Portuguese Foundation for the Development of Science and Technology, Ministry of Science, Technology and Higher Education, CIDEM, R&D, INEGI and LAETA, and CIETI—Center for Innovation in Engineering and Industrial Technology, funded by national funds through the FCT/MCTES (PIDDAC), Portugal, Base Funding—UIDB/04730/2020.

## References

ADENE (2013) Agência para a Energia: Guia Técnico—Soluções para melhorar os sistemas accionados por motores elétricos 2013:15. https://dpydhb3wsr746.cloudfront.net/sites/www.voltimum.pt/files/fields/attachment_file/pt/flipbooks/others/F/201203161976633.pdf

AMR (2027) Allied market research: hydraulic equipment market outlook. https://www.alliedmarketresearch.com/hydraulic-equipment-market-A06534. Last accessed 23 Apr 2022

Bosch Rexroth (2022) https://apps.boschrexroth.com/rexroth/en/connected-hydraulics/products/cytrobox/. Last accessed 23 Apr 2022

Bucher Hydraulics (2022) Axial piston pumps AX, https://www.bucherhydraulics.com/ax. Last accessed 23 Apr 2022

Costa C (2013) WEG—Evoluções Tecnológicas dos Motores Eléctricos, Eficiência Energética de um Sistema—Soluções. Jorn Técnicas, Ualg. https://www.ualg.pt/sites/ualg.pt/files/ise/Electrica/weg_cc_ualgarve.pdf. Last accessed 7 May 2022

Danfoss Power Solutions (2022). https://www.danfoss.com/en/about-danfoss/news/dps/danfoss-completes-full-acquisition-of-artemis-intelligent-power/. Last accessed 23 Apr 2022

De Almeida AT, Ferreira FJTE, Both D (2005) Technical and economical considerations in the application of variable speed drives with electric motor systems. In: Conference record of industrial and commercial power systems technical conference, vol 41(1), pp 136–144. https://doi.org/10.1109/icps.2004.1314992

De Almeida A, Fong J, Brunner CU, Werle R, Van Werkhoven M (2019) New technology trends and policy needs in energy efficient motor systems—A major opportunity for energy and carbon savings. Renew Sustain Energy Rev 115. https://doi.org/10.1016/j.rser.2019.109384

Doe (2004) Variable speed pumping—A guide to successful applications. Hydraulic Institute, Europump, and the U.S. Department of Energy's (DOE) Industrial Technologies Program. A Guide to Variable Speed Pumping, https://www1.eere.energy.gov/manufacturing/tech_assistance/pdfs/variable_speed_pumping.pdf. Last accessed 17 Apr 2022

Een (2018) Eficiência Energética—Informação de apoio às empresas. Enterprise Europe Network, LNEG—Laboratório Nacional de Energia e Geologia, I.P (2018). https://www.lneg.pt/wp-content/uploads/2020/06/Brochura-EE-LNEG.pdf. Last accessed 7 May 2022

European Union (2019) Commission regulation (EU) 2019/1781. https://eur-lex.europa.eu/legal-content/PT/TXT/?uri=CELEX:32019R1781. Last accessed 29 May 2022

Fernandes MC, Matos HA, Nunes CP et al (2016) Medidas transversais de eficiência energética para a indústria. Direção-Geral de Energia e Geologia (2016). ISBN 978-972-8268-41-1. https://meesi.pt/sites/default/files/ebook/19893-medidas-transversais-miolo-prova5.pdf

Heitzig S, Sgro S, Theissen H (2012) Energy efficiency of hydraulic systems with shared digital pumps. Int J Fluid Power 13(3):49–57. https://doi.org/10.1080/14399776.2012.10781060

IEA (2021a) International Energy Agency. World energy outlook 2021a. IEA, Paris, Dec 2021, https://www.iea.org/reports/world-energy-outlook-2021. Last accessed 23 Apr 2022

IEA (2021b) International Energy Agency. Minimum Energy Performance Standards (MEPS). https://www.iea.org/policies/333-minimum-energy-performance-standards-meps, Last accessed 27 Apr 2022

IEC 60034-30-1:2014 (2014) Rotating electrical machines—Part 30-1: efficiency classes of line operated AC motors (IE code). Int Electrotechnical Comm. https://webstore.iec.ch/publication/136. Last accessed 30 Apr 2022

Li Z, Su L, Lin T (2021) Overflow energy loss recovery system based on hydraulic motor-electric generator. Appl Sci 11(3):941. https://doi.org/10.3390/app11030941

Lyu L, Chen Z, Member S, Yao B, Member S (2019) Development of pump and valves combined hydraulic system for both high tracking precision and high energy efficiency. IEEE Trans Ind Electron 66(9):7189–7198. https://ieeexplore.ieee.org/abstract/document/8495014

Mahato AC, Ghoshal SK (2021) Energy-saving strategies on power hydraulic system: An overview. Proc Inst Mech Eng Part I J Syst Control Eng 235(2):147–169. https://doi.org/10.1177/0959651820931627

PORDATA (2022) Consumo de energia elétrica: total e por sector de atividade económica—Indústria transformadora. https://www.pordata.pt/Portugal/Consumo+de+energia+el%c3%a9trica+total+e+por+sector+de+atividade+econ%c3%b3mica-1125-9099. Last accessed 8 May 2022

Quan Z, Quan L, Zhang J (2014) Review of energy efficient direct pump controlled cylinder electro-hydraulic technology. Renew Sustain Energy Rev 35:336–346. https://doi.org/10.1016/j.rser.2014.04.036

Schmidt L, Hansen KV (2022) Electro-hydraulic variable-speed drive networks-idea, perspectives, and energy saving potentials. Energies 15(3):1228. https://doi.org/10.3390/en15031228

Scroggins R (2018) Variable-speed pump drive save energy, cut noise and heat. Hydraul Pneumatics (2018). https://cdn.baseplatform.io/files/base/ebm/hydraulicspneumatics/document/2019/03/hydraulicspneumatics_4474_variablespeedpumps.pdf. Last accessed 1 May 2022

Song Y, Hu Z, Ai C (2022) Fuzzy compensation and load disturbance adaptive control strategy for electro-hydraulic servo pump control system. Electronics 11(7):1159. https://doi.org/10.3390/electronics11071159

Volvo Construction Equipment (2020) Pioneering electro-hydraulic solution significantly improving fuel efficiency In: Construction equipment. https://www.volvoce.com/global/en/news-and-events/press-releases/2020/pioneering-electro-hydraulic-solution-significantly-improving-fuel-efficiency-in-construction-equipm/. Last accessed 23 Apr 2022

WEG (2021) Novo Regulamento UE para a Eficiência Energética 2019/1781—Motores Elétricos de Baixa Tensão e Variadores de Velocidade. https://static.weg.net/medias/downloadcenter/h7b/h0d/WEG_Brochura_Eficiencia_Energetica-Portugal.pdf

Xu B, Shen J, Liu S, Su Q, Zhang J (2020) Research and development of electro—hydraulic control valves oriented to industry 4.0: a review. Chin J Mech Eng 33(29):1–20. https://doi.org/10.1186/s10033-020-00446-2

Zagar P, Kogler H, Scheidl R, Winkler B (2020) Hydraulic switching control supplementing speed variable hydraulic drives. Actuators 9(4):1–13. https://doi.org/10.3390/act9040129

# Chapter 46
# Influence of Electric Vehicles on Urban Traffic Noise and Fuel Consumption

**R. Calejo Rodrigues**

**Abstract** For the common citizen electric vehicles are symbol of sustainability because of its high-energy efficiency. Nevertheless, other environmental factors should be attended and in this work the noise emission is considered together with energy for urban traffic average speed decisions. The research strategy, developed in theoretic field, is backgrounded by former studies that provided the relation between speed and energy both for electric and internal combustion vehicles. In addition, the noise reduction of electric vehicles is obtained from the revision of literature. The present study adapted and uniformed the referred information and developed representative equations. Using the ESdB criteria, that is the energy saved by decibel reduction as speed varies; a new nomogram is proposed that illustrates the importance of considering both noise and energy in urban traffic speed decisions. Theoretical simulations are presented considering different percentages of electric vehicles. These simulations allow conclusions on the effect of electric vehicles in urban traffic and on the optimum average speed to be considered as different percentages of electric vehicles are in service. The benefits of EV's are only interesting when over 90% of urban vehicles are electrical and average speed is 50 km/h.

**Keywords** Electric vehicles · Traffic energy · Urban mobility · Urban noise

R. C. Rodrigues (✉)
Porto University—FEUP—NI&DEA—GEQUALTEC—CONSTRUCT, Rua Dr. Roberto Frias s/n, 4200-465 Porto, Portugal
e-mail: calejo@fe.up.pt

 
N. S. Caetano and M. C. Felgueiras (eds.), *The 9th International Conference on Energy and Environment Research*, Environmental Science and Engineering,
https://doi.org/10.1007/978-3-031-43559-1_46

## 46.1 Introduction

## 46.2 Background and motivation

Sustainability of the planet depends on multiple factors but, recently, carbon footprint is in the center of environmental worries. Because $CO_2$ emissions of most energy sources have a key role in this discussion and as mobility is (still) driven by energy, decisions are taken to avoid mobility supported by natural, finite and $CO_2$ emitting sources.

According to the European Union roadmap on transports, only half of ICE (Internal Combustion Engine vehicles) should be in use in urban areas by 2030; with their total ban in cities by 2050; aiming to achieve essentially $CO_2$-free city logistics in major urban centers by 2030 (European Commission , 2011). Portugal has recently adopted legislation to banish the commerce of ICE till 2035, U.K. and France will banish new ICE from 2040 (Petroff 2017). Also in the USA, predictions indicate that by 2040 electrical vehicles (EV's) will be more than 50% of urban traffic (Randall 2016). EV's are more energy efficient than ICE ones. Around 75% of supplied energy is available on wheels of EV's while only an average of 25% energy is converted to power in wheels on ICE. Although EV's have in general more moving-weight, in fact they are more energy efficient than ICE.

The environmental meter for this change of source energy for the mobility is mostly based on carbon footprint.

Urban mobility implies physical displacement that results in the need of energy consumption to be carried out. From the environmental point of view, several studies are limited to consider only the source energy consumption as the only environmental meter. However, there are others such as air quality, raw resources availability, economy, cultural heritage, and so on, that should also be taken into consideration.

The background research of this work introduced the environmental evaluation of traffic speed considering both energy consumption and noise. It was concluded that the optimization of both parameters, considering the diversity of vehicles associated with mobility, could only be done using the Monte Carlo method since from the mathematical point of view one is faced with a random system of two non-normal variables. Without EV's results appointed for an optimum average urban speed of 41 km/h (Calejo 2022).

The increasing use of electricity-powered vehicles (EV's) with a lower sound emission intensity and much more optimized energy consumption changes the assumptions of the previous work. This fact gave rise to the development of the present research that aims to evaluate the influence of electric vehicles on the relationship between energy consumption and noise related to urban mobility solutions.

The research question of this work deals with the following challenge:

***What is the importance of the progressive introduction of EV's in urban mobility considering both the influence of noise and energy?***

### 46.2.1 Background and Motivation Literature Review

Several authors have been exhibiting research about sound emissions from electric vehicles. Iversen and Skov (2015) conducted comparative studies between two electric and two internal combustion vehicles and concluded that below 30 km/h EV's are the quietest, but they stated that the same is not observed at the remaining speeds, nevertheless EV's are globally quieter than ICE.

Verheijen and Jabben (2010) referred identical conclusions and announced a practical case of 3 dB(A) sound reduction in urban sound environment if all vehicles were electric. In their study, they present the expected sound reduction between EV's and ICE, whose curve is reproduced in Fig. 46.1. It can be observed that for speeds up to 50 km/h the sound reduction is significant, but above this speed the rolling noise overlaps that of the engine and the gains of an electric engine at these speeds are questionable.

Sobrino et al. (2016) developed a model to study fuel consumption reduction strategies named HERA (Highway EneRgy Assessment). Under this research it is proposed a fuel consumption nomogram considering speed and different European environmental standards for vehicles. Figure 46.2 obtained from Sobrino et al. (2016) states the imposed reduction in fuel consumption and the variation with speed.

Van Haaren (2009) studied and identified the relationship between required power and speed in electric vehicles, he identified the different energy consumption motivating factors, correlating them with speed. Figure 46.3, obtained from Van Haaren (2009) illustrates the respective conclusions, according to which for speeds up to about 80 km/h the energy consumption is due to the need to overcome the inertia of the transmission mechanics, but for higher speed, the power needs are generated to overcome aerodynamic resistance.

He and Wu (2018) developed a study resulting in Fig. 46.4 in which they compared the electrical energy consumption of an EV's in Wh/mile with the gasoline consumption of an ICE in g/mile. This study published in 2018 presented a nomogram that

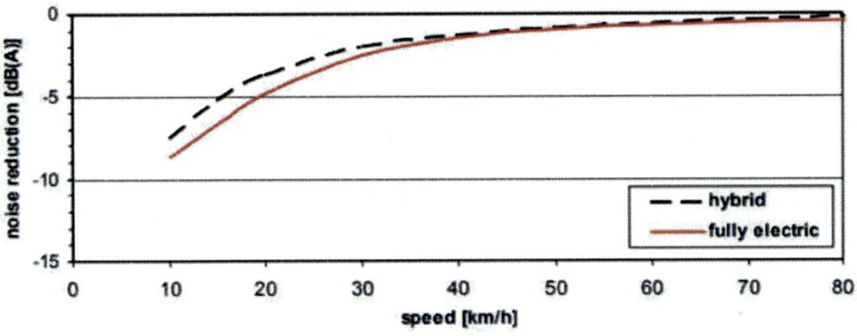


**Fig. 46.1** Noise reduction of hybrid and electric passenger cars compared to conventional passenger cars (Iversen and Skov 2015)

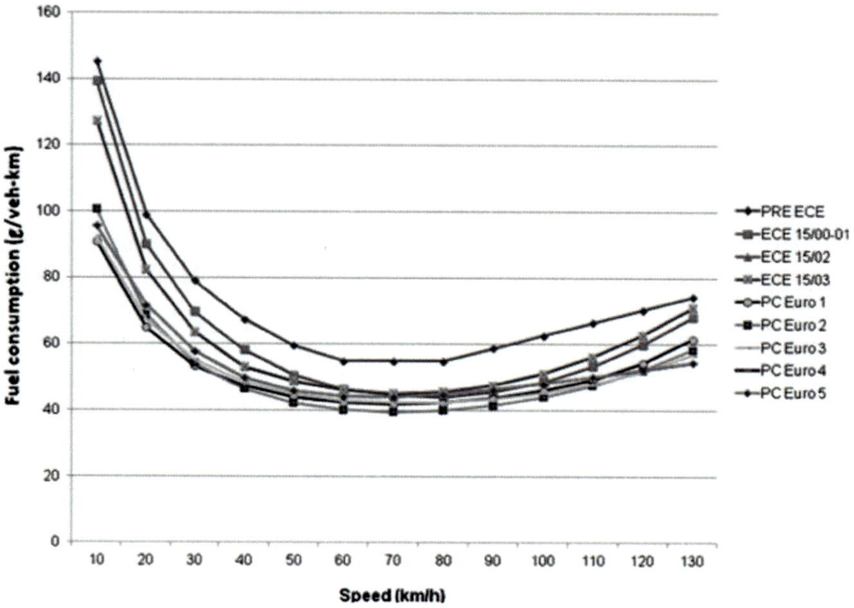


**Fig. 46.2** Speed-fuel consumption for gasoline passenger car versus technology (Sobrino et al. 2016)

could at the first glance show that electric vehicles, in this case the Tesla Roadster, consume much less power for low speeds, equal energy at 41 km/h and more energy after this speed. Contrarily, converting Fig. 46.4 for the same equivalent units (as presented on Fig. 46.7), energy consumption of EV's is always less than ICE. In line with the study by Verheijen and Jabben (2010); it is also at these low speeds that EV's emit less noise.

## 46.3 Research Methodology

### *46.3.1 Flowchart*

As mentioned, the research objective is to identify the influence of electric vehicles on urban traffic considering not only the reduction of energy consumption, but also the consequences in terms of noise. To this end, the following steps are proposed as presented in Fig. 46.5:

1. Update and adapt the present study in terms of units to the work of van Haaren (2009); He and Wu (2018); Iversen and Skov (2015); Sobrino et al. (2016)

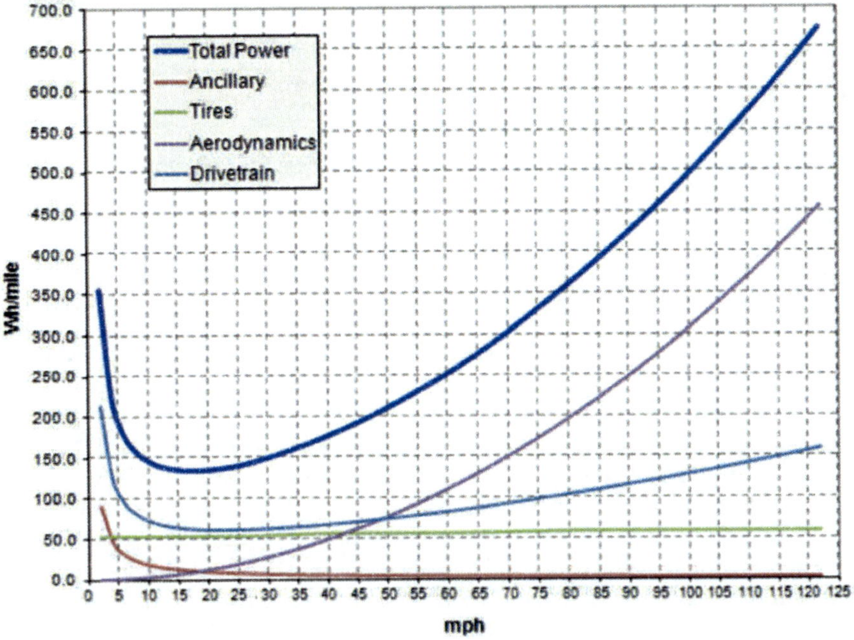


**Fig. 46.3** Relationship between power and speed (Van Haaren 2009)

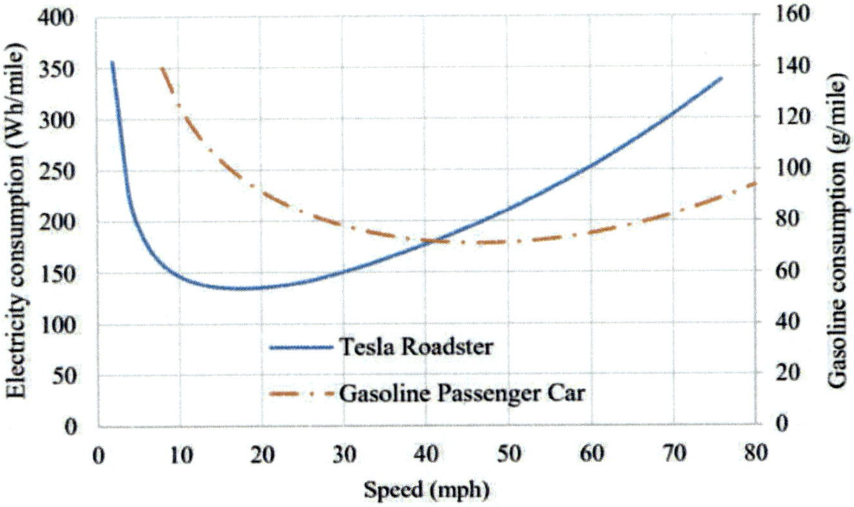


**Fig. 46.4** Energy consumption vs speed for EV's Tesla and ICE (He and Wu 2018)

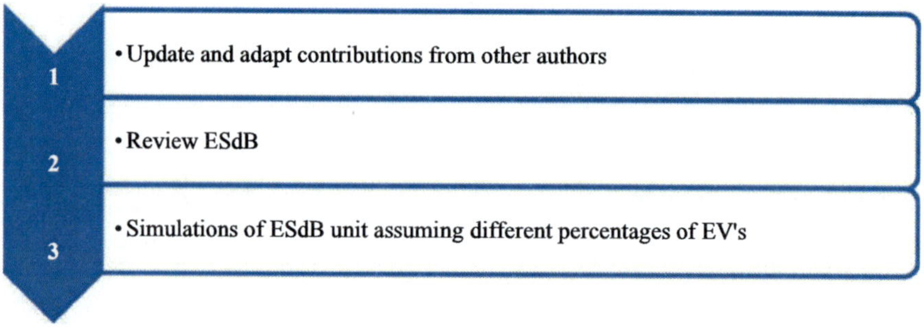


**Fig. 46.5** Research steps

2. Update the methodology proposed by the author (Calejo 2022) and according with that,modeling urban traffic noise dependence on energy, assisted with Monte Carlo simulation in order to use the evaluation unit proposed there, ESdB, that allows to identify the optimum speed to optimize the binomial energy consumption and noise.
3. Simulate different percentages of electric vehicles and identify the optimal speed trend.

### *46.3.2 Step 1—Conversion Background Contributions to ESdB Model*

Energy and noise variation with speed was obtained from former studies van Haaren (2009); He and Wu (2018); Iversen and Skov (2015); Sobrino et al. (2016). These studies were developed in different unit systems; therefore, to use their information a conversion must be done.

**Noise Evaluation**

The flowing graphic in Fig. 46.6 sets the evaluation between noise emission by ICE and EV's. As it could be seen no difference exists, let us say, above 50 km/h, but it is important for low speeds, for example a 5 dB(A) difference is identified at 20 km/h. This data will be considered to obtain ESdB.

**Energy Evaluation**

To obtain comparable data of energy consumption vs speed, both for EV's and ICE, the referred background studies were taken into account, but it was needed to standardize the units. Considering that 1 mpg = 2.3521 L/km, 1 Wh/mile = 33.705 mpg, and that the average gasoline density is 0.75 it was possible to construct the graph of Fig. 46.7.

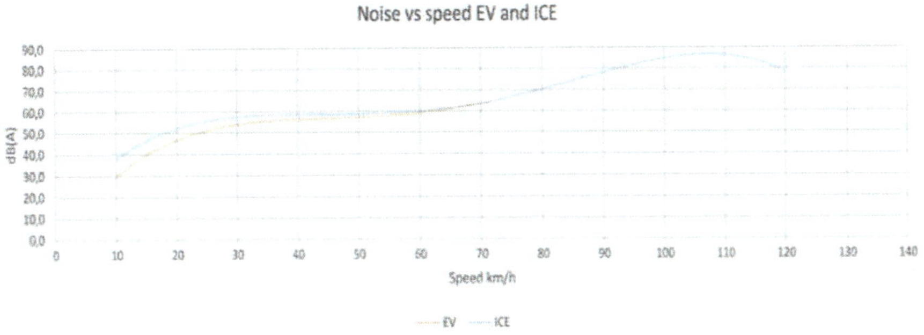


Fig. 46.6 Evaluation of noise emission between EV's and ICE

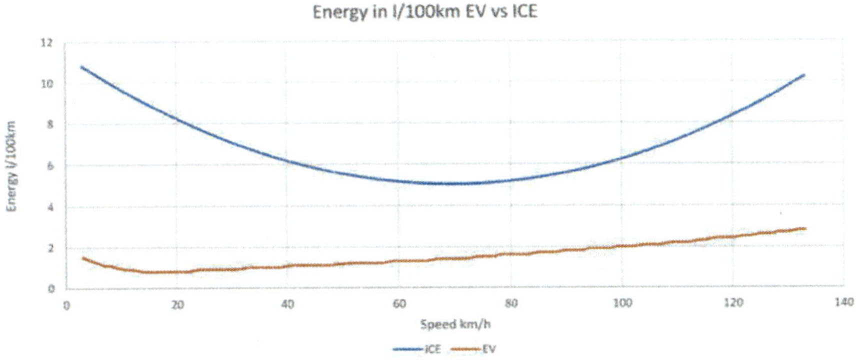


Fig. 46.7 Evaluation of energy vs speed for EV's and ICE

### 46.3.3 Step 2—Review ESdB

ESdB stands for Energy Savings by decibel that is the rate of energy consumption versus velocity. The parameter is obtained from de ratio of energy saved in liters of gasoline per one hundred kilometers over the noise increase as speed increases as seen in Eq. (46.1):

$$ESdB = \frac{\Delta energy}{\Delta noise} \tag{46.1}$$

where

*ESdB* Energy saved by noise variation as speed increases, in (L/100 km)/dB(A).
*Δ energy* Energy variation, in L/100 km.
*Δ noise* Noise variation in dB(A).

This parameter is numerically obtained from equations that support Figs. 46.6 and 46.7:

Noise emission driven by velocity for EV's, Eq. (46.2):

$$L = -0.000005 \times v^4 + 0.0013 \times v^3 - 0.1142 \times v^2 + 4.1698 \times v + 3.7898 \quad (46.2)$$

where

$L$ Noise level in dB(A)
$v$ Velocity in km/h.

Noise emission driven by velocity for ICE, Eq. (46.3)

$$L = -0.000005 \times v^4 + 0.0013 \times v^3 - 0.118 \times v^2 + 4.507 \times v + 7.278 \quad (46.3)$$

where

$L$ Noise level in dB(A).
$v$ Velocity in km/h.

Energy consumed by velocity for EV's, Eq. (46.4).

$$E = 0.0001 \times v^2 - 0.0012 \times v + 0.919 \quad (46.4)$$

where

$E$ Energy in L/100 km.
$v$ Velocity in km/h.

Energy consumed by velocity for ICE, Eq. (46.5):

$$E = 0.0012 \times v^2 - 0.1572 \times v + 10.53 \quad (46.5)$$

where

$E$ Energy in L/100 km.
$v$ Velocity in km/h.

### *46.3.4 Step 2—Simulations of ESdB Unit Assuming Different Percentages of EV's*

To simulate the different percentages of EV's and ICE the energy consumption was balanced according to the same percentage to be considered. Regarding the noise emission, it was balanced considering the Eq. (46.6):

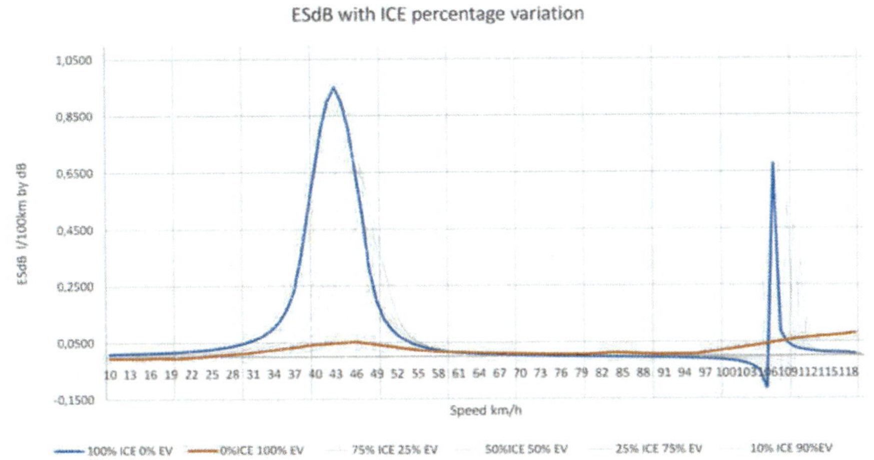


**Fig. 46.8** Variation of EsdB criteria with urban traffic speed for different mixes of EV's and ICE

$$L_u = 10 \times \log\left(\frac{1}{100} \times \left(\left(\%_{EV} \times 10^{L_{EV}}\right) + \left(\%_{ICE} \times 10^{L_{ICE}}\right)\right)\right) \quad (46.6)$$

where

| | |
|---|---|
| $L_u$ | Level of urban noise because of mixed vehicles. |
| $\%_{EV}$ | Percentage of EV's. |
| $L_{EV}$ | Noise of EV's. |
| $\%_{ICE}$ | Percentage of ICE. |
| $L_{ICE}$ | Noise of ICE. |

## 46.4 Results and Discussions

### *46.4.1 Results*

Figure 46.8 presents the global results aiming to answer to the research question regarding the environmental influence of EV's considering dual criteria, the energy consumption and noise emission.

### *46.4.2 Discussion*

ESdB is a complex parameter because no linearity is identified with speed variation. Two speed intervals are important to consider [20; 60] km/h and [95;120] km/h.

The first interval deals with the change of energy spent to beat drivetrain inertia to aerodynamics. It is clear that there is a reduction in energy when after the vehicle inertia has expired the necessary energy is used to overcome aerodynamic resistance that at low speed is not carried out. As noise production is also low, gains of around 1 L of petrol per 100 km in ICE are also generated. In EV's the gain is much lower because it is used much less energy, although the process is similar.

The second mentioned interval shows a different behavior between EV's and ICE. While ICE exhibit in this interval of velocity a gearshift for a high multiplication, EV's do not have this need.

## 46.5 Conclusion

The influence of EV's on traffic noise and energy consumption is important because on the one hand they produce less noise but on the other hand they consume less energy. However, the evidence of this conclusion is dependent on the number of EV's in circulation.

The results of this study allowed identifying that only for over 90% EV's in circulation, these conclusions are obvious. The obtained data allow identifying that if more EV's are in circulation, the optimum ESdB speed could be higher. In this case, it is about 50 km/h facing 42 km/h for 100% ICE.

Nevertheless, energy savings are lower for EV's than for ICE, partially because EV's are more energy efficient (almost 75%) compared to ICE (that have an average efficiency of 25%), so ICE when optimizing speed as by ESdB decision, are able to save more energy while at the same time optimize noise emission.

**Acknowledgements** The author thank NI&DEA Project collaborators for gathering basic information for the development of this work.

**Funding** This work was financially supported by: Base Funding—UIDB/04708/2020 and Programmatic Funding—UIDP/04708/2020 of the CONSTRUCT—Instituto de I&D em Estruturas e Construções—funded by national funds through the FCT/MCTES (PIDDAC).

## References

Calejo RR (2022) Modeling urban traffic noise dependence on energy, assisted with Monte Carlo simulation. Energy Rep 8:583–588. https://doi.org/10.1016/J.EGYR.2022.02.254

European Commission (2011) Roadmap to a Single European Transport Area—Towards a competitive and resource efficient transport system—European Environment Agency 2011. https://www.eea.europa.eu/policy-documents/roadmap-to-a-single-european (Accessed 10 July 2022)

He X, Wu X (2018) Eco-driving advisory strategies for a platoon of mixed gasoline and electric vehicles in a connected vehicle system. Transp Res Part D Transp Environ 63:907–922. https://doi.org/10.1016/J.TRD.2018.07.014

Iversen LM, Skov RSH (2015) Measurement of noise from electrical vehicles and internal combustion engine vehicles under urban driving conditions. Euronoise 2015:2129–2134

Petroff A (2022) These countries want to ditch gas and diesel cars 2017. https://money.cnn.com/2017/07/26/autos/countries-that-are-banning-gas-cars-for-electric/index.html (Accessed 10 July 2022)

Randall T (2022) Here's how electric cars will cause the next oil crisis 2016. https://www.bloomberg.com/features/2016-ev-oil-crisis/ (accessed 10 July 2022)

Sobrino N, Monzon A, Hernandez S (2016) Reduced carbon and energy footprint in highway operations: the Highway Energy Assessment (HERA) methodology. Netw Spat Econ 16:395–414. https://doi.org/10.1007/S11067-014-9225-Y/FIGURES/7

vanHaaren R (2009) Assessment of electric cars' range requirements and usage patterns based on driving behavior recorded in the national household travel survey of 2009 2011:1

Verheijen E, Jabben J (2010) Effect of electric cars on traffic noise and safety. Pub Health 2010:29

# Chapter 47
# Automation, Project and Installation of Photovoltaic System in a Rural Farm

**Filipe Pereira, Adriano A. Santos, António Ferreira da Silva, Nídia S. Caetano, and Carlos Felgueiras**

**Abstract** Rural farms are an important part of each country's consumption and typically include traditional energy installations, which are relatively reliable but energy-intensive and energy inefficient. In rural farms, the implementation of the Smart Farm concept with renewable energy and integrated resource management technologies has been slower than in the domestic and industrial sectors. This work describes a solution that was developed and implemented in a rural farm in Portugal which shows a significant reduction in energy consumption from the electricity grid, and the corresponding reduction in $CO_2$ emissions. The present work makes a technical description of the developed solution and a comparison with other published scientific case studies.

**Keywords** Photovoltaic · Renewable energy · Smart farm · Sustainable farms · SCADA · PLC

F. Pereira · N. S. Caetano · C. Felgueiras (✉)
CIETI—Centre of Innovation on Engineering and Industrial Technology/IPP-ISEP, School of Engineering, R. Dr. António Bernardino de Almeida 431, 4249-015 Porto, Portugal
e-mail: mcf@isep.ipp.pt

F. Pereira · A. A. Santos · A. F. da Silva
CIDEM, School of Engineering (ISEP), Polytechnic of Porto (P.Porto), Rua Dr. António Bernardino de Almeida 431, Porto, Portugal

A. A. Santos · A. F. da Silva
INEGI, Institute of Science and Innovation in Mechanical Engineering and Industrial Engineering, R. Dr. Roberto Frias S/N, Porto, Portugal

N. S. Caetano
LEPABE-Laboratory for Process Engineering, Environment, Biotechnology and Energy, Faculty of Engineering, University of Porto (FEUP), R. Dr. Roberto Frias, 4200-465 Porto, Portugal

ALiCE-Associate Laboratory in Chemical Engineering, Faculty of Engineering, University of Porto, Rua Dr. Roberto Frias, 4200-465 Porto, Portugal



N. S. Caetano and M. C. Felgueiras (eds.), *The 9th International Conference on Energy and Environment Research*, Environmental Science and Engineering,
https://doi.org/10.1007/978-3-031-43559-1_47

## 47.1 Introduction

Our economy is currently highly dependent on traditional energies. These have an impact that is detrimental to the ecological balance of the planet. The prices of oil, coal and natural gas continue to rise, as their reserves are increasingly dwindling. Portugal spends a lot of money on energy imports, making it difficult to invest in other areas that are much more beneficial to the country development.

Currently, the electricity sector faces several challenges, such as in terms of environmental sustainability, combined with the need to explore new energy sources as an alternative to fossil resources, the reliability and quality of supply, as well as the challenges arising from increased competitiveness in the power sector. The solutions to these challenges will allow an increase in the capacity to integrate in the electrical networks, distributed production systems based on renewable energy sources, which are of an intermittent and variable nature. Among the available sources of energy on the planet, solar is the main one, since it is a practically inexhaustible and constant resource. Compared to other European countries, Portugal is one of the countries that has the best conditions for taking advantage of this resource, as it has an average of 2200–3000 h per year of sunshine (mainland) and good temperatures with low humidity, while Germany, for example, has only between 1200 and 1700 h per year of sunshine (Pereira et al. 2022). The system presented in this paper addresses each of the following specific Sustainable Development Goals (SDGs) (United Nations 2017):

- SDG6—Drinking water and sanitation—The system has a 10,000 L tank at the bottom of the subsoil, whose capacity supplies sanitary and bathing consumption. This capacity depends on the amount of rain that the rural farm can store annually.
- SDG7—Affordable and clean energy—The rural farm is supplied based on clean alternative energy. It consists of a photovoltaic system, supported by a wind turbine and a mini-hydro. If any of the energy sources mentioned above is not available, the farm will have the Portuguese national electricity grid with an alternative energy source.
- SDG9—Industry, Innovation and Infrastructure—This system presents innovation, the rural farm is fully automated. The implemented system allows remote access to rural farm owners and users, where it is possible to monitor and control in real time the proper functioning of domestic and rural equipment associated with the rural house/farm.
- SDG11—Sustainable Cities and Communities—This project is associated with energy sustainability, which reduces the negative environmental impact per-capita in the area where it operates, being also adapted to climate change. This allows one to make it a smart-grid.

## 47.2 Sustainable Farms

The consumption of electric energy in a house represents one of the biggest monthly expenses of a family. Cultural factors also contribute to this cost—Portugal is one of the countries in the world where most of the energy consumed in home buildings is directed towards home cooking, and where it can reach values of 40% (Toshiba White Paper 2022). Buildings, in particular, account for 40% of total energy consumption in Europe (30% in Portugal). However, in many cases, this consumption could drop to less than half with the implementation of the right solutions. Several papers have been published and have a special relationship with the present work. Faria and Vale (2011), described an efficient form of Demand Response in electrical energy supply, through an optimal real-time pricing approach. Guerrero et al. (2010), discussed distributed generation as a new paradigm to produce locally good quality electrical power, in a way particularly interesting when different kinds of energy resources are available, such as photovoltaic (PV) panels, or wind turbines. Spanaki et al. (2021) discussed the use of Internet of Things (IoT) in the farming sector. This work and application are presented through a smart farm scenario, while they incrementally explored the data sharing in a so-called Agriculture 4.0. Yahya (2018), explored the concept of Agriculture 4.0 through a system that employed drones, robotics, IoT, vertical farms, Artificial Intelligence (AI), and solar energy. Through the integration of digital technology into farming practices, companies are able to increase yield, reduce costs, experience less crop damage, and minimize water, fuel, and fertilizer usage. Despite the high number of articles published about Smart Farms, only a limited set effectively present technical and scientific descriptions that allow the comparison of solutions.

In order to respond to these needs, this project emerged, which uses the capabilities of intelligent electrical energy management systems to promote the reduction of electrical energy consumption in a home, with the purpose of integrating the production of energy from renewable sources, in particular photovoltaic solar energy, and the use of automation equipment incorporated into the home.

Undoubtedly, the use of automation equipment is extremely important in the management of efficient, safe and comfortable energy for the user and household within the dwelling.

The solutions for Smart Farms are intrinsically complex and include several studies. In our previous work (Pereira et al. 2022), an economic analysis was presented. In the current work, technical solutions and details about the implemented system are presented.

## 47.3 Case Study

The rural farm is located in Portugal, and, in this work, it will be called System A. According to a previous study concerning consumption, the rural farm had very high monthly electricity costs (Pereira et al. 2022). This situation allowed for the development of a large-scale project that combined a source of renewable energy, several other sources of renewable energy as support systems and the automation of the farm's housing, in order to make it more energy efficient. The number of inhabitants of the rural farm is 8 people, who work mainly as farmers. This rural farm has an artesian well that supplies water to the farm and housing at the same time. The problem presented relies in the high values of energy consumption, leading to a high electricity bill. The house had a daily consumption of 89.43 kWh/day and an annual turnover of around €5,000. The developed system includes a photovoltaic generator, energy storage batteries, monitoring system and microgrid control. The project includes the dimensioning of the various components as well as the dimensioning of cables and protection systems.

### *47.3.1 Case Study Presentation*

The photovoltaic system that was implemented allows controlling loads and connecting to the existing photovoltaic system, or through the grid (in the event of a power failure). The control is done using three grid-connected inverters in which one of them (the master) gives orders to control the operation of the loads. For the photovoltaic generator, 72 photovoltaic modules were provided with six followers of an azimuthal axis with 15.28 $m^2$ of area of photovoltaic modules per follower, a total capture area of 91.6 $m^2$, inclined at 36°. Sunny Boy inverters were fixed to each follower. These modules were chosen due to their quality/price ratio, which allowed the presentation of an adequate budget in order to obtain a return on investment as soon as possible. The inverters in question have a 5-year warranty and an extension of another 10 years is possible. There is no longer any inverter brand on the market with the characteristics of Sunny Island 5.0. Figure 47.1 presents the block diagram of the implemented photovoltaic system (Multicluster Box and 6.3–11—Europe-SolarStore.com, 2021).

Alongside this, plans were also made to install a monitoring system interconnected with the inverter via an RS485 communication port, capable of monitoring the installation and making data available locally (via LCD), via the Web or through a smartphone application. The Sunny Island inverters were connected to a panel called Multicluster 6.3, through piped and buried cables, with the intention of using part of the manholes of the external electrical infrastructures. Regarding the Multicluster 6.3 board, there is nothing similar in the market with the quality/price ratio of the product in question, hence no more possible solutions are presented. In terms of the automation system, the system allows end users, among other actions, to:

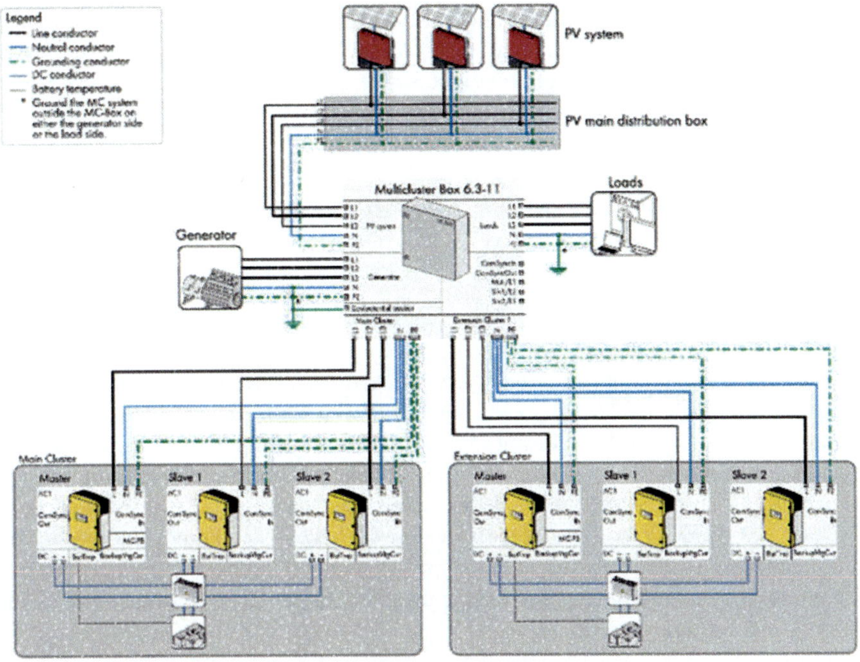


**Fig. 47.1** Block diagram of the implemented photovoltaic system

- Activate or deactivate loads in the house, such as the pump motor that allows the collection of water coming from the subsoil;
- Activate or deactivate loads, such as lighting, blinds, irrigation system, etc.;
- Control the operation of some equipment and/or appliances at home and abroad;
- Monitor the state of operation of the various pieces of equipment that make up the dwelling;
- Generate warning signals if any anomaly occurs;
- Send SMSs in case of fire or other type of anomaly.

### *47.3.2 Sizing the Photovoltaic Generator*

The autonomous system was dimensioned using the PVGIS software, having obtained an annual production in the order of 27 MWh. For the respective dimensioning, the following input parameters referring to the implemented photovoltaic system were adopted:

- Rated power of the photovoltaic generator: 13.0 kW (crystalline silicon)
- Estimated losses due to temperature and low irradiation: 10.8% (using local ambient temperature)

- Estimated loss due to angular reflection effects: 2.6%
- Other losses (cables, inverters, etc.): 14.0%
- Losses from the combined photovoltaic system: 25.3%
- Optimal tilt angle: 36°

### *47.3.3 Components of the Photovoltaic System Implemented in the Housing*

The dimensioned photovoltaic system consists of 6 solar trackers with 18 photovoltaic modules each and has 24 OPzS 2 V batteries each, as shown in Fig. 47.2.

The OPzS Solar range provides excellent results in medium and high-power industrial applications. These are lead-acid batteries with low maintenance liquid electrolyte. Thanks to their robustness and longevity they are ideally suited for use in solar and wind energy systems, as well as in telecommunications and emergency power supply applications. It should be noted that the system has a Multicluster MC-Box 6.3 control box to interface and control the loads and there is also control and monitoring through a PLC with a supervision application. Table 47.1 presents a summary of the equipment that makes up the photovoltaic system as well as its monitoring and control system.

**Fig. 47.2** The implemented photovoltaic system and the set of batteries associated with

**Table 47.1** List of equipment that make up the photovoltaic system

| Equipment list | Quantity |
|---|---|
| SILIKEN panels with 180 Wp peak power | 72 |
| Structure for fixing the panels to the ground, with ETTATRACK 1500 solar tracker system (for 12 panels each) | 6 |
| Sunny Boy 3800 V inverters (up to 95.6% efficiency) | 3 |
| Battery Inverter (off-grid) Sunny Islands 5048, (efficiency up to 95%) | 3 |
| (3 × 8) Batteries 8OPzS 800 Elem.2V-1166 Ah/C120 h | 24 |
| Battery shelf | 1 |
| Material earth protection system, modules, junction boxes, electrical wiring, circuit breakers, electrical panels, etc | 1 |
| Multicluster MC-Box 6.3 | 1 |
| Monitoring option—Sunny WebBox | 1 |
| Router | 1 |

### *47.3.4 Equipment that Composes the Photovoltaic System Implemented in the Housing*

A monitoring system was installed in the farm where all data was collected in real time (such as production in kWh, string currents, string voltages, DC power at the input and AC power at the inverter output, among others). An OMRON NS5 console was used, although another brand could have been used. As the company had it in stock, the decision was made to use it, thus optimizing the final cost of the installation. The same applies to the use of the PLC and current and voltage monitoring equipment of the photovoltaic generator. The PT-100 temperature probe with its transmitter could also have been chosen from other brands. In relation to temperature control, two PT-100 probes from OMRON were used together with two 0–10 V transmitters from the company Transducer LKM 224. One of the probes measured the temperature in the batteries of the photovoltaic system and the remaining six probes measured the temperature in the photovoltaic generator.

As photovoltaic installations must achieve the maximum energy efficiency in the shortest possible time, equipment from the manufacturer Phoenix Contact, called SOLARCHECK, was used. This equipment makes it possible to efficiently monitor the power losses of individual branches that could be caused, for example, by damaged panels or faulty contacts and cabling.

The solar check modular monitoring system consists of several pieces of equipment for measuring current and voltage and a corresponding communication module. The communication module allows for:

- Connecting and collecting measurement values from up to eight measurement modules;
- Preparation of data for transmission to higher commands;
- Supplying power to the connected measurement modules;

- Current measurement modules allow to/up to:
  - 8-channel current measurement up to 20 A DC;
  - Detection of reverse currents up to -1 A;
  - 4-channel add-on modules for 20 A DC;
- Digital input for monitoring, from remote signaling contacts of surge current protection modules;
- Power through the communication module;
- Voltage measurement modules allow to/up to:
  - Voltage measurement up to 1500 V DC in any earthed photovoltaic systems;
  - Connection and power supply normally via the intended analogue input of the 8-channel Solar check current measuring module;
  - Output of the voltage measurement value as an analog signal.

The communication module gathers the measurement values from the current measurement modules and forwards them to a higher hierarchical controller via RS485 MODBUS. A maximum of eight current measurement modules of any type can be connected to a communication module. The two-conductor communication cable is used to supply the current to the measurement modules at the same time. Therefore, no additional power supply is required. All measured values can be read through open registers from the communication module. The CP1L-EM PLC with built-in Ethernet port and two analog input expansion cards was used. An RS-485 communication card has also been incorporated on the front of the PLC to allow RS-485 communications between the devices.

### *47.3.5 Comparison with Similar Cases*

According to our research, there are few reported cases of smart rural farms with data that allow comparison with our case study—hereafter referred to as System A. Soufi et al. (2021), described the optimal sizing of the solar panel and battery in an autonomous photovoltaic (SPV) system to provide the necessary electricity for a rural farm located in Algeria—henceforth called System B. Salihu et al. (2020) presented an off-grid photovoltaic system for rural electrification in Nigeria—hereinafter called System C. Their work reported the design and implementation of the photovoltaic microgrid system carried out. Ibrik (2020), carried out a work that referred to the impact of the use of micro-grid connected solar photovoltaic (PV) systems at rural level in Palestine—hereinafter referred to as System D.

Table 47.2 summarizes some of the characteristics of the solutions proposed in the above-mentioned papers in comparison with the case study proposed in the present research.

Comparing all the solutions presented in Table 47.2, it can be concluded that not all of them respond to a set of important specifications.

**Table 47.2** Characteristics of the previously proposed solutions compared to the case study

| System characteristics | System A (Case study) | System B | System C | System D |
|---|---|---|---|---|
| Peak power of the installed photovoltaic system (Wp) | 13,000 | 56 | 24,000 | 10,000 |
| PV System—fixed or solar tracker | Solar tracker | Fixed | Fixed | Fixed |
| Energy consumption per day (kWh) | 89.43 | 121 | 54.64 | 36.2 |
| Installed battery size (Ah) | 1166 | 85 | 1202 | 1800 |
| Own production | Yes | Yes | Yes | Yes |
| Local and remote control/monitoring system ability | Yes | No | No | No |
| Prepared to include a wind generator or mini hydro | Yes | No | No | No |
| Automatic energy selection from own production or external grid network | Yes | No | No | No |
| Batteries autonomy in days for the installed power | 5/6 | n.a | n.a | n.a |
| Reduction energy from grid (%) | 83.24% | n.a | n.a | 86% |
| Percentage (%) of load rate Batteries autonomy for the installed power; | 86 | n.a | n.a | n.a |
| Type of automation (PLC or other) | Yes | No | No | No |
| System uses HMI or other equipment for monitoring and controlling the system | Yes | No | No | No |
| Automation able to optimize/reduce consumption | Yes | No | No | No |
| System must comply with the IEC 60364-7-712 standard | Yes | n.a | n.a | n.a |
| Peak power of the installed photovoltaic system (Wp) | 13,000 | 56 | 24,000 | 10,000 |
| PV System—fixed or solar tracker | Solar Tracker | Fixed | Fixed | Fixed |
| Energy consumption per day (kWh) | 89.43 | 121 | 54.64 | 36.2 |

The developed solution called System A has some differentiating factors, as follows:

- System A provides power for agricultural equipment and rural housing, while System D provides power only for rural housing. In addition, System A has a high degree of automation of the housing and the implemented photovoltaic system, which allows a significant cost reduction in the monthly energy bill;
- System A allows for the management of two alternative sources of renewable energy, one being a solar photovoltaic and another wind or mini-hydro. System D has a diesel generator, that is, a non-renewable energy source. In addition, System A allows the national electricity grid to be used as a source of energy support, should there be a failure in the supply of any of the renewable energy sources;

- System A will present, in the future, a lower cost for replacing batteries, as the installed capacity is lower than that of System D. In addition, System A has a higher capacity percentage (%) of the charge rate battery autonomy for the installed power of 5/6 days, which is very good;
- System A has control, local/remote monitoring for the photovoltaic system as well as for the automated equipment on the farm and rural housing. System D does not have a degree of automation, as well as remote access and integrated controls.

In summary, Systems A and D have some similarities, but System A stands out, not only for the level/degree of automation, but also for the type of control and monitoring, as well as for its versatility. The Support system allows the integration of other renewable energy sources as well as the automation of various agricultural equipment and allows remote control and access that are important for the sustainability, comfort and well-being of the residents. It also allows a degree of autonomy of 5–6 days relying only on batteries and has a supply capacity of up to 85% of the loads existing in the farm and rural housing.

## 47.4 Conclusions

This work describes the development of a photovoltaic system and a home automation system, capable of monitoring and controlling the installation and the equipment that make up the house in real time. The main focus is undoubtedly to reduce rural farm energy consumption and eliminate dependence on fossil energy sources, and on energy grid dependence. Display pages were created for real-time monitoring of some electrical quantities of the house and farm, which permits a quick analysis and verification of the photovoltaic system so that corrective measures can be implemented, in order to maximize the performance of the photovoltaic electric energy production system. In short, this paper summarizes and highlights the difficulty of comparison in terms of the implementation of this type of systems and/or solutions. The authors tried to explore the economic analysis part, however technical data about the installation or system is not always available. It is partially understandable, as it is often confidential data, but it makes the comparison of similar solutions more difficult.

In this work, we seek to present in a condensed way a list of technical requirements and specifications and the respective implementation details as a contribution to scientific dissemination.

**Acknowledgements** This work was financially supported by Base Funding—UIDB/04730/2020 of Center for Innovation in Engineering and Industrial Technology, Portugal, CIETI; LA/P/0045/2020 (ALiCE) and UIDB/00511/2020—UIDP/00511/2020 (LEPABE) funded by national funds through FCT/MCTES (PIDDAC), Portugal.

## References


Faria P, Vale Z (2011) Demand response in electrical energy supply: an optimal real time pricing approach. Energy 36:5374–5384. https://doi.org/10.1016/J.ENERGY.2011.06.049

Guerrero JM, Blaabjerg F, Zhelev Y, Hemmes K, Monmasson E, Jemeï S, Comech MP, Granadino R, Frau JI (2010) Distributed generation: toward a new energy paradigm. IEEE Ind Electron Mag 4:52–64. https://doi.org/10.1109/MIE.2010.935862

Ibrik I (2020) Micro-grid solar photovoltaic systems for rural development and sustainable agriculture in Palestine. Agron 10:1474. https://doi.org/10.3390/agronomy10101474

Multicluster Box 6.3–11—Europe-SolarStore.com (2021) Connection Overview—MULTI-CLUSTER BOX 6.3–11 (europe-solarstore.com)

Pereira F, Caetano NS, Felgueiras C (2022) Increasing energy efficiency with a smart farm—an economic evaluation. Energy Rep 8:454–461. https://doi.org/10.1016/J.EGYR.2022.01.074

Salihu TY, Akorede MF, Abdulkarim A, Abdullateef AI (2020) Off-grid photovoltaic microgrid development for rural electrification in Nigeria. Electr J 33:106765. https://doi.org/10.1016/J.TEJ.2020.106765

Soufi A, Chermitti A, Mostafa BM, Zehor A. Spanaki K, Karafili E, Despoudi S (2021) AI applications of data sharing in agriculture 4.0: a framework for role-based data access control. Int J Inf Manage 59:102350. https://doi.org/10.1016/j.ijinfomgt.2021.102350

Spanaki K, Karafili E, Despoudi S (2021) AI applications of data sharing in agriculture 4.0: a framework for role-based data access control. Int J Inf Manag 59:102350. https://doi.org/10.1016/j.ijinfomgt.2021.102350

Toshiba White Paper (2022) Induction heating: the technology driving efficiency in home cooking appliances. https://toshiba.semicon-storage.com/content/dam/toshiba-ss-v2/master/en/semiconductor/design-development/innovationcentre/whitepapers/TCM0542_GT20N135SRA.pdf. Accessed 24 May 2022

United Nations (2017) Resolution adopted by the General Assembly on 6 July 2017. In: Work of the Statistical Commission pertaining to the 2030 Agenda for Sustainable Development.

Yahya N (2018) Agricultural 4.0: its implementation toward future sustainability. Green Energy Technol. https://doi.org/10.1007/978-981-10-7578-0_5/COVER

# Chapter 48
# Indoor Radon Remediation in Highly Constrained Built Environments: Balancing Indoor Air Quality and Energy Efficiency Through Collaborative Sensing

**António Curado, Leonel J. R. Nunes, Joaquim P. Silva, Nuno Lopes, Rolando Azevedo, and Sérgio I. Lopes**

**Abstract** Energy Efficiency (EE) and Indoor Air Quality (IAQ) are mandatory features in the rehabilitation of historical buildings. Even though it is challenging to manage both sides, EE demands are fundamental to increasing buildings´ thermal comfort, and IAQ requirements are essential for improving occupants' wellness. There are, however, several constraints that limit the intervention in complex buildings, such as the installation of mechanical ventilation systems in heritage structures. This paper presents a case study where radon risk communication has been considered as a remediation measure in a highly constrained building—classified as National Architectural Patrimony—and whose recent retrofit had focused on EE improvement. Due to the intervention, the lack of significant air renovation can harm IAQ, and therefore, increase indoor radon concentration, which is a known problem in the building under study. This paper aims to evaluate the impact of proper radon

A. Curado (✉) · L. J. R. Nunes
proMetheus, Instituto Politécnico de Viana Do Castelo, Rua da Escola Industrial E Comercial Nun'Álvares, 4900-347 Viana Do Castelo, Portugal
e-mail: acurado@estg.ipvc.pt

J. P. Silva · N. Lopes
2Ai, School of Technology, IPCA, Barcelos, Portugal

R. Azevedo · S. I. Lopes
Centro de Interface Tecnológico E Industrial, Largo da Feira, 5, 4970-786 Arcos de Valdevez, Portugal

S. I. Lopes
ADiT-Lab, Instituto Politécnico de Viana Do Castelo, Rua da Escola Industrial E Comercial Nun'Álvares, 4900-347 Viana Do Castelo, Portugal

IT—Instituto de Telecomunicações, Campus Universitário de Santiago, 3810-193 Aveiro, Portugal

 
N. S. Caetano and M. C. Felgueiras (eds.), *The 9th International Conference on Energy and Environment Research*, Environmental Science and Engineering,
https://doi.org/10.1007/978-3-031-43559-1_48

risk communication in the actions undertaken to mitigate radon risk exposure. Moreover, it puts forward a conceptual framework that manages manual ventilation actions through collaborative sensing. Results have shown that under constrained retrofitting conditions, the fine-tuning between natural ventilation actions and thermal comfort can be the key to ensuring the building EE.

**Keywords** Energy Efficiency · Indoor Air Quality · Radon risk exposure · Thermal comfort · Building ventilation · Collaborative sensing

## 48.1 Introduction

An Energy Efficiency (EE) upgrade should be designed to improve buildings' indoor environment, enhancing Indoor Air Quality (IAQ) without putting down the overall energy performance (Aleixo and Curado 2022). However, if the retrofitting is implemented in a highly constrained building, classified as National Architectural Patrimony, the lack of options to install mechanical ventilation systems will adversely affect IAQ (Lopes et al. 2018). Under these circumstances, the balance between natural ventilation and thermal comfort is the path to guarantee the building's EE (Golshan et al. 2018).

A historic building operating as a school was recently retrofitted. Due to its high architectural quality, the building is listed as National Architectural Patrimony, and therefore the rehabilitation process has met unique challenges concerning façades, windows, and roof preservation to maintain the original design (Curado et al. 2018). Given the age, the building presents very thick external walls, which favor energy-saving characteristics and considerably improve the overall thermal inertia. The original envelope allows natural sources of lighting, heating, and ventilation, contributing, therefore, to assure reasonable EE (Lopes et al. 2018). The retrofitting works were focused on building historic preservation and major renovations were set on stylistic measures to preserve the building's original trace and character, therefore issues like IAQ in classrooms and cabinets for professors, researchers, and staff, were less attended to due to the architectural features of the building, a difficult issue to deal with (Azevedo et al. 2019). Recently, there have been some complaints that most classrooms, cabinets, and cabinets need improvements in ventilation and air renovation to avoid moisture concentration and to attenuate peaks of $CO_2$ production. All rooms are operated with natural ventilation, and to prevent leaky cracks and joints, weatherstripping was added to doors and windows so that the building's EE could be reinforced.

To evaluate the impact of proper radon risk communication in the actions undertaken to mitigate radon risk exposure and to assess the positive effect of the energy retrofitting, two adjacent cabinets that have been previously identified with long-term average radon concentrations above 300 Bq $m^{-3}$, as defined by the 2013/59/EURATOM EU directive, were selected for a 3 months experimental campaign to assess continuously indoor radon exposure, air temperature, and relative humidity

(Lopes et al. 2018, 2021). Despite sharing the same architecture, type of construction, and occupancy schedule, only one of the room occupants has been introduced to the radon problem in indoor environments, and therefore the ventilation routines of this occupant are established according to a predefined schedule. The effect of air renovation on IAQ and thermal comfort will be thoroughly analyzed for both rooms, and the conceptual framework of an IoT tool to manage manual ventilation actions through collaborative sensing by using a mobile app will be deployed. Results should reveal that the adoption of a manual ventilation schema priory scheduled to sustain EE and promote air renovation flow, must allow the radon gas concentration to remain below the legal threshold limit, without compromising the building EE. Under constrained retrofitting conditions, the fine-tuning between natural ventilation actions—performed by informed occupants—and thermal comfort can be the key to ensuring the building EE.

This document is organized as follows: Sect. 48.2 presents a set of related works on the subject; Sect. 48.3 describes the Materials and Methods by detailing the adopted methodology to implement the analysis, the case study under analysis considering its specific architectural constraints, and a brief reference to risk communication applied to the referred case study. In Sect. 48.4, the results are presented and discussed, and in Sect. 48.5, a conceptual framework to address the problem of high radon concentrations in highly constrained built environments is designed to be a user-based approach to remediate the radon risks. Finally, in Sect. 48.6 the conclusions are outlined.

## 48.2 Related Works

The positive impact of air renovation on IAQ has been stated by several research works. According to Azevedo et al. (2019), air exchange through ventilation has effective and immediate effects in reducing indoor radon concentration in buildings. In the same way, it is recognized that ventilation is the most used strategy for radon remediation in existing buildings (Khan et al. 2019). Building maintenance works to improve EE, motivated in part by greenhouse gas reduction targets, involve reducing uncontrolled ventilation following best practice guidelines. These changes in buildings have increased lung cancer risk with implications for quality of life and health care costs. Studies show that increasing the airtightness of dwellings in England, without compensatory ventilation, has increased the average indoor radon concentration by more than 50% (Milner et al. 2014). The work of Silva et al. (2018) presents a pre-diagnosis model for selecting a specified set of variables that will permit the evaluation of radon potential for occupant's risk exposure while considering IAQ and the building's EE.

The retrofitting of school buildings leads to a potential improvement in the building's energy consumption, which can achieve considerable gains. In the study presented by Elkhapery et al. (2021), energy consumption was reduced by 30% in

Dubai schools. Another study in German schools produced a reduction of 80% in energy consumption (Reiss 2014).

Nevertheless, the energy retrofitting of a school building by itself does not provide any guarantee of improvement on the IAQ of the building. The work of Becker et al. (2007) proposed to achieve a balance between IAQ and EE, noticing that the Mediterranean climate demands indoor air temperature control, however, an improved ventilation scheme can improve IAQ, leading at the same time to substantial energy savings.

Korsavia et al. (2020) studied the IAQ in 29 naturally ventilated classrooms in the United Kingdom (UK), across two seasons, and concluded that IAQ is affected mostly by the occupant's behavior, and inferred that even rooms with good natural ventilation potential may not provide adequate IAQ when users are not properly informed about the positive impact of air renovation on the healthiness of the classroom. Moreover, some studies state that even with comprehensive ventilation systems, the balance between ventilation and IAQ is difficult to get, concluding that most of the time emphasis is generally put on buildings' EE neglecting, therefore, IAQ (Golshan et al. 2018; Ramos et al. 2015). The work of Monge-Barrio et al. (2022) moves in the same direction by proposing a combination of natural and mechanical ventilation mechanisms to improve IAQ in Northern Spain schools during the COVID pandemic, which raised the awareness of providing a good and healthy indoor environment for students, extending far beyond the concept of sustainable buildings.

## 48.3 Materials and Methods

The adopted methodology consisted of selecting two adjacent cabinets that have been previously identified with long-term average radon concentrations above 300 Bq $m^{-3}$, as defined by the 2013/59/EURATOM EU directive, placed in a building that was transformed into a school and is still in operation today, after being recently subject to a very prudent rehabilitation process to improve the EE, and at the same time to conserve the original trace (Curado et al. 2019). The building taken as a case study was built over the centuries showing a wide variety of architectural styles, such as Renaissance, Neoclassical and Rococo, thus reflecting the time of its interventions, ranging from the twelth to eighteenth century (Lopes et al. 2018; Curado et al. 2018).

The retrofitting process was focused on the EE improvement, which tried to preserve the buildings' very thick and heat-retaining masonry façades built with granite stone and painted with a light color to reflect summer sun radiation, resulting in cooler interior spaces with improved energy performance. Some new energy-efficient improvements were implemented during the retrofitting, namely by adding thermal insulation to the attic and crawlspaces and applying weatherstripping to doors and windows to avoid air leakage and heat losses. Despite the implemented rehabilitation having enhanced energy savings, the adoption of new mechanical ventilation systems to improve IAQ is largely conditioned by the exceptional situation of the building, given its high architectural quality which allowed its classification as

National Architectural Patrimony. Therefore, rooms are operated with natural ventilation by windows opening which can significantly reduce indoor air renovation damaging air quality. Given the architectural restraints inherent to a historic school building listed as National Architectural Patrimonial, the balance between air quality and thermal comfort even though is difficult to get, is mandatory to assure the balance between EE and IAQ.

The selected cabinets were measured by using radon sensors with 1-h resolution installed distant from the laptops and other radiation sources and electronic devices and moved away from air intakes and air streams to avoid sudden air pressure drops. Both cabinets share the same architecture, type of construction, and occupancy schedule, have similar constructed areas, air volume, glazing area, identical equipment applied (light fittings and heating devices), and similar building options. The variation between both adjacent cabinets lies in the ventilation habits of the occupants: in the room where the occupant kept a predefined air renovation procedure, indoor pollutants were removed without compromising the EE, while in the other room no ventilation actions were performed, the IAQ was affected. The experimental work involved three months monitoring campaign implemented during the Winter and early Spring of 2022, carried out to assess IAQ, thermal comfort, and EE (Lopes et al. 2018).

It is known that the communication of radon risk exposure is a subject largely unfamiliar to most citizens. Besides transmitting information to the public, radon risk communication seeks to alert policymakers to indoor radon exposure and its importance on public health, which requires cooperation between organizations in order to lead to the creation of clear and objective messages that the public understands, and duly adjusted to the recipients (Lopes et al. 2021). For these reasons, it is essential to measure the level of knowledge on the subject by the public to adjust the communication of the radon risk exposure to the adequate capacity of understanding of the target audience, namely through the forms traditionally used, such as pamphlets, clarification sessions, use of social networks, among others. Based on this assumption, one of the cabinets' occupants has been introduced to the radon problem in indoor environments and trained to implement a pre-scheduled set of natural ventilation actions designed to balance IAQ and EE and taking into consideration the room occupancy and the resulting radon concentration over time (Lopes et al. 2021). On the other hand, the other cabinet occupant didn't receive specific training on the subject and therefore wasn't stimulated to promote a set of predefined natural ventilation actions to improve IAQ.

To facilitate radon risk communication, the use of the Internet of Things (IoT) and Big Data technologies will be applied for the development of a conceptual platform designed to enhance decision-making and thus trigger in-situ remediation actions, for a specific situation, always having in mind the balance between IAQ and EE. The deployment of IoT-based tools for Radon Risk Management (RRM), as presented by Lopes et al. (2021), can be used with the Indoor Radon Risk Exposure Indicator (IRREI), a simple and intuitive tool that enables effective—and online—risk communication with the ordinary citizens, and other energy indicators design to assess EE.

Moreover, it contributes to a better perception of the exposure risk through the immediate quantification of the doses to which the occupants may be exposed in each compartment. Additionally, it allows the evaluation of a set of radon risk-oriented metrics, such as the dosimetry limits presented by the International Commission on Radiological Protection (ICRP), the occupancy rate of buildings, which corresponds to the spatial component of the system, and the periods of occupancy of buildings, which corresponds to the temporal component. The combination of all these factors allows an assessment of the actual state of indoor radon concentration and the potential impacts on the occupants/users of the buildings, secondly, to quickly assess the level of the risk that these occupants/users may be exposed, and thirdly to promote a balance between IAQ and EE (Curado et al. 2018).

## 48.4 Results

By implementing an in situ long-term monitoring campaign, significant experimental data has been collected over three months in two adjacent teacher cabinets with the same occupancy schedule. The monitored parameters, namely, the indoor radon concentration, air temperature, pressure, and relative humidity, were monitored under identical measurement conditions. The volume of the cabinets (both with 105 $m^3$), doors and windows dimensions and accesses (door to an interior aisle and window to the outside of the building), and the type of construction are all the same, therefore both cabinets share similar indoor radon potential (Silva et al. 2018). Only one of the cabinet occupants has been introduced to the radon problem in indoor environments, and therefore the ventilation routines are established according to a predefined schedule to preserve IAQ without compromising EE. In the ventilated cabinet, the window was opened every day following the same procedure, as well as the cabinet door to promote air renovation. On the other hand, in the second cabinet, the room was rarely ventilated by windows opening to facilitate cross-air circulation.

The results indicate a difference between the radon concentration in the two cabinets since higher values are detected in the cabinet without ventilation habits than in the cabinet with a predefined air renovation procedure implemented, reaching a maximum value of 1275 Bq $m^{-3}$. In contrast, in the cabinet with a predefined air renovation procedure implemented, the highest value was 867 Bq $m^{-3}$. This difference in maximum values is confirmed in the analysis of the averages obtained, which were $309 \pm 198$ and $230 \pm 150$ Bq $m^{-3}$, respectively. The natural ventilation performed through the opening of the window seems to contribute significantly to maintaining radon concentration below the limit value of 300 Bq $m^{-3}$. In contrast, in the cabinet without ventilation habits, this value is above (Fig. 48.1).

However, if, on the one hand, a predefined ventilation schedule seems to mitigate high indoor radon concentration, it may, on the other hand, raise an additional problem related to thermal comfort, especially during the winter months, when both cabinets are heated by applying a central heating system fed by a biomass boiler. During the period in which the long-term measurement took place, the outdoor

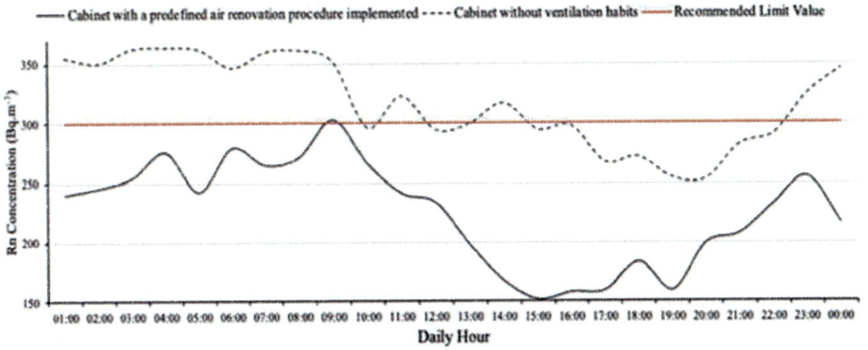


**Fig. 48.1** Evolution of the daily hourly average of radon concentration in the two analyzed cabinets

temperature fluctuated between a minimum of 1.3 °C and a maximum of 25.7 °C, as the measurement period took place during Winter and early Spring. The average outdoor temperature of the period was 12.2 ± 4.4 °C.

The analysis of indoor air temperature for both cabinets showed similar indoor air temperatures: the minimum indoor temperatures verified were 14.4 and 14.7 °C, respectively, for the cabinet without ventilation and the cabinet with a predefined air renovation procedure implemented, and the maximum temperatures recorded were, respectively, 20 °C and 22 °C. The average values are, respectively, 17.2 ± 1.1 °C and 18.9 ± 1.6 °C.

Based on the results, it seems that ventilation plays a decisive role in reducing indoor radon, contributing to maintaining the concentration below the limit value of 300 Bq $m^{-3}$. On the other hand, the indoor air temperature of the cabinet with a predefined air renovation procedure implemented does not seems to have a reduced thermal comfort when compared to the cabinet not ventilated, both under the same heating scheme.

## 48.5 Conceptual Framework

According to Sect. 48.4, the fine-tuning between natural ventilation actions and thermal comfort can be the key to ensuring the building EE. This approach can be implemented by applying a physical collaborative sensing methodology employing an infrastructure-based sensor network to collaboratively collect and analyze data from sensing devices. An IoT platform for monitoring IAQ parameters, namely, radon concentration, indoor temperature, relative humidity, and atmospheric pressure is being updated under the scope of RnHealth Project (Silva et al. 2018). The new architecture has the free and open-source Fiware platform at its core, which represents a modular approach to the development of smart applications and services (Azevedo

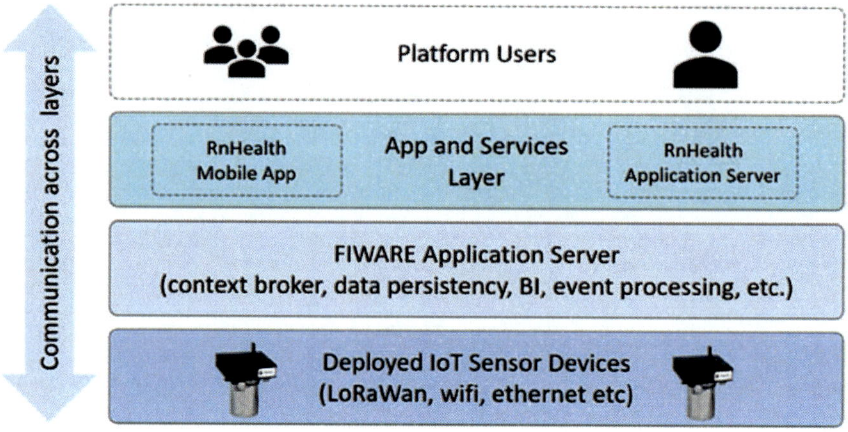


**Fig. 48.2** RnHealth high-level architecture for radon remediation

et al. 2019). Fiware supports the ability to easily add new IoT sensors, applications, or services to the existing platform.

To address the problem of high radon concentrations in highly constrained built environments, it was defined a user-based approach to remediate radon exposure risks (Fig. 48.2). The application server continuously gathers IAQ measurements from the deployed sensors and, based on the user occupancy profile, will trigger and push mobile app notifications to recommend remediation actions to the user.

The mobile app will provide the features for implementing a participatory-sensing approach (Fig. 48.3). The RnHealth platform users will receive notifications to inform about the ventilation duty cycle and the schedule of the actions to be executed. The ventilation duty cycle will be managed by the platform according to the building and compartment characteristics, the user occupancy profile, and the data received from the IoT sensors. The user will receive notifications to confirm the execution of the recommended mitigation measures. The duty cycle may change depending on the evolution of the IAQ conditions. The user will be able to check the active radon remediation prescription plan for the compartment. The Application Server will send alerts to comply with the predefined ventilation protocol. Moreover, the server will also be able to store the IAQ data provided by the user when a non-IoT-enabled radon sensor is used.

The radon mitigation protocol must balance the remediation of indoor radon exposure with thermal comfort to preserve EE. In this sense, the ventilation protocol should regulate the ventilation hours with the occupancy schedule, along with the variation of indoor air temperature and relative humidity, which vary throughout the year. In Winter, due to higher indoor-outdoor temperature differences, manual ventilation will guarantee a good air exchange, and the ventilation should occur during warmer day periods to favor EE. However, on hot days, manual ventilation

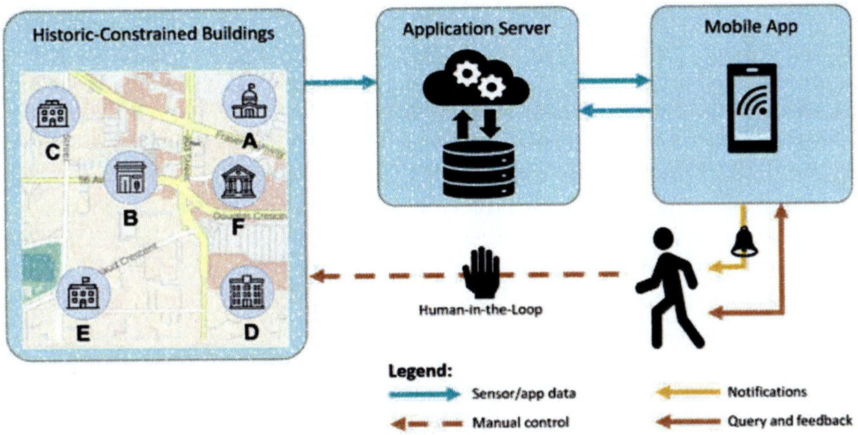


**Fig. 48.3** Conceptual RnHealth architecture with human-in-the-loop

should be done in the early morning or in the evening to cool the building to promote the building's energy performance.

## 48.6 Conclusions

The rehabilitation of a historic school building listed as National Architectural Patrimony was focused on preserving the building's original trace and character, by increasing, as far as it was possible, its EE and occupants' thermal comfort. However, given all architectural restraints related to the application of partitions and false ceilings, which didn't allow the installation of mechanical ventilation designed to take exhaust air from less ventilated classrooms and cabinets, issues like IAQ were less attended to. Under these restrictions, the improvement of natural cross-ventilation is mandatory not only to improve IAQ but also to preserve the energy performance of the building. To tackle the problem, after having been previously identified with long-term average radon concentrations above 300 Bq $m^{-3}$ according to 2013/59/EURATOM EU Directive, two adjacent cabinets of professors were taken as a case study and subject to a specific analysis to evaluate the impact of natural ventilation actions on IAQ, thermal comfort and EE. Only one of the cabinet occupants has been introduced to the radon problem in indoor environments and was stimulated to promote a set of predefined natural ventilation actions to improve IAQ. The other cabinet occupant did not receive specific training on the subject and therefore was not stimulated to undertake natural ventilation actions to preserve both IAQ and EE. The results show that the adoption of a manual ventilation schema priory scheduled to sustain EE and promote air renovation flow allows not only to decrease indoor radon exposure but also to reduce indoor moisture and $CO_2$ concentration, all without compromising the building EE. Under constrained retrofitting conditions, the fine-tuning between natural ventilation actions and thermal comfort can be the key to ensuring the building EE, by applying, within the short term, a physical collaborative sensing approach using an infrastructure-based sensor network to collaboratively collect and analyze data from sensing devices. This user-based approach to remediate radon risks will continuously monitor IAQ parameters from the deployed sensors and, based on the user occupancy profile, will trigger, and push mobile app notifications to recommend remediation measures based on natural ventilation actions. The continuous measurement of the indoor air temperature and relative humidity allows an optimal balance between and EE, by preserving the building energy performance and the occupants' thermal comfort.

**Acknowledgements** This research is a result of the project TECH—Technology, Environment, Creativity and Health, Norte-01-0145-FEDER-000043, supported by Norte Portugal Regional Operational Program (NORTE 2020), under the PORTUGAL 2020 Partnership Agreement, through the European Regional Development Fund (ERDF). LJRN was supported by proMetheus, Research Unit on Energy, Materials and Environment for Sustainability—UIDP/05975/2020, funded by national funds through FCT—Fundação para a Ciência e Tecnologia. RA was supported by operation NORTE-06-3559-FSE-000226, funded by Norte Portugal Regional Operational Program (NORTE 2020), under the PORTUGAL 2020 Partnership Agreement, through the European Social Fund (ESF). AC co-authored this work within the scope of the project proMetheus—Research Unit on Materials, Energy and Environment for Sustainability, FCT Ref. UID/05975/2020, financed by national funds through the FCT/MCTES.

## References

Aleixo K, Curado A (2022) Thermal discomfort assessment in school buildings: study based on short-term measurements. AIP Conf Proc 2425(1):200003

Azeved R, Silva JP, Lopes N, Curado A, Lopes SI (2019) Short-term indoor radon gas study in a granitic school building: a comparative analysis of occupation periods. In: International Summit Smart City 360°. Springer, Berlin

Becker R, Goldberger I, Paciuk M (2007) Improving energy performance of school buildings while ensuring indoor air quality ventilation. Build Environ 42(9):3261–3276

Curado A, Silva JP, Lopes SI (2018) Radon risk assessment in a low-energy consumption school building: a dosimetric approach for effective risk management. Energy Rep 6:897–902

Curado A, Silva JP, Lopes SI (2019) Radon risk analysis in a set of public buildings in Minho region, Portugal: from short-term monitoring to radon risk assessment. Proc Struct Integrity 22:386–392

Elkhapery B, Kianmehr P, Doczy R (2021) Benefits of retrofitting school buildings in accordance to LEED v4. J Build Eng 33:101798

Golshan M, Thoen H, Zeiler W (2018) Dutch sustainable schools towards energy positive. J Build Eng 19:161–171

Khan SM, Gomes J, Krewski DR (2019) Radon interventions around the globe: a systematic review. Heliyon 5(5):e01737

Korsavi SS, Montazami A, Mumovic D (2020) Indoor air quality (IAQ) in naturally-ventilated primary schools in the UK: occupant-related factors. Build Environ 180:106992

Lopes SI, Silva J, Antão A, Curado A (2018) Short-term characterization of the indoor air radon concentration in a XII century monastery converted into a school building. Energy Proc 153:303–308

Lopes SI, Nunes LJ, Curado A (2021) Designing an indoor radon risk exposure indicator (IRREI): an evaluation tool for risk management and communication in the IoT age. Int J Environ Res Public Health 18(15):7907

Milner J, Shrubsole C, Das P, Jones B, Ridley I, Chalabi Z (2014) Home energy efficiency and radon related risk of lung cancer: modelling study. BMJ 348:f7493

Monge-Barrio A, Bes-Rastrollo M, Dorregaray-Oyaregui S, González-Martínez P, Martin-Calvo N, López-Hernández D (2022) Encouraging natural ventilation to improve indoor environmental conditions at schools. Case studies in the north of Spain before and during COVID. Energy Build 254:111567

Ramos NM, Almeida RM, Curado A, Pereira PF, Manuel S, Maia J (2015) Airtightness and ventilation in a mild climate country rehabilitated social housing buildings—what users want and what they get. Build Environ 92:97–110

Reiss J (2014) Energy retrofitting of school buildings to achieve plus energy and 3-litre building standards. Energy Proc 48:1503–1511

Silva J, Lopes N, Curado A, Nunes LJ, Lopes SI (2018) A pre-diagnosis model for radon potential evaluation in buildings: A tool for balancing ventilation, indoor air quality and energy efficiency. Energy Rep 8:539–546

# Chapter 49
# Efficiency of Indian Cement Firms: A DEA Analysis of Large Cement Producers of PAT Cycle I and II

**Hena Oak**

**Abstract** Government of India launched the PAT scheme to improve the energy intensity of high energy consuming industries in India. The first PAT cycle ran from 2012–2013 to 2014–2015, and currently the country is implementing PAT-VI. Cement is one of the industries that has been a part of all the six PAT cycles. Although PAT-I and II had included large cement producers, they were dropped from the subsequent PAT cycles. The paper does an input-oriented DEA model to examine the efficiency scores of large cement firms that were a part of PAT-I and II, but not PAT-III and IV. Efficiency scores are first calculated using one energy input and two non-energy inputs, and then using only energy inputs for the years 2012–2013, 2016–2017, 2017–2018, and 2018–2019, where these years represent the first year of implementation of PAT Cycles-I, II, III and IV. Results show when all inputs are considered, more number of firms are on the frontier, exhibiting pure technical efficiency. But most of them are scale-inefficient. When purely energy efficiency scores are estimated, the number of firms on the frontier decline considerably for all the PAT years. The efficiency scores for top ten cement producers do improve over the various PAT cycles in the first model that can justify their exclusion from subsequent PAT cycles. But the low energy efficiency scores do not support the exclusion.

**Keywords** Cement industry · Data Envelopment Analysis · Energy efficiency scores · India · Perform–achieve–trade

## 49.1 Introduction

Energy consumption plays a vital role in any economy, especially in its industrial sector. This dependence on non-renewable energy needs to be reduced to control the rising carbon dioxide ($CO_2$) emissions. Government of India, in partnership

H. Oak (✉)
Miranda House College, University of Delhi, New Delhi 110007, India
e-mail: hena.oak@mirandahouse.ac.in

 
N. S. Caetano and M. C. Felgueiras (eds.), *The 9th International Conference on Energy and Environment Research*, Environmental Science and Engineering,
https://doi.org/10.1007/978-3-031-43559-1_49

with the Bureau of Energy Efficiency (BEE), has taken an important step in this direction by launching a market-based flagship programme called the PAT policy. It is defined as a market-based compliance mechanism to accelerate improvements in energy efficiency in energy intensive industries (Bureau of Energy Efficiency Ministry of Power 2022). PAT was one of the four initiatives started under the National Mission for Enhanced Energy Efficiency, with the objective to strengthen the market for energy efficiency through implementation of innovative business models in the energy efficiency sector (Bureau of Energy Efficiency Ministry of Power 2022).

PAT specifically targeted the Indian industrial sector because of its high consumption of non-renewable energy. In 2019–2020, 55.85% of the final energy consumption of the country went into the industrial sector (Government of India 2021). So, the aim was to gradually improve the energy intensity of the Indian industries, starting with the high energy consuming industries. The first PAT Cycle ran from 2012–2013 to 2014–2015, under which eight industries were selected. Within these eight industries, a set of plants, called designated consumers were identified for implementation. These plants had an energy consumption level that was higher than the industrial threshold level. Each of these plants were given energy intensity targets, that had to be achieved by the end of 2014–2015. The plants that successfully achieved the same were given energy saving certificates, that could be traded in the Indian Energy Exchange Limited and Power Exchange India Limited. The defaulters were required to purchase certificates to undertake production. One energy saving certificate is equivalent to 1 million tons of oil equivalent worth of energy consumption. PAT-I exceeded its energy saving target by 30% and helped to reduce emissions by 31 million tons of $CO_2$.

Cement industry was one of the eight industries selected under PAT-I, and continues to be a part of all the PAT cycles. Cement had 85 designated consumers under PAT-I, and 111 under PAT-II. However, the subsequent PAT cycles saw a sharp decline in designated consumers from this industry, with the numbers falling to 14, and 1 under PAT-III and IV respectively. The large cement producers of PAT-I and II were dropped from PAT-III and IV. The objective of this paper is to see if there was an improvement in efficiency that led to these large cement producers being excluded from the scheme's successive cycles. A sample of 32 cement firms, that includes the top 10 producers of cement, is used for analysis. All these firms were a part of PAT-I or II or both. Data Envelopment Analysis (DEA) models are used to estimate the efficiency scores of these firms, first by using both energy and non-energy inputs, and then by using only energy inputs. The paper estimates the efficiency scores of these firms in the first year of implementation of PAT Cycles I, II, III and IV.

Several studies in the literature have used DEA to examine the efficiency of various units. The energy use pattern of wheat production for the important cropping areas of Punjab in Pakistan is analyzed using DEA (Ilahi et al. 2019). DEA analysis helped to identify the inefficient energy inputs. A 3-stage DEA model is used to evaluate China's provincial energy efficiency during 2008–2016 (Zhao et al. 2019). 3-stage DEA model is also used to evaluate the industrial environmental regulation efficiency of 30 Chinese regions (Feng and Li 2020). For the Indian case, to the best of my knowledge, no study has been taken to evaluate the efficiency of designated consumers of PAT-I and II in the successive PAT cycles. The paper uses input oriented radial DEA model to fill the gap in literature by studying if the exclusion of these large cement firms had a link with their energy consumption and other inefficiencies.

Results show when all inputs are considered, a greater number of firms get an efficiency score of 1. However, when purely energy efficiency scores are estimated, the number of firms on the frontier decline considerably for all the PAT years. The efficiency scores for the top ten cement producers do improve over the various PAT cycles, that could have led to their exclusion from identified designated consumers under PAT-III and IV. The reference units for inefficient firms are mostly firms that do not belong to the top 10 group, and the reference units with the highest lambda value that form a benchmark for the inefficient firms to reach, are also not from the top 10 category. However, efficiency scores alone are not adequate to examine the reasons that caused the dropping of cement firms from PAT-III and IV.

## 49.2 Econometric Methodology, Data and Variables

Data Envelopment Analysis is a non-parametric method that is used to construct efficiency frontiers in economics and compare the relative efficiencies of similar decision making units or DMUs. The inefficient DMUs are also given efficiency targets that will help them reach the frontier. The two DEA models used in this paper are input-oriented CCR and BCC models, named after Charnes, Cooper, Rhodes, and Banker, Charnes, Cooper, respectively. The input-oriented model uses linear programming to measure the efficiency of a DMU by minimizing inputs while holding the output level constant. CCR model assumes that the efficiency frontier operates with constant returns to scale, while BCC model assumes variable returns to scale. The efficiency scores under CCR model reflect technical inefficiency, that includes both pure technical inefficiency which arises due to managerial inefficiency, and scale inefficiency which arises due to unsuitable size of the DMU.

In this paper, DMUs are the 32 cement firms that were identified for the implementation of PAT-I or PAT-II or both. The objective is to see if the efficiency of these firms improved or worsened at the time PAT-III and IV were launched, given that these firms were dropped from the next two PAT cycles. Scores are calculated for the years 2012–2013. 2016–2017, 2017–2018 and 2018–2019, where each year represents the beginning of the implementation period of PAT-I, II, III and IV respectively. There is homogeneity among the DMUs as they all belong to the same industry, and hence comparison of efficiency scores is possible. Efficiency scores are calculated first using energy and non-energy inputs, and then by using only energy inputs. Energy input is proxied by power and fuel expenditure in millions of rupees. It is defined as the cost of consumption of energy, like electricity, petroleum products, coal, etc., for carrying out the business of a company (Centre for Monitoring Indian Economy Private Limited 2022). The two non-energy inputs are raw material expenses (Rs. Million) that covers all raw materials used for production, and compensation to employees which is a proxy for labour and covers periodic payments made to the employees for the services rendered by them (Centre for Monitoring Indian Economy Private Limited 2022). Output is production of cement (Rs. Million) defined as the sum of sale of goods and change in stock of finished goods. Data on all the variables is taken from ProwessIQ. It provides time series data of Indian companies, and is built from annual reports, quarterly financial statements, and sources. The database is provided by Centre for Monitoring Indian Economy Private Limited (CMIE). The basic two-stage CCR model for the jth DMU can be written as follows.

$$\begin{aligned} &Stage\ I:\ Minimize\ \theta \\ &subject\ to\ \theta x_j - X\lambda \geq 0, \quad Y\lambda \geq y_j,\ \lambda \geq 0 \end{aligned} \tag{49.1}$$

where $X\lambda$ and $Y\lambda$ represent the best practice input and output respectively, such that the DMU is located on the efficiency frontier. $Y\lambda \geq y_j$ implies that the jth DMU is inefficient because there exists a DMU that can produce the same or more output with fewer inputs. The input slack given by $s^- = \theta x_j - X\lambda$ is the reduction in input to reach the frontier but without changing the output level. The output excess or output slack is given by $s^+ = Y\lambda - y_j$. Solving (49.1) gives the optimal solution $\theta^*$ for the efficiency score. If $\theta^* = 1$, then the DMU is said to be CCR efficient or Farrell efficient point. The value of $\theta^*$ is used to maximize the sum of input and output slacks in stage II.

$$\begin{aligned} &Maximize\ \epsilon s^- + \epsilon s^+\ where\ \epsilon = (1, \ldots, 1)\ is\ a\ vector\ of\ 1s \\ &subject\ to\ s^- = \theta^* x_j - X\lambda,\ s^+ = Y\lambda - y_j,\ \lambda \geq 0,\ s^- \geq 0,\ s^+ \geq 0 \end{aligned} \tag{49.2}$$

If a DMU has $\theta^* = 1$ and all slacks are 0, then it is said to satisfy Pareto–Koopmans strong efficiency condition (Cooper et al. 2007). The BCC model modifies the CCR model by adding a convexity condition, $\sum_{j=1}^{32} \lambda_j = 1$ and separates the technical efficiency into pure technical efficiency and scale efficiency under the assumption of variable returns to scale. If the optimal solution of the BCC model given by $(\theta^*, \lambda^*, s^{-*}, s^{+*})$ satisfies the condition that $\theta^* = 1$ and all slacks are 0, then the solution is BCC efficient.

CCR and BCC efficiency scores are estimated for the years 2012–2013, 2016–2017, 2017–2018, and 2018–2019. Finally, scale efficiency is calculated as the ratio of the CCR and BCC efficiency scores. Scale efficiency of 1 implies the DMU is operating at the optimal size.

## 49.3 Results and Discussion

Results show that out of 32 firms, 18.75% display technical efficiency as per the CCR model, and 40.63% display pure technical efficiency as per the BCC model with 0 slack values. Pure technical efficiency does not imply that the size is optimal. Out of the 13 firms with efficiency score of 1 under the BCC model, 6 firms have an efficiency score of 1 as per scale efficiency, and are thereby operating on their optimal size. 5 out of these 6 firms had a share of <2% in the total cement production, implying that more than 80% of the firms that are operating at optimal scale are not large cement producers. Detailed table has been omitted due to brevity.

In the year 2014–2015, the top ten cement producers of India were Ultratech Cement Limited, ACC Limited, Ambuja Cements Limited, Shree Cement, India Cement, Ramco Cement, Birla Corporation Limited, Dalmia Cement, J.K. Group, and Chettinad Cement (Government of India, Ministry of Mines, Indian Bureau of Mines 2013, 2016). Efficiency scores were calculated for all the 32 DMUs for the years 2012–2013, 2016–2017, 2017–2018 and 2018–2019, but Table 49.1 reports detailed analysis for the year 2012–2013 for the top 10 producers only due to brevity.

**Table 49.1** Efficiency analysis of top 10 cement producers in 2012–2013

| DMU | CCR | BCC | SE | Power and fuel exp | Raw material | CoE | Ref. group |
|---|---|---|---|---|---|---|---|
| ACC | 0.854 | 0.912 | 0.936 | 0 | 0 | 409.5 | B (0.119) C (0.382) **G(0.499)** |
| Ambuja | 0.862 | 0.926 | 0.930 | 645.7 | 0 | 0 | D (0.275) E (0.293) **G(0.432)** |
| Birla Corp | 0.693 | 0.709 | 0.977 | 0 | 0 | 0 | B (0.326) **C (0.447)** F (0.169) G (0.058) |
| Chettinad Cement | 0.785 | 0.809 | 0.970 | 0 | 0 | 0 | C (0.097) **D (0.442)** F (0.399) G (0.063) |
| Dalmia Cement | 0.794 | 0.918 | 0.865 | 1239.3 | 0 | 0 | A (0.021) **F (0.912)** G (0.067) |
| India Cement | 0.753 | 1 | 0.753 | 0 | 0 | 0 | |
| JK Cement* | 0.854 | 0.885 | 0.965 | 0 | 0 | 0 | C (0.114) D (0.385) **F(0.413)** G(0.089) |
| JK Lakshmi* | 1 | 1 | 1 | 0 | 0 | 0 | |
| Ramco Cement | 0.853 | 0.892 | 0.956 | 0 | 0 | 126 | **B (0.671)** C (0.225) G (0.104) |
| Shree Cement | 0.646 | 0.704 | 0.917 | 509.4 | 0 | 0 | A (0.048) **F (0.779)** G (0.172) |
| Ultratech Cement | 0.949 | 1 | 0.949 | 0 | 0 | 0 | |

A—India Cement, B—JK Lakshmi Cement, C—My Home Industries, D—Orient Cement, E—Penna Cement, F—Sanghi industries, G—Ultratech Cement

Columns 5, 6, and 7 report input slack values for Power and Fuel Expenditures, Raw Materials, and Compensation to Employees

*JK Cement and JK Lakshmi Cement together are a part of the JK Group, and are treated as one group

The values in bold are the reference units with the highest weight for a given DMU

As per Table 49.1, JK Lakshmi Cement alone operates on optimal size among the large cement producers, that makes it overall technically efficient. The other inefficient firms need to reduce their input consumption in order to be on the frontier. For example, to achieve pure technical efficiency in the year 2012–2013, the highest reduction in all its inputs was needed by Shree Cement at 29.55%, and the least reduction in inputs was needed by Ambuja Cement at 7.39%. In terms of input slack values reported in columns (5), (6), and (7), for ACC Limited a further reduction of compensation to employees by Rs. 409.49 million would have improved its efficiency, without worsening output. For Ambuja and Shree Cement Limited, a further reduction in power and fuel expenditure by Rs. 645.66 million and Rs. 509.36 million, respectively would have improved its efficiency, with no change in output. Column (8) reports the reference points of the inefficient DMUs. This is a point that an inefficient DMU targets to move from the Farrell efficient point to the Pareto–Koopman efficient point (Ji and Lee 2010). In case the inefficient DMU has more than one reference point, then the point with the largest weight (or lambda) will be selected as a benchmark (Cooper et al. 2011). The values in bold are the reference units with the highest weight for a given DMU. The efficient DMUs are already on the frontier, and hence do not have a reference unit. The benchmark for most of the inefficient firms is Wonder Cement as it had the highest lambda value the maximum number of times, followed by Sanghi Industries.

Table 49.2 reports the pure technical efficiency scores of the top cement firms only (the remaining firms have not been reported due to brevity). Results show that 37.5% of the DMUs are efficient in all the three PAT cycles. Most of the DMUs that were on the frontier under PAT-I continue to be so under PAT-II, III and IV. Ultratech Cement Limited, the largest producer of cement in recent years, has consistently reported efficiency score of 1, with 0 slacks. Compared to PAT-I, there is a rise in the minimum and average efficiency scores in these subsequent cycles. There is a fall in standard deviation that implies less dispersion in the data set and greater clustering around the mean efficiency score. Results from the calculation of scale efficiency shows that Deccan Cement and Wonder Cement have a scale efficiency of 1 in all the 4 years and 3 years, respectively. Therefore, these two firms alone operate at an optimal plant size. All the other firms, including the top 10 producers, recorded a decline in scale efficiency scores after the implementation period of PAT-II ended, implying a move away from their optimal size of production. In PAT-III except for Shree Cements and JK Lakshmi Cement Limited, none of the large cement firms were included. What is interesting is that other than these two firms, all the remaining designated consumers under PAT-III had a share of <1% in the total cement production for the year 2017–2018. PAT-IV had only one firm as designated consumer.

Factors like capacity utilization, age, PAT scheme, etc. can explain the inter-firm variations in efficiency scores. A higher capacity utilisation reduces the per unit cost of production due to economies of large-scale production. Indian cement industry's capacity utilization was well above 70% prior to 2015–2016 (Government of India, Ministry of Mines, Indian Bureau of Mines 2016), after which it came down to an average of 57.92% between 2015–2016 to 2018–2019. Since the cement industry's growth is closely intertwined with the economy, 2012–2018 was a period

**Table 49.2** Efficiency analysis of top 10 cement firms in 2016–2017, 2017–2018, and 2018–2019

| DMU | Efficiency scores as per BCC model | | |
|---|---|---|---|
| | PAT-II | PAT-III | PAT-IV |
| | 2016–2017 | 2017–2018 | 2018–2019 |
| ACC Ltd | 0.843 | 1 | 1 |
| Ambuja | 1 | 0.984 | 0.954 |
| Birla Corp | 0.776 | 0.815 | 0.884 |
| Chettinad Cement | 0.867 | 0.837 | 0.833 |
| Dalmia Cement | 0.985 | 0.913 | 1 |
| India Cement | 0.853 | 0.832 | 0.839 |
| JK Cement* | 0.962 | 0.938 | 0.891 |
| JK Lakshmi* | 0.879 | 0.882 | 0.793 |
| Ramco Cements | 1 | 0.923 | 0.798 |
| Shree Cement | 1 | 1 | 1 |
| Ultratech Cement | 1 | 1 | 1 |

*JK Cement and JK Lakshmi Cement together are a part of the JK Group, and are treated as one group

of deceleration for the Indian economy (Mohan 2019) that led to lesser production in general. But even during the slowdown, smaller firms seem to be doing better than the larger firms. Only 4 out of 12 firms with an efficiency score of 1 as per the BCC model in 2016–2017 belonged to the top 10 category. Similarly for the years 2017–2018 and 2018–2019, most of the firms operating above the average capacity utilisation of the industry, did not belong to the top 10 producers' group. Age of the firm is important, because older firms have more resources at their disposal to invest in productivity improvements of all inputs. 8 of the DMUs are among the most efficient in terms of their position on the frontier because they either reported efficiency score of 1 consistently in all the four years, or in 3 out of the four years. 5 out of these 8 firms were more than 40 years old. Finally, exclusion from the PAT scheme can also have an effect on all the 32 firms. After 2016–2017, when the implementation period of PAT-II ended, 9 firms showed a clear decline in efficiency scores, thereby highlighting the importance of PAT. 5 firms showed a rise in scores in 2017–2018, followed by a decline, and for the remaining firms efficiency scores either increased or remained static.

The paper also estimates efficiency scores for the top ten cement firms when only energy input is considered. Table 49.3 below reports the energy efficiency scores for these firms. Ultratech Cement is the only large cement firm that lies both on the efficiency frontier and energy efficiency frontier for all the PAT implementation years, and satisfies the Pareto–Koopmans strong efficiency condition. The other large cement firms report much lower energy efficiency scores and will have to reduce their input combinations to move up on the frontier. Also, they do not exhibit any clear pattern. Ultratech Cement, Kesoram Industries and Meghalaya Cement have been

**Table 49.3** Energy efficiency analysis of top 10 cement firms in PAT Cycles I, II, III and IV

| DMU | Efficiency scores as per BCC model | | | |
|---|---|---|---|---|
| | PAT-I | PAT-II | PAT-III | PAT-IV |
| | 2012–2013 | 2016–2017 | 2017–2018 | 2018–2019 |
| ACC Ltd | 0.886 | 0.787 | 0.919 | 0.994 |
| Ambuja | 0.794 | 0.769 | 0.821 | 0.817 |
| Birla Corp | 0.468 | 0.609 | 0.204 | 0.607 |
| Chettinad Cement | 0.383 | 0.648 | 0.178 | 0.236 |
| Dalmia Cement | 0.323 | 0.963 | 0.779 | 1 |
| India Cement | 0.527 | 0.639 | 0.372 | 0.444 |
| JK Cement | 0.498 | 0.802 | 0.434 | 0.758 |
| JK Lakshmi | 0.381 | 0.671 | 0.175 | 0.457 |
| Ramco Cements | 0.649 | 1 | 0.457 | 0.649 |
| Shree Cement | 0.459 | 0.821 | 0.794 | 0.777 |
| Ultratech Cement | 1 | 1 | 1 | 1 |

reference units for the inefficient firms, but only Kesoram Industries and Meghalaya Cement have been benchmark reference units.

## 49.4 Conclusion

India started the PAT scheme with the view to improve the energy intensity of high energy consuming industries, cement industry being one of the selected industries. However, after PAT-II, the number of designated consumers from this industry fell considerably and all the large producers were dropped from the successive cycles. This paper uses input-oriented DEA analysis to evaluate if the efficiency scores of these firms were improving overtime, that led to their omission from the scheme. BCC model shows that though in general efficiency scores were improving, especially for the 10 largest cement producers, the improvement in energy efficiency scores was not outstanding enough to cause an exclusion of these firms. Moreover, since the PAT scheme's primary objective was to improve the energy intensity of large energy consuming firms, its judicious to account for the low energy efficiency scores and include these firms in the upcoming PAT cycles. At the same time, allowance must be given for considerations other than efficiency scores that could have influenced the decision on whether to add a DMU within the policy or not.

## References

Bureau of Energy Efficiency Ministry of Power, Govt. of India (2022) https://beeindia.gov.in. Accessed August 10, 2022

Centre for Monitoring Indian Economy Private Limited (2022) https://prowessiq.cmie.com/. Accessed 14 Jul 2022.

Cooper W, Seiford L, Tone K (2007) Data envelopment analysis: a comprehensive text with models, applications, references and DEA-solver software, 2nd edn. Springer, USA

Cooper W, Seiford L, Zhu J (2011) Handbook on data envelopment analysis, 2nd edn. Springer, USA

Feng M, Li X (2020) Evaluating the efficiency of industrial environmental regulation in China: a three-stage data envelopment analysis approach. J Clean Prod 242:118535

Government of India, Ministry of Mines, Indian Bureau of Mines (2013) Indian Minerals Yearbook 2013: Cement, 52nd edition

Government of India, Ministry of Mines, Indian Bureau of Mines (2016) Indian Minerals Yearbook 2016: Cement, 55th edition

Government of India, Ministry of Statistics and Programme Implementation (2021) Central statistics office: energy statistics

Ilahi S, Wu Y, Raza M, Wei W, Imran M, Bayasgalankhuu L (2019) Optimization approach for improving energy efficiency and evaluation of greenhouse gas emission of wheat crop using data envelopment analysis. Sustainability 11:3409

Ji Y, Lee C (2010) Data envelopment analysis. Stand Genomic Sci 10(2):267–280

Mohan R (2019) Moving India to a new growth trajectory: need for a comprehensive big push. Brookings India Research Paper 072019

Zhao H, Guo S, Zhao H (2019) Provincial energy efficiency of China quantified by three-stage data envelopment analysis. Energy 166:99–107

# Chapter 50
# Circular Economy of Household Used Cooking Oil: Waste-to-Energy Potential Geospatial Mapping

**Jose Armando Hidalgo Crespo, Cesar Alvarez-Mendoza, C. M. Moreira, Manuel Soto, and Jorge Luis Amaya-Rivas**

**Abstract** Large amounts of edible oils are discarded each day from various sources, principally from households. When disposed wrongly into the nature, they can reach the ground, reducing the absorption of nutrients from plants, the aquifers, making them unusable, and finally, water surfaces, affecting the aquatic systems. From a circular economy perspective, one essential condition is looking for alternatives to reuse the used cooking oil (UCO). The objective of this paper was to discuss the application of Geographical Information Systems (GIS) to map the renewable energy potential for three possible waste-to-energy potential paths: biodiesel, anaerobic digestion, and direct combustion for the city of Guayaquil. For the cases of biodiesel (a), methane (b), and direct combustion (c), maximum values of 95, 88, and 95 MJ/kg/person, respectively, were found. Finally, Tarqui parish has much higher values than Ximena and Pascuales. Given its potential, the collection should start from this parish as a pilot project.

**Keywords** Waste-to-energy · Geospatial mapping · Circular economy · Biogas · Biodiesel · Combustion

J. A. Hidalgo Crespo (✉)
Carrera de Ingeniería Industrial, Universidad Politécnica Salesiana, Guayaquil 090150, Ecuador
e-mail: jhidalgo@ups.edu.ec

J. A. Hidalgo Crespo · M. Soto
Universidade da Coruña, 15008 Coruña, Spain

J. A. Hidalgo Crespo
G-Scop Laboratory, Polytechnic Institute of Grenoble, 46. Av. Félix Viallet, 38000 Grenoble, France

C. Alvarez-Mendoza
Grupo de Investigación Ambiental en el Desarrollo Sustentable GIADES, Carrera de Ingeniería Ambiental, Universidad Politécnica Salesiana, Quito 170517, Ecuador

C. M. Moreira · J. L. Amaya-Rivas
ESPOL Polytechnic University, Escuela Superior Politécnica del Litoral, ESPOL, Guayaquil 090150, Ecuador



N. S. Caetano and M. C. Felgueiras (eds.), *The 9th International Conference on Energy and Environment Research*, Environmental Science and Engineering,
https://doi.org/10.1007/978-3-031-43559-1_50

## 50.1 Introduction

Rapid population growth coupled with urbanization development of the city deteriorates the environment rapidly due to the increasing energy consumption and waste generation, which at the same time provokes an exponential increase in fossil fuel consumption. Sustainable development Goal 7 seeks to ensure access to affordable, reliable, sustainable, and modern energy for all by 2030.

Following this path, the circular economy (CE) expresses a new concept that focuses on maintaining the value of products, materials, and resources in the economy for as long as possible and minimizing waste generation (European Commission 2015). The focus of a circular economy is not ought to be only on maintaining materials in the market for as long as possible and minimizing the use of natural resources, but it should also keep in mind other aspects, such as health, economic and environmental, the three pillars of sustainability, when looking for the best waste treatment option. The best alternative to create a synergy between the economic and environmental development goals for any country, replacing the traditional take, make, consume and dispose of patterns, is the circular economy. In this context, waste-to-energy WtE technologies have started playing an essential role in this emerging environment as it helps to maximize resource efficiency. They refer to the energy recovery from waste materials into usable heat, electricity, or fuels (Tan et al. 2015). Many crucial environmental problems, such as the use of land and climate change are addressed by the circular economy paradigm. In fact, CE has grown to be a vital part of waste management, reducing the dependence of fossil-based fuels, while ensuring the security of energy supply (Xiao et al. 2020).

WtE fits the CE agenda correctly, and it's one prime example of closing the biological loop, enabling the reuse of waste for energy generation and the production of other environmental raw materials (Charter 2018; Garmulewicz et al. 2018). There are different technologies available for WtE: combustion, gasification, and anaerobic digestion (AD). By direct burning, UCO can also be processed to obtain pyrolytic oil, hydrogen gas, biodiesel, or electricity production.

In Ecuador, diesel is the petroleum-based derivate with the highest consumption, next are gasoline and liquefied petroleum gas (LPG). One reason for the high demand for diesel fuel is mainly related to the dependence of the transport sector for this fuel, which is responsible for 69% of the share according to the energy balance of 2016 (MICSE 2016) for the internal combustion engines. Calculations effectuated by Ribau-Teixeira et al. (2018) accounted that 32% of the 200 million tons of edible vegetable oil consumed worldwide become waste, meaning 64 million tons of UCO are available for further study usage. According to information from the Ministry of Production and Foreign Trade, Ecuador produces mainly palm oil, which is the country's seventh agricultural export product. Data from the Internal Revenue Service indicates there are 142 active businesses in the country dedicated to producing oils and fats of vegetable origin.

Using vegetation (crops) to generate biofuels causes adverse effects. The most serious one is the increase in the price of food-related products, followed by the desertification due to the excessive agricultural exploitation, and not always, the greenhouse gas emissions (GHG) are higher for biofuels than for fossil fuels, when performing the cradle-to-grave analysis (Carriquiry et al. 2011; Edenhofer 2011). Sugarcane-based ethanol combined with gasoline, and African palm-based biodiesel combined with diesel have been the latest venture of the country since 2012. Also, toxic and cancer-related components, such as polycyclic aromatic hydrocarbons are released during the frying process of any edible oil, due to the processes of hydrolysis, polymerization (Wang et al. 2016). For Europe, a total of 0.008 $m^3$ of UCO are produced annually per capita. For the case of Guayaquil, it was estimated in 2020 that household used cooking oil generation was 5.11 g (5.70 mL) daily per person (Hidalgo-Crespo et al. 2020a, b). Considering a total population of 2,654,433 inhabitants, Guayaquil has the potential to collect 5.52 thousand metric tons of UCO yearly, just from households alone.

One renewable energy concept refers to an energy resource, which is not lost when consumed, and that is renewed by nature. To ensure the energy supply of any country with limited fossil-fuel reserves, the search for renewable and non-polluting alternatives is imperative. It should be a top priority of every government. To ensure the region's sustainable development, it's essential to trace the predisposition of renewable energy sources like vegetation, livestock, and waste. The final disposition of UCO in landfills or into the drainage runs free for the inhabitants of the country, due to the lack of regulations and laws for its control. Indeed, 54% of people dispose of their used cooking oil in the garbage, 24% in drains, 21% use it as animal feed, and only 1% deposit it in a particular container or give it to a person in charge of the collection of this product. It has been estimated that one liter of UCO can affect the taste of one million liters of water (Milano 2008). Due to its slow degradation, it can cause severe problems after disposal. It can damage sewage systems (clogging) and wastewater treatment plants, increasing maintenance costs. The economic and energy costs of wastewater treatment plants due to UCO treatment are about 0.45 euros/kg (GESENU 2016) and 3 kWh/kg (Ripa et al. 2014). Additionally, when disposed wrongly into the nature, it can reach the ground, reducing the absorption of nutrients from plants, the aquifers, making them unusable, and finally, water surfaces, affecting the aquatic systems. From a circular economy perspective, one essential condition is looking for alternatives to reuse the used cooking oil (Loizides et al. 2019). This paper aims to discuss the application of Geographical Information Systems (GIS) to map the renewable energy potential for three possible waste-to-energy potential paths: biodiesel, anaerobic digestion, and direct combustion for the city of Guayaquil.

## 50.2 Materials and Methods

For this research, information on GPS coordinates (latitude and longitude WGS84) and household size were gathered from a total of 530 households in the city of Guayaquil. For a time-lapse of 2 weeks, they poured their UCO into a glass jar, allowing them to calculate the total daily per capita production later. After, samples

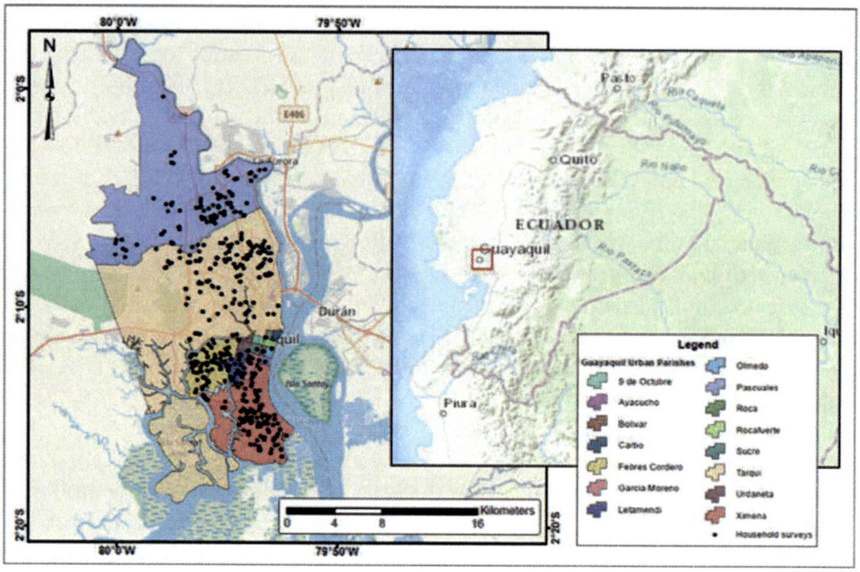


**Fig. 50.1** City of Guayaquil urban parishes

were taken to the Laboratory of Hydrocarbons of the Faculty of Natural Sciences and Mathematics of one local university. Finally, the results are processed in a Geographic Information System (GIS) to make a spatial analysis. Guayaquil is divided into parishes or districts, having 16, being the most populated Tarqui, Ximena, and Pascuales. Figure 50.1 shows the study area for Guayaquil Urban Parishes. Additionally, the 530 GPS coordinate points are mapped across the city.

### *50.2.1 Biodiesel*

Through the esterification reaction of ethanol or methanol, with the presence of some kind of catalysts like sodium or potassium hydroxide, and with glycerides, whether mono, di, or tri, from vegetable, or animal oils, biodiesel is produced (Barriga 2009). This biofuel can also be generated from residual edible oils, however, they require first to pass through a cleansing process constituted but not limited to filtration, and washing, to remove solid particulates and the content of humidity, before transesterification. One major company specializing in the biodiesel production from African palm oil has pointed out that their cost of production ($1.00/kg), is 53% higher than that of regular diesel (OLADE 2010), however this cost can be reduced if used cooking oil is transformed instead, reducing the cost to $0.52/kg (Elgharbawy and Sayed 2022).

Some advantages of biodiesel besides being renewable and biodegradable is the fact that it does not contain sulfur or any polynuclear compounds. Coming from a renewable resource such as biomass from crops, combustion of this fuel helps reduce the emissions of GHG, and particulate matter, compared to the emissions related to the combustion of petroleum-based diesel (Hasan and Rahman 2017). One research concluded that biodiesel derived from UCO could replace the diesel to help control air pollution and reduce the dependency on fossil fuel resources with a small sacrifice in the performance of an engine (Kumar et al. 2017). However, different blends of biodiesel in 10, 20, and 30% (B10, B20, B30) increase the specific fuel consumption and NOX and CO2 emissions than regular diesel (Abed et al. 2018). This research followed a yield of 96.55% for the biodiesel-making process and a calorific value of 37.27 MJ/kg (Hidalgo-Crespo et al. 2020a).

### *50.2.2 Anaerobic Digestion*

Anaerobic digestion allows the generation of the much-needed energy with a nutrient-rich for agricultural compost through the degradation of biological waste. As fats can give origin to elevate methane yields, used cooking oil seems a promising solution. However, some biological limits are present when using UCO for biogas production. Nevertheless, co-digestion with animal effluents, rich in water and nitrogen, can help mitigate them. One example is the use of UCO in the co-digestion with pig slurry, which significantly has proved to shorten the lag phase (by almost 58%), and increase the overall methane production (by more than 15%) when compared to that of pig slurry alone (Marchetti et al. 2019).

The introduction of UCO in the feeding mix of crop-based biogas plants could lead to up to 50% of land savings by reducing the energy crops and maintaining a stable energy production (Carnevale et al. 2017). Co-digestion can augment biogas production by 25–400% more than mono-digestion of the same substrates. It is beneficial due to the dilution of toxic chemicals, enhancing the balance of nutrients, and the synergistic effect of microorganisms. The co-substrates can supply micro-nutrients and alkalinity to overcome the disadvantages of food waste's single digestion.

Previous research on the subject showed a high lipids content of 99.19% for the used cooking oil in the city (Hidalgo-Crespo et al. 2020a). The methane potential generation of lipids is very high, making the recycling of edible and used oils an attractive alternative. In fact, the theoretical production of $CH_4$ is close to 1 L per kg of lipid. This value is more than two times higher than the methane generation from carbohydrates and proteins (Martínez et al. 2016; Oliveira et al. 2018). The calculations considered 0.657 kg/m$^3$ and 50 MJ/kg for the density and heating value.

### 50.2.3 *Direct Combustion*

Direct combustion of UCO can help achieve the efficient use of energy by eliminating unnecessary energy conversions, which imply certain losses as wasted heat, limiting the total efficiency of the process. The direct application of UCO in engines can prevent the treatment necessary for biodiesel generation (transesterification, neutralization, washing, and distillation), reducing the environmental impacts and its related costs as a consequence (Tsai et al. 2007; Knothe et al. 2015; Badvipour 2009). This application is much more feasible for stationary production of electrical and thermal energy (Capuano et al. 2017) rather than for automotive applications. As an alternative source for combining heat and power production (CHP), the use of UCO represents an intelligent way for waste valorization and circular economy promotion. To determine the calorific heating value (LHV), previously collected samples were passed through a calorimeter IKA C200, following the INEN2903 norm, obtaining 37.78 MJ/kg.

### 50.2.4 *Geospatial Mapping*

Several studies have tried to assess the geospatial availability of many biological waste, since it helps support decision-making when thinking to implement technologies to recover them. In this context, Geographic System Information (GIS) helps deliver spatial and temporal analyses of renewable resources. The most challenging task to carry out is the collection of household used cooking oil due to the dispersion of the population. However, this deficiency can be overcome by improving environmental education and the self-awareness increment as producers of this waste (Marchetti et al. 2020). Also, its collection and storing is very much possible due to its very uniform chemical and microbial composition, making it manageable with little cost of supervision. For this research, we assumed a collection rate of 20% (Qamar et al. 2020). The heat maps show the yearly MJ available for the city for the case of each technology: biodiesel, biogas, and direct burning. Moreover, we build a spatial model using the inverse distance weighting (IDW) interpolation using GPS coordinates.

## 50.3 Results

With the help of GIS software, the potential energy generation in MJ is shown in Fig. 50.2. This figure considers only a 20% collection efficiency, as stated in the previous section. For the cases of biodiesel (a), methane (b), and direct combustion (c), maximum values of 95, 88, and 95 MJ/kg/person, respectively, were found. The most promising points seem to be in tourist sites of the city, particularly in middle-town. Thus, the areas near the tourist sites have a higher waste-to-energy potential. Considering a total sample of 530 households, the three most populated parishes represented close to 75%: Pascuales (21%), Tarqui (30%), and Ximena (24%). Table 50.1 shows the average values for each waste-to-energy option for the entire sample and the three most populated parishes. As can be seen in the figure, Tarqui parish has much higher values than the other two parishes. Given its potential, the collection should start from this parish as a pilot project.

## 50.4 Conclusion and Discussion

This study aimed to geo-locate the potential renewable energy generation of used cooking oil converted to biodiesel, methane, or used directly for burning. For the cases of biodiesel (a), methane (b), and direct combustion (c), maximum values of 95, 88, and 95 MJ/kg/person, respectively, were found. Finally, Tarqui parish has much higher values than Ximena and Pascuales. Given its potential, the collection should start from this parish as a pilot project.

Businesses should capture value leakage in their everyday activities and their supply chain. The reuse of UCO prevents increasing the costs of edible oils, avoids the potential improper return for human consumption, and represents new job opportunities helping the local economy.

When comparing the environmental impacts of biodiesel and direct burning WtE pathways with the more common cradle-to-grave approach, one study found that the second use of UCO has lower has implications for both comparisons, being natural burning the best profile (Albuquerque et al. 2022). In one previous study, it was possible to evaluate the influence of waste cooking oil as a co-substrate of cow, pig, and chicken manure and also restaurant food waste in anaerobic co-digestion. As demonstrated, waste cooking oil ameliorated the values of 75% of the types of garbage. This study also indicated that the biogas production was comparatively higher in the co-digested bottles than in the mono-digested bottles. The main problem of digesting lipids from UCO is that this product is water repellent, which hinders biogas production and creates a two-phase system, with UCO floating over the water, and developing foaming. Another problem is the agglomeration of long-chained fatty acids, that are poisonous to the bacteria consortia, but this problem can be avoided if UCO co-digests with other substrates that are much more degradable, like a carbo-rich waste (Wang et al. 2016). The use of biodigesters to produce biogas through anaerobic digestion may play an essential role in local economies due to the opportunity to make renewable fuel from organic waste and alternative waste treatment.

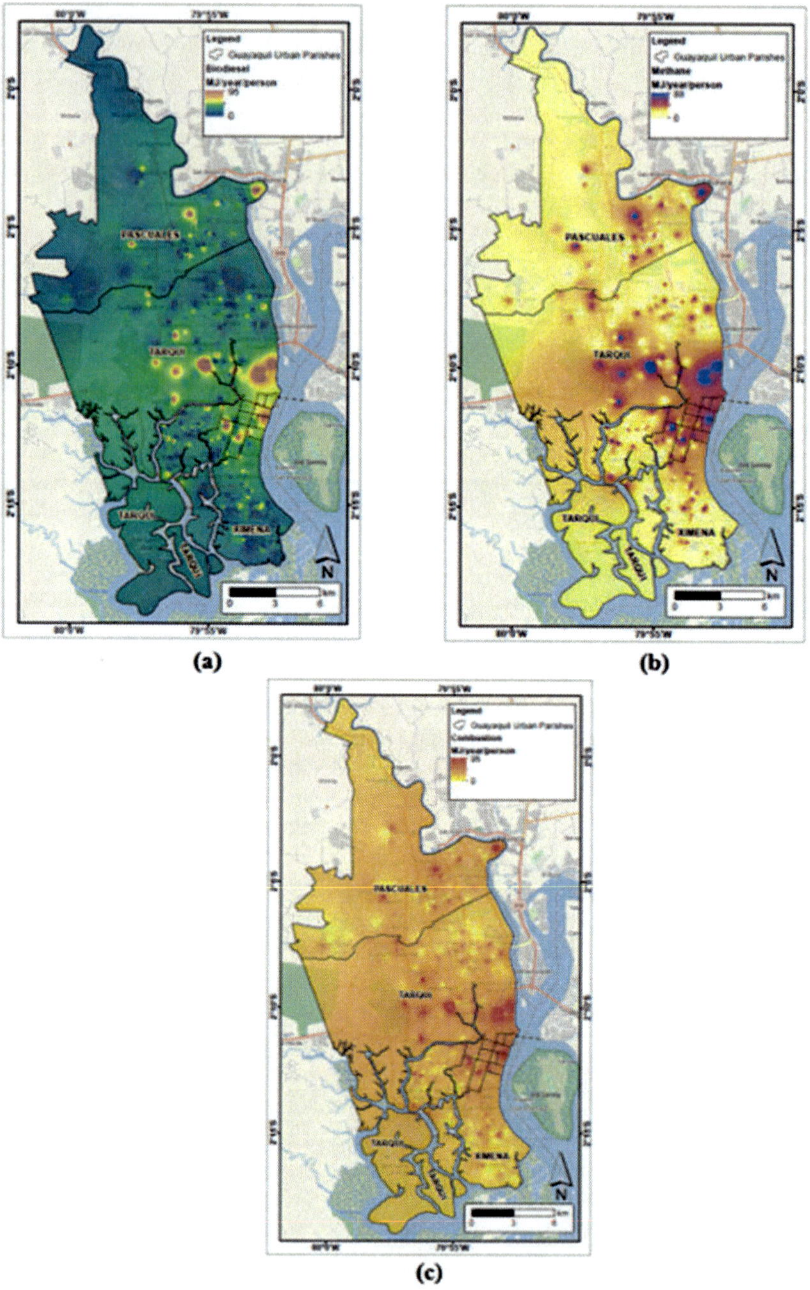


**Fig. 50.2** UCO waste-to-energy potential

**Table 50.1** Stratification of households' samples and parishes

| Parishes | Biodiesel | Methane | Combustion |
|---|---|---|---|
| Tarqui | 12.284 | 10.889 | 12.327 |
| Ximena | 10.305 | 9.136 | 10.342 |
| Pascuales | 11.170 | 9.902 | 11.209 |
| Total | 11.737 | 10.405 | 11.779 |

# References


Abed KA, El Morsi AK, Sayed MM, Shaib AA, El Gad MS (2018) Effect of waste cooking-oil biodiesel on performance and exhaust emissions of a diesel engine. Egypt J Petrol 2018:S111006211830014X. https://doi.org/10.1016/j.ejpe.2018.02.008

Albuquerque Landi F, Fabiani C, Castellani B, Cotana F, Pisello AL (2022) Environmental assessment of four waste cooking oil valorization pathways. Waste Manag 138:219–233. https://doi.org/10.1016/j.wasman.2021.11.037

Badvipour S (2009) Biodiesel production from waste cooking oil. Opportunities and challenges. In: Proceedings of the 2nd Waste-to-Fuel Conference, San Diego, CA

Barriga A (2009) Producción y Uso de Aceites Vegetales y Biodiesel en Ecuador

Capuano D, Costa M, Di Fraia S, Massarotti N, Vanoli L (2017) Direct use of waste vegetable oil in internal combustion engines. Renew Sustain Energy Rev 69:759–770. https://doi.org/10.1016/j.rser.2016.11.016

Carnevale E, Molari G, Vittuari M (2017) Used cooking oils in the biogas chain: a technical and economic assessment. Energies 10(2):192. https://doi.org/10.3390/en10020192

Carriquiry M, Du X, Timilsina G (2011) Second generation biofuels: economics and policies. Energy Policy 39(7):4222–4234

Charter M (ed) (2018) Designing for the circular economy. Routledge, London

European Commission (2015) Closing the loop—an EU action plan for the Circular Economy.COM 614. Brussels.

Edenhofer O (2011) IPCC special report on renewable energy sources and climate change mitigation. Prep. By Work. Gr. III Intergov. Panel Clim. Chang. Cambridge University Press, Cambridge

Elgharbawy A, Sayed AA (2022) Cost analysis for biodiesel production from waste cooking oil plant in Egypt. Int J Smart Grid 1(1):16–25

Garmulewicz A, Holweg M, Veldhuis H, Yang A (2018) Disruptive technology as an enabler of the circular economy: what potential does 3D printing hold? Calif Manage Rev 60(3):112–132. https://doi.org/10.1177/0008125617752695

GESENU (2016) Perugia: Attivi 4 nuovi punti di raccolta Olly. http://www.gesenu.it/articolo/perugia-attivi-4-nuovi-punti-di-raccolta-olly

Hasan MM, Rahman MM (2017) Performance and emission characteristics of biodiesel-diesel blend and environmental and economic impacts of biodiesel production: a review. Renew Sustain Energy Rev 74:938–948. https://doi.org/10.1016/j.rser.2017.03.045

Hidalgo-Crespo J, Coello-Pisco S, Crespo-Vaca T, López-Vargas A, Borja-Caicedo D, Martínez-Villacrés H (2020a) Domestic waste cooking oil generation in the city of Guayaquil and its relationship with social indicators. Paper presented at the Proceedings of the LACCEI International Multi-Conference for Engineering, Education and Technology. https://doi.org/10.18687/LACCEI2020.1.1.484

Hidalgo-Crespo J, Coello-Pisco S, Crespo-Vaca T, Amaya JL, Soto M, Jervis FX, Moreira CM (2020b) Waste to energy potential of domestic waste cooking oil in Guayaquil: a review. Paper presented at the Proceedings of the LACCEI International Multi-Conference for Engineering, Education and Technology. https://doi.org/10.18687/LACCEI2020.1.1.485

Knothe G, Krahl J, Van GJ (2015) The biodiesel handbook. Elsevier

Kumar SB, Sushma US, Chandrasagar L, Bheema Raju V, Devi V (2017) Use of waste frying oil as ci engine fuel—a review. Open Access Library J 4:e3958. https://doi.org/10.4236/oalib.1103958

Loizides MI, Loizidou XI, Orthodoxou DL, Petsa D (2019) Circular bioeconomy in action: collection and recycling of domestic used cooking oil through a social, reverse logistics system. Recycling 4:16. https://doi.org/10.3390/recycling4020016

Marchetti R, Vasmara C, Fiume F (2019) Pig slurry improves the anaerobic digestion of waste cooking oil. Appl Microbiol Biotechnol 103(19):8267–8279.https://doi.org/10.1007/s00253-019-10087-8

Marchetti R, Vasmara C, Bertin L, Fiume F (2020) Conversion of waste cooking oil into biogas: perspectives and limits. Appl Microbiol Biotechnol. https://doi.org/10.1007/s00253-020-10431-3

Martínez EJ, Gil MV, Fernández C, Rosas JG, Gómez X (2016) Anaerobic co-digestion of sludge: addition of butcher's fat waste as a cosubstrate for increasing biogas production. PLoS ONE 11:e0153139. https://doi.org/10.1371/journal.pone.0153139

MICSE (2016) Agenda Nacional de Energía 2016–2040. http://biblioteca.olade.org/opac-tmpl/Documentos/cg00362.pdf. Accessed 7 May 2022

Milano SR (2008) La produzione di biodiesel da olio vegetale esausto. Presented at BioDieNet, Milan

OLADE (2010) Atlas de la agro energía y los biocombustibles en las Américas

Oliveira JV, Duarte T, Costa JC, Cavaleiro AJ, Pereira MA, Alves MM (2018) Improvement of biomethane production from sewage sludge in co-digestion with glycerol and waste frying oil, using a design of experiments. Bioenergy Resourc 11:763–771. https://doi.org/10.1007/s12155-018-9933-0

Qamar MA, Liaquat R, Jamil U, Manor R, Azam S (2020) Techno-spatial assessment of waste cooking oil for biodiesel production in Pakistan. SN Appl Sci 2(5):901. https://doi.org/10.1007/s42452-020-2716-1

Ribau-Teixeira M, Nogueira R, Nunes LM (2018) Quantitative assessment of the valorization of used cooking oils in 23 countries. Waste Manage 78:611–620. https://doi.org/10.1016/j.wasman.2018.06.039

Ripa M, Buonaurio C, Mellino S, Fiorentino G, Ulgiati S (2014) Recycling waste cooking oil into biodiesel: a life cycle assessment. Int J Perform Eng 10(4):347–356

Tan ST, Ho WS, Hashim H, Lee CT, Taib MR, Ho CS (2015) Energy, economic and environmental (3E) analysis of waste-to-energy (WTE) strategies for municipal solid waste (MSW) management in Malaysia. Energy Convers Manage 102:111–120. https://doi.org/10.1016/j.enconman.2015.02.010

Tsai W-T, Lin C-C, Yeh C-W (2007) An analysis of biodiesel fuel from waste edible oil in Taiwan. Renew Sustain Energy Rev 11(5):838–857

Wang L, Csallany AS, Kerr BJ, Shurson GC, Chen C (2016) Kinetics of forming aldehydes in frying oils and their distribution in french fries revealed by LC-MS-based chemometrics. J Agric Food Chem 64:3881–3889. https://doi.org/10.1021/acs.jafc.6b01127

Xiao H, Li Z, Jia X, Ren J (2020) Waste to energy in a circular economy approach for better sustainability: a comprehensive review and SWOT analysis. Waste-to-Energy 2020:23–43. https://doi.org/10.1016/b978-0-12-816394-8.00002-1

# Chapter 51
# Towards a Circular Economy for Restaurant Waste in Guayaquil: Characterization and Energy Generation Potential

**Jose Armando Hidalgo Crespo, Manuel Soto, Jorge Luis Amaya-Rivas, L. Borja-Mora, R. Robles-Iglesias, and Leonardo Alvaro Banguera Arroyo**

**Abstract** The objective of this study was to characterize the solid wastes generated in 13 different restaurants in the city of Guayaquil, located in the province of Guayas, Ecuador, evaluating the potential of using the generated waste to produce energy through incineration in a power plant. Data was collected for a period of 7-days for five categories of restaurants going from 1 to 5 forks, which serve meals as their business models. A total of 817.30 kg of waste was collected from the 13 restaurants during the time-lapse of the study, and 19.57 kg of lixiviate, calculating a final solid waste of almost 800 kg. Important to mention that the 1 and 3 forks categories produce the most quantity of waste with 11.79 kg/day/r and 11.36 kg/day/r, and the unique variety the least with 5.87 kg/day/r). Food waste was the most representative percentage for all types, with the highest rates in mobile establishments. Further on, the most present inorganic waste was plastics, with percentages ranging from 6.70 to 9.30%. Data was extrapolated to the total number of restaurants for one year, yielding a possible production of 31 TJ of electricity, concluding an environmentally acceptable alternative.

**Keywords** Restaurant · Waste · Characterization · Energy valorization · Circular economy

J. A. Hidalgo Crespo (✉) · L. Borja-Mora · L. A. Banguera Arroyo
Carrera de Ingenieria Industrial, Universidad de Guayaquil, Guayaquil 090150, Ecuador
e-mail: jose.hidalgocr@ug.edu.ec

J. A. Hidalgo Crespo · M. Soto · R. Robles-Iglesias
Universidade da Coruña, 15008 Coruña, Spain

J. L. Amaya-Rivas
ESPOL Polytechnic University, Escuela Superior Politécnica del Litoral, ESPOL, Guayaquil 090150, Ecuador



N. S. Caetano and M. C. Felgueiras (eds.), *The 9th International Conference on Energy and Environment Research*, Environmental Science and Engineering,
https://doi.org/10.1007/978-3-031-43559-1_51

## 51.1 Introduction

Over the years, as the number of inhabitants increased in the country (over 14% in the last 10 years) and with the continuous development of society, the generation of waste is increasing, and the spaces to store become scarcer. Banguera Arroyo et al. (2020) analyzes the per capita generation of waste in Latin America and the Caribbean linked to the economic and social aspects that occur in cities. Besides, Hidalgo-Crespo et al. (2021) mentioned that each inhabitant produces 0.69 kg/day/c at the national level. Of this waste, 15.45% comes from the destruction of hotels, restaurants, catering services (HORECA), and industry sectors with similar characteristics to households. Several factors further aggravate this situation. One of them is the growth of single-use non-biodegradable materials, which negatively impact the environment due to the significant amount of time these elements take to decompose in nature or under landfill conditions. Wastes generated from restaurants contribute to the stress in our natural systems through impacts associated with the production and disposal, representing both a challenge and opportunity since the governmental units are responsible for the waste management of the entire process: collection, transportation, and final disposal. Tons of still-edible food is discarded every day at the end of the food chain, causing problems related to removing these wastes and financial costs (Fahrenkamp-Uppenbrink 2016).

Several research has been done regarding household solid waste for Guayaquil; however, little is known about restaurant waste. Studies should consider all waste, from households and the contribution from commercial sectors, including restaurants, as a significant waste generation source (Hogan et al. 2004; Christensen and Fruergaard 2010).

At the national level of the total waste generated, 96% is buried in various final disposal systems, while 4% is recovered by grassroots recyclers, mostly impoverished women who work in precarious conditions (Telegraph 2020). According to the information declared by the municipal decentralized autonomous governments (MDAG), 48.2% of municipalities dispose of urban solid waste in sanitary landfills and 51.8% in open dumps. In 2019, at the regional level, 58.1% of MDAG in the north of the country disposed of their solid waste in landfills, while 25.0% on the Coast (Ecuador in Figures 2020).

The circular economy concept has recently gained ground in European policy-making as a positive, solution-based perspective for achieving economic development within a growing environmental framework. One of the central pillars of a circular economy is to return materials to the economy and prevent waste from being sent to landfills or incinerated, thus capturing the value of materials as much as possible and reducing losses (Ribic et al. 2016). Waste must be treated effectively to start with a stable circular economy, and this means that it must suffer an alteration in its physical, chemical, or biological properties, allowing to improve the efficiency of operations and waste management systems, recover reusable and recyclable materials and produce energy in the form of heat and fuel biogas" (Gómez 2000).

**Table 51.1** Total number of establishments in Guayaquil

| Category | N | Percentage (%) |
|---|---|---|
| 5 forks | 2 | 0.13 |
| 4 forks | 587 | 36.73 |
| 3 forks | 496 | 31.04 |
| 2 forks | 361 | 22.59 |
| 1 fork | 152 | 9.51 |
| Total | 1598 | 100.00 |

According to the Ministry of Tourism, in their Ministerial Agreement No. 53, ·Tourist Regulation of Foods and Beverages defines restaurants as establishments where prepared foods are made and sold. In these establishments, alcoholic and non-alcoholic beverages can be sold. They can also offer cafeteria services and, depending on the category; they can have a self-service service. Their definition also includes establishments with a fast-food specialty. Tourist food and beverage establishments are categorized using a scoring system obtained according to compliance with the requirements established in ministerial regulations. The restaurants are categorized into: (5) five, (4) four, (3) three, (2) two, and (1) one fork, with (5) five forks being the highest category and one (1) fork the lowest category. Table 51.1 shows the number of establishments formally registered by the tourism ministry. The total number of active establishments in the city is 1598, including restaurants, food courts, and mobile establishments, grouped according to their category. As shown in Table 51.1, the highest percentages are between 3 and 4 forks with almost 31 and 36%, respectively, and the lowest ones are five forks (luxury) with <1%. Given that five forks restaurants were scarce in the city, it was decided to keep them out of the characterization. However, a new category was included named unique category, which was meant to quantify the waste from fast food carts in the city.

At the European level, the most sustainable approach for waste management should be based on an integrated hierarchical system with the following priority order: prevention, reuse, recycling, energy recovery, and final disposal (Directive 2008/98/EC). It's necessary to increase the knowledge of the quantities and physical and chemical characteristics of waste generated at varying types and operating modes of restaurants, covering from fast food carts to traditional restaurants. In this context, the objective of this study is to characterize the solid wastes generated in different restaurants in the city of Guayaquil, located in the province of Guayas, Ecuador, evaluating the potential of using the waste to generate energy through incineration.

## 51.2 Materials and Methods

### 51.2.1 *Collection of Restaurant Waste*

Sepulveda et al. (2017), after analyzing waste management systems, establish four essential steps for comprehensive solid waste management (gatekeeping, collection, sorting, and disposal) through a conceptual model of reverse logistics. Banguera et al. (2018) apply these essential steps for the comprehensive management of solid waste in a case study to collect out-of-use tires in Santiago, Chile. For this work, these steps in the direction of organic waste were considered in this sequence. Before the collection stage, we first contacted different establishments, giving them a brief introduction to the research objectives. We obtained the participation of 13 establishments dedicated to the food industry (2 of one and two forks, and 3 of 3, 4 divisions, and unique category). After collecting relevant information, such as the opening and closing hours, several days worked, etc. It was possible to cover the five categories mentioned in the previous section, from recognized franchises in the market, renowned restaurants, and mobile establishments, which belong to the fifth category or are called the single category (fast food trucks). The establishments that belonged to three and four forks categories had an internal waste treatment system, carrying out a pre-classification of the waste generated within their facilities. This is one of the characteristics that position them in said categories.

The solid waste collection in Guayaquil restaurants was carried out for 7 days, starting on Monday, August 16, until the night of Sunday, August 22, 2021, when normal activities occurred. This activity was carried out at three different times, and the owners approved each establishment at the beginning of the process. Jumbo bags were given the night before to each establishment with a code to recognize them and collect them the night later. After being taken to the University of Guayaquil, the waste was collected, sampled, and sorted the following morning. This process was repeated every day, as shown in Fig. 51.1 (1).

### 51.2.2 *Sampling and Sorting of Restaurant Waste*

Once the bags arrived at the university, they were grouped by categories, depending on their origin; the next step was to weigh each bag to keep a record of the initial daily weight generated (W1) (3) in each restaurant, then we proceeded to suspend the bags to a certain height (2), making a hole at the bottom, so that the lixiviates could exist. To keep track of the weights of the bags that contain solid waste from restaurants, a hanging scale was used, with a hook to hold the bag, which supported up to 50 kg. On the other hand, at the time of weight, the amount of lixiviate present in each pack, a digital scale was used to obtain a precise value of a said variable. The amount of lixiviated liquid was measured (5), and the bags were weighed again, receiving (W2). Once this operation was finished, the waste bags of each category were opened on the

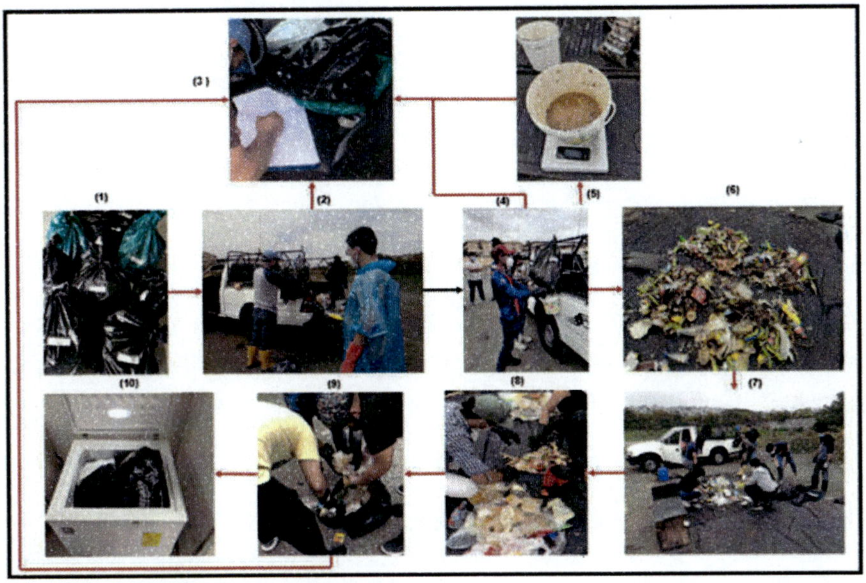

**Fig. 51.1** Restaurant organic waste characterization

surface of the sailcloth, and this process was repeated for each type. The quartering method was used to homogenize the waste and further sorting (6). The objective of this method was to have wasted with homogeneous characteristics, that is, with a similar physical–chemical composition. The solid already on the sailcloth was homogenized using shovels to carry out the quartering, and a 4-m diameter pile was formed. The bank was then divided into four equal parts A, B, C, and D, and opposite parts A and C, or B and D were eliminated. Continuing waste remaining was weighted. This process was made as many times as necessary to achieve around 5 kg for sorting and freezing. Once the sample was determined, it was classified manually, separating the waste according to its nature (7), forming two groups organic waste (OW) and inorganic waste (IW). These two were divided into subgroups (8): paper and cardboard, food waste, plastic, glass, and other inorganic waste (9). The next step after classification was weighing each group already classified (10). Finally, samples were frozen at –20 °C for further experimentation.

### *51.2.3 Energy Incineration Valorization*

Following the results of Hidalgo-Crespo et al. (2019, 2021), we considered low heating values of food, paper and cardboard, plastic, glass, and metal waste as 3.03, 16.55, 28.79, 29.13, and 25.93 MJ/kg. This was due to the waste's high moisture content, which accounts for 75%. To calculate the total energy potential through

**Table 51.2** Total waste collected after 7 days

| Category | W1 (kg) | Lixiviate (kg) | W2 (kg) | Daily generation (kg/day/r) |
|---|---|---|---|---|
| 1 Fork | 170.30 | 5.22 | 165.08 | 11.79 |
| 2 Forks | 111.70 | 0.26 | 111.44 | 7.96 |
| 3 Forks | 245.80 | 7.16 | 238.64 | 11.36 |
| 4 Forks | 164.40 | 5.12 | 159.28 | 7.58 |
| Unique | 125.10 | 1.81 | 123.29 | 5.87 |
| Total | 817.30 | 19.57 | 797.73 | 8.77 |

combustion for each category of waste, we followed the equation below:

$$LHV_{RC} = 3.03 * \%Food + 16.55 * \%Paper\ \&\ Cardboard + 28.79 * \%Plastic + 29.13 * \%Glass + 25.93 * \%Metal \quad (51.1)$$

where $LHV_{RC}$ is the low heating value per restaurant category.

## 51.3 Results

### *51.3.1 Waste Collection*

Table 51.2 shows the total weights per category during the seven days of the study, together with the quantity of lixiviates, both in kilograms. A total of 817.30 kg of waste was collected from the 13 restaurants during the time-lapse of the study, and 19.57 kg of lixiviate, calculating a final solid waste of almost 800 kg as shown in Table 51.2. Remarkably, the 1 and 3 forks categories produce the most quantity of waste with 11.79 kg/day/r and 11.36 kg/day/r, and the unique category the least with 5.87 kg/day/r).

### *51.3.2 Waste Characterization*

Table 51.3 shows the participation percentage of the different types of waste found in the samples provided by the restaurants in Guayaquil. After the quartering method was effectuated, we had an approximate value of 5.00 kg that were sorted into food, paper and cardboard, plastic, glass, and metal wastes. Each day, this process was done five times (one per category). Table 51.3 presents the total sample collected for each type after the seven days. It can be seen that food waste is the most representative of all categories, with the highest percentages in mobile establishments, followed by establishments with 3 and 4 forks, with values of 76.25%, 74.02%, and 71.34%.

Table 51.3 Waste characterization of each category

| Category | Sample (kg) | Food [kg (%)] | Paper/ Cardboard [kg (%)] | Plastic [kg (%)] | Glass [kg (%)] | Metals [kg (%)] |
|---|---|---|---|---|---|---|
| 1 Fork | 40.80 | 24.90 (61.03) | 6.10 (14.95) | 9.30 (22.79) | 0.50 (1.23) | 0.00 (0.00) |
| 2 Forks | 40.80 | 25.3 (62.01) | 3.80 (9.31) | 6.70 (16.42) | 4.9 (12.01) | 0.1 (0.25) |
| 3 Forks | 40.26 | 29.8 (74.02) | 1.20 (2.98) | 7.28 (18.08) | 1.80 (4.47) | 0.18 (0.45) |
| 4 Forks | 41.21 | 29.40 (71.34) | 2.60 (6.31) | 8.10 (19.66) | 1.10 (2.67) | 0.01 (0.02) |
| Unique | 48.76 | 37.18 (76.25) | 2.30 (4.72) | 8.20 (16.82) | 0.80 (1.64) | 0.28 (0.57) |
| Total | 211.83 | 146.58 (69.20) | 16.00 (7.55) | 40.03 (18.90) | 9.1 (4.30) | 0.57 (0.27) |

Further on, the most present inorganic waste is plastics, with percentages ranging from 6.70% to 9.30%. Restaurants in the one fork category generate the most plastic waste with 9.30%, while glass is more present in the two forks category restaurants with 12.01%. The presence of metals is insignificant for all categories.

### *51.3.3 Waste Energy Valorization*

Using Eq. 51.1, we can see in the second column of Table 51.4 the low heating values of each restaurant category. There are no official records of the number of fast-food carts in Guayaquil, but the Municipality reported around 1800 carts located in different parts of the city. Some serve only at night; others sell outside colleges and schools at noon (The Universe 2003). However, two forks restaurants have the highest low heating value. Even though they have the lowest value, unique category, they have the most daily energy potential with approximately 90 GJ due to the city's high presence of fast-food carts.

There is a total potential of almost 240 GJ of energy to be produced daily. For the case of thermal power plants, some authors show a thermal conversion efficiency of ~35% (Xu et al. 2019; Eboh et al. 2019), which would mean a possible daily electric production of 84 GJ, and a potential yearly generation of almost 31 TJ from waste incineration.

Table 51.4 Waste energy incineration valorization

| Category | LHV (MJ/kg) | Number of Restaurants | Daily waste generation (kg/day/r) | Daily energy potential (GJ/day) |
|---|---|---|---|---|
| 1. Fork | 11.24 | 152 | 11.79 | 20.14 |
| 2. Forks | 11.71 | 361 | 7.96 | 33.65 |
| 3. Forks | 9.36 | 496 | 11.36 | 52.74 |
| 4. Forks | 9.65 | 587 | 7.58 | 42.94 |
| unique | 8.56 | 1800 | 5.87 | 90.44 |

## 51.4 Conclusion

The collection, sorting, and characterization of restaurant waste was achieved in August 2021. Waste consisted of food, paper and cardboard, plastic, glass, and metal. The proportion of each waste was 69.20%, 7.55%, 18.90%, 4.30%, 0.27%. Total potential energy from incineration is around 31 TJ daily for the entire formal restaurants of the city. A total of 817.30 kg of waste was collected from the 13 restaurants during the time-lapse of the study, and 19.57 kg of lixiviate, calculating a final solid waste of almost 800 kg. Daily generation was found to be 8.77 kg/day.

Although the obtained data provided helpful information about the amount of waste produced by different types of restaurants, this research study should be recognized as a first step in studying this matter for the country. Further research could concentrate on obtaining the proximate and ultimate analysis, and measuring the potential of this waste to obtain biogas, another viable choice to diminish the amount of waste while providing energy.

## References


Banguera Arroyo LÁ, Sepúlveda Salas JM, Molestina Malta CJ, Vásquez Pérez OC, Carrasco Armijo R (2020) Marco De La Ley De Responsabilidad Extendida Del Productor (REP). Biblioteca Colloquium. Recuperado a partir de https://colloquiumbiblioteca.com/index.php/web/article/view/22

Banguera LA, Sepúlveda JM, Ternero R, Vargas M, Vásquez Óscar C (2018) Reverse logistics network design under extended producer responsibility: the case of out-of-use tires in the Gran Santiago city of Chile. Int J Prod Econ 205:193–200. https://doi.org/10.1016/j.ijpe.2018.09.006

Christensen TH, Fruergaard T (2010) Commercial and institutional waste. In: Christensen TH (ed) Solid waste technology and management, vol 1. John Wiley & Sons Ltd., Chichester, pp 97–99

Eboh FC, Ahlström P, Richards T (2019) Evaluating improvements in a waste to-energy combined heat and power plant. Case Stud Therm Eng 14:100476. https://doi.org/10.1016/j.csite.2019.100476

Ecuador in Figures (2020) Statistics of economic environmental information in autonomous decentralized municipal governments. https://www.ecuadorencifras.gob.ec/documentos/webinec/Encuestas_Ambientales/Municipios_2019/Residuos_solidos_2019/PRESENTACION RESIDUOS_2019.pptxV06.pdf. Accessed 3 May 2022

Fahrenkamp-Uppenbrink, J. (2016). Reducing food loss and waste. Science 352(6284):424–426. https://doi.org/10.1126/science.352.6284.424-p

Gómez CIS (2000) Problemática y gestión de residuos sólidos peligrosos en Colombia. Rev Innov J 15:84595. https://revistas.unal.edu.co/index.php/innovar/article/download/24163/24792/84595

Hidalgo-Crespo J, Amaya J, Jervis F, Moreira C, Crespo-Vaca T, Zabala-Ortiz G, Coello-Pisco S (2019) Waste-to-energy incineration: Evaluation of energy potential for urban domestic waste in guayaquil. RISTI 2019(E23):392–403

Hidalgo-Crespo J, Moreira CM, Jervis FX, Soto M, Amaya JL (2021) Development of sociodemographic indicators for modeling the household solid waste generation in Guayaquil (Ecuador): quantification, characterization and energy valorization. Paper presented at the European Biomass Conference and Exhibition Proceedings, pp 252–259

Hogan J, Cunnigham D, Finn J (2004) Characterisation of non-household municipal waste in Ireland and the development of an approach to tracking municipal waste composition. Final Report 2000-MS-7-M1, Environmental Protection Agency, Ireland

Bojan R, Voca N, Ilakovac B (2016) Concept of sustainable waste management in the City of Zagreb: towards the implementation of circular economy approach. J Air Waste Manage Assoc 67(2):10962247. https://doi.org/10.1080/10962247.2016.1229700

Sepulveda JM, Banguera L, Fuertes G, Carrasco R, Vargas M (2017) Reverse and inverse logistic models for solid waste management. South Afric J Indus Eng 28(4):1701. https://doi.org/10.7166/28-4-1701

The Telegraph (2020) Alianza Basura Cero promotes the sustainable management of solid waste. https://www.eltelegrafo.com.ec/noticias/sociedad/6/alianza-basura-cero-manejo sustentable-residuos-solidos. Accessed 3 May 2022

The Universe (2003) Fast food, a delicious street business. https://www.eluniverso.com/2003/10/12/0001/9/2F370B288AFD424CB3F5DBD999B36E27.html/. Accessed May 04, 2022

Xu H, Lin WL, Dal Magro F, Li T, Py X, Romagnoli A (2019) Towards higher energy efficiency in future waste-to-energy plants with novel latent heat storage-based thermal buffer system. Renew Sustain Energy Rev 112:324–337. https://doi.org/10.1016/j.rser.2019.05.009

# Chapter 52
# Agrivoltaics System as an Integral Part of Modern Farming

**Jiri Bim and Michaela Valentová**

**Abstract** Photovoltaics represents one of the key parts of the low carbon energy transition. The combination of producing electricity from photovoltaics and agriculture production is gaining momentum. Such combination, when set up correctly, brings synergy, i.e., much more than just electricity from a renewable source. Energy security and food security are in this new concept called agrivoltaics. The paper reviews current knowledge on agrivoltaic systems as a branch of decarbonisation, and decentralisation energy strategies. Agrivoltaics in general means a photovoltaic installation that improves the conditions for cultivated crops, or cattle breeding, while renewable electricity production is an integral part and both activities are operated together at the same time. Different countries are trying to implement agrivoltaics in their policy differently using the knowledge of (so far limited) pilot installations. Pilot installations demonstrate that levelized costs of electricity (LCOE) are higher than ground-mounted photovoltaics installations. However, this is generally without considering the additional benefits for crops. Agrivoltaics has many forms and types. It can be combined with permanent cultures like berries, apple farms or with conventional agriculture. LCOE will depend on the type of agrivoltaics because each type needs different constructions and percentages of overshading. Vertical agrivoltaic installations bring a new daily diagram curve in electricity production from photovoltaics, better fitting into the electricity consumption curve. Further research will focus on multiple impacts of agrivoltaics, connections between agrivoltaic electricity production and consumption and finding new synergies between electricity production and crops cultivation.

**Keywords** Agriculture · Agrivoltaics · Agrophotovoltaics · Clean electricity · Green transition · Photovoltaics

J. Bim (✉) · M. Valentová
Faculty of Electrical Engineering, Czech Technical University in Prague, Prague 16000, Czech Republic
e-mail: bimjiri@fel.cvut.cz; bim.jiri@gmail.com



N. S. Caetano and M. C. Felgueiras (eds.), *The 9th International Conference on Energy and Environment Research*, Environmental Science and Engineering,
https://doi.org/10.1007/978-3-031-43559-1_52

## 52.1 Introduction

The world population has been constantly growing in modern history and has increased from 2 billion in 1927 to 7.7 billion in 2021. The forecast for year 2100 is 10.9 billion in the United Nations projection (Johannessen and Shalina 2022). Growing population brings about increased needs for food security, and thus demands for spaces to agriculture and connected processes (Mahmood et al. 2021). Putting it simply, this means more land for citizens and more land for cultivating crops (Beek et al. 2010). Energy security is as important as food security. Electricity consumption is globally increasing. The energy sector is transforming to renewable energy sources, but at the same time it must be able to cover the consumption increase (Ahmad and Zhang 2020). The availability of both disponible land for agriculture and energy sources is limited. Combining the need for food security and clean energy security at the same time are at the core of current sustainable development requirements (Akpoti et al. 2018).

Climate change impacts differ geographically and depend on local conditions and capacity to adapt on climate change (Bonfante et al. 2018). In many places, conditions for cultivating crops are becoming worse. One of the solutions to both issues can be in dual use of land. The connection between farming, food security, and energy security is thus called agrivoltaics. The agrivoltaics combine electricity production and farming on the same land at the same time (Abidin et al. 2021). While generating green electricity, agrivoltaic systems can improve conditions for cultivated crops. The solutions are highly location-dependent (Leon and Ishihara 2018). Special features and types of agrivoltaics relate to location and its environmental, geographical, and weather conditions. Sun irradiance, altitude, or soil quality are examples of factors that influence types of agrivoltaics in different locations.

The agrivoltaics concept is a relatively new one (Goetzberger 1982) and therefore also the research on the topic has been rather scarce and scattered. The main aim of this article is therefore to provide one of the first thorough reviews of the existing knowledge on the concept of agrivoltaics. It discusses the different types of installations, and technical–economic aspects. It also tackles the multiple benefits of agrivoltaics systems. It concludes with the areas of further research.

## 52.2 Material and Methods

This article is based on a thorough literature review of available research articles in agrivoltaics. Because it is a new sector and it combines energy and agriculture there were 139 results in Science Direct, as the most relevant source of research articles. Other databases were also scanned, together with a general search in Google Scholar. Only circa 30 articles were exactly on agrivoltaic systems. The methods used to analyse the articles then include screening the results for definitions of agrivoltaics systems, to anchor the main idea of this new concept. The articles were then analysed to cover the other issues, such as technical–economic aspects and multiple benefits. Complex case studies and materials come primarily from Germany, France, the Netherlands, and Japan. Articles from these countries provide a good basis with relevant results. Because of differences between each country such as geographical or atmospheric impacts, there are difficulties in determining what exactly constitutes agrivoltaics (i.e., what still is agrivoltaics, and what is not) (Ramsebner et al. 2021).

The article identifies new trends in agrivoltaics topic. In this new trend, just a few case studies throughout the world were published, but nevertheless, they present a good basis and starting point for research, implementation, and expansion of agrivoltaics around the world. Despite the diversity of the worldwide geographical and socio-economic conditions, the general trends in agrivoltaics are likely to be similar worldwide, including construction, used photovoltaics panels, and water management.

Regarding the techno-economic aspects, the analysed case studies often used a slightly different methodology for economic calculations. Therefore, it is very difficult to compare them. Calculation of economic parameters of each different type can be made by levelized cost of electricity (LCOE) formula. LCOE in selected articles is compared in this paper. Total lifetime costs are divided by the total lifetime output (Nissen and Harfst 2019).

## 52.3 Agrivoltaics Review

### *52.3.1 Definition and Legislative Framework*

The article describes worldwide used types of agrivoltaics. So far, there are many types of agrivoltaics, from high construction above conventional crops to special construction for wine yards. France, Belgium, Germany, and Italy are among the first countries trying to implement agrivoltaics in national legislation. The core of the concept is similar in all countries: it is a combination of food production and renewable electricity production. Nevertheless, agrivoltaics has not had any formal definition. German case study defined agrivoltaics as a dual land use system, for agricultural outputs and at the same time secondary, not agricultural output (Feuerbacher et al. 2021). The study by the Fraunhofer institute defined agrivoltaics as an integrated food-energy system. In their definition, the agricultural process must be maintained or improved with agrivoltaics systems (Schindele et al. 2020). Italian study defined agrivoltaics with differences between Photovoltaics-ground mounted (PV-GM) and Agrivoltaics (AG-PV). In AG-PV systems, crops can be cultivated with conventional machinery under high-mounted photovoltaics panels (Agostini et al. 2020). German Institute for Standardization published a draft of agrivoltaics specification. The definition covers all of the above approaches stating that AG-PV means "a combined use of one and the same land area for agricultural production as a primary use and for electricity production by photovoltaic systems as a secondary use" (Spec and Spec 2021).

The most advanced legislation process is in France. The certification manual defines the agrivoltaics systems and what can be certified as agrivoltaics (Cedex 2021). The second most advanced country in the legislation process of agrivoltaics so far is Germany which has implemented a few additions in its renewable electricity sources (RES) law to cover agrivoltaics (Verordnung and Problem 2021).

### 52.3.2 Types of Agrivoltaics Installations

The combination of farming and electricity production has many forms. The main division of agrivoltaic installations is in vertical and horizontal forms. Bifacial photovoltaics panels are used for vertically mounted panels (Jain et al. 2021). The use of bifacial panels is becoming more and more common; they can produce more electricity and at the same time create better conditions for cultivated crops. Bifacial panels help to mitigate the effects of climate change better than monofacial panels (Tahir et al. 2022). Vertically mounted systems are typically east–west oriented. Half of the panels with front face to the east and half to the west, which gives a well-balanced diagram (Fraunhofer 2020). East–west oriented vertical panels start producing electricity early in the morning, around noon, there is a large production decrease, and in the evening electricity production increases again (Willockx et al. 2022). The second form is the horizontal agrivoltaic system. In this form, panels are mostly south-oriented. There are many types of constructions of agrivoltaics systems. Vertical agrivoltaic systems have a structure similar to conventional photovoltaic installation. Horizontal agrivoltaics systems typically take the form of greenhouses, high mounted panels above conventional crops, and construction for orchards, vineyards or foil tunnels for berries typically (Fraunhofer 2020). In the case of greenhouses with photovoltaics on rooftops, typically, special organic panels are used, that can use part of the solar spectrum, which is useless for photosynthesis processes in plants (Fernández 2022). Construction for conventional farming is developed for conventional farming methods, including heavy machinery. High construction is the biggest cost of this agrivoltaic type. In Germany, the project APV-RESOLA used 5-m vertical clearance, to keep the possibility of using conventional agriculture machinery (Trommsdorff et al. 2020).

Conversely, agrivoltaics installations in orchards and vineyards combine construction for cultivated trees and for photovoltaics panels. This decreases final costs. Agrivoltaics with permanent crops has advantages with similar life cycle periods. Installation in vineyards in France by Sun'Agri combined high mounted construction with special single-line construction for orchards (Fraunhofer 2020). Special foil tunnels with agrivoltaics are used for berries farms. For instance, in the Netherlands, special photovoltaic panels with larger sun throughput were used to allow for more sunshine for cultivated crops, in this case, raspberries (Schindele 2022).

Each type of the above agrivoltaics system has its own specifics. Vertical mounted east–west oriented installations and south oriented installation will make well-balanced curve of daily electricity production and better fit into the electricity consumption curve (Khatib and Deria 2022). The choice of agrivoltaic system will depend largely on atmospheric and geographical conditions.

Solar irradiation, usual weather and typical cultivated crops are variables which determine the key pros and cons of each agrivoltaic system type in the given country (and location). Nevertheless, there are several key advantages and disadvantages of agrivoltaic systems that are common. The common main advantages of the synergy between the agrivoltaic solution (and electricity production) and agriculture are shade formation for cultivated crops, lower water evaporation rate and physical protection. The level shade of shadow formation depends on the geographical location of the agrivoltaic system. Pilot installations across Europe are now trying to analyse the magnitude of the service of photovoltaic panels in different conditions (Ise 2020).

Some advantages and disadvantages intersect. Each agrivoltaic system will have a different light transmission depending on the location and the selected crop so that the shadow created has only a positive effect. Physical plant protection by photovoltaic panels, in turn, will be an obstacle to the use of mechanization. Therefore, the agrivoltaic system will always be built in accordance with the agricultural practices common to the crop (Schindele et al. 2020). From an energy point of view, the main advantage is the production of clean electricity from a renewable source. By consuming or selling this electricity, farmers can improve and balance their economic situation (Campana et al. 2021). Figure 52.1a shows a vertical agrivoltaics system in a pilot project in Vienna, Fig. 52.1b shows the agrivoltaics system above raspberries in the Netherlands.

**Fig. 52.1** **a** Vertical agrivoltaics system, **b** agrivoltaics system above raspberries (own author's)

### 52.3.3 Multiple Benefits of Agrivoltaics

The synergy between cultivating crops and producing "green" electricity brings both direct and indirect multiple benefits. The indirect, multiple benefits can be divided into four main groups: Technical, Economic/Financial, Social and Environmental with subcategories in each of them (Makešová and Valentová 2021). Agrivoltaics, a subcategory of photovoltaics, is a renewable energy source and, therefore, on its own, it may bring many co-benefits similar to other RES. In the Technical section, these can be distributed electricity production, diversification of electricity mix, energy savings, and decrease of fossil power generation. Economically, agrivoltaics can decrease electricity costs and bring additional economic profit/savings for farmers. Social section means sector development and local employment increase. From the environmental point of view, agrivoltaics systems produce "green" electricity, which can replace fossil fuels (Makešová and Valentová 2021). From the agricultural point of view, agrivoltaic systems can be a useful tool for solutions against the erosion of the fields (Borrelli 2021). Water and wind erosion can be decreased or stopped. Other positive impacts entail better soil conditions for cultivating crops, slowing down the decrease in soil quality or protection of cities and villages against fast floods (Tang et al. 2020). However, in general, multiple impacts are an area to be further explored and more in the agrivoltaics systems.

### 52.3.4 Technical–Economic Aspects

The conventional photovoltaic system and the agrivoltaic system differ in technical aspects, mainly in the construction that holds the panels and the types of panels used. The construction for agrivoltaics requires special harmony with the given type of cultivated crop. The design for vines and fruit trees will be different from that for growing berries such as currants or gooseberries or completely different from the design for conventional crops such as potatoes, cereals, and corn (Schindele et al. 2019). Similarly, the vertical construction is completely different. Only bifacial panels are used for this purpose, which increases the efficiency of the agrivoltaic system located in this way (Riaz et al. 2021).

The second difference is the type of photovoltaic panels used. Due to the dual use of land, the main aim and result is synergy, not maximizing the electricity produced. For this reason, bifacial or special photovoltaic panels are used. The main common feature of these panels is the light transmission (Campana et al. 2021). This can be determined according to the needs of cultivated crops, as well as the dependence on geographical and atmospheric conditions of implementation. From classic bifacial panels, we can also see transparent glass photovoltaic cells on pilot projects. The result should, therefore, be a synergy between sufficient light for the crops grown, corresponding to the maximum electricity produced, at the lowest possible economic cost.

**Table 52.1** Comparison of LCOE for agrivoltaic systems in different countries

| Country | Year | Installed power (KWP) | LCOE (EUR/kWh) |
|---|---|---|---|
| Germany | 2020 | 1038 | 0.0829 |
| Italy | 2021 | 1000 | 0.0754 |
| India | 2022 | 520 | 0.0400 |
| Netherlands | 2022 | 130 | 0.0713 |

Vertical installations can be seen mainly in Germany. Nex2Sun (Next2Sun GmbH 2020), together with Fraunhofer ISE, is investigating the impacts of such panels on areas with permanent grassland. The advantage of such an agrivoltaic solution is that it does not take up much agricultural land. Vertical mounted photovoltaic panels can also protect the land against wind erosion (Fraunhofer 2020). Vertically mounted systems have another practical advantage in dusty areas. Dust adheres to their surface much less than with horizontally placed panels. As a result, it is possible to use bifacial panels not only on permanent grassland but also to farm between the individual rows of panels in the classic way (Riaz et al. 2021). They have a similar advantage in winter when covered with snow. In this case, even due to the increased reflectivity of light from snow, these systems produce much more electricity (Kopecek et al. 2021). In the Table 52.1, agrivoltaics systems in each country are compared by LCOEs. The range of LCOE for photovoltaics in Fraunhofer ISE study is 0.0312 EUR/kWh and 0,1101 EUR/kWh in 2021(Kost et al. 2021). LCOE for agrivoltaics systems in pilot projects in Germany, Italy, India, and the Netherlands are between 0.040 EUR/kWh and 0.0829 EUR/kWh. The significantly lower LCOE value of the agrivoltaic system in India indicates a higher rate of solar exposure in this country compared to European countries.

Economic aspects are important, but sometimes hard to compare due to different methodologies in the individual case studies. The thorough economic analysis is therefore outside the scope of this paper but offers a possibility for further research. The key task to solve is to find balance in the economic situation of each farmer and maintain or improve their crop production. In agrivoltaics, food production must be in the first place; otherwise, it is a conventional photovoltaic powerplant.

### *52.3.5 Clean Electricity from Agrivoltaics System Utilization*

To present agrivoltaic systems as a whole concept, the usage of electricity is in close connection with electricity production. Agrivoltaics brings a unique opportunity to consume electricity in the place of production. Water pumps and the whole watering system can be powered by electricity from panels above cultivating crops. If the agrivoltaics solution has a large area of panels, water can be pulled out to the water reservoir (Li et al. 2022). Water pumps can again distribute water to crops. Agriculture machinery can be powered by electricity, and autonomous but also person-driven machinery can be charged on the field (Lombardi and Berni 2021). Most agricultural technical processes are done during the day, which means that it is necessary to install battery storage too, for charging agriculture machinery during the night, when there is no electricity production. A combination of battery storage, charging agricultural machinery or other technical processes on the field can consume electricity in the place of production. Additionally, the produced electricity can be used to charge electric cars (Heinisch et al. 2021). Roads often intersect individual fields, so the continuity of agrivoltaic systems and their further use of electricity to charge electric cars is logical. The battery storage, which would store the electricity produced in the sunshine, would be supplemented by a fast electric car charger. Such devices can be part of either existing petrol stations or newly created rest areas. So far, the most expensive use of excess electricity at the production site is the so-called power to X, which is the conversion of electricity into hydrogen, longer or the conversion of hydrogen into methane. In this way, we can acquire additional commodities that we can consume, sell, or store and use later. If there is the excess of electricity, the power plant must be connected to the grid and electricity sell it. Connection to the grid has many terms, and it depends on the robustness of the local grid. Again, such solutions are yet to be fully explored in the research field (Safarzyńska and Bergh 2018).

### 52.3.6 Further Research in Agrivoltaics

The review has shown there are several areas where further research is needed. First, the need to find new synergies between crops and electricity production will continue with the climate change impacts. Due to climate change and intensive agriculture, places with good agronomic conditions are changing into worse conditions, less quality soil, and thus lower cost of field (Tang et al. 2020) Second, in the sector coupling concept, individual energy and electricity sources are getting closer and composing smaller energy centres (Ramsebner et al. 2021). Thus, agrivoltaics contribute to decentralizing electricity sources and at the same time enhancing the cultivation of the crops. The interconnection between agrivoltaics and final electricity consumption is, therefore, yet to be researched. Within the sector coupling concept, the agrivoltaic system can be connected to small wind turbines or a hydroponic solution in the future. Agrivoltaics and the energy gained from it will also be an integral part of energy communities (Cheng et al. 2022). Third, many crops have not yet been tested in an agrivoltaic system. This means, among others, analysing the economic impacts and benefits for individual farmers as well as agricultural cooperatives. Agrivoltaic systems bring many options and opportunities for local production in combination with clean electricity (Ramsebner et al. 2021). There is a clear need to study agrivoltaics impacts on each type of crop and compare the economics of different types of agrivoltaics and their impact on farmers. Agrivoltaics is a complex and wide topic, with connections to many study fields, and their interconnections offer a vast space for further research.

## 52.4 Conclusions

The article mapped the current state of agrivoltaic systems, and their technical, economic, and legislative state. There are many forms of these solutions, always combining the need for dual land use. Some AG-PVs are almost conventional photovoltaic power plants. Other pilot projects show high structures with low installed power but a high share of light transmission leading to less favourable economic cost-effectiveness of such a solution. It is therefore very difficult to cross the line that can already be considered agrivoltaics. The agrivoltaic system is a part of decentralized electricity production. The synergy does not have to be only between the cultivation of crops and the production of electricity, but also the place of production and the consumption of electricity. The energy from the agrivoltaic system can be used as part of the community energy system of the given municipality or consumed for charging agricultural machines, or for pumping water for irrigation, for example. The agrivoltaic system is an integral part of the transformation of the energy sector towards renewable and emission-free sources of energy. Agrivoltaics has therefore a clear potential to contribute to achieving the goals of climate neutrality in Europe in 2050 and ensure food security at the same time.

The review has shown that the very definition of the agrivoltaic systems is still to be fully established. Countries such as France and Germany are very close agreeing on their definitions, but due to different geographical and climatic conditions, the replicability to other regions may be difficult. Specifically, the difference between conventional photovoltaics and Ag-PV is yet to be fully established.

In contrast, LCOEs are comparable for PV-GM and AG-PV. This testifies to the economic efficiency of agrivoltaics projects, without the calculation containing positive externalities such as better conditions for cultivated crops. Agrivoltaic systems can therefore play an important role in future, decarbonised energy mixes in many countries. It has been shown that some types of installations better cover the typical daily consumption diagram while needing less space on the land. However, the still higher price of such solutions (special panels and constructions) affects the economy of agrivoltaics. With the development of these panels and a lower price, the economy of agrivoltaics projects will improve. At the same time, the use of the electricity produced will of course greatly affect the economy of the whole project. The multiple impacts of the agrivoltaic systems are manifold (such as decentralized local electricity production, local employment increase or protection against erosion), but have so far been rarely studies and quantified.

Further research in the area of economics of agrivoltaics, their multiple impacts, and its contribution to sector coupling can contribute to the development of this concept and, thus, contribute to both mitigation and adaptation actions to climate change, including reducing greenhouse gas emissions, increasing energy security as well as food security.

**Acknowledgements** Research described in the paper was supervised by Prof. Jaroslav Knápek, FEE CTU in Prague and Dr. Michaela Valentová, FEE CTU in Prague and supported by Students grant competition at CTU, under grant number SGS21/117/OHK5/2T/13.

## References

Abidin MAZ, Mahyuddin MN, Zainuri MAAM (2021) Solar photovoltaic architecture and agronomic management in agrivoltaic system: a review. Sustain 13(14):18–21. https://doi.org/10.3390/su13147846

Agostini A, Colauzzi M, Amaducci S (2020) Innovative agrivoltaic systems to produce sustainable energy: an economic and environmental assessment. Appl Energy 281:116102. https://doi.org/10.1016/j.apenergy.2020.116102

Ahmad T, Zhang D (2020) A critical review of comparative global historical energy consumption and future demand: the story told so far. Energy Rep 6:1973–1991. https://doi.org/10.1016/j.egyr.2020.07.020

Akpoti K, Kabo-bah AT, Zwart SJ (2019) Agricultural land suitability analysis: state-of-the-art and outlooks for integration of climate change analysis. Agric Syst 173:172–208. https://doi.org/10.1016/j.agsy.2019.02.013.

Bonfante A et al (2018) A dynamic viticultural zoning to explore the resilience of terroir concept under climate change. Sci Total Environ 624:294–308. https://doi.org/10.1016/j.scitotenv.2017.12.035

Borrelli P et al (2021) Soil erosion modelling: a global review and statistical analysis. Sci Total Environ 780:146494. https://doi.org/10.1016/j.scitotenv.2021.146494

Campana PE, Stridh B, Amaducci S, Colauzzi M (2021a) Optimisation of vertically mounted agrivoltaic systems. J Clean Prod 325:129091. https://doi.org/10.1016/j.jclepro.2021.129091

Cedex LPS (2021) Label Projet Agrivoltaïque. 33:1–37

Cheng C, Blakers A, Stocks M, Lu B (2022) 100% renewable energy in Japan. Energy Convers Manag 255:115299. https://doi.org/10.1016/j.enconman.2022.115299

Fernández EF et al (2022) Global energy assessment of the potential of photovoltaics for greenhouse farming. Appl Energy 309:118474. https://doi.org/10.1016/j.apenergy.2021.118474

Feuerbacher A et al. (2021) An analytical framework to estimate the economics and adoption potential of dual land-use systems: the case of agrivoltaics. Agric Syst 192:103193. https://doi.org/10.1016/j.agsy.2021.103193

Fraunhofer ISE (2020) Agrivoltaics: Opportunities for agriculture and the energy transition—a guideline for Germany. pp 1–56

Goetzberger A (1982) On the coexistence of solar-energy conversion and plant cultivation. Appl Energy 1(1):55–69. https://doi.org/10.1080/01425918208909875.

Heinisch V, Göransson L, Erlandsson R, Hodel H, Johnsson F, Odenberger M (2021) Smart electric vehicle charging strategies for sectoral coupling in a city energy system. Appl Energy 288:116640. https://doi.org/10.1016/j.apenergy.2021.116640

Ise F (2020) Fraunhofer institute for solar energy systems Ise agrivoltaics : opportunities for agriculture and the energy transition a guideline for Germany.

Jain P, Raina G, Sinha S, Malik P, Mathur S (2021) Agrovoltaics: step towards sustainable energy-food combination. Bioresour Technol Rep 15:100766. https://doi.org/10.1016/j.biteb.2021.100766.

Johannessen OM, Shalina EV (2022) Perspective population increase impacts the climate, using the sensitive Arctic as an example. Atmos Ocean Sci Lett 15(2):100192. https://doi.org/10.1016/j.aosl.2022.100192

Khatib T, Deria R (2022) East-west oriented photovoltaic power systems: model, benefits and technical evaluation. Energy Convers Manag 266:115810. https://doi.org/10.1016/j.enconman.2022.115810

Kopecek R, Libal J (2021) Bifacial photovoltaics 2021: status, opportunities and challenges. Energies 14(8):en14082076. https://doi.org/10.3390/en14082076

Kost C, Shammugam S, Fluri V, Peper D, Memar AD, Schlegel T (2021) Levelized cost of electricity—renewable energy technologies, pp 1–45.

Leon A, Ishihara KN (2018) Assessment of new functional units for agrivoltaic systems. J Environ Manage 226(August):493–498. https://doi.org/10.1016/j.jenvman.2018.08.013

Li R, Wu M, Aleid S, Zhang C, Wang W, Wang P (2022) An integrated solar-driven system produces electricity with fresh water and crops in arid regions. Cell Reports Phys Sci 3(3):100781. https://doi.org/10.1016/j.xcrp.2022.100781

Lombardi GV, Berni R (2021) Renewable energy in agriculture: farmers willingness-to-pay for a photovoltaic electric farm tractor. J Clean Prod 313(May):127520. https://doi.org/10.1016/j.jclepro.2021.127520.

Mahmood F, Bicer Y, Al-Ansari T (2021) Design and thermodynamic assessment of a solar powered energy–food–water nexus driven multigeneration system. Energy Rep 7:3033–3049. https://doi.org/10.1016/j.egyr.2021.05.032

Makešová M, Valentová M (2021) The concept of multiple impacts of renewable energy sources: a critical review. Energies 14(11):3183. doi: https://doi.org/10.3390/en14113183

Next2Sun GmbH (2020) Bifacial solar fences. https://www.next2sun.de

Nissen U, Harfst N (2019) Shortcomings of the traditional 'levelized cost of energy' [LCOE] for the determination of grid parity. Energy 171:1009–1016. https://doi.org/10.1016/j.energy.2019.01.093

Ramsebner J, Haas R, Ajanovic A, Wietschel M (2021) The sector coupling concept: A critical review. Wiley Interdiscip. Rev. Energy Environ. 10(4):1–27. https://doi.org/10.1002/wene.396

Riaz MH, Imran H, Younas R, Butt NZ (2021) The optimization of vertical bifacial photovoltaic farms for efficient agrivoltaic systems. Sol Energy 230(October):1004–1012. https://doi.org/10.1016/j.solener.2021.10.051

Safarzyńska K, van den Bergh JCJM (2018) A higher rebound effect under bounded rationality: interactions between car mobility and electricity generation. Energy Econ 74:179–196. https://doi.org/10.1016/j.eneco.2018.06.006

Schindele S (2022) Baywa re. https://www.baywa-re.com/de/solar-projekte/agri-pv

Schindele S et al (2020) Implementation of agrophotovoltaics: techno-economic analysis of the price-performance ratio and its policy implications. Appl Energy 265(October):114737. https://doi.org/10.1016/j.apenergy.2020.114737

Spec DIN, Spec DIN (2021) 3264280. May:1–25

Tahir F, Baloch AAB, Al-Ghamdi SG (2022) Impact of climate change on solar monofacial and bifacial Photovoltaics (PV) potential in Qatar. Energy Rep 8:518–522. https://doi.org/10.1016/j.egyr.2022.02.197

Tang B, Jiao J, Zhang Y, Chen Y, Wang N, Bai L (2020) The magnitude of soil erosion on hillslopes with different land use patterns under an extreme rainstorm on the Northern Loess Plateau, China. Soil Tillage Res 204(26):104716. https://doi.org/10.1016/j.still.2020.104716

Trommsdorff M et al (2020) Combining food and energy production: design of an agrivoltaic system applied in arable and vegetable farming in Germany. Renew Sustain Energy Rev 140(April):2021. https://doi.org/10.1016/j.rser.2020.110694

van Beek CL, Meerburg BG, Schils RLM, Verhagen J, Kuikman PJ (2010) Feeding the world's increasing population while limiting climate change impacts: linking N2O and CH4 emissions from agriculture to population growth. Environ Sci Policy 13(2):89–96. https://doi.org/10.1016/j.envsci.2009.11.001

Verordnung E, Problem VA (2021) Referentenentwurf

Willockx B, Lavaert C, Cappelle J (2022) Geospatial assessment of elevated agrivoltaics on arable land in Europe to highlight the implications on design, land use and economic level. Energy Rep 8:8736–8751. https://doi.org/10.1016/j.egyr.2022.06.076

# Chapter 53
# Evaluation of Co-digested Biogas Production Using Waste Cooking Oil as a Co-substrate

**Jose Armando Hidalgo Crespo, N. M. Berrones-Rivera, A. F. Teran-Alvarado, Manuel Soto, and Jorge Luis Amaya-Rivas**

**Abstract** Nowadays, many countries suffer economic, social, and environmental problems due to the lack of proper final disposal systems for different types of residues, like municipal solid waste (MSW), agricultural waste (AG), and waste cooking oil (WCO). In this context, the controlled degradation of wastes through anaerobic digestion (AD) in the absence of oxygen with the help of different groups of bacteria offers a sustainable solution for otherwise unusable waste. The main objective of this study was to develop artisanal anaerobic biodigesters and water displacement gasometers, to measure the influence of waste cooking oil as co-substrate for the anaerobic co-digestion with four different types of waste: cow, pig, chicken, and food from restaurants. As demonstrated, waste cooking oil ameliorated the values of 75% of the types of garbage. This study also indicated that the biogas production was comparatively higher in the co-digested bottles than in the mono-digested bottles.

**Keywords** Cow · Pig · Chicken · Manure · Waste cooking oil · Anaerobic digestion

J. A. Hidalgo Crespo (✉) · N. M. Berrones-Rivera · A. F. Teran-Alvarado
Universidad Politécnica Salesiana, Guayaquil 090150, Ecuador
e-mail: jose-armando.hidalgo-crespo@grenoble-inp.fr

J. A. Hidalgo Crespo · M. Soto
Universidade da Coruña, 15008 Coruña, Spain

J. A. Hidalgo Crespo
G-Scop Laboratory, Polytechnic Institute of Grenoble, 46. Av. Félix Viallet, 38000 Grenoble, France

J. L. Amaya-Rivas
ESPOL Polytechnic University, Escuela Superior Politécnica del Litoral, ESPOL, Guayaquil 090150, Ecuador

 
N. S. Caetano and M. C. Felgueiras (eds.), *The 9th International Conference on Energy and Environment Research*, Environmental Science and Engineering,
https://doi.org/10.1007/978-3-031-43559-1_53

## 53.1 Introduction

The disposal of large amounts of residues is ubiquitous worldwide. One major problem that challenges many countries, particularly, third-world developing ones, is the generation of solid and liquid wastes from different sources, such as households, restaurants, industries, and agricultural farms. Their poor management causes environmental, social, and economic issues. Waste is disposed of in places that lack adequate management infrastructure, becoming troublesome areas (Carlini et al. 2017). The elevated costs of their collection, transport, and treatment make them challenging to manage for any government (Apetato et al. 1999).

One way to ease the living standards of developing countries is the production of affordable and adequate energy. Bioenergy presents a sustainable option to manage waste while producing energy (Ebner et al. 2015). In this context, biodegradation in the absence of oxygen in organic matter, together with the help of different sorts of bacteria to accelerate the microbial action, can be achieved through anaerobic digestion (AD). AD presents a promising solution with a well-established organic solid waste management technology, superior to traditional disposal methods, such as landfilling, incineration, and composting. However, some social and technological challenges need to be overcome first.

In the process of anaerobic decomposition of organic waste, anaerobic microbes convert various biomass and organic waste into biogas. Biogas is constituted primarily by (60–70%) methane ($CH_4$), followed by (30–40%) carbon dioxide ($CO_2$), and other gases, such as hydrosulfuric acid ($H_2S$), nitric oxide (NO), sulfur monoxide (SO), and water vapor. The biodegraded matter left after the process can serve for land amelioration due to its nutrient-rich nature (Sheets et al. 2015; Yang et al. 2015). Many of the organic materials ready for the AD process are carbohydrates, proteins, and lipids. These materials can be degraded to much simpler compounds following the AD stages: hydrolysis, acidogenesis, acetogenesis, and methanogenesis (Hagos et al. 2017).

Annually, enormous amounts of waste cooking oil (WCO) are discarded around the world. This liquid waste is primarily used for biodiesel production or to generate other chemicals and materials (Lin et al. 2013). WCO's composition is transformed due to the high temperatures of the frying process, and the elements that make part of the cooking meals, which can leave rests of solids and water. Logically, the quantity and quality of WCO will depend on the consumed quantities and types of edible oils. In fact, the global consumption of vegetable oils sums up to ~200 million tons annually, and one study found that of the total consumption 32% becomes waste (Ribau-Teixeira et al. 2018). In Ecuador, it is estimated that each year citizens produce 5.11 g of daily WCO. Considering a density of 0.8958 g/ml and a total population for the country of 2,654,433, it can be expected a total of 13.56 tons daily (15,141 L) (Hidalgo-Crespo et al. 2020).

Restaurant waste is mainly composed of food waste (RFW). Food waste is rich in lipids and is considered highly attractive for anaerobic digestion due to the methane potential of lipids (Labatut et al. 2011; Meng et al. 2015). Different types of farm animal manures, such as cow (CM), pig (PM), and chicken (CHM), are some essential substrates for anaerobic digestion because of their high water content and fibers. Even though anaerobic digestion is already a well-proven, evolved, and fully-applied technology in many treatments, such as wastewater, sewage sludge, food waste, and animal manure, adopting this technology for animal manure and food waste, using waste cooking oil as a co-substrate has not yet been adopted been well researched.

On the other hand, the use of co-substrates can increase methane generation production by between 25 and 400%, given that they can supply the necessary nutrients and alkalinity. These values are higher than single digestion of the same substrates (Shah et al. 2015). The main objective of this study was to develop artisanal anaerobic biodigesters and water displacement gasometers, to measure the influence of waste cooking oil as a co-substrate for the anaerobic digestion of the four different types of waste: CM, PM, CHM, and RFW.

## 53.2 Materials and Methods

### *53.2.1 Sources of CM, PM, CHM, WCO, and Restaurant Organic Waste*

Cow, pig, and chicken manure was collected from a farm located just outside the city on the outskirts of Guayaquil. For the case of waste cooking oil, the collection was made for two weeks from 532 households in the town, with the help of 650 ml glass jars, which were then poured into 20-L drums. Finally, for the case of restaurant waste, we collected various samples from 13 restaurants of 5 different categories in Guayaquil, according to the classification done by the ministry of tourism for a time of 1 week, with the quarter method waste was homogenized. Organic matter was separated from inorganic, and samples were taken and frozen under –20 °C. Restaurant organic waste and cooking oil samples were taken during other projects to quantify, characterize, and geolocate.

### *53.2.2 Preparation of Restaurant Organic Waste*

Samples of restaurant organic waste were first unfrozen. Pieces of paper and cardboard were removed, leaving only the food waste (FW). After, FW was ground with the help of a blender, getting a total of 1.00 kg, as shown in Fig. 53.1.

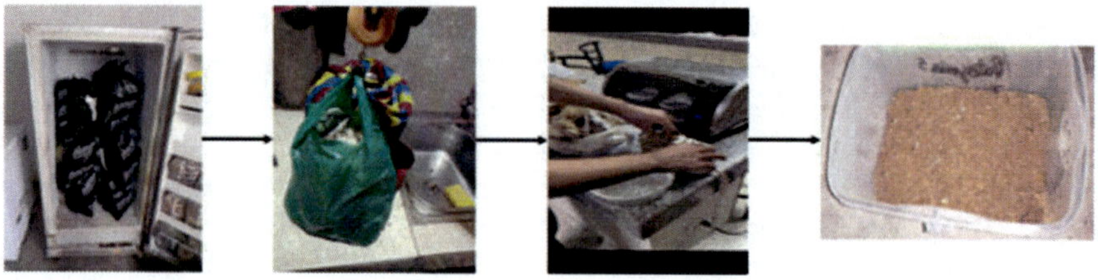

Fig. 53.1 Restaurant organic waste preparation

### 53.2.3 Assemble Artisanal Anaerobic Digesters

Laboratory-based batch digesters were made using 500 ml glass bottles (Fig. 53.2). A hole of 4 mm diameter was made in the center of each bottle's cap. One-liter test tube was submerged in a 5-L transparent box (28.00 × 20.00 × 14.00 $cm^3$) with a lid filled with water. The test tube was kept inverted and immersed in the box. The test tube was kept 2.00 cm from the bottom of the box using its lid to hold it in place. A disposable infusion set with a 2 feet length and 5 mm diameter PVC tube was used to connect the glass bottle with the test tube. The spike of the infusion set was inserted with pressure through the hole of the cap. The spike's surroundings and the infusion set's air vent were sealed with epoxy putty. The needle was removed from the infusion set, and the PVC tube was placed in the middle of the test tube. The PVC tube served as a conduit leading the generated gas from the glass bottle into the test tube. All generated gas displaced the water from the test tube, creating a headspace volume. This volume represented the generated biogas. Measurements were made daily for 21 days. The experiment followed the water displacement method previously tested in other research (Roman et al. 2021). The water displacement method applies to any gas which is not particularly soluble in water, including biogas and is still widely used to record the volume of methane generated.

### 53.2.4 Humidity Calculation

The humidity of each type of waste was obtained by sampling, weighing, and drying with an electric oven. Each sample rested for 3 h inside the range at 180 °C, and final weights were measured, as shown in Table 53.1.

### 53.2.5 Samples Preparation

To find the role and efficiency of waste cooking oil as a co-substrate, we filled each of the 12 glass bottles with different mixtures of residues. An average of 208 g of waste

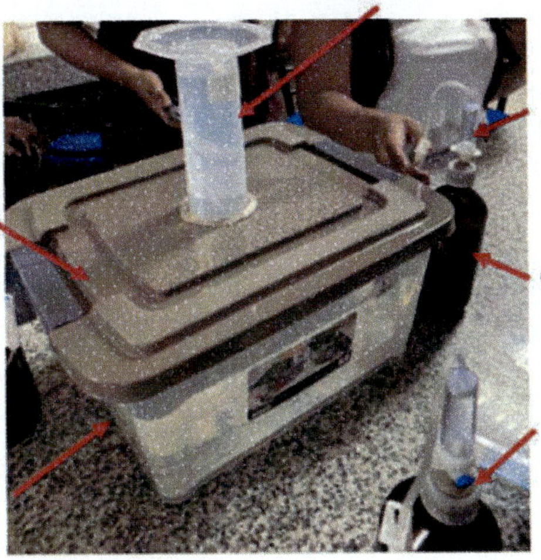


**Fig. 53.2** Experimental artisanal digesters set up

**Table 53.1** The humidity of each type of waste

| Sample | Wi (g) | Wf (g) | Humidity (%) |
|---|---|---|---|
| CM | 52.25 | 10.98 | 79.00 |
| PM | 42.14 | 12.95 | 71.40 |
| CHM | 86.87 | 37.29 | 57.07 |
| RFW | 110.96 | 55.55 | 49.94 |

were introduced in total. Percentages of the different types of waste allocated to each jar are shown in Table 53.2. The experiment was conducted at ambient temperature and lasted 21 days. Each pot was agitated for one minute to mix all co-substrates. Absolute humidity was calculated for each jar having values between 42.80% and 79%, with an average humidity value of 55.90%. Given the high humidity levels of each jar, additional water was not necessary. The experiment was conducted under the meteorological conditions of Salesian Polytechnic University in March 2022.

### *53.2.6 Pressure and Volume of Gas Calculation*

One method to calculate volumes of non-soluble water gases are the displacement gasometers. These devices require that readings are taken directly from the gas column (liquid levels, pressure) are used to calculate gas volumes. It is also necessary to consider the vapor content and correct any hydrostatic pressure on the gas (Pham et al. 2013). To calculate the total pressure of biogas produced ($P_g$), we followed

**Table 53.2** Mixing ratios of waste

| Bottle | FW (%) | CM (%) | CHM (%) | PM (%) | WCO (%) | w/w Humidity (%) |
|---|---|---|---|---|---|---|
| 1 | 100 | | | | | 49.94 |
| 2 | 83 | | | | 17 | 41.45 |
| 3 | | 100 | | | | 79.00 |
| 4 | | 76 | | | 24 | 60.04 |
| 5 | | | 100 | | | 57.07 |
| 6 | | | 75 | | 25 | 42.80 |
| 7 | | | | 100 | | 71.40 |
| 8 | | | | 74 | 26 | 52.84 |
| 9 | 70 | 30 | | | | 58.66 |
| 10 | 74 | | 26 | | | 52.30 |
| 11 | 75 | | | 25 | | 55.31 |
| 12 | 45 | 13 | 14 | 13 | 15 | 50.01 |

manometry rules, getting the following formula (Fig. 53.3):

$$P_{atm} = P_1 + \rho_{Agua} g(13.50 - h_g)/100 \quad (53.1)$$

$$P_1 = P_{biogas} + P_v \quad (53.2)$$

where $P_v$ is the vapor pressure. Vapor pressure is considered 3.56 kPa at room temperature of 300 K. Atmospheric pressure at sea level of 101.3 kPa was also considered. Finally, the biogas pressure is:

$$P_{biogas} = P_{atm} - \frac{\rho_{Agua} g\left(13.50 - h_g\right)}{100} - P_v \quad (53.3)$$

## 53.3 Results

Figure 53.4 shows the pressure calculated for 21 days. However, due to the artisanal nature of the experiment, biodigesters 1, 2, 6, 7, 11, and 12 started losing pressure after day 11, meaning the presence of leaks. The rest of the biodigesters continued for the total time of the project.

The highest and lowest yields in 11 days were BD10 (98.14 kPa) and BD4 (97.04 kPa), and in 21 days were BD3 (98.85 kPa) and BD4 (97.26 kPa). To determine whether or not waste cooking oil helps improve biogas production, we have to compare biodigesters with and without WCO. As a result, for 11 days, 3 out of 4 types of waste produced more biogas with the presence of biogas (BD2 > BD1, BD6

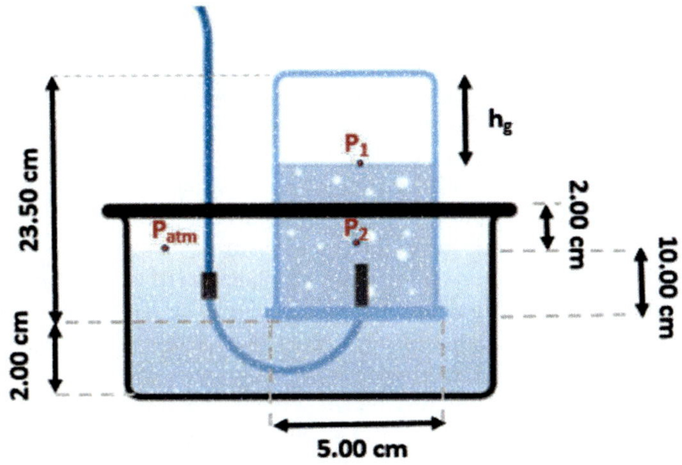


**Fig. 53.3** Pressure measurement

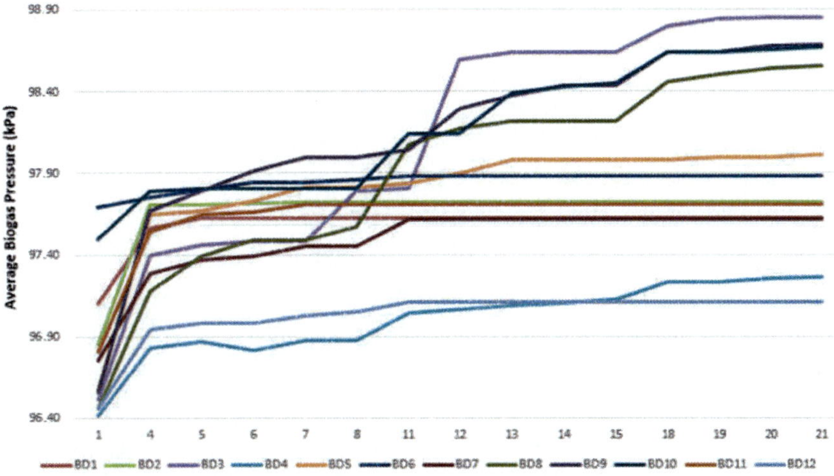


**Fig. 53.4** Cumulative biogas pressure (kPa) from all biodigesters

> BD5, BD8 > BD7), except for the case of cow manure. These results show that, indeed, waste cooking oil can serve as a co-substrate. This is not surprising due to the high lipid content of this liquid waste.

Another critical remark is that the study followed which manure and restaurant food waste generated the best biogas yield. Chicken manure was the best co-substrate, followed by cow manure. Also, mixing all four types of waste produced less than only working with a mix of two of them.

## 53.4 Conclusion

In this work, it was evaluated the influence of waste cooking oil as a co-substrate of cow, pig, and chicken manure and restaurant food waste in anaerobic co-digestion. As demonstrated, waste cooking oil ameliorated the values of 75% of the types of garbage. This study also indicated that the biogas production was comparatively higher in the co-digested bottles than in the mono-digested bottles.

One major problem to tackle in future studies is hydrophobicity due to the presence of fats. This produces floating foams and the accumulation of long-chain fatty acids, killing the microbial consortia. This inhibiting factor can be avoided by co-digestion of fats with substrates that are more easily degradable, like carbohydrates (Wan et al. 2011).

The use of biodigesters to produce biogas through anaerobic digestion may play an essential role in local economy of the country, due to the opportunity to make renewable fuel from organic waste and alternative waste treatment.

## References

Apetato MM, Nobre AM, Alves JC, Robalo GS, Ferreira F (1999) Taxa de Resíduos Urbanos: Deficiências e Soluções. In: Procedéis of the 6a Conferência Nacional sobre Qualidade do Ambiente, Lisboa, Portugal, Editora Plátano, Vol 3, pp 363–369

Carlini M, Mosconi EM, Castelluci S, Villarini M, Colantoni A (2017) An economical evaluation of anaerobic digestion plants fed with organic agro-industrial waste. Energies 10:1165. https://doi.org/10.3390/en10081165

Ebner JH, Labatut RA, Rankin MJ, Pronto JL, Gooch CA, Williamson AA, Trabold TA (2015) Lifecycle greenhouse gas analysis of an anaerobic codigestion facility processing dairy manure and industrial food waste. Environ Sci Technol 49(18):11199–11208. https://doi.org/10.1021/acs.est.5b01331

Hagos K, Zong J, Li D, Liu C, Lu X (2017) Anaerobic co-digestion process for biogas production: Progress, challenges and perspectives. Renew. Sust. Energ. Rev. 76:1485–1496. https://doi.org/10.1016/j.rser.2016.11.184

Hidalgo-Crespo J, Coello-Pisco S, Crespo-Vaca T, López-Vargas A, Borja-Caicedo D, Martínez-Villacrés H (2020) Domestic waste cooking oil generation in the city of Guayaquil and its relationship with social indicators. Paper presented at the Proceedings of the LACCEI International Multi-Conference for Engineering, Education and Technology. https://doi.org/10.18687/LACCEI2020.1.1.484

Labatut R, Angenent LT, Scott N (2011) Biochemical methane potential and biodegradability of complex organic substrates. Bioresour Technol 102(3):2255–2264. https://doi.org/10.1016/j.biortech.2010.10.035

Lin C, Pfaltzgraff L, Herrero-Davila L, Mubofu E, Abderrahim S, Clark J, Koutinas A, Kopsahelis N, Stamatelatou K, Dickson F, Thankappan S, Mohamed Z, Brocklesby R, Luque R (2013) Food waste as a valuable resource for the production of chemicals, materials and fuels. Current situation and global perspective. Energy Environ Sci 6:426–464. https://doi.org/10.1039/C2EE23440H

Meng Y, Li S, Yuan H, Zou D, Liu Y, Zhu B, Chufo A, Jaffar M, Li X (2015) Evaluating biomethane production from anaerobic mono- and co-digestion of food waste and floatable oil (FO) skimmed from food waste. Bioresour Technol 185:7–13. https://doi.org/10.1016/j.biortech.2015.02.036

Pham CH, Triolo JM, Cu TTT, Pedersen L, Sommer SG (2013) Validation and recommendation of methods to measure biogas production potential of animal manure. Asian-Australas J Anim Sci 26:864–873. https://doi.org/10.5713/ajas.2012.12623

Ribau-Teixeira M, Nogueira R, Nunes LM (2018) Quantitative assessment of the valorization of used cooking oils in 23 countries. Waste Manage 78:611–620. https://doi.org/10.1016/j.wasman.2018.06.039

Roman R, Heinz-Peter M, Humayun K, Mohummad R, Khasruzzaman A, Uddin M, Islam Md (2021) Evaluation of biogas production from mono and co-digested cow manure with household waste. J Chem Biol Phys Sci 11:168–177. https://doi.org/10.24214/jcbps.D.11.2.16877

Shah FA, Mahmood Q, Rashid N, Pervez A, Raja IA, Shah MM (2015) Co-digestion, pretreatment and digester design for enhanced methanogenesis. Renew Sustain Energy Rev 42:627–642. https://doi.org/10.1016/j.rser.2014.10.053

Sheets JP, Yang L, Ge X, Wang Z, Li Y (2015) Beyond land application: emerging technologies for the treatment and reuse of anaerobically digested agricultural and food waste. Waste Manag 44:94–115. https://doi.org/10.1016/j.wasman.2015.07.037

Wan C, Zhou Q, Fu G, Li Y (2011) Semi-continuous anaerobic codigestion of thickened waste activated sludge and fat, oil and grease. Waste Manage 31:1752–1758. https://doi.org/10.1016/j.wasman

Yang L, Xu F, Ge X, Li Y (2015) Challenges and strategies for solid-state anaerobic digestion of lignocellulosic biomass. Renew Sustain Energy Rev 44:824–834. https://doi.org/10.1016/j.rser.2015.01.002

# Chapter 54
# Electrical Energy Generation Through Microbial Fuel Cells Using *Pichia membranifaciens* Yeasts

**S. Rojas-Flores, M. De La Cruz-Noriega, R. Nazario-Naveda, Santiago M. Benites, D. Delfín-Narciso, Cecilia V. Romero, and F. Diaz**

**Abstract** The current world's significant problems are due to the dependence on the consumption of environmentally harmful materials for obtaining electrical energy where the search for new generating sources of this resource is a necessity. For this reason, this research tries to provide a novel way of generating electricity by using the *Pichia membranifaciens* yeast as a fuel source through microbial reactors manufactured at scale with Zn and Cu electrodes. The *Pichia membranifaciens* yeast was molecularly identified with 99.47% identity, managing to generate current and voltage peaks of 2.782 ± 0.241 mA and 0.865 ± 0.351 V, respectively, with an optimal operating pH of 7.519 ± 0.102 on the 7 days, where the values of electrical conductivity increased to 75.92 ± 3.54 mS/cm. The maximum values of the PD were 5.4534 ± 0.251 W/cm$^2$ at a CD of 361.71 mA/cm$^2$ for a voltage of 0.807 V. This investigation gives high values compared to other experiments using different types of substrates and generating a new natural aggregate to increase electrical values without using chemical compounds harmful to nature.

**Keywords** Current · Electrical energy · Generation · *Pichia membranifaciens* · Yeast

---

S. Rojas-Flores (✉) · M. De La Cruz-Noriega · R. Nazario-Naveda · S. M. Benites
Vicerrectorado de Investigación, Universidad Autónoma del Perú, Lima 15842, Peru
e-mail: segundo.rojas.89@gmail.com

D. Delfín-Narciso
Grupo de Investigación en Ciencias Aplicadas Y Nuevas Tecnologías, Universidad Privada del Norte, Trujillo 13007, Peru

C. V. Romero
Facultad de Medicina, Universidad Nacional de Trujillo, Trujillo 13001, Peru

F. Diaz
Universidad Tecnológica del Perú, Lima 15022, Peru

 
N. S. Caetano and M. C. Felgueiras (eds.), *The 9th International Conference on Energy and Environment Research*, Environmental Science and Engineering,
https://doi.org/10.1007/978-3-031-43559-1_54

## 54.1 Introduction

In recent years, factors such as accelerated industrialization and energy demand to meet humanity's social and economic development have led to a notable negative impact on the environment. In this sense, it is worth mentioning that the production of electrical energy is its primary source is fossil fuels, which, due to their combustion, have contributed in large proportion to the increase in $CO_2$ emissions and greenhouse gases (Opoku and Boachie 2020; Owusu and Asumadu-Sarkodie 2016). This has accelerated climate change and global warming, generating problems that the world faces today (Chakrabarti and Chakrabarti 2002). Due to this, there is a growing interest in the use of alternative energy sources; one of the alternatives that have emerged is microbial fuel cells (MFCs), which use microorganisms to convert chemical energy (Rojas-Flores et al. 2022a), which process is achieved when the bacteria transfer electrons from an anodic electrode to a cathodic one, within the MFC, where the bacteria oxidize and reduce organic molecules (Idris et al. 2016; Flores et al. 2022). There are various designs of MFCs, but all consist of two chambers, the anaerobic (anode) and the aerobic (cathode); these chambers are joined by a bridge called the cation exchange membrane (Ullah and Zeshan 2020). On the other hand, it is worth mentioning that MFCs use various substrates for electricity generation, such as agricultural, brewing, domestic wastewater, and food wastewater (Pandey et al. 2016), as well as a wide range of soluble or dissolved complex organic waste, lignocellulosic biomass among other substrates, serves as a source of carbon and energy, this being an essential factor in the efficiency of MFC (Pant et al. 2010).

Another factor of vital importance in the performance of the MFCs is the microorganisms present in the substrates, which biologically oxidize the organic matter and transfer electrons to the anode (Flores et al. 2020). In this sense, in recent years, a large number of microorganisms have been identified as possessing the capacity to generate an electric current in the MFCs; among the best-known bacteria, *Pseudomonas aeruginosa*, *Geobacter sulfurreducens*, *Escherichia coli*, and Yeasts of the genus Pichia (Logan 2009). The last species used in this research (yeast *Pichia membranifaciens*) is widely distributed in nature and can form as cospores. This microorganism has been isolated from various habitats, including spoilage of rotten fruit and other plant materials and a contaminant of fermented beverages (Kurtzman and Pichia 2011) and fresh meats, for which it is considered a food and beverage spoilage organism (Fleet 2011).

In this sense, various investigations have been reported in the literature where multiple species of the genus *Pichia* are used in MFCs; for example, Pal et al. (2019) explored the exoelectrogenic behavior of the species *Pichia fermentans* in a double-chamber and one-chamber MFC, where they used a mediator (methylene blue) managing to generate open-circuit voltage peaks of 0.40 and 0.397 V in a MFC-SC with and without a mediator, respectively. Likewise, a maximum power density of 1.64 $\mu W/cm^2$ was recorded in the presence of a mediator, concluding that this species can produce a high power density (Pal and Sharma 2020). In the same sense, Kumar et al. (2018) designed a double-chamber MFC to degrade wheat straw through various fungi in a cathodic chamber, while the species *Pichia fermentans* was used in the anodic compartment, evaluation of the MFC was carried out for 30 days, from which the increase in the concentrations of reducing and phenolic sugars in the catholyte was reported, as well as a maximum voltage of 0.504 and 0.496 V recorded on day 17 (Kumar et al. 2018). Therefore, MFC applications can be used for wastewater treatment, energy production, bioremediation, and environmental biosensors, for which MFC has excellent potential as an alternative for clean energy generation (Ochoa and Juárez 2004).

Main objective of this research is to evaluate the generation of electricity using the yeast *Pichia membranifaciens* (molecularly identified) as a substrate in the minimal salt medium in a single-chamber microbial fuel cell manufactured at low cost with Cooper-Zinc electrodes to which the values of conductivity, pH, voltage, current, current density, and power were monitored for 15 days. This research will give a potential use to the *Pichia membranifaciens* yeast for future applications as aggregates in MFCs to increase their electrical values.

## 54.2 Materials and Methods

Manufacturing of MFC-SC fuel cells: three microbial fuel cells were manufactured which were polyethylene containers, to which a 10 $cm^2$ hole was made at one end of a square shape of ~144 $cm^2$ to place a cathode electrode (Copper, Cu). The anodic electrode (Zinc, Zn) was placed inside the cell. Both electrodes were linked by an external circuit with an internal resistance of 100 Ω. Yeast isolation used an aliquot of grape juice sample fermented in Sabouraud Agar medium was added and streak-seeded, incubating at 30 °C for 48 h. After the time, isolated colonies were observed to which a Gram stain was performed to verify the purity of the strain and preserve it in an inclined tube with Sabouraud Dextrose Agar for later identification (Wang et al. 2018) (Table 54.1).

**Table 54.1** Results of the sequencing performed on the anodic electrode

| BLAST characterization | Length of consensus sequence (nt) | % Maximum identity | Accession number | Phylogeny |
|---|---|---|---|---|
| *Pichia membranifaciens* | 376 | 99.47 | MK898934.1 | Cellular organisms; Eukaryota; Opisthokonta; Fungi; Dikarya; Ascomycota; saccharomyceta; Saccharomycotina; Saccharomycetes; Saccharomycetales; Pichiaceae; *Pichia* |

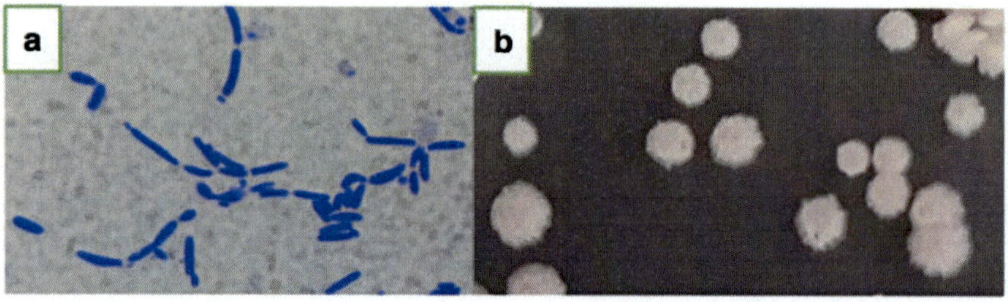


**Fig. 54.1** **a** Microscopy of *P.* by Gram staining (100×) and **b** Macroscopic view of colonies of *P. membranifaciens* in Sabouraud Agar using a dark background

While, identification at the genus and species level was performed, recording the microscopic characteristics of oval cells in the process of multipolar budding (see Fig. 54.1a), macroscopic white, opaque, rough colonies, normally centralized, with volcanic elevation, typical of the genus *Pichia* (see Fig. 54.1b) and using the molecular identification methodology based on the ITS regions of the rDNA of the yeast strain (Gustincich et al. 1991).

Molecular identification was carried out by the Analysis and Research Center of the "Biodes Laboratories". For this reason, a pure culture of the yeast was sent, which will undergo the extraction of genetic material (DNA) using the CTAB extraction method (Kumar et al. 2018), and then the PCR products were sequenced in the Macrogen laboratory (USA). To see more details, see the publication given by Rojas-Flores et al. (2023). Preparation of the inoculum for the Microbial Fuel Cell, from a 24-h culture of *Pichia membranifaciens*, a suspension was made in 20 ml of sterile physiological saline solution (SSFE) of 0.85%, adjusted with tube No. 3 of Mc Farland ($9 \times 10^8$ CFU/ml), and added to 80 ml of minimal mineral medium buffered with phosphate buffer, according to the following composition (g/L): $Na_2HPO_4$.6; $KH_2PO_4$, 3; $NH_4Cl$, 1; NaCl, 0.5; $MgSO_4{\cdot}7H_2O$, 0.246; $CaCl_2$, 0.01, was then homogenized and distributed evenly in the microbial combustion cells. The entire process was carried out at room temperature (20 ± 2 °C); it was performed in triplicate (Valenzuela-González et al. 2015). The voltage and current values (Rext. 100 Ω) were made using a multimeter (Prasek Premium PR-85) for 15 days. The method used for the measurement PDF and CD were made using the method used by Segundo et al. (2022). Monitoring of changes in conductivity (conductivity meter CD-4301) and pH (pH-meter 110 Series Oakton) was also measured.

## 54.3 Results and Discussion

Figure 54.2a shows the generated voltage values during the 15 days of monitoring, which increase slightly from the first day (0.751 ± 0.108 V) to the seventh day (0.865 ± 0.351 V) after falling until the last day, the increase in voltage is mainly

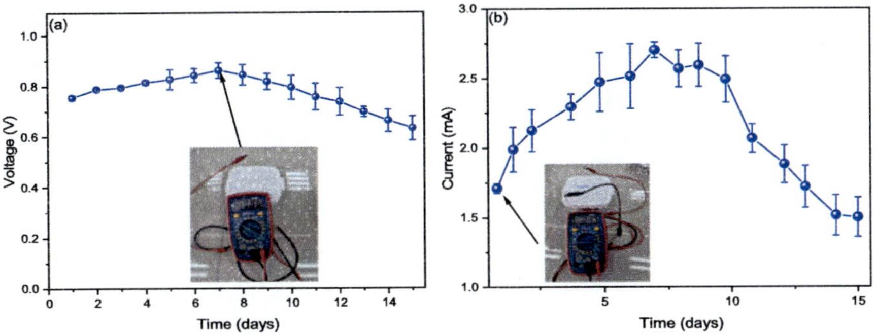


**Fig. 54.2** Values of **a** voltage and **b** current of the microbial fuel cells

due to the large amounts of initial nutrient broth in the cells, while the decrease in values is due to the antibacterial activity of the copper electrode used (Kabir et al. 2018; Noriega et al. 2021). Figure 54.2b shows the monitoring of the electric current values, being the maximum generation peak on the seventh day (2.782 ± 0.241 mA) and the minimum on day 15 (1.47 ± 0.325 mA). High current values during the monitoring period confirm the presence of an exoelectrogenic biofilm (bacterial active community) on the anodic (Magaly et al. 2020). These first results of the use of this yeast to obtain electrical energy will open new paths for the research groups due to the high values of voltage and current obtained, according to Gul et al. (2021), the key for the generation of electrons is part of the glucose available in the substrate used, and a poor performance in this part would lead to a limitation of the production of electric current.

Figure 54.3a shows the pH values obtained from the monitoring of the cells during the 15 days, which increase from the first day, going from a slightly acid level to a moderately alkaline one. Each substrate and specific cell design have an optimal pH in which it generates higher voltage and current values (Shanthi Sravan et al. 2021). In this investigation, the optimal pH would be 7.519 ± 0.102 belonging to the seventh day, which is corroborated by Fig. 54.2a, b showing higher values of electrical. This is because microorganisms have certain optimal environmental values for their growth and generation of electricity (Li et al. 2021; Halim et al. 2021). Figure 54.3b shows the electrical conductivity values of the substrate, achieving a maximum conductivity on the tenth day (75.92 ± 3.54 mS/cm) to then slowly decay until the last day of monitoring.

Figure 54.3c shows the values of the power density (DP) and voltage as a function of the current density (DC), as shown, a $DP_{max}$ was obtained of 5.4534 ± 0.251 W/cm$^2$ at a DC of 361.71 mA/cm$^2$ and a peak voltage of 0.807 V. These values are higher compared to those of other researchers using other substrates (fuel), for example, Zhang et al. (2016) used sewage sludge as substrate and carbon felt as the anode, obtaining $DP_{max}$ and DC values obtained were 20.4 mW/cm$^2$ and 25.86 mA/cm$^2$, which may be due to adjusting the substrate to pH > 8 (Zhang et al. 2016), this being not its optimal pH for its operation in microbial fuel cells (Rojas-Flores et al. 2022b).

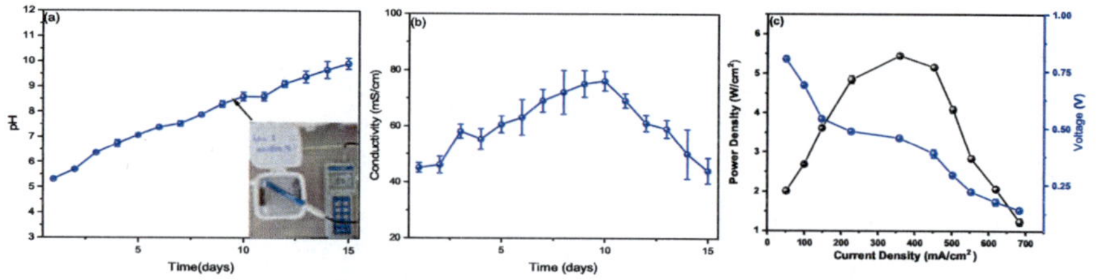

**Fig. 54.3** Monitoring of **a** pH, **b** conductivity and **c** values of power density, current density and voltage

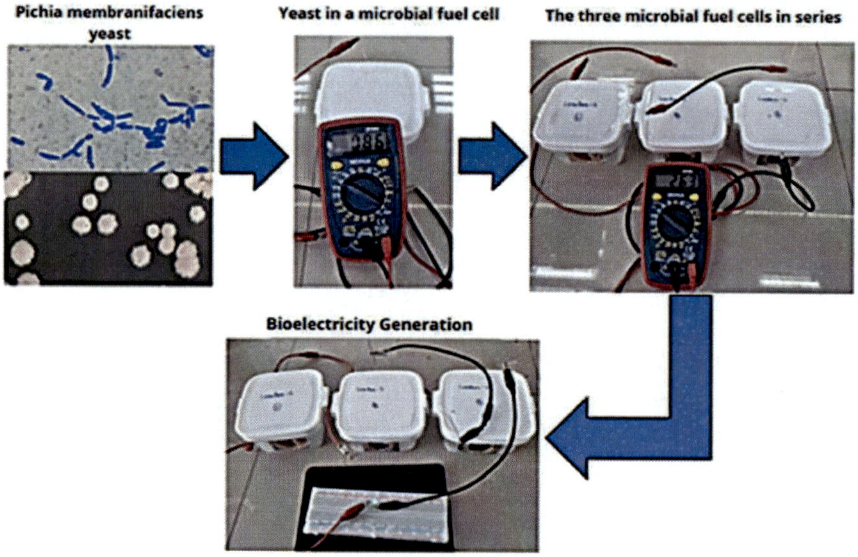


**Fig. 54.4** Schematization of the bioelectricity generation process

Recent studies show that power density values can be influenced by pH values (Rossi and Logan 2021; Jiang et al. 2016). In Fig. 54.4, the bioelectricity generation process of the MFCs is observed, in which it was placed in the individual cells (03 in total) to connect them in series, giving a value of 2.63 V, enough to turn on a led bulb of green color.

## 54.4 Conclusion

Bioelectricity was successfully generated using the *Pichia membranifaciens* yeast (molecularly identified) as fuel in low-cost, laboratory-scale MFC-SC. It managed to generate a maximum voltage and current in the seventh of 0.865 ± 0.351 V and

2.782 ± 0.241 mA, respectively. While the pH values went from slightly acidic to moderately alkaline, with its optimum pH being 7.519 ± 0.102 (seventh day), and its conductivity values increased to 75.92 ± 3.54 mS/cm on the tenth day and then decreased. The cells showed a maximum $PD_{max}$ of 5.4534 ± 0.251 W/cm$^2$ at a CD of 361.71 mA/cm$^2$ and a $V_{max}$ of 0.807 V. Finally, the three MFC-SC were connected in series, generating a voltage of 2.63 V and turning on a led. Green color. This research gives the beginnings for further investigations by varying the different parameters that microbial fuel cells have for better efficiency.

## References


Chakrabarti S, Chakrabarti S (2002) Rural electrification programme with solar energy in remote region—a case study in an island. Energy Policy 30(1):33–42

Fleet GH (2011) Yeast spoilage of foods and beverages. The Yeasts. Elsevier, pp 53–63

Flores SJR, Benites SM, Rosa ALRAL, Zoilita ALZAL, Luis ASL (2020) Using lime (*Citrus aurantiifolia*), orange (*Citrus sinensis*), and tangerine (*Citrus reticulata*) waste as a substrate for generating bioelectricity. Environ Res Eng Manag 76(3):24–34

Flores SR, Pérez-Delgado O, Naveda-Renny N, Benites SM, De La Cruz-Noriega M, Narciso DAD (2022) Generation of bioelectricity using molasses as fuel in microbial fuel cells. Environ Res Eng Manag 78(2):19–27

Gul H, Raza W, Lee J, Azam M, Ashraf M, Kim K-H (2021) Progress in microbial fuel cell technology for wastewater treatment and energy harvesting. Chemosphere 281(130828):130828. https://doi.org/10.1016/j.chemosphere.2021.130828

Gustincich S, Manfioletti G, Del Sal G, Schneider C, Carninci P (1991) A fast method for high-quality genomic DNA extraction from whole human blood. Biotechniques 11(3):298–300, 302.

Halim MA, Rahman MO, Ibrahim M, Kundu R, Biswas B (2021) Study of the effect of pH on the performance of microbial fuel cell for generation of bioelectricity. Res Square. https://doi.org/10.21203/rs.3.rs-151072/v

Idris SA, Esat FN, Abd Rahim AA, Zahin Rizzqi WA, Ruzlee W, Zyaid Razali WM (2016) Electricity generation from the mud by using microbial fuel cell. MATEC Web Conf 69:02001. https://doi.org/10.1051/matecconf/20166902001

Jiang Y-B, Zhong W-H, Han C, Deng H (2016) Characterization of electricity generated by soil in microbial fuel cells and the isolation of soil source exoelectrogenic bacteria. Front Microbiol 7:1776. https://doi.org/10.3389/fmicb.2016.01776

Kabir MM, Fakhruddin ANM, Chowdhury MAZ, Pramanik MK, Fardous Z (2018) Isolation and characterization of chromium(VI)-reducing bacteria from tannery effluents and solid wastes. World J Microbiol Biotechnol 34(9):126. https://doi.org/10.1007/s11274-018-2510-z

Kumar R, Singh L, Zularisam AW, Hai FI (2018) Microbial fuel cell is emerging as a versatile technology: a review on its possible applications, challenges and strategies to improve the performances: microbial fuel cell is emerging as a versatile technology. Int J Energy Res 42(2):369–394. https://doi.org/10.1002/er.3780

Kurtzman CP, Pichia EC (2011) The yeasts. Elsevier. pp 685–707

Li X, Lu Y, Luo H, Liu G, Torres CI, Zhang R (2021) Effect of pH on bacterial distributions within cathodic biofilm of the microbial fuel cell with maltodextrin as the substrate. Chemosphere 265(129088):129088. https://doi.org/10.1016/j.chemosphere.2020.129088

Logan BE (2009) Exoelectrogenic bacteria that power microbial fuel cells. Nat Rev Microbiol 7(5):375–381. https://doi.org/10.1038/nrmicro2113

Magaly DLCN, Otiniano Garcia NME, Rojas Flores SJ, Silva Palacios F, Angelats Silva L, Benites Castillo SM, et al (2020) Bioelectricidad a partir de la levadura Saccharomyces cerevisiae a través de celdas de combustible microbiana de bajo costo. In: Proceedings of the 18th LACCEI International Multi-Conference for Engineering, Education, and Technology: Engineering, Integration, and Alliances for A Sustainable Development. Hemispheric Cooperation for Competitiveness and Prosperity on A Knowledge-Based Economy. Latin American and Caribbean Consortium of Engineering Institutions

Noriega DLC, Rojas-Flores M, Benites S, Otiniano SM, Cabanillas-Chirinos NM, Rodriguez-Yupanqui L et al (2021) Generación bioelectricidad a partir de aguas residuales mediante celdas de combustible. LACCEI Inc.

Ochoa JL, Juárez RV (2004) Las levaduras marinas como herramientas científica y biotecnológica. Universidad y Ciencia. 1:39–50

Opoku EEO, Boachie MK (2020) The environmental impact of industrialization and foreign direct investment. Energy Policy 137(111178):111178. https://doi.org/10.1016/j.enpol.2019.111178

Owusu PA, Asumadu-Sarkodie S (2016) A review of renewable energy sources, sustainability issues and climate change mitigation. Cogent Eng 3(1):1167990. https://doi.org/10.1080/23311916.2016.1167990

Pal M, Sharma RK (2019) Exoelectrogenic response of *Pichia fermentans* influenced by mediator and reactor design. J Biosci Bioeng 127(6):714–720. https://doi.org/10.1016/j.jbiosc.2018.11.004

Pal M, Sharma RK (2020) Development of wheat straw based catholyte for power generation in microbial fuel cell. Biomass Bioenergy 138(105591):105591. https://doi.org/10.1016/j.biombioe.2020.105591

Pandey P, Shinde VN, Deopurkar RL, Kale SP, Patil SA, Pant D (2016) Recent advances in the use of different substrates in microbial fuel cells toward wastewater treatment and simultaneous energy recovery. Appl Energy 168:706–723. https://doi.org/10.1016/j.apenergy.2016.01.056

Pant D, Van Bogaert G, Diels L, Vanbroekhoven K (2010) A review of the substrates used in microbial fuel cells (MFCs) for sustainable energy production. Bioresour Technol 101(6):1533–1543. https://doi.org/10.1016/j.biortech.2009.10.017

Rojas-Flores S, La Cruz-Noriega D, Nazario-Naveda R, Benites SM, Delfín-Narciso D, Angelats-Silva L, Murga-Torres E (2022a) Use of banana waste as a source for bioelectricity generation. Processes 10(5):942

Rojas-Flores S, De La Cruz-Noriega M, Nazario-Naveda R, Benites SM, Delfín-Narciso D, Rojas-Villacorta W, Romero CV (2022b) Bioelectricity through microbial fuel cells using avocado waste. Energy Rep 8:376–382

Rojas-Flores S, De La Cruz-Noriega M, Cabanillas-Chirinos L, Benites SM, Nazario-Naveda R, Delfín-Narciso D et al (2023) Use of kiwi waste as fuel in MFC and its potential for use as renewable energy. Fermentation 9(5):446

Rossi R, Logan BE (2021) Using an anion exchange membrane for effective hydroxide ion transport enables high power densities in microbial fuel cells. Chem Eng J 422(130150):130150. https://doi.org/10.1016/j.cej.2021.130150

Segundo R-F, De La Cruz-Noriega M, Milly Otiniano N, Benites SM, Esparza M, Nazario-Naveda R (2022) Use of onion waste as fuel for the generation of bioelectricity. Molecules 27(3):625. https://doi.org/10.3390/molecules27030625

Shanthi Sravan J, Tharak A, Annie Modestra J, Seop Chang I, Venkata Mohan S (2021) Emerging trends in microbial fuel cell diversification—critical analysis. Bioresour Technol 326(124676):124676. https://doi.org/10.1016/j.biortech.2021.124676

Ullah Z, Zeshan S (2020) Effect of substrate type and concentration on the performance of a double chamber microbial fuel cell. Water Sci Technol 81(7):1336–1344. https://doi.org/10.2166/wst.2019.387

Valenzuela-González F, Casillas-Hernández R, Villalpando E, Vargas-Albores F (2015) The 16S rRNA gene in the study of marine microbial communities. Cienc Mar 41(4):297–313. https://doi.org/10.7773/cm.v41i4.2492

Wang Y, Zhao Y-C, Fan L-L, Xia X-D, Li Y-H, Zhou J-Z (2018) Identification and characterization of *Pichia membranifaciens* Hmp-1 isolated from spoilage blackberry wine. J Integr Agric 17(9):2126–2136. https://doi.org/10.1016/s2095-3119(18)62027-1

Zhang C, Liang P, Yang X, Jiang Y, Bian Y, Chen C (2016) Binder-free graphene and manganese oxide coated carbon felt anode for high-performance microbial fuel cell. Biosens Bioelectron 81:32–38. https://doi.org/10.1016/j.bios.2016.02.051

# Part VII
# Renewable Energy

# Chapter 55
# Energy Transition in a Business Company—Solar PV for a Car Fleet

**Paulo Silva, Nídia S. Caetano, and Carlos Felgueiras**

**Abstract** Fossil fuels are increasingly limited in today's world, causing an energy crisis due to external factors, increasing prices in international markets. To solve this global problem, the energy transition related to mobility in companies that oversee their car fleets is highlighted. This transition to electric mobility influences several economic, technical, and social aspects, thus it becomes crucial for companies to adapt their infrastructure and dynamics to have more sustainable practices. According to the 2021–2022 EIB Climate Survey, 55% of Portuguese young people consider climate change when looking for a job. Furthermore, when asked about future car purchases, 84% of Portuguese car buyers say they will purchase either a hybrid or electric car, making Portugal the EU No. 1 country in terms of intentions to purchase electric cars. These statistics show the urgency for companies to adapt to future needs, as well as align with the European goals of reducing greenhouse gases emissions to 45% by 2030 and to zero by 2050.

**Keywords** Energy efficiency · Energy markets · Renewable energy · Sustainable cities

P. Silva · N. S. Caetano · C. Felgueiras (✉)
CIETI—Centre of Innovation on Engineering and Industrial Technology/IPP-ISEP, School of Engineering, R. Dr. António Bernardino de Almeida 431, 4249-015 Porto, Portugal
e-mail: mcf@isep.ipp.pt

N. S. Caetano
LEPABE—Laboratory for Process Engineering, Environment, Biotechnology and Energy, Faculty of Engineering, University of Porto, Rua Dr. Roberto Frias, 4200-465 Porto, Portugal

ALiCE—Associate Laboratory in Chemical Engineering, Faculty of Engineering, University of Porto, Rua Dr. Roberto Frias, 4200-465 Porto, Portugal

 
N. S. Caetano and M. C. Felgueiras (eds.), *The 9th International Conference on Energy and Environment Research*, Environmental Science and Engineering,
https://doi.org/10.1007/978-3-031-43559-1_55

## 55.1 Introduction

Due to population growth and industrial development, the need for burning of fossil fuels to produce electrical energy has increased, leading to a drastic increasing level of pollution on Earth, resulting in effects that are directly linked to climate change and to the phenomenon known as global warming. Over the years, the development of technology allows, more and more, to obtain alternatives to polluting energies. Obtaining electrical energy through alternative technologies, such as solar energy, contributes not only to environmental savings, but also, in some cases, to capital savings by those who opt for these alternatives, such as companies. The use of renewable energy can diversify in their application, from consumption in buildings, to sale to distribution entities, as well as application in own infrastructures, as is the example of charging stations for electric/hybrid vehicles. As a matter of fact, global electric car sales broke records in 2021, more than doubling the values achieved in the previous year, even in face of supply chain bottlenecks (IEA 2022). Ambitious government policy announcements, strategies and budgetary commitments also characterized electric vehicles (EVs) developments in 2021.

In general, businesses succeed when they can meet people's needs with solutions that people want, and when a rewarding business model can be established around that interaction (Pedersen 2018). Encouraging and implementing measures on intrinsic internal dynamics of companies will allow progress towards economic, social, and environmental sustainability. Among such measures, the implementation of photovoltaic (PV) charging stations for EVs stands out, stimulating the transition to electric mobility, not only on an individual basis, but also as a business (Pardo-Bosch et al. 2021). The creation of these infrastructures will contribute to the decarbonization of the economy, progress in infrastructure modernization, reduction of energy dependence, reduction of financial charges related to fuels, reduction in carbon footprint and increased energy efficiency for transportation (Yang et al. 2021). It is based on these principles that companies should act, carrying out energy efficiency interventions and investing in infrastructures that reduce energy dependence. The commitment to these actions will allow companies not only to reduce their annual energy bill, but also offer their employees a more sustainable and healthier lifestyle.

Adding a Vehicle-to-Grid (V2G) configuration to the system results in a bidirectional flow of energy between an EV and the charging station. In addition to offering complete control over the charging process and load-balancing, V2G charging systems can reduce energy bills for businesses and individuals by feeding energy back to the grid at peak times whilst charging during cheap periods. The first remarkable advantage of using this solution is that the dynamic load can be placed on a local substation by EV charging—for instance, in the early evening peak. The second one solves an energy storage problem, using the vehicles as mobile energy storage systems that can be charged off-peak and discharged at peak times (Lauinger et al. 2017). Furthermore, due to the current geopolitical situation and the energy crisis, it is increasingly necessary to seek solutions that respond to the SGDs, increase energy independence, and solve energy and environmental obstacles for society in cohesion.

Thus, the present study aims to represent the clear benefits for companies that this type of infrastructure provides, having a strong economic and social impact. After the study is presented, some considerations and iterations will be made on some possibilities to optimize results. An approach will also be made to estimate the possible impact at national level resulting from the use of this type of technologies and methodologies.

## 55.2 Photovoltaic Stations for EVs Charging

Charging with self-generated electricity from a photovoltaic system is becoming increasingly interesting in a business context, bringing benefits in its operating dynamics (Ala et al. 2021). Nowadays, even electric company fleets can be intelligently controlled and charged with photovoltaic electricity in a very cost-efficient manner, as shown in Fig. 55.1.

As mentioned, the purchase of EVs by the population has been increasing over the years, as has been expected. Sales of EVs reached another record high peak in 2021 despite the COVID-19 pandemic and supply chain challenges, including semiconductor chip shortages. Looking back, about 120,000 electric cars were sold worldwide in 2012. In 2021, that many were sold in a week. After increasing in 2020 despite a depressed car market, sales of electric cars nearly doubled year-on-year to 6.6 million in 2021. This brought the total number of electric cars on roads to over 16.5 million. As in previous years, battery electric vehicles (BEVs) accounted for most of the increase (about 70%) (IEA 2022).

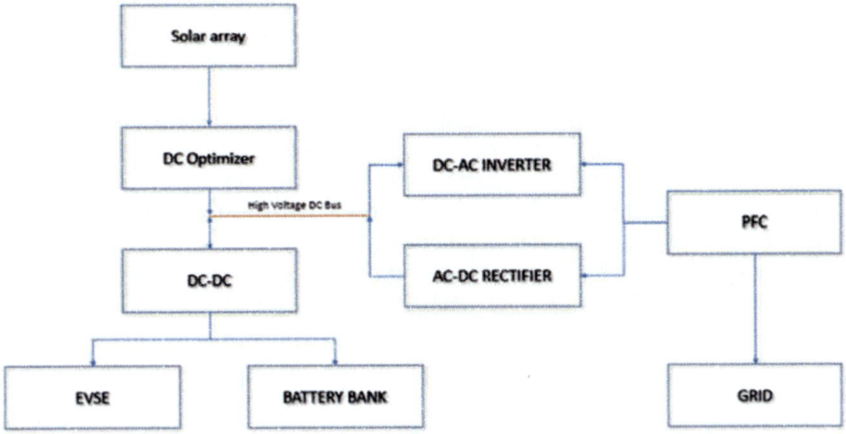


**Fig. 55.1** V2G power architecture

The possibility of having at home the energy needed to power the vehicle with very affordable conditions attracts more and more people to this trend. Furthermore, this solution is seen as environmentally friendly, which attracts most of the population concerned about the environment. As for companies, for example, when buying a car with an acquisition cost of 50,000 €, adding VAT, the price of the car would immediately rise to 61,500 €. For an electric car, VAT is deductible, which means that the company would not have to pay this tax, paying only the 50,000 €.

To the value of the car, in Portugal, the autonomous taxation must also be added, a tax independent from the Corporate Income Tax (IRC) which, for diesel or gasoline passenger vehicles, can range from a minimum value of 10% to a maximum rate of 35% on the value of the company with the car. Given the cost of the vehicle in this example (50,000 €), autonomous taxation would correspond to the maximum rate, that is, you would still have to pay 35% of the value of depreciation and other expenses associated with the vehicle, such as tolls, maintenance costs, insurance, fuel, etc.

As for electric vehicles, they are again exempt from this tax, and the company will not need to pay autonomous taxation on any charge it has with the car. Due to this increase of EVs within the population and companies, as also benefits and advantages, the number of investments in charging stations has been increasing significantly (Pande and Report 2022). In 2021, there were roughly 375,900 public charging stations for electric vehicles in Europe. The number of these infrastructures grew consecutively between 2010 and 2021, with exponential increases observed in 2011, 2012, and 2016 (Statista 2022).

## 55.3 Case Study

The company in this case study is active in water distribution, wastewater drainage and treatment, rainwater drainage, waterline management, seafront management, energy management and promotion of environmental education and sustainability in the city of Porto—Portugal. From the point of view of environmental sustainability, the company concluded in 2018 an electricity production project from renewable energy that is worth 330,000 €, more specifically, installed a photovoltaic plant on the rooftop of its headquarters. The estimated annual energy production is 462,396 kWh, of which 72.7% will be used for self-consumption (included for EVs charging stations, as represented in Fig. 55.2), with the remaining 159,162 kWh being sold and injected into the national grid. The carbon dioxide emissions ($CO_2$) avoided represent 217 tons $CO_2$ per year. A 50% reduction in costs associated with the portion of active energy consumed at the company' Headquarters in the first year of production was expected, which allowed the amount invested to be amortized in about 8 years.

This energy efficiency project resulted in annual savings between 38,000 € and 67,000 €. In 25 years, the estimated useful life for the photovoltaic park, under the current consumption scenario, the company expects to save about 740,000 € in

**Fig. 55.2** Electric vehicle charging stations

**Fig. 55.3** Overview of the implemented PV plant

electricity costs. To promote electric mobility, the company acquired 84 vehicles on a leasing plan, of which 52 are fully electric, 8 hybrids and the rest combustion, totaling more than 1,500,000 €. It is estimated a 30% reduction, in liters, in the fuel consumption and, consequently, a significant decrease in $CO_2$ emissions.

The photovoltaic plant was installed on the rooftop of the park where the company's fleet is parked, as shown in Fig. 55.3. The inverter and DC/AC switchboard are installed in the technical area, minimizing losses. The interconnection from the photovoltaic system to the existing electrical system is low voltage. The characteristics of the main equipment are described in Table 55.1.

## 55.4 Economic and Environmental Analysis

Using the licensed software, PV*SOL, and according to the data presented, the photovoltaic plant displays the values listed in Table 55.2.

As mentioned earlier, the case study company sells the energy produced as surplus to the distributor. Therefore, estimating that about 75% is consumed by electric

**Table 55.1** PV power plant main features

| PV power plant technical features | |
|---|---|
| Location | Porto, Portugal |
| Number of modules Q-Cells. Plus L-G4.1 345 | 357 |
| Total installed power (kWp) | 123 |
| Rated output power on the inverter (kVA) | 25 |
| Number of SMA 250,000 TL inverters | 5 |
| Orientation modules | 235° |
| PV slope | 8° |

**Table 55.2** PV power plant main energetic features

| PV power plant energetic features | |
|---|---|
| Energy produced by the PV power plant (kWh/year) | 167,277 |
| $CO_2$ emissions avoided (kg/year) | 78,620 |

vehicles for charging, it is possible to conclude that about 41,820 kWh are injected into the grid. Adding to the 159,162 kWh surplus from the previous project, it results in the injection and sale of an estimated 200,982 kWh, resulting in an annual financial return of 7034.37 € (estimated sale value of 0.035€ per kWh). The combination of energy efficiency projects results in an estimated reduction of ~295.6 tons of $CO_2$ emissions avoided per year.

### *55.4.1 Impact on $CO_2$ Emissions*

According to European statistical data, the total $CO_2$ emissions attributed to the Portuguese road transport sector in 2020 was around 800 Mton $CO_2$ (European Environment Agency 2019). Assuming that 3% of the cars registered in Portugal that run on fossil fuel belong to car fleets of companies that have invested in the energy transition, traveled the same annual distance in number of kilometers as the European average (estimated 14,000 km) (European Environment Agency 2019) and that they would make the same investment in photovoltaic park infrastructure for charging vehicles, this would mean an annual reduction of 451,151 or 647,665 ton $CO_2$, depending on whether the fleet consumes gasoline or diesel, respectively, representing between 6 and 8% of total $CO_2$ emissions in Portugal transportation sector in 2019.

### 55.4.2 Contribution to Achieve Sustainable Development Goals

This work brings an original contribution, with a practical case, as it presents a couple solutions that address some specific Sustainable Development Goals (SDGs), as follows:

- *SDG 3—Good health and well-being*—The improvement in the facilities of this type of infrastructure promote the quality of the services provided by the companies, reducing the exposure of its employees to polluting agents, as well as creating a welcoming space that promotes values which benefit the environment, contributing to mental well-being;
- *SDG 7—Affordable and clean energy*—The creation of a charging infrastructure will allow to obtain clean energy, with investments that reflect favorable returns in the short/medium term. The main objectives of this infrastructure were to reduce energy dependence, to reduce the annual energy bill associated with fossil fuels in the vehicle fleet and to reduce annual $CO_2$ emissions associated with the use of vehicles belonging to the fleet;
- *SDG 8—Decent Work and economic growth*—The modernization of infrastructures will allow the improvement of working conditions for company employees. For companies, this modernization will mean a competitive advantage and advancement in the market compared to other companies;
- *SDG 9—Industry, innovation, and infrastructure*—Industry investment in new infrastructures, technologies, strategies, and renewable energy paves the way for progress and increased competitiveness in the market. These strategies can be implemented at the national level, promoting innovation, through investment in new technologies and infrastructures;
- *SDG 11—Sustainable cities and communities*—This project is associated with energy sustainability, which reduces the energy footprint, and consequently, the negative environmental impacts per capita, and is also adapted to climate change;
- *SDG 12—Responsible consumption and production*—Highlights the importance of clean energy production and responsible consumption. In the present case, the excess production is sold to the distributor so that it can be integrated and distributed in the national electricity grid;
- *SGD 13—Climate action*—The investment made in electric mobility by the case study company will significantly reduce GHG emissions associated with the use of its vehicle fleet. This decrease, although not very significant at a global level, can serve as a stimulus for other companies to follow the same path, making a joint effort in the investment in electric mobility and the positive effects become quite significant at a national level, as well as at a global level;
- *SGD 17—Partnership for the goals*—The case study company's strategy focuses not only on the individual aspect as a company, but as a government partner to achieve the European goals outlined for 2030 and 2050. This company aims

to increase the supply of EVs' charging points, being able to include articulation with energy production infrastructures and Vehicle-to-Grid (V2G)/Grid-to-Vehicle (G2V) configurations. Highlighting the Vehicle-to-grid (V2G) configuration, this operation describes a system in which plug-in electric vehicles communicate with the power grid to sell demand response services by either returning electricity to the grid or by throttling their charging rate. Since at any given time 95% of cars are parked, the batteries in electric vehicles could be used to let electricity flow from the car to the electric distribution network and back (Texas Instruments 2016).

## 55.5 Conclusions

The implementation of the proposed measures, combined with a previous energy efficiency project, had a very considerable impact, representing a saving in the annual energy bill estimated between 45,034 € and 74,034 € and an estimated reduction of 295 tons of $CO_2$ emissions per year. It was also possible to conclude that the injection of surplus energy into the national grid, although not quite profitable compared to the investment in the short or medium term, reinforces the robustness of the project payback time, reducing the annual energy bill and having a direct impact in the national and European goals outlined for 2030 and 2050, decreasing energy dependence and $CO_2$ emissions associated by its energy and fossil fuels consumption. In addition, it increases the amount of green energy produced in the country, contributing favorably to the energy mix of electricity distributed on the national grid, lowering the $CO_2$ emission factor per kWh. Additionally, V2G configurations will allow an increase in the energy efficiency of the system, resulting in an increase in the annual savings of the energy bill.

**Acknowledgements** This work was financially supported by Base Funding—UIDB/04730/2020 of Center for Innovation in Engineering and Industrial Technology, Portugal, CIETI—funded by national funds through the FCT/MCTES (PIDDAC), Portugal; LA/P/0045/2020 (ALiCE) and UIDB/00511/2020–UIDP/00511/2020 (LEPABE) funded by national funds through FCT/MCTES (PIDDAC).

## References

Ala G, Colak I, Di Filippo G et al (2021) Electric mobility in Portugal: Current situation and forecasts for fuel cell vehicles. Energies. https://doi.org/10.3390/en14237945

European Environment Agency (2019) Sectoral profile-transport energy consumption

IEA (2022) Global EV Outlook 2022 Securing supplies for an electric future, p 221

Lauinger D, Vuille F, Kuhn D (2017) European battery, hybrid and fuel cell electric vehicle congress: a review of the state of research on vehicle-to-grid (V2G): progress and barriers to deployment. Eur Batter Hybrid Fuel Cell Electr Veh Congr, pp 1–8

Pande MS, Report IM (2022) Electric vehicle charging station market share, size, growth and industry overview 2022–20288
Pardo-Bosch F, Pujadas P, Morton C, Cervera C (2021) Sustainable deployment of an electric vehicle public charging infrastructure network from a city business model perspective. Sustain Cities Soc 71:102957. https://doi.org/10.1016/J.SCS.2021.102957
Pedersen CS (2018) The UN sustainable development goals (SDGs) are a great gift to business! Proc CIRP 69:21–24. https://doi.org/10.1016/j.procir.2018.01.003
Statista (2022) Number of electric vehicle charging stations in Europe from 2010 to 2021
Texas Instruments (2016) Eletric vehicles (EV) charging station—grid infrastructure industrial systems, p 53
Yang M, Zhang L, Zhao Z, Wang L (2021) Comprehensive benefits analysis of electric vehicle charging station integrated photovoltaic and energy storage. J Clean Prod 302:126967. https://doi.org/10.1016/J.JCLEPRO.2021.126967

# Chapter 56
# A Preliminary Study on the Ignition of Some Ligneous Biomass Pellets Inside a Traveling Grate Furnace

**Tânia Ferreira, Edmundo Marques, João Monney Paiva, and Carlos Pinho**

**Abstract** With the growing interest in using biomass as an alternative energy source, the scientific study of its behavior becomes even more relevant. Furthermore, the heterogeneous nature of biomass makes the process more challenging. A laboratory-scale traveling bed furnace was used to study the ignition of some ligneous biomass pellets, specifically the measurement of the ignition time of volatiles released by different types of wood pellets. The experimental ignition tests were carried out at five different corrected furnace temperatures, namely between 359 and 443 °C, and using the same mass flow rate of the supplied air. The results show that for the studied pellets, the ignition time is linear furnace temperature dependent; as expected, for all species, the pellets ignition time was higher at 443 °C. In general, it was verified that the pellets species presented some variableness in the ignition time, being the *Pinus pinaster* the specie that obtained a higher ignition time.

**Keywords** Biomass · Ignition · Pellets · Traveling grate

## 56.1 Introduction

The furnace used in this work was designed to carry out experiments for burning small batches of char pellets in a traveling bed. With the kinetic and diffusive data previously determined in laboratory tests of burning batches in fluidized and fixed bed (Ferreira et al. 2020), the evaluation of the applicability of these data to a situation closer to

T. Ferreira (✉) · E. Marques · J. M. Paiva
Escola Superior de Tecnologia e Gestão, Instituto Politécnico de Viseu, Campus Politécnico, 3504-510 Viseu, Portugal
e-mail: taniaferreira@estgv.ipv.pt

C. Pinho
CEFT-DEMEC—Faculdade de Engenharia, Universidade Do Porto, Rua Dr. Roberto Frias, S/N, 4200-465 Porto, Portugal



N. S. Caetano and M. C. Felgueiras (eds.), *The 9th International Conference on Energy and Environment Research*, Environmental Science and Engineering,
https://doi.org/10.1007/978-3-031-43559-1_56

the real situations of biomass particulate burning on a circulating grid is intended. However, in the initial installation procedures of this new laboratory reactor and its start-up tests, which were carried out not only with carbonized pellets but also with pellets just as they had been produced, and therefore not yet carbonized, an absence of any publications that referred to results of the ignition times of volatiles that are released from woody pellets during their travel on a traveling grate was found from previous reading of the scientific and technical literature. As such, given the challenge posed by the lack of experimental data regarding the ignition of wood pellets on their way through the traveling grate, a series of tests measuring the ignition of pellets from six different biomasses was carried out. By pellet ignition, in this particular context, the ignition of volatiles released from the pellets in the initial phase of their path on the traveling grate is meant.

The knowledge of the ignition behavior of biofuels during their handling, transport and storage is of high importance in order to avoid serious incidents (Torrent et al. 2012). Even though the literature concerning the characterization of woody pellet ignition is practically non-existent, some experimental studies concerning the ignition of different fuels have been found. Tognotti et al. (1989) experimentally determined the ignition temperature of a South African coal using two distinct cylindrical beds. The data obtained were compared with thermal ignition models. Also Grotkjaer et al. (2003) experimentally determined the ignition temperature of biomasses, specifically wheat straw, at 21% (v/v) oxygen. Ferrero et al. (2009) developed a mathematical model that predicts the heating of wood chip piles in open air to prevent the auto ignition of these materials during storage. These authors found that microbial processes are responsible for the highest heat production in the early stages of storage. Jones et al. (2015) evaluated the ignition risk of seven biomasses and classified them according to ignition risk using kinetic parameters. These authors found that slow combustion occurred, ignition only occurred in the carbonaceous residue, and combustion of the released volatiles was not observed.

## 56.2 Materials and Methods

### *56.2.1 Pellets Preparation*

For the present work six wood species were selected: *Pinus pinaster*, *Acacia dealbata*, *Cytisus scoparius*, *Cistus ladanifer*, *Paulownia cotevisa* and *Eucalyptus globulus*. With the exception of *Pinus pinaster*, the other species were used as raw material for the production of pellets used in the tests, at the laboratories of the Polytechnic Institute of Viseu.

The *Pinus pinaster* pellets were acquired to a Portuguese company, while the species *Acacia dealbata*, *Cytisus scoparius*, *Cistus ladanifer*, *Paulownia cotevisa* and *Eucalyptus globulus* were collected in a forest surrounding Viseu and dried in a solar kiln, with the moisture content monitored throughout the process. A knife mill, Retsch

SM 100, was used to ground the raw material using a sieve with an average diameter of 7 mm, in order to obtain the necessary raw material for the pellets production. After characterization, regarding particle size and moisture content, 6 mm pellets diameter were produced using a Tojaltec pelletizer press machine. Subsequently, the pellets were cut manually into smaller and uniform particles with average lengths between 14 and 15 mm. The pellets exact dimensions, length and diameter, were determined using a digital caliper with a resolution of 0.1 mm. The mass of each pellet, which allowed the calculation of the average particle density, was determined using a Sartorius BP 310 P digital balance. Figure 56.1 illustrates the pellets used and Table 56.1 shows their average dimensions, length and diameter, and particle density.

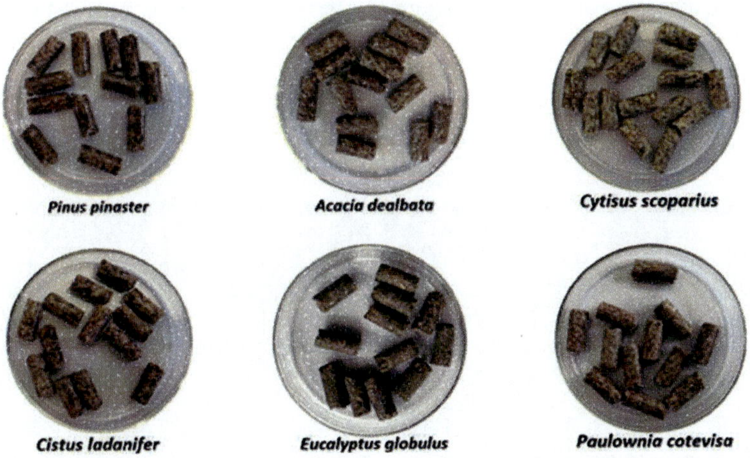


**Fig. 56.1** Photographs of the tested pellets

**Table 56.1** Dimensions and particle density of the tested pellets

| | Length (mm) | Diameter (mm) | Particle density (kg/m$^3$) |
|---|---|---|---|
| *Pinus pinaster* | 14.62 | 6.09 | 1162.77 |
| *Acacia dealbata* | 14.48 | 6.22 | 1062.33 |
| *Cytisus scoparius* | 14.49 | 6.13 | 997.04 |
| *Cistus ladanifer* | 14.61 | 6.08 | 1093.45 |
| *Eucalyptus globulus* | 14.42 | 6.56 | 1014.27 |
| *Paulownia cotevisa* | 14.50 | 6.10 | 1173.95 |

### *56.2.2 Experimental Setup*

The experimental setup used in the pellet ignition tests consisted of a traveling grate furnace, Fig. 56.2, and a set of instruments that allowed the measurement and control of the furnace temperature, as well as the air mass flow rate used in the tests.

The furnace was constructed of ST37 steel with 610 mm length and 320 mm width internal dimensions. The front part had a high temperature glass door with 23 mm long and 9 mm wide. The fuel supply was fulfilled from the left side, through a gutter, and the ashes exit was on the right side. The traveling grate was made from a refractory stainless steel, with 30 cm long and 9 cm wide (Fig. 56.3).

Regarding the furnace heating system, it consisted of an electric resistance made from 1.14 mm diameter Kanthal wire with a power of 4.5 kW, inserted in refractory bricks and placed around the traveling grate. The furnace was also thermally insulated with Kaowool ceramic blanket. The experimental setup is schematically represented in Fig. 56.4.

The furnace temperature was measured by 2 K-type thermocouples, which were coated with stainless steel sheaths with 3 mm diameter and 1500 mm long, positioned

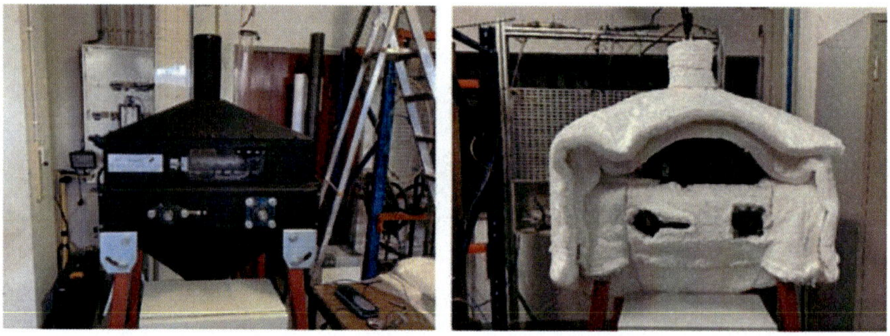

**Fig. 56.2** Experimental installation (without and with thermal insulation)

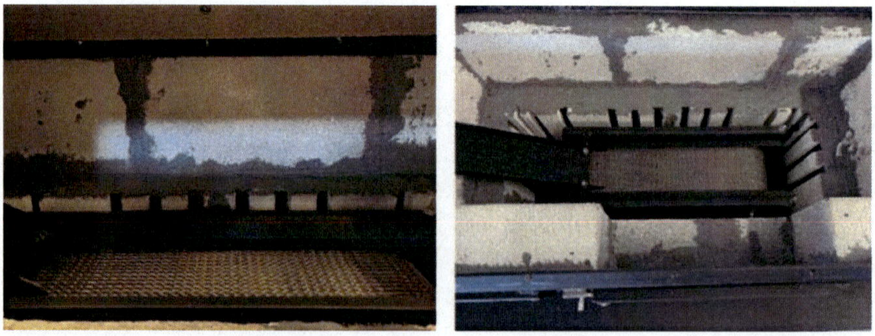

**Fig. 56.3** Traveling grate (front view and top view)

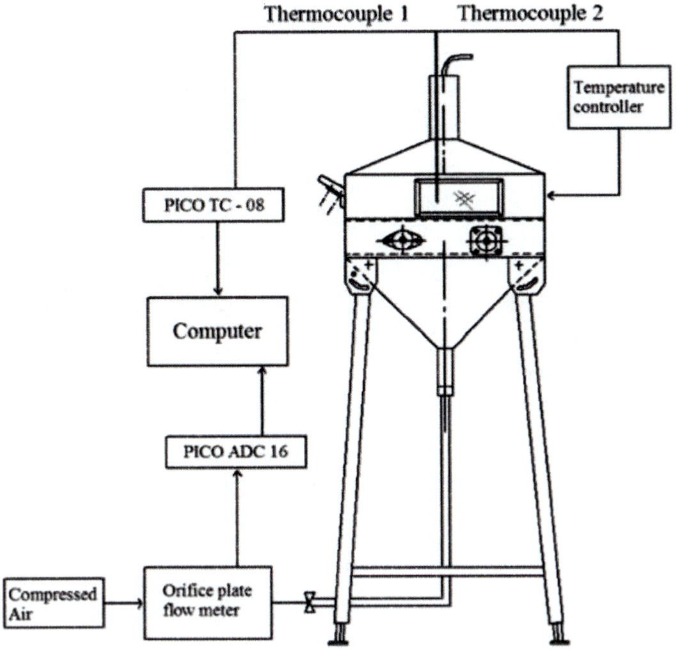


**Fig. 56.4** Schematic representation of the experimental setup

in the middle of the traveling grate, at a height of 10 mm. One of the thermocouples was connected to the temperature controller, an Omron ESCSV, and the other one was connected to a Pico TC-08 data logger. The mass flow rate of the air, coming from the compressed air network, properly filtered and dehumidified, was measured with an orifice plate flow meter, using both a U-tube water pressure manometer and a differential pressure transducer, Omega PX143-2,5BD5V, connected to the PicoLog Recorder through an ADC 16 data logger. The used sampling rate was 1.0 Hz.

#### 56.2.2.1 Experimental Procedure

6 g pellet batches were burned in the traveling grate furnace at set point temperatures of 360, 385, 410, 435 and 460 °C. For all experiments the mass flow rate of the supplied air was ~43 mg/s. For each ignition experiment, the evolution of the furnace temperature and the air mass flow rate were continuously measured and registered.

The ignition experiments were started when the furnace temperature was stabilized. Each pellets batch was placed in the feeder, Fig. 56.5, and introduced, through the feed zone, into the furnace. To do this, the ceramic blanket cover, which sealed the burner entrance, had to be removed and the feeder was quickly introduced into the furnace, which was opened to deposit the pellet batch at the beginning of the

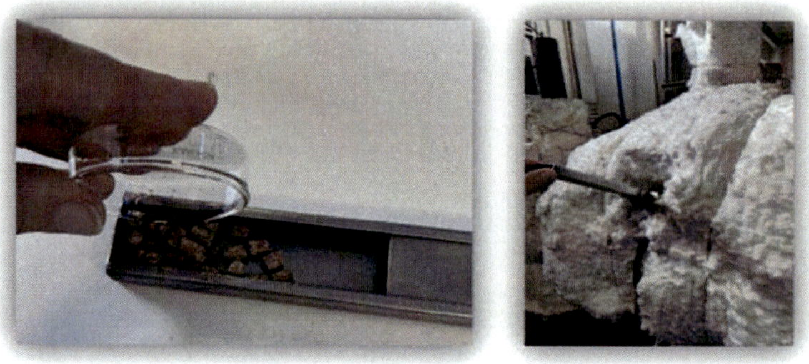

**Fig. 56.5** Batches feeding procedure

traveling grate. As soon as the pellets fell, a chronometer was activated, the feeder was removed from inside the furnace, and the burner feed was covered again. At the instant the pellet batch ignited, the chronometer was stopped and the time was recorded.

## 56.3 Results and Discussion

Since the thermocouples positioned inside the burner were in a gaseous environment and the temperature measured by them was the radiation equilibrium temperature and not the real temperature of this environment, a temperature correction according to Holman (2012) was necessary. A Testo 835-T2 high-temperature infrared thermometer was therefore used to measure the furnace walls temperatures in the range measured during the ignition tests. This procedure was carried out by three persons simultaneously, so that it was done as quickly and accurately as possible. Thus, when the furnace door was opened, the temperature measured by the thermocouple was recorded and, with the high-temperature infrared thermometer, the temperature of the furnace walls was also measured. Table 56.2 shows the corrected temperatures determined for the temperature ranges used in the experiments.

**Table 56.2** Corrected furnace temperature

| Temperature read by the thermocouple (°C) | Particle density (°C) |
|---|---|
| 360 | 359 |
| 385 | 381 |
| 410 | 403 |
| 435 | 424 |
| 460 | 443 |

Although the traveling bed furnace set point temperatures chosen for the experiment temperatures were, as referred before, 360, 385, 410, 435 and 460 °C, taking into account the corrected temperatures for each case, the actual ignition tests temperatures are respectively 359, 381, 403, 424 and 443 °C.

As previously mentioned, the ignition tests consisted in introducing batches of pellets, with 6 g of mass, inside the traveling grate furnace that was heated to the desired temperature, to determine the time between the introduction instant and the moment of ignition of volatiles released from that batch of pellets (Fig. 56.6).

The furnace temperature evolution during the ignition tests was continuously recorded. Figure 56.7 shows graphically the temperature evolution curves during two ignition tests performed with *Pinus pinaster*, for the furnace at 359 and 443 °C, respectively.

From the analysis of the previous figures, the temperature evolution presents a similar trend. There is a slight drop in temperature when the batch is introduced into the burner and maximum temperature reached is higher in tests performed at

**Fig. 56.6** Moment of ignition of a batch of *Pinus pinaster* pellets

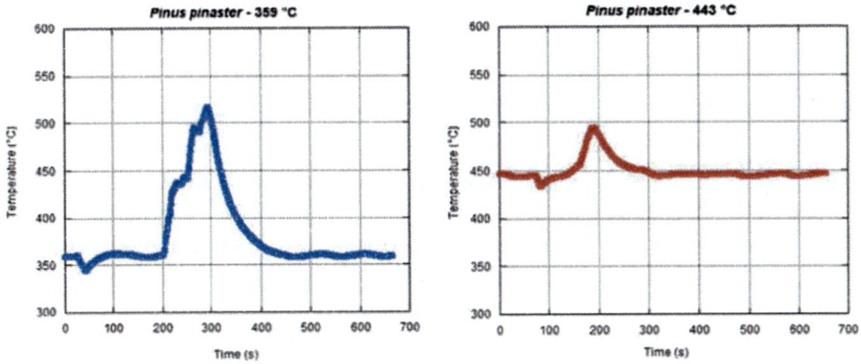


**Fig. 56.7** Furnace temperature evolution during tests performed with *Pinus pinaster* pellets

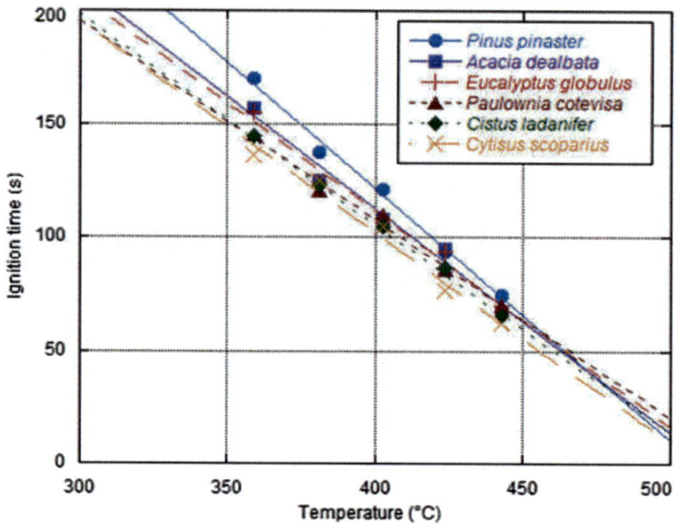


**Fig. 56.8** Pellet ignition time as a function of furnace temperature

359 °C. This is related to the fact that at 359 °C the batch takes longer to ignite and therefore, its position, on the traveling grate at the time of ignition, will be closer to the thermocouple. The results obtained for the ignition time are graphically represented in Fig. 56.8, for the six species and the five temperatures used.

From the previous figure it can be observed that as the furnace temperature increases, there is a decrease in the pellet ignition time. In general, the evolution of the ignition time presents a linear trend. *Pinus pinaster* was the species that presented a higher ignition time, being this difference more evident with the tests performed at 359 °C, where the average ignition time of *Pinus pinaster* was about 170 s. *Cytisus scoparius*, for the same conditions, was the species that took the least time to ignite. With the tests performed at 359 °C, the average ignition time for this species was 136 s.

## 56.4 Conclusion

In this study, the ignition time of six types of wood pellets made from *Pinus pinaster*, *Acacia dealbata*, *Cytisus scoparius*, *Cistus ladanifer*, *Paulownia cotevisa* and *Eucalyptus globulus* were determined under five different furnace temperatures, 359, 381, 403, 424 and 443 °C. For the studied pellets, there are significant differences in the ignition time, particularly at the lowest temperature, 359 °C. For furnace temperatures above 443 °C the time ignition was so short that they could not be determined with a minimum of precision and, as such, those results were discarded.

**Funding** This work was supported by the Portuguese Foundation for Science and Technology (FCT), grant reference SFRH/BD/137170/2018, awarded to Tânia Ferreira.

## References

Ferreira T, Paiva JM, Pinho C (2020) Comparative analysis of fluidized and fixed beds to obtain data on the char pellet's combustion regime. Int J Energy Clean Environ 21:237–268

Ferrero F, Lohren C, Schmidt B, Noll M, Malow M (2009) A mathematical model to predict the heating-up of large-scale wood pile. J Loss Prev Process Ind 22:439–448

Grotkjær T, Johansen K, Jensen A, Glarborg P (2003) An experimental study of biomass ignition. Fuel 82:825–833

Holman JP (2012) Experimental methods for engineers, 8th edn. McGraw-Hill, New York

Jones M, Saddawi A, Dooley B, Mitchell E, Werner J, Waldron D (2015) Low temperature ignition of biomass. Fuel Process Technol 134:372–377

Tognotti L, Petarca L, Zanelli S (1989) Spontaneous combustion in beds of cool particles. Sym-Posium Combust 22:201–210

Torrent J, Gómez Á, Arogón E, Olmedo C, Pejic L (2012) Determination of the risk of self-ignition of coals and biomass materials. J Hazard Mater 213–214:230–235

# Chapter 57
# Evaluation of the Conversion Potential of Maize Stover from Soil Phytoremediation to Bioethanol

**Nídia S. Caetano, Mariana Santos, and Ana P. Marques**

**Abstract** This work aimed to evaluate the conversion potential of maize stover (MS) from phytoremediation of heavy metals contaminated soil to bioethanol. Thus, MS was submitted to an acid pretreatment with 3% (v/v) $H_2SO_4$, HCl, $HNO_3$ or $CH_3COOH$ at 85 °C for 48 h. An enzymatic hydrolysis step with *Accellerase* or *Ultraflo* was applied at 50 °C for 13 h. Finally, *Saccharomyces cerevisiae* was used to ferment the glucose at 37 °C, followed by distillation to recover ethanol. The average yield in ethanol for the MS produced in the two contaminated soils was 0.51 and 0.32 $g_{ethanol}/g_{MS}$ for the MS treated with HCl and *Accellerase* and 0.39 and 0.27 $g_{ethanol}/g_{MS}$ for the MS treated with $HNO_3$ and *Ultraflo*, respectively. For the MS produced in the control soil, the yield was 0.37 and 0.44 $g_{ethanol}/g_{MS}$ for the treatment with $HNO_3$ and *Ultraflo* and HCl and *Accellerase*, respectively, being the differences in ethanol yield assigned to the different cellulose/hemicellulose content of the MS samples. No metals were detected in the ethanol recovered. This research demonstrated the feasibility of valorization of MS from heavy metals contaminated soil phytoremediation through ethanol production, contributing to a more sustainable process of soil phytoremediation.

**Keywords** Biofuel · Cadmium · Heavy metal contaminated soil · Phytoremediation · Soil decontamination · Zinc

---

N. S. Caetano (✉)
LEPABE—Laboratory for Process Engineering, Environment, Biotechnology and Energy, Faculty of Engineering, University of Porto, R. Dr. Roberto Frias S/N, 4200-465 Porto, Portugal
e-mail: nsc@isep.ipp.pt

ALiCE—Associate Laboratory in Chemical Engineering, Faculty of Engineering, University of Porto, Rua Dr. Roberto Frias, 4200-465 Porto, Portugal

N. S. Caetano · M. Santos
CIETI, School of Engineering (ISEP), Polytechnic Institute of Porto (P.Porto), R. Dr. António Bernardino de Almeida 431, 4249-015 Porto, Portugal

A. P. Marques
CBQF – Centro de Biotecnologia e Química Fina – Laboratório Associado, Escola Superior de Biotecnologia, Universidade Católica Portuguesa, Rua de Diogo Botelho 1327, 4169-005 Porto, Portugal

 
N. S. Caetano and M. C. Felgueiras (eds.), *The 9th International Conference on Energy and Environment Research*, Environmental Science and Engineering,
https://doi.org/10.1007/978-3-031-43559-1_57

## 57.1 Introduction

Technological development has been driven by the high population growth over the years and by the growing demand for new goods and appliances, with the consequent increase in materials and energy consumption (BP 2019).The consequent depletion of natural resources and the intensive use of energy, mostly derived from fossil sources, caused significant negative environmental impacts that are on the basis of climate changes, causing great concern not only to scientists, but also to the general public that is getting increasingly aware of sustainability issues. This state of affairs has been the engine of modern society commitment to sustainable development, making it fundamental to seek new and more viable solutions for society and the environment, that reduce the dependence on fossil fuels and contribute to the GHG reduction, minimizing the potential for global warming (Oliveira et al. 2020). Thus, different techniques have been developed to use renewable energy sources. For the transportation sector, the use of biofuels produced from biomass feedstock represents the most immediate and interesting alternative, as currently most renewable energy technologies for this purpose are still under development (REN21 2018). Bioethanol is presently the most widely used liquid biofuel alternative to gasoline (Mata et al. 2022).

Increasing soil contamination due to intensive agriculture and to industrial activities such as mining, which introduces considerable quantities of heavy metals into agricultural soil, has been a growing concern, as the presence of heavy metals in soil makes it unusable for food agriculture (Silva et al. 2019). This has motivated the need to develop sustainable techniques for the extraction of these contaminants. Currently, phytoremediation is one of the most promising used processes since it has the advantage of being a treatment performed at the contaminated site (in situ). Therefore, phytoremediation plays an important role in soils remediation, making them suitable again for cultivation by removing contaminants (Coutinho and Barbosa 2007). However, biomass resulting from phytoremediation, that may contain high levels of contaminants such as heavy metals, is usually composted, landfilled or sometimes submitted to thermal processes, with a significant loss of valuable resources in such process (Marques et al. 2020). To produce second generation bioethanol, it is necessary to convert polysaccharides from lignocellulosic material. Lignocellulose is a material belonging to the wall constitution of plant cells and refers to the material composed of lignin, cellulose and hemicelluloses. Its composition varies with the type of plant, age and environment where it grows. Generally, plants have in their composition 45–25% of lignin, 40–50% cellulose and 25–30% of hemicellulose (Costa et al. 2019; Carpita and McCann 2020; Brethauer et al. 2020).

Maize (*Zea mays* L.) has been reported as potentially effective in Cd, Zn, Cr, Cu and Pb removal from contaminated soil (Hong et al. 2009; Jiang et al. 2010; Soudek et al. 2010). Thus, in the present work, maize was used to perform the Cd and Zn contaminated soil phytoremediation. In order to further valorize this organic matter of lignocellulosic nature that would otherwise be discarded, and aiming to reduce the competition with the agricultural food sector, it was proposed the possibility of transforming lignocellulosic material (maize stover) resulting from phytoremediation of soils contaminateds with heavy metals into bioethanol. Therefore, the present study investigated the feasibility of using maize stover (MS) from phytoremediation of soils contaminated with Cd and Zn for the production of bioethanol, to be used as a liquid biofuel, and compared it with similar material grown in uncontaminated soil (control). The effect of the potential presence of these heavy metals was assessed not only on alcohol yield but also on the presence of heavy metals in the biofuel produced.

## 57.2 Materials and Methods

### *57.2.1 Sample Preparation*

Maize stover (MS) was produced from the lignocellulosic parts (roots, stem and leaves) of the plants of *Zea mays* L. cultivated in three different soils: an agricultural soil, one industrial soil from Estarreja, both from the North of Portugal, and a third mining soil from Panasqueira, in the inner center of Portugal, as described in a previous research by Paulo et al. (2022). The plants were collected 120 days after seeding, sundried, cleaned from the attached soil and the separated parts were ground with a Fritsch grinder, to particles of about 1 × 5 mm. The stem and leaves fraction was selected for the rest of the process.

### *57.2.2 Acid Pretreatment*

As previously described, the main function of pretreatment is to break down the cellulose and hemicellulose polymers present in lignin, thus increasing the accessibility to enzymes in the enzymatic hydrolysis process. In the present research an acid pretreatment was applied. Thus, about 20 g of maize stover were weighed into a 500 ml glass bottle. Subsequently, 150 ml of one of each of the acids was added to each of the flasks used—$H_2SO_4$, $HNO_3$, HCl or $CH_3COOH$, at 3% (v/v) concentration (Fig. 57.1a).

The pretreatment was performed for 48 h in a thermostatic water bath shaking at 100 rpm (Julabo, SW22) at 85 °C. Each treatment was performed in duplicates.

### *57.2.3 Enzymatic Hydrolysis*

The enzymes used in this study were kindly offered by their respective supplier. Thus, it was used *Accellerase* 1000, a liquid enzyme preparation that contains 5–10% fungal cellulase and <1% β-glucosidase (supplied by *Genencor*, a *Danisco*

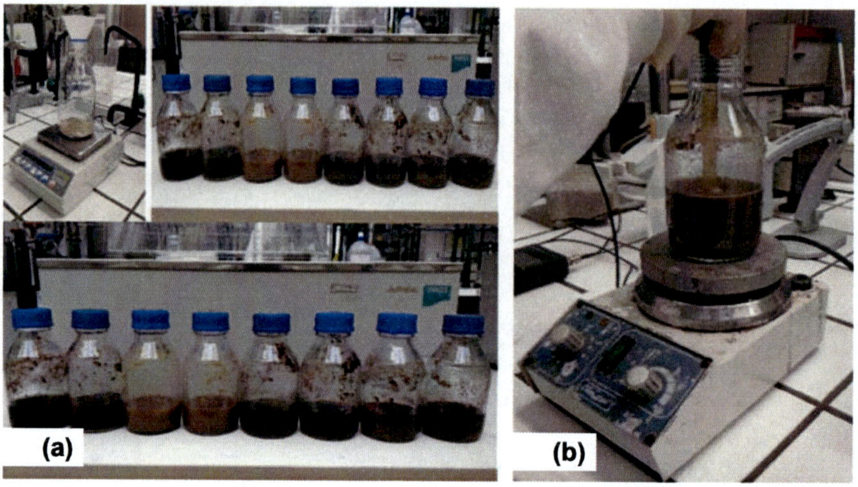


**Fig. 57.1** **a** Top Left: Weighing maize straw in glass bottle; Top Right: Samples before pre-treatment. Bottom: Samples after pre-treatment; **b** pH regulation after pre-treatment

division) and *Ultraflo*, a liquid enzyme preparation that contains up to 15% endo-1,3(4)-β-glucanase (supplied by *Novozymes*).

To achieve the maximum efficiency in the enzymatic hydrolysis process, after the end of pre-treatment, the glass bottles were cooled and the pH value was measured, and adjusted to 5, through small additions of NaOH at 40% (w/v) or 0.1 M when closer to the desired pH (Fig. 57.1b).

After pH regulation, 2 ml of *Accellerase* 1000 (*Danisco*) or *Ultraflo* (*Novozymes*) enzyme were added to each one of the bottles, stirring to homogenize. The bottles were inserted again in the water bath with agitation (Julabo, SW22) at a temperature of 50 °C for 13 h and the hydrolysate was used for fermentation (Fig. 57.2).

**Fig. 57.2** Left: Samples before the hydrolysis process; Right: Samples after hydrolysis process

**Fig. 57.3** **a** Applied yeast and fermentation in a thermostatic bath with slow shaking; **b** Vacuum filtration and ethanol recovery using a rotary evaporator

### *57.2.4 Fermentation*

Commercial *Saccharomyces cerevisiae*, currently used in the bakery process, was used for the fermentation process. 40 g of yeast was diluted in 200 ml of deionized water and after homogenization, 25 ml of inoculum was added to each of the bottles. The mixture was stirred and the bottles were covered with porous stopper, placed in the thermostatic water bath with agitation at 60 rpm at 37 °C for 11 days (Fig. 57.3a). After fermentation, the samples were filtered under vacuum, to separate the liquid phase from the sludge. Ethanol was recovered from the filtered fermentation broth in a rotary evaporator (Buchi, R-210) at 120 mbar and 60 °C (Fig. 57.3b).

### *57.2.5 5-Dinitrosalicylic Acid Method (DNS) for Sugar Quantification*

The reducing sugars content in the samples collected during the different stages of ethanol production was evaluated according to the methodology described by Miller (1959), in which the DNS is reduced to 3-amino, 5-nitrosalicylic acid with the simultaneous oxidation of the free carbonyl group present in reducing sugars.

The concentration of reducing sugars was determined reading the absorbance at $\lambda = 500$ nm using a UV–Vis spectrophotometer (Shimadzu, UV-1700 Pharma Spec; Shimadzu, V260) (Fig. 57.4) and a calibration curve previously prepared.

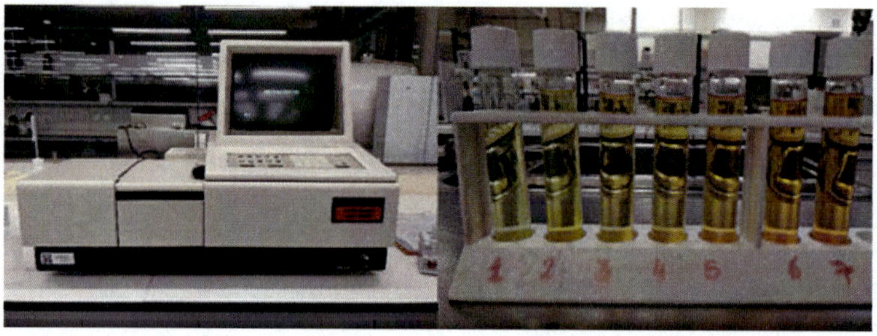

**Fig. 57.4** UV–Vis spectrophotometer (left) and the standards used to prepare the calibration curve (right)

## 57.3 Results and Discussion

### 57.3.1 *Identification of the Most Promising Acids for Biomass Pretreatment*

For the samples collected from the supernatant of maize stover from non-contaminated soil, after the pre-treatment and enzymatic hydrolysis, the amount of sugar released was evaluated using the DNS method, as previously described. Figure 57.5a shows the total reducing sugars concentration for each combination acid + enzyme, which allowed to identify the combinations that were the most effective in the conversion of polysaccharides to simple sugars. Figure 57.5a shows that the liquor produced from the pretreatment with acetic acid, is the one that has the lowest concentration of sugars for both enzymes, which shows that it is the least effective in pretreatment and therefore promotes lower conversion of sugars. This conclusion is consistent with the fact that acetic acid is the weakest one, and therefore was not able to effectively break the lignin molecules. These results are consistent with those obtained by Gundupalli and Bhattacharyya (2023) who tested the effects of sulphuric, hydrochloric and nitric acids on coconut coir, and concluded that nitric acid yielded the highest reducing sugar amount.

### 57.3.2 *Optimization of Pretreatment and Enzymatic Hydrolysis Conditions*

After analyzing the results obtained in the preliminary tests (Fig. 57.5a) and verifying the most effective combinations of acid and enzyme, the same steps described above were performed applying the two most effective combinations to biomass collected from all of the three soils. The tests applying the best operating conditions to biomass from non-contaminated soil were performed again to serve as a

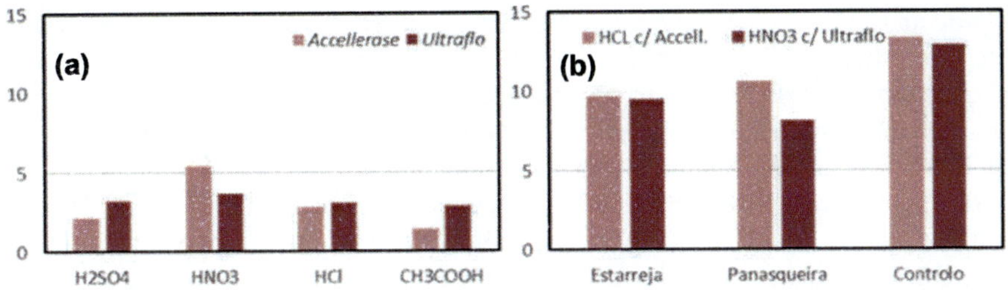


**Fig. 57.5** Reducing sugar concentration **a** after pretreatment followed by enzymatic hydrolysis of biomass from control soil, in the preliminary experiments; **b** using the two best combinations of acid pretreatment followed by enzymatic hydrolysis applied to all three biomass samples

control, simultaneously to the other maize stover samples from the phytoremediation of the contaminated soils from Estarreja and Panasqueira. The results of total reducing sugar concentration obtained in these experiments are shown in Fig. 57.5b. The results obtained for both biomasses under the same processing conditions (Fig. 57.5b), show the presence of different concentrations of contaminants in the soil can influence the production of reducing sugars, probably because the presence of Zn and Cd in the soil affects the composition of the plant parts that were used in the study (Paulo et al. 2022).

### 57.3.3 *Results After Fermentation*

Samples were collected at two different stages of fermentation (6th day and end), and the total reducing sugars concentration were read using the spectrophotometer. Figure 57.6 shows the results obtained for total reducing sugars concentration during the different stages of ethanol production.

Through the analysis of the total reducing sugar concentration along the fermentation process (Fig. 57.6), it was concluded that at all stages, the pretreatment with $HNO_3$ and the enzymatic hydrolysis with *Ultraflo* provided a rather constant amount

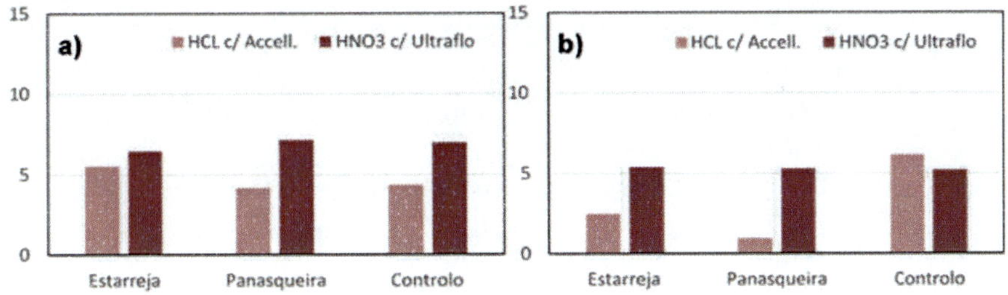


**Fig. 57.6** Concentration of total reducing sugars **a** on the 6th day of fermentation; **b** at the end of fermentation

of reducing sugars. For the treatment with HCl and *Accellerase* 1000, it is concluded that there was a high consumption of the sugars present in the liquor, which may indicate that this combination contributes to higher consumption of glucose by the yeasts in the fermentation process, probably due to the presence of lower amount of inhibitors that resulted from the $HNO_3$ pretreatment.

It was also concluded that the final concentration of reducing sugars for the biomass from control soil is higher. It was hypothesized that: (1) the contaminants accumulated in the biomass during phytoremediation may have a positive effect on the conversion of sugars, or (2) the contaminants present in soil affected the biomass composition, contributing to increase the contents in glucose of the corresponding biomass. The second hypothesis was further confirmed by the compositional analysis of the three biomasses (Almeida et al. 2022), that showed that the cellulose content was 27.1, 32.5 and 24.0%, and the hemicellulose content was 32.0, 31.3 and 23.5%, respectively, for the biomasses from control, Panasqueira and Estarreja soils. For the same biomasses, the lignin content was 20.9, 18.6 and 23.6%, and the ash content was 13.9, 9.0 and 21.8%, respectively.

Table 57.1 summarizes the results of the sugar and ethanol concentration and recovery obtained for the biomass from contaminated and control soil.

Biomass from the control soil yielded the highest amount of total sugars, of 13.4 g/L when the pretreatment was performed using HCl followed by an enzymatic hydrolysis with *Accellerase* 1000. However, the yield in ethanol was also lower for this biomass, probably due to the fact that it contained more hemicellulose that release C5 sugars, that are non-fermentable by *Saccharomyces cerevisiae*. The biomass from

**Table 57.1** Summary of total reducing sugar and ethanol concentration for biomass from contaminated and control soil

| Soil | Pretreatment + enzyme | Reducing sugar concentration after hydrolysis (g/l) | Reducing sugar concentration remaining after fermentation (g/l) | Volume of ethanol solution recovered (ml) | Volume of ethanol (ml) |
|---|---|---|---|---|---|
| Estarreja | HCl + *Accellerase* | 9.74 | 2.47 | 338.75 | 8.05 |
| | $HNO_3$ + *Ultraflo* | 9.47 | 5.40 | 288.22 | 5.13 |
| Panasqueira | HCl + *Accellerase* | 10.61 | 0.94 | 330.50 | 6.22 |
| | $HNO_3$ + *Ultraflo* | 8.15 | 5.30 | 253.04 | 4.25 |
| Control | HCl + *Accellerase* | 13.38 | 6.17 | 355.24 | 7.02 |
| | $HNO_3$ + *Ultraflo* | 12.91 | 5.20 | 293.00 | 5.79 |

Estarreja soil, which is a Petrochemical site, yielded the highest amount of ethanol, probably because the biggest fraction of released sugars was glucose.

Hydrochloric acid pretreatment followed by enzymatic hydrolysis with *Accellerase* yielded the best results in our study. Consequently, no acetic acid and no sulfuric acid were yielding the best results (data not shown). The pretreatment with hydrochloric acid followed by enzymatic hydrolysis with *Accellerase* and pretreatment with nitric acid followed by enzymatic hydrolysis with *Ultraflo* yielded the best results in ethanol production from maize stover, compared with the pretreatment with any of the sulfuric or acetic acids. Sulfuric acid is a strong acid that probably originates fermentation inhibitory compounds, while acetic acid is a week acid that does not sufficiently damage the lignin molecules, thus impeding further polysaccharides hydrolysis.

Maize stover from contaminated soils yielded more ethanol while pre-treated with hydrochloric acid followed by hydrolysis with *Accellerase*—and it was 32–36% higher—than when the pre-treatment was performed with $HNO_3$ and the enzymatic hydrolysis with *Ultraflo*. On the other hand, maize stover from contaminated soils yielded 12% higher and 11% lower ethanol concentration, as compared to the control soil, for the same conditions.

In this work, the maize stover pretreated with $HNO_3$ followed by enzymatic hydrolysis with *Ultraflo* yielded on average 290, 210 and 255 L ethanol per 1 ton of dry biomass, respectively for the control, Panasqueira and Estarreja soil. Interestingly, when HCl was used in the pretreatment and *Accellerase* in the enzymatic hydrolysis, the yield was significantly higher, namely 350, 310 and 405 L ethanol per 1 ton of dry-biomass, respectively for the control, Panasqueira and Estarreja soil.

Table 57.2 lists the ethanol yields expected for various biomass feedstock.

Comparing the data in Table 57.2 with the estimated average yield of the biomass in the current work, it is observed that under the best conditions tested, the ethanol

**Table 57.2** Biomass and its potential ethanol yield

| Biomass feedstock | Potential ethanol yield (L/ton dry biomass) | References |
|---|---|---|
| Bagasse | 437 | Yusoff et al. (2015) |
| Barley | 410 | Aikawa et al. (2013) |
| Corn grain | 470 | Yusoff et al. (2015) |
| Corn stover | 428 | Yusoff et al. (2015) |
| Cotton gin trash | 215 | Yusoff et al. (2015) |
| Forest thinning | 309 | Yusoff et al. (2015) |
| Hardwood sawdust | 382 | Yusoff et al. (2015) |
| Mixed paper | 440 | Yusoff et al. (2015) |
| Oat | 410 | Aikawa et al. (2013) |
| Rice straw | 416 | Yusoff et al. (2015) |
| Sweet sorgum | 80 | Towler et al. (2004) |
| Wheat | 430 | Aikawa et al. (2013) |

yields obtained are quite close to those found in the literature, particularly for the biomass grown in Estarreja soil. Thus, it is concluded that the conversion of maize stover from contaminated soil phytoremediation is of potential interest for bioethanol production.

Concerning the potential contamination with the target heavy metals of the bioethanol produced in the current work, no zinc or cadmium was detected in none of the samples of ethanol recovered from our assays, making it suitable for biofuel or for sanitary uses.

Although the ethanol yield achieved is interesting, it could be increased if a pentose fermenting yeast had been used.

## 57.4 Conclusion

In this work, maize stover from plants used in the phytoremediation of heavy metal contaminated soils was used as a lignocellulosic biomass feedstock for bioethanol production. The biomass was submitted to acid pretreatment, enzymatic hydrolysis followed by fermentation with Saccharomyces cerevisiae yeast. To conclude, the yield in ethanol using hydrochloric acid and *Accellerase* 1000 or nitric acid and *Ultraflo* was of 7.0 ml and 5.8 ml ethanol per 20 g biomass. Based on the results obtained, the yield in alcohol for maize stover produced in both contaminated soils was 0.5, 0.39 g ethanol per 1 g of maize stover grown in the Estarreja's soil, and 0.32 and 0.27 g ethanol per 1 g of maize stover grown in the Panasqueira's soil, while for the maize stover produced in non-contaminated soil it was 0.44 and 0.37 g ethanol per 1 g of maize stover when hydrochloric acid and nitric acids were used, respectively. It would probably be recommended to use another yeast that could also ferment the C5 sugars to yield an higher amount of ethanol, particularly when converting the maize stover from the control and the Estarreja soils, as the remaining unfermented sugar concentration was still high at the end of the fermentation. Therefore, it can be concluded that ethanol production from biomass of maize plants from phytoremediation of heavy metal contaminated soils is technically feasible, allowing to further valorise an underexploited feedstock for bioethanol production.

**Funding** This work was financially supported by: LA/P/0045/2020 (ALiCE) and UIDB/00511/2020—UIDP/00511/2020 (LEPABE) funded by national funds through FCT/MCTES (PIDDAC), and UIDB/04730/2020—Center for Innovation in Engineering and Industrial Technology (CIETI), funded by national funds through FCT/MCTES (PIDDAC); project PhytoEnergy PTDC/BTA–BTA/28761/2017—POCI–01–0145–FEDER–028761 funded by FEDER funds through COMPETE2020—Programa Operacional Competitividade e Internacionalização (POCI) and by national funds through FCT/MCTES (PIDDAC); scientific collaboration under the FCT project UID/Multi/50016/2013.

## References


Aikawa S, Joseph A, Yamada R et al (2013) Direct conversion of *Spirulina* to ethanol without pretreatment or enzymatic hydrolysis processes. Energy Environ Sci 6:1844–1849. https://doi.org/10.1039/c3ee40305j

Almeida F, Júnior W, Marques AP, Caetano NS (2022) Optimization of pretreatment and enzymatic hydrolysis steps towards valorization of maize stover from phytoremediation of heavy metals contaminated soil for bioethanol production (submitted)

BP (2019) BP Energy Outlook 2019—China, p 73

Brethauer S, Shahab RL, Studer MH (2020) Impacts of biofilms on the conversion of cellulose. Appl Microbiol Biotechnol 104:5201–5212. https://doi.org/10.1007/s00253-020-10595-y

Carpita NC, McCann MC (2020) Redesigning plant cell walls for the biomass-based bioeconomy. J Biol Chem 295:15144–15157. https://doi.org/10.1074/jbc.REV120.014561

Coutinho HD, Barbosa AR (2007) Phytoremediation: general topics and methods for utilization. Silva Lusit 15:103–117 (in Portuguese)

Da Costa RMF, Pattathil S, Avci U et al (2019) Desirable plant cell wall traits for higher-quality *Miscanthus* lignocellulosic biomass. Biotechnol Biofuels 12:1–18. https://doi.org/10.1186/s13068-019-1426-7

Da Silva TJ, Hansted FAS, Tonello PS, Goveia D (2019) Phytoremediation of soils contaminatedwith metals: current outlook and prospects of use of forest species. Rev Virtual Quim 11:18–34 (in Portuguese)

Hong CO, Gutierrez J, Yun SW et al (2009) Heavy metal contamination of arable soil and corn plant in the vicinity of a Zinc smelting factory and stabilization by liming. Arch Environ Contam Toxicol 56:190–200. https://doi.org/10.1007/s00244-008-9195-5

Jiang C, Wu QT, Sterckeman T, et al (2010) Co-planting can phytoextract similar amounts of cadmium and zinc to mono-cropping from contaminated soils. Ecol Eng 36:391–395. https://doi.org/10.1016/j.ecoleng.2009.11.005

Marques APGC, Caetano NSC, Castro PML (2020) Chapter 12. Strategies for enhancing soil phytoremediation and biomass valorization. In: The Handbook of Environmental Remediation. Royal Society of Chemistry, pp 331–356

Mata TM, Rodrigues S, Caetano NS, Martins AA (2022) Life cycle assessment of bioethanol from corn stover from soil phytoremediation. Energy Rep 8:468–474. https://doi.org/10.1016/j.egyr.2022.01.059

Miller GL (1959) Use of dinitrosalicylic acid reagent for determination of reducing sugar. Anal Chem 31:426–428. https://doi.org/10.1021/ac60147a030

Oliveira GM, Caetano N, Mata TM, Martins AA (2020) Biofixation of $CO_2$ emissions from natural gas combined cycle power plant. Energy Rep 6:140–146. https://doi.org/10.1016/j.egyr.2019.08.032

Paulo A, Caetano NS, Castro PML, Marques APGC (2022) Phytomanagement of Zn and Cd contaminated soil: *Helianthus annuus* biomass production and metal remediation abilities with plant growth promoting microbiota assistance. Authorea. https://www.authorea.com/doi/full/https://doi.org/10.22541/au.168130633.32498772/v1

Paulraj Gundupalli M, Bhattacharyya D (2023) Effect of different mineral acids on coconut coir for recovery of reducing sugar: process optimization using response surface methodology (RSM). Mater Today Proc 80:2260–2267. https://doi.org/10.1016/J.MATPR.2021.06.225

REN21 (2018) REN21. Renewables 2018-Global Status Report, Paris, REN21 Secretariate

Soudek P, Petrová Š, Benešová D, Vaněk T (2010) Phytoextraction of toxic metals by sunflower and corn plants. J Food Agric Environ 8:383–390

Towler GP, Oroskar AR, Smith SE (2004) Development of a sustainable liquid fuels infrastructure based on biomass. Environ Prog 23:334–341. https://doi.org/10.1002/EP.10052

Yusoff MNAM, Zulkifli NWM, Masum BM, Masjuki HH (2015) Feasibility of bioethanol and biobutanol as transportation fuel in spark-ignition engine: a review. RSC Adv 5:100184–100211. https://doi.org/10.1039/C5RA12735A

# Chapter 58
# Bioelectricity from the Yeast *Candida boidinii*

**S. Rojas-Flores, M. De La Cruz-Noriega, R. Nazario-Naveda, Santiago M. Benites, D. Delfín-Narciso, Cecilia V. Romero, and F. Diaz**

**Abstract** Due to the great changes that the environment is undergoing due to the excessive use of fuels to obtain electrical energy, in the search for new ways to generate this type of energy, this research gives in a new way the use of *Candida boidinii* yeast as fuel for the generation of electricity through MFC at laboratory scale. Managing to generate maximum values of electric current and voltage of 3.3904 ± 0.0916 mA and 0.70342 ± 0.04726 V, whose values were obtained at a pH of ~6.5848 ± 0.2174 and with a value of 63.701 ± 0.911 mS/cm of electrical conductivity. The cells managed to generate a maximum power density of 7.839 ± 0.159 W/cm$^2$ whose current density was 449.657 mA/cm$^2$. Finally, the cells were connected in series, managing to turn on a green LED, thus demonstrating its effectiveness in the generation of bioelectricity.

**Keywords** *Candida boidinii* · Current · Generation · MFC-SCs · Voltage · Yeast

## 58.1 Introduction

Demand for energy to meet the needs of society to develop multiple development activities in various sectors, this has implied a negative environmental impact of great magnitude since the great part of this energy is produced from fossil fuels

S. Rojas-Flores (✉) · M. De La Cruz-Noriega · R. Nazario-Naveda · S. M. Benites
Vicerrectorado de Investigación, Universidad Autónoma del Perú, Lima 15842, Peru
e-mail: segundo.rojas.89@gmail.com

D. Delfín-Narciso
Grupo de Investigación en Ciencias Aplicadas Y Nuevas Tecnologías, Universidad Privada del Norte, Trujillo 13007, Peru

C. V. Romero
Facultad de Medicina, Universidad Nacional de Trujillo, Trujillo 13001, Peru

F. Diaz
Universidad Tecnológica del Perú, Lima, Peru



N. S. Caetano and M. C. Felgueiras (eds.), *The 9th International Conference on Energy and Environment Research*, Environmental Science and Engineering,
https://doi.org/10.1007/978-3-031-43559-1_58

which generate high emissions of greenhouse gases and/or toxic particles, producing climate change effects (Hanif 2018; Shindell and Smith 2019). On the other hand, the limited availability of fossil fuel reserves in the future is considered, since these are not renewable (Wang et al. 2020). In response to this type of environmental problem, various technologies have emerged, among them, one of the most innovative are MFC-SCs (MFCs), which have great application potential to solve both the energy crisis and environmental pollution (Hoang et al. 2022). These bioelectrochemical systems can generate electricity from the interaction of microorganisms, by the oxidation of organic matter such as wastewater, organic waste, among other substrates, as an alternative strategy for the production of clean energy (Leung et al. 2020). Regarding their design, MFCs are commonly composed for two chambers (anodic and cathodic) which are separated by proton exchange membranes (almost always) (Walter et al. 2022). Within these systems, various substrates are usually used, such as acetate and glucose, wastewater from distilleries based on molasses, agricultural residues, among others, rich in the organic matter (Pant et al. 2010).

Another factor of vital importance in the performance of the MFC is the microorganisms since they carry out the transport of electrons in the processing of various substrates, so the literature on this subject has shown the findings and isolations of microorganisms, including bacteria and yeasts, the latter being an ideal biocatalyst for MFC applications, since most of the strains are not pathogenic, can metabolize a wide range of substrates and are easily manipulated (Sayed and Abdelkareem 2017). One of the most used species is the yeast *Saccharomyces cerevisiae, Hansenula anomala*, as well as species of *Candida*, among others (Sayed and Abdelkareem 2017), such as the yeast *Candida boidinni*, which belongs to the fungal division Ascomycota, class Saccharomycetes, order Saccharomycetales; This microorganism is widespread in various environments, so it has been isolated from wine fermentation, olive production, among other substances, and from natural environments such as soil, seawater, and sap flows from many trees (Camiolo et al. 2017). Likewise, this non-infectious asporogenic yeast has also been isolated from plant surfaces (Oku et al. 2014). Linked to the above, the species of *Candida* sp. are closely associated with various diseases present in humans, around the world this fungus is an important cause of morbidity and mortality (Rodrigues et al. 2019). Within this context, it is worth mentioning that the genus *Candida* is methylotrophic yeasts, microorganisms that have particular characteristics such as the ability to use methanol as the only source of carbon and energy, so these yeasts oxidize methanol with an alcohol oxidase (AO) using molecular oxygen as electron acceptor (Yurimoto and Sakai 2019). In this regard, a study reported the identification of the yeast *Candida boidinni* present in the MFC-SCs used in said research for the generation of electricity from blueberry waste (Rojas-Flores et al. 2021).

Likewise, in another investigation the elaboration and application of a metabolic engineering strategy were studied to design and direct the metabolic flux for the improvement of the regeneration of NADH (Nicotinamide adenine dinucleotide) (Feng et al. 2018), a coenzyme that serves as the intracellular reducer. In this way, the increase in NADH regeneration can lead to an increase in the efficiency of intracellular electron transfer (EET), which within bioelectrochemical systems such as

MFC will have a favorable influence since it facilitates the degradation of substrates and energy generation by microorganisms (Segundo et al. 2022). In this sense, in said research they studied 16 genes, of which 4 were assembled, these caused an increase in the intracellular reserve of NADH, leading to an increase in the maximum power in the MFC, it is worth mentioning that within all of them an exogenous gene was identified of the species *Candida boidinii*, which increased the generation of NADH (Flores et al. 2022). Based on this, this *Candida boidinii* species has a clear biotechnological potential for applications such as biomedical use, since it can release butyric acid, a compound relevant to human health. In addition to this, this yeast has strains that show a phenotypic and genetic variation, as well as important metabolic characteristics for its use in food industries (Corbu et al. 2019). Likewise, it is worth mentioning that the MFC by themselves have various applications, among which stand out for the treatment of wastewater, environmental bioremediation, the development of biosensors, the production of biohydrogen, the microbial capture of carbon and the production of energy, among others. Others (Prathiba et al. 2022).

For this reason, the main objective of this research is to evaluate the generation of electricity through the use of MFC-SCs on a laboratory scale, using the yeast *Candida boidinii* as a substrate (fuel) in a minimum salt medium, monitoring the values of electric current, voltage, pH, electrical conductivity, current density and power density, for 15 days. Giving in this way a new and innovative way of generating electric current using this yeast.

## 58.2 Materials and Methods

### *58.2.1 Single-Chamber MFC-SC*

MFC-SCs were manufactured (03) on a laboratory scale, in polyethylene containers, to which a 10 $cm^2$ hole was made at one end with an area of ~144 $cm^2$ to place the cathodic electrode (Copper, Cu), and the anodic electrode (Zinc, Zn); which were placed inside the cell. Both electrodes were linked by an external circuit with an internal resistance of 100 Ω.

### *58.2.2 Reactivation of* Candida boidinii *Yeast*

A molecularly identified yeast strain isolated from a cranberry juice sample by Rojas-Flores et al. (2021) was used. The regions sequenced and analyzed in the BLAST program obtained an identity percentage of 99.86%, which corresponds to the species *Candida boidinii* (Table 58.1). The strain was reactivated in sterile 10% peptone water for 30 min at 30 °C. Then it was planted in tubes with Sabouraud Agar inclined at 30 °C for 24 h to obtain axenic cultures.

**Table 58.1** rDNA sequence of yeast isolated from the MFC anode electrode

| BLAST characterization | Consensus sequence length (nt) | % coverage | % maximum identity | Accession number | Phylogeny Filogenia |
|---|---|---|---|---|---|
| *Candida boidinii* | 713 | 96% | 99.86% | KY101980.1 | Cell organism; eukaryota; Opisthokonta; fungi; dikarya; Ascomy bound; Saccharomyceta; Saccharomycotina; Saccharomycetes; Saccharom ycetales; Pichiaceae; Og ataea; Ogataea/*Candida clade* |

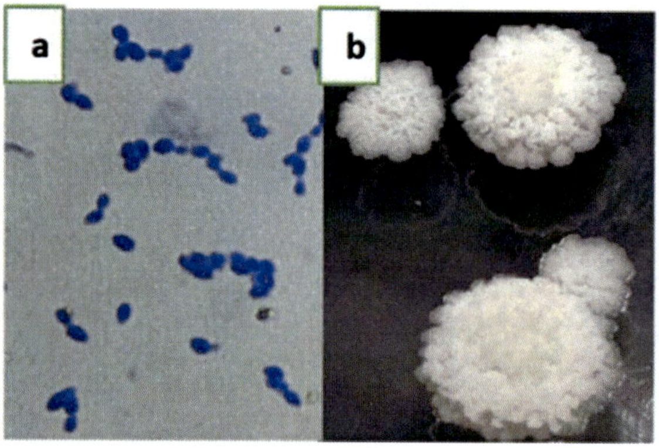


**Fig. 58.1** **a** Microscopy of *Candida boidinii* by Gram stain (100×) and **b** Macroscopic view of *Candida boidinii* colonies in Sabouraud Agar using a dark background

## 58.2.3 *Macroscopic and Microscopic Observation of* Candida boidinii

A Gram stain was performed to observe the microscopic characteristics: cylindrical oval cells in the process of budding (see Fig. 58.1a), macroscopic white, opaque colonies, irregular edges, with pulpinous elevation, curly margin typical of *Candida boidinii* (see Fig. 58.1b) (Cuervo Mulet et al. 2015).

## 58.2.4 *Preparation of the Inoculum for the MFC-SC*

A pure culture of *Candida boidinii* activated for 24 h was suspended in 20 ml of sterile physiological saline solution (SSFE) of 0.85%, adjusting the turbidity with tube No. 3 of Mc Farland ($9 \times 10^8$ CFU/ml) incorporating 80 ml of minimal mineral medium buffered with phosphate buffer. The process was carried out at room temperature (20 ± 2 °C). It was performed in triplicate (Kabir et al. 2018).

## 58.3 Results and Discussion

In Fig. 58.2a, the electric potential values of the MFC-SC are observed during the 15 days of monitoring. The cells managed to create a voltage from the first day (0.371 ± 0.002 V) to its maximum peak on the tenth (0.70342 ± 0.04726 V) day of monitoring and then decay until the 15 day (0.628 ± 0.0475 V The values obtained are being influenced by the microbiota present on the surface of the electrodes, which create a potential difference, according to Miran et al. (2016) has observed that the thickness of the particles are of vital importance for the generation of electrical potential because they can retard the hydrolysis processes that are generated inside the cells, limiting the generation of voltage (Miran et al. 2016). In Fig. 58.2, can be seen the electric current values obtained during the monitoring period, achieving maximum values of 3.3904 ± 0.0916 mA on the twelfth day; these values decreased slightly in the fifteenth day (3.0599 ± 0.0844 mA). The electrical current values obtained are possibly due to the activity of microorganisms that perform the function of fermenting, converting the substrate into fuel, for example, glucose into organic acids (small chain), which cause electrical energy (Takahashi et al. 2016; Lovley and Nevin 2008).

In Fig. 58.3a the monitored pH values are shown, being able to observe values from slightly acidic (on the first days, 3.75) to slightly basic (on the last day, 7.99 ± 0.28); being its optimum pH of 6.5848 ± 0.2174 according to its current and voltage values generated. It has been reported that the groups of microorganisms are diverse, that they can grow in the substrate at different environmental conditions; but the pH variations are capable of affecting the ability of these to generate electrical energy, all this because there is an optimal operating pH for MFC and each substrate used (Puig

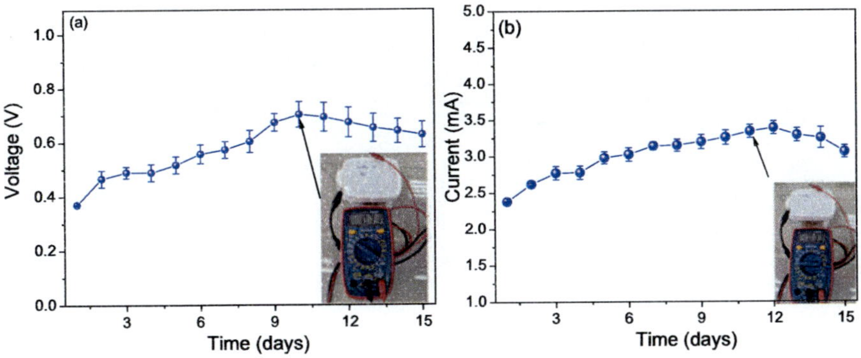


**Fig. 58.2** **a** Voltage data and **b** Current data of the MFC-SCs

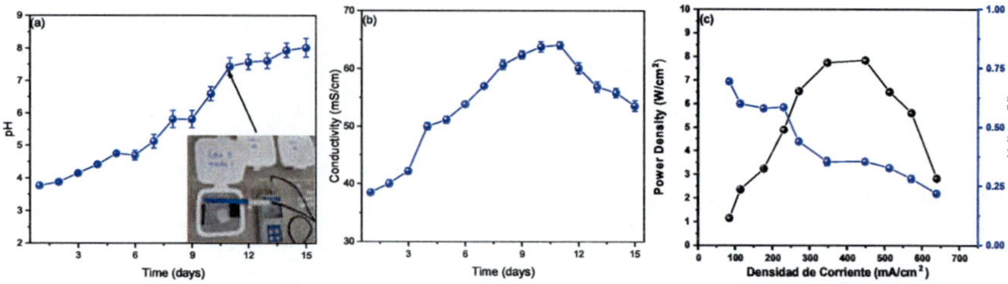


**Fig. 58.3** **a** pH data, **b** conductivity data and **c** data of power density vs current density and voltage

et al. 2010). The values of electrical conductivity have been reported in Fig. 58.3b, where the values increase from the first day (38.446 mS/cm) of operation to the tenth day (63.701 $\pm$ 0.911 mS/cm) and then decrease slightly until the 15 day (53.480 $\pm$ 0.921 mS/cm).

In Fig. 58.3c be able to observe the data of the density of power (PD) in function of the density of corriente (DC), whete the $DP_{MAX.}$ shown was 7.839 $\pm$ 0.159 W/cm$^2$ at a DC of 449.657 mA/cm$^2$ and a maximum voltage of 0.692 $\pm$ 0.0598 V. However, the DP values of *Candida boidinii* yeast are higher compared to the data reported in the literature where other similar substrates were used. Zhang et al. (2016) used sewage sludge as fuel (substrate) and carbon felt as anode electrode, managing to generate $DP_{max}$ and DC peaks of 20.4 mW/cm$^2$ and 25.86 mA/cm$^2$, due to adjusting the pH to 8 (Zhang et al. 2016). Similarly, Jiang et al. (2016) reported an 80% increase in its PD, that is, from 0.36 W/m$^2$ at pH 6 to 0.66 W/m$^2$ for pH 9.5 (Zhang et al. 2016). In Fig. 58.4 the bioelectricity generation process can be observed using the yeast *Candida boidinii* as fuel in the MFC-SC, managing to generate a peak of 2.57 V connecting the three MFCs in series; managing to turn on a green led.

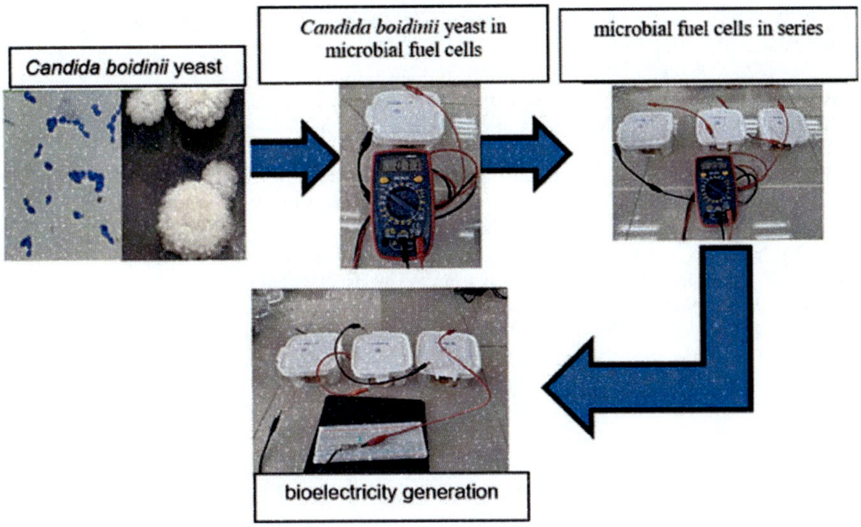


**Fig. 58.4** Schematization of the bioelectricity generation process through MFC-SCs

## 58.4 Conclusion

Bioelectricity was successfully obtained through MFC-SC at laboratory scale using yeast *Candida boidinii* as fuel. Maximum values of voltage and current of 0.70342 $\pm$ 0.04726 V and 3.3904 $\pm$ 0.0916 mA were shown, respectively, which were generated at an optimum pH of 6.5848 $\pm$ 0.2174 on the tenth day. While the substrate showed electrical conductivity values of 63.701 $\pm$ 0.911 mS/cm on the tenth day. While the highest observed power density values were 7.839 $\pm$ 0.159 W/cm$^2$ whose current density was 449.657 mA/cm$^2$. Finally, the MFC-SC with the substrate inside (*Candida boidinii* yeast) were connected in series, generating 2.57 V, turning on a green LED. In this way, a novel and ecological use is given to this type of yeast by observing its electrical current that generates energy.

## References

Camiolo S, Porru C, Benítez-Cabello A, Rodríguez-Gómez F, Calero-Delgado B, Porceddu A, Budroni M, Mannazzu I, Jiménez-Díaz R, Arroyo-López FN (2017) Genome overview of eight Candida boidinii strains isolated from human activities and wild environments. Stand Genomic Sci. https://doi.org/10.1186/s40793-017-0281-z

Corbu VM, Vassu T, Istim A, Csutak O (2019) Biodiversity and biotechnology of *Candida boidinii* industrial strains. J Biotechnol 305:S81. https://doi.org/10.1016/j.jbiotec.2019.05.280

Cuervo Mulet RA, Argote Vega FE, Fernández Daza FF, Osorio Cadavid E (2015) Hongos levaduriformes aislados de frutas y de suelo de la región del Valle del Cauca, Colombia.

Feng L, Yuanxiu L, Liming S, Xiaoli C, Xingjuan A, Changji Y, Yingxiu C, Hui W, Hao S (2018) Modular engineering intracellular NADH regeneration boosts extracellular electron transfer of *Shewanella oneidensis* MR-1. ACS Synth Biol. https://doi.org/10.1021/acssynbio.7b00390

Flores SR, Pérez-Delgado O, Naveda-Renny N, Benites SM, De La Cruz-Noriega M, Narciso DA (2022) Generation of bioelectricity using molasses as fuel in MFC-SCs. Environ Res Eng Manag 78(2):19–27

Hanif I (2018) Impact of fossil fuels energy consumption, energy policies, and urban sprawl on carbon emissions in East Asia and the Pacific: a panel investigation. Energ Strat Rev 21:16–24. https://doi.org/10.1016/j.esr.2018.04.006

Hoang AT, Nižetić S, Ng KH, Papadopoulos AM, Le AT, Kumar S, Hadiyanto H, Pham VV (2022) MFC-SCs for bioelectricity production from waste as sustainable prospect of future energy sector. Chemosphere 287:132285. https://doi.org/10.1016/j.chemosphere.2021.132285

Kabir MM, Fakhruddin A, Chowdhury M, Pramanik MK, Fardous Z (2018) Isolation and characterization of chromium (VI)-reducing bacteria from tannery effluents and solid wastes. World J Microbiol Biotechnol 34(9):126. https://doi.org/10.1007/s11274-018-2510-z

Leung DHL, Lim YS, Uma K, Pan G-T, Lin J-H, Chong S, Yang TC-K (2020) Engineering *S. oneidensis* for performance improvement of MFC-SC—a mini review. *Appl Biochem Biotechnol* 193(4):1170–1186. https://doi.org/10.1007/s12010-020-03469-6

Lovley DR, Nevin KP (2008) Electricity production with electricigens. In: Bioenergy, pp 295–306. American Society of Microbiology

Miran W, Nawaz M, Jang J, Lee DS (2016) Conversion of orange peel waste biomass to bioelectricity using a mediator-less MFC-SC. Sci Total Environ 547:197–205

Oku M, Takano Y, Sakai Y (2014) The emerging role of autophagy in peroxisome dynamics and lipid metabolism of phyllosphere microorganisms. Front Plant Sci. https://doi.org/10.3389/fpls.2014.00081

Pant D, Van Bogaert G, Diels L, Vanbroekhoven K (2010) A review of the substrates used in MFC-SCs (MFC-SC) for sustainable energy production. Biores Technol 101(6):1533–1543. https://doi.org/10.1016/j.biortech.2009.10.017

Prathiba S, Kumar PS, Vo D-VN (2022) Recent advancements in MFC-SCs: a review on its electron transfer mechanisms, microbial community, types of substrates and design for bio-electrochemical treatment. Chemosphere 286:131856. https://doi.org/10.1016/j.chemosphere.2021.131856

Puig S, Serra M, Coma M, Cabré M, Balaguer MD, Colprim J (2010) Effect of pH on nutrient dynamics and electricity production using MFC-SCs. Biores Technol 101(24):9594–9599

Rodrigues C, Rodrigues M, Henriques M (2019) *Candida* sp. infections in patients with diabetes mellitus. J Clin Med 8(1):76. https://doi.org/10.3390/jcm8010076

Rojas-Flores S, Benites SM, La Cruz-Noriega D, Cabanillas-Chirinos L, Valdiviezo-Dominguez F, Quezada Álvarez MA, et al (2021) Bioelectricity production from blueberry waste. *Processes* 9(8):1301. https://doi.org/10.3390/pr9081301

Sayed ET, Abdelkareem MA (2017) Yeast as a biocatalyst in MFC-SC. Old Yeasts New Quest. https://doi.org/10.5772/intechopen.70402

Segundo RF, La Cruz-Noriega D, Nazario-Naveda R, Benites SM, Delfín-Narciso D, Angelats-Silva L, Díaz F (2022) Golden berry waste for electricity generation. Fermentation 8(6):256

Shindell D, Smith CJ (2019) Climate and air-quality benefits of a realistic phase-out of fossil fuels. Nature 573(7774):408–411. https://doi.org/10.1038/s41586-019-1554-z

Takahashi S, Miyahara M, Kouzuma A, Watanabe K (2016) Electricity generation from rice bran in MFC-SCs. Bioresour Bioprocess 3(1):1–5

Walter XA, Madrid E, Gajda I, Greenman J, Ieropoulos I (2022) MFC-SC scale-up options: performance evaluation of membrane (c-MFC) and membrane-less (s-MFC) systems under different feeding regimes. J Power Sources 520:230875. https://doi.org/10.1016/j.jpowsour.2021.230875

Wang H, Wang G, Qi J, Schandl H, Li Y, Feng C, Yang X, Wang Y, Wang X, Liang S (2020) Scarcity-weighted fossil fuel footprint of China at the provincial level. Appl Energy 258:114081. https://doi.org/10.1016/j.apenergy.2019.114081

Yurimoto H, Sakai Y (2019) Methylotrophic yeasts: current understanding of their C1-metabolism and its regulation by sensing methanol for survival on plant leaves. Curr Issues Mol Biol 33:197–210. https://doi.org/10.21775/cimb.033.197

Zhang C, Liang P, Yang X, Jiang Y, Bian Y, Chen C et al (2016) Binder-free graphene and manganese oxide coated carbon felt anode for high-performance MFC-SC. Biosens Bioelectron 81:32–38

# Chapter 59
# Mixing Effect on Anaerobic Digestion of Wine Vinasse Wastewater for Energy Production

**Andreia D. Santos, João R. Silva, F. A. Nuno, Rosa M. Quinta-Ferreira, and Luís M. Castro**

**Abstract** The wine distillery wastewater, called wine vinasse, is an aqueous by-product of wine distillation containing the non-volatile fraction of the distilled wine and other organic waste material. The wine distillation industry generates a high volume of wastewater (10–20 L per liter of produced ethanol). This wastewater is characterized by its low pH values and high organic matter concentration. This work studies the effect of mechanical mixing of an anaerobic digester reactor to produce biogas from wine vinasse wastewater in batch operation. When a mechanical mixing system at 8 rpm was employed, biogas volume increased by 20.6% and methane volume by 21.9%, evidencing the better performance under continuous mixing of the anaerobic reactor. The batch assay reached a specific methane production of 1.03 NL $CH_4$/g-VS. The wastewater treatment shows a COD removal efficiency of about 90%. The kinetic evaluation confirmed the experimental results as the assay with the mixing system did not describe a lag phase contrary to the system without mixing. The biogas produced during the anaerobic digestion process can be converted into electrical energy or used as an energy source in the distillation process (reducing its environmental impact), with a power of 2.54 kWh/kg-VS and 8.26 kWh/kg-VS, respectively.

**Keywords** Anaerobic digestion · Biochemical methane potential · Circular economy · Energy production · Kinetic study · Wine vinasse wastewater

---

A. D. Santos (✉) · J. R. Silva · R. M. Quinta-Ferreira · L. M. Castro
Department of Chemical Engineering, Faculty of Sciences and Technology, CIEPQPF—Chemical Engineering Processes and Forest Products Research Center, University of Coimbra, Coimbra, Portugal
e-mail: duartesantos@uc.pt

A. D. Santos · J. R. Silva · L. M. Castro
Department of Chemical and Biological Engineering, Polytechnic of Coimbra, Coimbra Institute of Engineering, Rua Pedro Nunes – Quinta da Nora, 3030-199 Coimbra, Portugal

F. A. Nuno
Quinta das Bágeiras - Mário Sérgio Alves Nuno, Rua Principal nº598, Fogueira, 3780-523 Sangalhos, Portugal

 
N. S. Caetano and M. C. Felgueiras (eds.), *The 9th International Conference on Energy and Environment Research*, Environmental Science and Engineering,
https://doi.org/10.1007/978-3-031-43559-1_59

## 59.1 Introduction

World alcohol consumption has been increasing. According to World Health Organization, during the year 2019, 5.8 L of pure alcohol were consumed per person, whereas in 2000, it was only 5.2. The major consumers were Europe, Russia, and Australia (Chowdhary et al. 2017; World Health Organization 2022). In Portugal, the number of alcoholic beverages available per capita between 2016 and 2020 registered an increase of around 14% compared with 2012–2015. The consumption of spiritual drinks is lower than wine consumption. However, the consumption of spiritual beverages is increasing (Instituto Nacional de Estatística 2021). Spirit drinks for human consumption follow specific guidelines and organoleptic properties and contain at least 15% ethyl alcohol. These drinks are produced by fermentation of several raw materials, such as wine, grains, fruits, vegetables, or sugar, followed by distillation to increase the alcoholic content (Wiśniewska et al. 2015; Botelho et al. 2020). *Aguardente* falls in the category of a spiritual drink and is well-known and consumed in Portugal. The beverage is obtained from wine distillation and is usually consumed pure or used in liqueur production.

When discharged into soils or water mediums, distillery wastewater causes severe pollution problems. In soils, it causes inhibition of seed germination and vegetation depletion inhibition by reducing soil alkalinity and manganese availability. Water bodies reduce sunlight penetration, decreasing photosynthetic activity and dissolved oxygen content (Chowdhary et al. 2017). Wine distillery or wine vinasse wastewater (WVW) is an aqueous by-product of wine distillation with a very high organic content resulting from its non-volatile fraction. This industry produces a high volume of wastewater within a range of 10–20 L per liter of ethanol generated (Melamane et al. 2007; Tena et al. 2020). The vinasse wastewater is seasonal, which raises specific problems such as the volume produced and its composition. These wastewaters are acidic, with a pH around 3.5–5, present a dark brown color, and a very high Chemical Oxygen Demand (COD) of 50–150 g $O_2$/L, as well as Biochemical Oxygen Demand ($BOD_5$) of 5.5–20 g $O_2$/L. It is also characterized by high levels of phenolic compounds: 29–474 mg/L (Melamane et al. 2007). These wastewater characteristics demand a treatment adapted to the load and the start-up, the shutting down, and periods of inactivity. Vinasse wastewaters are commonly treated with physicochemical methods capable of color and organic load reduction. However, these methods present high costs and generate a large amount of sludge as secondary pollutants (Chowdhary et al. 2018).

The energy consumed during the alcohol distillation is another important aspect that must be considered. Generating power from the waste produced during the beverage process follows the circular economy strategy. The best waste treatment and utilization are promoted to achieve a minimum waste production, with all residues transformed into fertilizers, fuel, or energy (UNESCO World Water Assessment Programme 2009). In the Circular Economy, Anaerobic digestion (AD) is strongly supported once it provides job opportunities and adequate biowaste treatment (Nattassha et al. 2020). Resource recovery for vinasse wastewater is highly valuable for producing value-added resources such as volatile fatty acids, biohydrogen, and biomethane from AD (Silva et al. 2021). According to Kharayat's study, the waste from distillery industries can produce biogas. However, the anaerobic process is quite vulnerable to the organic load, low pH, and the lower growth rate of the anaerobic microorganism, requiring longer HRT (Kharayat 2012). Borja et al. (1993a) studied the biomethanation of wine vinasse wastewater, founding that it can be easily biodegradable by anaerobic digestion, reaching a chemical oxygen demand (COD) removal of 90%.

This paper aims to optimize energy production, which, according to an economic circularity strategy, can be used to minimize the environmental impact of the distillation process. For this purpose, the mixing effect in biogas production from wine vinasse wastewater was evaluated, performing a Biochemical Methane Potential (BMP) study with and without mixing. Specific Biogas Production (SBP—the volume of biogas produced per mass of Volatile Solids (VS) fed to the reactor) and Specific Methane Production (SMP—the volume of methane produced per mass of VS fed to the reactor) were determined to evaluate the BMP performance.

## 59.2 Materials and Methods

### *59.2.1 Substrate and Inoculum*

The wastewater used in this study was collected from a distillery in the central region of Portugal, preserved at pH = 3.5, and stored in a dry environment until the experimental test was used. The anaerobic sludge came from an anaerobic reactor of a Municipal Wastewater Treatment Plant.

### *59.2.2 Experimental Setup*

The experimental procedure was conducted in the AD Reactor, as specified by Castro et al. (2020), with a working volume of 9.0 L. Two assays were performed: one continuously mixed at 8 rpm and another without mixing. The gas volume was determined and converted to standard temperature and pressure conditions (normal liter—NL). The temperature was set to 37.5 ± 1 °C, and the Substrate Inoculum Ratio (SIR), based on VS, was 0.5. The biogas production and methane content were evaluated daily until gas production ceased.

### 59.2.3 Analytical Methods

Chemical Oxygen Demand (COD), total ($COD_t$) and soluble (CODs) were determined by the standard Closed Reflux Colorimetric Method 5220 D. Total Solids (TS), Volatile Solids (VS), Total Suspended Solids (TSS), and Volatile Suspended Solids (VSS) were calculated using standard gravimetric methods 2540 B, D, and E (APHA, Awwa, WEF 2017). The Methane content was evaluated using the absorption of $CO_2$ in the alkaline liquid method proposed by Abdel-Hadi (2008).

### 59.2.4 Kinetic Study

For the kinetic evaluation of the BMP assay, the following empirical kinetic models, first-order, Gompertz, Logistic, Feller, Cone, Chen-Hashimoto, Weibull, France, and Cauchy, were employed as described by Santos et al. (2022). The selection of the most suitable model was assessed with the determination of the second-order Akaike Information Criterion (AIC) test (Eq. 59.1), along with the Bayesian Information Criterion (BIC) test (Eq. 59.2).

$$AIC = \begin{cases} N \ln\left(\frac{RSS}{N}\right) + 2K, & when\ \frac{K}{N} \geq 40 \\ N \times \ln\left(\frac{RSS}{N}\right) + 2K + \frac{2K(K+1)}{N-K-1}, & when\ \frac{K}{N} < 40 \end{cases} \tag{59.1}$$

$$BIC = N \times \ln\left(\frac{RSS}{N}\right) + K \ln(N) \tag{59.2}$$

where the RSS is the Residual Sum of Squares, N is the number of data points, and K is the number of parameters the model estimates.

### 59.2.5 The Electrical Energy Potential of the WVW

To predict the energetic potential of the WVW through the methane produced during the BMP assay, Eq. 59.3 is employed (Suhartini et al. 2019).

$$Energy\left(\frac{\text{kWh}}{\text{g-VS}}\right) = \frac{calorificvalue_{biogas} \times Volume_{biogas}}{mass_{VS}} \times Efficiency \tag{59.3}$$

where the calorific value of biogas is the lower calorific power (PCI) of biogas (PCI = 50 MJ/kg) (Haynes et al. 2016), assuming an efficiency of conversion to the electricity of 35% in the electrical conversion and 82% heat conversion (Hakawati et al. 2017).

## 59.3 Results and Discussion

### *59.3.1 BMP Performance*

A BMP was performed in both conditions to evaluate the mixing effect in biogas production. Table 59.1 presents the characterization of the Inoculum and Substrate used in the BMP assay.

It was possible to verify that the mixing effect on AD performance influences the volume of biogas and methane produced. Additionally, the time required to reach maximum biogas production is shorter. Figure 59.1 presents the cumulative biogas and methane production without and with mixing. The cumulative biogas production was 46.20 NL resulting in 34.77 NL of methane in the assay without mixing, corresponding to a biogas production with an average of 78 $\pm$ 8% methane content. The mixed BMP produced 58.18 NL and 44.51 NL of biogas and methane, respectively, corresponding to an average methane content of 80 $\pm$ 7%, with an increase of 20.59 and 21.88% in biogas and methane production.

The SBP and the SMP determined the BMP performance for each operational condition. An SBP of 1.27 NL/g-VS and an SMP of 0.95 NL $CH_4$/g-VS were achieved for the assay without mixing. The mixed assay obtained 1.35 NL/g-VS and 1.03 NL $CH_4$/g-VS for SBP and SMP, respectively. Other authors have studied the anaerobic digestion of vinasse. Belhagj et al. (2013) looked at the AD of vinasse from a production line of ethyl alcohol at a lab-scale reactor with magnetic stirring and mesophilic temperatures. The authors reached a SBP of 0.7 NL/g-VS, lower values than the ones achieved in this study (Belhagj et al. 2013). Wolmarans and De Villiers show in their research the high efficiency of Upflow Anaerobic Sludge Blanket (UASB) technology. Besides having reached 90% of COD removal efficiency, treated wastewater with a COD concentration of 30,000 mg $O_2$/L, the authors obtained a biogas production of 0.43 $m^3$/kg-VS, which, once again, is lower than the presented results (Wolmarans and Villiers 2002).

Borja et al. evaluated the AD of red wine vinasse using a lab-scale reactor with magnetic stirring at mesophilic temperature. They reached a methane yield of 0.32 NL $CH_4$/g-$COD_{added}$, higher than the values presented in this study: 0.25 and 0.28 NL $CH_4$/g-$COD_{added}$ (Borja et al. 1993b). However, Borja and co-authors used immobilized microorganisms to improve fluidization, facilitating the degradation of organic matter and enhancing biogas production (Borja et al. 1993b). Likewise, Lalov et al. studied the methane production of the WVW with a similar setup, reaching 0.27 NL $CH_4$/g-$COD_{removed}$ (Lalov et al. 2001). In this study, the higher methane production per gram of removed COD obtained with mixing corresponded to 0.28 NL $CH_4$/g-$COD_{removed}$. Compared to Lalov et al. experiments, this study presents more promising results than the fixed-film reactor used by the authors (Lalov et al. 2001).

From the BMP assay, it is possible to predict the technical digestion time (TDT), which Kafle and Chen presented as the required time to produce 80–90% of the total biogas (Kafle and Chen 2016). That is a critical parameter when changing to a semi or continuous operation. The TDT can be used as Hydraulic Retention Time

**Table 59.1** Characterization of the substrate and the inoculum used in the BMP assay regarding CODt, CODs, TS, VS, TSS, and VSS

| Parameter (g/L) | $COD_t$ | $COD_s$ | TS | VS | TSS | VSS |
|---|---|---|---|---|---|---|
| Inoculum | 40.62 ± 5.79 | 2.278 ± 0.034 | 41.33 ± 0.152 | 20.79 ± 0.148 | 24.63 ± 1.62 | 12.76 ± 0.817 |
| Substrate | 57.19 ± 3.52 | 15.57 ± 0.352 | 21.43 ± 0.890 | 15.36 ± 0.539 | 1.097 ± 0.971 | 1.003 ± 0.886 |

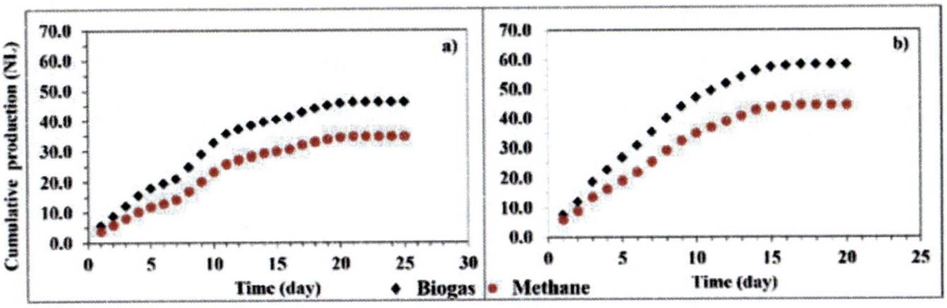


**Fig. 59.1** Cumulative biogas and methane production for **a** no mechanical and **b** with mechanical mixing

(HRT). The TDT was found to be between 13 and 17 days for the unmixed assay, whereas in the stirring process was 11–13 days. The reduction of TDT shows that the mechanical mixing of the AD reactor allows reducing the time required to attain the maximum production of biogas. The stirring provides better contact between the organic matter and the microorganisms. It helps the gas bubbles flow to the liquid surface, promoting a better mass transference from the liquid to the gas phase and increasing AD efficiency. Concerning the wastewater treatment performance, COD removals efficiencies were 91% and 94%, with and without mixing, proving that the WVW is easily biodegradable, as previously stated by Borja et al. (1993b).

### 59.3.2 *Kinetic Study*

The kinetic study was performed in MATLAB® software to predict the kinetic parameters and assess the AIC and BIC criteria with the obtained values presented in Table 59.2.

The lowest value of the AIC and BIC must be considered to select the best model. It has been found that the empirical-kinetic models that better describe cumulative biogas production are the Modified Gompertz (Eq. 59.4) for the assay where there was no mechanical mixing. For the one with mixing, the Weibull (Eq. 59.5) model enabled a more accurate prediction of the parameters.

$$BMP(t) = BMP_0 \times e^{-e^{-\frac{e \times \mu_{max}}{BMP_0}\left(t_{lag} - t\right)+1}} \quad (59.4)$$

$$BMP(t) = BMP_0 \times \left[1 - e^{-\left(k\left(t - t_{lag}\right)\right)^{\gamma}}\right] \quad (59.5)$$

where BMP(t) is the cumulative methane yield (NL $CH_4$/g-VS), $BMP_0$ is the methane potential of the substrate (NL $CH_4$/g-VS), k is the hydrolysis rate constant (1/d), t

**Table 59.2** RSS, AIC and BIC values for the assay without stirring and stirring at 8 rpm

| | Model | RSS | AIC | Δ AIC | Akaike weight | BIC | Δ BIC |
|---|---|---|---|---|---|---|---|
| No stirring | First-order | 0.03627 | – 92.641 | – 18.887 | 3.48E+05 | – 90.647 | – 18.198 |
| | Gompertz | 0.01589 | – 111.53 | 0 | 6.29E+01 | – 108.84 | 0 |
| | Logistic | 0.01835 | – 107.78 | – 3.7461 | 3.55E+02 | – 105.1 | – 3.7461 |
| | Feller | 0.01948 | – 106.32 | – 5.2119 | 6.98E+02 | – 103.63 | – 5.2119 |
| | Cone | 0.03068 | – 94.424 | – 17.104 | 1.69E+05 | – 91.74 | – 17.104 |
| | Chen e Hashimoto | 0.04315 | – 85.553 | – 25.975 | 1.01E+07 | – 82.87 | – 25.975 |
| | Weibull | 5.8243 | 44.792 | – 156.32 | 1.51E+33 | 47.92 | – 156.76 |
| | France | 0.02975 | – 92.408 | – 19.121 | 4.77E+05 | – 89.28 | – 19.565 |
| | Cauchy | 0.03026 | – 94.785 | – 16.743 | 1.43E+05 | – 92.102 | – 16.743 |
| With stirring | First-order | 0.02070 | – 81.099 | – 12.298 | 2.26E+04 | – 79.677 | – 12.042 |
| | Gompertz | 0.01177 | – 90.22 | – 3.1771 | 4.16E+02 | – 88.498 | – 3.2208 |
| | Logistic | 0.01943 | – 79.687 | – 13.71 | 4.88E+04 | – 77.966 | – 13.753 |
| | Feller | 0.02235 | – 76.743 | – 16.654 | 1.85E+05 | – 75.022 | – 16.697 |
| | Cone | 0.01983 | – 79.261 | – 14.136 | 5.92E+04 | – 77.539 | – 14.18 |
| | Chen e Hashimoto | 0.02743 | – 72.447 | – 20.95 | 1.29E+06 | – 70.725 | – 20.994 |
| | Weibull | 0.00873 | – 93.397 | 0 | 1.15E+02 | – 91.719 | 0 |
| | France | 0.01729 | – 79.046 | – 14.351 | 7.56E+04 | – 77.368 | – 14.351 |
| | Cauchy | 0.01774 | – 81.599 | – 11.798 | 2.06E+04 | – 79.877 | – 11.842 |

is the time (d), $\mu$max the maximum methane production rate (NL $CH_4$/g-VS.d), $t_{lag}$ the lag phase (d), and e is the Neper number.

By employing Eqs. 59.4 and 59.5 to the experimental data in the MATLAB® software, one can predict the kinetic parameters and visualize the fitting of the model for each assay, as presented in Fig. 59.2.

From Fig. 59.2 is possible to verify the accuracy of the fitting between the experimental and the predicted SMP values of each assay.

The kinetic parameters are presented in Table 59.3.

The coefficient of determination $R^2$ = is 0.994 and 0.996, respectively, for the assay without and with mixing. No lag phase or adaptation time was registered compared to the other assay, which predicted a lag phase of 0.3126 d. As the hydrolysis rate constant and the maximum methane production rate are correlated following Eq. 59.6, it is possible to estimate one parameter or another depending on the ones given by the kinetic model (Pererva et al. 2020).

$$k = \frac{\mu_{max}}{BMP_0} \tag{59.6}$$

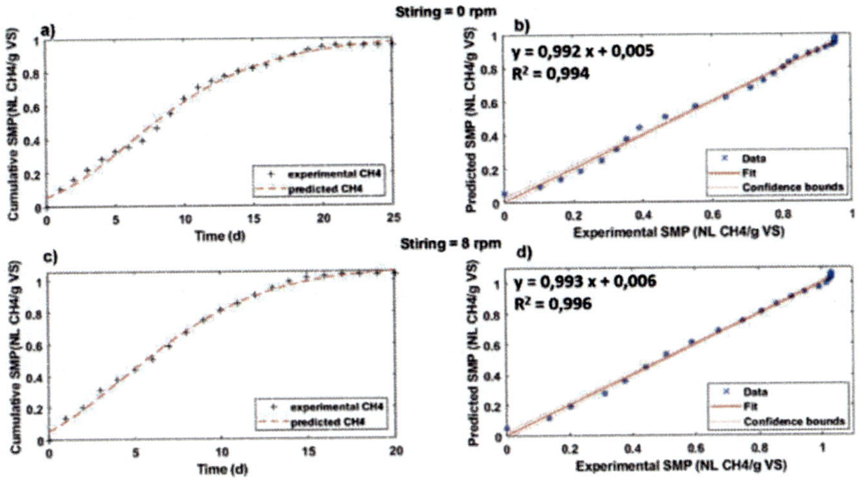


**Fig. 59.2** Fitting of the empirical-kinetic models, **a** modified Gompertz fitting for the assay without mixing, **b** plot of experimental and predicted data for stirring = 0 rpm, **c** Weibull fitting for the mixed assay, **d** plot of experimental and predicted data for stirring = 8 rpm

**Table 59.3** Kinetic parameters $BMP_0$, μmax, k and $t_{lag}$ of the best fitting models for assay without stirring and stirring at 8 rpm

| Parameter | $BMP_0$ (NL $CH_4$/g-VS) | $\mu_{max}$ (NL $CH_4$/g-VS d) | k (1/d) | $t_{lag}$ (d) |
|---|---|---|---|---|
| Modified Gompertz (no mixing) | 1.009 | 0.0666 | – | 0.3126 |
| Weibull (with mixing) | 1.075 | – | 0.104 | 0.000 |

It was calculated a $\mu_{max}$ of 0.1118 NL $CH_4$/g-VS d for the mixed assay, compared with one with no mixing ($\mu_{max}$ = 0.0666 NL $CH_4$/g-VS), showing an improvement of 40% in the methane production rate, proving that stirring increases the biogas and methane production.

### *59.3.3 Energy Production from the AD of WVW*

This work aims to contribute to the circularity of the wine distillation process through the recovery by anaerobic digestion of a residue, producing heat or electricity that can be used in the distillation process. As the PCI of methane is 50.0 MJ/kg (Hakawati et al. 2017), each $Nm^3$ of biogas (with an average composition of 80% molar in methane) generated in the anaerobic digestion of wine vinasse has a calorific value of 28.6 MJ/$Nm^3$. Assuming an efficiency of 35% in the electrical conversion and 82% in heat conversion (Belhagj et al. 2013), one has that 1 $Nm^3$ of biogas generates

2.8 kWh in electricity and 6.5 kWh for heating purposes, meaning that by Eq. 59.3, it is possible to generate 2.54 kWh/kg-VS for power generation and 8.26 kWh/kg-VS for heating. These values indicate the potential to produce electrical energy from the waste used in the facility as electricity. However, the most promising application of biogas can be as a heating source for the distillation unit. According to [27], in a traditional wine batch distillation in a copper pot heated directly by natural gas combustion, the heat consumption is 0.375 kWh/kg of distilled wine. Since each kg of distilled wine produces about 0.8 kg of vinasse with a VS concentration of 15.5 g-VS/kg, the produced heat might be 0.10 kWh/kg of distilled wine, representing energy savings and a reduction in greenhouse gases emissions of 27%. Thus, with the energy recovery of a residue generated during the production process, the level of contamination of the generated liquid effluent can be reduced, and simultaneously supply part of the energy needs for the distillation unit.

## 59.4 Conclusion

The Anaerobic Digestion proved to be a reliable circular economy system allowing for the wine vinasse wastewater treatment and biomethane production. It achieved COD removals efficiencies higher than 90%. The WVW can be a reliable energy source by producing 16 NL of methane per liter of vinasse digested with mechanical stirring. The effect of the mechanical stirring was also noticeable in the time required to produce 80–90% of the total biogas from 17 to 13 days. Additionally, in the kinetic evaluation of the BMP, the experiment without mixing does not have an identifiable lag phase. The biogas produced from the WVW can be used as a reliable heating source, as it can produce 8.26 kWh of calorific energy per kilogram of VS fed to the reactor. Thus, with the energy recovery of a residue from the production process, 0.10 kWh/kg of distilled wine. The degree of contamination of the generated liquid effluent can be reduced, and simultaneously supply part of the energy needs necessary to proceed with the distillation of the wine. Anaerobic digestion is a reliable mechanism allowing for energy savings and a reduction in greenhouse gases emissions of 27%.

**Acknowledgements** The authors thank the company Quinta das Bágeiras - Mário Sérgio Alves Nuno for providing the wine vinasse samples.

**Funding** This work was developed under the TratVinhaça Research and Development project (Vale R&D Project, approved by IAPMEI under the Portugal 2020 program, Notice 12/SI/2015—phase 4), Proj. No. 18356 in partnership with the company Mário Sérgio Alves Nuno, Sociedade, and Project SAFE—Water recovery in cellars using photosensitized oxidation, POCI-01-0247-FEDER-039708.

## References


Abdel-Hadi MA (2008) A simple apparatus for biogas quality determination. Misr J Ag Eng 25:1055–1066

APHA, AWWA, WEF (2017) Standard methods for the examination of water and wastewater, 23rd edn. American Public Health Association, American Water Works Association, Water Environment Federation

Belhagj S, Karouach F, El Bari H, Joute Y (2013) The biogas production from mesophilic anaerobic digestion of vinasse. IOSR J Environ Sci Toxicol Food Technol 5:72–77. https://doi.org/10.9790/2402-0567277

Borja R, Martín A, Luque M, Durán MM (1993a) Kinetic study of anaerobic digestion of wine distillery wastewater. Process Biochem 28:83–90. https://doi.org/10.1016/0032-9592(93)80011-5

Borja R, Martín A, Durán MM et al (1993b) Kinetic study of anaerobic digestion of brewery wastewater. Process Biochem 29:645–650. https://doi.org/10.1016/0032-9592(94)80041-3

Botelho G, Anjos O, Estevinho LM, Caldeira I (2020) Methanol in grape-derived fruit and honey spirits: a critical review on source, quality control, and legal limits. Processes 8:1–21. https://doi.org/10.3390/pr8121609

Castro LM, Santos A, Silva J, Quinta-Ferreira R (2020) Floating drum anaerobic digester with mechanical stirring

Chowdhary P, Yadav A, Kaithwas G, Bharagava RN (2017) Distillery wastewater: a major source of environmental pollution and its biological treatment for environmental safety. Green technologies and environmental sustainability. Springer, Cham, pp 409–435

Chowdhary P, Raj A, Bharagava RN (2018) Environmental pollution and health hazards from distillery wastewater and treatment approaches to combat the environmental threats: a review. Chemosphere 194:229–246. https://doi.org/10.1016/j.chemosphere.2017.11.163

Hakawati R, Smyth BM, McCullough G et al (2017) What is the most energy-efficient route for biogas utilization: heat, electricity or transport? Appl Energy 206:1076–1087. https://doi.org/10.1016/j.apenergy.2017.08.068

Haynes WM, Lide DR, Bruno TJ (2016) CRC handbook of chemistry and physics: a ready-reference book of chemical and physical data, 2016–2017th edn. CRC Press, Boca Raton, FL

Instituto Nacional de Estatística (2021) Recenseamento Agrícola. Análise dos principais resultados: 2019. Lisboa

Kafle GK, Chen L (2016) Comparison on batch anaerobic digestion of five different livestock manures and prediction of biochemical methane potential (BMP) using different statistical models. Waste Manag 48:492–502. https://doi.org/10.1016/j.wasman.2015.10.021

Kharayat Y (2012) Distillery wastewater: bioremediation approaches. J Integr Environ Sci 9:69–91. https://doi.org/10.1080/1943815X.2012.688056

Lalov IG, Krysteva MA, Phelouzat JL (2001) Improvement of biogas production from vinasse via covalently immobilized methanogens. Bioresour Technol 79:83–85. https://doi.org/10.1016/S0960-8524(01)00045-1

Melamane XL, Strong PJ, Burgess JE (2007) Treatment of wine distillery wastewater: a review with emphasis on anaerobic membrane reactors. S Afr J Enol Viticult 28:25–36. https://doi.org/10.21548/28-1-1456

Nattassha R, Handayati Y, Simatupang TM, Siallagan M (2020) Understanding circular economy implementation in the agri-food supply chain: the case of an Indonesian organic fertiliser producer. Agric Food Secur 9:1–16. https://doi.org/10.1186/s40066-020-00264-8

Pererva Y, Miller CD, Sims RC (2020) Existing empirical kinetic models in biochemical methane potential (BMP) testing, their selection and numerical solution. Water (basel) 12:1831. https://doi.org/10.3390/w12061831

Santos AD, Silva JR, Castro LM, Quinta-Ferreira RM (2022) Kinetic prediction of biochemical methane potential of pig slurry. Energy Rep 8:159–165. https://doi.org/10.1016/j.egyr.2022.01.128

Silva AFR, Brasil YL, Koch K, Amaral MCS (2021) Resource recovery from sugarcane vinasse by anaerobic digestion—a review. J Environ Manag 295:113137. https://doi.org/10.1016/j.jenvman.2021.113137

Suhartini S, Lestari YP, Nurika I (2019) Estimation of methane and electricity potential from canteen food waste. IOP Conf Ser Earth Environ Sci 230. https://doi.org/10.1088/1755-1315/230/1/012075

Tena M, Luque B, Perez M, Solera R (2020) Enhanced hydrogen production from sewage sludge by co-fermentation with wine vinasse. Int J Hydrogen Energy 45:15977–15984. https://doi.org/10.1016/j.ijhydene.2020.04.075

UNESCO World Water Assessment Programme (2009) Water in a world changing

Wiśniewska P, Śliwińska M, Dymerski T et al (2015) Application of gas chromatography to analysis of spirit-based alcoholic beverages. Crit Rev Anal Chem 45:201–225. https://doi.org/10.1080/10408347.2014.904732

Wolmarans B, De Villiers GH (2002) Start-up of a UASB effluent treatment plant on distillery wastewater. Water SA 28:63–68. https://doi.org/10.4314/wsa.v28i1.4869

World Health Organization (2022) Alcohol, total per capita (15+) consumption (in liters of pure alcohol) (SDG Indicator 3.5.2). https://www.who.int/data/gho/data/indicators/indicator-details/GHO/total-(recorded-unrecorded)-alcohol-per-capita-(15-)-consumption

# Chapter 60
# Renewable Energy from Agro-industrial Residues: Potato Peels as a Case Study

**P. V. Almeida, F. S. Henriques, L. M. Gando-Ferreira, and M. J. Quina**

**Abstract** The potato chip is an important and widespread food industry that generates large quantities of organic residues, which can be valorized through energy recovery. This work aims to assess methane yield obtained through the anaerobic digestion of potato peel (PP) and compare the energy recovered with alternative processes such as ethanolic fermentation, and pyrolysis. Anaerobic digestion assays were performed to determine the biochemical methane potential of potato peel. A maximum of 177 ± 39 L $CH_4$/kg (on a volatile solid basis—VS) was attained while the modeling through a first-kinetic order indicated 187 L $CH_4$/kg. Considering the experimental value (177 L $CH_4$/kg VS), the anaerobic digestion allows for generating 6.2 MJ/kg of PP (dry basis). Theoretically, anaerobic digestion seems to present a higher energy recovery than ethanolic fermentation and pyrolysis. Overall, 7.6 × $10^3$ GJ/year can be recovered through anaerobic digestion of PP produced by a potato chip industry with a production capacity of 26.3 kton/year.

**Keywords** Anaerobic digestion · Ethanolic fermentation · Pyrolysis

## 60.1 Introduction

The food processing industries generate a huge amount of organic residues that can be used for energy production. The potato chip industry is among the most relevant activities in the EU. The main process units in the potato chip industry are presented in Fig. 60.1, using 1 ton of potato as a basis. Throughout this process, different residues are generated, such as potato peels (PP), defective potatoes, leftovers from the potato cut, defective chips, and liquid effluent from the washing stages. PP has been highlighted as an abundant residue with no proper valorization treatment. About 10–12% of the 89.4 kton of potato processed per year corresponds to PP, representing

P. V. Almeida (✉) · F. S. Henriques · L. M. Gando-Ferreira · M. J. Quina
Department of Chemical Engineering, University of Coimbra, CIEPQPF, Rua Sílvio Lima, Pólo II—Pinhal de Marrocos, 3030-790 Coimbra, Portugal
e-mail: patriciav@eq.uc.pt



N. S. Caetano and M. C. Felgueiras (eds.), *The 9th International Conference on Energy and Environment Research*, Environmental Science and Engineering,
https://doi.org/10.1007/978-3-031-43559-1_60

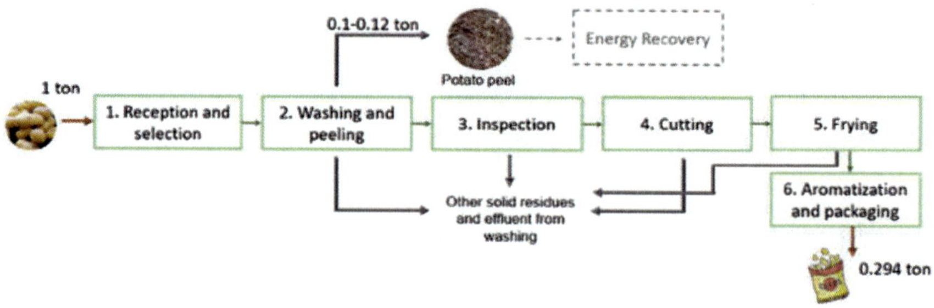


**Fig. 60.1** Schematic representation of a potato chips industry

8.9–10.7 kton/year. This study will be focused on PP due to its abundance and rapid decomposition, and the objective is the assessment of energy recovery.

Anaerobic digestion and alcoholic fermentation are both complex biological processes that generate biogas and bioethanol, respectively. The former uses microorganisms to break down the organic matter and produce methane over four different stages in the absence of oxygen. The maximum methane production is commonly referred to as the biochemical methane potential (BMP—L $CH_4$/kg) of a specific substrate. Before the ethanolic fermentation an acid, alkaline or enzymatic treatment must be done to convert the lignocellulosic structure into monomers such as glucose (saccharification). Then, the monomers are converted into ethanol and carbon dioxide by enzymes. Pyrolysis is a thermochemical process also usually used to convert biomass into three different products: syngas, bio-oil, and biochar. In order to select the most promising treatment pathway, it is relevant to compare the energy recovery of each technology. PP has been considered a substrate for anaerobic digestion (Achinas et al. 2019; Pathak et al. 2018), ethanolic fermentation (Barampouti et al. 2021; Chohan et al. 2020; Mazaheri and Pirouzi 2020), and pyrolysis (Daimary et al. 2022; Liang et al. 2015). However, there is a lack of studies in the literature that compare these three processes for the valorization of PP residues.

This work aims to evaluate the possibility of recovering energy through anaerobic digestion as a strategy to manage the main residue produced in the potato chips industry—potato peel. A specific industry is used as a case study to analyze the production capacity and estimate the residues generation. A comparison among the pathways of energy recovery from potato peel will also be performed. This work stands out because uses data from industry and compares the net calorific value of the fuels produced through anaerobic digestion, ethanolic fermentation, and pyrolysis.

## 60.2 Materials and Methods

### *60.2.1 Residue Collection and Characterization*

Two samples of potato peels were collected from a potato chip industry located in Portugal in April 2021 (S1) and April 2022 (S2). The samples were frozen until anaerobic digestion assays were performed. Total solids (TS) and volatile solids (VS) were determined by drying the sample respectively, in an oven at 105 °C and in a furnace at 550 °C until constant weight. For characterization purposes, the PP was dried in an oven at 40 °C until constant weight and then it was ground to reduce the particle size. An Analyzer NA 2500 equipment (Xylem, New York, NY, USA) was used to determine the ultimate analysis (CHNS, %wt). The difference between VS and the sum of CHNS was considered the oxygen content. The National Renewable Energy Laboratory (NREL) procedure was followed to determine the content of lignin (LG) with some modifications (Sluiter et al. 2008). A suspension obtained at a liquid to solid ratio of 10 (mL/g) was left over 2 h in agitation to determine the natural pH of the solid sample. The determinations were performed in triplicates and the results are presented as mean ± standard error (SD). The theoretical higher heating value ($HHV_{elem}$) and lower heating value ($LHV_{elem}$) were calculated using the elemental analysis (Almeida et al. 2022).

### *60.2.2 Data Collection and Treatment*

The potato chips industry provided the quantity of raw potatoes processed and chips produced. The amount of PP was estimated considering that about 10–12% of the potato weight corresponds to PP (Pathak et al. 2018). Data regarding ethanolic fermentation and pyrolysis were collected from the literature selecting the ones in which PP presented similar properties. The LHV (MJ/kg PP) for ethanolic fermentation was calculated according to Eq. 60.1, where [EtOH] (g/L) corresponds to the concentration of ethanol produced, D is the dosage (g/L) of PP in the fermentation and 26.7 MJ/kg is the net calorific value of ethanol.

$$LHV\ (\text{MJ/kg PP}) = \left(\frac{[EtOH]}{D}\right) \times 26.7 \tag{60.1}$$

Regarding the pyrolysis of PP, the LHV was obtained by the sum of the yield of products ($Y_{bio\text{-}oil}$, $Y_{biochar}$, and $Y_{syngas}$—g/g) multiplied by the respective calorific value ($CV_{bio\text{-}oil}$, $CV_{biochar}$, and $CV_{syngas}$—MJ/kg)—Eq. 60.2.

$$LHV(\text{MJ/kg PP}) = Y_{bio\text{-}oil}CV_{bio-oil} + Y_{biochar}CV_{biochar} + Y_{syngas}CV_{syngas} \tag{60.2}$$

The energy conversion efficiency related to all technologies was calculated to compare the LHV of the residue obtained experimentally ($LHV_{elem}$) with the LHV of the fuel products—Eq. 60.3.

$$Conversion\ efficiency\ (\%) = \frac{LHV}{LHV_{elem}} \times 100 \tag{60.3}$$

### 60.2.3 Anaerobic Digestion

The anaerobic digestion sludge (inoculum) used in the experimental assays was collected from a wastewater treatment plant (WWTP) that treats urban effluents located in Portugal.

The BMP was assessed in duplicate using 5 L reactors under mesophilic conditions (37 °C). The substrate to inoculum ratio (S/I) was 0.5 and the total concentration of volatile solids in the reactors was 20 g/L. Nutrient solutions were added to the reaction mixture (Angelidaki et al. 2009). The biogas volume was estimated through the displacement of a graduated gasometer sealed with a solution of NaCl 60% at pH 2. The methane content was measured by GAS DATA (GMF 406) equipment. The trials were stopped when the biogas production was $< 5\%$ for consecutive days to save time. The BMP results were normalized to the standard conditions.

A first-order kinetic model (Eq. 60.4) was used to fit the experimental data,

$$SMP(t) = BMP_{\infty}\left(1 - e^{-kt}\right) \tag{60.4}$$

where SMP (L $CH_4$/kg VS) corresponds to the specific methane potential in a determined time, $BMP_{\infty}$ (L $CH_4$/kg VS) is the methane yield when time tends to infinite, k (1/d) is the rate degradation constant, and t (d) corresponds to time. The non-linear regression was performed in the SigmaPlot V11.0 software.

For comparison purposes, the theoretical BMP ($BMP_{theo}$—mL $CH_4$/g VS) was also calculated based on the elemental composition, using the Buswell and Muller method (Buswell and Mueller 1952).

The LHV of the biogas produced was calculated considering its composition and the LHV of methane (35.8 MJ/$Nm^3$).

**Table 60.1** Properties (mean ± SD) of two samples of PP

| | S1 | S2 |
|---|---|---|
| pH | 6.96 ± 0.10 | 7.19 ± 0.07 |
| TS (%) | 8.90 ± 0.16 | 13.7 ± 0.3 |
| VS (%TS) | 90.2 ± 0.35 | 97.9 ± 0.2 |
| C (%TS) | 45.9 ± 0.6 | 47.1 ± 0.9 |
| H (%TS) | 6.03 ± 0.29 | 6.01 ± 0.4 |
| N (%TS) | 2.52 ± 0.06 | 3.50 ± 0.10 |
| O (%TS) | 35.7 ± 0.8 | 42.0 ± 0.7 |
| LG (%TS) | 34.8 ± 0.6 | 21.2 ± 1.1 |
| $LHV_{elem}$ (MJ/kg TS) | 17.9 ± 0.4 | 17.6 ± 0.6 |

## 60.3 Results and Discussion

### *60.3.1 Potato Peel Characterization*

The main physical and chemical properties of the two PP samples are summarized in Table 60.1. PP is characterized by a natural pH around neutrality and presents low content of total solids (about 11%). In particular, the total solids are mainly composed of organic matter (> 90% TS), which indicates that PP is a potential good substrate for anaerobic digestion. PP is mainly composed of C and O and presents a C/N ratio of 24, which is within the range preferred to assure anaerobic digestion stability (Jain et al. 2015). Similar results are found in the literature for elemental analysis of PP (Daimary et al. 2022; Liang et al. 2015; Liang and McDonald 2014). PP presents a lignin content ranging from 20 to 35%, which can hinder the biodegradability of the residue by microorganisms since lignin exhibits resistance to biodegradation. The variability of PP noted in Table 60.1 may be related to the processing of potatoes from different regions (presenting different growing conditions) and multiple varieties. The theoretical lower heating value of PP was also calculated to evaluate the maximum energy that can be recovered by complete combustion. PP presents an $LHV_{elem}$ around 18 MJ/kg TS.

### *60.3.2 Anaerobic Digestion of Potato Peel*

A preliminary assessment of the potential to recover energy from PP through anaerobic digestion was performed using S2. The operating conditions used in the reactors were selected according to the recommendations found in the literature and the most used by other authors (Achinas et al. 2019; Holliger et al. 2016; Liang and McDonald 2015). The experimental assays with PP led to a BMP of 177 ± 39 L $CH_4$/kg VS, corresponding to almost 55% of the theoretical BMP—Table 60.2. The high lignin content of S2 (Table 60.2) can explain this result. Chandler et al. (1980) proposed a correlation often used in the literature, between the biodegradability of residues and the lignin content applied to residues with a maximum of 20% (VS) of lignin. Although PP does not meet the requirements to apply this equation, they stated that at most 83% of a substrate is degraded. Therefore, a methane yield of 272 L $CH_4$/kg VS is the maximum that can be expected for this substrate. Slightly higher values of BMP for PP are found in the literature (239 and 267 L $CH_4$/kg VS)—Table 60.2 (Gunaseelan 2004; Liang and McDonald 2015). The difference can be related to the operating conditions and chemical properties of the residues. Achinas et al. (2019) reported the maximum methane production (218 L $CH_4$/kg VS) for S/I ratio of 0.5 and this operating condition is the one that affects the most the BMP. Moreover, they obtained a biogas production (384 L/kg VS) similar to the result achieved in this work when comparing assays with an equal S/I ratio.

Figure 60.2 presents the cumulative methane yield over time for PP. The ultimate methane yield is reached in 15 days, as reported by Achinas et al. (2019).

A first-order kinetic model was applied to the experimental results and the parameters estimated are summarized in Table 60.3. The degradation kinetic rate is 0.198 1/d, higher than the kinetic constant found in the literature—0.090 1/d (Gunaseelan 2004). $BMP_\infty$ is 187 L $CH_4$/kg VS, proving that after 15 days the increment in biogas production is residual (< 6%).

**Table 60.2** Results from the anaerobic digestion of PP obtained experimentally (biogas and $BMP_{exp}$), theoretical ($BMP_{theo}$), and reported in the literature ($BMP_{lit}$)

| Biogas (L/kg VS) | $BMP_{exp}$ (L $CH_4$/kg VS) | $BMP_{theo}$ (L $CH_4$/kg VS) | $BMP_{lit}$ (L $CH_4$/kg VS) | | |
|---|---|---|---|---|---|
| 381 ± 78 | 177 ± 39 | 327 | 218[1] | 267[2] | 239[3] |

1—Achinas et al. (2019); 2—Liang and McDonald (2015); 3—Gunaseelan (2004)

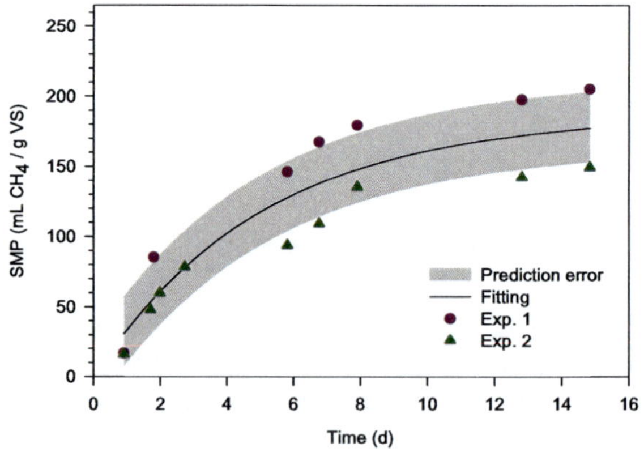


**Fig. 60.2** Cumulative methane production (SMP) of 2 trials and corresponding fitting curve and prediction error

**Table 60.3** First order kinetic parameter for the methane production of PP

| | |
|---|---|
| $BMP_{\infty}$ (L $CH_4$/kg VS) | 187 |
| k (1/d) | 0.198 |
| $R^2$ | 0.84 |
| RMSE | 25 |

### *60.3.3 Assessment of Energy Recovery*

Energy recovery from industrial residues can be performed by alternative processes to anaerobic digestion, such as ethanolic fermentation and pyrolysis. The LHV of the products obtained from the three different processes was compared, aiming to determine which process maximizes the conversion of the feedstock to fuel products. Table 60.4 summarizes the LHV calculated as described in Sect. 2.2. Considering all the processes, the LHV ranges between 4.5 and 11.2 MJ/kg PP. However, it is important to highlight that the energetic contribution of the syngas generated through pyrolysis (about 35–50% w/w of the yield products) was not considered since its composition was not evaluated by the authors (Daimary et al. 2022; Liang et al. 2015). Moreover, the LHV of 6.2 MJ/kg PP corresponds to the heating value of biochar (30.5% w/w) (Liang et al. 2015). If all the pyrolysis products were considered, the conversion efficiency would be higher than in biological processes. However, it is important to note that PP contains high moisture content, and thus water in the biomass must be driven off before the first stage of pyrolysis can take place. As removing water requires energy (the latent heat of evaporation at 100 °C is 2256 kJ/kg), the overall pyrolysis efficiency is reduced. In the literature, the pyrolysis of PP has mostly been assessed to produce biochar with catalytic and sorbent properties (Daimary et al. 2022; Liang et al. 2015). Comparing the biogas obtained through anaerobic digestion and the bioethanol produced by ethanolic fermentation, it is possible to note that slightly better conversion efficiencies were reached with anaerobic digestion.

Overall, only about 25–50% of PP is converted to biofuel. Some strategies can be adopted to increase the energetic conversion efficiency. For example, Barampouti et al. (2021) proposed to treat PP with an ethanolic fermentation followed by anaerobic digestion. They reported an increase of 6.6% and 25% of the energy recovered when compared to anaerobic digestion and alcoholic fermentation, respectively. However, in addition to the conversion efficiency, the input energy, the chemicals introduced in each process, the investment and maintenance costs are also important aspects to consider when selecting the best management route. So, an economic analysis is needed to determine which treatment pathway is more suitable. For example, the investment and maintenance cost of a treatment process with fermentation followed by anaerobic digestion may not compensate the energy revenues. Another alternative to improve energy recovery may be co-digestion. Achinas et al. (2019) reported an increase of 10% in the BMP when PP is mixed with cow manure. Nevertheless, the co-digestion of PP with other potato residues is an interesting study to be conducted in the future to assess the global potential of valorization of all the solid residues of the potato chips industry.

**Table 60.4** Lower heating value (LHV) of the products of anaerobic digestion, alcoholic fermentation, and pyrolysis and conversion efficiency

| | Anaerobic digestion | | | | Alcoholic fermentation | | | Pyrolysis | |
|---|---|---|---|---|---|---|---|---|---|
| | This work | Achinas et al. (2019) | Gunaseelan (2004) | Liang and McDonald (2015) | Barampouti et al. (2021) | Chohan et al. (2020) | Mazaheri and Pirouzi (2020) | Daimary et al. (2022) | Liang et al. (2015) |
| LHV (MJ/kg PP) | 6.2 | 7.6 | 9.6 | 7.0 | 6.6 | 4.8 | 4.5 | 11.2 | 6.2 |
| Conversion efficiency (%) | 35 | 42 | 53 | 39 | 37 | 27 | 25 | 62 | 34 |

In summary, operating anaerobic digestion with a methane yield of 177 L $CH_4$/kg VS generates about $7.6 \times 10^3$ GJ of energy when the PP produced over a year in the potato chips industry is treated. It is relevant to mention that industrial PP is produced regularly throughout the year, and thus anaerobic digestion can operate in this case without the seasonal stress observed for other agro-residues.

## 60.4 Conclusion

In this work, the energy recovered from potato peel through anaerobic digestion was evaluated. Potato peel is the most abundant and critical residue produced by the potato chip industry, accounting for about 8.9 kton/year for a production of 26.3 kton/year. Anaerobic digestion of potato peel was assessed to determine the biochemical methane production. A maximum methane yield of $177 \pm 39$ L $CH_4$/kg (on a volatile solid basis) was achieved in 15 days. The cumulative specific methane production was evaluated through a first-order kinetic model to determine the kinetic parameters. In particular, the ultimate methane production was 187 L $CH_4$/kg (on a volatile solid basis) and the kinetic degradation rate was 0.198 1/d. At the end, a comparison of the net calorific value recovered by anaerobic digestion, alcoholic fermentation, and pyrolysis was performed. Although pyrolysis may present the highest energy recovery, the gain probably does not compensate for the energy required to reduce the moisture of potato peel. Therefore, anaerobic digestion seems to be a promising process to recover energy from potato peel. Additional studies are recommended to maximize the methane yield, namely testing co-digestion with other residues from the same industry or even from other sectors.

**Acknowledgements** This work was financially supported by Fundação para a Ciência e Tecnologia (FCT, Portugal) that provided a Ph.D. Grant (2020.08445.BD) to Patrícia V. Almeida.

## References

Achinas S, Li Y, Achinas V, Euverink GJW (2019) Biogas potential from the anaerobic digestion of potato peels: process performance and kinetics evaluation. Energies (Basel) 12:2311

Almeida PV, Rodrigues RP, Slezak R, Quina MJ (2022) Effect of phenolic compound recovery from agro-industrial residues on the performance of pyrolysis process. Biomass Convers Biorefinery 12:4257–4269

Angelidaki I, Alves M, Bolzonella D, Borzacconi L, Campos JL, Guwy AJ et al (2009) Defining the biomethane potential (BMP) of solid organic wastes and energy crops: a proposed protocol for batch assays. Water Sci Technol 59:927–934

Barampouti EM, Christofi A, Malamis D, Mai S (2021) A sustainable approach to valorize potato peel waste towards biofuel production. Biomass Convers Biorefinery

Buswell AM, Mueller HF (1952) Mechanism of methane fermentation. Ind Eng Chem 44:550–552

Chandler JA, Jewell WJ, Gossett JM, van Soest PJ, Robertson JB (1980) Predicting methane fermentation biodegradability. Biotechnol Bioeng Symp 10

Chohan NA, Aruwajoye GS, Sewsynker-Sukai Y, Gueguim Kana EB (2020) Valorisation of potato peel wastes for bioethanol production using simultaneous saccharification and fermentation: process optimization and kinetic assessment. Renew Energy 146:1031–1040

Daimary N, Eldiehy KSH, Boruah P, Deka D, Bora U, Kakati BK (2022) Potato peels as a sustainable source for biochar, bio-oil and a green heterogeneous catalyst for biodiesel production. J Environ Chem Eng 10:107108

Gunaseelan VN (2004) Biochemical methane potential of fruits and vegetable solid waste feedstocks. Biomass Bioenergy 26:389–399

Holliger C, Alves M, Andrade D, Angelidaki I, Astals S, Baier U, Bougrier C, Buffière P, Carballa M, De Wilde V, Ebertseder F, Fernández B, Ficara E, Fotidis I, Frigon J, De Laclos H, Ghasimi D, Hack G, Hartel M, Heerenklage J, Horvath I, Jenicek P, Koch K, Krautwald J, Lizasoain J, Liu J, Mosberger L, Nistor M, Oechsner H, Oliveira J, Paterson M, Pauss A, Pommier S, Porqueddu I, Raposo F, Ribeiro T, Pfund F, Strömberg S, Torrijos M, Van Eekert M, Van Lier J, Wedwitschka H, Wierinck (2016) Towards a standardization of biomethane potential tests. Water Sci Technol 74:2515–2522

Jain S, Jain S, Wolf IT, Lee J, Tong YW (2015) A comprehensive review on operating parameters and different pretreatment methodologies for anaerobic digestion of municipal solid waste. Renew Sustain Energy Rev 52:142–154

Liang S, McDonald AG (2014) Chemical and thermal characterization of potato peel waste and its fermentation residue as potential resources for biofuel and bioproducts production. J Agric Food Chem 62:8421–8429

Liang S, McDonald AG (2015) Anaerobic digestion of pre-fermented potato peel wastes for methane production. Waste Manag 46:197–200

Liang S, Han Y, Wei L, McDonald AG (2015) Production and characterization of bio-oil and bio-char from pyrolysis of potato peel wastes. Biomass Convers Biorefinery 5:237–246

Mazaheri D, Pirouzi A (2020) Valorization of *Zymomonas mobilis* for bioethanol production from potato peel: fermentation process optimization. Biomass Convers Biorefinery

Pathak PD, Mandavgane SA, Puranik NM, Jambhulkar SJ, Kulkarni BD (2018) Valorization of potato peel: a biorefinery approach. Crit Rev Biotechnol 38:218–230

Sluiter A, Hames B, Ruiz R, Scarlata C, Sluiter J, Templeton D et al (2008) Determination of structural carbohydrates and lignin in biomass. Laboratory Analytical Procedure

# Chapter 61
# Enhanced Biogas Production from Press Mud Using Molasses-Based Distillery Wastewater as Co-substrates Through an Immobilized Anaerobic Digestion

**Michelle Almendrala, Zhane Ann Tizon, Bonifacio Doma, and Ralph Carlo Evidente**

**Abstract** As a promising tool in converting waste to energy, anaerobic digestion (AD) was utilized in this study. Two experiments were performed by (1) co-digestion of press mud (PM), and molasses-based distillery wastewater (DWW) and (2) immobilized AD of DWW. Experiment 1 was done with variation of volume of DWW, PM and tap water. The highest methane yield was 61.3% and 78.23% for PM mixed with DWW, with or without added nutrients, respectively. With added nutrients, a 300:600 (DWW to PM) ratio produced the highest methane content. Experiment 2 was carried out in a thermophilic 2-L Erlenmeyer flask inside an orbital shaker using a K1 moving bed biofilm carriers to immobilize DWW with addition of nutrients and glucose. Batch 5 had a BOD/COD ratio of 0.23, yielding the highest volume of 2-L in 1 day. Since there was a niche for bacterial growth, DWW mixed with nutrients, glucose, and media performed the best in terms of biodegradability. Both setups include a homogenous feed had negligible amount of methane due to the less carbon present. The results demonstrated that added nutrients can improve AD efficiency using plastic carriers with immobilization and increasing volume of the substrates.

**Keywords** AD · Co-digestion · DWW · Immobilization · Moving bed biofilm reactor · PM

M. Almendrala (✉) · Z. A. Tizon · B. Doma
Mapua University, 1002 Manila, Philippines
e-mail: mcalmendrala@mapua.edu.ph

R. C. Evidente
POSTECH, Pohang, Gyeongbuk 37673, South Korea



N. S. Caetano and M. C. Felgueiras (eds.), *The 9th International Conference on Energy and Environment Research*, Environmental Science and Engineering,
https://doi.org/10.1007/978-3-031-43559-1_61

## 61.1 Introduction

From industrialization to the technological era, the world still faces one of the most challenging issues, the formation of organic wastes. By further treatment of biowastes through anaerobic digestion (AD), these can be valorized by converting into a renewable energy source, biogas, and produced digestate that acts as a fertilizer. Biomethane potential yield from anaerobic digestion alone is considered insufficient due to the limited supply of homogenous feedstocks. To alleviate this concern and ease the fluctuations of compositions in feedstocks, co-digestion can accommodate a wide range of feedstocks with different homogeneity and humidity in one digester (AgStar 2012). An optimal amount of carbon (C), nitrogen (N), phosphorus (P), and sulfur (S) is required in feedstocks to maximize AD efficiency fully. An inadequate ratio of C/N in substrates results in incomplete conversion of carbon since it cannot produce enzymes or even capture ammonia from the conversion of urea (Gerardi 2003). Thus, a moving bed biofilm reactor can preserve the microbiome through biofilm growth in the plastic media to enhance AD while being cost-effective. The purpose of this paper is to investigate the effects of co-digestion of press mud with molasses-based distillery wastewater and biofilm formation on biogas yield enhancement using distillery wastewater.

## 61.2 Materials and Methods

### 61.2.1 Substrates Used and Pretreatment

In this work, the distillery wastewater (DWW) used for co-digestion came from Central Azucarera de Tarlac, along with press mud and bagasse. In contrast, the distillery wastewater samples for the biofilm formation were obtained from Batangas' Absolut Distillers, Inc, whereas cow manure was freshly taken from YGGACC HAI farms in Laguna. Batch experiments were performed under mesophilic conditions and halted after 42 days for co-digestion and four (4) weeks in thermophilic conditions for the immobilized AD. The press mud was treated by two-step hydrolysis before mixing with the solution. Press mud was about 1108.8 g, submerged in 1 L 62.0 mEq/L of $Ca(OH)_2$ for 15 h. The dilution of alkali hydrolysate with water was taken up to 1200 mL, then heated until the solution boils for 20 min. Before experimentation, the treated press mud was cooled for seven (7) days. Sequentially, cow manure was sealed and incubated at a temperature of 40 °C through an anaerobic process for seven (7) days before experimenting.

### *61.2.2 Experimental Design for Co-digestion of Press Mud and DWW*

The main focus of this experiment is the dilution of distillery wastewater (DWW) with the same amount of press mud and the variation of volume of both DWW and press mud. Batch anaerobic digestion was conducted in this experiment in 10 2-L Erlenmeyer flask. The first batch added 300 mL of pretreated press mud into 200 mL DWW solution and 100 $cm^3$ of bagasse. A pure DWW was used in the preparation of Sample 1. The batch contained remaining samples and were diluted with tap water in a volume ratio of 3:2 and 2:3, and with or without the presence of micronutrients. Glucose, yeast, and macronutrients were added to each sample for about 20 g/L and 5 g/L, respectively. In addition, each media was added a 200 $cm^3$ of inoculum. In Table 61.1, the operating conditions of Batch 1 samples are presented.

Based on a study (Menon et al. 2017), the amount of micronutrients utilized in the experiment, presented in Table 61.2, is twice as the suggested amount in the literature. Each media was purged with nitrogen gas for 15 min to secure the absence of oxygen in AD. A 2-L urine bag was connected to each media to accumulate the gas presence shown in Fig. 61.1. The digestion took up to 42 days and was carried out at a room temperature.

Table 61.3 illustrates the volumetric ratios for mixed feedstocks, DWW, and press mud, of Batch 2 samples were adjusted discretely by 1:0, 1:1, 1:1, 2:1, and 1:2. Samples 7 and 8 had the same DWW and press mud ratio but differed in pretreatment. Sample 7 utilized thermal dilution, while Sample 8 did undergo alkali pretreatment. Consequently, the amount of inoculum and nutrients added in each sample in Batch 1 were not varied. However, the total volumetric amounts of DWW and press mud

**Table 61.1** Controlled parameters used in Batch 1

| Sample ID | DWW (mL) | $H_2O$ (mL) | Nutrient |
|---|---|---|---|
| 1 | 200.0 | 0.0 | ✓ |
| 2 | 120.0 | 80.0 | ✓ |
| 3 | 80.0 | 120.0 | ✓ |
| 4 | 120.0 | 80.0 | × |
| 5 | 80.0 | 120.0 | × |

**Table 61.2** Volumetric amounts of micronutrients used

| Nutrients | Amount (g/L) |
|---|---|
| $NH_4Cl$ | 4.00 |
| $KH_2PO_4$ | 1.00 |
| $MgSO_4 \cdot 7H_2O$ | 1.20 |
| $MnSO_4 \cdot 7H_2O$ | 0.04 |
| $FeSO_4 \cdot 7H_2O$ | 0.04 |

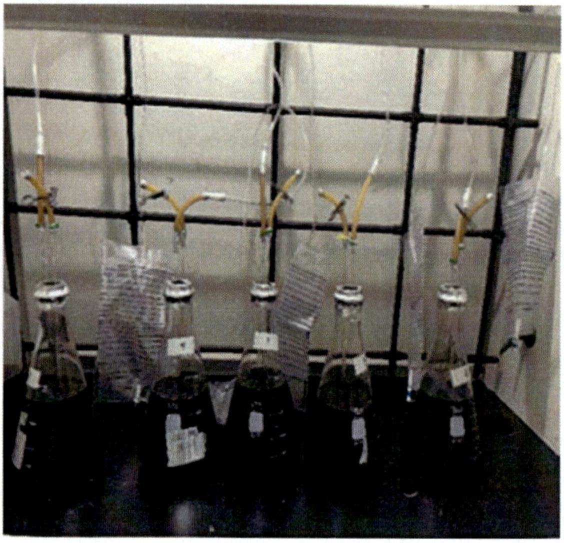

**Fig. 61.1** Experimental setup for lab-scale anaerobic digestion

**Table 61.3** Controlled parameters used in Batch 2

| Sample ID | DWW (mL) | Press mud solution (mL) | Pretreated press mud |
|---|---|---|---|
| 6 | 0.0 | 1200.0 | ✓ |
| 7 | 600.0 | 600.0 | × |
| 8 | 600.0 | 600.0 | ✓ |
| 9 | 600.0 | 300.0 | ✓ |
| 10 | 300.0 | 600.0 | ✓ |

were increased by 2.4 compared to Batch 1. This was done to counteract the excessive micronutrients present in Batch 1.

### *61.2.3 Experimental Design for Biofilm Formation in an Immobilized AD of DWW*

To further enhance AD due to the lack of immobilized substrates in the prior experiment, using a moving bed biofilm reactor with 10 mm diameter, 7 mm length, a density of 0.96–0.98 g/cm$^3$, and a surface area of > 800 m$^2$/m$^3$ in DWW was conducted in this set-up. The volumetric ratios of DWW and water are depicted in Table 61.4, including the added micronutrients, glucose, and media carrier. The required amount of nutrients used was adapted from literature (Menon et al. 2017) shown in Table 61.5.

**Table 61.4** Various controlled parameters in AD

| Sample ID | DWW (mL) | $H_2O$ (mL) | Nutrient | Glucose (g/200 mL) | Carriers |
|---|---|---|---|---|---|
| 1 | 200.0 | 0.0 | × | × | × |
| 2 | 120.0 | 80.0 | ✓ | ✓ | × |
| 3 | 80.0 | 120.0 | ✓ | ✓ | ✓ |
| 4 | 120.0 | 80.0 | × | × | ✓ |
| 5 | 80.0 | 120.0 | ✓ | ✓ | ✓ |

**Table 61.5** List of the used macronutrients and micronutrients in AD

| Nutrients | Required amount based on the literature (g/L) | Weighed amount (g) |
|---|---|---|
| Glucose | 30 | 42.0364 |
| $KH_2PO_4$ | 0.5 | 0.7033 |
| $MgSO_4 \cdot 7H_2O$ | 0.2 | 0.2863 |
| $MnSO_4 \cdot 7H_2O$ | 0.01 | 0.0163 |
| $FeSO_4 \cdot 7H_2O$ | 0.01 | 0.0146 |
| NaCl | 0.01 | 0.0152 |
| $CuSO_4 \cdot 5H_2O$ | 0.1 | 0.1437 |
| $CoCl_2 \cdot 6H_2O$ | 0.1 | 0.1444 |
| $ZnSO_4 \cdot 7H_2O$ | 0.02 | 0.0317 |

Hence, removing oxygen and gas collection from the first study was done similarly in this experiment. However, some operating conditions were changed. To eliminate the oxygen present, nitrogen gas was utilized by purging through a glass tube for each media bottle for 10 min. A 2-L urine bag was connected to each media bottle with rubber tubes to collect the biogas produced as shown in Fig. 61.1. These bottles were placed in an orbital shaker at 120 rotations per minute and were exposed to a constant temperature of 37 °C for 4 weeks.

### 61.2.4 Analytical Methods

The physico-chemical properties of the collected biogas, including pH, COD (Chemical Oxygen Demand), BOD (Biological Oxygen Demand), TSS (Total Suspended Solids), VS (Volatile Solids), amount of Phosphorus, Total Carbon, and Total Nitrogen, were determined in their initial and final values. The volume of biogas was calculated using water displacement method. For concentration of methane, it was determined in the laboratory using a gas chromatograph thermal conductivity detector (GC-TCD) in accordance with the American Standard Test method ASTM D2504-88(1998).

## 61.3 Results and Discussion

### *61.3.1 Methane Yield from Co-digestion of Press Mud and DWW*

The highest methane yield from the samples of Batch 1 gave 61.3% and 78.23% (v/v) for press mud mixed with DWW, with or without the added nutrients, respectively, as shown in Fig. 2a. Accordingly, acidogenesis and acetogenesis occur to produce various forms of fatty acids and acetate during anaerobic digestion (den Boer and den Boer 2021; Achinas et al. 2020). The generation of these products affects the pH of all the samples throughout the digestion as well as the methanogens produced. Sample 2 had the lowest methane yield and a low initial value of pH. Due to the stable pH of 5.00 in Sample 5, it yielded the highest methane content. The percent BOD/COD ratio is determined by the capacity for degradation of an organic matter that varies from 0 to 100%. Samples 3 and 5 exhibited an optimum % (BOD/COD) that ranges from 38.46 to 55.55% due to the press mud containing sugar products (Radjaram and Saravanane 2011). In relation to the C/N ratio, Samples 1 and 4 had nonsignificant methane yield but performed the lowest and highest C/N ratios, respectively. The addition of micronutrients to the samples remarkably affects the C/N ratio values to obtain optimal values compared to the C/N ratios of 30.0 for DWW, 25.0 for press mud (PM), and 24.7 for PM (25%) + bagasse (75%). An excessive amount of micronutrients could inhibit the methanogenesis's bacterial activity. Substantially, there was a direct proportionality between the dilution ratio and methane yield, as the sample with the highest dilution ratio produced a significant amount of methane.

For Batch 2 samples, an increase in the volumetric amount of DWW and PM solution had been found to reduce the high concentrations of nutrients. Sample 10 yielded the highest methane content, having a ratio of 300:600 (DWW to PM) with additional nutrients, as illustrated in Fig. 2b. The change in % pH was the lowest in Sample 10 due to a stable acidic pH value for the sludge; thus, anaerobic digestion

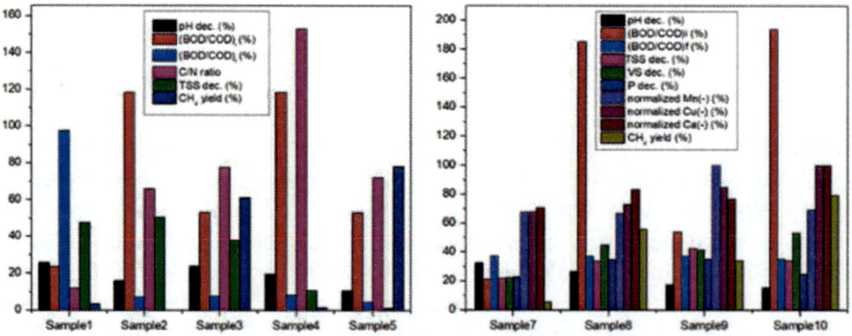


**Fig. 61.2** Summary of physico-chemical data (left) from Batch 1 and (right) Batch 2

**Table 61.6** Comparison of $CH_4$ yields from different substrates

| Substrate | $CH_4$ yield (%) | Source |
|---|---|---|
| Bagasse (B) | 26.00 | Vargas et al. (2016) |
| Press mud (PM) | 45.00 | Gonzales et al. (2017) |
| DWW | 50.00 | Kumar and Aklilu (2016) |
| PM + vinasse | 51.60 | Gonzales et al. (2017) |
| FW + sludge + glycerol | 77.10 | Silva et al. (2018) |
| DWW + PM + B + nutrients | 3.45 | Evidente et al. (2021) |
| DWW/$H_2O$ (120:80) + PM + B + nutrients | 0.016 | |
| DWW/$H_2O$ (80:120) + PM + B + nutrients | 61.30 | |
| DWW/$H_2O$ (120:80) + PM + B | 1.36 | |
| DWW/$H_2O$ (80:120) + PM + B | 78.20 | |
| DWW/PM (600:600) + nutrients (not pretreated) | 5.88 | |
| DWW/PM (600:600) + nutrients | 55.79 | |
| DWW/PM (600:300) + nutrients | 34.40 | |
| DWW/PM (300:600) + nutrients | 79.43 | |

possibly occurred. The correlation of BOD/COD had been inversely proportional with methane yield. As there was a dramatic decrease of BOD/COD for both Samples 8 and 10, the methane yield exhibited in these samples was high enough. Sample 6 did not have a significant methane yield since it only contained press mud during digestion. The efficiency of AD was greatly affected by the trace elements found in micronutrients such as Ca, Cu, and Mn present in Samples 3, 4, and 5. In Table 61.6, the effect of dilution in AD of press mud using DWW as co-substrates in sample nutrients. The pretreatment using alkali hydrolysis had an exceptional methane yield than the thermal water hydrolysis method.

### *61.3.2 Biogas Yield from Immobilized AD of DWW in Biofilm Formation*

A pure DWW as a homogenous feed produced the most negligible methane yield, as shown in Fig. 3a, similar to the previous experiment, due to the less carbon utilized by anaerobic organisms. It had been found that Batch 2 and Batch 5 produced the most significant volume of 2-L within 1 day. During Day 27, Batch 1 diffused in volume from 2000 to 1850 mL. Regardless, both batches were added nutrients and glucose for optimal intake of anaerobes, resulting in a shorter duration of AD to perform.

Figure 3b shows that Sample 4 had the lowest decrease pH, yielding a small amount of biogas, while the change in pH of other batch samples remained constant. In regard to BOD/COD ratio, samples of DWW mixed with nutrients, glucose, and

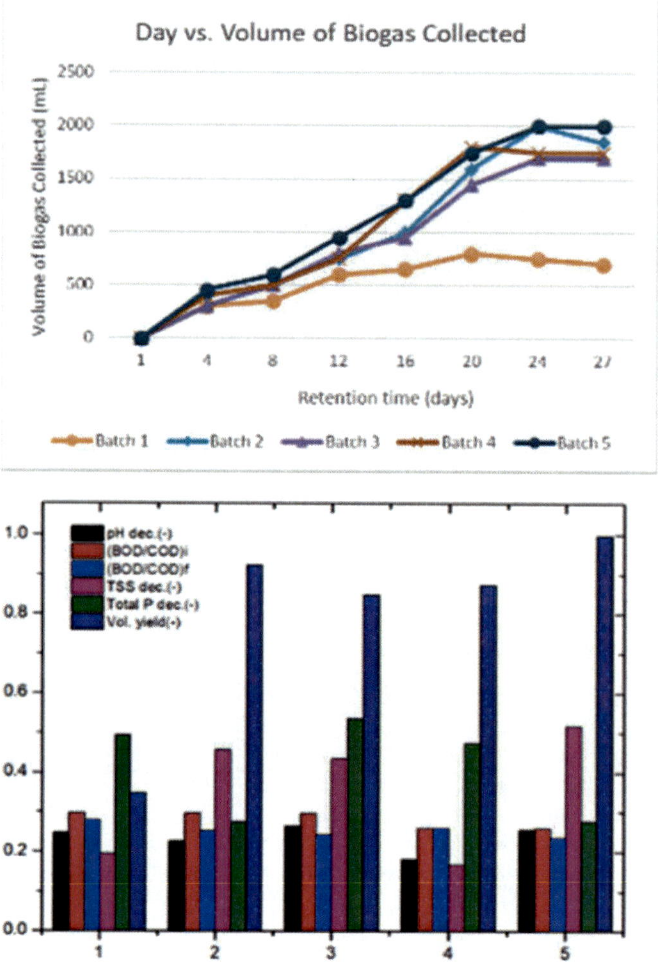


**Fig. 61.3** Volumetric profile of biogas for 27-day period (top); summary of data per batch (bottom)

media executed the most biodegradability since there was an occurrence of niche for bacterial growth. Sample 5 demonstrated a BOD/COD ratio that approaches zero; thereby, degradation occurred virtually. Changes in BOD to COD ratio favored the principle of AD due to the significant difference between the initial and final values. Sample 5 also performed a higher % reduction value of TSS, suggesting that AD was executed significantly with biodegradation in its BOD/COD ratio. Therefore, the greater reduction of suspended solids could result in a significant volume yield of biogas. This experiment suggested that to generate a larger volume of biogas yield, the optimum range of TSS is from 938 to 1456 mg/L.

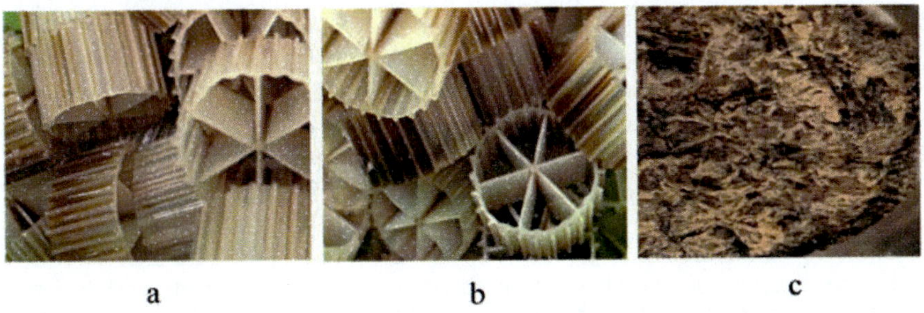


**Fig. 61.4** Comparison of plastic carriers of **a** Batch 3; **b** Batch 4; and **c** Batch 5 in AD

**Table 61.7** Summary of the volume of biogas produced by substrate

| Substrate | Feed (L) | Biogas yield (L) |
|---|---|---|
| Pure DWW with nutrients and glucose | 1.4 | 1.85 |
| Pure DWW with nutrients, glucose, and carriers | 1.4 | 1.7 |
| 49:21 (DWW to $H_2O$) with carriers | 1.4 | 1.75 |
| 49:21 (DWW to $H_2O$) with nutrients, glucose, and carriers | 1.4 | 2 |
| Pure DWW with nutrients and glucose | 1.4 | 1.85 |

As shown in Fig. 4c, Batch 5 was the most remarkable sample observed in relation to biofilm formation. This phenomenon could improve the metabolic processes due to the highly coordinated interactions between each microbe that grows in one microbial culture. Nutrients such as phosphorus and nitrogen-containing complexes were digested in the cultures of mixed microbes in the biofilm during AD. Overall, the quantitative biogas output from the original volumetric amount of substrates used is compared in Table 61.7.

## 61.4 Conclusion

The objectives of the study were achieved to determine the effects of dilution ratio in co-digestion and to compare the BOD/COD ratio, TSS, and pH from biogas yield in immobilized AD. The microorganisms in AD are highly sensitive to pH, and a significant yield of methane was observed in all batch samples. The rate of conversion was also analyzed based on BOD/COD ratio, C/N ratio, and the pretreatment methods. For Batch 1 and 2 in Experiment 1, Sample 5 with a 3:2 dilution ratio yielded higher methane of 78.23% (v/v), and 1:2 of DWW and PM had 79.43% methane yield (v/v), respectively. In Experiment 2, Batch 5, with a dilution ratio of 7:3 of DWW and pure water generated the highest biogas yield of 2-L, and a BOD/COD ratio suggested a value that is closest to zero. Meanwhile, Batch 2 of pure DWW with no plastic media yielded 1850 L. There had been no correlation found between the reduction of %TSS and methane yield and total phosphorus with the biogas yield. As micronutrients remarkably affect the C/N ratio, an excessive amount of micronutrients added should be considered as it could cause toxicity. The AD efficiency is driven by catalysts such as cobalt or iron and plastic media. Ultimately, this study provides new insights to further understand the generation of biogas and using plastic carriers consisting of microbial culture in distillery wastewater from sugar refineries.

**Acknowledgements** The authors extend their gratitude to the Center for Renewable Bioenergy Research and the School of Chemical, Biological, Materials Engineering and Sciences of Mapúa University.

## References

Achinas S, Achinas V, Euverink GJW (2020) Microbiology and biochemistry of anaerobic digesters: an overview. Bioreactors. https://doi.org/10.1016/B978-0-12-821264-6.00002-4

AgStar (2012) Increasing anaerobic digester performance with codigestion. https://www.epa.gov

den Boer E, den Boer J (2021) Production of volatile fatty acids in biorefineries. Waste biorefinery: value addition through resource utilization, pp 159–179. https://doi.org/10.1016/B978-0-12-821879-2.00006-5

Evidente RC, Almendrala M, Villaflor SM, Doma B (2021) Anaerobic co-digestion of press mud and molasses-based distillery wastewater for enhanced biogas production. IOP Conf Ser: Earth Environ Sci. https://doi.org/10.1088/1755-1315/943/1/012017

Gerardi MH (2003) The microbiology of anaerobic digesters, chap 7. Wiley, pp 51–59

Gonzales LM, Reyes IP, Romero OR (2017) Anaerobic co-digestion of sugarcane press mud with vinasse on methane yield. Waste Manag 68:139–145

Kumar MR, Aklilu EG (2016) Production of biogas fuel from alcohol distillery plant. Int J Sci Res Publ 6(12). ISSN: 2250-3153

Menon A, Wang JY, Giannis A (2017) Optimization of micronutrient supplement for enhancing biogas production from food waste in two-phase thermophilic anaerobic digestion. Waste Manag 59:465–475. https://doi.org/10.1016/j.wasman.2016.10.017

Radjaram B, Saravanane R (2011) Assessment of optimum dilution ratio for biohydrogen production by anaerobic co-digestion of press mud with sewage and water. Biores Technol 102(3):2773–2780

Silva FM, Mahler CF, Oliveira LB, Bassin JP (2018) Hydrogen and methane production in a two-stage anaerobic digestion system by co-digestion of food waste, sewage sludge, and glycerol. Waste Manag 76:339–349

Vargas JA, Larios AF, Alvarez VG, Gonzales RI, Acosta HO (2016) Single and two-stage anaerobic digestion for hydrogen and methane production from acid and enzymatic hydrolysates of *Agave tequilana* bagasse. Int J Hydrogen Energy 41(2):897–904

# Chapter 62
# Biochemical Methane Potential Enhancement Through Biomass Fly Ash Addition

**R. P. Rodrigues, P. V. Almeida, C. M. O. Martinho, L. M. Gando-Ferreira, and M. J. Quina**

**Abstract** The bioenergy production from alternative sources such as biomass has been leading to an increase of biomass fly ash (BFA) that need proper management. In this study, the performance and optimization of the biochemical methane potential of tomato residues were studied through the addition of BFA. The results demonstrated that higher methane production can be achieved when BFA is added in the concentration of 2500 and 10,000 mg/L instead of the use of a synthetic nutrient medium. The optimization of the substrate-to-inoculum ratio (SIR) and the BFA concentration demonstrated that a higher SIR can be applied in the reactor due to the higher buffering capacity of the system, which allows the digestion of higher substrate quantities. The optimum conditions determined were a SIR equal to 1.0 and a BFA concentration of 5000 mg/L. Moreover, the kinetic study revealed that the BFA can be added to the anaerobic digestion process without compromising the rate of biogas formation. Overall, this study demonstrated that biomass fly ash can act as a buffering agent as well as a nutrient supplier in the anaerobic digestion process.

**Keywords** Anaerobic digestion · Nutrient substitute · Tomato residue

## 62.1 Introduction

Bioenergy production from alternative sources such as biomass has been widely used due to the need for fossil fuel substitution and because of the renewable nature of biomass (Alavi-Borazjani et al. 2021). However, high quantities of ash are generated resulting from biomass combustion. According to Alavi-Borazjani et al. (2021), about 68.6 million tons of ash are generated per 1 billion tons of biomass fully converted into energy. The high amount of biomass ash produced can lead to a

R. P. Rodrigues (✉) · P. V. Almeida · C. M. O. Martinho · L. M. Gando-Ferreira · M. J. Quina
Department of Chemical Engineering, University of Coimbra, CIEPQPF, Rua Sílvio Lima, Pólo II—Pinhal de Marrocos, 3030-790 Coimbra, Portugal
e-mail: rafaelar@eq.uc.pt



N. S. Caetano and M. C. Felgueiras (eds.), *The 9th International Conference on Energy and Environment Research*, Environmental Science and Engineering,
https://doi.org/10.1007/978-3-031-43559-1_62

serious environmental problem if not managed properly. Besides the disposal issues related to ash, the depletion of natural sources of biomass has been a major concern if the utilization rate is not adequate.

In order to overcome these problems, it is necessary to find new alternatives for valorizing these residues. Anaerobic digestion (AD) is a biological process that can convert biomass into bioenergy in the absence of oxygen. Indeed, the biogas (a mixture of $CH_4$ and $CO_2$) formed in AD can be further valorized as electrical energy, heat, or fuel gas (Rodrigues et al. 2022; Xiao et al. 2020). AD is a multi-step biological process that takes place through four successive phases: hydrolysis, acidogenesis, acetogenesis, and methanogenesis (Zhang et al. 2014). Hydrolysis is a complex phenomenon mediated by extracellular enzymes in which organic polymers (e.g. proteins, carbohydrates, lipids, etc.) are degraded into low molecular weight compounds. These small molecules are then degraded through fermentative bacteria resulting mainly in volatile fatty acid (VFA)—acidogenesis. The acidogenesis products are further converted, through the action of acetogenic bacteria into acetate, hydrogen, and carbon dioxide—acetogenesis. In the final stage of AD, methanogenesis, the methanogenic bacteria metabolize formic acid, acetic acid, methanol, carbon monoxide, carbon dioxide, and hydrogen into methane. It is well known that the accumulated VFA during acidogenesis can inhibit the formation of methane (Xiao et al. 2021). Therefore, it is necessary to ensure the stability of the process to be efficient and profitable.

Due to the nature of the AD process, in addition to providing bioenergy using residues as substrates, it has become an attractive and promising destination for the use of biomass ash (Alavi-Borazjani et al. 2021). In fact, biomass ash has been reported as a good nutrient supplier and buffering agent in AD processes (Alavi-Borazjani et al. 2021; Podmirseg et al. 2013). The tomato industry is one of the most relevant socio-economic activities in Europe (Almeida et al. 2022). However, a high amount of residues are generated during all process stages (Gaspar et al. 2019). In particular, the tomato industry generates a significant amount of tomatoes unfit for consumption (high maturation stage). This represents an environmental problem that needs to be assessed properly. Several studies point out AD as an effective process to valorize wastes from the tomato industry (Almeida et al. 2021a, b; Rodrigues et al. 2019).

In this context, this study aims to evaluate and optimize the biochemical methane potential (BMP) of tomato residues (acidic substrate) through the addition of biomass fly ash (BFA).

## 62.2 Material and Methods

### 62.2.1 *Materials*

The tomato residue was simulated by leaving tomatoes (*Solanum lycopersicum*) collected in a local supermarket at room temperature for two weeks to increase the state of maturation (unfit for consumption). The whole tomato fruit (peels, skins, and seeds) was crushed to ensure its homogeneity. The anaerobic inoculum was collected in a local wastewater treatment plant in the central region of Portugal. The sample of BFA was collected from a biomass power plant located in the central region of Portugal, that burns forestry residues (about 85% eucalyptus).

### 62.2.2 *Materials Characterization*

The natural pH and the electric conductivity (EC) of substrates were determined in a water suspension using a liquid-to-solid ratio (L/S) of 10 L/kg. The alkalinity of the BFA was determined through a titration method with HCl (0.1 N) using an L/S of 10 L/kg. The total solid (TS) content was assessed by drying the samples at 40 $\pm$ 5 °C until constant weight. The volatile solids (VS) were obtained by calcinating the material in a furnace at 550 °C for 2 h. Total chemical oxygen demand (COD) was determined according to standard methods. The elemental analysis of the tomato was conducted using Elemental Analyzer NA 2500 equipment, and the elemental composition of BFA was determined by X-ray fluorescence (Bruker TXRF S2 PICOFOX).

### 62.2.3 *Biochemical Methane Potential*

Biochemical methane potential (BMP) assays were performed in reactors with a working volume of 100 mL at 37 °C, in duplicate. The VS concentration was set at 34.5 g/L. The flasks were flushed with $N_2$ gas to assure an inert atmosphere. The biogas volume was estimated by manometric measurements and the composition was measured using GAS DATA (GMF 406) equipment. Two different and independent studies were conducted. In order to evaluate the nutrient capacity of the BFA, a substrate-to-inoculum ratio (SIR) of 0.5 was used, a synthetic nutrient medium indicated by Angelidaki et al. (2009) was added in half of the experiments, and a concentration of BFA between 0 and 10,000 mg/L was tested. Moreover, a simple Tuckey test (with a confidence level of 95%) was performed on five sets of samples to evaluate the effect of adding BFA under the same concentration conditions. The optimization of the BMP using BFA as an additive (no synthetic medium was used in these experiments) was performed through a design of experiments (DoE) using JMP Pro 16 software. A full factorial design was selected, while three levels of SIR

(0.5, 1.0, and 1.5) and four concentrations of BFA (0, 2500, 5000, and 1000 mg/L) were tested. The optimum conditions were obtained through the maximization of the desirability function and based on the surface response plot obtained through the construction of a second-degree polynomial model. Moreover, the methane formation kinetics was also evaluated through the fitting of the first-order kinetic model described by Eq. 62.1, as suggested by Browne and Murphy (2013).

$$BMP(t) = BMP_{\infty} \times (1 - e^{-kt}) \tag{62.1}$$

where BMP ($NmL_{CH4}/gVS$) corresponds to the cumulative specific methane production over time, and $BMP_{\infty}$ ($NmL_{CH4}/g_{VS}$) is the maximum volume of methane produced.

## 62.3 Results and Discussion

### *62.3.1 Materials Characterization*

This study was conducted with tomato residues as a substrate for the evaluation of the anaerobic digestion performance with the addition of biomass fly ash (BFA). The tomato residue and the BFA were characterized, and the main physical and chemical properties are summarized in Tables 62.1 and 62.2, respectively.

The tomato residue is characterized by a high water content (around 83%) and high organic matter (around 95%TS of volatile solids), as stated by Almeida et al. (2021a).

According to Jain et al. (2015), the C/N ratio of the substrates must be between 20 and 30 to have a proper anaerobic digestion performance. According to the results obtained, this substrate presents a low C/N ratio (= 18.1). In the case of low C/N, an increase in ammonia formation can be observed and consequently the inhibition of the methane formation. Moreover, the acidic nature of the tomato residues can lead to an acidification of the reactor, causing a detrimental effect on methanogenic microorganisms.

To overcome the possibility of system instability, it is necessary to ensure a proper buffering capacity of the system. BFA is known for its high pH and buffering capacity.

**Table 62.1** Characterization of tomato residue

| Parameter | Value | Parameter | Value |
|---|---|---|---|
| pH | $4.19 \pm 0.01$ | C (%, db) | $44.03 \pm 0.13$ |
| TS (%) | $16.82 \pm 1.96$ | N (%, db) | $2.43 \pm 0.04$ |
| VS (%, db) | $92.76 \pm 0.53$ | O (%, db) | $39.81 \pm 0.20$ |
| COD (mg $O_2/g_{VS}$) | $1517 \pm 19$ | H (%, db) | $6.49 \pm 0.35$ |

*db*—dry basis

**Table 62.2** Characterization of biomass fly ash

| Parameter | Value | Element | Value (g/kg, db) | Element | Value (mg/kg, db) |
|---|---|---|---|---|---|
| pH | 12.84 ± 0.05 | Ca | 60.56 ± 6.66 | Zn | 148.41 ± 9.55 |
| TS (%) | 98.90 ± 0.10 | K | 23.35 ± 2.83 | | |
| VS (%, db) | 4.33 ± 0.11 | Cl | 18.05 ± 3.56 | Br | 78.36 ± 15.19 |
| EC (mS/cm) | 12.53 ± 0.42 | S | 7.12 ± 1.10 | Cu | 28.55 ± 1.68 |
| Alkalinity ($mg_{CaCo3}$/kg, db) | 250.2 ± 0.5 | Fe | 6.80 ± 0.50 | Co | 16.03 ± 1.58 |
| | | Mn | 1.66 ± 0.10 | Ni | 13.66 ± 0.76 |
| | | P | 0.92 ± 0.22 | Cr | 12.61 ± 0.78 |

*db*—dry basis

In fact, this is verified through the experimental characterization of the BFA used in this study. The BFA tested in the present study revealed a pH of 12.84 and alkalinity of 250.2 $mg_{CaCO3}/kg_{db}$, which can be beneficial for the stabilization of the process (Alavi-Borazjani et al. 2021).

BFA can also act as a nutrient substitute due to the presence of some key elements essential for the AD process. Takashima et al. (2009) referred that P, K, and S are essential elements for the activity of methanogenic bacteria. Moreover, elements such as Ca, Fe, Zn, Ni, Co, and Cu were also reported as stimulants for the anaerobic activity of the bacteria.

### 62.3.2 *Biomass Fly Ash as a Nutrient Supplier*

The potential of BFA to be used as a nutrient supplier was evaluated by performing some tests using external nutrient media and different quantities of ash. The BMP obtained can be seen in Fig. 62.1, and the results show that in the case where BFA was not added (0 mg/L) the removal of the external nutrient media caused a slight decrease in the BMP of tomato residue (from 172 to 124 $NmL_{CH4}/g_{VS}$).

This is an indication of nutrient deficiency in the system. The addition of the BFA in the experiments where synthetic nutrient media was used did not have a significant impact since a low variation of the BMP value was observed. However, when no nutrient media was applied to the reactor, the addition of BFA caused a positive impact on the BMP. In fact, the high EC observed in Table 62.2 for the BFA is an indication of nutrient solubilization on an aqueous media. With the addition of BFA in the concentrations of 2500 and 10,000 mg/L, it was possible to obtain higher BMP values (234 and 261 $NmL_{CH4}/g_{VS}$, respectively) than those obtained using a standard nutrient media. In fact, through the Tuckey test, it is possible to observe that only in these two concentrations a significant difference between the use or not of

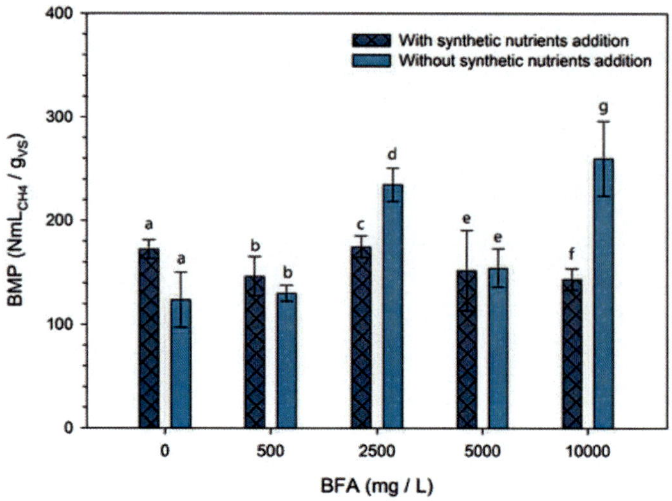


**Fig. 62.1** Effect of biomass fly ash addition in the biochemical methane potential

a synthetic nutrient media is achieved. This study indicates the potential beneficial use of biomass fly ash as a substitute for synthetic nutrient medium.

### *62.3.3 Optimization of Biochemical Methane Potential*

According to Almeida et al. (2021b), the substrate-to-inoculum ratio was found to be the key operational condition for tomato residues processed by anaerobic digestion. To optimize the BMP, in this study, three SIR (0.5, 1.0, and 1.5) and four BFA concentrations (0, 2500, 5000, and 1000 mg/L) were evaluated. The results of the optimization study are presented in Fig. 62.2, as well as the first-order kinetic parameters obtained from fittings which are summarized in Table 62.3.

The results demonstrate that, as suggested by Almeida et al. (2021b), the SIR plays an important role in the formation of methane from tomato residues. Indeed, as can be seen in Fig. 2a), a slight variation in the SIR causes a high variation in the BMP value. As in the case of SIR, the concentration of BFA added to the AD system can also affect the methane potential yield. The effect of two variables studied on the BMP value can also be observed in Fig. 2b), where the surface response plot of the second-order polynomial model obtained is presented. Higher BMP values were obtained for intermediate BFA concentration and SIR conditions. Moreover, the individual BMP values revealed that higher values are achieved for the SIR ratio of 1.0 (for the BFA concentrations of 5000 mg/L) and 1.5 (for the BFA concentrations of 2500 mg/L).

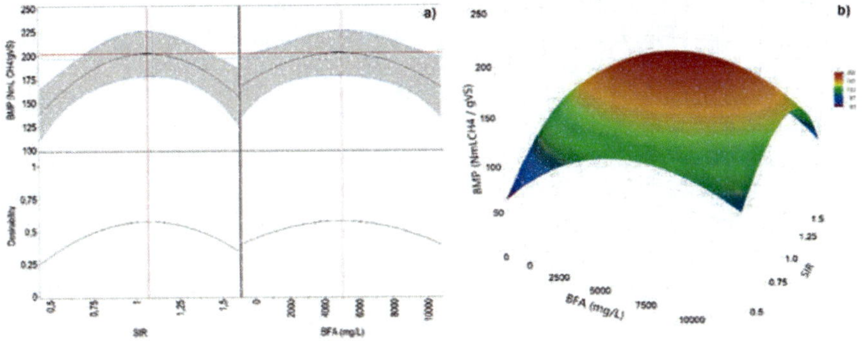


**Fig. 62.2** Optimization of the BMP tests: **a** desirability function; **b** surface response plot

**Table 62.3** Biochemical methane potential optimization kinetic parameters

| SIR | BFA (mg/L) | $BMP_\infty$ ($NmL_{CH4}$/ gvs) | k (1/d) | $R^2$ | $R^2_{Adj}$ | RMSE |
|---|---|---|---|---|---|---|
| 0.5 | 0 | 119.6 ± 1.4 | 0.91 ± 0.06 | 0.982 | 0.981 | 5.18 |
| 0.5 | 2500 | 130.0 ± 2.9 | 0.77 ± 0.09 | 0.942 | 0.939 | 10.26 |
| 0.5 | 5000 | 143.9 ± 7.0 | 0.61 ± 0.14 | 0.788 | 0.777 | 24.35 |
| 0.5 | 10,000 | 138.1 ± 1.1 | 0.55 ± 0.02 | 0.994 | 0.993 | 3.68 |
| 1.0 | 0 | 164.8 ± 3.3 | 0.65 ± 0.06 | 0.958 | 0.956 | 11.45 |
| 1.0 | 2500 | 179.6 ± 2.9 | 0.57 ± 0.04 | 0.975 | 0.973 | 9.88 |
| 1.0 | 5000 | 238.7 ± 5.8 | 0.44 ± 0.04 | 0.950 | 0.948 | 18.93 |
| 1.0 | 10,000 | 170.0 ± 1.8 | 0.51 ± 0.02 | 0.989 | 0.988 | 6.10 |
| 1.5 | 0 | 155.9 ± 2.4 | 0.35 ± 0.02 | 0.981 | 0.980 | 7.57 |
| 1.5 | 2500 | 206.0 ± 4.9 | 0.35 ± 0.03 | 0.960 | 0.958 | 15.08 |
| 1.5 | 5000 | 135.0 ± 2.4 | 0.48 ± 0.04 | 0.970 | 0.968 | 8.11 |
| 1.5 | 10,000 | 141.5 ± 1.4 | 0.53 ± 0.02 | 0.990 | 0.989 | 4.88 |

RMSE—root mean square error; $R^2$—determination coefficient; $R^2_{Adj}$—adjusted determination coefficient

In previous studies using tomato residues as an AD substrate, the process showed the best performance when lower SIR, 0.5, was used. Higher SIR can lead to acidification of the reactor and methane inhibition (Almeida et al. 2021b). However, in the present study, the addition of the BFA allowed the utilization of a higher SIR, which can be an indication of a better level of the buffering capacity of the system. In fact, the addition of BFA demonstrated a positive impact on the BMP value in all tests except for the experiment performed at SIR 1.5 with 5000 and 10,000 mg/L of BFA. These results indicate the benefit of incorporating BFA in the AD process.

The optimum conditions obtained through the maximization of the desirability function present in Fig. 2a), are a SIR of 1.0 with the addition of BFA at a concentration of 5000 mg/L. A first-order kinetic model was used to evaluate the $CH_4$ formation in the optimization tests (Table 62.3). Overall, the kinetics of the $CH_4$ formation is described by the chosen model.

Moreover, it is possible to observe that the addition of the BFA did not have a significant impact on the kinetic constant, which demonstrates that the ash can be used without compromising the production performance. It is important to note that after the AD process, the BFA remains in the digestate fraction. However, the presence of the BFA in this sludge does not represent an additional problem for further valorization in agronomical applications. In fact, BFA is often used directly in the soil as an amendment and/or fertilizer (Alavi-Borazjani et al. 2021).

## 62.4 Conclusions

The results demonstrated that the addition of BFA at concentrations of 2500 and 10,000 mg/L allowed for higher methane production when compared to the values obtained when a standard external nutrient medium was used. The optimization of the substrate-to-inoculum ratio and the BFA concentration led to the conclusion that this addition of BFA allows using a higher SIR, which may be due to the increased buffering capacity of the system. The optimum conditions were obtained for a SIR of 1.0 and a BFA concentration of 5000 mg/L, with a methane production of 237.9 $NmL_{CH4}/g_{VS}$. The kinetic study indicated that the BFA can be added to the system without compromising the system performance in terms of biogas production rate. Overall, biomass fly ash can act as a buffering agent as well as a nutrient supplier in anaerobic digestion systems.

**Acknowledgements** R.P. Rodrigues acknowledges the Fundação para a Ciência e Tecnologia (FCT) for the Ph.D. grant (SFRH/BD/145694/2019) and the fellowship from the project MATIS, Fundo Europeu de Desenvolvimento Regional (FEDER), através do Programa Operacional Regional do Centro (CENTRO2020) CENTRO-01-0145-FEDER-000014. Thanks are due to the Strategic Project of CIEPQPF (UIDB/00102/2020), financed by FCT through national funds. P.V. Almeida acknowledges the Fundação para a Ciência e Tecnologia (FCT) for the Ph.D. grant (2020.08445.BD).

## References

Alavi-Borazjani SA, Tarelho LAC, Capela I (2021) A brief overview on the utilization of biomass ash in biogas production and purification. Waste Biomass Valoriz 12:6375–6388

Almeida PV, Rodrigues RP, Gaspar MC, Braga MEM, Quina MJ (2021a) Integrated management of residues from tomato production: recovery of value-added compounds and biogas production in the biorefinery context. J Environ Manag 299:113505

Almeida PV, Rodrigues RP, Teixeira LM, Santos AF, Martins RC, Quina MJ (2021b) Bioenergy production through mono and co-digestion of tomato residues. Energies 14:5563
Almeida PV, Rodrigues RP, Slezak R, Quina MJ (2022) Assessment of NIR spectroscopy for predicting biochemical methane potential of agro-residues—a biorefinery approach. Biomass Bioenergy 151:106169
Angelidaki I, Alves M, Bolzonella D, Borzacconi L, Campos JL, Guwy AJ, Kalyuzhnyi S, Jenicek P, Van Lier JB (2009) Defining the biomethane potential (BMP) of solid organic wastes and energy crops: a proposed protocol for batch assays. Water Sci Technol 59:927–934
Browne JD, Murphy JD (2013) Assessment of the resource associated with biomethane from food waste. Appl Energy 104:170–177
Gaspar MC, Mendes CVT, Pinela S, Moreira R, Carvalho MGVS, Quina MJ, Braga MEM, Portugal A (2019) Assessment of agroforestry residues: their potential within the biorefinery context. ACS Sustain Chem Eng 7:17154–17165
Jain S, Jain S, Wolf IT, Lee J, Tong YW (2015) A comprehensive review on operating parameters and different pretreatment methodologies for anaerobic digestion of municipal solid waste. Renew Sustain Energy Rev 52:142–154
Podmirseg SM, Seewald MSA, Knapp BA, Bouzid O, Biderre-Petit C, Peyret P, Insam H (2013) Wood ash amendment to biogas reactors as an alternative to landfilling? A preliminary study on changes in process chemistry and biology. Waste Manag Res 31:829–842
Rodrigues RP, Klepacz-Smolka A, Martins R, Quina M (2019) Comparative analysis of methods and models for predicting biochemical methane potential of various organic substrates. Sci Total Environ 649:1599–1608
Rodrigues RP, Gando-Ferreira LM, Quina MJ (2022) Increasing value of winery residues through integrated biorefinery processes: a review. Molecules 27:4709
Takashima M, Speece RE, Parkin GF (2009) Mineral requirements for methane fermentation. Crit Rev Environ Control 19:465–479
Xiao Q, Chen W, Tian D, Shen F, Hu J, Long L, Zeng Y, Yang G, Deng S (2020) Integrating the bottom ash residue from biomass power generation into anaerobic digestion to improve biogas production from lignocellulosic biomass. Energy Fuels 34:1101–1110
Xiao Q, Hu J, Huang M, Shen F, Tian F, Zeng Y, Jang M-K (2021) Valorizing the waste bottom ash for improving anaerobic digestion performances towards a "Win-Win" strategy between biomass power generation and biomethane production". J Clean Prod 295:126508
Zhang C, Su H, Baeyens J, Tan T (2014) Reviewing the anaerobic digestion of food waste for biogas production. Renew Sustain Energy 38:383–392

# Part VIII
# Energy Policy, Economics, Planning, and Regulation

# Chapter 63
# Levelized Cost of Storage of Second-Life Battery Applications in Flanders, Belgium

**Dominik Huber, Maeva Lavigne Philippot, Daniele Costa, Jelle Smekens, and Maarten Messagie**

**Abstract** Batteries from electric vehicles can still provide around 80% of initial capacity, being suitable for stationary applications. Electricity storage is a solution for integrating renewable energies, as the electrification of the transportation fleet will make used batteries available. However, it is unclear if second-life batteries (SLB) will offer a more sustainable alternative to stationary new batteries (SNB). Therefore, this study investigates the life cycle economic impacts of future SLB in Flanders, Belgium. It focuses on collecting, dismantling, repurposing, using them in a second application, and recycling those batteries. Hence, a levelized cost of storage (LCOS) calculation is carried out from cradle-to-grave. Three use cases are assessed: (a) a residential use case, (b) an industrial use case, and (c) a utility use case. Cost data represent the geographical scope of Flanders, Belgium. The calculated LCOS of SNB for the three use cases were found to be between 71.77 €/MWh and 202.25 €/MWh in 2040, whereas the LCOS of SLB vary between 64.99 €/MWh and 211.10 €/MWh. For SLB, dismantling and repurposing costs dominate the first years economically. Towards 2040 battery purchase and charging costs will be major contributors. Overall, the utility use case proposes the lowest LCOS. The cost-competitiveness amongst SLB and SNB varies over the different years. This study highlights the importance of the application. Furthermore, it stresses that the cost of extending the EV batteries lifetime can be considerable.

**Keywords** Belgium · Levelized cost of storage (lcos) · Li-ion battery · Second-life battery · Stationary energy storage

D. Huber (✉) · M. Lavigne Philippot · J. Smekens · M. Messagie
Electric Vehicle and Energy Research Group (EVERGI), Mobility, Logistics and Automotive Technology Research Centre (MOBI), Department of Electrical Engineering and Energy Technology, Vrije Universiteit Brussel, Pleinlaan 2, 1050 Brussels, Belgium
e-mail: Dominik.Huber@vub.be

D. Costa
VITO—EnergyVille, Unit Smart Energy and Built Environment (SEB), Thor Park 8310, 3600 Genk, Belgium

 
N. S. Caetano and M. C. Felgueiras (eds.), *The 9th International Conference on Energy and Environment Research*, Environmental Science and Engineering,
https://doi.org/10.1007/978-3-031-43559-1_63

## 63.1 Introduction

With the increase in the electrification of the transport sector, more and more retired traction batteries will become available in the following years. As the capacity of such devices after use in a vehicle ranges typically between 70 and 80% of the initial capacity, they still could serve other applications in second-life. Even though extending the first life of batteries is not new, its costs remain uncertain. Various levelized cost of storage (LCOS) studies addressing different research directions are available in the scientific literature [9, 13, 18]. So far, only two studies have evaluated the LCOS of second-life batteries (SLB), both from the perspective of the United States of America [14, 20]. Both studies compare LCOS of SLB with stationary new batteries (SNB), pointing out different results. Thus, this study identifies the LCOS of future SLB for a particular geographical region in Belgium, called Flanders, considering 3 different use cases: a residential, an industrial and a utility use case. Data are obtained from the Re2Live project, dedicated to determining the potential of SLB in Flanders.

## 63.2 Materials and Methods

After batteries have been utilized in battery electric vehicles (BEV), additional value chain steps are required to obtain a SLB: collection, dismantling, repurposing and, after serving as stationary storage, dismantling and recycling (Fig. 63.1). Sections 63.2.1 to 63.2.3 present the methodology, the use cases and the cost data, respectively.

### *63.2.1 Levelized Cost of Storage*

The LCOS (Eq. 63.1) is the ratio of the battery installation, operation and storage device disposal costs and the cumulated discharged electricity over its lifetime. The lifetime costs for SLB are calculated by using Eq. 63.2.

$$\mathrm{LCOS}\,\frac{\mathrm{EUR}}{\mathrm{MWh}} = \frac{\text{Lifetime costs (EUR)}}{\text{Cumulated, discharged electricity (MWh)}} \tag{63.1}$$

$$\text{Lifetime costs}_{SLB}(€) = \begin{cases} t_0 = (CAPEX_{NB} * C_{NB})/(Eld_{SLB}/Eld_{total}) \\ t_N = \sum_n^N \frac{T_c+D_c+R_c}{(1+r)^N} + \frac{O\&M_c+a_{deg}}{(1+r)^n} \\ t_{N+1} = \frac{EoL_c}{(1+r)^{N+1}}/(Eld_{SLB}/Eld_{total}) \end{cases} \tag{63.2}$$

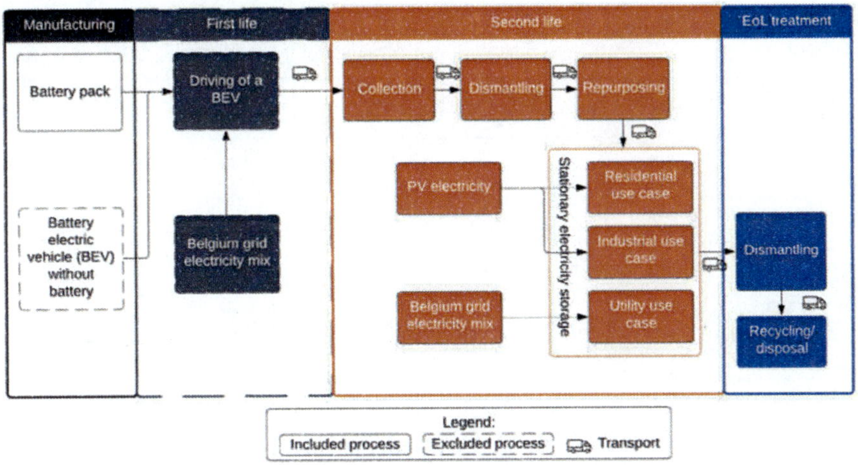


**Fig. 63.1** Scope of the economic evaluation of second-life batteries (BEV = battery electric vehicle, PV = photovoltaic, EoL = end of life)

The application of LCOS for SLB claims a standardized approach, reflecting, among others, the consideration of SLB-specific parameters, such as initial state of health ($SoH$), replacements, repurposing and new battery module costs [20]. The LCOS calculation should reflect additional costs required to extend the battery's lifetime and the additional discharged electric energy. The first application is excluded due to the focus on the second-life. Degradation, maintenance, operating, transport and repurposing costs can be fully allocated to the time where the second-life batteries are used ($t_N$). However, other costs occur regardless of a second application, e.g., when acquiring the batteries ($t_0$) or after its use ($t_{N+1}$): the initial purchase, the battery removal, the end of life ($EoL$)—treatment and the dismantling costs. Hence, this second group of costs is allocated based on the discharged electricity of the first and second applications [1]. In Eq. 63.2, the costs are grouped according to their occurrence: capital expenditure of new batteries ($CAPEX_{NB}$), the initial capacity of new batteries ($C_{NB}$), discharged electricity for SLB and total ($Eld_{SLB}$ and $Eld_{total}$) are used to determine the cost at the purchasing time. Transportation, dismantling, removal, operating and maintenance, and degradation costs ($T_c$, $D_c$, $R_c$, $O\&M_c$, $a_{deg}$) are also considered during the second-life. Costs for EoL treatment ($EoL_c$) are reflected as well. Similarly to costs, only discharged electricity of the second-life application is considered, neglecting the discharged electricity of the first life. Use case-specific discharged electricity has been evaluated by multiplying the different battery capacities, full equivalent cycles, round-trip efficiencies, depth-of-discharges and operating days [10, 11, 19].

**Table 63.1** Use cases

| Use case | First life | Residential | Industrial | Utility |
|---|---|---|---|---|
| Battery installation capacity (MWh) | 0.041 | 0.0082 | 1.50 | 20 |
| Utilized number of battery packs (number) | 1 | 0.25 | 38 | 548 |
| Lifetime battery (years) | 9 | 12 | 12 | 12 |
| Total delivered electricity battery (MWh) | 23.39 | 22.48 | 3800 | 21,000 |
| PV capacity (kWp) | n.a. | 4 | 1000 | n.a. |
| Electricity price for charging (€/kWh) | n.a. | 0.16 | 0.12 | n.a. |
| Maintenance costs (€/year) | n.a. | 37.86 | 1426.10 | 19,696.20 |
| CAPEX of mobile battery (€/kWh) | 97.68 | 97.68 | 97.68 | 97.68 |

Adapted from [4, 8, 12, 21, 22]

### *63.2.2 Use Cases*

The assessed SLB is a battery with lithium nickel manganese cobalt oxide (NMC111) obtained after the first life in a 2017 Renault Zoe ZE40, consuming 175 Wh/km under the worldwide harmonized light vehicles test procedure (WLTP) cycles [2]. The total mileage is 133,668 km [7]. Table 63.1 provides an overview of all use cases.

The case studies and their main assumptions are as follows:

- Residential use case: a 2017 Renault Zoe ZE40 battery of 41 kWh capacity is dismantled and 3 of its modules are used to store solar energy for one Belgian household. At this household, an average domestic photovoltaic (PV) installation with the corresponding storage system in Belgium is in place. As charging costs, the levelized cost of electricity (LCOE) is applied, representing an average of European residential PV installations [12]. Maintenance costs include electricity for capacity and visual checks, besides labor costs [21].
- Industrial-scale use case: this installation is set up to provide behind-the-meter services, such as peak shaving or an uninterrupted power supply. The SLBs are grouped in a sea freight container in racks with a cooling system [19]. Charging costs are evaluated by the LCOE of industrial PV installations [12]. Maintenance costs were upscaled to the system size.
- Utility-scale use case: a battery system is set up to participate in the secondary reserve market. The packs are mounted in containers with racks and a cooling system [19]. By participation in the secondary reserve market, a certain capacity of the batteries is contracted [3]. Afterwards, the installation owner is not responsible for charging and discharging. Hence, charging costs are assumed to be zero. Due to the bigger system capacity, the maintenance is expected to take 16 h per year.

### 63.2.3 Cost Data

If not indicated otherwise, all data are primary and obtained from the project consortium. A price of 111 $/kWh for an electric vehicle (EV) NMC111 battery is utilized and translated into 97.68 €/kWh per battery pack [5, 16]. The purchase price is allocated between the first and second-life according to the discharged electricity in each life. A linear degradation over the lifetime is included based on the purchase costs accounting for the difference between initial and final SoH. Additionally, 117.00 €/pack is considered for removing the EV battery after its first life [17] and allocated to both first and second-life. The collection costs are the same for all use cases and consider transportation between the different locations and the required packaging. The costs are considered from end-of-first-life to the dismantling facility, from the dismantling to the repurposing and second-life application and from dismantling to a recycling facility. Collection costs from the end of the first life directly to the second-life application are excluded. The included SNB collection costs are: transporting batteries and packaging from the end of the first life to the dismantling facility and from dismantling to a recycling facility. Depending on the use cases, dismantling costs are assessed. For the residential use case, dismantling up to the module level is required, while dismantling up to the pack level is assumed for the industrial and utility use case. Dismantling costs include the infrastructure, the electricity for dismantling the packs and the transportation to the facility. In 2022, the manual dismantling of the packs is assumed, followed by the slow uptake of automation levels over the years. Dismantling costs per pack are then aligned to each use case. Repurposing costs before the second use include investments for the facility, labor costs for the repurposing and variable costs, such as electricity, a new battery management system, a thermal management system, other safety features and disposal costs. The value of materials after the batteries have been recycled is evaluated. Therefore, the LCOS calculation considers the recycling value as a negative cost. To calculate such benefits, the material composition of the modeled NMC111 battery is multiplied by the expected value for different materials. This approach is adjusted for the other two use cases by adjusting the weight of all components to the required battery capacity. Recycling benefits are allocated to first and second-life according to the discharged electricity. For comparison, a new stationary benchmark battery is included (with the same chemistry as the SLB). Furthermore, a sensitivity analysis is carried out, in which one parameter at a time is changed while the others remain constant. The sensitivity analysis of this study is limited to 3 parameters: lifetime, allocation rule and discount rate. The values used in the initial calculations for the lifetime of SNB are obtained from literature, while the lifetime of SLB originate from the Re2Live project [20]. The 10-year yield curve spot rate for government bonds, rated with triple-A for the Euro area, is used as indication for the discount rate [6, 22]. Table 63.2 outlines the parameters for SNB and the sensitivity analysis.

Table 63.2 Parameter modification for sensitivity analysis 1–3

| No | Parameter | SLB: initial | SLB: adapted | SNB: initial | SNB: adapted |
|---|---|---|---|---|---|
| 1 | Lifetime (years) | 12 | 20 | 15 | 30 |
| 2 | Allocation rule (%) | 62–81 | 50 | n.a. | n.a. |
| 3 | Discount rate (% p.a.) | 0.10 | 5.00 | 0.10 | 5.00 |

SLB = second-life batteries, SNB = stationary new batteries

## 63.3 Results and Discussion

### 63.3.1 *Residential Use Case*

The LCOS of SLB is 428.71 €/MWh for a residential use case in Belgium in 2022, while the LCOS for SNB in the same use case is 315.85 €/MWh. Dismantling and repurposing costs account for over 50% of the LCOS in the first year. This is reasonable as these two value chain steps have to be set up, requiring significant investment, while only low battery pack throughput can be obtained. In 2025, a cut-back of dismantling and repurposing costs by almost 90% results in LCOS of 237.92 €/MWh. Until 2040, the LCOS for SLB diminishes only marginally due to changes in the charging costs. Towards 2040, a shift in the cost drivers is observed, where operating costs represent the greatest contributors. For the SNB, a slight but constant decrease in the LCOS over time can be observed. Even though the initial purchase prices are much higher for SNB than for SLB, the LCOS of SNB is still lower than SLB. The constant decrease of LCOS over time is due to the decline of the purchase price for SNB over time and the corresponding degradation costs. The main cost driver for SNB is the operation costs, similar to the SLB. In 2040, the LCOS of a SLB is 211.10 €/MWh and the LCOS of a SNB is 202.25 €/MWh.

### 63.3.2 *Industrial Use Case*

Regarding the LCOS of SLB in an industrial use case, the initial investment for dismantling and repurposing facilities drives up the LCOS in the first years. Therefore, LCOS for SLB will be 351.10 €/MWh in 2022. The dismantling and repurposing costs in the industrial use case make up almost 60% of the LCOS in 2022. High repurposing costs per pack are due to a low battery pack throughput, resulting in high repurposing costs for all required packs. Analogous operating costs represent about 35% of the industrial use case in 2022. In this use case, lower operating costs resulted from lower charging costs than in the residential use case. Furthermore, there is a considerable difference in LCOS of SLB and SNB in 2022, which dwindles towards 2040. The LCOS of SLB in 2040 is 149.86 €/MWh, while the LCOS of SNB is 145.61 €/MWh. The lower LCOS of SNB compared to a residential use case can mainly be assigned to a greater amount of delivered electricity and a lower price for charging the batteries.

### 63.3.3 Utility Use Case

The smallest LCOS for SLB can be observed in the utility use case, as no charging costs are considered. However, this does not hold for 2022, where the LCOS of SLB is 594.66 €/MWh. Once more, dismantling and repurposing costs are significant in 2022. However, these costs will level out over time by 2030. From 2030 until 2040, the purchase price will account for at least 60% of the LCOS for SLB. Other costs contribute around 10% to the LCOS in these years. For the SNB, the main cost driver is the purchase costs, which account for 80% or more. The LCOS of SLB in 2040 is 64.99 €/MWh, whereas the LCOS for SNB in 2040 is 71.77 €/MWh.

### 63.3.4 Benchmarking

The LCOS of SLB in 2022 for the residential and the industrial use cases are in the same range as the LCOS in similar studies [15, 20] (Fig. 63.2). On the contrary, LCOS for SLB and SNB in 2022 for the utility use case are well above comparable study results [15], even though no charging costs are included here. However, considering LCOS evolution by 2040, all results are well below the LCOS of both consulted studies.

### 63.3.5 Sensitivity Analysis

Overall, extending the lifetime will reduce the LCOS of all use cases. The lifetime extension shows a substantial decrease in 2022, while the decrease in 2040 is divergent. These divergent tendencies reflect the difference in LCOS amongst the use cases. For the second sensitivity, the allocation rule was changed. In all use cases, the LCOS hardly responds to another allocation rule (0.00% up to −9.64% changes in LCOS). The most significant changes are revealed for the industrial use case, followed by the residential use case. The utility use case does show only in 2022 some changes due to another allocation rule, while over the other years, no implications are observed. In the third sensitivity, the real discount rate of 0.10% increased to 5.00%. Again, this sensitivity was in the same range amongst all use cases, except in 2040 for a utility use case.

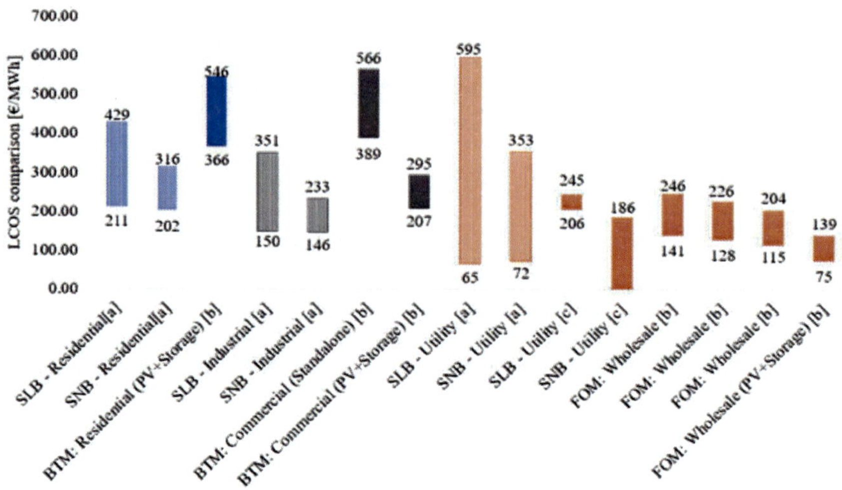


**Fig. 63.2** Levelized cost of storage of different system sizes (€/MWh) ([a] = own calculation, [b] = [15], [c] = [20]. Legend: LCOS = levelized cost of storage, BTM = behind the meter, FOM = in front of the meter, PV = photovoltaic, SLB = second-life battery, light blue = own calculations of the residential use case, dark blue = results of the residential use case of [15], light gray = own calculations for the industrial use case, dark grey = results of the industrial use case of [15], light orange = own calculations for the utility use case, dark orange = results of the utility use case of [15, 20])

## 63.4 Results

The LCOS of NMC111 SLB installed in different use cases (residential, industry and utility) is studied. Besides purchase, degradation, operation and maintenance costs, the assessment also includes removal, dismantling, transport and repurposing costs of SLB. The utility use case reveals the lowest LCOS. The main drivers are charging costs combined with the high electricity throughput. This result underlines the importance of electricity prices to charge the batteries. Next to operating costs, purchase costs are essential. Compared to SNB, the SLB currently profits from the price difference for mobile and stationary pack prices. However, the NMC111 battery pack price is expected to decrease linearly to 75.00 €/kWh by 2040 [16], representing stationary and mobile battery pack prices. Unfortunately, this price advantage of SLB is diminished by the high dismantling and repurposing costs. These costs are particularly high while upscaling the SLB market. Until 2040, the cost competitiveness between SLB and SNB varied, reaching almost equal LCOS in 2040. Conducting sensitivity analysis uncovered that adjusting the allocation rule hardly changed results, while extending the lifetime had a more significant influence. Despite, this study is subject to limitations. For example, the calculated results represent only NMC batteries. Furthermore, separate value chain steps are not entirely harmonized and incorporated into each other, neglecting the full potential of SLB. Finally, social and environmental impacts also should be used to determine the integration of SLB.

**Acknowledgements** This work was supported by the Agency for Innovation and Entrepreneurship (VLAVIO) [grant number HBC.2019.0125].

## References


1. Cusenza MA et al (2019) Energy and environmental benefits of circular economy strategies: the case study of reusing used batteries from electric vehicles. J Energ Storage 25:100845. ISSN 2352-152X. https://doi.org/10.1016/j.est.2019.100845, https://www.sciencedirect.com/science/article/pii/S2352152X19302944
2. Electric Vehicle Database. Renault Zoe ZE40 R110 price and specifications 202. https://ev-database.org/car/1236/Renault-Zoe-ZE40-R110 (visited on 2019). Accessed: 27.05.2022
3. ELIA. Auction Results 2022. https://www.elia.be/en/griddata/balancing/capacity-auction-results. Accessed: 21.03.2022
4. EU Science Hub. Photovoltaic Geographical Information System (PVGIS) 2022. https://re.jrc.ec.europa.eu/pvg_tools/en/. Accessed: 11.05.2022
5. European Central Bank. US dollar (USD) 2021. https://www.ecbeuropa.eu/stats/policy_and_exchange_rates/euro_reference_exchange_rates/html/eurofxref-graph-usd.en.html. Accessed: 24.04.2021
6. European Central Bank. Yield curve spot rate, 10-year maturity—Government bond, nominal, all issuers whose rating is triple A—Euro area (changing composition). https://sdw.ecb.europa.eu/quickviewdo?SERIES_KEY=165.YC.B.U2.EUR.4F.G_N_A.SV_C_YM.SR_10Y. Accessed: 07.03.2022
7. FEBIAC. Datadigest 2020. http://www.febiac.be/public/statistics.aspx?FID=23
8. Frischknecht R et al (2020) Life cycle inventories and life cycle assessments of photovoltaic systems. Technical report Version Report T12-19:2020. International Energy Agency (IEA)
9. Gupta R et al (2020) Levelized cost of solar photovoltaics and wind supported by storage technologies to supply firm electricity. J Energ Storage 27:101027. ISSN 2352-152X. https://doi.org/10.1016/j.est.2019.101027, https://www.sciencedirect.com/science/article/pii/S2352152X19302749
10. Hesse HC et al (2017) Lithium-ion battery storage for the grid—a review of stationary battery storage system design tailored for applications in modern power grids. Energies 10(12). ISSN 1996-1073. https://doi.org/10.3390/en10122107, https://www.mdpi.com/1996-1073/10/12/2107
11. Hosen MdS et al (2021) Twin-model framework development for a comprehensive battery lifetime prediction validated with a realistic driving profile. Energ Sci Eng 9(11):2191–2201. https://doi.org/10.1002/ese3.973 , eprint https://onlinelibrary.wiley.com/doi/pdf/10.1002/ese3.973, https://onlinelibrary.wiley.com/doi/abs/10.1002/ese3.973
12. International Renewable Energy Agency (IRENA) (2021) Renewable power generation costs in 2020. Technical report, International Renewable Energy Agency (IRENA), Abu Dhabi. https://www.irena.org/-/media/Files/IRENA/Agency/Publication/2021/Jun/IRENA_Power_Generation_Costs_2020.pdf?rev=c9e8dfcd1b2048e2b4d30fef671a5b84
13. Jülch V (2016) Comparison of electricity storage options using levelized cost of storage (LCOS) method. Appl Energ 183:1594–1606. ISSN 0306-2619, https://doi.org/10.1016/j.apenergy2016.08.165, https://www.sciencedirect.com/science/article/pii/S0306261916312740
14. Kamath D et al (2020) Evaluating the cost and carbon footprint of secondlife electric vehicle batteries in residential and utility-level applications. Waste Manage 113:497–507. ISSN 0956-053X. https://doi.org/10.1016/j.wasman.2020.05.034, https://www.sciencedirect.com/science/article/pii/S0956053X20302749

15. Lazard. Lazard's Levelized Cost of Storage—Version 7.0 2021:1-35. https://www.lazard.com/media/451882/lazards-levelized-costof-storage-version-70-vf.pdf. Accessed: 27.05.2022
16. Mauler L et al (2021) Battery cost forecasting: a review of methods and results with an outlook to 2050. Energ Environ Sci 14:4712–4739. https://doi.org/10.1039/D1EE01530C
17. Rallo H et al (2020) Economic analysis of the disassembling activities to the reuse of electric vehicles Li-ion batteries. Resour Conserv Recycl 159:104785. ISSN 0921-3449. https://doi.org/10.1016/j.resconrec.2020.104785, https://www.sciencedirect.com/science/article/pii/S0921344920301063
18. Schmidt O et al (2019) Projecting the future levelized cost of electricity storage technologies. Joule 3:81–100. ISSN 2542-4785. https://doi.org/10.1016/j.joule.2018.12.008
19. Schulz-Mönninghoff M et al (2021) Integration of energy flow modelling in life cycle assessment of electric vehicle battery repurposing: evaluation of multi-use cases and comparison of circular business models. Resour Conserv Recycl 174:105773. ISSN 0921-3449. https://doi.org/10.1016/j.resconrec.2021.105773, https://www.sciencedirect.com/science/article/pii/S0921344921003827
20. Steckel T, Kendall A, Ambrose H (2021) Applying levelized cost of storage methodology to utility-scale second-life lithium-ion battery energy storage systems. Appl Energ 300:117309. ISSN 0306-2619. https://doi.org/10.1016/j.apenergy.2021.117309. https://www.sciencedirect.com/science/article/pii/S0306261921007212
21. Vlaamse Regulator van de Elektriciteits- en Gasmarkt. Gemiddlede electriciteitsprijs 2022. https://dashboard.vreg.be/report/DMR_Prijzen_elektriciteit.html. Accessed: 01.03.2022
22. Wernet G et al (2016) The ecoinvent database version 3 (part I): overview and methodology. Int J Life Cycle Assess 21:1218–1230. ISSN 1614-7502. https://doi.org/10.1007/s11367-016-1087-8

# Chapter 64
# Wind Farms End-of-Life: An Economic Evaluation for Climate Neutrality Through a Literature Review

**Gisela Mello, Marta Ferreira Dias, and Margarita Robaina**

**Abstract** One of the major challenges for the European Union is to achieve climate neutrality by 2050. The European directives and action plans such as the Green Deal are preparing the economic sector for climate neutrality. The EU has improved its electricity and gas markets, moreover, has fomented energy efficiency, renewable energy deployment, reduction of greenhouse gas emissions and a stronger carbon price. This transition involves the decarbonization of the energy industry, and the energy economy sector leading to changes in the business models and the labour market. Adopting the principles of a circular economy allows for changing the current economic system into a more sustainable one. In this context, generating power through renewable sources, like wind, is a viable way to reach the energy transition. At the same time, the first wind farms are reaching their end of life. Therefore, this work analyses the economic aspects related to the end-of-life of wind farms based on a systematic literature review of life cycle assessments to verify the challenges and advantages. As observed, studies are required to equalize some issues such as reducing the co-dependency of raw materials and developing secondary markets to treat waste and components for reuse and repowering strategy.

**Keywords** Climate neutrality · Economic aspects · End-of-life · Life cycle assessment

## 64.1 Introduction

Currently, coal is responsible, globally, for 44% of $CO_2$ emissions and 40% of installed power generation capacity. However, 75% of this capacity belongs to emerging markets and developing economies (EMDEs) where the energy mix is

G. Mello (✉) · M. F. Dias · M. Robaina
Research Unit on Governance, Competitiveness and Public Policies (GOVCOPP), Department of Economics, Management, Industrial Engineering and Tourism (DEGEIT), University of Aveiro, Campus Universitário de Santiago, 3810-193 Aveiro, Portugal
e-mail: gisela.mello@ua.pt



N. S. Caetano and M. C. Felgueiras (eds.), *The 9th International Conference on Energy and Environment Research*, Environmental Science and Engineering,
https://doi.org/10.1007/978-3-031-43559-1_64

based on coal. While the EMDEs have a prediction of growth of more than 500GW of coal capacity in the next 10–15 years (Stark et al. 2022), the European Union (EU) seeks to reach zero emissions by 2050.

According to Broom (2022) reaching climate neutrality requires a global investment of $3.5 trillion a year by governments, businesses and individuals in energy and land-use systems. Thus, the priority investments to get net-zero aim to clean electrification which may eliminate 73.2% of global emissions However, it is necessary for a complete energy transition from fossil fuels to carbon-free of power grids (Bhattacharya et al. 2021). Moreover, energy transition requires a change to renewable sources or other zero-emissions technology (Pennington 2022), which does not only depend on the low emissions technologies but also on how it will impact the involved stakeholders, such as CEOs, employees, and the communities. This coal transition will have social advantages as well as: land space, interconnection lines, generators, synchronous condensers, and substations (Stark et al. 2022).

By 2030, it is predicted for the energy sector: the settlement of global carbon prices, the highest investments into renewable sources, Europe becoming the primary producer of renewable energy, a significant reduction of the fossil fuel lobby and the growth of public engagement due to community energy ownership (Burston 2030).

According to the International Renewable Energy Agency, by 2050, all countries may increase the renewable energy proportion significantly in their mixes, reaching up to 60% or more of many countries' total final energy consumption. Moreover, it is suggested that the decarbonization of the power sector, through a considerable increase of renewable sources in the energy mix, is the path to a sustainable energy future. However, this transformation requires new approaches to energy sector planning and operation, the functioning of the market, but also new regulation and public policies (IRENA 2050).

Thus, to encourage the insertion of renewable sources in the European Union mix, in 2009, Directive 2009/28/EC of the European Parliament and the Council established to the horizon 2020, target of 20% of the energy supply coming from renewable sources (Mello et al. 2020). An update was released in 2018, with a new target of a minimum of 32% of final energy consumption for 2030 on the Renewable Energy Directive (EU) 2018/2001, being a part of the Clean Energy for all Europeans package to promote the shift from fossil fuels towards cleaner energy sources. All this path is aimed at, reducing greenhouse gas (GHG) emissions negotiated in the Paris Agreement (Arvesen and Hertwich 2012) and climate neutrality with net-zero emissions in the Continent by 2050 (Wood 2050).

For (Pianta et al. 2021) the COVID-19 pandemic became the evaluation of the climate impact associated with the public policies implemented uncertain. Moreover, governmental decisions, in the next few years, regarding economic recovery and climate policies will show if the pandemic boosted the efforts for decarbonization and green recovery.

## 64.2 Literature Review

### *64.2.1 Current European Scenario*

The restrictions and lockdowns due to the COVID-19 pandemic provoked a serious economic crisis which directly influenced the energy sector (IEA 2021). Globally, in 2020, the demand for fossil fuels has dropped (especially oil and coal). Compared to 2019, in European Union, the annual $CO_2$ emissions decreased by 10%, due to the lower electricity demand and the growth of renewable generation. These factors contributed to the decline in coal-fired power generation which lead to an increase in the share of renewables generation to 39% in 2020 (IEA 2021).

According to the 2021 report of Eurostat, the primary energy consumption falls to 1.236 million tons of oil equivalent (Mtoe) in 2020. This represents 5.8% better than the efficiency expected target for 2020, however, is 9.6% away from the 2030 goal. Moreover, compared with the 2017–2019 average, there was a reduction of 9.9%, namely due to the pandemic restrictions (Eurostat 2021).

These rates are important if we consider that more than 130 countries are engaged or considering reaching net-zero emissions by 2050. As described by Bhutada (2022) to attain this goal, globally, it will be necessary to invest in climate measures of $125 trillion by 2050. Nonetheless, only $755 billion were invested in 2021 in deploying low-carbon energy technologies (IEA 2020).

With the pandemic framework in mind, in 2020 there was a 7–10% economic downturn in the EU. And the energy sector was also affected. It was registered a decrease in energy demand and supply, which lead to a fall in $CO_2$ emissions but also in air pollution (IEA 2020).

Moreover, due to the restrictions, in the first quarter of 2020 in the EU, there was a reduction of 20% in coal demand, but also a fall in $CO_2$ emissions by 8%. In this period, it was also registered a reduction in generation from coal, gas and nuclear, but also a record in renewable generation. Despite these trends which may allow the EU to meet its targets for renewable and energy efficiency, the investment in renewable sources ad a historic decline in 2020 compared to 2019 (IEA 2020).

By 2022–26, it is expected the installation of 116 GW of new wind farms, which means building on average 18 GW of new wind farms in EU-27. However, to achieve the EU's new 40% renewable energy goal, it should be built 32 GW a year. Nonetheless, this prediction may be reduced to less than 89 GW, if issues related to permitting issues, less effective strategies for repowering, but also new restrictions due to the pandemic were not solved by Governments (WINDEUROPE 2022).

At the same time that raises the wind power share in the energy mix, the number of wind farms reaching their end of service also increases. Notably, the end of their life cycle occurs after typically 20 years of activity (Mello et al. 2020). In 2021, Europe decommissioned 396 MW of wind capacity and commissioned 515 MW from repowering. In the next five years, it is expected a decommissioning of 11.4 GW of which only 2.9 GW should be repowered. The remaining 8.6 GW will be removed from the system (WINDEUROPE 2022).

### 64.2.2 *Life Cycle Assessment*

As pointed out by Ziegler et al. (2018) the industry needs to be prepared, at the end of the life cycle of wind farms to address some of the issues of this phase such as the guarantee of structural device integrity and the decision to extend the useful life of the farms and the windmills. In other words, investments into processes, design and new technologies and materials improvements, to manage more sustainable projects, may supply the energy demands without compromising the environmental quality and the power supply efficiency. The main challenges of the wind energy sector are related to the ageing fleets, structural integrity and lifetime extension or decommissioning strategies (Ziegler et al. 2018; Topham et al. 2019).

Therefore, the life cycle assessment (LCA) is a methodology widely used to qualify the main challenges, but also the economic and socio-environmental impacts related to a process or product by analyzing its entire life cycle (Ji and Chen 2016). Thus, as highlighted by Yang et al. (2018) LCA is a decision-making tool to identify inputs resources and environmental impacts of renewable energy technologies.

### 64.2.3 *Economic Assessment of the Wind Farms Lifecycle*

The economic assessment allows for estimating the costs and incentives necessary to make any project market competitive and to attract possible investors (Al-Behadilia and El-Osta Wedad 2015). The costs associated with LCA can be broken down by each of the main stages of the process (Castro-Santos and Diaz-Casas 2014). In the evaluations made by Yang and Chen (2013) it is shown that of the total investment for the installation of a wind farm, 69.77% corresponds to the manufacture of turbines, 6.71% to the substation, 14.69% to construction work and 8.83% for other activities associated with the installation.

During the planning phase, the cost of the project (offshore wind resources, area conditions, geotechnical conditions) and legislative factors (such as socio-environmental aspects and mandatory authorizations) are determined (Castro-Santos and Diaz-Casas 2014; Yang and Chen 2013; IWEA 2019). In general, the definition of costs at this stage will determine what will happen throughout the rest of the project's life cycle. In the design phase, the management and engineering costs are considered, as well as other factors such as the number of turbines which depends on the energy generation demand, the distance between wind turbines and transmission lines, dimensioning of substation electrical cables and calculation of structures weight (Castro-Santos and Diaz-Casas 2014).

During the installation stage, the costs associated with the execution of civil works, assembly of structures, electrical connections, transport of structures, commissioning, and necessary equipment for assembly are counted. Regarding offshore wind farms, the costs of the platform assembly, mooring and anchoring installation, port activities, and maritime transportation must also be considered. References

Ortegon et al. (2013) and Nian et al. (2019) consider the location of the port and the shipyard up to the wind farm's final site as a factor that directly affects the costs of transport and the number of required connection cables. Nian et al. (2019) add that sites farthest from the material supply centres and the substation register higher energy and emission values due to the need to use more cables and transportation. Reference Castro-Santos and Diaz-Casas (2014) stressed the necessity of specific vessels for transportation and installation, besides their equipment such as cranes. For the working phase, in general, the following aspects are considered in the cost calculation: taxes, energy delivery fees, exploration management and operation and maintenance costs (Castro-Santos and Diaz-Casas 2014). It's important to highlight that the distance between the port and the site may also influence the maintenance costs.

For offshore wind farms, the tendency is the installation of new projects more distant (far from 200 km of the coast, for example) and in deeper waters, as well as the increasing of the nominal rated power generated due to the use of larger rotors, swept area and transmission lines (Gomes et al. 2018). This requires equipment manufacturing more robust, bigger and heavier, which may directly influence the amount of material for building the components. Regarding the foundations, the monopile types are frequently used in Europe, for adapting to different seabed conditions and deepwater depths up to 50 m, but also, they are easy to build and cost-competitive (Topham et al. 2019). Reference Topham et al. (2019) also highlight that, as the primary material of the monopiles is steel, due to its strength, flexibility and resistance to marine environments, it has become a fundamental part of the circular economy because it is 100% recyclable.

### *64.2.4 Economic Assessment of the End-of-Life Phase*

As pointed out by Athanasios and Martínez-Luengo (2016) the two possibilities at the end of life are extended or not a lifetime. This means repowering (through the reuse of part of the original infrastructure to reduce the cost of the new one) or decommissioning (which leads to returning the current wind farm to the same state before the installation activities).

So, each option has challenges and costs, and the definition of the solution to be adopted will be based on aspects such as the profitability facing the reduction in reliability and performance; the cost–benefit ratio compared with the total decommissioning and the expected profits with the life extension.

For (Ji and Chen 2016) there is a lack of data about the capital input for the end-of-service phase, and for that reason, it is considered that the cost of decommissioning a wind farm is the same as the activities in the construction phase. Therefore, the waste material transportation costs are considered the same as the construction phase.

Ortegon et al. 2013) point out that the end-of-life, costs are associated with crane mobilization and demobilization, the removal and disassembly of overall components, transportation, foundation demolition, and soil recovery.

In this sense, almost half of the total costs of this phase are related to the removal of the foundations due to their weight, the complex techniques and the specialized equipment involved. The type of foundation and the location of the wind farm influence directly in the costs' calculation. For instance, the estimated costs for offshore wind farms are about 2–3% of the original total capital cost (Topham and McMillan 2017). Another factor that will affect the costs is the decision regarding the destination of the farm. Nevertheless, regardless of the solution adopted, some costs should be accounted for: such as transportation, mobilization of workers and equipment, the destination of components and environmental recovery.

Within this frame of reference, there are two possible ways to recycle the materials: through the partial and complete decomposition of the base material (Andersen 2015). At the same time that there is a significant amount of waste generated, there is also a lack of treatment centres and demand for them. The low motivation for sending the equipment to recycling may be explained by the transport costs and the process itself since there are few specialized locations and sometimes they are very distant from the park site (Andersen 2015). With the decommissioning of the first wind farms, there is an opportunity to develop secondary markets for the treatment of equipment. This new market should be considered, not only for those who see them as a simple waste but also for remanufacturing and reusing on the same site of the original farm, with guarantees of the quality of these components (Ortegon et al. 2013). The growth of those markets may be dependent on factors such as prices, timeless and warranty and expected performance balance between the new equipment and the remanufactured ones (Ortegon et al. 2013).

Topham et al. (2019), estimate that if the offshore wind turbines materials were recycled approximately 9% of the decommissioning costs could be paid, and including the monopile foundations these values rise to 13%, namely, due to the steel scrap prices, while the scrap values for cast iron and copper were considered minor influencers in this reduction. Thus, high scrap metal values, namely steel, may increase the recoverable value from recycling, but at the same time reduces the decommissioning costs, which is a good strategy to encourage the materials to be recycled.

In this context, the landfill destination should be the last solution, prioritizing the reuse and recycling of the materials. Between 80 and 90% of a wind turbine's weight may be recycled, namely the construction materials and metals (Topham et al. 2019).

Considering the new challenges brought up by the end of the life cycle, many researchers have been looking for recycling solutions, namely for the composite materials present in wind blades, to reuse the equipment in the original function or even reuse it in other ways. For example, one of these uses is a 1200 $m^2$ renovated playground called Wikado located in Rotterdam designed by 2012Architecten for KinderparadijsMeidoorn using 5 discarded rotor blades from wind turbines (Beauson and Brøndsted 2016). Another possibility is to build furniture. However, the process of remanufacturing for these two usages is complex due to the structure and the geometry of the wind blades which turns it difficult to use them on a large scale in the industry (Beauson and Brøndsted 2016).

The Directive 2008/98/EC establishes the basic concepts and definitions regarding waste management, in terms of destination, recycling and reuse, also including guidelines for sending the waste to landfill. Its implementation by each State Member led to a rise in taxes connected to sending the waste to a landfill and also a restriction by Germany on using this measure (Beauson and Brøndsted 2016). While in Sweden, the disposal of organic waste or fuels in landfills is prohibited (Andersen 2015). Since certain components, such as shovels, cannot be reused, Germany required them to be recycled and supplied to secondary material markets. With a generation between 30.000 in 2020 and 40.000 t in 2040 (with a peak of 52.000 t in 2045) of wind turbine waste per year, some companies have identified this scenario as a potential market offering solutions for recycling (Volk et al. 2021). However, these markets need improvement to ensure the quality of remanufactured equipment, which may be damaged by the disassembly, reprocessing and assembly (Ortegon et al. 2013).

Given the impossibility of adopting the reuse or remanufacturing techniques for all equipment, certain basic components, such as *aluminium* for electric cables or steel for floating platforms, may be sold to reduce the costs at this stage (Castro-Santos and Diaz-Casas 2014; Yang and Chen 2013; IWEA 2019; Ortegon et al. 2013). In addition, if it considers all processes involved in demobilizing, the selling of each component may result in significant reductions in total cost (Ortegon et al. 2013; Nian et al. 2019; Gomes et al. 2018; Athanasios and Martínez-Luengo 2016; Topham and McMillan 2017; Andersen 2015; Beauson and Brøndsted 2016; Volk et al. 2021; Sun et al. 2017). Ortegon et al. (2013), expressed concern about the consumption of materials, such as copper and *aluminium*, which is higher than the quantity produced and about the geographic source of the supply of these inputs.

In addition, Ortegon et al. (2013) highlight that the adoption of strategies for reusing and remanufacturing system components critical materials, such as steel, copper, rare earth magnets and fibreglass may reduce the total consumption of resources as well as the supply of unexpected changes in the market (a supply shortage, a dependence on exporters and an increase in added value). As it is estimated by 2050, an increase in the cost of 30% of producing steel and 45% in cement-making (Broom 2022).

The transition to clean energy needs other technologies which require a significant amount of critical minerals such as lithium, cobalt, and rare earth materials (Pennington 2022). The International Energy Agency (IEA) estimates an increase of six-fold in mineral input by 2040 to reach climate neutrality. Other metals, such as lithium may need an input of over 40 times while nickel and cobalt more than 20-fold. These additional metal inputs lead to a rise in prices. For instance, in February 2021, lithium prices reach a record of $50.000 per ton up from $10,000 in 2020 (Pennington 2022).

Moreover, the dependence on these materials may generate volatility in the market and an increase in prices and the project as a whole. As also pointed out by Ortegon et al. (2013) small quantities of rare earth elements, such as dysprosium, praseodymium, and neodymium are recovered from wind turbines rely on. Additionally, the dominance of rare earth metals production, namely from China, may also affect the international market prices. The development of a secondary manufacturing market for these critical materials and attractive costs may boost the adoption of a remanufacturing long-term and sustainable strategy for wind farms at the end of service. However, meeting industry-specific requirements and small quantities of material (namely rare earth) recovered are challenges to establishing a secondary market for wind farms' components (Ortegon et al. 2013).

## 64.3 Conclusion

Decarbonization of the energy sector is a key for the European Union to reach its goal of climate emergency challenge. In a scenario of digital transformation, post-Covid recovery and the recent conflict between Russia and Ukraine the effects on the energy sector are uncertain. And the impacts on the established goals of emission reductions and zero emissions are also unknown.

However, this transition to a green economy and green energy should be kept, and this process requires a massive investment in the deployment of renewable sources or other zero-emissions technology and reduce coal dependence.

Regarding the wind source, the challenges of the end-of-service for recycling may result in a higher carbon footprint and a secondary market for the treatment of the components is not well established. Moreover, the co-dependency of raw materials exportation and the continuation of the efforts towards an energy transition are challenges for the energy sector. Today, few quantities of materials, namely rare earth, are recovered from those components, increasing the dependence of the producer countries which leads to volatility in global prices. Additionally, the repowering option may reduce costs with maintenance and due to advanced technologies, the efficiency and performance will increase, with a reduction of the overall capital costs for the installation and the creation of new job opportunities.

In conclusion, long-term strategies for the end of service of energy projects should be considered as they may help to reach the energy mix decarbonization and improve the adoption of the circular and green economy concepts.

**Acknowledgements** This work was financially supported by the research unit on Governance, Competitiveness and Public Policy (UIDB/04058/2020) + (UIDP/04058/2020), funded by national funds through FCT—Fundação para a Ciência e a Tecnologia.

## References

Al-Behadilia SH, El-Osta Wedad B (2015) Life Cycle Assessment of Dernah (Libya) wind farm. Renew Energy 2013:1227–1233

Andersen N (2015) Wind turbine end-of-life: characterisation of waste material. Master's thesis, University of Gävle

Arvesen A, Hertwich EG (2012) Assessing the life cycle environmental impacts of wind power: a review of present knowledge and research needs. Renew Sustain Energy Rev 16(8):5994–6006. https://doi.org/10.1016/j.rser.2012.06.023

Athanasios K, Martínez-Luengo M (2016) The end of the line for today's wind turbines. Renew Energy Focus 17(3):109–111. ISSN: 1755-0084. https://doi.org/10.1016/j.ref.2016.05.003

Beauson J, Brøndsted P (2016) Wind turbine blades: an end of life perspective. MARE-WINT, pp 421–432

Bhattacharya S, Fthenakis V, Kammen D (2021) The role of offshore wind farms in the decarbonisation of energy systems to tackle climate change. Acad Lett. https://doi.org/10.20935/AL4416

Bhutada G (2022) These are the top 10 countries by energy transition investment. World Econ Forum

Broom D (2022) Europe is beating its renewable energy targets. Which countries are leading the charge? World Econ Forum

Burston J (2016) 5 predictions for energy in 2030. World Econ Forum

Castro-Santos L, Diaz-Casas V (2014) Life-cycle cost analysis of floating offshore wind farms. Renew Energy 66:41–48. https://doi.org/10.1016/j.renene.2013.12.002

Eurostat (2021) Energy efficiency statistics

Gomes M, Moris V, Nunes A (2018) Avaliação de ciclo de vida da energia eólica offshore: Uma revisão da literatura. Revista Brasileira De Energias Renováveis 7(2):199–213. https://doi.org/10.5380/rber.v7i2.58259

IEA (2020) Global $CO_2$ emissions in 2019—analysis

IEA (2021) Global energy review: $CO_2$ emissions in 2020. IEA, Paris

IRENA (2018) Global energy transformation: a roadmap to 2050

IWEA (2019) Life-cycle of an onshore wind farm

Ji S, Chen B (2016) Carbon footprint accounting of a typical wind farm in China. Appl Energy 180:416–423

Mello G, Ferreira Dias M, Robaina M (2020) Wind farms life cycle assessment review: $CO_2$ emissions and climate change. Energy Rep 6(8):214–219. https://doi.org/10.1016/j.egyr.2020.11.104

Nian V, Liu Y, Zhong S (2019) Life cycle cost-benefit analysis of offshore wind energy under the climatic conditions in Southeast Asia—setting the bottom-line for deployment. Appl Energy 233–234:1003–1014

Ortegon K, Nies L, Sutherland JW (2013) Preparing for end of service life of wind turbines. J Clean Prod 39:191–199. https://doi.org/10.1016/j.jclepro.2012.08.022

Pennington J (2022) 3 ways the circular economy is vital for the energy transition. World Econ Forum

Pianta S, Brutschin E, van Ruijven B, Bosetti V (2021) Faster or slower decarbonization? Policymaker and stakeholder expectations on the effect of the COVID-19 pandemic on the global energy transition. Energy Res Soc Sci 76:102025. ISSN: 2214-6296. https://doi.org/10.1016/j.erss.2021.102025

Stark M, Cataudella F, Mayer J (2022) A cleaner future for coal power plants and coal-reliant communities

Sun H, Yang H, Gao X (2017) Study on offshore wind farm layout optimization based on decommissioning strategy. Energy Procedia 143:566–571. https://doi.org/10.1016/j.egypro.2017.12.728

Topham E, McMillan D (2017) Sustainable decommissioning of an offshore wind farm. Renew Energy 102:470–480

Topham E, McMillan D, Bradley S, Hart E (2019) Recycling offshore wind farms at decommissioning stage. Energy Policy 129:698–709. https://doi.org/10.1016/j.enpol.2019.01.072

Volk R, Stallkamp C, Herbst M, Schultmann F (2021) Regional rotor blade waste quantification in Germany until 2040. Resour Conserv Recycl 172:105667. https://doi.org/10.1016/j.resconrec.2021.105667

WINDEUROPE (2022) Wind energy in Europe 2021 Statistics and the outlook for 2022–2026

Wood J (2020) Renewable energy could power the world by 2050. Here's what that future might look like. World Econ Forum

Yang J, Chen B (2013) Integrated evaluation of embodied energy, greenhouse gas emission and economic performance of a typical wind farm in China. Renew Sustain Energy Rev 27:559–568. https://doi.org/10.1016/j.rser.2013.07.024

Yang J, Chang Y, Zhang L, Hao Y, Yan Q, Wang C (2018) The life-cycle energy and environmental emissions of a typical offshore wind farm in China. J Clean Prod 180:316–324. https://doi.org/10.1016/j.jclepro.2018.01.082

Ziegler L, Gonzalez E, Rubert T, Smolka U, Melero J (2018) Lifetime extension of onshore wind turbines: a review covering Germany, Spain, Denmark, and the UK. Renew Sustain Energy Rev 82:1261–1271

# Chapter 65
# Use Cases for Contextual Load Flexibility Remuneration Strategies

**Débora de São José, Fernando Lezama, Pedro Faria, and Zita Vale**

**Abstract** In the new paradigm of smart grids, load flexibility remuneration strategies play a key role in defining innovative business models and market frameworks using distributed energy resources and demand response. This article establishes a set of use cases used to support the application of remuneration strategies considering contextual load flexibility. In other words, the various new tariffs and remuneration plans consider the context provided by solar and wind power, different times of the day and week, temperature, and electricity market prices. Also, depending on the circumstances, dynamically specified consumer clusters are envisaged for a realistic validation of results while also analysing higher smart grid efficiency, decreased pollutant emissions, and cheaper energy prices. The proposed use cases are proposed in the context of COLORS (COntextual LOad flexibility Remuneration Strategies), a project investigating business models and market frameworks with a particular emphasis on demand response.

**Keywords** Demand response · Load flexibility · Remuneration

## 65.1 Introduction

Environmental concerns around energy generation and consumption have pushed the power grid to change quickly in the last decades. Such transformation is also driving the adoption of renewable and distributed generation, the introduction of new actors as the prosumer, the reduction and even deactivation of non-renewable/fossil fuels-based power plants, among other related aspects. In that context, demand response

D. de São José · F. Lezama · P. Faria (✉) · Z. Vale
Research Group on Intelligent Engineering and Computing for Advanced Innovation and Development (GECAD), Intelligent Systems Associated Laboratory (LASI), Polytechnic of Porto (P.PORTO), Rua Dr. António Bernardino de Almeida 431, 4200-072 Porto, Portugal
e-mail: pnsfaria@gmail.com

Z. Vale
e-mail: zav@isep.ipp.pt

 
N. S. Caetano and M. C. Felgueiras (eds.), *The 9th International Conference on Energy and Environment Research*, Environmental Science and Engineering,
https://doi.org/10.1007/978-3-031-43559-1_65

(DR) represents a solution to increase the flexibility needed to make this change possible and reliable and plays a key role to deal with power generation uncertainty and load demand fluctuations (O'Connell et al. 2014; Faria and Vale 2022).

Besides that, DR also offers many benefits to consumers, prosumers, network operators, electricity markets and aggregators (Faria and Vale 2022; Siano and Sarno 2016). The benefits include, for instance, the reduction in prices, outages, and price volatility, increase in customer participation, diversification of resources, and billing savings. Also, the control and coordination of home appliances is an interesting source of flexibility for demand response, calling for efficient and scalable algorithms than can handle thousands of devices (Limmer et al. 2021). Finally, DR can contribute to grid regulation and support network operators in different grid services (Elma and Selamogullari 2017). For those reasons, DR is considered an asset supporting energy conservation and efficiency and an essential component of demand-side management (Ghadi et al. 2019; Groppi et al. 2021). Thus, in this article, a set of use cases are presented and defined with special emphasis on contextual load flexibility remuneration strategies. The proposed use cases are also within the scope of COLORS (COntextual LOad flexibility Remuneration Strategies), a Portuguese 2020 project investigating business models and market frameworks with a particular emphasis on demand response.

This paper is organized as follows: after the introduction in Sect. 65.1, Sect. 65.2 addresses the definition of use cases for contextual load flexibility remuneration strategies. Section 65.3 develops further the three proposed use cases. Finally, Sect. 65.4 presents the conclusions of this work.

## 65.2 Definition of Use Cases

In a first step, the value proposition for the use cases is identified according to three different perspectives: wholesale and energy system market value (i.e., use case 1); energy community value (i.e., use case 2); and grid value (i.e., use case 3). As previously mentioned, with the increasing integration of renewable generation, several advantages and goals in economic and environmental terms can be achieved. However, specific challenges are associated with the large penetration of distributed generation and DR. For instance, the actual response of consumers to implicit and explicit DR events has some associated uncertainty, which is not negligible. In this way, it is important to incentivize consumers to provide full load flexibility potential to contribute to better integration of renewable resources. Thus, to overcome this, it is necessary to enable improved remuneration strategies for the consumers' participation in DR programs, preferably prepared to adapt to different contexts or situations. For instance, consumers and load can be adequately modelled using knowledge discovery in data concerning consumers' needs (Antonopoulos et al. 2020). Therefore, an accurate and adequate remuneration strategy will take full advantage of the consumers' participation in DR programs. In addition, suitable key performance indicators should be established to measure the effectiveness of use cases (Faria et al. 2021). Each use case is described in the following section. For more detailed information, the reader can be referred to http://www.gecad.isep.ipp.pt/COLORS/home/.

## 65.3 Use Cases

The use cases proposed in this work were defined after several iterations of planning, resulting in the following three use cases:

- UC-1 Energy community load flexibility for wholesale and local market value: use case targeting the usage of flexibility and DR of the local community for the wholesale and energy system market value.
- UC-2 Energy community load flexibility for internal use: use case targeting the use of flexibility and DR for the benefit of the local market.
- UC-3 Energy community load flexibility for distribution network operator (DSO): use case targeting the usage of local community distributed flexibility for DR, including available storage systems and energy/power services to enable a more efficient renewable integration at a low voltage level, with a reserved capacity to serve grid needs and mitigate technical constraints.

The detailed definition of each of those use cases is presented in the following subsections.

### *65.3.1 Use Case 1—Local Community Flexibility and Demand Response for Wholesale and Energy System Market Value*

The third use case targets the usage of local community distributed flexibility for DR, including available storage systems and energy/power services to enable a more efficient renewable integration at low voltage level, with reserved capacity to serve grid needs and mitigate technical constraints. This use case is envisaged to clarify the definition and implementation of technical grid support services, in different timeframes, regarding voltage control and congestion management within the local low voltage grid, taking also into account that some of the constraints can have an origin in the medium voltage grid (medium voltage constraints due to abnormal low voltage loads).

The scope considers the interaction between a flexibility procurer and local users/resources that may provide the required flexibility. These interactions run through a local market designed to provide adequate signals and market processes to regulate the use of flexibility for both grid and commercial applications (see Fig. 65.1).

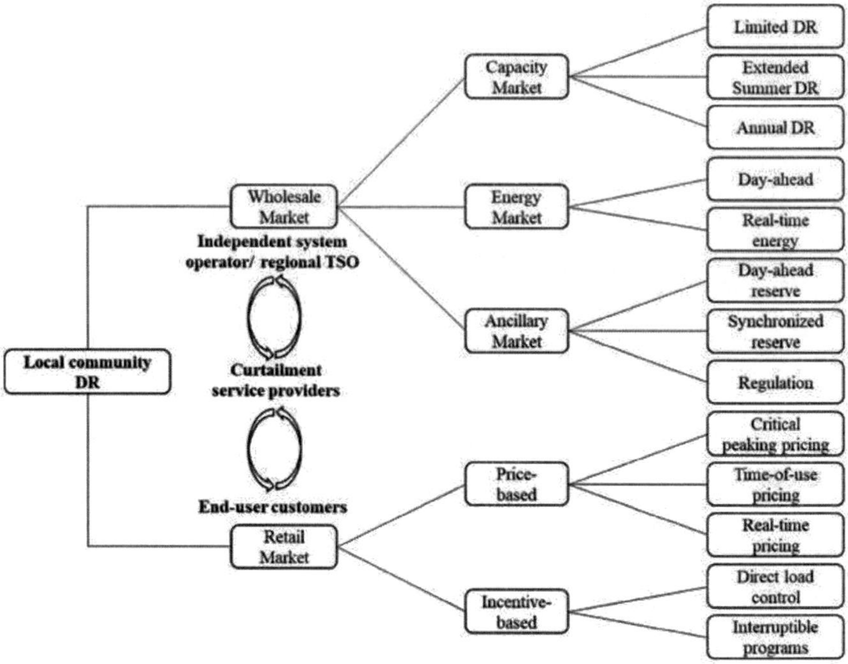


**Fig. 65.1** Use case 1: local community flexibility and demand response for wholesale and energy system market value

A set of rules can be defined with the goal of benefiting consumers who voluntarily reduce demand during needed periods (e.g., during system emergencies) by eliminating a group of charges; assessing and reporting obstacles to DR's equal treatment; and making the market price mirror the energy value during operating reserve shortages periods. The wholesale market value can be effectively realized by involving the resources in the purchase and sale of energy of a retailer or an aggregator. Furthermore, the value of the resources can be included in the bidding process of the retailer with the help of the local market, which aggregates the flexibility of the resources and offers it to the retailer/aggregator. The bidding can then be operated on the day-ahead and intra-day markets. It is considered that multiple retailers are present in the ecosystem. Thus, end-consumers and aggregators can present flexibility for the purposes of their respective retailers.

Three different scenarios are considered within the use case: (1) prosumers selling balancing services is a scenario where the prosumers directly offer the flexibility of their resources to the local market where it can then be allocated to aggregators for sale; (2) aggregators selling balancing services is a scenario where an aggregator offers the flexibility of its managed resources to the local market; and (3) retail services where multiple retailers can acquire flexibility from the market for managing their market positions. For more information, the reader can be referred to http://www.gecad.isep.ipp.pt/COLORS/home/.

### 65.3.2 *Use Case 2—Local Community Market with Flexibility and Demand Response for Energy Community Value*

The use case 2 targets the use of flexibility and DR for the benefit of the local market. The definition of this use case is motivated by the fact that conventional households cannot participate in electric markets actively (Mengelkamp et al. 2018). This problem discourages people from investing in electric resources for self-consumption and storage and creates a mistrusting environment. Considering the increase of residential solar generation, several initiatives have proposed local market environment where prosumers and consumers are empowered (Lezama et al. 2021), making it easier to get information on the status of the electric system to respond better and adapt to the price/demand curve fluctuation. An environment where end users can perform an active role in the market can lead to a more reliable, efficient, and balanced system and every agent involved in the market could benefit from that. However, this is possible only if information and communication technologies that integrate local markets can handle the communication between all the elements in the market. Therefore, it is assumed that this local market process is managed by an energy community service provider, which acts as a community manager (see for instance Fig. 65.2).

Thus, this use case aims to validate the community manager's activities, whose goal is to manage the local market creating value for all the relevant actors, especially for the community.

Besides that, it also investigates the functioning of community energy markets managed by a community manager, who will take advantage of the local flexibility and DR programs to create economic value for all the participants, namely the energy clients.

First, the community manager runs the forecast tools to understand the load and generation profiles for the day ahead. It is assumed that each client's flexibility and DR level are also known or predicted with a certain degree of accuracy. Based on the aggregated amount of flexibility, the flexibility value is commutated and generates the DR set points. Customers must decide whether these orders are accepted or not and inform back the community manager. Then, the community manager buys the electricity accordingly to maintain the local market running without constraints.

Furthermore, for the intraday scenario, the story is similar. However, instead of running the forecast tools, the community manager continuously runs a tool to verify the existence of deviations from the plan designed the day before and then initiates the collection of available flexibility and sends DR set points.

Making use of the DR and other necessary platforms and systems, the inclusion of prosumers and consumers into the market and allowing them to make decisions

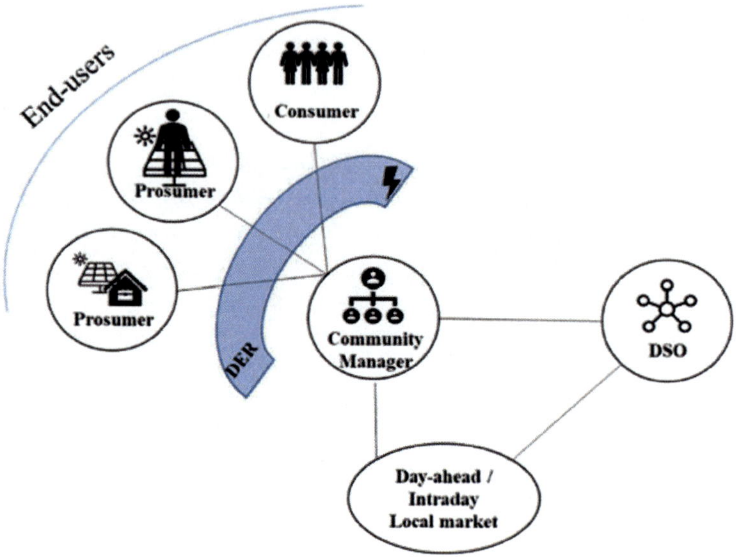


**Fig. 65.2** Use case 2: local community market with flexibility and demand response for energy community value

could lead to a more efficient and independent electric market with more intensive and efficient use of distributed energy resources (DER).

The reader is referred to http://www.gecad.isep.ipp.pt/COLORS/home/ for more details on the design and assessment of this use case.

### *65.3.3 Use Use Case 3—Local Market Flexibility and Demand Response for Grid Value*

The third use case targets the usage of local community distributed flexibility for DR, including available storage systems and energy/power services to enable a more efficient renewable integration at low voltage level, with reserved capacity to serve grid needs and mitigate technical constraints. This use case is envisaged to clarify the definition and implementation of technical grid support services, in different timeframes, regarding voltage control and congestion management within the local low voltage grid, taking also into account that some of the constraints can have an origin in the medium voltage grid (medium voltage constraints due to abnormal low voltage loads).

The scope considers the interaction between a flexibility procurer and local users/ resources that may provide the required flexibility. These interactions run through a

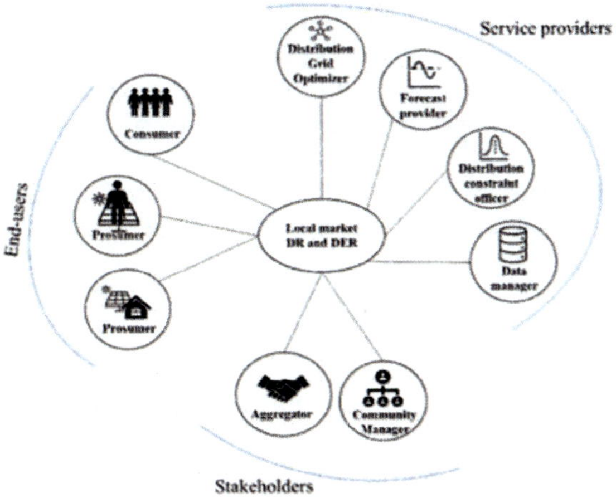


**Fig. 65.3** Use case 3: local market flexibility and demand response for grid value

local market designed to provide adequate signals and market processes to regulate the use of flexibility for both grid and commercial applications (see Fig. 65.3).

The goal of the user case is focused on the optimization of distribution network operation using local flexibility driven by DR, namely: efficient planning, monitoring and operation of distribution grid; anticipate and solve distribution grid contingencies by managing available flexibilities with DR to mitigate constraints and optimize network operation, deploy optimal solutions for grid management, integrate DERs in the most cost-effective way, and ensure high levels of quality of service.

Furthermore, power grids require a system and local balance, which is to say that certain technical needs and criteria must be guaranteed, ensuring compliance with nominal parameters. So, it is key to ensure efficient and resilient management of new DERs connected to the distribution network, considering a multi-period approach and DR. This means that the distribution grid operator may search and use novel approaches to manage its network and solve technical constraints. As such, the implementation of a local data platform that integrates and exchanges data from the different energy sector players (e.g., available flexibility from DERs in different timeframes) will enable not only new market functionalities but also an advanced approach for distribution management, by using decision-making algorithms which can ascertain the validity of the transactions fostering an advanced grid operation.

For that, the rollout of smart meters enables players to take advantage of three possible channels to obtain data on their consumption. The first one is the DSO, by using the smart meter communication topology to collect data (e.g., loads, voltage profiles) that will foster market facilitation as well as the optimization of network operation, mainly in low voltage levels. The second one is the prosumer, by single smart appliances connected to the home area network (HAN) port of the smart meter to extract consumption/production data allowing new functionalities (e.g., energy efficiency services; active participation in ancillary services), and the last one is the market entities fostering new market services and products that can be provided to prosumers/DER. In that way, DR will use this data to evaluate flexibility to help consumers to make better decisions that will prevent technical issues at the local grid level considering all scenarios calculated and, in the local energy market hub a rule-based system will flag all operation as valid or invalid when a transaction takes place, such as selling energy to the wholesale market, P2P transactions, procurement/offer of flexibility, islanding a grid for local management, and new DER integration.

The reader is referred to the COLORS website http://www.gecad.isep.ipp.pt/COLORS/home/ for more details.

## 65.4 Conclusions

The adequate handling of DR in electricity markets and distribution networks requires studies on how the flexibility can be obtained from the consumers, adequately dealing with the consumers remuneration for the provided flexibility. In fact, in different situations, providing a certain amount of flexibility can cause a different impact on consumer comfort. Thus, defining standard use cases for testing DR initiatives becomes relevant. The three use cases included in this article result from a close and carefully executed analysis of contextual load flexibility remuneration strategies proposed in the context of Portugal project COLORS. The use cases definition helped to condensate previous analyses on the matter, providing a starting point for the further development of research in the project. The three resulting use cases covered different viewpoints to be considered in the scope of DR, local markets, and distribution network operation, ensuring that all actors and stakeholders, from end-users to the wholesale market, could benefit from DR. In several contexts, this can serve as a tool for demand response comparison and the assessment of innovative business models and projects.

**Acknowledgements** The present work has received funding from European Regional Development Fund through COMPETE 2020—Operational Programme for Competitiveness and Internationalisation through the P2020 Project TIoCPS (ANI|P2020 POCI-01-0247-FEDER-046182) and has been developed under the EUREKA—ITEA3 Project TIoCPS (ITEA-18008), we also acknowledge the work facilities and equipment provided by GECAD research center (UIDB/00760/2020) to the project team.

## References

Antonopoulos I et al (2020) Artificial intelligence and machine learning approaches to energy demand-side response: a systematic review. Renew Sustain Energy Rev 130:109899

Elma O, Selamogullari US (2017) An overview of demand response applications under smart grid concept. In: 2017 4th international conference on electrical and electronic engineering (ICEEE), pp 104–107

Faria P, Vale Z (2022) Application of distinct demand response program during the ramping and sustained response period. Energy Rep 8:411–416

Faria P, Lezama F, Vale Z, Khorram M (2021) A methodology for energy key performance indicators analysis. Energy Inform 4(1):6

Ghadi MJ, Rajabi A, Ghavidel S, Azizivahed A, Li L, Zhang J (2019) From active distribution systems to decentralized microgrids: a review on regulations and planning approaches based on operational factors. Appl Energy 253:113543

Groppi D, Pfeifer A, Garcia DA, Krajačić G, Duić N (2021) A review on energy storage and demand side management solutions in smart energy islands. Renew Sustain Energy Rev 135:110183

Lezama F, Pinto T, Vale Z, Santos G, Widergren S (2021) From the smart grid to the local electricity market. Local electricity markets. Elsevier, pp 63–76

Limmer S, Lezama F, Soares J, Vale Z (2021) Coordination of home appliances for demand response: an improved optimization model and approach. IEEE Access 9:146183–146194

Mengelkamp E, Bose S, Kremers E, Eberbach J, Hoffmann B, Weinhardt C (2018) Increasing the efficiency of local energy markets through residential demand response. Energy Inform 1(1):11

O'Connell N, Pinson P, Madsen H, O'Malley M (2014) Benefits and challenges of electrical demand response: a critical review. Renew Sustain Energy Rev 39:686–699

Siano P, Sarno D (2016) Assessing the benefits of residential demand response in a real time distribution energy market. Appl Energy 161:533–551

# Chapter 66
# European Union Electricity Production and Air Pollution Emissions

**Florinda F. Martins and Nídia S. Caetano**

**Abstract** There is a growing concern about global issues such as Global Warming. However, other problems concerning air emission pollutants such as sulphur dioxide, particles, and heavy metals are also of great importance, since they are responsible for many health and sustainability problems. Various energy studies give much relevance to greenhouse gas emissions (GEE) leaving behind the other air pollutants. In this work, several scenarios for the year of 2020 were considered and for each one the air pollutants emissions were estimated. It was possible to conclude that if all fossil fuels were replaced by natural gas there would be a significant reduction in all pollutants except CO and non-methane volatile organic compounds (NMVOC). Heavy metals would present a reduction of almost 100%. The other scenarios considered the replacement of fossil fuels, the solid ones first, then liquid and finally natural gas with photovoltaic and/or wind technology. As expected, reduction obtained for total renewables scenarios was of 100% for all pollutants except for $CO_2$ that presented reductions between 90 and 100% (only wind). It would have been avoided the emission of 240 Mt of GEE and 2.32 Mt of other pollutants which is rather important considering the global warming and health problems they caused.

**Keywords** Air pollutants · Energy scenarios · Environment · Health · Sustainability

F. F. Martins (✉) · N. S. Caetano
School of Engineering (ISEP), Polytechnic of Porto (P.Porto), Rua Dr. António Bernardino de Almeida, 431, 4249-015 Porto, Portugal
e-mail: ffm@isep.ipp.pt

N. S. Caetano
LEPABE—Laboratory for Process Engineering, Environment, Biotechnology and Energy, Faculty of Engineering, University of Porto (FEUP), R. Dr. Roberto Frias s/n, 4200-465 Porto, Portugal

ALiCE—Associate Laboratory in Chemical Engineering, Faculty of Engineering, University of Porto, Rua Dr. Roberto Frias, 4200-465 Porto, Portugal



N. S. Caetano and M. C. Felgueiras (eds.), *The 9th International Conference on Energy and Environment Research*, Environmental Science and Engineering,
https://doi.org/10.1007/978-3-031-43559-1_66

## 66.1 Introduction

There are many global problems that must be addressed in a holistic way, bringing together the efforts of several parties, in order to be effective in the solution of problems. Global warming and consequently climate change are one of those challenges that requires such an approach, as well as other problems. This is known since the first conferences on Sustainable Development (SD) of the United Nations (UN). There have been many efforts to engage countries and organizations in the quest for sustainable solutions and potentiate synergistically the actions of each party, sometimes with more success than others.

Global warming has been addressed in many different ways, such as for example studying peoples' attitudes (Pitiruţ et al. 2022). Other authors studied global warming consequences which is also very important (Adler et al. 2022; Feroz et al. 2009). Air pollution is often seen as a more localized problem although it affects thousands of people because pollutants cause health problems, that sometimes can result in death (Chen et al. 2022; Gelb and Apparicio 2022). But as previously seen, the solution of global problems can also be the way to solve more local issues.

Renewables are seen as a key strategy to solve global warming but they are also important to reduce air pollution. Electricity production is a source of air pollutants from greenhouse gas emissions (GEE) to sulphur dioxides, heavy metals, etc. Due to electricity production and to the industrial sector activity, many cities face air pollution and that may lead to the shutdown of facilities to decrease air pollution to acceptable levels. To change the electricity mix shifting to reduced usage of fossil fuels, following a strategy that substitutes first solid fuels, then liquid and finally gas, can be valuable in the reduction of air pollutant emissions with an improvement in health and sustainability, while contributing to decarbonization. In this work the reduction of air pollutants was studied considering several scenarios for the electricity production mix for European Union for the year of 2020, following the above-mentioned strategy.

## 66.2 Base Case and Scenarios

In order to structure the work a base case for 2020 was first considered, that corresponds to actual electricity production mix for that year. Then, the replacement of fossil fuels was considered, substituting first solid fuels by photovoltaic and wind (equal parts), then solid and liquid fuels by renewables (photovoltaic and wind) and finally substituting all fossil fuels including natural gas. This strategy lead to 3 different scenarios and air pollutant emissions were estimated as well as the reduction achieved for each pollutant. Additionally, it was considered a fourth and fifth scenario where fossil fuels were replaced with photovoltaic only or wind technology only to verify the differences obtained in the emissions and in the air pollutants

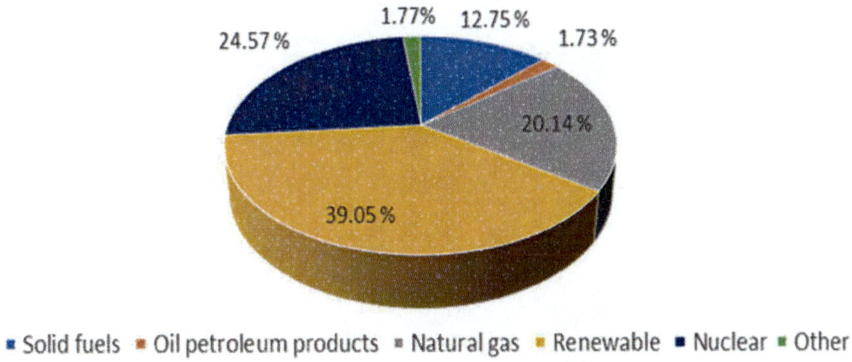


**Fig. 66.1** EU electricity mix for 2020

reduction. Finally, it was considered a scenario where solid and liquid fossil fuels were replaced by natural gas.

Two major categories of solid fuels were considered, namely hard coal (bituminous coal, etc.) and brown coal (lignite, etc.) and for liquid fossil fuels two categories were also considered, namely fuel and light oil, that were used for selecting the emission factors.

Tier default approach was used. Pollutant reduction calculations were based on base case total fossil fuel emissions. It was considered that the amount of electricity produced was the same for each scenario.

Emission factors for GEE and other pollutants collected from EMEP/EEA (2019) and IPPC (2006). Data for base case from EUROSTAT (2022) and emissions factors for photovoltaic (Multi Si) and wind technologies from Tawalbeh et al. (2021) and Yang et al. (2018).

Electricity mix for 2020 is presented in Fig. 66.1 showing that fossil fuels represent 34.6% of electricity production.

Figure 66.2 presents in detail the renewable contribution to electricity production, showing that major contributors are wind, hydro and solar photovoltaic.

The six scenarios considered in this study are listed in Table 66.1. Briefly, they considered the replacement of increasing amount of fossil energy sources with renewable energy, starting with hard and brown coal only, then adding fuel oil and light oil, and finally natural gas. As for the renewable energy, it was considered equal share of photovoltaic and wind energy, photovoltaic only and wind only. A final scenario where the solid and liquid fossil energy sources were replaced totally with natural gas was also considered.

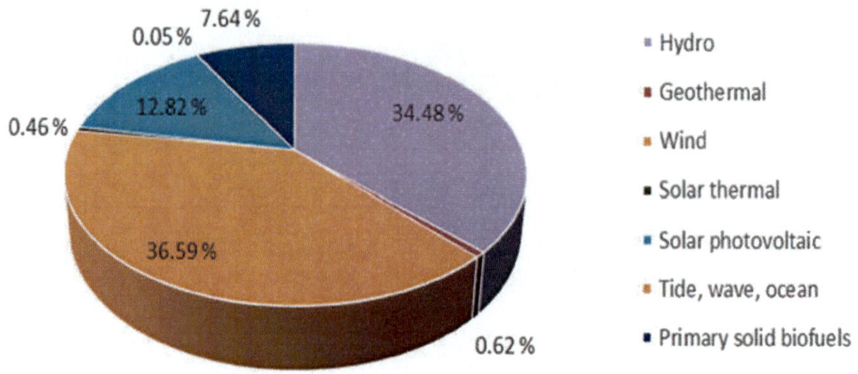


**Fig. 66.2** Renewable contribution to electricity mix EU2020

**Table 66.1** Energy scenarios

| Scenario | Characteristics |
|---|---|
| Scenario 1 | Hard and brown coal replaced by photovoltaic and wind (equal parts) |
| Scenario 2 | Hard coal, brown coal, fuel oil and light oil replaced by photovoltaic and wind (equal parts) |
| Scenario 3 | Hard coal, brown coal, fuel oil, light oil and natural gas replaced by photovoltaic and wind (equal parts) |
| Scenario 4 | Hard coal, brown coal, fuel oil, light oil and natural gas replaced by photovoltaic |
| Scenario 5 | Hard coal, brown coal, fuel oil, light oil and natural gas replaced by wind |
| Scenario 6 | Hard coal, brown coal, fuel oil and light oil replaced by natural gas |

## 66.3 Results and Discussion

The results were calculated for all the proposed scenarios and Table 66.2 presents the results obtained for pollutant reduction. If solid fossil fuels were replaced by renewable sources (photovoltaic and wind) there would be a significant reduction in all pollutants being the values between 12 and 100%.

If the other fossil fuels such as fuel oil and light oil were replaced, the reduction would increase, remaining only three pollutants below 50% (namely methane, carbon monoxide and NMVOC (non-methane volatile organic compounds)) values ranging from 14.9 to 100%.

If natural gas were replaced with renewables (photovoltaic and wind) it would be possible to conclude that the reduction would go to 100%, except for carbon dioxide reduction that would range between 90 and 100.0% (wind only).

**Table 66.2** Air pollutants reduction for the different scenarios

| | Scenario | | | | | |
|---|---|---|---|---|---|---|
| | 1 | 2 | 3 | 4 | 5 | 6 |
| | % reduction | | | | | |
| $CO_2$ | 48.2 | 53.0 | 95.4 | 90.8 | 100.0 | 22.6 |
| $CH_4$ | 27.3 | 42.2 | 100.0 | 100.0 | 100.0 | 9.1 |
| $N_2O$ | 85.8 | 90.6 | 100.0 | 100.0 | 100.0 | 84.4 |
| $NO_x$ | 58.1 | 62.0 | 100.0 | 100.0 | 99.9 | 37.5 |
| CO | 12.0 | 14.9 | 100.0 | 100.0 | 100.0 | – 46.2 |
| NMVOC | 22.0 | 26.3 | 100.0 | 100.0 | 100.0 | – 26.7 |
| $SO_x$ | 96.0 | 99.3 | 100.0 | 100.0 | 100.0 | 99.9 |
| TSP | 70.7 | 91.4 | 100.0 | 100.0 | 100.0 | 85.3 |
| $PM_{10}$ | 67.6 | 87.9 | 100.0 | 100.0 | 100.0 | 79.1 |
| $PM_{2.5}$ | 51.1 | 77.9 | 100.0 | 100.0 | 99.2 | 62.4 |
| Pb | 95.1 | 100.0 | 100.0 | 100.0 | 100.0 | 100.0 |
| Cd | 89.2 | 100.0 | 100.0 | 100.0 | 100.0 | 100.0 |
| Hg | 89.9 | 93.7 | 100.0 | 100.0 | 100.0 | 89.1 |
| As | 94.7 | 98.4 | 100.0 | 100.0 | 100.0 | 97.2 |
| Cr | 96.1 | 100.0 | 100.0 | 100.0 | 100.0 | 100.0 |
| Cu | 87.0 | 100.0 | 100.0 | 100.0 | 100.0 | 100.0 |
| Ni | 96.3 | 100.0 | 100.0 | 100.0 | 100.0 | 100.0 |
| Se | 98.5 | 100.0 | 100.0 | 100.0 | 100.0 | 99.9 |
| Zn | 63.1 | 100.0 | 100.0 | 100.0 | 100.0 | 100.0 |
| Benzo(a)pyrene | 54.0 | 54.0 | 100.0 | 100.0 | 100.0 | 20.9 |
| Benzo(b) fluoranthene | 95.6 | 96.6 | 100.0 | 100.0 | 100.0 | 94.1 |
| Benzo(k) fluoranthene | 94.4 | 95.7 | 100.0 | 100.0 | 100.0 | 92.6 |
| Indenol (1.2.3-cd)pyrene | 42.4 | 66.2 | 100.0 | 100.0 | 100.0 | 41.9 |
| HCB | 100.0 | 100.0 | 100.0 | 100.0 | 100.0 | 100.0 |

In Scenario 5 there are two pollutants for which the reduction would not be 100%, but it would be very near it (nitrogen oxides, $NO_X$ and particles, $PM_{2.5}$). Fossil fuels contribute to the emission of almost every pollutant considered in this study. On the other hand, renewables such as wind and photovoltaic contribute mainly to $CO_2$ emissions, but at a much lower level.

If all fossil fuels were replaced by natural gas there would be a significant reduction in all pollutants except CO and NMVOC, that would increase 42,712 t and 1896 t, respectively. Natural gas emission factors for those pollutants are higher, especially for CO, being in some cases slightly above fourfold the value of the other emissions factors (ex.: solids fuels). Heavy metals present a reduction of almost 100%.

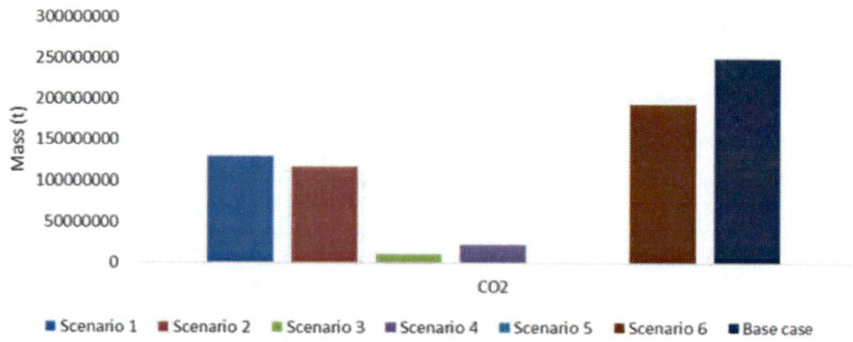


**Fig. 66.3** Carbon dioxide emissions

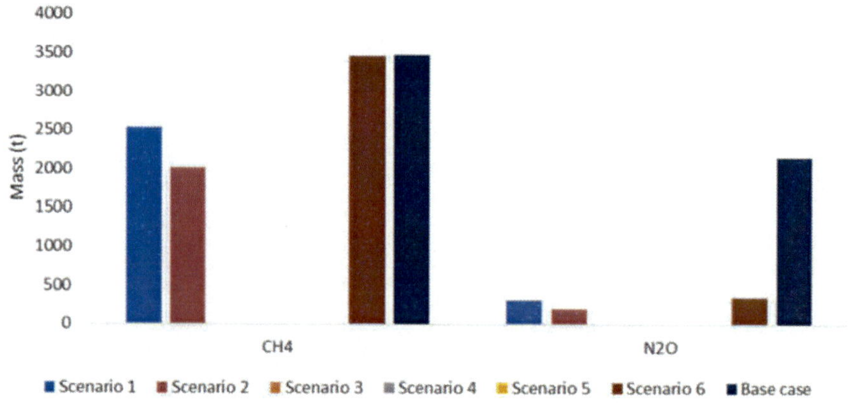


**Fig. 66.4** Other GEE emissions

Figures 66.3 and 66.4 show the GEE emissions for the several scenarios being the Scenario 6 the one that considers that natural gas replaces all other fossil fuels.

Figures 66.5 and 66.6 represent the other pollutant emissions for the several scenarios. Scenarios 1, 2 and 6 show the highest contributions.

As mentioned before, natural gas lead to the reduction of heavy metals and other pollutants. However, its usage causes the increase of CO and NMVOC emissions and given the actual circumstances, the renewables become even more attractive since they can solve climate change and reduce all pollutant emissions, being available in almost every country, promoting the reduction of countries energy dependency.

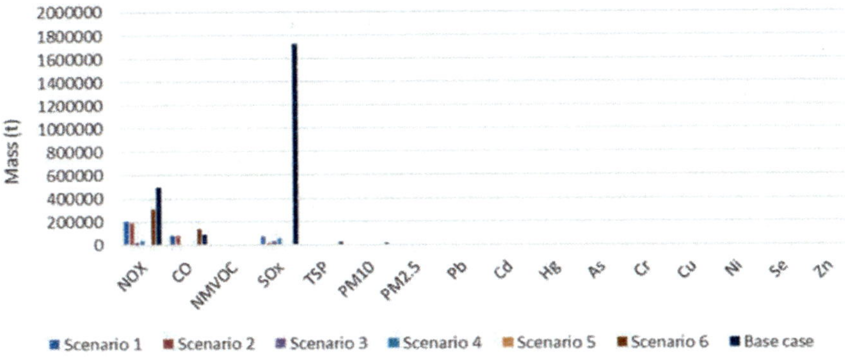


**Fig. 66.5** Nitrogen oxides, sulphur dioxide, particulate matter, heavy metals

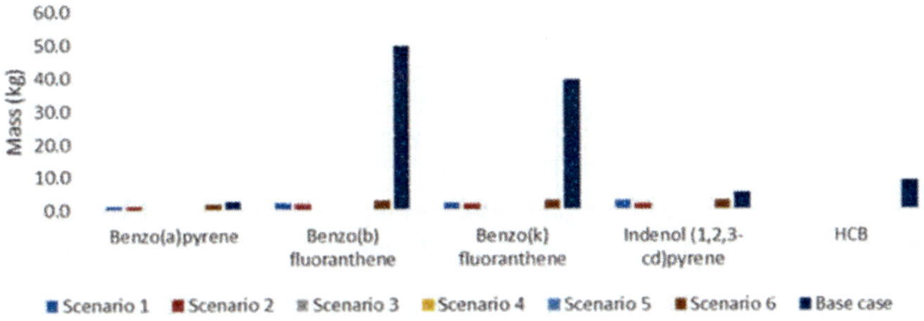


**Fig. 66.6** Remaining air pollutants emissions

## 66.4 Conclusion

Energy is a core and important issue to achieve sustainable solutions for global warming and pollution problems. This work considered six scenarios for electricity production in the European Union for 2020 where the pollutant emissions were estimated and replacement of fossil fuels with renewables, namely photovoltaic and wind was considered. The replacement of fossil fuels, as expected, conducted to the pollutant emission reduction. If solid fossil fuels were replaced, there would be a significant reduction in all pollutants, a trend that is emphasized if fuel oil and light oil were also replaced, in which case only 3 pollutants would show a reduction below 50% (namely methane, carbon monoxide and NMVOC).

When natural gas is replaced with renewables, it is possible to conclude that the reduction for all pollutants is equal to 100% with the exception of carbon dioxide reduction, that ranges between 90 and 100%. If all fossil fuels were replaced by natural gas there would a significant reduction in all pollutants except CO and NMVOC, that would increase. Heavy metals present a reduction of almost 100%. It is possible to say that the increase of renewables (photovoltaic and wind) can solve the pollution problems with advantages to environment and people's heath, increasing sustainability. As demonstrated, it is also important to climate change and to decrease energy dependency because those resources, solar and wind, are available in most countries.

**Acknowledgements** This work was financially supported by: LA/P/0045/2020 (ALiCE) and UIDB/00511/2020—UIDP/00511/2020 (LEPABE) funded by national funds through FCT/MCTES (PIDDAC), and UIDB/04730/2020 of the Center for Innovation in Engineering and Industrial Technology (CIETI), funded by national funds through FCT/MCTES (PIDDAC).

## References

Adler C, Athanassiou C, Carvalho MO, Emekci M, Gvozdenac S, Hamel D, Riudavets J, Stejskal V, Trdan S, Trematerra P (2022) Changes in the distribution and pest risk of stored product insects in Europe due to global warming: need for pan-European pest monitoring and improved food-safety. J Stored Prod Res 97. https://doi.org/10.1016/J.JSPR.2022.101977

Chen Y, Yang R, Wong CWY, Ji J, Miao X (2022) Efficiency and productivity of air pollution control in Chinese cities. Sustain Cities Soc 76. https://doi.org/10.1016/J.SCS.2021.103423

EMEP/EEA (2019) Air pollutant emission inventory guidebook 2019

EUROSTAT (2022) No Title. https://ec

Feroz EH, Raab RL, Ulleberg GT, Alsharif K (2009) Global warming and environmental production efficiency ranking of the Kyoto Protocol nations. J Environ Manag 90:1178–1183. https://doi.org/10.1016/J.JENVMAN.2008.05.006

Gelb J, Apparicio P (2022) Cyclists' exposure to air and noise pollution, comparative approach in seven cities. Transp Res Interdiscip Perspect 14:100619. https://doi.org/10.1016/j.trip.2022.100619

IPPC (2006) Guidelines for national greenhouse gas

Pitiruţ B, Ogunbode C, Enea V (2022) Attitudes towards global warming: the role of anticipated guilt and the Dark Triad traits. Pers Individ Dif 185. https://doi.org/10.1016/J.PAID.2021.111285

Tawalbeh M, Al-Othman A, Kafiah F, Abdelsalam E, Almomani F, Alkasrawi M (2021) Environmental impacts of solar photovoltaic systems: a critical review of recent progress and future outlook. Sci Total Environ 759:143528. https://doi.org/10.1016/J.SCITOTENV.2020.143528

Yang J, Chang Y, Zhang L, Hao Y, Yan Q, Wang C (2018) The life-cycle energy and environmental emissions of a typical offshore wind farm in China. J Clean Prod 180:316–324. https://doi.org/10.1016/J.JCLEPRO.2018.01.082

# Chapter 67
# Examining the Financial Performance of Renewable Energy Companies Through a Hybrid Multi-criteria Decision Making Model

**Ali Cilesiz and Faruk Dayi**

**Abstract** This study compared the financial performance of renewable energy companies. The study focused on 2019–2021 data from the top ten companies in the S&P Global Clean Energy Index. The research model consisted of the current ratio, quick ratio, accounts receivables turnover, asset turnover, stock turnover, liability ratio, debt to equity ratio, fixed assets to constant capital ratio, return on assets, and return on equity. The data were analysed using the Best Worst Method (BWM)-based TOPSIS method. The financial ratio group with the highest weight in the renewable energy sector is the profitability ratio (42%). There is no significant change in the annual performance ranking of the companies.

**Keywords** Renewable energy · S&P Clean Energy Index · Financial performance · Best–worst method · TOPSIS

## 67.1 Introduction

In recent years, there has been a growing demand for energy due to rapid industrialization, global population growth, and the escalating consumption of electricity. Governments seek new sources of energy to address the challenges of energy management and procurement and achieve energy independence. Sources of clean energy are increasing prevalent due to environmental concerns such as fossil fuels. The share of renewable energy in total energy production worldwide increased from one percent in 2010 to six percent in 2020. However, most states have failed to meet their targets for renewable energy despite many commitments (KPMG 2022). In the coming years, renewable energy will have a larger share in the power generation portfolio for several reasons. First, it becomes cheaper with advances in science

A. Cilesiz (✉) · F. Dayi
Kastamonu University, Kastamonu 37150, Turkey
e-mail: acilesiz@kastamonu.edu.tr



N. S. Caetano and M. C. Felgueiras (eds.), *The 9th International Conference on Energy and Environment Research*, Environmental Science and Engineering,
https://doi.org/10.1007/978-3-031-43559-1_67

and technology. Second, people become more aware of environmental problems and adopt eco-friendly lifestyles. Third, governments introduce "green" policies to achieve sustainable economic growth. For these reasons, it is very important to evaluate financial performance for renewable energy companies for the growth of the energy sector and the elimination of threats.

Company performance is defined as how well a company uses its assets to achieve its financial objectives (Barney 2014). Financial performance is a measure of outcomes resulting from changes in a company's financial structure (Carton 2004). Companies that want to survive and thrive in the increasingly competitive marketplace must assess their financial performance. The renewable energy sector is growing exponentially. Therefore, renewable energy companies must measure their financial performance to develop the sector.

Multi-criteria decision-making (MCDM) methods are popular methods that allow researchers to solve choice (among alternatives), sequencing, and classification problems. Some of the MCDM methods are the Analytic Hierarchy Process (AHP), Analytic Network Process (ANP), Best Worst Method (BWM), the Multi-Attribute Utility Theory/Utility Additive Method (MAUT/UTA), Measuring Attractiveness by a Categorical Based Evaluation Technique (MACBETH), Preference Ranking Organization Method for Enrichment Evaluation (PROMETHEE), Elimination et Choix Traduisant la REalité (ÉLECTRE), Technique for Order of Preference by Similarity to Ideal Solution (TOPSIS), and Višekriterijumsko Kompromisno Rangiranje (VIKOR). Decision-makers use one or more of these methods to solve decision problems (Ishizaka and Nemery 2013).

This study investigated the financial performance of the top ten renewable energy companies in the S&P Global Clean Energy Index. Financial ratios are used as variables in the BWM-based TOPSIS method. This study employed the method to assess the financial condition of the companies. According to the analysis results, the companies were ranked according to their financial performance.

This study consisted of five sections. The second section was a literature review. The third section addressed the methodology. The fourth section presented the findings. The fifth section was the conclusion.

## 67.2 Literature Review

Liquidity, financial structure, activity, and profitability ratios are often used to measure performance (Bunea et al. 2019; Ghadikolaei et al. 2014; Molina-Azorín et al. 2009; Wu et al. 2009). MCDM methods are used to evaluate financial performance. Research on the energy sector has used various MCDM methods for analysis.

Ozdemir and Parmaksiz (2022) used the TOPSIS and the Evaluation Based on Distance from Average Solution (EDAS) to analyze the financial ratios of 16 energy companies. They reported minor differences in performance rankings between the two methods. Yang et al. (2022) used the TOPSIS method to assess China's energy security. They found that economic growth, urbanization, and other driving forces put enormous pressure on energy security. Agyekum et al. (2021) used the BWM method to weight criteria for opportunities and challenges in the renewable energy sector. They determined that the great opportunity was exporting renewable energy, whereas the great challenge was the government's low interest in renewable energy. Dong et al. (2021) used the ANP and Matter-Element Extension Cloud Model (MEECM) to examine the trade performance of five power generation companies operating in China. They concluded that the companies should carry out market and support technology activities to pay more attention to the criteria of trading ability and behavior. Ecer (2021) employed the BWM to weight the criteria used to evaluate the sustainability performance of wind farms. The main criterion group with the highest weight was the environmental dimension, followed by the economic and social dimensions. Arsu (2021) used the Entropy-Based Additive Ratio Assessment (ARAS) method to evaluate the financial performance of companies in the Borsa Istanbul (BIST) Electricity, Gas, and Steam sector. They reported that the most critical ratios were equity turnover rate, debt coefficient, and asset turnover. They also found that the three companies with the highest financial performance were Enerjisa (ENJSA), Aksa Enerji (AKSEN), and Zorlu Enerji (ZOREN).

Alizadeh et al. (2020) used the Benefits, Opportunities, Costs, and Risks (BOCR) method to weigh the barriers to using renewable resources in order of importance. They predicted that the most preferred renewable energy source in Iran would be solar energy. Mercan and Cetin (2020) used the Complex Proportional Assessment (COPRAS) and VIKOR methods to examine the financial performance of seven electric power companies. They found that the financial performance rankings of the companies did not change in both methods.

Kayahan-Karakul and Ozaydin (2019) used the TOPSIS and VIKOR methods to analyze the financial performances of eight electricity companies based on seven financial ratios. They reported apparent differences in performance rankings because the two methods used different algorithms. Lee and Chang (2018) employed the WSM, VIKOR, TOPSIS, and ELECTRE methods to compare renewable energy sources in Taiwan. They concluded that the best alternative was hydroelectric power, followed by solar, wind, biomass, and geothermal energy. Halkos and Tzeremes (2012) used the Data Envelopment Analysis (DEA) to assess the financial performance of 78 companies operating in the renewable energy sector. They found that high return on assets, return on equity ratio, and low debt to equity ratio positively affected financial performance.

**Table 67.1** Companies list

| Company name | |
|---|---|
| Enphase Energy Inc. | Plug Power Inc. |
| Vestas Wind Systems AS | Scottish & Southern Energy |
| Consolidated Edison Inc. | Energias de Portugal SA |
| Orsted A/S | Iberdrola SA |
| SolarEdge Technologies Inc | First Solar Inc. |

## 67.3 Methodology and Data

### *67.3.1 Data and Variables*

This study aimed to compare the financial performance of companies operating in the renewable energy sector. The sample consisted of the top ten companies in the S&P Global Clean Energy Index as of 31.03.2022 according to their weight (S&P Global 2022). Financial data was collected from independently audited consolidated financial statements in company web pages. Table 67.1 shows the companies list.

Variables were determined to examine the financial performance of the companies. The variables were popular variables based on literature (Bunea et al. 2019; Ghadikolaei et al. 2014; Molina-Azorín et al. 2009; Wu et al. 2009). Table 67.2 shows the variables considered.

### *67.3.2 Methods*

MCDM methods systematically solve decision problems in many categories. Those methods allow researchers to make decisions about decision problems involving multiple conflicting factors (Belton and Stewart 2002). Academic studies on renewable energy generally use the WSM, WPM, AHP, ELECTRE, TOPSIS, VIKOR, PROMETHEE, and MAUT methods (Kumar et al. 2017). This study employed the BWM and TOPSIS methods to evaluate the financial performance of companies operating in the renewable energy sector.

**Technique for order of preference by similarity to ideal solution (TOPSIS).** TOPSIS allows us to focus on criteria and identify alternatives to be preferred or ranked. TOPSIS steps are below (Hwang and Yoon 1981):

Step 1. A normalized decision matrix is generated: $a_{ij}$ shows the alternatives.

$$n_{ij} = \frac{a_{ij}}{\sqrt{\sum_{i=1}^{m} a_{ij}^2}} (i = 1, \ldots, m \ and \ j = 1, \ldots, n) \tag{67.1}$$

**Table 67.2** Variables

| Main variable | Sub-variable | Definition |
|---|---|---|
| Liquidity ratios | Current ratio | Current ratio is a ratio that reveals the ability of a company to pay its short-term debts. It is obtained by dividing current assets by current liabilities |
| | Quick ratio | Quick ratio is obtained by subtracting inventories from current assets while calculating the current ratio |
| Operating ratios | Accounts receivables turnover | Account receivables turnover is the ratio that expresses the collection period of the receivables arising from the sales of the company as a result of its activities |
| | Asset turnover | Asset turnover the ratio that indicates how many times more of its assets the company sales |
| | Stock turnover | Stock turnover is the ratio that shows how quickly stocks turn into cash |
| Profitability Ratios | Return on assets | Return on assets is the ratio that shows how much profit the company makes with one unit of money with its assets |
| | Return on equity | Return on equity is the ratio that shows how much profit the company makes with one unit of money with its equity |
| Financial structure ratios | Liability ratio | Liability ratio is the ratio that shows what percentage of assets are financed by liabilities |
| | Debt to equity ratio | Debt to equity ratio shows the capital structure of the company. It is the ratio obtained by dividing the total debts of the company by the equity |
| | Fixed asset/constant capital | Fixed asset/constant capital shows how much of the operating fixed assets are financed by constant capital |

Step 2. A weighted normalized decision matrix is generated.

$$V_{ij} = n_{ij} w_{ij} \quad (i = 1, \ldots, m \; and \; j = 1, \ldots, n) \tag{67.2}$$

Step 3. Positive ideal and negative ideal solution values are obtained.

$$A^{+} = \left\{v_i^{+}, v_i^{+}, \ldots, v_n^{+}\right\} = \left\{\max_{j} v_{ij} | j = 1, \ldots, p; i = 1, \ldots, n\right\} \tag{67.3}$$

$$A^{-} = \left\{v_i^{-}, v_i^{-}, \ldots, v_n^{-}\right\} = \left\{\min_{j} v_{ij} | j = 1, \ldots, p; i = 1, \ldots, n\right\} \tag{67.4}$$

Step 4. Distances to the positive ideal and negative ideal values are determined.

$$S_i^+ = \sqrt{\sum_{j=1}^{n} \left(v_{ij} - v_j^+\right)^2} \quad (67.5)$$

$$S_i^- = \sqrt{\sum_{j=1}^{n} \left(v_{ij} - v_j^-\right)^2} \quad (67.6)$$

Step 5. The relative closeness to the ideal solution is calculated.

$$C_i^* = \frac{S_i^-}{S_i^-, S_i^*} \quad (67.7)$$

Step 6. Alternatives are ranked according to the $C_i^*$ value.

**Best Worst Method (BWM).** BWM was developed by Rezaei (2015). The method is based on weighting via only two comparison vectors instead of a pairwise comparison of all criteria. Matrices expressed as vectors are matrices from best to others and others to worst. Comparing only two vectors yields more consistent results (Salimi and Rezaei 2018). The method weighs the criteria to evaluate alternatives. BWM steps are below (Rezaei 2015):

Step 1. A set of criteria $\{C_1, C_2, \ldots, C_n\}$ is determined to choose among alternatives.
Step 2. The best and worst criteria are selected.
Step 3. The best criterion is compared with a number between 1 and 9 according to the level of preference with all criteria. The comparison yields the Best-Others ($A_B$) vector.

$$A_B = (a_{B1}, a_{B2}, \ldots, a_{Bn}) \quad (67.8)$$

Step 4. The worst criterion is compared with a number between 1 and 9 according to the level of preference with all criteria. The comparison yields the Other-Worst ($A_W$) vector.

$$A_W = (a_{1W}, a_{2W}, \ldots, a_{nW})^T \quad (67.9)$$

Step 5. Optimum weights $\left(w_1^*, w_2^*, \ldots, w_n^*\right)$ are determined for each criterion.

The objective of Step 5 is to determine the optimal weights of all criteria to create maximum absolute differences. The optimum weight for the criteria is $w_B/w_j = a_{B_j}$ and $w_j/w_w = a_{j_w}$ for each $w_B/w_j$ and $w_j/w_w$ pair, respectively. There must be a j value of $\left|\frac{w_B}{w_j} - a_{Bj}\right|$ and $\left|\frac{w_j}{w_w} - a_{jw}\right|$ at which the maximum absolute differences are minimized, which is translated into the following min–max model:

Under the constraints of

$$\min \max\{|w_B - a_{Bj}w_j|, |w_j - a_{jw}w_w|\}j \tag{67.10}$$

$$\sum_j w_j = 1$$

$$w_j \geq 0, \quad \text{for all } j$$

## 67.4 Results

An appropriate weighting of variables is critical for realistic financial performance analysis. This study focused on the objective weights of the variables. The variables were weighted using the BWM based on experts' opinions in the renewable energy sector and finance. Table 67.3 shows the weights of the variables.

Ten variables under four main variable groups were weighed using the BWM method. The main variable groups "profitability ratios" and "liquidity ratios" had the highest (42%) and lowest weights (13%), respectively. The main variable groups "operating ratios" and "financial structure ratios" weighted 22.5%. The variable "return on assets" and "return on equity" had the highest weight (21%), whereas the variable "stock turnover" had the lowest weight (3.2%).

In 2019, Enphase Energy had the highest score (0.905), indicating the best financial performance. SolarEdge Technologies (0.638) and Scottish & Southern Energy (0.592) ranked second and third, respectively. Plug Power had the lowest score (0.123), suggesting the worst financial performance.

In 2020, Enphase Energy had the highest score (0.848), indicating the best financial performance. SolarEdge Technologies (0.778) and Vestas Wind Systems (0.765)

**Table 67.3** Weights of variables

| Variables | Weight | Sub-variable | Weight |
|---|---|---|---|
| Liquidity ratios | 0.130 | Current ratio | 0.065 |
| | | Ouick ratio | 0.065 |
| Operating ratios | 0.225 | Accounts receivables turnover rate | 0.052 |
| | | Asset turnover | 0.141 |
| | | Stock turnover | 0.032 |
| Profitability ratios | 0.420 | Return on assets | 0.210 |
| | | Return on equity | 0.210 |
| Financial structure ratios | 0.225 | Liability ratio | 0.075 |
| | | Debt to equity ratio | 0.075 |
| | | Fixed asset/constant capital | 0.075 |

**Table 67.4** BWM-based TOPSIS results

| | 2019 | | 2020 | | 2021 | |
|---|---|---|---|---|---|---|
| | Score | Rank | Score | Rank | Score | Rank |
| Enphase Energy Inc. | 0.905 | 1 | 0.848 | 1 | 0.776 | 1 |
| Vestas Wind Systems AS | 0.578 | 4 | 0.765 | 3 | 0.436 | 8 |
| Consolidated Edison Inc. | 0.482 | 7 | 0.655 | 8 | 0.444 | 7 |
| Orsted A/S | 0.497 | 5 | 0.750 | 4 | 0.527 | 5 |
| SolarEdge Technologies Inc. | 0.638 | 2 | 0.778 | 2 | 0.615 | 3 |
| Plug Power Inc. | 0.123 | 10 | 0.149 | 10 | 0.204 | 10 |
| Scottish & Southern Energy | 0.592 | 3 | 0.607 | 9 | 0.764 | 2 |
| Energias de Portugal SA | 0.458 | 8 | 0.668 | 7 | 0.417 | 9 |
| Iberdrola SA | 0.495 | 6 | 0.696 | 6 | 0.483 | 6 |
| First Solar Inc. | 0.403 | 9 | 0.716 | 5 | 0.571 | 4 |

ranked second and third, respectively. Plug Power had the lowest score (0.149), suggesting the worst financial performance.

In 2021, Enphase Energy had the highest score (0.776), indicating the best financial performance. Scottish & Southern Energy (0.764) and SolarEdge Technologies (0.615) ranked second and third, respectively. Plug Power had the lowest score (0.123), suggesting the worst financial performance.

## 67.5 Conclusion

This study employed the BWM-based TOPSIS method to analyze the 2019–2021 financial performance of the top ten companies in the S&P Global Clean Energy Index. The main variable group, "profitability ratios," had the highest weight (42%). Operating ratios, financial structure ratios, and liquidity ratios weighted 22.5%, 22.5%, and 13%, respectively. Enphase Energy had the best financial performance between 2019 and 2021, probably due to high profitability, favorable liquidity, and operating structure. Plug Power had the worst financial performance between 2019 and 2021, probably because it constantly makes a loss and has problems in its operational structure. There was no significant difference in rankings between 2019, 2020, and 2021. The greatest difference was observed in Vestas Wind System. The company, which had ranked third in 2020, ranked eighth in 2021. This is probably due to the decrease in return on assets and returns on equity by around 75%. There was also a reduction in its liquidity ratios. The second greatest difference was observed in Scottish & Southern Energy. The company, which had ranked ninth in 2020, ranked second in 2021. This is because the company, which made a loss in 2020, reported a significant amount of profit in 2021. This study focused on the financial performance of the top ten companies in the S&P Global Clean Energy Index. Researchers should use the BWM-based TOPSIS method to analyze the financial performance of different renewable energy companies.

## References


Agyekum EB, Kumar NM, Mehmood U, Panjwani MK, Haes-Alhelou H, Adebayo TS (2021) Decarbonize Russia—a Best-Worst Method approach for assessing the renewable energy potentials, opportunities and challenges. Energy Rep 7:498–515

Alizadeh R, Soltanisehat L, Lund PD, Zamanisabzi H (2020) Improving renewable energy policy planning and decision-making through a hybrid MCDM method. Energy Policy 37:111174–111191

Arsu T (2021) Finansal Performanslarin Entropi Tabanli ARAS yontemi ile degerlendirilmesi: BIST Elektrik, Gaz ve Buhar sektorundeki isletmeler uzerine bir uygulama. Hacettepe Universitesi Iktisadi Ve Idari Bilim Fakultesi Dergisi 39(1):15–32

Barney J (2014) Gaining and sustaining competitive advantage, 4th edn. Pearson

Belton V, Stewart TJ (2002) Multiple criteria decision analysis. Springer

Bunea OI, Corbos RA, Popescu RI (2019) Influence of some financial indicators on return on equity ratio in the Romanian energy sector—a competitive approach using a DuPont-based analysis. Energy 189:1–10

Carton RB (2004) Measuring organizational performance: an exploratory study. Doctoral dissertation, University of Georgia

Dong J, Liu D, Liu Y, Huo H, Dou X, Bao A (2021) Trading performance evaluation for traditional power generation group based on an integrated matter-element extension cloud model. Energy Rep 7:3074–3089

Ecer F (2021) Sustainability assessment of existing onshore wind plants in the context of triple bottom line: a best-worst method (BWM) based MCDM framework. Environ Sci Pollut Res 28(16):19677–19693

Ghadikolaei AS, Esbouei SK, Antucheviciene J (2014) Applying fuzzy MCDM for financial performance evaluation of Iranian companies. Technol Econ Dev Econ 20(2):274–291

Halkos GE, Tzeremes NG (2012) Analyzing the Greek renewable energy sector: a Data Envelopment Analysis approach. Renew Sustain Energy Rev 16:2884–2893

Hwang C, Yoon K (1981) Multiple attribute decision making methods and applications a state-of-the-art survey. Springer, Berlin, Heidelberg

Ishizaka A, Nemery P (2013) Multi-criteria decision analysis: methods and software. Wiley

Kayahan-Karakul A, Ozaydin G (2019) TOPSIS ve VIKOR Yontemleri ile Finansal Performans Degerlendirmesi: XELKT Uzerinde Bir Uygulama. Dumlupinar Universitesi Sosyal Bilimler Dergisi 60:68–86

KPMG: Enerji Sektörel Bakış. https://assets.kpmg/content/dam/kpmg/tr/pdf/2022/04/enerji-sektorel-bakis.pdf. Accessed 16 May 2022

Kumar A, Sah B, Singh AR, Deng Y, He X, Kumar P (2017) A review of multi criteria decision making (MCDM) towards sustainable renewable energy development. Renew Sustain Energy Rev 69:596–609

Lee HC, Chang C (2018) Comparative analysis of MCDM methods for ranking renewable energy sources in Taiwan. Renew Sustain Energy Rev 92:883–896

Mercan Y, Cetin O (2020) COPRAS ve VIKOR Yontemleri ile BIST Elektrik Endeksindeki Firmalarin Finansal Performans Analizi. Uluslararasi Afro-Avrasya Arastirmalari Dergisi 5:123–139

Molina-Azorín JF, Claver-Cortés E, López-Gamero MD, Tarí JJ (2009) Green management and financial performance: a literature review. Manag Decis 47(7):1080–1100

Ozdemir O, Parmaksiz S (2022) BIST Enerji Isletmelerinin Finansal Performanslarinin Cok Kriterli Karar Verme Teknikleri ile Karsilastirilmasi: TOPSIS ve EDAS Yontemleri ile Analiz. Baskent Universitesi Ticari Bilimler Fakültesi Dergisi 6(1):34–56
Rezaei J (2015) Best-worst multi-criteria decision-making method. Omega 53:49–57
S&P Global (2022) Global Clean Energy Index
Salimi N, Rezaei J (2018) Evaluating firms' R&D performance using best worst method. Eval Program Plan 66:147–155
Wu HY, Tzeng GH, Chen YH (2009) A fuzzy MCDM approach for evaluating banking performance based on Balanced Scorecard. Expert Syst Appl 36(6):10135–10147
Yang B, Ding L, Zhan X, Tao X, Peng F (2022) Evaluation and analysis of energy security in China based on the DPSIR model. Energy Rep 8:607–615

# Chapter 68
# Natural Gas and $H_2$: The Role of the Iberian Countries to EU Supply Diversification and Decarbonization

**João Moura and Isabel Soares**

**Abstract** The main goal of this article is to identify and to discuss the significant potential of the Iberian countries to EU decarbonization strategy and security of supply, due to their natural gas infrastructures and green hydrogen ($H_2$) projects implementation. In line with the renewable energy sources (RES) expansion, $H_2$ has emerged as a promising energy source. Notwithstanding, there are still major economic and technological constraints that can explain why $H_2$ represents only 2% of the European energy consumption. Energy transition costs are widely recognized as a major constraint, while security of supply also remains a strategic driver. Among others, the inclusion of a low percentage (10–15%) of green $H_2$ in natural gas grid appears to be a reliable contribution to the European decarbonization process while allowing for an economically consistent strategy along time.

**Keywords** Decarbonization · European Union · $H_2$ · Iberia · Natural gas

## 68.1 Introduction

According to some authors, a gradual and smooth global transition to clean energy seems unrealistic as the energy system, which is the central element of the international economy—thus supporting geopolitical order, will be completely transformed (Bordoff and O'Sullivan Meghan 2022). Notwithstanding, the European process towards a clean energy future continues subject to specific countries profile besides dealing with short-medium term options and constraints. Dependence on fossil fuels is still a reality for most developed countries, even for economies with good performance regarding climate change indicators, such as the

J. Moura (✉) · I. Soares
Faculdade de Economia do Porto and CEFUP, Rua Dr. Roberto Frias, 4400-464 Porto, Portugal
e-mail: up200706386@up.pt

I. Soares
e-mail: isoares@fep.up.pt



N. S. Caetano and M. C. Felgueiras (eds.), *The 9th International Conference on Energy and Environment Research*, Environmental Science and Engineering,
https://doi.org/10.1007/978-3-031-43559-1_68

United Kingdom (2° in climate change action, ranking with 403,076 $CO_2$ thousand ton/year) or France (4° with 344,891.76 $CO_2$ thousand ton/year (Environmental Performance Index Homepage: https://epi.yale.edu/downloads/epi2022report06062022.pdf; Oliveira et al. 2021a; Eurostat Homepage: https://ec.europa.eu/eurostat/databrowser/view/ENV_AIR_GGE/default/table?lang=en). In this context, energy-intensive industries are facing unprecedented pressures to lower carbon emissions at the same time they seek innovative strategies to reduce production costs (Liu et al. 2021). Hydrogen ($H_2$) has been identified as a new viable alternative fuel because of its neutral greenhouse gas emissions (Gielen et al. 2019) as well as natural gas cannot be totally discarded despite current geostrategic problems with Russia (putting a dramatic pressure on prices and security of supply), due to its $CO_2$ low impact and world availability.

The Iberian Gas Market (MIBGAS) was created in June 2014 to reinforce security of supply, competitiveness and obliged a revision of both regulatory framework of Portugal and Spain, to prompt efficiency on the new liberalized activities. With electric and gas markets forming a Single Iberian market, Portugal and Spain are benefiting from a fully integrated regional energy market, that presents lower risks in terms of supply comparing with other European regional markets. This specially accounts for the natural gas current situation: while most member states of EU strive for short-term alternatives due to the Ukrainian war, the Iberian countries are still facing a relatively stable situation because of their wide gas suppliers, with alternatives amongst Africa, North America and Asia. Despite this, the low interconnection capacity with France creates a huge bottleneck that makes unfeasible the gas flow between Iberian Peninsula and the rest of Europe, negatively contributing for a general security of supply and harshening the probability of geostrategic risk.

In terms of green $H_2$, Spanish and Portuguese governments have already turned public their commitment to $H_2$ roadmaps (European Commission Homepage: https://commission.europa.eu/documents_en?f%5B0%5D=document_title%3Aspain%2C%20H2; Ministério do Ambiente e da Ação Climática (MAAC). https://www.portugal.gov.pt/download-ficheiros/ficheiro.aspx?v=%3d%3dBQAAAB%2bLCAAAAAAABAAzNDCyMAUArNeLzAUAAAA%3d). Some large green projects are expected to be operational until 2025, such as GreenH2Atlantic in Sines with a 100 MW electrolyzer capacity (Energias de Portugal Homepage (EDP): https://www.edp.com/en/news/2021/12/21/renewable-H2-production-sines-advances-through-greenh2atlantic-project), the Green Hysland project in Mallorca with 300 tones/year of renewable $H_2$ production (Clean H2 Partnership Homepage: https://www.clean-H2.europa.eu/media/news/green-hysland-inauguration-first-renewable-H2-industrial-plant-mallorca-2022-03-14_en) or the Catalina Project in Aragon with a 500 MW electrolyzer (Power Technology Homepage: https://www.power-technology.com/news/cip-project-catalina-spain/). Recently, HyDeal introduced the largest green $H_2$ hub project in Asturias, with an installed capacity to reach 9.5 GW of solar power and 7.4 GW of electrolyzers by 2030 (HyDeal España Homepage: https://www.hydeal.com/copie-de-hydeal-ambition). This is, not only, the largest project for planned operation in the peninsula, but it is also the world's largest renewable $H_2$ project (The International Renewable Energy

Agency (IRENA) Homepage: https://irena.org/-/media/Files/IRENA/Agency/Publication/2022/Jan/IRENA_Geopolitics_H2_2022.pdf). The Iberian Peninsula renewable energy potential (wind and photovoltaic) greatly induces general interest to bring together RES and green $H_2$ generation, thus, it is expected that heavy industry will be provided with close or onsite green $H_2$ production in the medium term. The European Commission $H_2$ roadmap also considers $H_2$ blending with natural gas in an early phase (European Commission Homepage: https://eur-lex.europa.eu/legal-content/EN/TXT/?uri=CELEX:52020DC0301), which makes current European gas infrastructure an important key to promote the $H_2$ markets. Considering this, the limited interconnection between Spain and France constitutes a serious barrier.

The main goal of this article is to identify the significant potential of the Iberian countries to contribute to EU decarbonization strategy and security of supply, due to their natural gas infrastructures and green $H_2$ projects implementation. After this introduction, this article is organized as follows: Sect. 68.2—Materials and Methods identifies $H_2$ as a tool for energy and economic change and sustains the importance of natural gas as a transition fuel; in this section, a critical analysis of both countries' current situation in terms of natural gas infrastructures strategies and green $H_2$ initiatives is proposed. Results and discussion are stated in Sect. 68.3 followed by conclusions in Sect. 68.4.

## 68.2 Materials and Methods

We follow a methodology that starts by identifying both $H_2$ and natural gas current situation in EU, namely $H_2$ initiatives as well as natural gas/LNG infrastructures in the Iberian countries. This will be followed by the presentation of the main reasons that sustain this article proposition. Our aim is presenting and discussing our findings in a reproducible transparent and objective way.

### 68.2.1 *$H_2$ as a Tool for Energy Transition and Economic Change*

An economy based on $H_2$ and sustainable power generation is comparable to a new major "industrial revolution", promoting innovative expertise and technological leadership as well as a better environment for future generations (Trattner et al. 2022). Besides being an optimal energy carrier, it can also help surpass various critical challenges of the energy transition. For example, $H_2$ can be used to decarbonize energy intensive sectors where electricity is not an option to consider, such as long-distance transport or even heavy industry (refining, chemical, metallurgical, cement, mining, ceramics, glass industries, steelmaking) (Portuguese Government Homepage: https://files.dre.pt/1s/2020/08/15800/0000700088.pdf; Cheng and Lee 2022).

$H_2$ is also versatile in terms of supply and use, since it is generated from several sources and it will unlock renewables potential through the storage of electricity over time (Gielen et al. 2019). Considering that energy sector is facing the growth of energy demand at the same time there are crucial environmental concerns, $H_2$ can be the perfect "missing link" in the decarbonization roadmap (Nicita et al. 2020). However, it was only recently that interest by the scientific community and the International Energy Agency (IEA) gained a massive support considering the key role $H_2$ can play in decarbonisation (Kovač et al. 2021). Being recognized as a versatile energy vector with high energy storage potential, $H_2$ has several potential applications in the power, buildings, transport and industrial sectors and it will be replacing natural gas as a carbon free green fuel (Kovač et al. 2021; Papadis and Tsatsaronis 2020). It is one of the most promising clean energy alternatives in the twenty-first century because of its production diversity, high calorific value, good thermal conductivity, and high reaction rate (Kakoulaki et al. 2021). Nevertheless, $H_2$ represents only 2% of Europe's energy consumption (European Commission Homepage: https://eur-lex.europa.eu/legal-content/EN/TXT/?uri=CELEX:52020DC0301). It is mostly used in chemical industry and produced through natural gas (approximately, 95%), with captive production representing 64% of its total production, while 15% are originated from merchant production and 21% from by-product processes (Pawelec et al. 2020). Having a low overall impact on actual energy systems, current $H_2$ production is far from being clean. About 95% of $H_2$ production is produced through highly carbon-intensive processes, such as steam methane reforming (SMR) or autothermal reforming (ATR), representing a fossil-based process—Grey $H_2$ (Cihlar et al. 2021). To minimize pollution, these methods can be coupled with carbon capture, usage, and storage (CCUS) systems in order to reduce $CO_2$ emissions—Blue $H_2$ (Cihlar et al. 2021). Despite not being clean, blue $H_2$ is an important alternative to promote decarbonization at short-term, since it is produced in larger volumes at lower cost (Damman et al. 2021). Similarity with previous color-band terminologies, when $H_2$ and oxygen are broken together through water electrolysis using renewable power, the production process is free from greenhouse gas emissions—Green $H_2$ (Oliveira et al. 2021a; Cheng and Lee 2022). This last type is the ultimate choice to tackle the energy sector change, since electrolysis has the highest potential for decarbonization for a carbon-free economy in the long-term. Thus, the EU Commission recognizes $H_2$ investment as one of the main priorities for achieving carbon neutrality by 2050, having decreed that at least 6 GW of electrolyzers powered by renewable energy will be installed between 2020 and 2024 (European Commission Homepage: https://eur-lex.europa.eu/legal-content/EN/TXT/?uri=CELEX:52020DC0301). Until then, green $H_2$ technology must develop lower overall costs of processes, to scale up properly (Kovač et al. 2021).

### 68.2.2 Importance of Natural Gas as a Transition Fuel

A transition fuel is a temporary substitute for heavier conventional fossil fuels using low-carbon alternative compounds, functioning as a bridge until the dominance of renewable energy is a reality, producing secure and sustained energy (Aguilera and Aguilera 2020; Gürsan and Gooyert 2021). Natural gas is currently one of the most used fossil fuels with a well-defined growing demand per year, accounting for about 23% of primary energy demand and being responsible for 25% of electricity generation in the world (International Energy Agency (IEA) Homepage: https://www.iea.org/fuels-and-technologies/gas). Although not carbon- free, natural gas has an important role of optimizing the energy mix in the ideal energy structure at near-term, being capable of dealing with the intermittency of increased RES (Safari et al. 2019). Specifically, natural gas is an energy source immune to intermittency, has a high flexibility and has a good potential to correspond to high peak demands, making it one of the preferred reliable resources all over the world and the main candidate as a great facilitator of energy transition (Gürsan and Gooyert 2021). Nevertheless, there is also controversy about natural gas impact on energy transition, causing an open debate since investment and financing on a low carbon substitute might have a negative effect by delaying innovation on zero-carbon technologies (Gürsan and Gooyert 2021). Regarding GHG emissions, this fact seems to reduce mitigation efforts in the long-term, due to accommodation of strategies based on the existing gas infrastructure, such as maintaining gas power plants with an advanced useful life through carbon capture technologies (Woollacott 2020). Despite providing a low cost and a low carbon solution in comparison to oil or coal, natural gas role must consist of backup instead of being the "round-the-clock" main supplier. In the case of Europe, natural gas share on energy mix is only second to total petroleum products, representing 23.7% of overall data when considering the EU-27, with some of its members reaching even higher values: 38% for Netherlands) or 40% for Italy (Eurostat Homepage: https://ec.europa.eu/eurostat/web/products-interactive-publications/-/ks-fw-22-002). With integrated trans-European transmission and a wide distribution pipeline network, natural gas imports are mostly dependent on Russia, the main supplier of crude oil, natural gas and solid fossil fuels for the last decades (European Network of Transmission System Operators for Gas (ENTSOG) Homepage: https://transparency.entsog.eu/#/map). However, with recent events related with post-Covid pandemics allied with the Ukrainian-Russian war, gas prices for both industrial and household consumers raised drastically in the first quarter of 2022.

### 68.2.3 Iberian Natural Gas Infrastructures Strategies and Green $H_2$ Initiatives

**Natural gas infrastructures**

**Fig. 68.1** Iberian natural gas pipelines. *Source* European Network of Transmission System Operators for Gas (ENTSOG, 2022)

All Iberian grid scaled up after completion of the Maghreb-Europe Gas Pipeline (MEG) in 1996, a project meant to gasify Portugal and Spain through Africa. All the infrastructure seems to satisfy overall own demand, with the grid having two main interconnections with both France (in the Pyrenees) and Africa (Fig. 68.1).

The current Iberian natural gas infrastructure situation has slightly changed over the last decade, despite the unification of energy markets, with only a couple of new projects going underway (Fig. 68.1—dashed lines). Adding to this, eight liquefied natural gas (LNG) terminals support the Iberian coast with liquefaction operations. Supply through seas is still having a great impact on Iberian gas system, being the main cause of different pricing systems when compared with other neighbors (Heather 2019). Table 68.1 shows a comparison of LNG terminals technical data in the Iberian Peninsula and some of the main western EU countries (Gas Infrastructure Europe Homepage: https://www.gie.eu/transparency/databases/lng-database/).

Table 68.1 considers aggregated data from LNG conversion annual capacity ($m^3$ (N)/year), LNG storage capacity ($m^3$), the maximum ship class size receivable ($m^3$ LNG) and total number of tanks. Only operational terminals and imports were considered on this analysis. The higher numbers presented by the Iberian Peninsula stand out easily, reflecting the importance of LNG in the Iberian gas market. Table 68.2 also presents the data at desegregated level for both Iberian countries.

**Green $H_2$ initiatives**

**Table 68.1** EU LNG technical data considering imports. Aggregated data (countries comparison)

| LNG terminal | Nom. annual cap (billion $m^3$(N)/ year) | LNG storage capacity ($m^3$ LNG) | Max. ship class size receivable ($m^3$ LNG) | Number of tanks |
|---|---|---|---|---|
| Iberian Peninsula | 68 | 8,212,800 | 760,000 | 28 |
| France | 33 | 1,370,000 | 267,000 | 10 |
| Italy | 16 | 487,500 | 217,000 | 8 |
| Netherlands | 12 | 540,000 | 266,000 | 3 |
| Greece | 7 | 225,000 | 260,000 | 3 |
| United Kingdom | 48 | 2,095,000 | 266,000 | 15 |

**Table 68.2** EU LNG technical data considering imports

| LNG terminal | Nom. annual cap (billion $m^3$ (N)/ year) | LNG storage capacity ($m^3$ LNG) | Max. ship class size receivable ($m^3$ LNG) | Number of tanks |
|---|---|---|---|---|
| Sines | 7.60 | 1,350,000 | 390,000 | 3 |
| Barcelona | 17.10 | 1,950,000 | 760,000 | 6 |
| Bilbao | 7.00 | 800,000 | 450,000 | 3 |
| Cartagena | 11.80 | 1,350,000 | 587,000 | 5 |
| Huelva | 11.80 | 1,350,000 | 619,500 | 5 |
| Mugardos | 3.60 | 412,800 | 300,000 | 2 |

Portugal's main industry hub is in Sines, which will house two of the most important national projects (Sapo Homepage: https://eco.sapo.pt/2022/05/16/portugal-pode-receber-projeto-de-hidrogenio-verde-de-23-mil-milhoes/). The first one, Green$H_2$Atlantic, by Energias de Portugal (EDP), consist in the installation of a 100 MW electrolyzer in the old coal-fired power plant. The second one, MadoquaPower2X, a project born of a partnership between Portuguese developer Madoqua Renewables, Power2x (Dutch energy transition project developer) and the Copenhagen Infrastructure Partners' Energy Transition Fund (CIP), will have an electrolysis capacity of 500 MW. A third project in Sines could be underway by the Vargas group, $H_2$ Green Steel, consisting of a 1GW eletrolyzer to generate $H_2$ that will be used as fuel for a steel direct reduction furnace, while other possible locations are also being investigated (Sapo Homepage: https://eco.sapo.pt/2022/05/16/portugal-pode-receber-projeto-de-hidrogenio-verde-de-23-mil-milhoes/). Spain has a vast array of projects under development. Figure 68.2 shows the regions receiving major $H_2$ projects (Flanders Investment & Trade Homepage: https://www.flandersinvestmentandtrade.com/export/sites/trade/files/market_studies/2021-Spanje-The%20green%20H2%20energy%20in%20Spain-Website.pdf). From the public information available, most projects consist of green $H_2$ production or synthesis of other chemical products (fertilizers, renewable fuels).

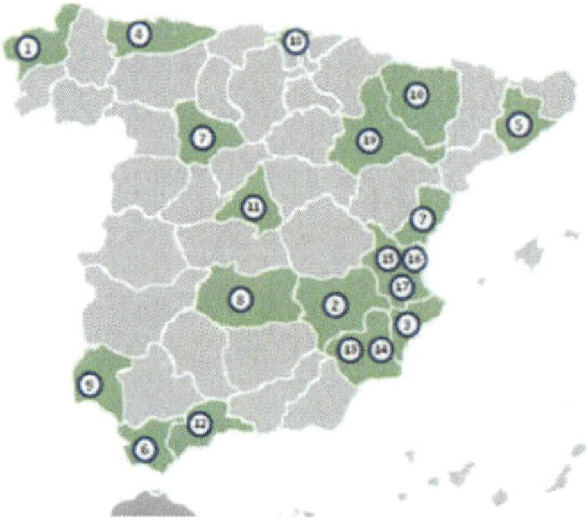


**Fig. 68.2** Regions in Spain with green $H_2$ proposals. *Source* Vermeulen and Campuzano (2021)

For example, in La Coruña, green $H_2$ is planned to be used for industrial purposes, to achieve a significant reduction in local emission of greenhouse gases, applied to transport of heavy vehicles and port machinery through fuel cells (H2bulletin Homepage: https://www.h2bulletin.com/enerfin-to-develop-a-green-H2-plant-in-the-port-of-a-coruna-spain/). Enerfín company leads building and operation of this project. Asturias will be probably one of the main regions considering green $H_2$ generation. *Energias de Portugal* (EDP) is going to develop projects worth 470 million euros, in Soto de Ribera and Aboño, as well as a major floating wind farm in Spain (Energias de Portugal Homepage (EDP): https://www.edp.com/en/news/2021/05/21/edp-presents-investment-470-million-asturias-energy-transition). In addition, the world's largest $H_2$ hub promoted by HyDeal, will take place in Asturias, with production purposes of green steel, green ammonia, green fertilizers amongst other low-carbon industrial products (ICEX Trade and Investment Homepage: https://www.investinspain.org/content/icex-invest/en/noticias-main/2022/hydeal-asturias.html). Considering both countries and their LNG terminals potential, there are no current projects involving adaptation of infrastructures to $H_2$ liquefaction.

## 68.3 Results and Discussion

The Iberian Peninsula benefits from an excellent geographic location, making it one of Europe's best regions to harvest and scale up RES, namely wind, solar and photovoltaic technologies (Carreno-Madinabeitia et al. 2021; Wetzel et al. 2022). Since it is a general assumption that green $H_2$ represents an advantage for decarbonisation by reducing RES intermittency (Oliveira et al. 2021b), it will be expected that the Iberian countries will become major players when considering its own generation potential. Regarding the European strategy of renewable gases blending into the existing gas grid as a first step (European Commission Homepage: https://eur-lex.europa.eu/legal-content/EN/TXT/?uri=CELEX:52020DC0301), it is necessary to understand some of the advantages and constraints. By adopting current natural gas infrastructure, overall investment costs will be lower (Quintino et al. 2021). On the other side, negative effects can arise considering infrastructure materials, potential gas leakages, safety issues considering the wider flammability of $H_2$, gas flows detrimental effects, wrong regulatory equipment measures, different degrees of severity depending on the user final application and also problems related to underground storage (Romeo et al. 2022). Considering this, the European $H_2$ roadmap seems to acknowledge that gas mixing can be accomplished to a certain extent, but that planning a $H_2$-based infrastructure should start as soon as possible (European Commission Homepage: https://eur-lex.europa.eu/legal-content/EN/TXT/?uri=CELEX:52020DC0301). In the Iberian case, the limited natural gas interconnection with France imposes an additional barrier when considering $H_2$ blending. Being limited to around 7 billion cubic meter/year (European Commission Homepage. https://ec.europa.eu/commission/presscorner/detail/de/MEMO_18_4622), these are values which contrast greatly with future European ambitions in terms of energy supply and security. Even with recent speculation about the construction of a third Spain-France gas link (Reuters Homepage: https://www.reuters.com/business/energy/spain-says-gas-pipeline-france-possible-8-9-months-2022-08-12/), it is not probable that Portugal and Spain could export significant amounts of green $H_2$ through the grid, since initial production will be confined to intensive industry hubs supported by local distribution. In theory, the diversity of LNG terminals along the Iberian coast present good potential in terms of EU future energy security, but there are many constraints to consider in terms of $H_2$ storage and transport. In fact, there is currently a controversy about the cost-effective shipping methods, with several options still open to debate: from conventional liquid $H_2$ to the use of other $H_2$ carriers, such as ammonia ($NH_3$), LNG, methylcyclohexane (MCH) or liquid organic $H_2$ carriers (LOHCs) (Johnston et al. 2022). Due to $H_2$ chemical properties, transport at ambient conditions is not possible due to low volumetric density (Niermann et al. 2021), which represent an additional technical difficulty since cargo shipping is mostly adapted to LNG properties. Practically, every form of carrying $H_2$ has a large margin to improve performance, since each of them have technical constraints to surpass. While liquefied $H_2$ deals with both problems of requiring a huge energy consumption during liquefaction process and $H_2$ losses by evaporation during storage, MCH requires a large amount of energy in dehydrogenation, as well as $NH_3$, which is also energy intensive in both synthesis and decomposition (Wijayanta et al. 2019). From all alternatives pointed above, some authors conclude that conventional liquid $H_2$ has the higher costs (\$2.09/kg$H_2$), followed by LOHCs (\$1.37/kg$H_2$), while using carriers such as MCH (\$0.68/kg$H_2$) and ammonia (\$0.56/kg$H_2$) are probably the cheapest ways to carry $H_2$ over the seas (Johnston et al. 2022). Finally, the shipping capacity problem remains. Currently, the largest operational ship for liquefied $H_2$ ($LH_2$) is manufactured by Kawasaki Heavy Industries, presenting a capacity of 1250 $m^3$. However, the cost of larger $LH_2$ ships remains uncertain.

## 68.4 Conclusions

Considering own demand and liquefaction capacity combined with a gas "web grid" supported by the installation of LNG terminals (Heather 2019), the Iberian Peninsula is relatively safe in terms of its own energy supply, benefiting from a diverse, resilient, and relatively isolated natural gas structure. However, it is obvious that currently, the Iberia has limited capacity to deliver significant amounts of natural gas. The two main interconnection gas pipelines in the border at Larrau and Biriatou are not enough to guarantee a stable and alternative way to enhance trade flows with Europe through France. Notwithstanding, this situation is likely to change in the next future, mainly due to recent geopolitical events. Both Portugal and Spain have diversified natural gas supply sources and good LNG terminals that are under expansion plans. As far as $H_2$ is concerned, Iberia also demonstrates solid potential and coherent roadmaps. Therefore, technology maturing together with adequate financing and political support, will be essential to the Iberian role in the European energy transition costs and risks.

**Acknowledgements** This paper is a result of the project "HyGreen&LowEmissions—Tackling Climate Change Impacts: the role of Green $H_2$ production, storage and use, together with low emissions energy systems", with the reference NORTE-01-0145-FEDER-000077, supported by Norte Portugal Regional Operational Programme (NORTE 2020), under the PORTUGAL 2020 Partnership Agreement, through the European Regional Development Fund (ERDF).

## References

Aguilera RF, Aguilera R (2020) Revisiting the role of natural gas as a transition fuel. Miner Econ 33(1):73–80

Bordoff J, O'Sullivan Meghan L (2022) Green upheaval: the new geopolitics of energy. Foreign Aff 101

Carreno-Madinabeitia S, Ibarra-Berastegi G, Sáenz J, Ulazia A (2021) Long-term changes in offshore wind power density and wind turbine capacity factor in the Iberian Peninsula (1900–2010). Energy 226:120364

Cheng W, Lee S (2022) How green are the national $H_2$ strategies? Sustainability 14(3):1930

Cihlar J, Lejarreta AV, Wang A, Melgar F, Jens J, Rio P (2021) $H_2$ generation in Europe: overview of costs and key benefits

Clean H2 Partnership Homepage. https://www.clean-H2.europa.eu/media/news/green-hysland-inauguration-first-renewable-H2-industrial-plant-mallorca-2022-03-14_en. Accessed 12 June 2023

Damman S, Sandberg E, Rosenberg E, Pisciella P, Graabak I (2021) A hybrid perspective on energy transition pathways: is $H_2$ the key for Norway? Energy Res Soc Sci 78:102116

de Oliveira AR, Collado JV, Saraiva JT, Doménech S, Campos FA (2021b) Electricity cost of green $H_2$ generation in the Iberian electricity market. In: 2021 IEEE Madrid PowerTech, pp 1–6

Energias de Portugal Homepage (EDP). https://www.edp.com/en/news/2021/12/21/renewable-H2-production-sines-advances-through-greenh2atlantic-project. Accessed 12 June 2023

Energias de Portugal Homepage (EDP). https://www.edp.com/en/news/2021/05/21/edp-presents-investment-470-million-asturias-energy-transition. Accessed 12 June 2023

Environmental Performance Index Homepage. https://epi.yale.edu/downloads/epi2022report06062022.pdf. Accessed 12 June 2023
European Commission Homepage. https://eur-lex.europa.eu/legal-content/EN/TXT/?uri=CELEX:52020DC0301. Accessed 12 June 2023
European Commission Homepage. https://commission.europa.eu/documents_en?f%5B0%5D=document_title%3Aspain%2C%20H2. Accessed 12 June 2023
European Commission Homepage. https://ec.europa.eu/commission/presscorner/detail/de/MEMO_18_4622. Accessed 12 June 2023
European Network of Transmission System Operators for Gas (ENTSOG) Homepage. https://transparency.entsog.eu/#/map. Accessed 12 June 2023
Eurostat Homepage. https://ec.europa.eu/eurostat/web/products-interactive-publications/-/ks-fw-22-002. Accessed 12 June 2023
Eurostat Homepage. https://ec.europa.eu/eurostat/databrowser/view/ENV_AIR_GGE/default/table?lang=en. Accessed 12 June 2023
Flanders Investment & Trade Homepage. https://www.flandersinvestmentandtrade.com/export/sites/trade/files/market_studies/2021-Spanje-The%20green%20H2%20energy%20in%20Spain-Website.pdf. Accessed 12 June 2023
Gas Infrastructure Europe Homepage. https://www.gie.eu/transparency/databases/lng-database/. Accessed 12 June 2023
Gielen D, Taibi E, Miranda R (2019) $H_2$: a renewable energy perspective. International Renewable Energy Agency, Abu Dhabi
Gürsan C, de Gooyert V (2021) The systemic impact of a transition fuel: does natural gas help or hinder the energy transition? Renew Sustain Energy Rev 138:110552
H2bulletin Homepage. https://www.h2bulletin.com/enerfin-to-develop-a-green-H2-plant-in-the-port-of-a-coruna-spain/. Accessed 12 June 2023
Heather P (2019) "A Hub for Europe": the Iberian promise? OIES Paper NG143
HyDeal España Homepage. https://www.hydeal.com/copie-de-hydeal-ambition. Accessed 12 June 2023
ICEX Trade and Investment Homepage. https://www.investinspain.org/content/icex-invest/en/noticias-main/2022/hydeal-asturias.html. Accessed 12 June 2023
International Energy Agency (IEA) Homepage. https://www.iea.org/fuels-and-technologies/gas. Accessed 12 June 2023
Johnston C, Khan MHA, Amal R, Daiyan R, MacGill I (2022) Shipping the sunshine: an open-source model for costing renewable $H_2$ transport from Australia. Int J $H_2$ Energy 47(47):20362–20377
Kakoulaki G, Kakoulaki G, Kougias I, Taylor N, Dolci F, Moya J, Jäger-Waldau A (2021) Green $H_2$ in Europe—a regional assessment: substituting existing production with electrolysis powered by renewables. Energy Convers Manag 228:113649
Kovač A, Paranos M, Marciuš D (2021) $H_2$ in energy transition: a review. Int J $H_2$ Energy 46(16):10016–10035
Liu W, Zuo H, Wang J, Xue Q, Ren B, Yang F (2021) The production and application of $H_2$ in steel industry. Int J $H_2$ Energy 46(17):10548–10569
Ministério do Ambiente e da Ação Climática (MAAC). https://www.portugal.gov.pt/download-ficheiros/ficheiro.aspx?v=%3d%3dBQAAAB%2bLCAAAAAAABAAzNDCyMAUArNeLzAUAAAA%3d. Accessed 12 June 2023
Nicita A, Maggio G, Andaloro APF, Squadrito GJIJ (2020) Green $H_2$ as feedstock: financial analysis of a photovoltaic-powered electrolysis plant. Int J $H_2$ Energy 45(20):11395–11408
Niermann M, Timmerberg S, Drünert S, Kaltschmitt M (2021) Liquid organic $H_2$ carriers and alternatives for international transport of renewable $H_2$. Renew Sustain Energy Rev 135:110171
Oliveira AM, Beswick RR, Yan Y (2021a) A green $H_2$ economy for a renewable energy society. Curr Opin Chem Eng 33:100701
Papadis E, Tsatsaronis G (2020) Challenges in the decarbonization of the energy sector. Energy 205:118025

Pawelec G, Muron M, Bracht J, Bonnet-Cantalloube B, Floristean A, Brahy N (2020) Clean H2 monitor

Portuguese Government Homepage. https://files.dre.pt/1s/2020/08/15800/0000700088.pdf. Accessed 12 June 2023

Power Technology Homepage. https://www.power-technology.com/news/cip-project-catalina-spain/. Accessed 12 June 2023

Quintino FM, Nascimento N, Fernandes EC (2021) Aspects of $H_2$ and biomethane introduction in natural gas infrastructure and equipment. H2 2(3):301–318

Reuters Homepage. https://www.reuters.com/business/energy/spain-says-gas-pipeline-france-possible-8-9-months-2022-08-12/. Accessed 12 June 2023

Romeo LM, Cavana M, Bailera M, Leone P, Peña B, Lisbona P (2022) Non-stoichiometric methanation as strategy to overcome the limitations of green $H_2$ injection into the natural gas grid. Appl Energy 309:118462

Safari A, Das N, Langhelle O, Roy J, Assadi M (2019) Natural gas: a transition fuel for sustainable energy system transformation? Energy Sci Eng 7(4):1075–1094

Sapo Homepage. https://eco.sapo.pt/2022/05/16/portugal-pode-receber-projeto-de-hidrogenio-verde-de-23-mil-milhoes/. Accessed 12 June 2023

The International Renewable Energy Agency (IRENA) Homepage. https://irena.org/-/media/Files/IRENA/Agency/Publication/2022/Jan/IRENA_Geopolitics_H2_2022.pdf. Accessed 12 June 2023

Trattner A, Klell M, Radner F (2022) Sustainable $H_2$ society—vision, findings and development of a $H_2$ economy using the example of Austria. Int J $H_2$ Energy 47(4):2059–2079

Wetzel M, Gils HC, Bertsch V (2022) Green energy carriers and energy sovereignty in a climate neutral European energy system. Renew Energy 210:591–603

Wijayanta AT, Oda T, Purnomo CW, Kashiwagi T, Aziz M (2019) Liquid $H_2$, methylcyclohexane, and ammonia as potential $H_2$ storage: comparison review. Int J $H_2$ Energy 44(29):15026–15044

Woollacott J (2020) A bridge too far? The role of natural gas electricity generation in US climate policy. Energy Policy 147:111867

# Chapter 69
# Proportions of the Relationship Between Economic Growth Rates and Energy Resources Consumption

**I. V. Filimonova, I. V. Provornaya, A. O. Haikina, and E. A. Kuznetsova**

**Abstract** The fuel and energy complex occupies one of the leading roles in the world economy, ensuring the dynamics and quality of economic growth in commodity-producing countries. At present, achieving high growth rates can create serious environmental problems, since the use of energy sources causes significant damage to the environment and determines climate change. In the work, the authors established the relationship between energy consumption and some factors for 89 countries, which are divided into seven macro regions. It was possible to determine the impact of consumption of oil, gas, coal and alternative energy sources, as well as the population, economic structure and energy intensity of macro regions on total energy consumption. The calculations confirm the fact that many countries have begun the transition to the use of alternative energy sources, namely: the countries of North America, Europe, South and Central America and the Asia–Pacific region. However, the proportions of the energy transition are significantly different.

**Keywords** Energy resources · Energy consumption · Climate change · Economic growth · Energy intensity · Population size · LMDI method

## 69.1 Introduction

For almost three centuries, energy has been the driving force of global industrialization. Speaking about the consumption of energy resources as a factor determining the growth rates of countries' economies, the most important assumption is the following

I. V. Filimonova (✉) · I. V. Provornaya · A. O. Haikina · E. A. Kuznetsova
Novosibirsk State University, 1, Pirogov St., 630090 Novosibirsk, Russia
e-mail: FilimonovaIV@list.ru

I. V. Filimonova · I. V. Provornaya · E. A. Kuznetsova
Trofimuk Institute of Petroleum Geology and Geophysics, 2, Koptug Av., 630090 Novosibirsk, Russia

 
N. S. Caetano and M. C. Felgueiras (eds.), *The 9th International Conference on Energy and Environment Research*, Environmental Science and Engineering,
https://doi.org/10.1007/978-3-031-43559-1_69

statement: the physical energy that a person spends on creating a social product is extremely small in comparison with the energy spent by machinery. That is why we can talk about the direct dependence of energy consumption and social wealth. As a consequence, a direct proportion between energy consumption and economic growth is also evident (Vickers 2017). Over time, energy consumption only increased, which caused environmental problems and, as a consequence, the issue of the insolvency of the further use of those energy sources, the use of which leads to emissions into the atmosphere (Wang and Zhang 2022). The problems of environmental pollution and global climate change have led to the need to conclude the Paris Agreement and the Kyoto Treaty on Reducing the Use of Energy Sources, the operation of which entails large-scale greenhouse gas emissions (Sun et al. 2022).

The global energy transition creates significant problems for sustainable economic growth, and therefore more and more authors are finding a new direction of research related to the analysis of changes in the structure of energy consumption and its impact on economic growth (Wang and Jia 2022). Technologies that are clean and safe for the environment are really necessary to achieve sustainable economic growth without negative consequences (Shahbaz et al. 2022). Thus, the topic of the future of the energy industry is being updated against the background of "green" energy transition and inter-fuel competition (Adebayo et al. 2021).

The aim of the study is to assess the impact of certain factors on energy consumption in the world and to search for suitable theories that allow us to get the most accurate answer to the question "what is the future of the energy industry taking into account the impact of climate change?".

In accordance with the purpose of the work, the authors set and solved the following tasks: (1) choosing a theory that allows assessing the impact of consumption of various energy resources, population, energy intensity and economic structure on energy consumption in the world; (2) creating a model that can give adequate and meaningful results; (3) assessment of the relevance of the approach and the results obtained by the author, justification of recommendations and prospects for the development of the energy complex.

The object of the research is the energy industry, the subject is indicators and methods for assessing the state of the energy complex and the impact of economic, environmental and institutional factors.

The empirical basis of the study is based on index theory, including the LMDI decomposition method, which in turn is an effective option for analysing the driving factors of energy consumption and greenhouse gas emissions (Wang et al. 2017). The most accurate and detailed description of this method in the framework of the analysis of the energy industry is presented in the works of B. W. Ang. In addition, the LMDI method is used to analyze energy consumption, energy efficiency, emissions into the atmosphere due to energy consumption, etc. extremely popular among foreign authors: Wei Zhang, Nan Wang, Chin Hao Chong, Jie Yang, Mohammad Maruf Hasan, Minda Ma, etc.

Thus, the results obtained by the author allow not only to assess the factors of inter-fuel competition, but also to determine the main prospects for its development in modern conditions.

## 69.2 Materials and Methods

### *69.2.1 Materials*

The study uses statistics from the official website of The World Bank: population and GDP PPP (in 2017 prices) of 89 countries in the period from 1995 to 2020 British Petroleum Statistics (Statistical Review): energy consumption, consumption of coal, oil, gas and other energy sources.

In the course of the work, countries were grouped into seven macro-regions: North America (Canada, Mexico, USA), South and Central America (Argentina, Brazil, Chile, Colombia, Ecuador, Peru, Trinidad and Tobago, Venezuela), Europe (Austria, Belgium, Bulgaria, Croatia, Cyprus, Czech Republic, Denmark, Estonia, Finland, France, Germany, Greece, Hungary, Iceland, Ireland, Italy, Latvia, Lithuania, Luxembourg, Netherlands, North Macedonia, Norway, Poland, Portugal, Romania, Slovakia, Slovenia, Spain, Sweden, Switzerland, Turkey, Ukraine, United Kingdom), CIS (Azerbaijan, Belarus, Kazakhstan, Russian Federation, Turkmenistan, Uzbekistan), Middle East (Iran, Iraq, Israel, Kuwait, Oman, Qatar, Saudi Arabia, United Arab Emirates), Africa (Algeria, Egypt, Morocco, South Africa), Asia–Pacific region (Australia, Bangladesh, China, China SAR, Hong Kong, India, Indonesia, Japan, Malaysia, New Zealand, Pakistan, Philippines, Singapore, Sri Lanka, Vietnam).

Currently, the largest regions in terms of energy consumption are the countries of the Asia–Pacific region, North America and Europe, using more than 78% of the world's energy (Fig. 69.1). In these regions, the largest amount of each type of energy is consumed. The current regional structure of primary energy use reflects the trends that have occurred in the global economy under the influence of economic, technological, resource, institutional and other factors (Filimonova et al. 2020; Eder et al. 2018).

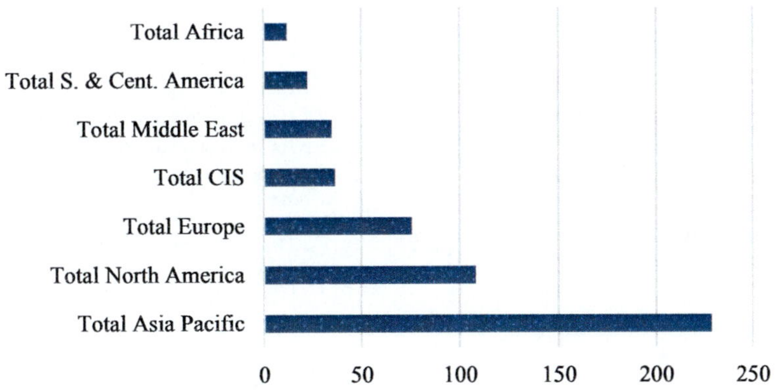


**Fig. 69.1** Energy consumption by macro regions in 2020, exajoule

### 69.2.2 Methods

In modern economic theory, there are many approaches to modelling economic growth and assessing the impact of various factors on it: optimization, intersectoral, balance models, etc. However, nowadays more and more researchers interested in the energy industry pay special attention to the method of the arithmetic mean division index, or LMDI. This method, since the 2000s, gradually began to gain popularity among authors. In addition, the trends of modern research indicate that the LMDI method is likely to further strengthen its dominance in research over time.

Energy consumption in the macro region was used as a dependent variable. The defining variables of the analysis were: (a) consumption of coal, oil, gas, alternative energy sources in the total energy consumption in the macro region; (b) the energy intensity of the macro region; (c) the economic structure of the macro region; (d) the population of the macro region.

In this paper, the authors used a model that allows us to determine the influence of certain factors on energy consumption:

$$E_j = \sum\nolimits_i \frac{E_{ij}}{Ej} \times \frac{Ej}{GDP_j} \times \frac{GDP_j}{Pj} \times Pj \tag{69.1}$$

An explanation of the designations used in the model is presented in Table 69.1.

Thus, it is possible to determine a ready-made formula for analysing energy consumption by macro-regions:

$$E_j = SE_j \times EI_j \times ES_j \times Pj \tag{69.2}$$

where $SE_j$—is the share of the i-th energy source in the total energy consumption of the macro region j, $EI_j$—is the energy intensity of the macro region j, $ES_j$—is

**Table 69.1** Model indicators, their formulas and designations

| Indicator | Formula | Name |
|---|---|---|
| The share of the energy source in total energy consumption ($SE_j$) | $\frac{E_{ij}}{Ej}$ | $E_{ij}$—consumption of the i-th energy source in the macro region j<br>$E_j$—energy consumption in the macro region j |
| Energy intensity ($EI_j$) | $\frac{Ej}{GDP_j}$ | $E_j$—energy consumption in the macro region j<br>$GDP_j$—gross domestic product of the macro region j |
| Economic structure ($ES_j$) | $\frac{GDP_j}{Pj}$ | $GDP_j$—gross domestic product of the macro region j<br>$Pj$—population of the macro region j |
| Population ($Pj$) | $Pj$ | $Pj$—population of the macro region j |

the economic structure of the macro region j, $Pj$—is the population of the macro region j.

For each country, as well as for each of the seven selected macro-regions, the influence of factors in the period 1995–2020 was calculated.

Decomposition of energy consumption $\Delta E_i$ with the help of this model determines the contribution of all selected factors: $\Delta E_i(SE_i)$—the effect of changing the share of the energy source, $\Delta E_i(EI_i)$—effect of changing energy intensity, $\Delta E_i(ES_i)$—the effect of changing the economic structure and $\Delta E_i(P)$—effect of population change.

$$\Delta E_i = E_i^t - E_i^{t-1} = \Delta E_i(SE_i) + \Delta E_i(EI_i) + \Delta E_i(ES_i) + \Delta E_i(A) + \Delta E_i(P) \quad (69.3)$$

To obtain the necessary estimate, the authors use the LMD method, which is expressed as follows on the example of the influence of the share of the i-th energy source in total consumption with the remaining factors unchanged:

$$\Delta E_i(SE_i) = \frac{E_i^t - E_i^{t-1}}{\ln(E_i^t) - \ln(E_i^{t-1})} \times \ln\frac{SE_i^t}{SE_i^{t-1}} \quad (69.4)$$

Similarly, the influence of the other three factors on the change in energy consumption was calculated. After receiving the results, dynamics were determined for each macro region, which reflects how much this or that factor matters for changes in energy consumption over time.

## 69.3 Results and Discussion

First of all, the authors presented the dynamics of the indicators determining energy consumption and made some conclusions.

It is determined that the dynamics of the energy intensity of the economy decreases over time, which cannot be said about the countries of the Middle East: for this macro-region, the energy intensity indicator has been increasing since 2000. The greatest importance of the energy intensity of the economy is observed in the CIS countries. However, since 1995, a fairly rapid decrease in energy intensity can be observed. The dynamics of the economic structure is positive for most macro-regions (the decline in this indicator in 2020 for the Middle East and Africa is primarily due to the crisis of the COVID-19 pandemic, so these emissions can be considered accidental). The leaders in this indicator are such macro-regions as North America, Europe and the Middle East. The CIS countries occupy only the 4th place in this "rating".

The size of the population undoubtedly has an impact on energy consumption, since the use of various energy sources is primarily associated with providing basic human needs. After the Asia–Pacific region, Europe remains the leader in terms of

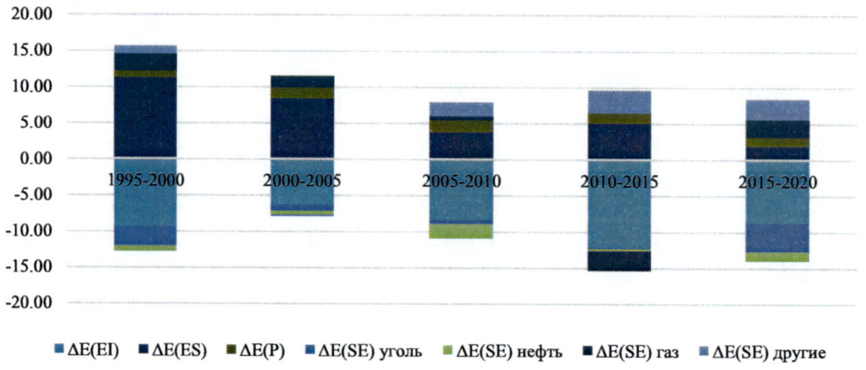


**Fig. 69.2** Dynamics of the influence of factors determining energy consumption in Europe

population, followed by North America. As for the CIS countries, the population has not changed significantly in the period from 2005 to 2020.

A comparison of the results obtained for all groups of countries shows a complete picture of the dependence of energy consumption on various factors.

Further, as a result of the study, it was possible to identify several trends for all seven macro-regions.

For example, Fig. 69.2 shows the dynamics of changes in shares by the degree of influence of factors on energy consumption in European countries. The following situation can be observed: over time, there is an increase in the share of the influence of the consumption of alternative energy sources, as well as a gradual decrease in the share of the influence of the factor of economic structure. In addition, in the period 2015–2020, the impact of coal consumption has increased significantly compared to previous periods.

Figure 69.3 shows how the shares of influence of one or another factor on the consumption of CIS energy changed over time. It can be seen that the influence of determining factors—economic structure and energy intensity, decreases with time. Moreover, recently we can observe an increase in the influence of the population on energy consumption, which is similar to the situation in Africa and the Middle East. It can also be seen that the impact of consumption of alternative energy sources has increased over the period 2010–2020.

In addition, based on the results obtained, the authors made the following important conclusions of the analysis: (1) the countries of North America and Europe can be grouped into one group based on the fact that the energy intensity of the economy is a factor determining energy consumption over a long period. In addition, these macro-regions are characterized by very similar dynamics of influencing factors; (2) The Middle East and African countries also have the same determining factor—population size. Moreover, the dynamics of the factors on which energy consumption in these countries depends is almost identical for the period 1995–2020; (3) South and Central America, the CIS countries and the Asia–Pacific region are also grouped into one group: the determining factor is the economic structure; (4) In the last years

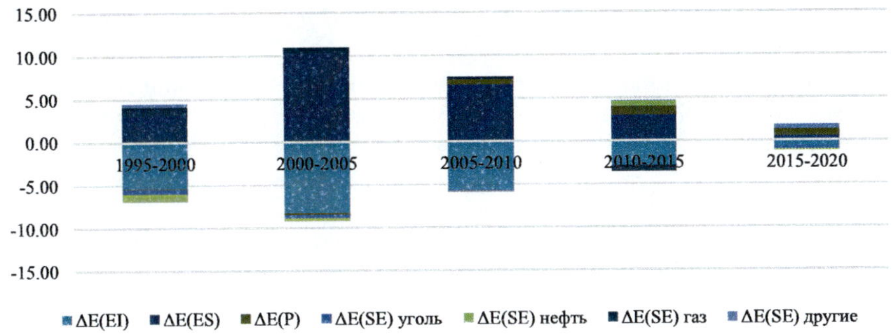


**Fig. 69.3** Dynamics of the influence of factors determining energy consumption in the CIS

of the period under review, there has been an increasing trend towards the use of alternative energy sources in the countries of North America, Europe, South and Central America, the CIS and the Asia–Pacific region; (5) In the countries of North America and Europe, the influence of coal consumption on the dependent variable has increased significantly.

## 69.4 Conclusion

Developed countries, as a rule, increase energy consumption primarily in order to achieve a high level of GDP, which in turn leads to an increase in energy consumption. Developing countries, which have a fairly significant amount of energy resources in their hands, are also accelerating economic growth through the use of energy (Mutumba et al. 2021). However, the most important task today remains to mitigate the effects of climate change in pursuit of high economic indicators.

The problem of climate change really seriously affects the energy industry and, as a result, economic growth, which is confirmed by the analysis carried out during the work (Wang et al. 2022). Statistics show that the volume of emissions into the atmosphere due to the use of energy sources such as coal, oil, gas is growing every year, and the results of the work are confirmation that more and more countries are beginning a gradual transition to more environmentally friendly energy sources.

As a result of the analysis of the period 1995–2020, it was possible to draw the following conclusions: (1) for the countries of North America and Europe, the determining factor is the energy intensity of EI; (2) the countries of South and Central America, the CIS and the Asia–Pacific region have united into one group, since the most influential factor on energy consumption is the economic structure of ES, or GDP per capita (3) for the countries of the Middle East and Africa, the determining factor was the population factor P; (4) for the countries of the Asia–Pacific region and North America, the first factor in terms of influence was the factor of coal consumption. So, to provide some recommendations for the coal industry of the

Russian Federation, we can give advice to focus on exports to the Asia–Pacific region and North America. (5) for the countries of Europe, the CIS, the Middle East and Africa, the most influential factor is the factor of gas consumption; (6) for the countries of South and Central America, the first factor in terms of influence is the factor of consumption of alternative energy sources, which indicates the gradual transition of this macro-region to more environmentally friendly energy sources.

**Acknowledgements** This study was carried out with the financial support of the Russian Science Foundation within the framework of the project №. 22-18-00424.

## References

Adebayo TS, Awosusi AA, Bekun FV, Altuntaş M (2021) Coal energy consumption beat renewable energy consumption in South Africa: developing policy framework for sustainable development. Renew Energy 175:1012–1024

Eder LV, Provornaya IV, Filimonova IV, Kozhevin VD, Komarova AV (2018) World energy market in the conditions of low oil prices, the role of renewable energy sources. Energy Procedia 153:112–117

Filimonova I, Provornaya I, Kozhevin V (2020) Identification of factors affecting renewable energy consumption by country groups. E3S Web Conf 157

Mutumba GS, Odongo T, Okurut NF, Bagire V (2021) A survey of literature on energy consumption and economic growth. Energy Rep 7:9150–9239

Shahbaz M, Song M, Ahmad S, Vo XV (2022) Does economic growth stimulate energy consumption? The role of human capital and R&D expenditures in China. Energy Econ 105:105662

Sun RS, Gao X, Deng LC, Wang C (2022) Is the Paris rulebook sufficient for effective implementation of Paris agreement? Adv Clim Chang Res 13(4):600–611

Vickers NJ (2017) Animal communication: when i'm calling you, will you answer too? Curr Biol 27(14):R713–R715

Wang Z, Jia X (2022) Analysis of energy consumption structure on $CO_2$ emission and economic sustainable growth. Energy Rep 8:1667–1679

Wang F, Zhang ZX (2022) Decoupling economic growth from energy consumption in top five energy consumer economies: a technological and urbanization perspective. J Clean Prod 357:131890

Wang Z, Pham TLH, Sun K, Wang B, Bui Q, Hashemizadeh A (2022) The moderating role of financial development in the renewable energy consumption-$CO_2$ emissions linkage: the case study of next-11 countries. Energy 254:124386

Wang H, Ang BW, Su B (2017) Assessing drivers of economy-wide energy use and emissions: IDA versus SDA. Energy Policy 107:585–599

# Chapter 70
# Oil Price Fluctuation Effects Over the Timor-Leste Economy

**Fernando Anuno, Mara Madaleno, and Elisabete Vieira**

**Abstract** This study examines the impact of oil price fluctuations (OP) on Timor-Leste's macroeconomic variables, i.e., the consumer price index (CPI), the exchange rate (ER), gross domestic product (GDP), the interest rate (IR), and money supply (M2). We used monthly data from 2003 to 2019 and the vector autoregressive (VAR) model. This study confirms that oil price fluctuations are associated with a significant positive and negative effect on the exchange rate and have a significant positive effect only in the second period, while the effects of the other variables are not statistically significant. The impulse response function results show that: (i) a positive OP fluctuation leads to an increase in CPI; (ii) a favorable OP reduces GDP in the short run and increases it in the long run; (iii) an OP fluctuation has a positive effect on ER; and (iv) a positive (negative) OP fluctuation affects the interest rate in different periods. Moreover, a negative OP fluctuation affects M2 in the short run, while a positive OP fluctuation affects M2 in the long run. The empirical results of the Granger causality tests show bidirectional causality between ER and CPI.

**Keywords** Oil price · Macroeconomic variables · VAR approach · Timor-Leste

F. Anuno (✉)
Faculty of Economics and Management, National University of Timor Lorosa'e (UNTL), Avenida Cidade de Lisboa, Díli, Timor-Leste
e-mail: fernando.anuno@ua.pt

F. Anuno · M. Madaleno
Research Unit On Governance, Competitiveness and Public Policies (GOVCOPP), Department of Economics, Management, Industrial Engineering and Tourism (DEGEIT), University of Aveiro, Campus Universitário de Santiago, 3810-193 Aveiro, Portugal

E. Vieira
GOVCOPP, ISCA—Higher Institute for Accountancy and Administration of Aveiro, University of Aveiro, Campus Universitário de Santiago, Aveiro, Portugal

 
N. S. Caetano and M. C. Felgueiras (eds.), *The 9th International Conference on Energy and Environment Research*, Environmental Science and Engineering,
https://doi.org/10.1007/978-3-031-43559-1_70

## 70.1 Introduction

Timor-Leste is a new country in the millennium with a government heavily dependent on oil revenue for development financing. The developed oil wealth can generate oil revenue. Oil revenue accounts for 70% of the country's GDP, and government revenue is 90% (Doraisami 2017). Therefore, oil revenue can be channeled in the form of an oil fund which is used as a government expenditure in the form of government revenue and an expenditure budget and asset investment in international financial markets. According to the Oil Fund Law, the transfer rate of the state budget follows the estimated sustainable income of 3% of total oil assets with the approval of the National Parliament (Timor-Leste Ministry of Finance 2019). Oil is tradable and is a product of natural resources that drives macroeconomic activity. Therefore, the cost of oil becomes a key element in determining the value of profit between bidders and demanders in the global economy and financial markets. Oil price fluctuations are the most important indicator for managing investment risks, prices, and business strategies aimed at obtaining the maximum benefit (Aimer 2016; Liu et al. 2019). Oil price fluctuations affect the instability of macroeconomic activity in both oil-exporting and oil-importing countries. Therefore, the highly volatile prices compared to other commodities are difficult to predict. An oil price increase has a positive effect on wealth transfer from oil-importing countries to oil-exporting countries, which leads to a decrease in the income of the oil-importing countries (Abdelsalam 2020).

Despite various studies conducted in oil-exporting and importing countries, the empirical results point to the above gaps. Several other studies have been conducted in different groups of countries such as The Association of Southeast Asian Nations (ASEAN), composed of Brunei, Cambodia, Indonesia, Laos, Malaysia, Myanmar, Philippines, Singapore, Thailand, and Vietnam (Aziz and Dahalan 2015; Nusair and Olson 2021), and the top five developing countries in the world, Brazil, Russia, India, China, and South Africa (BRICS) (Balcilar and Usman 2021; Li and Guo 2022; Salisu et al. 2021). There are other studies applied in other contexts, such as for GCC (The Gulf Cooperation Council) countries (Albaity and Mustafa 2018), the US (Hamilton 1983; Kilian and Vigfusson 2017), and the G7 countries (Wen et al. 2020). For this reason, this study analyses Timor-Leste, which has just restored its independence in 2002, wishing to contribute to the literature exploring economic activities whose main source is oil and gas wealth. Other studies that used the same economy have contributed to the exploration of economic activities through Timor-Leste's natural oil and gas resources, but they differ from the research methodology of the countries mentioned above. (John et al. 2020) explore Timor-Leste's dependence on petroleum resources, concluding that it serves to avoid a resource curse during a political and economic crisis. (Drysdale 2008) notes that countries dependent on oil and gas must adhere to the five principles of natural resource revenue management. These are accountability in managing oil revenues, acceptance of all-natural resources received by the state, wise investment of natural resource revenues, transparent management of natural resource revenues, and ensuring that natural resource revenues benefit future generations to achieve sustainable development. Similarly,

(Doraisami 2017) explains the use of sovereign wealth funds (SWFs) for sustainable future generations. (Scheiner 2021), for example, points to the need to diversify Timor-Leste's economic context, which is heavily dependent on oil and gas, through large-scale sunrise investments and petroleum infrastructure projects. From the contributions of this literature, it may be inferred that the authors do not use an econometric model, although they study the issue of petroleum funds, which have a great impact on economic activity through the state budget in each fiscal year. In addition, they only use data from the Ministry of Finance and the International Monetary Fund (IMF), the World Bank and the Asian Development Bank (ADB), and other sources to provide quantitative explanations using secondary data, and do not focus on macroeconomic variables towards oil price changes to determine Timor-Leste's economic growth. Consequently, as far as we are aware, no studies have been conducted by other authors to examine the macroeconomic impact of oil price fluctuations in Timor-Leste using the VAR model. Based on previous empirical evidence, our analysis, therefore, aims to examine the impact of oil price shocks on Timor-Leste's macroeconomic variables, namely CPI, ER, GDP, IR, and M2. We used monthly data from 2003 to 2019 and a VAR model. Moreover, as one of the youngest sovereign countries of the millennium, Timor-Leste is still dependent on the management of its natural resources, not having full control over these as many other economies in the world.

## 70.2 Data and Methodology

In this study, we use monthly data from January 2003 to December 2019 to analyze the impact of oil price fluctuation on Timor-Leste's macroeconomic variables, which are explained in detail in Table 70.1. In this study, six macroeconomic variables are used to analyze the case of Timor-Leste: Oil prices (OP), GDP growth (GDP), money supply (M2), interest rates (IR), exchange rates (ER), and consumer price index (CPI). The oil price and money supply variables are converted to natural logarithms. Other variables such as GDP growth, IR, and CPI are divided by 100. The exchange rate variable is converted to natural logarithms, IDR/USD the Indonesian exchange rate since Timor-Leste has adopted the US currency as its official currency and relies on imports of goods from Indonesia. Timor-Leste is an oil-exporting country and uses the West Texas Intermediate (WTI) crude oil price, an important benchmark for global oil prices. In addition, GDP growth (%) data is only available from the World Bank and other agencies in the form of annual data. To balance all the data of each variable, annual GDP growth (%) was converted into monthly data using EViews 11 Student Version Lite software (low to high-frequency method, quadratic).

The VAR model is a statistical method used to analyze relationships between time series data variables. (Sims 1980) developed the VAR in 1980 to augment the autoregressive process (AR) in time series data with one variable. In addition, (Enders 2019) showed two basic analyses that are possible with the model VAR: the variance decomposition and the impulse response function. The variance decomposition

**Table 70.1** Description of variables

| Variables | Symbol | Measurement Unit | Source |
|---|---|---|---|
| Oil prices | (OP) | Dollars per barrel | US Energy Information Administration (EIA) |
| GDP growth | (GDP) | Annual % | The World Bank |
| Money supply | (M2) | US$ | The Central Bank of Timor-Leste |
| Interest rates | (IR) | % | The Central Bank of Timor-Leste |
| Exchange rates | (ER) | IDR/USD | Investing |
| Consumer price index | (CPI) | 2010 = 100 | IMF |

*Note* Own elaboration based on the literature review performed

shows how significant the change in a group of variables is for other autoregression variables. The exogenous shock can explain the variance of each variable relative to the endogenous variable in a model. In addition, the impulse response function allows us to observe how the variables in a given system respond when each variable receives an impulse over the period, allowing us to analyze the relationship among different variables in different periods.

We have followed (Stock and Watson 2007) for Granger Causality. The Granger causality statistic is the F-statistic that tests if the coefficients of all the variables or one of the variables in the equation are zero in the null hypothesis.

## 70.3 Empirical Results

Due to space restrictions in this part of the article, we present only the Granger causality results in Table 70.2. The Granger causality test determines whether the value of past period variables predicts the changes in current variables in the VAR model (Granger. 1969). Table 70.2 shows the results of the VAR Granger causality test for Timor-Leste's macroeconomic variables. Table 70.2 shows that the exchange rate variable and CPI have bidirectional Granger causality with a significance level of 1%. This means that ER and CPI influence each other. However, unidirectional causality was found by Delgado et al. (2018); Charles and Marie 2020), who showed that ER has a positive effect on CPI. In (Antwi et al. 2020) it is found that CPI is not a Granger cause of ER in Ghana but indirectly influences ER. It is verified that there is unidirectional Granger causality between the IR and CPI variables at a 5% level of significance. This means that IR affects CPI, but not vice versa. This result is consistent with studies on evidence from emerging countries (Bui and Gábor 2021). Similarly, the M2 and the CPI variables exhibit unidirectional Granger causality at a 5% level of significance. This means that M2 influences CPI, but not the other way around. This study is similar to the studies of Bui and Gábor (2021), Batarseh (2021). However, this result contradicts a study by Kpagih (2022) mentioning that inflation is not caused by oil or gas price fluctuations.

**Table 70.2** Granger causality results

| Null Hypothesis | F-Statistic | Prob. | Decision | Null Hypothesis | F-Statistic | Prob. | Decision |
|---|---|---|---|---|---|---|---|
| ER does not granger cause CPI | 6.47768 | 0.0019*** | Causality | ER does not granger cause M2 | 1.65310 | 0.1941 | No causality |
| CPI does not granger cause ER | 4.55534 | 0.0116** | Causality | OP does not granger cause ER | 2.77472 | 0.0648* | Causality |
| GDP does not granger cause CPI | 1.63396 | 0.1979 | No causality | ER does not granger cause OP | 4.19193 | 0.0165** | Causality |
| CPI does not granger cause GDP | 0.49817 | 0.6084 | No causality | IR does not granger cause GDP | 1.77354 | 0.1726 | No causality |
| IR does not granger cause CPI | 3.10761 | 0.0469** | Causality | GDP does not granger cause IR | 1.50611 | 0.2245 | No causality |
| CPI does not granger cause IR | 1.08361 | 0.3404 | No causality | M2 does not granger cause GDP | 0.24204 | 0.7853 | No causality |
| M2 does not granger cause CPI | 3.05535 | 0.0493** | Causality | GDP does not granger cause M2 | 0.76029 | 0.4690 | No causality |
| CPI does not granger cause M2 | 0.30469 | 0.7377 | No causality | OP does not granger cause GDP | 2.73497 | 0.0675* | Causality |
| OP does not granger cause CPI | 4.01275 | 0.0196** | Causality | GDP does not granger cause OP | 0.95109 | 0.3882 | No causality |
| CPI does not granger cause OP | 2.37660 | 0.0955* | Causality | M2 does not granger cause IR | 0.49060 | 0.6130 | No causality |
| GDP does not granger cause ER | 0.02153 | 0.9787 | No causality | IR does not granger cause M2 | 0.51538 | 0.5981 | No causality |
| ER does not granger cause GDP | 1.20319 | 0.3026 | No causality | OP does not granger cause IR | 0.02534 | 0.9750 | No causality |
| IR does not granger cause ER | 0.07470 | 0.9280 | No causality | IR does not granger cause OP | 1.21940 | 0.2976 | No causality |

(continued)

**Table 70.2** (continued)

| Null Hypothesis | F-Statistic | Prob. | Decision | Null Hypothesis | F-Statistic | Prob. | Decision |
|---|---|---|---|---|---|---|---|
| ER does not granger cause IR | 0.20838 | 0.8121 | No causality | OP does not granger cause M2 | 0.40338 | 0.6686 | No causality |
| M2 does not granger cause ER | 3.30506 | 0.0387** | Causality | M2 does not granger cause OP | 1.91136 | 0.1506 | No causality |

*Source* Own elaboration. ***, ** (*) denotes significant causal relationship at 1, 5 (10%) significant level

Also, there is bidirectional Granger causality between the variables OP and CPI at a 5% and 10% level of significance, respectively. This implies that OP and CPI influence each other. Our result is consistent with that of Zakaria et al. (2021), who found that OP cause CPI via a Granger effect. Furthermore, there is unidirectional Granger causality between variables M2 and ER at a 5% significance level. This means that variable M2 influences variable ER and vice versa does not. In addition, there is a bidirectional Granger causality between the variables OP and ER with a significance level of 10% and 5%, respectively. This implies that oil prices and the exchange rate affect each other. Our results are inconsistent with a previous study by Olayungbo (2019), for Nigeria which found no Granger causal relationship between OP and ER. However, this result is consistent with the findings of Albulescu and Ajmi (2020), Beckmann et al. (2020).

There is unidirectional Granger causality between OP and GDP variables at a 10% level of significance. This means that OP affects GDP, but not the other way around. After the global oil crisis, the pioneer of macroeconomic research, (Hamilton 1983), proved that the Granger OP affects the economic performance of the US economy. Moreover, similar results by Awunyo-Vitor et al. (2018), show that fluctuations in OP cause, in the Granger sense, economic growth, but by Granger, economic growth does not cause changes in oil prices. The results of our study have several important implications for the government and policymakers in Timor-Leste. We believe that our empirical results can make a scholarly contribution to dealing with fluctuations in world oil prices and the government's use of oil revenues to fund public infrastructure as economic activities change, which depend not only on oil revenues, in this case, the Oil Fund, but also on government policies aimed at diversifying Timor-Leste's economy away from oil and gas.

The empirical findings can help Timor-Leste policymakers better shape the economic policies needed to achieve sustainable economic development in line with Timor-Leste's National Strategic Development Plan 2011–2030. The implications and proposed recommendations of the empirical results are presented below. The positive (negative) impact of oil prices can lead to wealth transfers to oil-exporting countries such as Timor-Leste, which can increase (decrease) the revenue balance of the Oil Fund. Therefore, when oil prices are high, the government of Timor-Leste should adopt expansionary macroeconomic policies that include increasing available funds for investment in priority sectors as per the Strategic Development Plan (SDP) 2011–2030 such as agriculture, tourism, and petroleum industry to increase the income of the non-oil sector and achieve the goal of increasing middle to upper income per Timor-Leste's Vision 2030.

Furthermore, global oil prices are constantly changing, affecting production costs and financial markets (Enwereuzoh and Odei-Mensah 2020; Liu et al. 2020) as well as other macroeconomic variables such as economic growth, income, production costs, interest rates, inflation, and consumer confidence (Hashmi et al. 2020). The consequences of production costs lead to inflation (Babuga and Naseem 2021). Thus, when the goods produced and exported to Timor-Leste are adjusted to the exchange rate of the national currency. Moreover, in cross-border trade, especially in different countries, the exchange rate of the local currency against the US dollar plays a very influential role, as evidenced by the fact that cash flow is affected by the volatility of the exchange rate (Tian et al. 2020). Since Timor-Leste uses the US dollars as legal tender, the competitiveness assessment requires the calculation of the trade-weighted average of the bilateral exchange rate as the trading partner country, in this case, Timor-Leste's neighboring country Indonesia. It can also be explained that the real effective exchange rate increased in the second quarter of 2020 mainly due to the appreciation of the US dollar (World Bank 2020). On the other hand, the real exchange rate depreciated slightly despite rising prices for domestic goods (World Bank Group 2021).

Although oil price shocks have significant positive and negative effects on exchange rates in the international oil market, this is beneficial for Timor-Leste as a new country that does not yet have its currency. However, Timor-Leste officially adopted the US dollar currency in 2000 under the United Nations Transitional Administration in Timor-Leste (UNTAET). The country has a floating exchange rate system for international payments and remittances to maximize revenue when oil prices rise and vice versa. In addition, an increase in the money supply by the central bank causes inflation and leads to economic uncertainty. Therefore, this is considered a "causality" between inflation and economic growth (Hussain and Zafar 2018). Moreover, the increase in inflation is the result of increasing dollarization (Mengesha and Holmes 2015). Furthermore, financial dollarization becomes an inflation target (Fabris and Vujanovic 2017). In this context, an increase in the money supply leads to an increase in inflation in the long run, while the money supply does not cause inflation in the short run (Doan 2020).

## 70.4 Conclusion

According to the estimation results of the VAR model, we describe only the impact of oil prices on macroeconomic variables. Thus, the VAR model shows that oil price shocks have a significant positive and negative effect on the exchange rate and a significant positive effect only in the second period, while the effects of the other variables are statistically insignificant. Overall, our findings indicate that macroeconomic conditions in Timor-Leste are susceptible to changes in oil prices. However, Timor-Leste achieved double-digit economic growth rates in 2007–2009. This sensitivity shows that economic activity is still declining and is highly dependent on oil and other mineral resources, and there is no economic diversification. In this condition, the Government is still concentrating on infrastructure development and human resource capacity.

**Acknowledgements** The support of the Research Unit on Governance, Competitiveness and Public Policy (UIDB/04058/2020) + (UIDP/04058/2020), funded by national funds through FCT—Fundação para a Ciência e a Tecnologia is greatly acknowledged.

## References

Abdelsalam MAM (2020) Oil price fluctuations and economic growth: the case of MENA countries. J Rev Econ Polit Sci

Aimer NMM (2016) Conditional correlations and volatility spillovers between crude oil and stock index returns of middle east countries. Oalib 03(12):1–23

Albaity M, Mustafa H (2018) International and macroeconomic determinants of oil price: evidence from gulf cooperation council countries. J Int Energy Econ Policy 8(1):69–81

Albulescu CT, Ajmi AN (2021) Oil price and US dollar exchange rate: change detection of bi-directional causal impact. J Energy Econ 100:105385

Antwi S, Issah M, Patience A, Antwi S (2020) The effect of macroeconomic variables on exchange rate: evidence from Ghana. J Cogent Econ Financ 8(1)

Awunyo-Vitor D, Samanhyia S, Bonney EA (2018) Do oil prices influence economic growth in Ghana? An empirical analysis. J Cogent Econ Financ 6(1):1–14

Aziz MIA, Dahalan J (2015) Oil price shocks and macroeconomic activities in asean-5 countries: a panel VAR approach. J Eurasian Bus Econ 8(16):101–120

Babuga UT, Naseem NAM (2021) Asymmetric effect of oil price change on inflation: evidence from sub Saharan Africa countries. J Int Energy Econ Policy 11(1):448–458

Balcilar M, Usman O (2021) Exchange rate and oil price pass-through in the BRICS countries: evidence from the spillover index and rolling-sample analysis. J Energy 229:120666

Batarseh A (2021) The nature of the relationship between the money supply and inflation in the Jordanian economy (1980–2019). J Banks Bank Syst 16(2):38–46

Beckmann J, Czudaj RL, Arora V (2020) The relationship between oil prices and exchange rates: revisiting theory and evidence. J Energy Econ 88:104772

Bui TT, Gábor KD (2021) Measuring monetary policy by money supply and interest rate: evidence from emerging economies. J Rev Econ Perspect 21(3):347–367

Charles S, Marie J (2020) A note on the competing causes of high inflation in Bulgaria during the 1990s: money supply or exchange rate? J Rev Polit Econ 32(3):433–443

Delgado NAB, Delgado EB, Saucedo E (2018) The relationship between oil prices, the stock market and the exchange rate: evidence from Mexico. J North Am Econ Financ 45:266–275

Doraisami A (2018) The Timor Leste petroleum fund, veterans and white elephants: fostering intergenerational equity? J Res Pol 58:250–256

Drysdale J (2008) Five principles for the management of natural resource revenue: the case of Timor-Leste's petroleum revenue. J Energy Nat Resour Law 26(1):151–174

Enders W (2019) Applied econometric time series, 4th edn., vol 53. Wiley. United States of America, 1–498

Enwereuzoh PA, Odei-Mensah J, Junior PO (2021) Crude oil shocks and African stock markets. J Res Int Bus Financ 55:101346

Fabris N, Vujanovic N (2017) The impact of financial dollarization on inflation targeting: empirical evidence from Serbia. J Cent Bank Theory Pract 6(2):23–43

Granger (1969) Investigating causal relations by econometric models and cross-spectral methods, 424–438

Hamilton JD (1983) Oil and the macroeconomy since world war II. J Polit Econ 91(2):228–248

Hashmi SM, Chang BH, Bhutto NA (2021) Asymmetric effect of oil prices on stock market prices: new evidence from oil-exporting and oil-importing countries. J Resour Policy 70:101946

Hussain MI, Zafar T (2018) The interrelationship between money supply, inflation, public expenditure and economic growth. J Eur Online Nat Soc Sci 7(1):1–24

John S, Papyrakis E, Tasciotti L (2020) Is there a resource curse in Timor-Leste? A critical review of recent evidence. J Dev Stud Res 7(1):141–152

Kilian L, Vigfusson RJ (2017) The role of oil price shocks in causing U.S. recessions. J Money Credit Bank 49(8):1747–1776

Kpagih LL (2022) Energy price fluctuation and inflation in Nigeria: a granger causality analysis. J Econ Financ Manag Stud 05(03)

Li Y, Guo J (2022) The asymmetric impacts of oil price and shocks on inflation in BRICS: a multiple threshold nonlinear ARDL model. Appl Econ: 1377–1395

Liu D, Meng L, Wang Y (2020) Oil price shocks and Chinese economy revisited: new evidence from SVAR model with sign restrictions. J Int Rev Econ Financ 69:20–32

Liu R, Chen J, Wen F (2021) The nonlinear effect of oil price shocks on financial stress: evidence from China. J North Am Econ Financ 55:101317

Mengesha LG, Holmes MJ (2015) Does dollarization reduce or produce inflation? J Econ Stud 42(3):358–376

Nusair SA, Olson D (2021) Asymmetric oil price and Asian economies: a nonlinear ARDL approach. J Energy 219:119594

Olayungbo D (2019) Effects of global oil price on exchange rate, trade balance, and reserves in Nigeria: a frequency domain causality approach. J Risk Financ Manag 12(1):43

Salisu AA, Cuñado J, Isah K, Gupta R (2021) Oil price and exchange rate behaviour of the BRICS. J Emerg Mark Financ Trade 57(7):2042–2051

Scheiner C (2021) Timor-Leste economic survey: the end of petroleum income. J Asia Pacific Policy Stud 8(2):253–279

Sims CA (1980) Macroeconomics and reality. J Econom Soc: 1–48

Stock JH, Watson MW (2007) Introduction to econometrics, 3rd edn. Pearson Education, Inc., 1–829

Tian M, Li W, Wen F (2021) The dynamic impact of oil price shocks on the stock market and the USD/RMB exchange rate: evidence from implied volatility indices. J North Am Econ Financ 55:101310

Timor-Leste Ministry of Finance (2019) Timor-Leste petroleum fund—annual report. https://www.mof.gov.tl/wp-content/uploads/2018/08/PF-FINAL-REPORT-2017.pdf. Accessed on 26 Jan 2022

Van Doan D (2020) Money supply and inflation impact on economic growth. J Financ J Econ Policy 12(1):121–136

Wen F, Zhang K, Gong X (2021) The effects of oil price shocks on inflation in the G7 countries. J North Am Econ Financ 57:101391

World Bank Group (2021) Timor-Leste economic report: steadying the ship, 1–47. Available from https://openknowledge.worldbank.org/bitstream/handle/10986/36733/TLER_December2021.pdf?sequence=5&isAllowed=y

World Bank (2020) Timor-Leste economic report: towards a sustained recovery, 1–56, Oct 2020. https://openknowledge.worldbank.org/handle/10986/34748. Accessed on 11 Jan 2021

Zakaria M, Khiam S, Mahmood H (2021) Influence of oil prices on inflation in South Asia: some new evidence. J Resour Policy 71:102014

# Chapter 71
# Oil Price Volatility Impacts Over the Timor-Leste Economy

**Fernando Anuno, Mara Madaleno, and Elisabete Vieira**

**Abstract** The present study applies time-series volatility models to study oil price volatility and the effect of macroeconomic variables, such as the Gross Domestic Product (GDP), Consumer Price Index (CPI), interest rate, exchange rates, and money supply on the economic growth of Timor-Leste during 2003M01-2019M12, as well as the reverse effect of GDP growth over oil prices. Considering the Autoregressive Conditional Heteroskedasticity ARCH (1) and the Generalized GARCH (1,1) models for the oil price volatility, results show that the interest rate can lower the oil price. Contrarily, the CPI, exchange rate, GDP, and money supply increase oil prices. Meanwhile, the CPI, exchange rate, and interest rate drive negatively GDP, reducing economic growth. Thus, oil prices and the money supply could influence Timor-Leste's economic growth. Exponential EGARCH (1,1) estimation of the oil price shows that interest rates could depress the oil price. Symmetrical and asymmetrical techniques have shown that oil prices and economic growth in the Timor-Leste region are volatile. This study is important for Timor-Leste policymakers to consider the impact of oil price volatility and macroeconomic variables to be significant to economic growth and prevent their negative effects.

**Keywords** Macroeconomic variables · Oil price · Time series · Volatility models · Timor-Leste

F. Anuno (✉)
Faculty of Economics and Management, National University of Timor Lorosa'e (UNTL), GOVCOPP, University of Aveiro, Avenida Cidade de Lisboa, Díli, Timor-Leste
e-mail: fernando.anuno@ua.pt

F. Anuno · M. Madaleno
Research Unit On Governance, Competitiveness and Public Policies (GOVCOPP), Department of Economics, Management, Industrial Engineering and Tourism (DEGEIT), University of Aveiro, Campus Universitário de Santiago, 3810-193 Aveiro, Portugal

E. Vieira
GOVCOPP, ISCA—Higher Institute for Accountancy and Administration of Aveiro, University of Aveiro, Campus Universitário de Santiago, Aveiro, Portugal

 
N. S. Caetano and M. C. Felgueiras (eds.), *The 9th International Conference on Energy and Environment Research*, Environmental Science and Engineering,
https://doi.org/10.1007/978-3-031-43559-1_71

## 71.1 Introduction

Crude oil is the major energy source of government revenue in many countries, able to significantly affect the real economy and financial markets (Álvarez-Díaz 2020). Oil is considered one of the macroeconomic factors, and oil prices determine economic growth (Awunyo-Vitor et al. 2018; Chiweza and Aye 2018; Mukhtarov et al. 2020). However, oil price fluctuations affect the income of oil-importing and exporting countries (Mukhtarov et al. 2020; Naser 2016). When the price of oil fluctuates, there is a redistribution of income from the oil-importing country to the oil-exporting country. This condition affects the real exchange rate and is a Dutch disease phenomenon. In addition, the effects of oil price changes affect the stock market and public finances (Liu et al. 2019; Śmiech et al. 2021). Oil price volatility also affects production costs and influences inflation (Akinsola and Odhiambo 2020). Based on the concept of the income effect, an increase in oil income improves the trade balance, consumption, and investment (Rafiq et al. 2016). However, a fall in oil prices improves the balance of payments and trade conditions of importing countries, increasing their disposable income and demand for other commodities (Akinsola and Odhiambo 2020). Therefore, these economic activities correlate with macroeconomic variables, such as inflation, interest rates, economic growth GDP, exchange rate, and money supply. As well, these variables correlate with oil prices. This phenomenon attracted the attention of macroeconomic researchers like (Hamilton 1983), a pioneer in the aftermath of the global oil crisis.

To the best of our knowledge, there is no specific literature that discusses the impact of oil prices on macroeconomic variables using econometric models. Moreover, macroeconomic activity is a very important measure of a country's position that determines whether it is classified as a poor country, a developing country, or a developed country. For this reason, Timor-Leste has been geographically located in Southeast Asia since it became a sovereign country in 2002. Although it has been registered as a member of the Association of Southeast Asian Nations (ASEAN) since 2011, it is not yet officially a member of ASEAN. Therefore, one of the requirements for ASEAN membership is to fulfill the economic pillar. To meet this pillar, Timor-Leste, which is currently heavily dependent on oil wealth, needs to diversify its economic activities outside the oil and gas sector, as the country has an economic market advantage, but limited oil reserves (Doraisami 2017). In addition, Timor-Leste is keen to increase investment to progress economically in line with the Strategic Development Plan 2011–2030 and transform Timor-Leste from a low-income country to a middle-income country with good health, education, and security by 2030. Finally, volatility models like ARCH-GARCH-EGARCH are useful to capture the macroeconomic impact of oil price volatility. This research contributes to the literature for several reasons. First, by investigating the impact of oil prices on macroeconomic variables, with an emphasis on Timor-Leste. The oil sector is a large market, but its oil wealth may be exhausted. Thus, governments should diversify their economies, using agriculture and tourism to boost their productivity, based on the Timor-Leste economic diversification and Strategic Development Plan 2011–2030. Second, by modeling the oil price volatility and the impact of macroeconomic variables on Timor-Leste's economic growth using monthly data between 2003 and 2019. A thorough search of the literature confirms that this study is the first to examine the oil price volatility and the impact of macroeconomic variables, such as oil price, GDP, CPI, interest rate, exchange rate, and money supply, on economic growth using time series techniques. Moreover, we simultaneously model the effect of oil prices on economic growth and growth in oil prices, not commonly done provided usually studies do not care as well for reverse causalities concurrently. Third, we found evidence of some stylized facts: (i) through the oil price volatility model, the interest rate reveals a negative impact on the oil price. In contrast, CPI, exchange rate, GDP, and money supply have positive effects; (ii) the effect of CPI, exchange rate, and interest rate on the GDP economic growth is negative. However, oil price and money supply have positive coefficients; (iii) the CPI, exchange rate, GDP, and money supply suggest that an increase in CPI, exchange rate, GDP, and money supply increases the oil price. The negative effect on the interest rate, based on its negative values, implies that the price of oil could fall; (iv) CPI, exchange rate, and interest rate have negative values, suggesting a negative effect on GDP economic growth. However, oil prices and money supply have a positive effect, and; (v) EGARCH(1,1) model results reveal negative effects of the interest rate, leading to a fall in the oil price, as well as the negative effect of CPI, exchange rate, interest rate, and money supply, all depressing economic growth; (vi) Finally, symmetrical and asymmetrical techniques showed that oil prices and economic growth in Timor-Leste are volatile.

## 71.2 Data and Methodology

We apply monthly time series data, from January 2003 to December 2019, to study the effect of oil prices on the Timor-Leste macroeconomic activity. This study used six macroeconomic variables to analyze the Timor-Leste case: oil prices (OP), GDP growth (GDP), money supply (M2), interest rates (IR), exchange rates (ER), and consumer price index (CPI). Oil price and money supply variables are converted into natural logs. Other variables such as GDP growth, IR, and CPI are divided by 100. The exchange rate variable is divided by the Indonesian currency exchange rate because Timor-Leste has adopted the US currency as the official currency and relies on imports of goods from Indonesia. Timor-Leste uses the WTI crude oil price, an essential benchmark for world oil prices. WTI price data were collected from the Energy Information Administration website. We accessed the CPI data from the Monetary International Fund website and currency exchange rates. Our empirical analysis uses GDP growth (annual %) data provided by the World Bank. Timor-Leste GDP monthly data is unavailable, but the annual data is available. To balance all data of each variable, annual GDP growth (%) is converted into monthly data using the Eviews 11 Student Version Lite Software (low to a high-frequency method, quadratic). In addition, interest rates and the money supply in Timor-Leste data are accessed on Banco Central. The ARCH model is a data variance to measure time series. Developed by Engle (1982), the ARCH model measures the volatility of financial and economic variables, such as share prices, oil prices, interest rates, exchange rates, and GDP. The model helps to predict future changes in volatility. (Bollerslev 1986) developed the GARCH model, which offered better results than the ARCH model. The GARCH model helps to develop the conditional variance, dependent on its previous delay so that the dependent variance equation is in the simplest case according to Brooks (2014). Nelson (1991) introduced the exponential GARCH-EGARCH model and it was later proposed by Reza et al. (2018) to explore oil price impacts. The model tests the asymmetric effect of time series data from good and bad news on conditional variance.

Table 71.1 describes the summary statistics of macroeconomic variables and oil prices for 2003–2019. The results show that the mean oil price is US$4.1537 per barrel, and the standard deviation is 0.3626. The GDP growth rate reached an average of 3.82% when the money supply was US$5.5045 million, with an average interest rate of 13.47%. Likewise, the mean of the consumer price index is 1.1225. Thus, the highest oil price is US$4.8969 per barrel, and the lowest price is US$3.3361 per barrel. In terms of the variable GDP growth, the lowest rate was − 4.11% and the highest growth was 11.34%. Timor-Leste uses the US$ as its official currency, and each transaction translates US$1 into each currency. The money supply reached US$6.7436 million, and the lowest performance was US$3.7932 million. Thus, the interest rate variable was high at 17.36% and low at 10.47%. Furthermore, the consumer price index peaked at US$1.4708 and the lowest at US$0.6985. Almost all variables are slightly distorted. Except for the oil price variable, the data is uneven, and the CPI is symmetrical. Kurtosis of all variables is greater than + 1, indicating that the distribution is too high.

**Table 71.1** Summary statistics

| | CPI | GDP | ER | IR | M2 | OP |
|---|---|---|---|---|---|---|
| Mean | 1.1225 | 0.0382 | − 9.2710 | 0.1347 | 5.5045 | 4.1537 |
| Median | 1.1314 | 0.0330 | − 9.1750 | 0.1293 | 5.6725 | 4.1523 |
| Maximum | 1.4708 | 0.1134 | − 9.0180 | 0.1736 | 6.7436 | 4.8969 |
| Minimum | 0.6985 | − 0.0411 | − 9.6291 | 0.1047 | 3.7932 | 3.3361 |
| Std. Dev. | 0.2899 | 0.0425 | 0.1859 | 0.0196 | 0.8892 | 0.3626 |
| Skewness | − 0.1324 | 0.2032 | − 0.4525 | 0.5233 | − 0.4022 | − 0.2502 |
| Kurtosis | 1.3250 | 2.0532 | 1.5696 | 2.1056 | 1.8349 | 2.3037 |
| Observations | 204 | 193* | 204 | 204 | 204 | 204 |

*Note* *The GDP growth variable sample did not reach a total sample of 204 like the total sample of other variables because the GDP growth data for Timor-Leste is only available annual GDP growth data, so the conversion of annual data to monthly data using the EVIEWS software only reached 193 samples. Oil prices (OP), GDP growth (GDP), money supply (M2), interest rates (IR), exchange rates (ER), and consumer price index (CPI)

## 71.3 Empirical Results

The ARCH model determines the ARCH effect before calculating and determining the GARCH model. The ARCH (1) model estimates the oil price volatility model as follows:

$$\begin{aligned} OP = & \ 18.8303 + 0.8890CPI + 1.7440ER \\ & + 2.2489GDP - 2.1783IR + 0.1161M2 \end{aligned} \qquad (71.1)$$

The interest rate negatively influences the oil price. However, CPI, exchange rate, GDP, and money supply have positive coefficients. This result suggests that every increase in CPI, exchange rate, GDP, and money supply increases the oil price. ARCH (1) effect is statistically significant, with a 1% significance value. However, the money supply is significant with a probability value of less than 10%. The model ARCH(1), as a model of economic growth, is estimated as in Eq. (71.2).

$$\begin{aligned} GDP = & -0.5236 - 0.1388CPI - 0.0673ER \\ & - 1.1270IR + 0.0125M2 + 0.0426OP \end{aligned} \qquad (71.2)$$

The CPI, exchange rate, and interest rate drive negatively the GDP economic growth. However, oil price and money supply have positive coefficients, suggesting that oil price and money supply have a positive relationship with the GDP level. The ARCH(1) is statistically significant, with a 1% significance level.

We also model the oil price equation using the GARCH and EGARCH models. Based on the results, the equation of the GARCH(1,1) model is provided in Eq. (71.3).

$$\begin{aligned} OP = & \ 18.6076 + 0.8736CPI + 1.7107ER \\ & + 2.1090GDP - 2.4077IR + 0.1110M2 \end{aligned} \qquad (71.3)$$

From the equation, the GARCH(1,1) model reveals the relationships between the variables. The interest rate is negative, suggesting a negative effect on the oil price. However, the CPI, exchange rate, GDP, and money supply have a positive effect. The CPI, exchange rate, GDP, and money supply have a positive effect on the CPI, exchange rate, GDP, and money supply leading to an increase in the oil price. However, the negative effect of the interest rate guided by its negative values implies that a decrease in the interest rate lowers the oil price. In addition, the coefficients of the $\alpha$ and $\beta$ parameters are 1.0119 and 0.0481, respectively. The p-value of $\alpha$ proves to be statistically significant. However, the p-value of $\beta$ is statistically insignificant, revealing ARCH and GARCH effects. The sum of the ARCH ($\alpha$) and GARCH ($\beta$) coefficients is significant > 1, showing that the oil price in Timor- Leste is volatile. Based on the results of the GARCH (1,1) model, the equation of variance of the GARCH (1,1) model is as follows.

$$\sigma_t^2 = 0.0027 + 1.0119\varepsilon_{t-1}^2 + 0.0481\sigma_{t-1}^2 \quad (71.4)$$

The EGARCH model is formulated as in Eq. (71.5).

$$\begin{aligned} \text{OP} = 19.4816 + 0.7480\text{CPI} + 1.8168\text{ER} \\ + 1.8463\text{GDP} - 4.5197\text{IR} + 0.1586\text{M2} \end{aligned} \quad (71.5)$$

The model EGARCH(1,1) shows a negative effect of the interest rate on the oil price, showing that an increase in CPI, exchange rate, GDP, and money supply increases the oil price. The p-value of α and β are statistically significant. The sum of α and β coefficients is > 1, which means that the oil price in Timor-Leste is volatile.

Based on the GARCH(1,1) model results, the GARCH(1,1) model equation describes the economic growth GDP as presented in Eq. (71.6).

$$\begin{aligned} \text{GDP} = -0.6634 - 0.1230\text{CPI} - 0.1028\text{ER} \\ - 1.5913\text{IR} + 0.0091\text{M2} + 0.0133\text{OP} \end{aligned} \quad (71.6)$$

In Eq. (71.6), the GARCH(1,1) model determines the relationships between the variables. According to the results, CPI, exchange rate, and interest rate have negative values that suggest a negative effect on economic growth. However, oil prices and money supply suggest a positive effect. An increase in money supply and oil price increases the GDP's economic growth. However, the negative effects on the CPI, exchange rate, and interest rate are aligned with their negative values, implying that the interest rates, exchange rates, and CPI affect negatively GDP growth. Furthermore, the coefficients of the parameters α and β are 0.8926 and 0.3102, respectively. The p-value of α and β are statistically significant and reveal the presence of both ARCH and GARCH effects. The sum of the ARCH (α) and GARCH (β) coefficients is higher than one for the GARCH model, showing that the economic growth in Timor-Leste is volatile. Based on the results of the GARCH (1,1) model, the variance equation of the GARCH (1,1) model is as described in Eq. (71.7).

$$\sigma_t^2 = 0.000001 + 0.8926\varepsilon_{t-1}^2 + 0.3102\sigma_{t-1}^2 \quad (71.7)$$

The EGARCH model is formulated following Eq. (71.8).

$$\begin{aligned} \text{GDP} = -0.6857 - 0.1070\text{CPI} - 0.1029\text{ER} \\ - 1.6433\text{IR} - 0.0100\text{M2} + 0.0397\text{OP} \end{aligned} \quad (71.8)$$

The EGARCH(1,1) model shows a negative effect on CPI, exchange rate, interest rate, and money supply on economic growth, showing that a decrease in CPI, exchange rate, interest rate, and money supply by 1% reduce the economic growth. The p-value of α and β are statistically significant. The sum of α and β coefficients is > 1, showing that the economic growth in Timor- Leste is volatile. Previous studies focused on the fluctuation of the economic impact of oil prices in the US, European countries, OPEC, MENA countries, and Gulf Cooperation Council (GCC) countries. This study is almost identical to the earlier study by Reza et al. (2018) on modeling oil price volatility and macroeconomic variables in South Africa using five variables such as oil price, interest rate, inflation, GDP, and exchange rate. This study is of interest to Timor-Leste. As a new country, this study uses six macroeconomic variables such as oil price, GDP, CPI, interest rates, exchange rates, and money supply. However, what makes this study different from previous studies is the money supply. The results of our study for ARCH (1) with the dependent variable oil prices are consistent with the results of Reza et al. (2018) for CPI, GDP, and IR, being the inconsistent variable ER. But for the M2 variable, our results are consistent with the results of Sekati et al. (2020) that money supply (M2) has a positive effect on crude oil prices as a proxy indicator for China's monetary policy. Moreover, our results confirm findings from previous work in the five South Asian Association for Regional Cooperation (SAARC) countries (Wen et al. 2019), which show that oil price shocks have significant effects on the economy in both the short and long run. However, each country deals with each oil price shock differently, for example, through macroeconomic policies, sector formation, and their heterogeneity across countries. The finding of the positive role of oil price volatility is consistent with recent studies by Chiweza and Aye (2018), Reza et al. (2018) but contradicts the findings of Ahmed et al. (2018). Essentially, the study shows that oil shocks affect the macroeconomic activity and concludes that measuring volatility based on daily crude oil futures prices has a negative and significant effect on future GDP growth. Moreover, (Ito 2010) concludes that the economic damage caused by oil price volatility is considered necessary to diversify the sector and increase the competitiveness of firms outside the energy sector. However, the results show a short-run asymmetric effect in the presence of negative and positive shocks to international oil prices. For example, the positive effect of an increase in relative crude oil prices tends to harm output and employment because the increase acts like a consumption tax. Moreover, as firms face higher costs, rising oil prices tend to increase inflation (Guo and Kliesen 2005). Our results are also consistent with the findings of Jawadi and Ftiti (2019), who found that oil price has a positive and significant impact on the dynamics of economic growth in Saudi Arabia. Moreover, our results are significantly negative, so unlike (Polbin et al. 2020) real exchange rate movements also affect the dynamics of real GDP. This is because oil price dynamics are the main source of revenue contributing to real GDP and real exchange rate movements. Likewise, the exchange rate is an important factor in determining international transactions. This is because macroeconomic policies can determine economic changes in performance internally and externally (Davari and Kamalian 2017). Finally, our results are consistent with those of Zulfigarov and Neuenkirch (2020) provided it was found that a decline in the exchange rate affects GDP and, subsequently, a decline in interest rates affects GDP.

## 71.4 Conclusion

Results showed that the interest rate negatively influences the oil price. However, CPI, exchange rate, GDP, and money supply have positive coefficients, suggesting that an increase in the GDP, money supply, exchange rate, and consumer price index is associated with an increase in the oil price. Furthermore, the ARCH(1) model for the estimated economic growth model evidenced that the CPI, exchange rate, and interest rate have negative effects on the GDP. However, oil price and money supply have positive coefficients. This result suggests that increases in the oil price and money supply lead to an increase in the GDP. The GARCH(1,1) model showed that CPI, exchange rate, and interest rate have negative values on economic growth. At the same time, the oil price and money supply indicate a positive effect. Oil price and money supply results suggest that this variable impacts positively economic growth. However, the CPI, the exchange rate, and the interest rates are negatively related to economic growth (GDP). The EGARCH(1,1) model, estimated at the oil price, showed that the interest rate negatively impacts the oil price. However, an increase in the CPI, exchange rate, GDP, and money supply increase the oil price. Similarly, the EGARCH(1,1) model reveals a negative effect of CPI, exchange rate, interest rate, and money supply on economic growth. Since this study is a new one for Timor-Leste, the results can be applied to other countries, especially those that rely heavily on oil revenues to diversify their economies. Similarly, it is hoped that similar results in countries with similar characteristics to Timor-Leste can translate these results to other realities and force the consideration of country characteristics. In addition, this study has limitations that can be easily modified in future research directions. For example, future research is needed to examine the impact of post-November 19 oil prices on oil importing and exporting countries. Since the 1973 oil crisis, several major historical events have occurred that have affected the oil crisis, including the global financial crisis, the Iraq War, the first Gulf War, the Covid-19 outbreak, and more recently the war between Ukraine and Russia.

**Acknowledgements** The support of the Research Unit on Governance, Competitiveness and Public Policy (UIDB/04058/2020) + (UIDP/04058/2020), funded by national funds through FCT—Fundação para a Ciência e a Tecnologia is greatly acknowledged.

## References

Ahmed K, Bhutto NA, Kalhoro MR (2019) Decomposing the links between oil price shocks and macroeconomic indicators: evidence from SAARC region. J Resour Policy 61:423–432

Akinsola MO, Odhiambo NM (2020) Asymmetric effect of oil price on economic growth: panel analysis of low-income oil-importing countries. J Energy Rep 6:1057–1066

Álvarez-Díaz M (2020) Is it possible to accurately forecast the evolution of Brent crude oil prices? An answer based on parametric and nonparametric forecasting methods. J Empir Econ 59(3):1285–1305

Awunyo-Vitor D, Samanhyia S, Bonney EA (2018) Do oil prices influence economic growth in Ghana? An empirical analysis. J Cogent Econ Financ 6(1):1–14

Bollerslev T (1986) A generalized least absolute deviation method for parameter estimation of autoregressive signals. IEEE Trans Neural Netw 31(3):307–327

Brooks C (2014) Introductory econometrics for finance, 3rd edn. Cambridge University Press, United Kingdom, pp 1–744

Chiweza JT, Aye GC (2018) The effects of oil price uncertainty on economic activities in South Africa. J Cogent Econ Financ 6(1):1–17

Davari H, Kamalian A (2017) Study short term and long term impact of effective real exchange rate on oil price growth in Iran. J Int Energy Econ Policy 7(6):159–163

Doraisami A (2018) The Timor Leste petroleum fund, veterans and white elephants: fostering intergenerational equity? J Resour Policy 58:250–256

Engle RF (1982) Autoregressive conditional heteroscedasticity with estimates of variance of United Kingdom Inflation. J Eco 50(4):987–1008

Guo H, Kliesen KL (2005) Oil price volatility and U.S. macroeconomic activity. J Fed Reserv Bank St Louis Rev 87(6):669–683

Hamilton JD (1983) Oil and the macroeconomy since world war II. J Polit Econ 91(2):228–248

Ito K (2010) The impact of oil price volatility on macroeconomic activity in Russia. Econ Anal Work Pap 9(05):1–10

Jawadi F, Ftiti Z (2019) Oil price collapse and challenges to economic transformation of Saudi Arabia: a time-series analysis. J Energy Econ 80:12–19

Liu Z, Ding Z, Lv T, Wu JS, Qiang W (2019) Financial factors affecting oil price change and oil-stock interactions: a review and future perspectives. J Nat Hazards 95(1–2):207–225

Mukhtarov S, Aliyev S, Zeynalov J (2020) The effect of oil prices on macroeconomic variables: evidence from Azerbaijan. J Int Energy Econ Policy 10(1):72–80

Naser H (2016) Estimating and forecasting the real prices of crude oil: a data rich model using a dynamic model averaging (DMA) approach. J Energy Econ 56:75–87

Nelson D (1991) Conditional Heteroskedasticity in asset returns: a new approach, 347–370

Polbin A, Skrobotov A, Zubarev A (2020) How the oil price and other factors of real exchange rate dynamics affect real GDP in Russia. J Emerg Mark Financ Trade 56(15):3732–3745

Rafiq S, Sgro P, Apergis N (2016) Asymmetric oil shocks and external balances of major oil exporting and importing countries. Energy Econ 56:42–50

Reza R, Tularam GA, Li B (2018) Returns and volatility of water investments. Cogent Econ Financ 6(1)

Sekati BNY, Tsoku JT, Metsileng LD (2020) Modelling the oil price volatility and macroeconomic variables in South Africa using the symmetric and asymmetric GARCH models. Cogent Econ Financ 8(1)

Śmiech S, Papież M, Rubaszek M, Snarska M (2021) The role of oil price uncertainty shocks on oil-exporting countries. J Energy Econ 93

Wen F, Min F, Zhang YJ, Yang C (2019) Crude oil price shocks, monetary policy, and China's economy. J Int Financ Econ 24(2):812–827

Zulfigarov F, Neuenkirch M (2020) The impact of oil price changes on selected macroeconomic indicators in Azerbaijan. J Econ Syst 44(4):100814

# Chapter 72
# PPE Waste Generation During COVID-19 Pandemic in Guayaquil: Geospatial Distribution and Thermochemical Valorization

**Jose Armando Hidalgo Crespo, Andrés Velastegui-Montoya, Manuel Soto, Jorge Luis Amaya-Rivas, Leonardo Alvaro Banguera Arroyo, Marcos Santos-Méndez, and Yomar Alexander González Cañizales**

**Abstract** Millions of tons of polluted personal protective equipment, such as face masks, face shields, and gloves related to the COVID-19 disease, are becoming dangerous infectious waste. Suppose PPE is littered in the open environment, whether aquatic or terrestrial. It will most likely produce the blockage of sewages in towns and cities and could also percolate to water vessels and prevent the aeration of agricultural land. Hence, the quantification, geospatial spreading, and thermochemical valorization of personal protective equipment waste were performed for the city of Guayaquil. The overall daily face masks, pairs of gloves, and face protectors waste generation in the town reach 816,673.25 masks, 43,173.34 glove pairs, and 29,148.88 face protectors. Considering that the standard weight of one mask is 5 g on average, this number of face masks turns into approximately four tons of waste daily (1.5

J. A. Hidalgo Crespo (✉) · L. A. Banguera Arroyo · M. Santos-Méndez · Y. A. González Cañizales
Universidad de Guayaquil, Carrera de Ingeniería Industrial, Guayaquil, Ecuador
e-mail: jose.hidalgocr@ug.edu.ec

J. A. Hidalgo Crespo · M. Soto
Facultad de Ciencias, Universidade da Coruña, Coruña, Spain

A. Velastegui-Montoya
Facultad de Ingeniería en Ciencias de la Tierra, ESPOL Polytechnic University, Guayaquil, Ecuador

Centro de Investigaciones y Proyectos Aplicados a las Ciencias de la Tierra, ESPOL Polytechnic University, Guayaqui, Ecuador

J. L. Amaya-Rivas
Facultad de Ingeniería Mecánica y Ciencias de la Producción, ESPOL Polytechnic University, Guayaquil, Ecuador

J. A. Hidalgo Crespo
G-Scop Laboratory, University of Grenoble Alpes, 46 Av. Félix Viallet, 38031 Grenoble, France

 
N. S. Caetano and M. C. Felgueiras (eds.), *The 9th International Conference on Energy and Environment Research*, Environmental Science and Engineering,
https://doi.org/10.1007/978-3-031-43559-1_72

thousand tons yearly). Three-ply surgical disposal (3PFM) is the most wasted for face masks, accounting for 52.31%, followed by KN/FPP masks with 15.36%. 3PFM is primarily wasted in El Fortin, La Florida, Trinitaria, and Guasmo neighborhoods, all urban-marginal of the city. Finally, when talking about final disposition, incineration is preferred over landfilling. Given that the vast majority of thermal electricity production plants are based on the combustion of fuels, whether liquid or solid, this is the most immediate solution for the PPE waste problem.

**Keywords** PPE · Waste · Generation · Geospatial · Thermochemical · COVID-19

## 72.1 Introduction

Millions of tons of polluted personal protective equipment, such as face masks, face shields, and gloves related to the COVID-19 disease, are becoming dangerous infectious waste (Sangkham 2020). Many studies have proven that the viral load of COVID-19 can survive several days on plastic and PPE materials (Nghiem et al. 2020), deteriorating the environment and endangering the wellness of the population if not disposed of correctly (Nzediegwu and Chang 2020; Dharmaraj et al. 2021). These products are mostly made of long-lasting plastic materials. They are classified as single use with concise service life, threatening the land and marine environment with immense loads of plastic waste. Their use is enforced by most governments around the world, going from indoors to the outdoors exacerbating severely the quantity of garbage generated daily. In many developing countries such as Ecuador, the lack of waste management imposes a severe contamination menace on the general population, particularly ours, which still relies on open landfills for its waste management. Suppose PPE is littered in a relaxed environment, whether aquatic or terrestrial, such as urban-marginal areas. In that case, it will most likely produce the blockage of sewages in towns and cities and could also percolate to water vessels and prevent the aeration of agricultural land (Patrício Silva et al. 2021). Some authors have studied that reusing previously collected and treated PPE waste can diminish the amount of waste by more than 90% and the natural resource depletion by more than 25% (Haque et al. 2021). However, reconditioning PPE waste for a second use is challenging due to the high number of related processes and the water and electricity necessary to sanitize them (Alcaraz et al. 2022). Another reason which makes recycling face masks almost impossible is their multilayer constitution which augments the probability of cross-contamination (Jung et al. 2021). One critical step toward formulating an efficient waste management system is the information available on the quantity of PPE waste generated, such as face masks, face shields, and gloves (Kumar et al. 2018). In addition, knowing how they are scattered in the city can help the planning of collection systems. However, the current streams of these PPE wastes remain unknown to the country. This research aims to estimate the current PPE waste generation for the country, using the city of Guayaquil as a case study in the context of the ongoing COVID-19 pandemic. Guayaquil was chosen due to its poor waste management and lack of necessary infrastructure to manage dangerous wastes. Additionally, a geospatial reference for the presence of the different PPE in the city and a review analysis for the potential combustion and pyrolysis of these wastes as an alternative method has been conducted.

## 72.2 Materials and Methods

### 72.2.1 *Survey Design and Data Gathering*

The ArcGIS Survey123 application (Mora-Araus et al. 2021) was used. A form was designed to record geolocated data of the inhabitants of Guayaquil, the most populated city of Ecuador, with around 2.645 million inhabitants. The survey was administered to a total of 797 household heads. To do so, 115 students in the subject of Energy and Environment from two local universities participated in collecting information. Each of them had to gather information from 7 household heads. The participants could be their neighbors or family members that didn´t live with them, as long as all were from the city. The survey was delivered between February and March 2022. The first part of the questionnaire contained the personal attributes of the interviewed person, such as gender, age, level of education, and general information such as GPS location and household size. The second part contained 13 questions (Qi) to pursue the study's specific aims, as stated at the end of the introductory section. In particular, Q1, Q2, Q3, Q4, and Q5 wanted to verify the quantities of weekly discarded masks. Q6, Q7, Q8, and Q9 were aimed to investigate the amounts of monthly discarded gloves per household. Finally, Q10 to Q13 were submitted to people participating in the research to examine the use of protecting face shields. An illustration was available on the survey to distinguish the different types of face masks, gloves, and face protectors (see Fig. 72.1; Table 72.1).

### 72.2.2 *Estimation of PPE Generation*

According to the data collected in the last census of 2010 for the country, the total amount of people was approximately 2.64 million (INEC 2010). To estimate the number of daily face masks (DFM), gloves (DGL), and face protectors (DFP) waste generation, the following equations were used based on the calculations previously stated by Torres and De la Torre (Torres and De-la-Torre 2021) with modifications:

$$DFM = \frac{\begin{array}{c}\sum T_P * (1 * \%GL_1 + 2 * \%GL_2 + 3 * \%GL_3 + 4 * \%GL_4 \\ +5 * \%GL_5 + 6 * \%GL_6 + 7 * GL_{>6})\end{array}}{7} \quad (72.1)$$

**Fig. 72.1** Surveyed types of PPE waste per household

$$DGL = \frac{\sum T_P * (1 * \%GL_1 + 2 * \%GL_2 + 3 * \%GL_3 + 4 * \%GL_4 + 5 * \%GL_5 + 6 * \%GL_6 + 7 * \%GL_{>6})}{30} \quad (72.2)$$

$$DFP = \frac{\sum T_P * (1 * \%FP_1 + 2 * \%FP_2 + 3 * \%FP_3 + 4 * \%FP_4 + 5 * \%FP_5 + 6 * \%FP_6 + 7 * \%FP_{>6})}{30} \quad (72.3)$$

Tp is the complete number of households in the city, %FM, %GL, and %FP are the different percentages of wasted PPE per option.

**Table 72.1** The submitted household questionnaire (English translation and adaptation)

| No. | Question | Answers |
|---|---|---|
| | *Gender* | Male; female |
| | *Age of household head* | < 26; 26–36; 37–47; 48–58; 59–69; > 69 |
| | *Level of education of household head* | Primary School; High School, Specialization/University; Master/Ph.D.; None |
| | *GPS Coordinates* | GIS location |
| | *Household Size* | 1; 2; 3; 4; 5–6; > 6 |
| Q1 | How many industrial face masks does your family dispose of weekly? | None; 1–2; 3–4; 5–6; 7–8; 9–10; 11–12; > 12 |
| Q2 | How many 3PFM face masks does your family dispose of weekly? | |
| Q3 | How many fabric face masks does your family dispose of weekly? | |
| Q4 | How many hypoallergenic face masks does your family dispose of weekly? | |
| Q5 | How many KN/FPP face masks does your family dispose of weekly? | |
| Q6 | How many pairs of latex gloves does your family dispose of monthly? | None; 1; 2; 3; 4; 5; 6; > 6 |
| Q7 | How many pairs of nitrile gloves does your family dispose of monthly? | |
| Q8 | How many pairs of polyethylene gloves does your family dispose of monthly? | |
| Q9 | How many pairs of vinyl gloves does your family dispose of monthly? | |
| Q10 | How many pairs of safety goggles does your family dispose of monthly? | None; 1; 2; 3; 4; 5; 6; > 6 |
| Q11 | How many face shields does your family dispose of monthly? | |
| Q12 | How many full-face shields does your family dispose of monthly? | |
| Q13 | How many mounted shields does your family dispose of monthly? | |

Italics represent general socio-demographic information

### *72.2.3 GIS Mapping of Waste PPE*

The data collected was processed in the ArcGIS Pro program (Environmental Systems Research Institute 2021). The Kernel Density tool was used to associate and fit a smoothly of Greater Guayaquil surface. This spatial analyst was made for each type of face mask, gloves, and face protector.

**Table 72.2** Characteristics of the household sample (N = 797)

| Demographic Value | N | Percentage (%) |
|---|---|---|
| *Gender* | 354 | 44.42 |
| Male | 443 | 55.58 |
| Female | | |
| *Age* | 103 | 12.92 |
| < 26 | 192 | 24.09 |
| 26–36 | 227 | 28.48 |
| 37–47 | 169 | 21.20 |
| 48–58 | 90 | 11.29 |
| 59–69 | 16 | 2.02 |
| > 69 | | |
| *Level of education* | 9 | I |
| None | 97 | 1.13 |
| Primary School | 430 | 12.17 |
| High School | 237 | 53.95 |
| Specialization/University | 24 | 29.74 |
| Master/Ph.D. | | 3.01 |
| *Household size* | 18 | 2.26 |
| 1 | 90 | 11.29 |
| 2 | 191 | 23.96 |
| 3 | 257 | 32.25 |
| 4 | 198 | 24.84 |
| 5–6 | 43 | 5.40 |
| > 6 | | |

## 72.3 Results

### *72.3.1 Population Characteristics*

Knowledge of the characteristics of the participating households is shown in Table 72.2. Of the 797 respondents, 55.58% were female. Approximately 29% of the household heads were aged between 37 and 47 years old. Most household heads have a minimum of a high school degree, with about 54%. Most households have four family members (32.25%), followed by households with 5 to 6 family members (24.84%).

### *72.3.2 PPE Waste Generation*

Table 72.3 shows the quantification of the total PPE waste projected for the city, considering a population of 505,769 urban households. The daily generation of face masks, pairs of gloves, and face protectors is approximately 817 thousand, 43 thousand, and 29 thousand. Three-ply surgical disposal (3PFM) is the most wasted for face masks, accounting for 52.31%, followed by KN/FPP masks with 15.36%. The

**Table 72.3** Quantification of total PPE waste generated

| Face masks | | Gloves | | Face protectors | |
|---|---|---|---|---|---|
| Industrial | 77,918.71 | Latex | 19,080.04 | Safety Goggles | 6,472.83 |
| 3PFM | 427,215.70 | Nitrile | 9,603.48 | Face shield | 9,815.01 |
| Fabric | 113,455.80 | Polyethylene | 6,938.19 | Full face shield | 7,953.54 |
| Hypoallergenic | 72,660.67 | Vinyl | 7,551.63 | Mounted shield | 4,907.50 |
| KN/FPP | 125,422.37 | | | | |
| Total | 816,673.25 | | 43,173.34 | | 29,148.88 |

sampled population prefers latex for the gloves, followed by nitrile with 44.19% and 22.24%, respectively. When using face protectors, the population prefers the novel face shields, followed by the full face shield, which covers the whole face with 33.67%, and 27.29%, respectively.

### *72.3.3 Geospatial Mapping*

Figure 72.2, 72.3 and 72.4 show the geospatial distribution of the different PPE items. According to Fig. 72.2, in Guayaquil, a greater volume of industrial masks was generated in the Pascuales industrial subdivision and the private urbanization sector of Metropolis (see Fig. 2a). The 3PFM masks was primarily generated in El Fortin, La Florida, Trinitaria, and Guasmo (see Fig. 2b), the Fabric masks in La Florida, Trinitaria, and Guasmo (see Fig. 2c), the Hypoallergenic masks in Metropolis and Trinitaria (see Fig. 2d) and, the KN FPP masks in Metropolis, La Florida, and Trinitaria (see Fig. 2e).

According to Fig. 72.3, in Guayaquil, a greater volume of industrial latex gloves was generated in Trinitaria and La Florida (see Fig. 3a). The Nitrile gloves were generated in Metropolis, Trinitaria, and Durán (see Fig. 3b), the Polyethylene gloves in Metropolis, La Florida, and Durán (see Fig. 3c), and the Vinyl gloves in Metropolis and La Florida (see Fig. 3d).

According to Fig. 72.4, in Guayaquil, a greater volume of Metropolis and Durán safety goggles was generated (see Fig. 4a). The Face shield was generated in Trinataria (see Fig. 4b), the Full shield phase in Metropolis, La Florida, and Durán (see Fig. 4c), and the Mounted shield in Metropolis, La Florida, and Durán (see Fig. 4d).

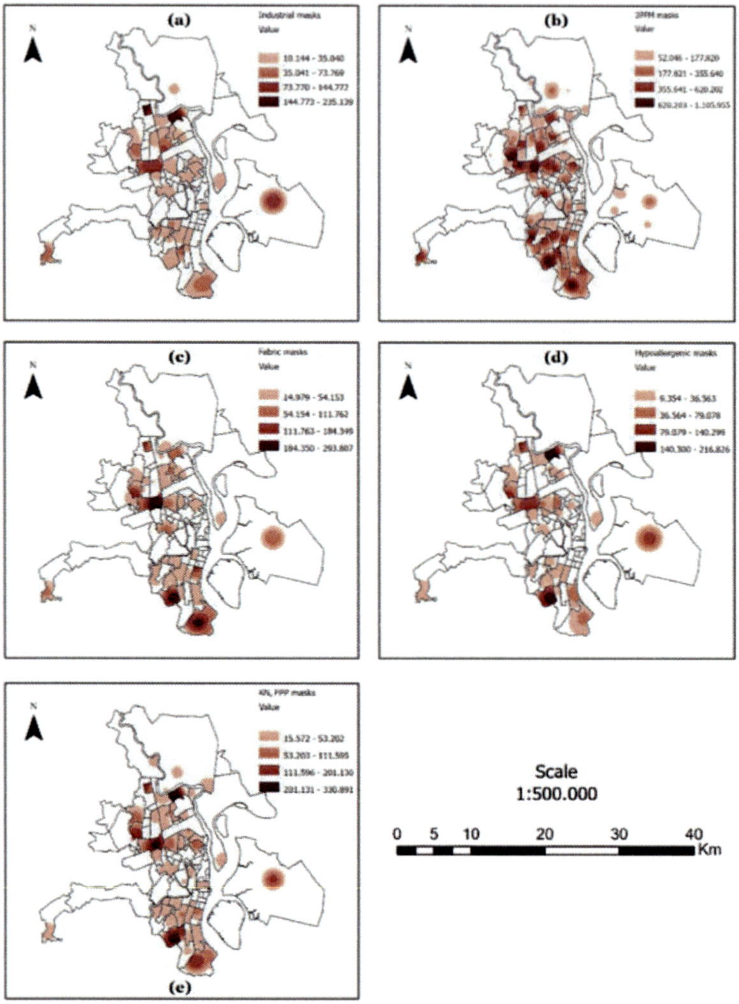


**Fig. 72.2** Quantity of face masks yearly geospatial distribution by type

## 72.4 Thermochemical Valorization of PPE Waste

Personal protective equipment is considered the primary source of protection against the COVID-19 virus reducing its spread to the general public; however, landfilling them can release microplastics and chemical additives via leachate that may contaminate the soil and groundwater. Affected settings must have the tools and systems required to destroy PPE effectively and efficiently to help protect those most susceptible to contracting the virus. In this context, combustion and pyrolysis are promising thermochemical methods for the final disposition of infectious waste materials while

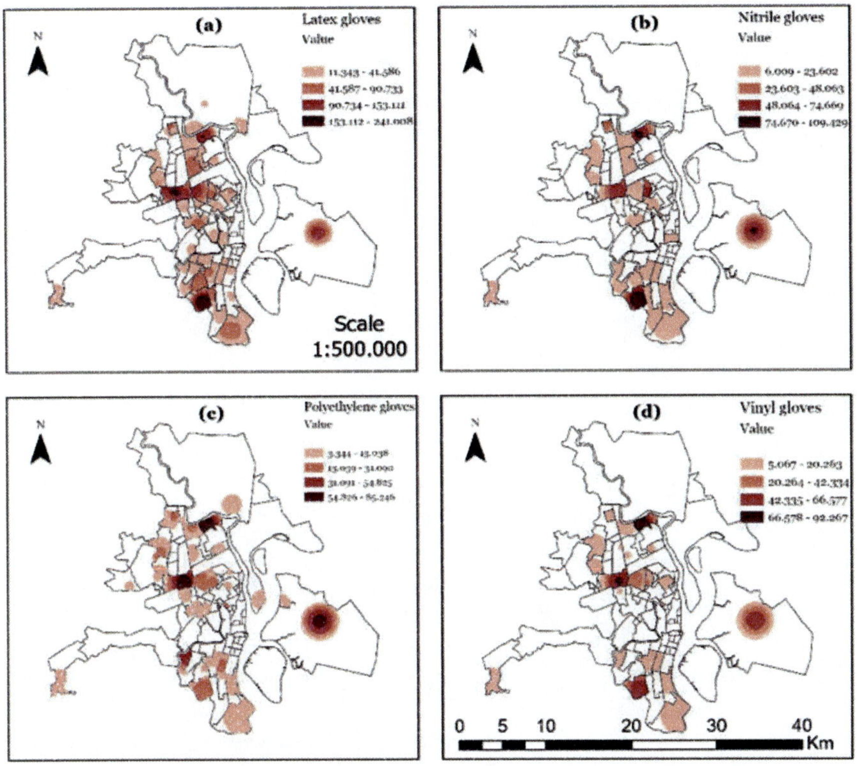


**Fig. 72.3** Quantity of gloves yearly geospatial distribution by type

converting them into valuable products. The World Health Organization (WHO) directs the incineration of PPE waste and other infectious wastes, especially those made from plastics (WHO 2020a; b). The incineration process of PPE, whether centralized or decentralized, showed a lower overall impact in the six categories studied through a life cycle assessment compared to landfilling (Kumar et al. 2021). From all PPE waste, 3-ply surgical disposable face masks (3PFM) represent a significant fraction. They are composed mainly of polypropylene and melt blow filters due to their low price, perfect filtration capacity, hypoallergenic feature, meager breathing resistance, resistance to fluids, and antiviral function. 3PFM face masks are rich in the volatile matter (97% according to Salema et al. (2022)) and can be decomposed in three stages in the range of 360–500 °C, with a mass loss between 67 and 96% (Yousef et al. 2021). Some studies have analyzed both the combustion and pyrolysis of this waste. Their heat of combustion and thermogravimetry show that most heat is generated above 450 °C, obtaining a heating value of 45 kJ/g. Therefore, the combustion of ·PFM waste should be carried out at 450–750 °C (Szefer et al. 2021). Pyrolysis of 3PFM waste produces a carbon-rich and oxygen-deficient liquid oil with a high heating value (HHV) of 43.5 kJ/g.

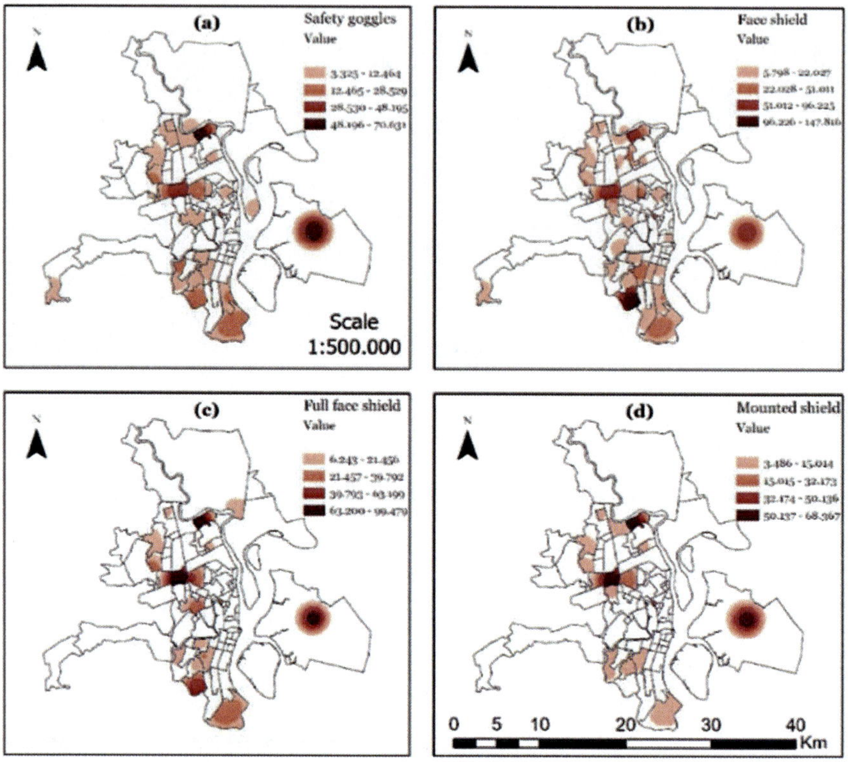


**Fig. 72.4** Quantity of face protectors yearly geospatial distribution by type

## 72.5 Conclusions

The quantification, geospatial spreading, and thermochemical valorization of personal protective equipment waste were performed for the city of Guayaquil. The overall daily face masks, pairs of gloves, and face protectors waste generation for the town reach 816,673.25 masks, 43,173.34 glove pairs, and 29,148.88 face protectors. Considering that the standard weight of one conventional face mask is five grams on average, this number of face masks becomes approximately four tons of daily waste daily, or 1.5 thousand tons each year. The daily generation of face masks, pairs of gloves, and face protectors is approximately 817 thousand, 43 thousand, and 29 thousand. Three-ply surgical disposal (3PFM) is the most wasted for face masks, accounting for 52.31%, followed by KN/FPP masks with 15.36%. 3PFM in El Fortin, La Florida, Trinitaria, and Guasmo. Given that the vast majority of thermal electricity production plants are based on the combustion of fuels, whether liquid or solid, this is the most immediate solution for the PPE waste problem (Salema et al. 2022).

## References


Alcaraz JP, Le Coq L, Pourchez J et al (2022) Reuse of medical face masks in domestic and community settings without sacrificing safety: ecological and economical lessons from the Covid-19 pandemic. Chemosphere. https://doi.org/10.1016/j.chemosphere.2021.132364

Dharmaraj S, Ashokkumar V, Hariharan S, Manibharathi A, Show PL, Chong CT, Ngamcharussrivichai C (2021) The COVID-19 pandemic face mask waste: a blooming threat to the marine environment. Chemosphere 272:129601

Environmental Systems Research Institute (2021) ArcGIS Pro. https://www.esri.com/en-us/arcgis/products/arcgis-pro/overview

Haque MS, Sharif S, Masnoon A, Rashid E (2021) SARS-CoV-2 pandemic-induced PPE and single-use plastic waste generation scenario. Waste Manag Res 39:3–17

INEC (2010) Resultados del Censo 2010 de población y vivienda en el Ecuador: Fascículo Provincial Guayas. https://www.ecuadorencifras.gob.ec/wp-content/descargas/Manu-lateral/Resultados-provinciales/guayas.pdf

Jung S, Lee S, Dou X, Kwon EE (2021) Valorization of disposable COVID-19 mask through the thermo-chemical process. Chem Eng J 405:126658

Kumar A, Samadder SR, Kumar N, Singh C (2018) Estimation of the generation rate of different types of plastic wastes and possible revenue recovery from informal recycling. Waste Manag 79:781–790

Kumar H, Azad A, Gupta A, Sharma J, Bherwani H, Labhsetwar NK, Kumar R (2021) COVID-19 Creating another problem? Sustainable solution for PPE disposal through LCA approach. Environ Dev Sustain 23:9418–9432

Mora-Araus M, Velastegui-Montoya A, Jaramillo-Lindao Y, Apolo H (2021) Mapping the sound landscape during social isolation due to COVID-19. In: 2021 IEEE international geoscience remote Sensing symposium. IGARSS. IEEE, Brussels, pp 8340–8343

Nghiem LD, Morgan B, Donner E, Short MD (2020) The COVID-19 pandemic: considerations for the waste and wastewater services sector. Case Stud Chem Environ Eng 1:100006

Nzediegwu C, Chang SX (2020) Improper solid waste management increases potential for COVID-19 spread in developing countries. Resour Conserv Recycl 161:104947

Patrício Silva AL, Prata JC, Walker TR, Duarte AC, Ouyang W, Barcelò D, Rocha-Santos T (2021) Increased plastic pollution due to COVID-19 pandemic: challenges and recommendations. Chem Eng J 405:126683

Salema AA, Mohd Zaifullizan Y, Wong WH (2022) Pyrolysis and combustion kinetics of disposable surgical face mask produced during Covid-19 pandemic. Energy Sour Part A Recover Util Environ Eff 44:566–576

Sangkham S (2020) Face mask and medical waste disposal during the novel COVID-19 pandemic in Asia. Case Stud Chem Environ Eng 2:100052

Szefer EM, Majka TM, Pielichowski K (2021) Characterization and combustion behavior of single-use masks used during covid-19 pandemic. Materials. https://doi.org/10.3390/ma14133501

Torres FG, De-la-Torre GE (2021) Face mask waste generation and management during the COVID-19 pandemic: an overview and the Peruvian case. Sci Total Environ 786:147628

WHO (2020a) Statement regarding cluster of pneumonia cases in Wuhan, China. https://www.who.int/china/news/detail/09-01-2020-who-statement-regarding-cluster-of-pneumonia-cases-in-wuhan-china. Accessed 19 May 2020

WHO (2020b) Report: Laboratory testing of human suspected cases of novel coronavirus (nCoV) infection. https://apps.who.int/iris/bitstream/handle/10665/330374/WHO-2019-nCoV-laboratory-2020.1-eng.pdf. Accessed 19 May 2020

Yousef S, Eimontas J, Striūgas N, Abdelnaby MA (2021) Pyrolysis kinetic behaviour and TG-FTIR-GC–MS analysis of coronavirus face masks. J Anal Appl Pyrolysis. https://doi.org/10.1016/j.jaap.2021.105118

# Chapter 73
# A Brief Systematic Review of the Literature on the Barriers and Solutions of Renewable Energy Acceleration in Malawi

**Sylvester William Chisale and Han Soo Lee**

**Abstract** Malawi's low electrification rate (18%) is pitiful, given the country's abundant energy resources. Several barriers to renewable energy installation were identified through a review of the literature. The barriers were classified as sociocultural, economic, market, technical, environmental, ecological, and geographical, as well as political-governmental barriers. The barriers were further discussed with some government and non-governmental organizations initiatives to overcome them. The study also discussed proposed solutions to the barriers and suggested further research into these issues. Thus, if the solutions are implemented substantially, the country's access to electricity will improve. Globally, Malawi will also be a part of the worldwide trend toward the use of renewables and the fight against climate change and global warming caused by greenhouse gas emissions.

**Keywords** Renewable energy · Barriers · Electricity · Malawi

S. W. Chisale (✉) · H. S. Lee
Transdisciplinary Science and Engineering, Graduate School of Advanced Science and Engineering, Hiroshima University, 1-5-1 Kagamiyama, Higashi-Hiroshima, Hiroshima 739-8529, Japan
e-mail: schisale@must.ac.mw

S. W. Chisale
Department of Applied Studies, Malawi University of Science and Technology (MUST), P. O. Box 5196, Limbe, Malawi

H. S. Lee
Center for Planetary Health and Innovation Science (PHIS), The IDEC Institute, Hiroshima University, Hiroshima, Japan

 
N. S. Caetano and M. C. Felgueiras (eds.), *The 9th International Conference on Energy and Environment Research*, Environmental Science and Engineering,
https://doi.org/10.1007/978-3-031-43559-1_73

## 73.1 Introduction

Malawi is a landlocked country in Africa's southern hemisphere. It covers a total area of 94,280 km$^2$. Malawi has a population of roughly 19 million people (World Population Review 2021). Malawi's electricity rate has recently increased from 18%, with only 12% of the population connected to the grid. According to the 2018 census, only 37% of the urban and 2% of the rural population have access to electricity (Mkumba et al. 2021).

The total installed capacity for Malawi is about 630.55 MW as of 2021. Hydropower has a huge contribution towards electricity in Malawi with total installed capacity of 372 MW. Big solar PV power plants have just been operationalized with installed capacity of 60 MW and 20 MW in Salima and Dedza, respectively (Tisheva 2021). Malawi chose to build diesel peaking generators in its cities as a quick solution to regular power disruptions. This resulted in a total installed capacity of 141.5 MW for diesel-related electricity and other sources. Biomass additionally contributes 18.5 MW to the total capacity deployed (USAID 2021). Figure 73.1 shows the electricity installed capacity, capacity utilization and net capacity change.

Malawi's low electrification rate is driving the government to implement a rural electrification initiative to ensure universal energy access. Malawi's grid, on the other hand, is quite old, and the government is tasked to maintain the aging national grid. Furthermore, several power plants are obsolete, and several attempts are currently ongoing to rehabilitate them (CESET Team 2020). For the past decade, Malawi experience frequent power outages which is more persistent during dry season due to dwindling water levels (ESCOM 2015). On the other hand, Malawi has vast energy resources that can be used to improve electrification rate. These resources include solar, wind, hydropower potential, bioenergy, and uranium.

Malawi has increased its usage of renewable energy technologies (RETs) for electricity generation in response to rising electricity demand brought on by population growth and urbanization, as shown in Fig. 73.2 for the years 2011–2016 (Macrotrends 2022). However, due to continued rising energy demand, non-renewable energy use, such as diesel, increased in 2017 and 2018, as shown in Fig. 73.1c, bringing the overall contribution to 22%, as shown in Fig. 73.1a. However, electrification rate

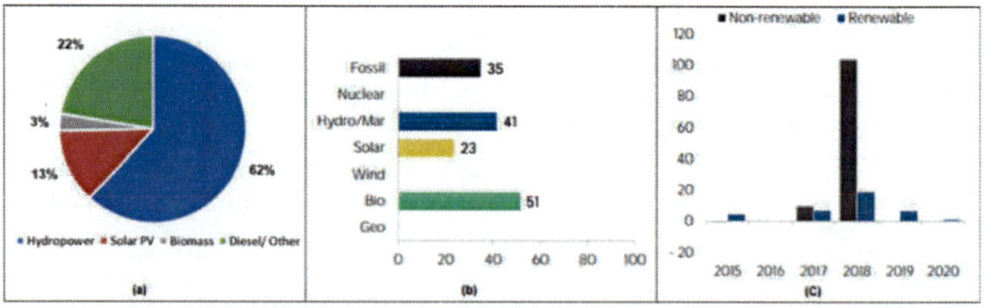


**Fig. 73.1** **a** Electricity installed capacity; **b** capacity utilization in 2019 (%); **c** net capacity change (MW) in Malawi (IRENA 2021)

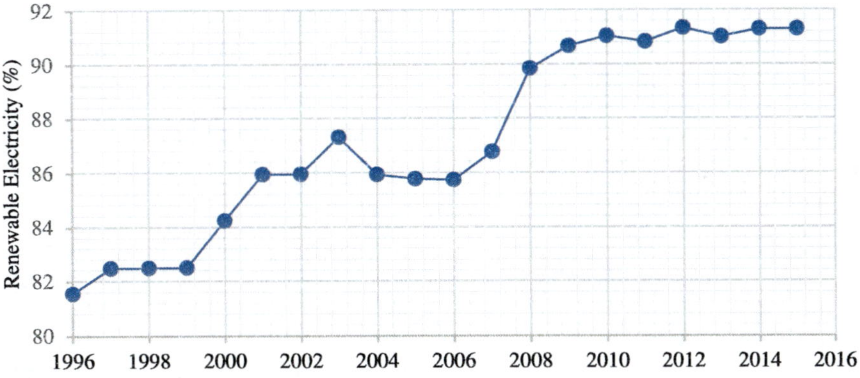


**Fig. 73.2** Electricity generation share from renewables (1996–2015) including hydropower, biomass, and solar PV

is very low (18%) regardless of Malawi facing energy hunger and having abundant renewable energy (RE) resources. Furthermore, as depicted in Fig. 73.1b, other RE resources such as solar, hydropower, and wind are still underutilized. Therefore, this study discusses the barriers and solutions to deploying RE in Malawi.

## 73.2 Barriers to Renewable Energy Installation

The literature review examines the challenges associated with the implementation of REs in Malawi and other developing countries in Africa. These barriers have been categorized into sociocultural, economic, market, technical, environmental, ecological, and geographical, and political-governmental barriers.

### *73.2.1 Economic Barriers*

*Lack of funding*: Over the years Malawi government has been funding its energy sector through budget allocation, grants, and concessional financing (Borgstein et al. 2019). As a result, the funding gap has grown, making it insufficient to meet future needs. Among the major non-governmental organizations that have contributed to the funding of many projects are: United Nations Development Programme (UNDP), Department for International Development (DfID), African Enterprise Challenge Fund (AECF), United States Agency for International Development (USAID) and the World Bank (UNDP 2020).

*High investment cost*: Despite rapid cost reductions around the world, RE developers in Africa continue to face high upfront costs with 20–25 years of payback period.

This, however, differs from country to country (IRENA, GIZ, and KfW Development Bank 2020; Eales et al. 2020a). The maintenance costs of several RE technologies are high because the equipment is not manufactured in the country. Furthermore, Malawi relies on expatriates to carry out most major maintenance activities. All these factors contribute to a high overall investment cost (CESET Team 2020). Malawi's government also emphasized the high level of investment in most RE projects (Government of Malawi 2003).

*Lack of financing mechanisms and incentives*: Malawi on its own has fewer financing institutions, grants, and incentives. However, in its capacity as a low-income country, it introduced zero rate VAT and removed import duty and excise tax on some RE related products (MRA 2022, 2019). It also imposed a 4.5% levy on energy sales such as electricity and petroleum products to fund the Malawi rural electrification program (MAREP) through RE (UNDP 2020). As an incentive to investors, government introduced a policy on Feed-in tariff (MERA 2012). These kind of challenges in most African countries triggered international organizations to provide support programs. For instance, USAID launched an off-grid facility through its Power Africa program to catalyze the market. Similarly, the World Banks' Off-Grid Market Development Fund supports a growing off-grid market sector in Malawi through financing components such as a mini-grid facility, resolving working capital loans and Recreational Boating Facilities (RBF) grant facility (UNDP 2020).

### 73.2.2 Technical Barriers

*Lack of infrastructure and technology*: Technologically, Malawi also lacks manufacturing capacity for components of RETs, resulting in high import costs (Government of Malawi 2003). Good physical and social infrastructure, such as roads, grid network, and communication systems, is one of the factors that attract investment in RE (Chirambo 2016). Household location has a clear impact on access to electricity. The most obvious reason is a lack of distribution infrastructure requirements and a poor road network. Thus, there is need to modernize the grid and mini-grid infrastructure to ensure flexibility and stability (IRENA, GIZ, and KfW Development Bank 2020). Furthermore, people in rural areas tend to be low-income, making it difficult for utilities to expand their infrastructure (UNDP 2020).

*Intermittency nature of energy*: RE generation is, to some extent, intermittent. Solar resources, for example, vary daily and seasonally, as does water volume for hydropower, and the availability of biomass, such as maize husk, varies with the harvest season (Numata et al. 2020). A power backup system is critical for intermittent energy sources. For example, a battery is required for a solar system because solar power is unavailable at night and on cloudy days. (Chisale and Mangani 2021). However, battery have short life span and expensive in Malawi because they are imported making it difficult to people to access the REs.

*Lack of skilled human resource*: Malawi has a scarcity of skilled labour in the installation and maintenance of RETs such as solar PV, which has clearly resulted in the failure of most installations (Eales et al. 2020a). Furthermore, human resources for the development and implementation of microgrids are scarce at all levels (Chirambo 2016).

*Lack of spare parts*: Energy system supply and maintenance are likely to be determined by the presence of a local market for spare and replacement products. Most of the work on renewable energy deployment has been concentrated in rural areas because most spare parts are found in urban areas, making it more difficult and costly (CESET Team 2020).

### *73.2.3 Market Barriers*

*Small market and Low demand*: Most communities in Malawi are low-income with little or no high energy consuming appliances resulting in small energy market (IRENA and AfDB 2022). The financial viability of solar microgrids is a major challenge particularly for low-income communities because the market for solar products in Malawi is less developed than in other countries in Africa (CESET Team 2020).

*Lack of consumer paying capacity and poverty*: Widespread poverty and income inequality result in consumers failing to pay for electricity (Zalengera et al. 2015). However, global dialogues consider electricity as a right. Hence, rural areas must also have access to electricity. Unfortunately, in Malawi lack of consumer paying capacity hinders ESCOM to expand the national grid to the rural areas (Gondwe 2007). This challenge in Malawi is mainly due to lack of economic incentives and low ability to pay (McCauley et al. 2022).

*Lack of R&D*: Malawi lacks market and technical data on various RETs. Research into the RE can assist investors in understanding the market situation and making an informed investment decision (Eales et al. 2020a). The Malawi government recognizes the value of R&D in supporting a diverse sector, which includes novel energy generation technologies, new off-grid solutions, energy storage, interconnection, and grid management (Government of Malawi 2017). Lack of RE R&D at all levels of society, including policymakers, decision makers, and the public, prevents further investments in RE (Government of Malawi 2003).

### *73.2.4 Political-Governmental Barriers*

*Lack of political commitment*: Malawi has tried to put in place measures to increase RE absorption through various policy reviews and introducing programs (Zalengera et al. 2014). However, when in power, political leaders implement development agendas that they will present to the people to be re-elected (Zalengera et al. 2014). On the other hand, there has been an increase in the number of projects that have been neglected because of leadership changes. This has an indirect impact on large-scale renewable energy deployment. However, as literature suggests, a strong political will is required to start addressing the existing measures, overcome rigidity, and increase electricity generation through renewables (Chirambo 2016; O'Reilly et al. 2015).

*Corruption, nepotism, and favoritism*: Most government departments now require people to bribe to obtain necessary approvals, documents, and certificates. It is difficult for stakeholders to invest in and obtain a project unless they have committed to corrupt government officials (Anders et al. 2020). Furthermore, most projects are assigned to people who have received special favors from leaders and are affiliated with a ruling party.

*Lack of coordination among local institutions*: There are several government departments and non-governmental organizations carrying out RE projects in Malawi. Additionally, different programs have been introduced to curb the challenges in energy sector such as MAREP and barrier removal for renewable energy in Malawi (BARREM) project and department of energy affairs have all actively participated. However, regulatory bodies such as MERA, MBS and other government departments have failed to coordinate to ensure quality standards and conducive environment for investors (Zalengera et al. 2014). Currently, existing microgrid systems are managed, operated, and maintained by the developer institutions in the absence of a relevant regulatory framework (Eales et al. 2020a). Furthermore, because of the ambiguity of microgrid ownership and governance, accountability and transparency procedures are not clearly defined, making regulation difficult. Malawi's regulatory environment is not yet conducive to private sector investment in mini grids despite efforts to rectify it (UNDP 2020).

### 73.2.5 Sociocultural Barriers

*Public awareness and information barrier*: A country's renewable energy adoption can be accelerated if the public understands renewables in terms of their desires. Lack of knowledge about renewable energy technologies and their financial benefits has a negative impact on RE adoption. Malawians rely heavily on biomass for cooking (Zalengera et al. 2014). People, on the other hand, find it difficult to use other technologies, such as solar, due to a lack of information on the environmental and health effects, as well as its low cost (Government of Malawi 2003; IRENA and AfDB 2022; Government of Malawi and UNDP 2020).

*Poverty*: Many Malawians live in rural areas and are impoverished, posing a barrier to wider RE adoption (Gamula et al. 2013; Kaunda 2013). Poverty prevents people from purchasing alternatives to their way of life such as solar system over kerosine.

Additionally, for mini grid, income levels have a bearing on bill payments (Jessel et al. 2019). However, energy sector growth has been suggested to be a critical prerequisite for the country's development and poverty alleviation. (Chirambo 2016).

*Cultural and gender issues*: Community projects such as mini grid requires community acceptance to ensure sustainability and implementation of the project (Numata et al. 2020). On the other hand, innovations like biogas generation from a pit toilet were rejected in Malawian society due to strong cultural concerns (Government of Malawi 2003). Additionally, women have traditionally performed most household energy-related tasks. Hence for biogas digester, women oversee management of the digester.

### 73.2.6 Ecological and Geographical Barriers

*Country's location*: Most renewable energy products come from China and other countries. Malawi, unfortunately, is a landlocked country with no access to a seaport. Thus, cargo charges, port fees, and transportation from the port to Malawi all contribute to the high overall cost of technologies (WorldBank 2008).

*Transport problems*: Most renewable energy projects, such as solar PV-based mini grids, are in rural areas. Unfortunately, rural areas have very poor road network, and transportation costs are relatively high (Ghimire and Kim 2018; Eales et al. 2020b). These difficulties can have a direct impact on product delivery, making maintenance and installation more expensive (Government of Malawi 2003).

*Scattered houses*: In rural areas of Malawi, most people live according to their relations making houses scattered. This makes it difficult for technologies that requires grid connection (Solangi et al. 2021).

## 73.3 Suggested Solutions to the Barriers

*Financial incentives*: Loans, grants, subsidies, and other financial incentives have been used around the world to attract RE investors (IRENA and AfDB 2022). Capital subsidies maybe applied at the initial stage of project while others are performance-based ones (Solangi et al. 2021). Renewables, on the other hand, are currently being deployed to help people meet their basic energy needs using subsidies (Government of Malawi 2017). As a result, it is recommended that subsidies be applied to countries that have two-thirds of the world's poorest people (IRENA and AfDB 2022). For instance, Malawi's mini grids are subsidized for rural people to afford electricity (Borgstein et al. 2019). For accelerated RE development, the government should expand methods of financing RE and energy efficiency programs such as grants, tax credits and loans (IRENA, GIZ, and KfW Development Bank 2020). With the introduction of the fuel levy for MAREP, there is a need for additional financing mechanisms for RE (Government of Malawi 2017).

*Coordination among Institutions*: Malawi has several departments and institutions involved in the energy sector. Coordination is required, for example, between MERA and MBS for the registration and certification process. Similarly, there must be coordination among government departments such as the DoEA, research institutions, EGENCO, ESCOM and Government itself and other development partners (IFC 2020).

*Direct, Enabling, and Integrating Policies*: Malawi has a lot of RE resources, so there's a need for direct, enabling, and integrating policies. These policies reassure the public and investors about the country's commitment to renewable energy (IRENA and AfDB 2022; Government of Malawi 2017). It is also necessary to integrate policies for specific energy consumers such as transportation, heating, cooling, and the power sector (Solangi et al. 2021).

*Carbon pricing*: Malawi can also use carbon pricing as a climate change mitigation policy, which will help absorb more renewables (IRENA and AfDB 2022; Solangi et al. 2021). For instance, countries like South Africa and China have introduced carbon pricing which can be used in RE projects (Eales et al. 2020a). However, there is a need to strike a balance between carbon pricing and encouraging investors to bring in more industries for the country's development.

*Public Awareness-Raising campaign*: Regardless of Malawi's energy challenges, most people are unaware of the financial and technical benefits of various RETs. As a result, the government must step up its public awareness campaign to promote the benefits of RETs (IRENA, GIZ, and KfW Development Bank 2020; IRENA and AfDB 2022; Government of Malawi 2017).

*Net metering*: Malawi implemented a Feed-in Tariff to ensure the purchase of renewable energy at a predetermined fixed tariff for a given period (MERA 2012). Net metering, on the other hand, has proven to be an effective method of increasing renewable electricity injection into the grid. Several African countries, including Rwanda, Senegal, South Africa, and Tanzania, have implemented net metering policies (IRENA and AfDB 2022). Furthermore, net metering allows small-scale producers with excess capacity to make electricity available to other users (Chisale et al. 2022).

*RE Targets, Quotas, and Mandates*: Regulatory policies, such as quotas, mandates, and targets, motivate the country's investments in the power sector. As a result, there is a need to establish goals and mandates for RE, which should be communicated to stakeholders and the private sector (Solangi et al. 2021).

*Training and Capacity building*: Malawi must increase RE human resources by expanding existing universities, technical colleges, and special training programs (Government of Malawi 2017). The primary goal of capacity building is to develop strong leadership, business modeling, marketing, partnership and collaboration, and governance skills among employees (Solangi et al. 2021). Capacity building in RE is required at all levels, including operation, maintenance, installation, management, and policy development (IRENA, GIZ, and KfW Development Bank 2020; IRENA and AfDB 2022). Malawian universities and technical colleges are currently implementing renewable energy-related programs (Eales et al. 2020a).

*Research and knowledge sharing*: Research is required in all aspects of RE, including the technical and economic levels. Universities and other research institutions must increase their research efforts to provide investors and the public with the necessary information about RE (IRENA and AfDB 2022; Government of Malawi 2017; Solangi et al. 2021).

## 73.4 Conclusion

Malawi has recently increased its use of fossil fuels for electricity generation. Nonetheless, the capacity utilization of several energy resources such as hydropower, solar, wind, and geothermal remains very low. However, because of government and international organization efforts to overcome barriers, the country's use of RE resources has increased proportionally. Therefore, more research into the barriers and solutions to RE deployment is required to increase the installation of RETs, which will improve Malawi's access to electricity. Globally, Malawi will also be a part of the worldwide trend toward the use of renewables and the fight against climate change and global warming caused by greenhouse gas emissions.

## References

Anders G, Kanyongolo FE, Seim B (2020) Corruption and the impact of law enforcement: insights from a mixed-methods study in Malawi. J Mod Afr Stud 58(3):315–336. https://doi.org/10.1017/S0022278X2000021X

Borgstein E, Santana S, Li B, Wade K, Wanless E (2019) Malawi sustainable energy investment study: summary for decision makers [online]. Available https://rmi.org/insight/malawi-study/

CESET Team (2020) Community energy in Malawi: an annotated bibliography. Sheffield, CESET, Sheffield

Chirambo D (2016) Addressing the renewable energy financing gap in Africa to promote universal energy access: integrated renewable energy financing in Malawi. Renew Sustain Energy Rev 62:793–803. https://doi.org/10.1016/j.rser.2016.05.046

Chisale SW, Mangani P (2021) Energy audit and feasibility of solar PV energy system: case of a commercial building. J Energy 2021:5544664. https://doi.org/10.1155/2021/5544664

Chisale SW, Chisanu L, Macheso PSB, Chikabvumbwa SR, Sibale D (2022) Investigation of net metering as a tool for increasing electricity access in Malawi. Int J Renew Energy Technol 13(1):66–83. https://doi.org/10.1504/IJRET.2022.120334

Eales A, Alsop A, Frame D, Strachan S, Galloway S (2020a) Assessing the market for solar photovoltaic (PV) microgrids in Malawi. J Sustain Res 2(1). https://doi.org/10.20900/jsr20200008

Eales A, Frame D, Coley W, Bayani E, Galloway S (2020b) Sustainable delivery models for achieving SDG7: lessons from an energy services social enterprise in Malawi. In: 2020b IEEE global humanitarian technology conference (GHTC), pp 1–8.https://doi.org/10.1109/GHTC46280.2020.9342877

ESCOM (2015) Water levels and the energy situation. Electricity Supply Corporation of Malawi (ESCOM)

Gamula GET, Hui L, Peng W (2013) An overview of the energy sector in Malawi. Energy Power Eng 5:8–17. https://doi.org/10.4236/epe.2013.51002

Ghimire LP, Kim Y (2018) An analysis on barriers to renewable energy development in the context of Nepal using AHP. Renew Energy 129:446–456. https://doi.org/10.1016/j.renene.2018.06.011

Gondwe K (2007) Barrier removal to renewable energy in Malawi (MLW99G31) draft mid term review. Lilongwe, June 2007. Accessed 03 Mar 2022 [online]. Available https://www.eartheval.org/sites/ceval/files/evaluations/44%20Barrier%20Removal%20to%20Sustainable%20Energy%20Malawi.pdf

Government of Malawi (2003) Malawi's climate technology transfer and needs assessment under united nations framework convention on climate change (UNFCCC)—expedited phase II. Lilongwe

Government of Malawi and UNDP (2020) Access to clean and renewable energy (ACRE) project 2020–2023. Accessed 01 Mar 2022 [online]. Available https://erc.undp.org/evaluation/managementresponses/keyaction/documents/download/6496

Government of Malawi (2017) Malawi renewable energy strategy. Accessed 08 Mar 2022 [online]. Available http://conrema.org/wp-content/uploads/2019/01/Malawi-Renewable-Energy-Strategy-_Final-.pdf

IFC (2022) Regulatory and tariff review for distributed generation in the commercial and industrial sectors in Southern Africa, Washington. Accessed 10 Mar 2022 [online]. Available https://www.ifc.org/wps/wcm/connect/090c58a2-2b98-482e-8c6d-b5931ed793e2/202006-Regulatory-Tariff-Review-Southern-Africa.pdf?MOD=AJPERES&CVID=nbDqlVa

IRENA (2022) Energy profile—Malawi. Masdar City, Abu Dhabi, 2021. Accessed 12 May 2022 [online]. Available https://www.irena.org/IRENADocuments/Statistical_Profiles/Africa/Malawi_Africa_RE_SP.pdf

IRENA and AfDB (2022) Renewable energy market analysis: Africa and its region. Abu Dhabi and Abidjan. Accessed 02 Mar 2022 [online]. Available https://irena.org/-/media/Files/IRENA/Agency/Publication/2022/Jan/IRENA_Market_Africa_2022.pdf?la=en&hash=BC8DEB8130CF9CC1C28FFE87ECBA519B32076013

IRENA, GIZ, and KfW Development Bank (2020) The renewable energy transition in Africa: powering access, resilience and prosperity

Jessel S, Sawyer S, Hernández D (2019) Energy, poverty, and health in climate change: a comprehensive review of an emerging literature. Front Public Health 7 [online]. Available https://doi.org/10.3389/fpubh.2019.00357

Kaunda CS (2013) Energy situation, potential and application status of small-scale hydropower systems in Malawi. Renew Sustain Energy Rev. https://doi.org/10.1016/j.rser.2013.05.034

Macrotrends (2022) Malawi renewable energy 1990–2022. https://www.macrotrends.net/countries/MWI/malawi/renewable-energy-statistics. Accessed 12 May 2022

McCauley D, Grant R, Mwathunga E (2022) Achieving energy justice in Malawi: from key challenges to policy recommendations. Clim Change 170(3–4):1–22. https://doi.org/10.1007/s10584-022-03314-1

MERA (2012) Malawi feed-in tariff policy: renewable energy resource generated electricity in Malawi. Accessed 01 Mar 2022 [online]. Available http://conrema.org/wp-content/uploads/2019/01/Malawi-feed-in-tariff-policy-final.pdf

Mkumba F, Phillips E, Sosis K, Kahinga E (2021) Stand alone solar (SAS): market update Malawi, Nairobi [online]. Available https://www.ace-taf.org/wp-content/uploads/2021/03/Stand-Alone-Solar-SAS-Market-Update-Malawi.pdf

MRA, "Malawi Revenue Authority," (2019) Amendments to the customs and excise (TARIFFS) ORDER, 12 Sept 2019. https://www.mra.mw/press-releases/amendments-to-the-customs-and-excise-tariffs-order. Accessed 01 Mar 2022

MRA, "Malawi Revenue Authority," (2022) No duty on solar lamps, sanitary pads, 22 Feb 2022. https://www.mra.mw/news/no-duty-on-solar-lamps-sanitary-pads. Accessed 01 Mar 2022

Numata M, Sugiyama M, Mogi G (2020) Barrier analysis for the deployment of renewable-based mini-grids in Myanmar using the analytic hierarchy process (AHP). Energies 16(3). https://doi.org/10.3390/en13061400

O'Reilly S, Young EW (2015) MREAP: policy options to support energy for development (dedicated study). Sheffield, Mar 2015. Accessed 07 Mar 2022 [online]. Available http://www.strath.ac.uk/eee/energymalawi/

Solangi YA, Longsheng C, Shah SAA (2021) Assessing and overcoming the renewable energy barriers for sustainable development in Pakistan: an integrated AHP and fuzzy TOPSIS approach. Renew Energy 173:209–222. https://doi.org/10.1016/j.renene.2021.03.141

Tisheva P (2021) RenewableNow. Malawi switches on 60-MW maiden solar plant. https://renewablesnow.com/news/malawi-switches-on-60-mw-maiden-solar-plant-761885/. Accessed 23 Dec 2021

UNDP (2020) Making access possible: energy and the poor: unpacking the investment case for clean energy in Malawi, New York [online]. Available https://www.undp.org/sites/g/files/zskgke326/files/publications/UNDP-UNCDF-Malawi-Energy-and-the-Poor.pdf

USAID (2021) Malawi power Africa fact sheet. Malawi Energy Sector Rev. https://www.usaid.gov/powerafrica/malawi. Accessed 23 Dec 2021

World Population Review (2021) Malawi population 2021 (live). https://worldpopulationreview.com/countries/malawi-population. Accessed 29 Dec 2021

WorldBank (2008) The World Bank. Landlocked countries: higher transport costs, delays, less trade, 16 June 2008. https://www.worldbank.org/en/news/feature/2008/06/16/landlocked-countries-higher-transport-costs-delays-less-trade. Accessed 08 Mar 2022

Zalengera C, Blanchard RE, Eames PC, Juma AM, Chitawo ML, Gondwe KT (2014) Overview of the Malawi energy situation and A PESTLE analysis for sustainable development of renewable energy. Renew Sustain Energy Rev 38:335–347. https://doi.org/10.1016/j.rser.2014.05.050

Zalengera C, Blanchard R, Eames PC (2015) Putting the end-user first: towards addressing contesting values in renewable energy systems deployment for low-income households—a case chapter 9. Springer International Publishing. https://doi.org/10.1007/978-3-319-20209-9

# Part IX
# Education for Sustainable Development

# Chapter 74
# Perceptions of Domestic Gas Consumption: Effects on the Economy, Urbanization Process and Environmental Proposal

**Silvia Magdalena Coello Pisco, Jose Armando Hidalgo Crespo, Benigno Antonio Rodriguez Gomez, Yomar Alexander González Cañizales, and Leonardo Alvaro Banguera Arroyo**

**Abstract** The purpose of this study was to explore, through the semi-structured survey technique, the perception of LPG gas consumption for domestic use in the canton of Guayaquil–Ecuador, with high rates of non-renewable energy consumption. Through a bibliographic study and relying on urban sustainability and the data obtained from the ministries of energy and statistics of Ecuador, the relationships between the increase in domestic gas consumption, its effect on the country's economy and on the urbanization process. In the discussion of the results, we indicate that the families of the Guayaquil canton perceive that the highest energy cost is found in the electricity sector; the dominant energy option (use of fossil energy) with many negative effects on society and the environment.

**Keywords** Survey of perception · Consumption of domestic gas · Energy cost · Urbanization process · Urban sustainability

S. M. Coello Pisco (✉) · J. A. Hidalgo Crespo · Y. A. González Cañizales · L. A. Banguera Arroyo
Facultad de Ingeniería Industrial, Universidad de Guayaquil, Av. Dr. Jiménez Lince y Av. Juan Tanca Marengo, 090501 Guayaquil, Ecuador
e-mail: silvia.coellop@ug.edu.ec

S. M. Coello Pisco · B. A. Rodriguez Gomez
Universidade da Coruña, Galicia, Spain



N. S. Caetano and M. C. Felgueiras (eds.), *The 9th International Conference on Energy and Environment Research*, Environmental Science and Engineering,
https://doi.org/10.1007/978-3-031-43559-1_74

## 74.1 Introduction

Ecuador for five years has been experiencing a progressive demand for consumption of liquefied petroleum gas (LPG), which represents a growing concern for the country's economy and the urbanization process (Biller and Nabi 2013). The Ecuadorian population essentially depends on gas of 15 kg where sometimes there is speculation with this supply of conventional energy, in certain marginal urban and rural sectors. According to the Statistical Bulletin of the Agency for the Regulation and Control of Hydrocarbons, the consumption of LPG gas for domestic use increased radically (DACE 2019). In such a way that over time, instead of decreasing said consumption, it increases. Compared to 2019, Petroecuador consigned more than 1198 million kilograms of liquefied petroleum gas (LPG) to the domestic (urban and rural), industrial, agro-industrial and vehicular sectors, which represented an increase of 3.5% compared to the previous year, that is, per year it dispatched 309,011,220 kg/year (DACE 2019; Vistazo Magazine 2020). The use of domestic gas is one of the main energy consumption problems in Ecuador, and the sector that demanded this product the most was domestic.

According to statistics from Petroecuador and DACE for this sector, 1063 million kilograms were shipped, which represents 88.64% of national consumption (Daily Trade 2020; Ministry of Hydrocarbons 2014). On the other hand, there is the problem of population growth in the main cities of the country. In addition, the reason for confinement due to the Sars-Covd-2 virus is added, families being confined to their homes increase the consumption of electricity (hydraulic) and gas (non-renewable energy) (Carreño et al. 2018; Carreño and Alfonso 2018). In other words, the expansion of economic development is a main factor that transfers: increase in urban waste, energy loss, environmental pollution as well as restriction in sustainable urban development (Burneo et al. 2020; Jia et al. 2017).

However, green technology is presented as a viable and economical solution, mainly in developing countries. Although, Ecuador is characterized by having good natural water resources; the processes of generating electricity from fossil resources do not include certain environmental costs, mainly in what refers to the phase of extraction of the raw material (transport, phase of electricity generation). With such background, the National Plan for Good Living is framed as a new model whose main focus is the satisfaction of needs that replaces the dominant paradigm of maximization of interests, based on perpetual economic growth. This has caused an overexploitation of natural wealth, generating: poverty, inequality and exclusion of certain sectors of the population. This improvement paradigm, supported by mass production, the waste of large amounts of raw material and energy, managed to separate the reciprocity between people and the cycles of the biosphere (Acosta and Martínez 2019).

With this in mind, Good Living requires promoting a joint regional perspective related to technology transfer and capacity building in favor of the Ecuadorian population and developing sectors, emphasizing clean technologies that are adapted to the national reality, which should contribute to the reduction of non-renewable resources, environmental impacts and improvement of sustainable development (EC CELAC

2012). Continuing with the idea, Carreño and Alfonso (2018) describe that the rapid demand for urban growth, its possible effects on the environment, the diversification of the industry and the market affect the increase in gross domestic product (GDP). Therefore, the consequences are the emissions of pollution indices that can negatively affect the urban sector (EC INEC National Institute of Statistics and Censuses INEC 2020). Based on these results, plus the problem of the Covid-19 pandemic that continues worldwide; it makes us think about looking for solutions that help us to gradually reduce dependence on the use of fossil energy. These events, which have been taking place for the last five years in the country, clearly show us these market failures (inefficient decisions), both on the part of producers and consumers, which have brought negative consequences such as: increase in waste in landfills, increased consumption of fossil energy, mainly domestic gas, generating an economic crisis in the energy sector, diseases, deterioration of the landscape, decrease in quality and/or quantity of certain productions and even in preventive expenses or mitigation costs, among other factors (Alfonso and Pardo 2012; Coello).

## 74.2 Materials and Methods

Given the qualitative nature of our inquiry, we opted for a descriptive-bibliographic type of research, with an investigative dimension by extension and a mixed measurement; that is, a non- experimental design. Secondary information was obtained through personalized interviews. The survey was validated by professionals familiar with the subject in the area of "energy techniques" through the "Expert Judgment" statistic and to measure its internal consistency, construct validity and reliability of the survey, the following statisticians were applied: V. Aiken and Alpha Cronbach (López and Fachelli 2015; Arquer 1994). The primary source was obtained from the Ministries of Energy and the Environment, as well as from the Ecuadorian Institute of Censuses and Statistics (INEC), to make a comparison with the data obtained in the consumer trends and perceptions survey; which allowed knowing the current state and the existing projections on the subject studied (EC INEC National Institute of Statistics and Censuses (INEC) 2019b). To demonstrate our assumption about: "LPG gas consumer trends and perceptions", we applied the Chi-Square Automatic Interaction Detector (CHAID) technique for market segmentation studies (Kinnear 1994; Calderón 2022; Díaz et al. 2020). Here is our assumption for the study.

### *74.2.1 Study Participants*

The consumption of domestic gas in the sectors of the Guayaquil canton has been affected in the last four years by the energy cost in the electricity sector, which means a negative contribution to the country's economy and the urbanization process.

An online survey of "Google forms" was developed between January 2 and February 12, 2021 to 1800 people. The sample error was 0.7%, only 1030 people answered the survey, discarding 60 participants because they had unanswered questions. There were a total of 970 responses from families, which represents 0.15% of the population of households in the city of Guayaquil. The characteristics of the families surveyed are described in Table 74.1. The questionnaire was structured according to the segmentation criteria (Table 74.2) and following the literature on consumer perception of energy, the environment, and pro-environmental attitudes. The questionnaire had 23 items (dichotomous questions and battery type); the dimensions are based on the context of the social phenomenon to be studied.

**Table 74.1** Characteristics of the sample of participants interviewed within the study of perception of LPG gas consumption in the canton of Guayaquil

| Gender | | Age | | | Level of education | | |
|---|---|---|---|---|---|---|---|
| Man | Woman | 20–35 | 35–50 | 50–65 | Basic | Middle | Higher |
| 124 | 852 | 167 | 680 | 123 | 347 | 526 | 97 |

**Table 74.2** Distribution of the questionnaire

| Structure | | Segmentation variable (criterion-objective) | |
|---|---|---|---|
| | | General | Specific |
| Consumption | Criterion | Direct get | Product usage, spending level |
| | Objective | Socioeconomic | Family economic income |
| | Subjective | Opinions and interests | Advantage, benefits, brand loyalty |
| Production | Objective | Social class and life cycle | Brand preference |
| | Subjective | Perceptions and inference (inferred) | Habits, benefits sought, latent attitudes (inferred) |
| Future scenario | Objective | Demographic | Social stratum, life cycle |
| | Subjective | Activities psychographic characteristics | Interest in the brand, device used, hours of consumption (inferred) |
| Characterization of the home | Objective | Demographic geographic | Household size, place of residence (inferred) |
| | Subjective | Psychographic, opinions and values | Lifestyle and behavior (inferred) |

Adapted by the author.
*Source* Principles of Marketing and Market Research (Díaz et al. 2020; Águeda and Madariaga 2008; Gordon and Langmaid 2022)

### *74.2.2 Structure of the Questionnaire*

The review of the last census on energy consumption, environmental economy and sustainable behavior of the Province of Guayas was carried out (EC INEC National Institute of Statistics and Censuses (INEC) 2019b). An analysis of the data obtained from the Guayaquil canton was carried out, which constitutes the basis for the elaboration of the questionnaire. The statistical instrument was adapted according to the social phenomenon to be investigated, using the "a priori market segmentation" technique (Águeda and Madariaga 2008). Due to the need to have simple and clear capture instruments, a structured questionnaire was designed through an integrated set of 23 questions on the perception, tendency and attitudes that people have about: the economy of the home and the country, habits personal energy consumption, environment, sustainable conduct. Such an instrument is elaborated in content and form, which allows us to obtain a high rate of response and acceptance at the household level. The type of questions was formalized based on different objective and subjective segmentation classification criteria (Table 74.2). There are closed questions, with a single answer, easy to formulate and tabulate; it also provides response alternatives, allowing the informant to respond freely to any of the options. The wording of the questionnaire is simple and consistent, avoiding distorting the answers and confusing the informant; rather, it allows the categories to be adapted to the knowledge and cultural level of the people providing the information (Águeda and Madariaga 2008; Fryn 2019). Next, we explain what each structure or chosen block of the survey consists of.

(a) Consumption structure: it is about inquiring about the use of domestic gas and electricity consumption (saving measures, knowledge of the price, taxes), identification of renewable and non-renewable energy sources.
(b) Production structure: the knowledge about the weight of the consumption of LPG gas and electricity in homes, the price of electricity for residential users, the identification of renewable and non-renewable energy sources and the problems associated with this and domestic gas production.
(c) Structure of the Future Scenario: the perception of Guayaquil families about the possibility of substituting part of the production of LPG gas (fossil source) and using a new device with the generation of organic bio-waste (food remains) and finally in the characterization of the Home: the importance of applying a new friendly technology (Biomass) is analyzed.
(d) Housing characterization structure: investigates society's perceptions of energy consumption problems and the advantages of using renewable energy for the urban environment compared to the use of a 15-kilo LPG gas cylinder (fossil fuels). Different socioeconomic data are collected from the interviewees (income, studies, composition of the family unit, etc.) and questions on environmental behavior (habits, beliefs and patterns of sustainable consumption) are included. This will allow an analysis of the structure of consumption and production of energy (gas and electricity) in the home.

## 74.3 Results and Discussion

In this study, four observations were made with respect to household gas and electricity consumption, which we describe below.

### 74.3.1 *Analysis of the Structure Characterization of the Home*

To carry out the validity of the construct and check the consistency of the instrument together with the assigned segmentations, we verify that if: "the expert criteria is: consistent", to apply the diagnostic survey on: "the perceptions of gas consumers domestic". The variables with which we are working, being qualitative (attitudes or perceptions), cannot be measured directly. That is, they must be assessed through indicators. It is now about examining the degree to which the defined indicators adequately measure the concept (construct) that is to be measured. Through the statisticians described in Table 74.3, an average score of 0.985 is obtained, which is a strong score. Such value indicates that our questionnaire is consistent, reliable and meets the parameters that we want to measure in the social phenomenon.

Regarding the characterization of the household in terms of psychographic segmentation, opinions and values, the data of the participating households was tabulated, that is, of the 970 families. In it we detail that 23.7% state they have a sustainable behavior in their homes; 36.5% of the interviewed households state that they practice some environmental program but only at the school type (age 7–16 years) but not in the community, that is, there are no practices of "green" consumption habits (see Table 74.4), that 138 families state that they have consumption habits based on clean energy, which represents 14.2%; what we can assess from now on in relation to your demand for LPG gas consumption, energy saving measures, knowledge of the advantages and disadvantages of using energy excessively.

Therefore, we can see that the knowledge about friendly technologies for the benefit of the environment is very low approximately 85.8% of the interviewees indicate that they do not know these types of friendly technologies to use in their homes. This may be in line with the record made by the Institute of Statistics and Censuses (INEC) which indicates that at the national level, 61.53% of Ecuadorian households classify or separate some type of waste (organic, paper, plastic, metal,

**Table 74.3** Qualitative validation of the personalized survey

| Validity | Punctuation | Statistician |
|---|---|---|
| Contents | 1.00 | Expert judgment |
| Criterion | 0.970 | Alfa Cronbach |
| Construct | 0.986 | V. Aiken |
| Full validity | 0.985 | |

**Table 74.4** Analysis of consumer perception (habits-sustainable behavior-knowledge). Guidelines for sustainable consumption and household habits

| Sustainable behavior | | | | Knowledge of clean energy | | Consumption habits (clean energy) | |
|---|---|---|---|---|---|---|---|
| Home | | School | | | | | |
| Yes | No | Yes | No | Yes | No | Yes | No |
| 230 | 740 | 354 | 616 | 108 | 862 | 138 | 832 |

etc.), which means that approximately six out of ten households do so (EC INEC National Institute of Statistics and Censuses (INEC) 2019b).

### 74.3.2 *Analysis of the Consumption Structure*

Guayaquil families used both energy resources (electric and gas); for example, only the gas cooker was used in 90.43% and the induction cooker in 9.67%. However, when asked how many times both appliances were used, it was found that 322 households used both appliances (86.74%); while 13.26% indicated that they used one or the other from time to time. The frequency of use of the 15 kg cylinder, used once a month is 73.2, 0.23% use 2 cylinders per month and 25.9% use up to three cylinders each month. Regarding the time of use for cooking food, it is concentrated in the peak hour with a time of 1 h (50.06%) with a consumption demand of 7.23% for the induction cooker; for consumption of 2 h there is 37.46% gas and 2.21% induction.

### 74.3.3 *Analysis of the Structure of the Future Scenario and Perception of the Home*

With regard to knowledge of other energy sources besides those based on fossil fuels, most of the respondents only know hydraulic energy as a source of electricity generation. Especially relevant for our case study is the widespread ignorance they have about domestic gas by not identifying it as a 100% non-renewable source. Therefore, 84.95% have a lack of knowledge about friendly clean energies; their ignorance about the use of Biomass as a renewable energy source also fits into a 100% ignorance. On the other hand, biomass and LPG gas are the only energy source that presents difficulties when it comes to being classified as renewable or non-renewable, perhaps due to lack of knowledge about it. These responses by the heads of households lead us to carry out an analysis of the structure of the future scenario of energy consumption of marginal urban families based on their study orientation as potential consumers.

Therefore, we show three questions that we consider interesting from our study that can be seen Fig. 74.1. These are classified according to market segmentation.

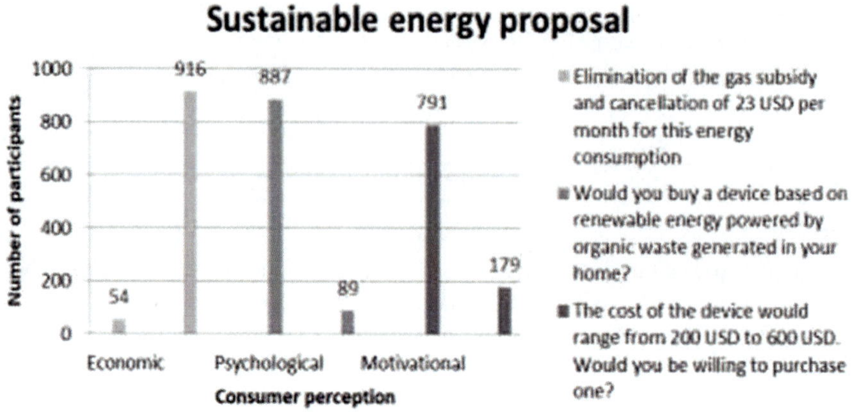


**Fig. 74.1** Analysis of consumer perception (motivational, psychological, emotional)

The results indicate that: if the domestic gas subsidy was eliminated, 82.03% would not be willing to pay more than 23 USD for a cylinder of LPG gas (economic segmentation). Families consider that such a price would negatively affect food consumption (basic family basket). 89.61% indicate that if there were a device that feeds on the waste they generate, they would agree to use it (psychological segmentation); while 79.11% would agree to pay a price of 200–600 USD for such substitution equipment and adaptation to their domestic kitchen (motivational segmentation). According to the technique "Automatic Interaction Detector with Chi-Square" (CHAID) for A priori segmentation studies, it shows us the corresponding tree diagram, where it can be seen that families of low social status (41.55%) are willing to acquire an anaerobic digester and belong to the group with a medium educational level (high school education) corresponding to 61.38%, where 51.03% have six people in the household. In these families, the people interviewed corresponds to 87.22% of the female sex (see Fig. 74.2).

## 74.4 Discussion

According to the results, it can be delimited that a large part of the participating families from the marginal urban area perceives that the highest energy cost is in the electricity sector; the dominant energy option (use of fossil energy) with many negative effects on society and the environment. Therefore, we wonder if the population of Guayaquil would be willing to compensate for these costs through a scenario of partial replacement of fossil fuels (household LPG gas) by renewables (residual biomass) in our energy generation processes. The devices with the highest demand are: electric stoves, induction, electric ovens, and refrigerators, among others. This, in part, responds to the fact that the Government decided to lower the tax on special

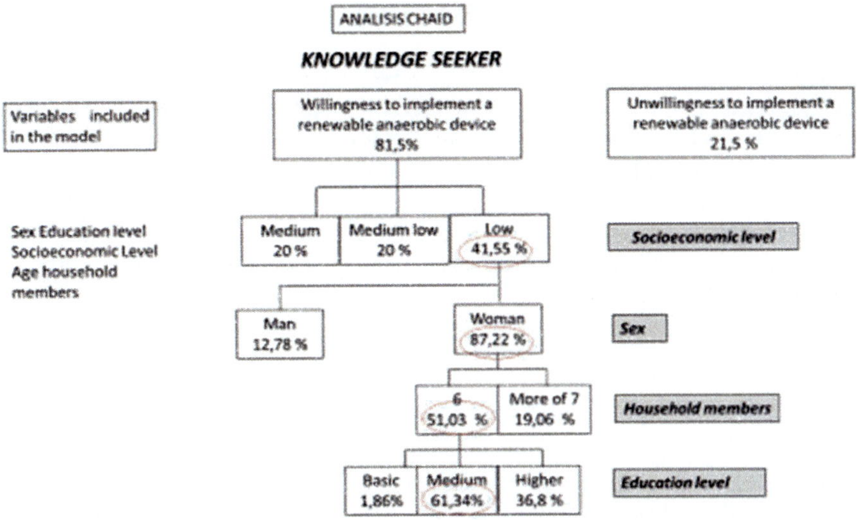


**Fig. 74.2** Profile of those who liked a new concept, product, based on renewable energy

consumption of this device (induction cookers) to 0%. Faced with this government decision, the president of the College of Electrical and Electronic Engineers of Pichincha, Ing. Salinas Fernando, stated: "When politicians take the risk of withdrawing the domestic gas subsidy, people will make the decision to use induction cookers and this way the State will save resources." According to our study, reality points in another direction (Trade 2020). It should be noted that the fuel subsidy has a very high economic cost for the country (Vinueza 2015); the results are foreign exchange leaks in a dollarized economy. However, some studies indicate that the most feasible way to respond to this increase in gas is to achieve the equilibrium point for the country under the same conditions of demand, that is, the retail price (PVP), setting it at 10 $0.04 (Raffino 2021). The author Villavicencio points out that: "changing the price of the cylinder above 10 USD, will have a decrease of 22% in the demand considered in border smuggling, which would cease to be attractive and profitable for smugglers; additional savings for the country due to the decrease in resources destined to control the illegal exit of LPG" (Villavicencio 2019). So, would this type of cost proposed by the author Villavicencio y Salinas (Villavicencio 2019) be a good option? To answer this question, we turn to the results of graph 1. Mainly, in the question about: the construction of a device that allows replacing the 15 kg domestic gas cylinder or can; for one that suits the domestic kitchen. This question is based on the psychographic dimension of the segmentation variable criteria and its study orientation is psychological segmentation (Table 74.1). This variable has a lot to do with the interest of a user regarding a brand, or development of a product or device (Raffino 2021). The results show us that the families were interested in this friendly technological proposal. Therefore, we would be giving a better offer than raising the price of gas to 10 USD as described and proposed by Villavicencio (Villavicencio

2019) in their study. The profit generated from the sale of a device designed for the urban area for domestic use can be used to cover the costs of the targeted subsidy. In what sense? In that the use of LPG gas would be gradually being eliminated and also reducing waste in the sanitary patios, it also contributes to the improvement of the budget in the basic family basket of Guayaquil households. The key is in the strategic management to be applied, in such a way that it starts with new energy projects based on friendly technology, circular economy through energy valuation and following the sustainable development objectives (SDG), improvement of sustainable urban development, etc., the application of such strategies would generate new jobs and improve the country's economy.

## 74.5 Conclusion

In this study, trends and perceptions regarding energy consumption were analyzed. It was found that urban households have experienced a high increase in their energy consumption needs over time (2016–2019), derived from technological advances and the presence of the COVID-19 Pandemic (2020–2022) due to the home confinement. Added to this is the increase in electricity consumption in homes; that is given by teleworking and the presence of family members who occupy distractions with technological equipment that feeds the electrical network. In the second instance, we have revealed how the families of the marginal urban sector perceive the reality of LPG gas consumption, in order to analyze the beliefs, habits and consumption patterns through the market segmentation technique and the use of a new friendly technology based on methane fermentation. Due to the high cost of consuming domestic gas without the subsidy granted by the Ecuadorian Government; to promote the use of urban biomass as a substitute for LPG gas is to encourage sustainable behavior among citizens. So far, the consumer's perceptive point of view based on "market segmentation techniques" has been discussed. At this point, it is recognized that in Ecuador the processes of hydrocarbon operations include activities of exploration, development and exploitation of hydrocarbons, as well as their phases of transportation, storage, refining, industrialization and production of oil and natural gas through natural resources. natural non- renewable fossils. It should also be noted that electricity generation is essentially fueled by fossil fuels. Insecurity of supply, price shocks, resource depletion and the negative effects of electricity production from fossil fuels (particularly coal and oil derivatives) have led to a profound debate on the insecurity of our energy future. In this context, the proposal is to develop a broader study at the national level on how to build anaerobic devices that are fed with organic bio-waste from urban areas. Thus, we contribute to caring for the environment, applying the criteria of the Circular Economy through the energy assessment of the substrate and finally we would be in tune with the 2030 Agenda and the sustainable development goals (SDG). In other words, the purpose is to produce biogas as a two-way energy source: either for the purpose of reducing domestic gas consumption or for electricity generation. These future strategies are based on the reduction of the use of fossil fuel inputs and the greater presence in the energy market of LPG gas based on green technologies, that is, that generated with renewable energy sources. Therefore, it is vital that the authorities on duty are interested in energy projects where new methods of environmentally friendly technologies are used. It should be noted that this research has shown that Guayaquil families would support this type of renewable energy and that they would do so at intensity no lower than the average of the countries in our community environment.

## References

Acosta A, Martínez E (2019) The good living. A way for development compilers. Abya-Ayala editions. Quito

Águeda E, Madariaga J (2008) Principle of marketing, 3rd edn. ESIC. Editorial. ISBN 978-84-7356-572-1

Alfonso W, Pardo C (2012) The Suburbanization Process in Bogotá D.C. and Municipalities of the Savanna of Bogotá 1998–2010. En M. Czerny y G. Hoyos (eds.), Suburbanization versus Peripheral Sustainability of Rural-Urban Areas Fringes. Nova Science Publishers Inc., Nueva York, pp. 103–122

Arquer M (1994) NTP 401: human reliability: methods of quantification, expert judgment [online document]. Available http://www.mtas.es/insht/ntp/ntp_401.htm

Biller D, Nabi I (2013) Investing in infrastructure. Harnessing its potential for growth in Sri Lanka. BAMCP Mundial, International Bank for Reconstruction and Development, Washington

Burneo D, Cansino JM, Yñiguez R (2020) Environmental and socioeconomic impacts of urban waste recycling as part of circular economy. The case of Cuenca (Ecuador). Sustainability 12:3406. https://doi.org/10.3390/su12083406

Calderón CA (2022) Study of circular models as a sustainable development alternative for the production of plastic. Doctoral dissertation, University of Guayaquil. Faculty of Industrial Engineering. Industrial Engineering Career

Carreño C, Alfonso P, William H (2018) Relationship between urbanization processes, international trade and its impact on urban sustainability. Housing Urbanism Notebooks 11(22):1–10

EC CELAC. Community of Latin American and Caribbean States (2012b) Declaration of Quito. XVIII meeting of the forum of ministers of environment of Latin America and the Caribbean and first meeting of ministers of environment of CELAC. Quito

Coello S. Energy recovery of organic waste from the city of Guayaquil for the production of biogas through a biodigester [Ph.D. thesis]. University of Coruña, Spain

DACE (2019) Statistical bulletin of the hydrocarbons regulation and control agency. Statistical Bulletin. Hydrocarbons Regulation and Control Agency (DACE). Available https://n9.cl/knu4z

Daily Trade (2020) A subsidy will be maintained for induction cookers. Business-News-Actuality. https://n9.cl/0oznt. 22 Dec 2019

Díaz F, García C, Fyall A (2020) The use of the CHAID algorithm for determining tourism segmentation: a purposeful outcome. Heliyon 6(7):e04256

EC INEC National Institute of Statistics and Censuses INEC (2019a) Environmental information in homes ESPND 2020. https://n9.cl/1gt9a

EC INEC National Institute of Statistics and Censuses (INEC) (2019b) Cantonal population projections. https://n9.cl/vtn5q

Fryn F (2019) Qualitative vs quantitative research—what is what? Imotion. https://n9.cl/252h0

Gordon W, Langmaid R (2022) Qualitative market research: a practitioner's and buyer's guide. Routledge

Jia S, Wang C, Li Y, Zhang F, Wei L, Jia S (2017) The urbanization efficiency in Chengdu city: an estimation based on a three-stage DEA model. Phys Chem Earth: 1–11

Kinnear TJ (1994) Market research. McGraw-Hill Interamericana S.A, Colombia

López P, Fachelli S (2015) Quantitative social research methodology, 1st edn. Editorial Creative Common, Barcelona-Spain. Available at https://n9.cl/xh86

Ministry of Hydrocarbons (2014) Management report 2014. Retrieved from https://n9.cl/zbq39

Raffino M (2020) Administrative process—what it is, its characteristics and stages. Accessed Mar 2021. Available at https://n9.cl/ta2uwr

Vistazo Magazine (2020) National gas production is not enough to cover the country's demand. Available https://n9.cl/8i3z. 15 June 2020

Villavicencio M (2019) Effect of eliminating gas subsidy for domestic use in Ecuador. Available at https://n9.cl/4d1jk

Vinueza RAL (2015) Economic analysis of the change in the energy matrix and its impact on the Ecuadorian economy, focused on the productive matrix and the National Plan for Good Living 2009–2013 (Graduation work prior to obtaining the title of Commercial Engineering, Pontifical Universidad Catholica of Ecuador). Retrieved from https://n9.cl/v58vg

# Chapter 75
# Energy Literacy Scale (ELS): Validated Survey Instrument to Measure Energy Knowledge, Attitude, and Behaviour

**Annie Feba Varghese and Divya Chandrasenan**

**Abstract** The Energy Literacy Scale (ELS) was created to assess students' energy-related knowledge and awareness of the implications of energy production and consumption, everyday energy use, and the adoption of energy-saving behaviors. The energy literacy scale was drafted and pilot tested among elementary school students across Kerala, India. Initial exploration of the measure yielded promising results: Cronbach's reliability coefficients for cognitive, emotional, and behavioral subscales varied from 0.68 to 0.78, while average discrimination indices ranged from 0.28 to 0.43. Factor Analysis was used to select appropriate questions for the Energy Literacy Scale with due importance being given to each of the Knowledge, Attitudinal and Behavioral domains. The field-tested ELS includes three knowledge factors namely (1) Energy sources, Efficiency and Conservation, (2) Energy Use and Implications and (3) Basic Energy Concepts. Three behaviour and two attitude dimension sub-scales are included in the accepted instrument. The ELS is particularly useful for determining the baseline energy literacy skills of potential responders and evaluating the broader effects of educational initiatives.

**Keywords** Energy literacy · Energy knowledge · Attitude · Behaviour · Energy literacy scale

## 75.1 Introduction

Awareness of the nature and role of energy in the world and in our everyday lives, as well as the capacity to utilize this knowledge to solve issues, constitute energy literacy. It is the capacity to have access to and make sensible use of energy. Literacy implies knowledge or competence in a particular area, but as far as Energy Literacy is concerned it takes into account the Knowledge, Attitude, and Behaviour of an

A. F. Varghese (✉) · D. Chandrasenan
Department of Education, University of Kerala, Thiruvananthapuram, India
e-mail: divyacsenan@kerelauniversity.ac.in

 
N. S. Caetano and M. C. Felgueiras (eds.), *The 9th International Conference on Energy and Environment Research*, Environmental Science and Engineering,
https://doi.org/10.1007/978-3-031-43559-1_75

individual. Energy is an essential good in today's world. The surge in the prices of the energy resources in our country is the talk of the common man now. As we go forward into a future with dwindling reserves of fossil fuel and deteriorating environmental circumstances, our society will be forced to define new paths with regard to the use of energy, the availability of resources, and its level of autonomy. Informed people who are energy-literate are more likely to be involved in the decision-making process. Furthermore, individuals who know more about energy will be better able to make intelligent and responsible energy decisions, choices, and actions because they will have more information to draw from DeWaters and Powers (2013).

A concentration on energy literacy in the next five to seven years will put India in an excellent position to meet the commitments it set at COP26, well ahead of schedule. The challenge is to produce more energy in a sustainable manner. An energy literacy movement from the grassroots is the only solution to meet this challenge. In the Indian context, energy literacy especially at the grassroots is still in its infancy. Awareness and a positive attitude to the prudent use of energy and sensitivity towards sustainable living and renewable energy will prove to be the foundation of a sustainable usage pattern of energy.

Only by understanding what energy is, its sources and methods of production, its applications and conservation measures can inform judgments regarding the most effective use of energy be made. However, knowledge alone does not ensure energy conservation. It is vital for individuals to be able to apply this information in real ways. Energy literacy is a critical tool that empowers individuals by enabling them to make more informed energy choices. It encompasses residents' energy knowledge, attitudes, and intentions, as well as citizens' energy-related habits and behaviour (Martins et al. 2019). Energy-literate individuals, should be able to trace energy flows and think in terms of energy systems, understand the purpose of the energy they consume, the amount they consume and the source, evaluate the veracity of energy-related information, and communicate effectively about energy and its use (U.S. 2017). Apart from that, the individual must be capable of making intelligent energy choices. When developing an Energy Literacy Scale, it is necessary to analyze the fundamental principles and concepts that an individual should understand about energy, its supply, usage, and future concerns.

## 75.2 Motivation

It is an undeniable fact that our fossil fuel reserves are depleting fast and though there are alternative sources for the same, it remains a question as to how much a common man has awareness or understanding of the same. The consequences of not using energy in a conscious way and the utilization of prospects available for wise investments in energy still need to be practiced by the common man. There lies the importance of assessing the Energy Literacy of an individual. Energy literacy should concentrate on critical thinking, leading to a positive attitude towards energy which then is reflected through actions. It thus becomes essential to determine the energy literacy levels as it gives an idea of where each individual stands with regard to energy issues at hand. Educational programs on familiarizing the students with energy issues have been initiated by many countries around the world and there have been curriculum revisions and relevant strategies adopted for the same. Children being the future citizens of a nation are surely going to frame how a nation acts on energy issues in near future. Schools are the ideal places to learn about energy because people's perception of energy is mostly formed in their formative years (Chandrasenan et al. 2021). Thus, it is of the utmost importance to determine the energy literacy levels of pupils, who will serve as the foundation for a robust and optimistic view of energy consumption. An extensive review of literature on related studies in India revealed that no such scale of the manner which can be used to assess the energy literacy of an individual is available in the country. To address this gap, this project seeks to construct a validated survey instrument to examine primary school students' knowledge, attitudes, and behaviours regarding energy literacy.

## 75.3 The Instrument Development Framework

### *75.3.1 Item Generation and Review*

Keeping in view that Energy Literacy embraces the Cognitive (Knowledge), Affective (Attitude), and Psychomotor (Behaviour) domains of a person, an Energy Literacy Scale was designed to measure the energy literacy of primary school students. The approach includes establishing the energy literacy concept and constructing a pool of survey items, as well as preparing, delivering, and assessing the pilot test (item pilot) needed to identify suitable survey items. Following a thorough review of the present curriculum for primary school kids and the researcher's own experiences, survey questions were developed. During the item/survey assessment process, a diverse, ten-member validity panel comprised a balance of energy and education experts.

### *75.3.2 Item Pilot Results*

A sample of 303 primary school pupils in grades 5–7 were given the item pilot form, which included 20 questions from the cognitive and behavioural elements and 10 questions from the attitudinal domain. The responses were gathered. The Item difficulty value was 0.40 for the cognitive scale and so within the range (Fleetwood and Hounshell 1976). The average discrimination indices were all well within acceptable limits (0.43, 0.35, and 0.28) for the cognitive, attitude, and behavioural subscales respectively (Hills 1976). Cronbach's alpha internal consistency reliability coefficient, $\alpha$. values were 0.68, 0.67, and 0.78 for the cognitive, attitude, and behavioural subscales respectively (Linn and Gronlund 2000; Qaqish 2006) (Table 75.1).

**Table 75.1** Subscale summaries, pilot study results

| | Knowledge | Behaviour | Attitude |
|---|---|---|---|
| N | 303 | 303 | 303 |
| Average item difficulty | 0.40 | – | – |
| Average discrimination index | 0.43 | 0.28 | 0.35 |
| Reliability | 0.676 | 0.776 | 0.683 |

## 75.4 Scale Validation

The sampling adequacy test was conducted using Kaiser–Meyer–Olkin (KMO) statistics. The KMO values of the subscales were computed to be 0.744, 0.770, and 0.861, and the Bartlett Sphericity test ($p < 0.001$) revealed that there was a significant difference between the data. Due to the fact that the KMO value was larger than 0.70 and the Bartlett Sphericity test indicated a significant difference, it was determined that factor analysis could be done on these data (Leech et al. 2005; Tavsancil and Yelboga 2010) (Table 75.2).

Exploratory Factor analysis was conducted to reduce the components using principal component analysis.

Three prominent knowledge components were figured out based on eigenvalues that are greater than 0.30. The values were greater than 0.30, falling in the range of 0.369–0.680, suggesting that the data set was appropriate (Stewart 1981).

From the scree plot (Fig. 75.1) and extensive review of literature, the investigator came to the conclusion that the cognitive subscale consisted of three factors namely Energy sources, Efficiency and Conservation, Energy Use and Implications, and Basic Energy Concepts (National Energy Foundation 2017). A total of 19 items from the 20 items given for factor loading were retained in the cognitive subscale. According to the factor loading values and the scree plot (Fig. 75.2), it is evident that the attitudinal subscale included two prominent factors. The values are greater than 0.3, falling in the range of 0.490–0.701, suggesting that the data set was appropriate (Stewart 1981). The subscales are namely Personal energy-related values identified in comparison with Energy Literacy Scale developed by Jan De Waters et al. in 2013 and Personal beliefs on energy use identified through extensive reading on the topic. Factor loading in the behavioural subscale and the scree plot (Fig. 75.3) revealed

**Table 75.2** KMO value and Bartlett's test of sphericity

| | | Knowledge | Attitude | Behavioural |
|---|---|---|---|---|
| Kaiser-Meyer-Olkin measure of sampling adequacy | | 0.744 | 0.861 | 0.770 |
| Bartlett's test of sphericity | Approx. Chi-square | 755.143 | 591.734 | 970.184 |
| | Df | 190 | 45 | 190 |
| | Sig. | 0.000 | 0.000 | 0.000 |

three prominent factors, namely Family and personal adjustments to save energy, Energy consumption behaviours, and Purchase of energy-saving equipment.

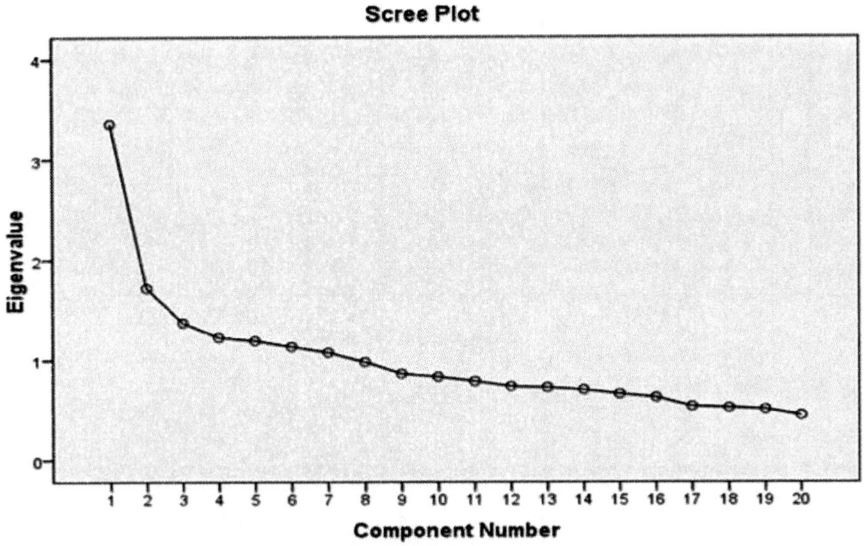


**Fig. 75.1** Scree plot diagram for cognitive scale

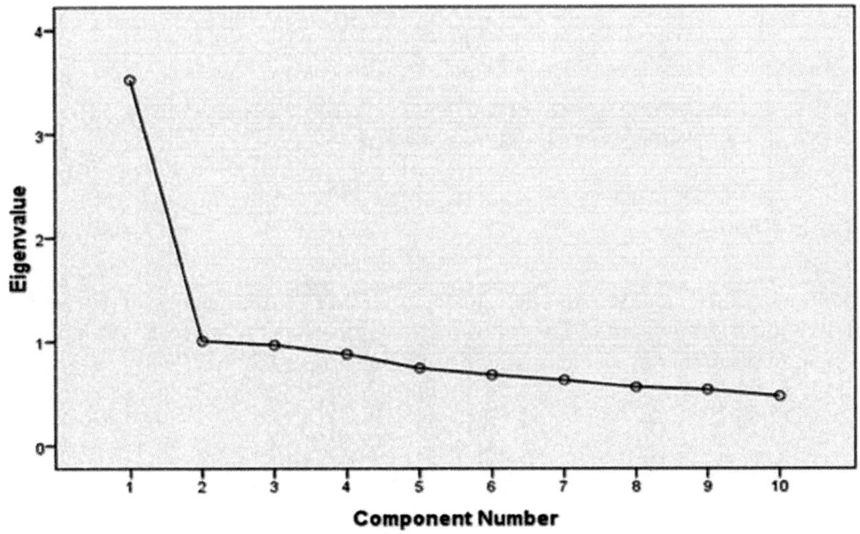


**Fig. 75.2** Scree plot diagram for attitudinal scale

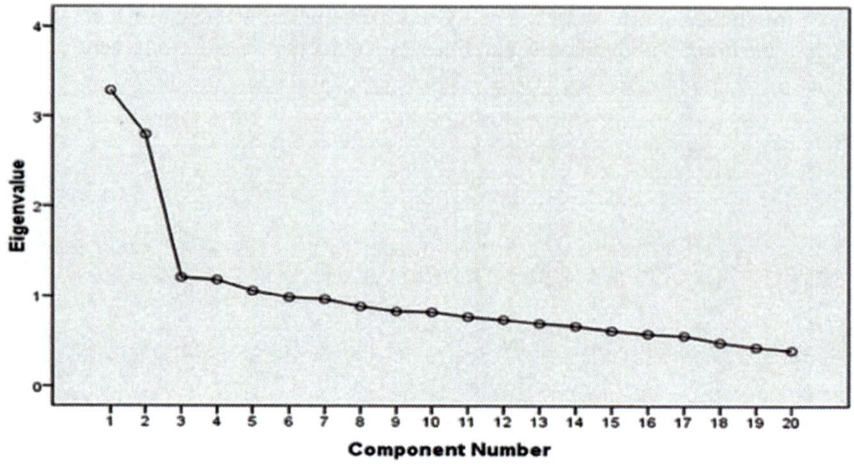


**Fig. 75.3** Scree plot diagram for behavioural scale

**Table 75.3** Factors of the subscale and number of items that fall in each factor

| Subscale | Factors | Number of items |
|---|---|---|
| Knowledge | Energy sources, efficiency and conservation | 12 |
| | Energy use and implications | 4 |
| | Basic energy concepts | 3 |
| Attitude | Personal energy-related values | 8 |
| | Personal beliefs on energy use | 2 |
| Behaviour | Family and personal adjustments to save energy | 7 |
| | Energy consumption behaviours | 11 |
| | Purchase of energy-saving equipment | 2 |

## 75.5 Result

Exploratory factor analysis was done to determine the 49-item energy literacy scale's underlying factor structure. The following table displays the subscale's component size and the number of elements inside each component (Table 75.3).

## 75.6 Discussion

The objective of the study was to design a multidimensional scale for measuring elementary-level energy literacy. The survey instrument examines students' knowledge, attitudes, and behaviour on energy usage, conservation, and management.

Energy Literacy Scale (ELS), with a 49-item, was validated through exploratory factor analysis and test samples of elementary school students across Kerala, India. The knowledge scale consists of 19 items spread across three factors focusing on areas like Energy Sources—Efficiency and Conservation, Energy Use and Implications, and Basic Energy Concepts. The scale tries to find the extent of knowledge in these areas. The attitudinal scale with ten questions deals with values and beliefs on energy use whereas the 20-item behavioural scale focuses on energy saving and consumption behaviour.

Akitsu et al. (2017) modified the energy literacy framework and instrument developed by DeWaters and Powers (2011) to study energy literacy among students in Japan. Guven Yildirim and Onder (2021) designed a 20-item literacy scale for renewable energy resources to determine pre-service science teachers' literacy levels for renewable energy resources, which comprised items that examined cognitive, attitudinal, and behavioural elements. Energy Literacy studies have been carried out in various countries which resound to assess the literacy levels in basically the three domains of knowledge, attitude, and behaviour. Studies show how important it is for people to have the right information and attitudes about energy in order to make good decisions in new situations. Students who know more about energy are more optimistic about it, and the opposite is true, too (Chandrasenan et al. 2022). The energy literacy scale developed in the study corresponds very much to the aspects considered in similar studies which makes it a valid and reliable tool for assessing energy literacy.

**Acknowledgements** This work was financially supported by Indian Council for Social Science Research (No. RFD/21-22/Short-Term/GEN/15), funded by Ministry of Education, Government of India. The findings and opinions presented here do not necessarily reflect the opinions of the funding agency. Also, we thank all the faculty experts who reviewed the scale.

## References

Akitsu Y, Ishihara KN, Okumura H, Yamasue E (2017) Investigating energy literacy and its structural model for lower secondary students in Japan. Int J Environ Sci Educ 12(5):1067–1095

Chandrasenan D, Mammen J, Yesodharan V (2021) Energy literacy of university graduate students: a multidimensional assessment in terms of content knowledge, attitude and behavior. In: Proceedings of the 7th international conference on advances in energy research. Springer, pp 879–889

Chandrasenan D, Kuleenan R, Yesodharan V, Varghese AF (2022) Clustering and exploring university students' knowledge and attitude towards energy sustainability. Energy Rep 8(3):608–613

DeWaters JE, Powers SE (2011) Energy literacy of secondary students in New York State (USA): a measure of knowledge, affect, and behavior. Energy Policy 39(3):1699–1710

DeWaters JE, Powers SE (2013) Establishing measurement criteria for an energy literacy questionnaire. J Environ Educ 44(1):38–55

Fleetwood GR, Hounshell PB (1976) Assessing cognitive and affective outcomes of environmental education. J Res Sci Teach 13(1):29–36

Guven Yildirim E, Onder AN (2021) Developing a literacy scale for renewable energy resources and identifying the literacy levels of pre-service science teachers. Online Sci Educ J 6(1):70–83
Hills JR (1976) Measurement and evaluation in the classroom. Merrill Publishing Company
Leech NL, Barrett KC, Morgan GA (2005) SPSS for intermediate statistics, use and interpretation, 2nd edn. Lawrence Erlbaum Associates Inc.
Linn RL, Gronlund NE (2000) Measurement and assessment in teaching, 8th edn. Prentice-Hall
Martins A, Madaleno M, Ferreira M (2019) Energy literacy: what is out there to know. Energy Rep 6:454–459
National Energy Foundation (2017) National energy literacy survey. https://nef1.org/survey/
Qaqish B (2006) Developing multiple-choice tests for social work training. Train Dev Human Serv 3(1):45–57
Stewart D (1981) The application and misapplication of factor analysis in marketing research. J Market Res 18(1)
Tavsancil E, Yelboga A (2010) The examination of reliability according to classical test and generalizability on a job performance scale. Educ Sci Theor Pract 10(3):1847–1854
U.S. Department of Energy (2017) Energy literacy: essential principles and fundamental concepts for energy education. https://www.energy.gov/eere/education/energy-literacy

# Chapter 76
# The Embodied Energy of Building Envelopes: Filling the Environmental Gap in Energy Performance Certificates

**Alexandre Soares dos Reis, Marta Ferreira Dias, and Alice Tavares**

**Abstract** The Energy performance certificates (EPCs) are an integral part of the Energy Performance of Buildings Directive (EPBD) and should have an essential role in enhancing the energy performance of buildings. For this reason, further improvements to EPCs might be necessary. Nowadays, contractors can refurbish building envelopes without providing environmental benefits. This paper emphasizes the need to track the environmental issues previously and during the rehabilitation of buildings envelopes. Due to the low heating practices of Portuguese families, in many cases, building materials' embodied energy may surpass the active energy demand for heating and cooling in a cost–benefit analysis. Some measures, like high insulation thicknesses and top-performance windows, may generate similar or even higher embodied energy than the potential reductions during operation. Hence, paradoxically, the global energy and Green House Gas (GHG) emissions might increase to reach Nearly Zero Energy Buildings (NZEBs), as the embodied energy will increase using higher insulation thicknesses. A new approach should provide a sustainable and more reliable service to end-users, informing them about the environmental impacts in the whole life cycle, including the embodied energy of the building envelopes components like thermal insulation systems and windows.

**Keywords** Building envelope · Embodied energy · Energy performance certificates · Environmental impacts

---

A. S. dos Reis (✉) · M. Ferreira Dias
Research Unit on Governance, Competitiveness and Public Policies (GOVCOPP), Department of Economics, Management, Industrial Engineering and Tourism (DEGEIT), University of Aveiro, Campus Universitário de Santiago, 3810-193 Aveiro, Portugal
e-mail: alexandre.soares.reis@ua.pt

A. Tavares
Centre for Research in Ceramics and Composite Materials (CICECO), Department of Materials and Ceramic Engineering, University of Aveiro, Campus Universitário de Santiago, 3810-193 Aveiro, Portugal

 
N. S. Caetano and M. C. Felgueiras (eds.), *The 9th International Conference on Energy and Environment Research*, Environmental Science and Engineering,
https://doi.org/10.1007/978-3-031-43559-1_76

## 76.1 Introduction

A building's thermal envelope, a physical barrier between the conditioned and unconditioned spaces consisting of opaque and transparent elements, plays an essential role in the occupant's comfort and well-being. First, it protects against heat, wind, rain, or noise. Secondly, it reduces the thermal losses from indoors to outdoors. Lastly, it is crucial for daylight, ventilation, and the outside view. Thermal insulation systems and windows are essential components of the building envelope. Architects, engineers, builders, and building owners should choose the proper solutions to minimize buildings' operational and embodied energy consumption (Zilberberg et al. 2021).

Embodied energy accounts for the necessary energy needed in all stages of a product. Hence, it includes the energy in producing, transporting, and assembling the product on-site (Reilly et al. 2020). Architects and engineers regularly assess systems and certifications in the design plan step. However, they should do these tasks before committing to the project. Reducing energy consumption during service life and embodied energy (Tushar et al. 2021) in buildings is critical to reducing greenhouse gas (GHG) emissions (Koh and Kraniotis 2020), and life cycle assessment (LCA) may provide helpful information for achieving sustainability (Yung et al. 2013). Usually, during the project appraisal stage, no concern is given to selecting environmentally friendly solutions with low embodied carbon (Topalović et al. 2018), where they could be best incorporated (Ding 2008).

Energy saving is crucial for buildings with a considerable potential for improvement by energy retrofitting. Thus, it is essential to evaluate and optimize the energy performance of the existing building stock (Simhachalam et al. 2021). European Union (EU) Member States (MS) reinforced thermal regulatory guidances to the current building stock. Still, the poor envelope of many houses requires a high level of rehabilitation to minimize the occupant's thermal discomfort. The global costs of such interventions may not become affordable, even with financial assistance (Matos et al. 2022).

On the other hand, embodied energy will increase as higher insulation thicknesses are used to reach NZEBs (Su et al. 2020). Thermal insulation solutions lack studies regarding environmental impacts like embodied energy and $CO_2$ emissions directly related to their production. These impacts account for 11% of global $CO_2$ emissions throughout the building life cycle (Gadallah and Aboulnaga 2020).

The main objective of this study is to review the literature about the embodied energy of the building envelope's main components—thermal insulation and windows. The final goal is to highlight the need to assess the trade-offs between operational and embodied impacts in EPCs, to achieve a more efficient methodology.

The structure of this paper is as follows. First, Sect. 76.1 introduces the problem. Secondly, in Sect. 76.2, the literature review is analyzed. The conclusions follow this section.

## 76.2 Literature Review

While we previously argued that EPCs have the potential to track the global environmental impact of building components (Reis and Dias 2020), this present study reviews the literature about the embodied energy of building envelopes.

Previous studies primarily select the optimum insulation materials with little or no consideration of embodied energy (Kumar et al. 2020). Many authors have studied the energy consumption of buildings considering the operational phase but ignoring the crucial influence of embodied energy to quantify the environmental impacts (Tushar et al. 2019). Luo and Oyedele (2021) stated that the main focus of traditional retrofitting is to reduce operational costs. Nevertheless, this strategy might increase embodied carbon and energy. In NZEBs, the embodied effects can be relevant compared with the active phase. However, building performance assessments focus mainly on the primary energy of the building use stage (Cusenza et al. 2022). Yet, embodied carbon needs to be considered, and develop methodologies to optimize the trade-off between incorporated and operational stages (Shadram et al. 2019).

It is essential to have a holistic approach (International Active House Alliance 2011), respecting each type of building and local climate conditions (Cristiano and Gonella 2019), considering energy-use behavior (Li et al. 2021), and looking beyond energy in the operational phase (Arent et al. 2020). In many cases, building materials' embodied energy (Gadallah and Aboulnaga 2020) may surpass the active energy demand for heating and cooling (Monteiro et al. 2021) in a cost–benefit analysis. Hence, excluding it from the whole building energy efficiency system might not be adequate (Chang and Wang 2011), requiring a balance between the potential energy savings and the material impact (Landuyt et al. 2021). In Germany, fossil and mineral insulations dominate the market despite numerous LCA showing that bio-based insulations can offer environmental benefits (Schulte et al. 2021). Additional criteria may be considered, such as the ecological impacts of the components used to enhance the thermal insulation of the building envelope (Schulte et al. 2021), local or non-local sourcing of building materials (Opher 2021), including the induced transport to and from the site (Crawley and Aho 1999). As thermal insulation systems and windows are essential components of the building envelope, it is crucial to study their embodied energy (Menoufi et al. 2012). Some measures, like high insulation thicknesses and top-performance windows, may generate similar or even higher embodied energy than the potential reductions during operation (Rivera et al. 2021). Energy-efficient windows can be costly and have high embodied environmental impacts (Rodrigues and Freire 2014). The building's insulation structure is a crucial factor in controlling the temperature in the building and minimizing the energy loss from the building to the outdoors as possible. Thus, any retrofit project must consider the importance of this component and choose the appropriate energy efficiency technologies to enhance the building's insulation and achieve positive and satisfactory results (Al-Habaibeh et al. 2022). Concerning thermal insulation thickness, the benefits of the energy bill for heating and cooling do not increase indefinitely. After a certain point, it is useless to keep thickening the insulation as it generates more embodied emissions than the reduction obtained in the operational phase (Rodrigues and Freire 2014). Environmental Product Declarations (EPDs) may help identify the most impacting materials and processes from an ecological point of view (Asdrubali et al. 2021). Nevertheless, EPDs are still scarce in Portugal.

According to the EPBD, MS must provide information to the end-users on energy performance certificates (EPCs), including the expected energy needs of their houses, generally through a sliding rating scale providing summarized information. However, there is often a difference between the theoretical and the actual energy consumption, the so-called energy gap. EPCs aim to provide owners with reliable information about the energy performance of their houses. They are supposed to convince them to assume investments in energy efficiency due to the expected increase in the building's value. Nevertheless, the theoretical concept of an economic valuation associated with higher energy classes is inconclusive (Olaussen et al. 2017). In Portugal, EPCs rate a building from A+ (most efficient) to F (least efficient), and they are a legal requirement when an owner promotes a house for selling or renting. However, the Portuguese methodology does not consider the specific habits of Southern European countries where a permanent heating practice does not exist. Hence, they are still an empiric statement of a building's energy efficiency and show no relationship with in-use energy performance (Pike 2020). Still, EPCs might be more helpful, as they can track additional information on environmental issues (Ahmed et al. 2020).

Using Building Information Model (BIM), Tushar et al. (2021) conducted a study about the building life cycle phases of a typical detached residential dwelling in the suburban area of Melbourne City, Australia. According to their findings, ceiling, and wall insulation were more critical influential factors in energy reduction than passive design strategies. However, the energy reduction was not linear to the applied thicknesses of insulation. Topalović et al. (2018) studied the consequences of thermal insulation systems on the environment due to implementing the new regulations on the energy efficiency of buildings. They concluded that it is necessary to consider embodied carbon and the whole life carbon to estimate the impact of a building on the environment. Tagliabue et al. (2018) stated that EPC calculation excludes the environmental effects of energy used for materials' production. Yet, few official sources provide accredited values for embodied energy, like EPDs. It is supposed that an NZEB reduces its environmental impact in the operational phase but incorporated energy claims an increasing contribution to the total life cycle of the building. Hence, architects, designers, and engineers should favor and endorse components with less potential embodied energy. Feng et al. (2022) designed an optimization framework based on a Whole Building Life Cycle Assessment (WBLCA) to evaluate and improve the environmental performance at the building level. However, to ensure that the WBLCA was comprehensive and reliable, they needed the EPDs of the materials.

O'Hegarty et al. (2021) reviewed past studies about in-situ U [$(W/(m^2{\cdot}K)$] monitoring of building envelopes and monitored the performance of ten highly insulated buildings in Ireland. Results show that nine out of the ten buildings underperform, with an average deviation of more than 100%. According to their conclusions, high insulation thicknesses are redundant and increase the embodied energy of the building without offering a beneficial decrease in the U [(W/($m^2$·K)] values. Miljan et al. (2020) concluded that studying the energy needs throughout the building life cycle, including the embodied energy of building materials, shows a more realistic overview of the energy efficiency of the building. Chen et al. (2020) proposed a BIM-based envelope thermal insulation optimization tool to balance the embodied energy and the envelope's thermal performance based on a case study of a dwelling in Sydney, Australia. They found out that improving the thermal performance by 10% has, as a consequence, increased the embodied energy by more than 50 times.

Veludo and Rato (2020) analyzed 90 solutions for heavyweight external walls based on the Portuguese common practice, proposing three indices. A sustainability index, ISG, is based on an environmental indicator, ISE, and a functional indicator, ISF. ISE results in a weighted average of the superficial embodied energy, EES, and the superficial embodied carbon, ECS. ISF is the result of a weighted average of U [(W/($m^2$·K)] and the net superficial thermal mass, Mtsu [J/($m^2$·K)]. Alla et al. (2020) developed a method that considers the embodied energy of thermal insulation materials, allowing them to estimate the potential energy savings in existing buildings. On the other hand, Vighnesh (2022) developed a process of selecting building envelope components based on total embodied energy (Vighnesh 2022).

## 76.3 Conclusions

EPCs focus mainly on primary energy needs in the operational stage, not considering the embodied energy of thermal insulation systems and windows. Thus, there is a need to balance the potential energy savings in the active phase and the embodied energy in producing, transporting, and assembling the building envelope's components on-site. Paradoxically, the global energy and GHG emissions might increase to reach NZEBs as the embodied energy will increase using higher insulation thicknesses.

Due to the habits of countries like Portugal, where a permanent heating practice does not exist, EPCs are only an empiric statement of a building's energy efficiency, showing no relationship with actual energy consumption for heating and cooling. But to reach NZEBs, embodied energy will increase. High insulation thicknesses might be redundant, increasing the embodied energy of the building without offering a beneficial increase in occupants' thermal comfort and well-being. Hence, assessing the trade-offs between operational and embodied impacts is crucial. Authors who have performed an LCA needed access to EPDs. However, there is still a lack of these certifications in the market.

Future developments should consider a non-complicated approach to tracking the environmental impacts of the building construction sector.

**Acknowledgements** GOVCOPP supported this work (project POCI-01-0145-FEDER-008540), financed by FEDER funds, through COMPETE2020—Competitiveness, and Internationalization

Operational Program (POCI), and by national funds through the Foundation for Science and Technology (FCT).

The author Alice Tavares thanks the support of FCT (2021.03830.CEECIND) and of the project CICECO-Aveiro Institute of Materials, UIDB/50011/2020, UIDP/50011/2020 and LA/P/0006/2020, financed by national funds through the FCT/MCTES (PIDDAC).

## References

Ahmed K, Hajian H, Hasu T, Kurnitski J (2020) Kouvola housing fair NZEB houses energy, cost and carbon analyses. Presented at the E3S web of conferences. https://doi.org/10.1051/e3sconf/202017213001

Al-Habaibeh A, Hawas A, Hamadeh L, Medjdoub B, Marsh J, Sen A (2022) Enhancing the sustainability and energy conservation in heritage buildings: the case of Nottingham playhouse. Front Archit Res 11(1):142–160. https://doi.org/10.1016/j.foar.2021.09.001

Arent J, Athalye R, Taylor S (2020) Clearing the path to ZNE with energy codes. Presented at the ASHRAE transactions, pp 47–54

Asdrubali F, Roncone M, Grazieschi G (2021) Embodied energy and embodied gwp of windows: a critical review. Energies 14(13). https://doi.org/10.3390/en14133788

Alla SA, Bianco V, Scarpa F, Tagliafico LA (2020) Retrofitting for improving energy efficiency: the embodied energy relevance for buildings' thermal insulation. Presented at the ASME 2020 14th international conference on energy sustainability, ES 2020. https://doi.org/10.1115/ES2020-1628

Chang Y, Wang Y (2011) Analysis of building embodied energy and atmosphere impacts in China based on economic input-output life-cycle assessment model. Tumu Gongcheng Xuebao/china Civ Eng J 44(5):136–143

Chen Z, Hammad AWA, Kamardeen I, Akbarnezhad A (2020) Optimising embodied energy and thermal performance of thermal insulation in building envelopes via an automated building information modelling (BIM) tool. Buildings 10(12):1–23. https://doi.org/10.3390/buildings10120218

Crawley D, Aho I (1999) Building environmental assessment methods: applications and development trends. Build Res Inform 27:300–308. https://doi.org/10.1080/096132199369417

Cristiano S, Gonella F (2019) Learning from hybrid innovative-vernacular solutions in building design: emergy analysis of sudanese energy-saving technologies. J Environ Account Manage 7(2):213–227. https://doi.org/10.5890/JEAM.2019.06.007

Cusenza MA, Guarino F, Longo S, Cellura M (2022) An integrated energy simulation and life cycle assessment to measure the operational and embodied energy of a Mediterranean net zero energy building. Energy Build 254. https://doi.org/10.1016/j.enbuild.2021.111558

Ding GKC (2008) Sustainable construction—the role of environmental assessment tools. J Environ Manage 86(3):451–464. https://doi.org/10.1016/j.jenvman.2006.12.025

dos Reis AS, Dias MF (2020) Cost-optimal levels and energy performance certificates: filling the gaps. Energy Rep 6:358–363. https://doi.org/10.1016/j.egyr.2020.11.172

Feng H, Kassem M, Greenwood D, Doukari O (2022) Whole building life cycle assessment at the design stage: a BIM-based framework using environmental product declaration. Int J Build Pathol Adapt. https://doi.org/10.1108/IJBPA-06-2021-0091

Gadallah S, Aboulnaga M (2020) Climate action and sdgs' attainment: Insulation materials' impacts assessment. Sustain Mediterranean Constr 2020(11):76–80

International Active House Alliance (2011) The active house principles—comfort, energy, environment—were defined in 2011, by a global group of likeminded visionary leaders, who strongly believed in a holistic approach to building design. https://www.activehouse.info/about/about-active-house/. Accessed 17 July 2021

Koh CHA, Kraniotis D (2020) A review of material properties and performance of straw bale as building material. Constr Build Mater 259. https://doi.org/10.1016/j.conbuildmat.2020.120385

Kumar D, Alam M, Zou PXW, Sanjayan JG, Memon RA (2020) Comparative analysis of building insulation material properties and performance. Renew Sustain Energy Rev 131. https://doi.org/10.1016/j.rser.2020.110038

Landuyt L, De Turck S, Laverge J, Steeman M, Van Den Bossche N (2021) Balancing environmental impact, energy use and thermal comfort: optimizing insulation levels for the mobble with standard HVAC and personal comfort systems. Build Environ 206. https://doi.org/10.1016/j.buildenv.2021.108307

Li L, Wang Y, Wang M, Hu W, Sun Y (2021) Impacts of multiple factors on energy consumption of aging residential buildings based on a system dynamics model—taking Northwest China as an example. J Build Eng 44. https://doi.org/10.1016/j.jobe.2021.102595

Luo XJ, Oyedele LO (2021) Assessment and optimisation of life cycle environment, economy and energy for building retrofitting. Energy Sustain Dev 65:77–100. https://doi.org/10.1016/j.esd.2021.10.002

Matos AM, Delgado JMPQ, Guimarães AS (2022) Linking energy poverty with thermal building regulations and energy efficiency policies in Portugal. Energies 15(1). https://doi.org/10.3390/en15010329

Menoufi K, Castell A, Navarro L, Pérez G, Boer D, Cabeza LF (2012) Evaluation of the environmental impact of experimental cubicles using life cycle assessment: a highlight on the manufacturing phase. Appl Energy 92:534–544. https://doi.org/10.1016/j.apenergy.2011.11.020

Miljan M, Miljan M-J, Keskküla K, Miljan J (2020) The combined impact of energy efficiency and embodied energy of external wall over 30 years of life cycle. Agron Res 18(3):2148–2155. https://doi.org/10.15159/AR.20.216

Monteiro H, Freire F, Soares N (2021) Life cycle assessment of a south European house addressing building design options for orientation, window sizing and building shape. J Build Eng 39. https://doi.org/10.1016/j.jobe.2021.102276

O'Hegarty R, Kinnane O, Lennon D, Colclough S (2021) In-situ U-value monitoring of highly insulated building envelopes: review and experimental investigation. Energy Build 252. https://doi.org/10.1016/j.enbuild.2021.111447

Olaussen JO, Oust A, Solstad JT (2017) Energy performance certificates—informing the informed or the indifferent? Energy Policy 111:246–254. https://doi.org/10.1016/j.enpol.2017.09.029

Opher T et al (2021) Life cycle GHG assessment of a building restoration: case study of a heritage industrial building in Toronto, Canada. J Clean Prod 279. https://doi.org/10.1016/j.jclepro.2020.123819

Pike J (2020) The future of sustainable real estate investments in a post-COVID-19 world. J Euro Real Estate Res 13(3):455–460. https://doi.org/10.1108/JERER-07-2020-0042

Reilly A, Kinnane O, O'hegarty R (2020) Energy embodied in, and transmitted through, walls of different types when accounting for the dynamic effects of thermal mass. J Green Build 15(4):43–66. https://doi.org/10.3992/jgb.15.4.43

Rivera ML, MacLean HL, McCabe B (2021) Implications of passive energy efficiency measures on life cycle greenhouse gas emissions of high-rise residential building envelopes. Energy Build 249. https://doi.org/10.1016/j.enbuild.2021.111202

Rodrigues C, Freire F (2014) Integrated life-cycle assessment and thermal dynamic simulation of alternative scenarios for the roof retrofit of a house. Build Environ 81:204–215. https://doi.org/10.1016/j.buildenv.2014.07.001

Schulte M, Lewandowski I, Pude R, Wagner M (2021) Comparative life cycle assessment of bio-based insulation materials: environmental and economic performances. GCB Bioenergy 13(6):979–998. https://doi.org/10.1111/gcbb.12825

Shadram F, Mukkavaara J, Schade J, Sandberg M, Olofsson T (2019) Trade-off optimization of embodied versus operational carbon impact for insulation and window to wall ratio design choices: a case study. Smart Innov Syst Technol 131:12–20. https://doi.org/10.1007/978-3-030-04293-6_2

Simhachalam V, Wang T, Liu Y, Wamelink H, Montenegro L, van Gorp G (2021) Accelerating building energy retrofitting with BIM-enabled BREEAM-NL assessment. Energies 14(24). https://doi.org/10.3390/en14248225

Su X, Tian S, Shao X, Zhao X (2020) Embodied and operational energy and carbon emissions of passive building in HSCW zone in China: a case study. Energy Build 222. https://doi.org/10.1016/j.enbuild.2020.110090

Tagliabue LC, Di Giuda GM, Villa V, De Angelis E, Ciribini ALC (2018) Techno-economical analysis based on a parametric computational evaluation for decision process on envelope technologies and configurations evaluation for decision process of envelope technologies and configurations. Energy Build 158:736–749. https://doi.org/10.1016/j.enbuild.2017.10.004

Topalović MN, Stanković M, Ćirović G, Pamučar D (2018) Comparison of the applied measures on the simulated scenarios for the sustainable building construction through carbon footprint emissions-case study of building construction in Serbia. Sustainability 10(12). https://doi.org/10.3390/su10124688

Tushar Q, Bhuiyan M, Sandanayake M, Zhang G (2019) Optimizing the energy consumption in a residential building at different climate zones: towards sustainable decision making. J Clean Prod 233:634–649. https://doi.org/10.1016/j.jclepro.2019.06.093

Tushar Q, Bhuiyan MA, Zhang G, Maqsood T (2021) An integrated approach of BIM-enabled LCA and energy simulation: the optimized solution towards sustainable development. J Clean Prod 289. https://doi.org/10.1016/j.jclepro.2020.125622

Veludo S, Rato V (2020) Performance-based selection of sustainable construction solutions for external walls. Presented at the WIT transactions on the built environment, pp 113–124. https://doi.org/10.2495/GD170101

Vighnesh R (2022) Use of parametric software for selecting building materials based on embodied energy. Lecture Notes Civ Eng 171:25–36. https://doi.org/10.1007/978-3-030-80312-4_3

Yung P, Lam KC, Yu C (2013) An audit of life cycle energy analyses of buildings. Habitat Int 39:43–54. https://doi.org/10.1016/j.habitatint.2012.10.003

Zilberberg E, Trapper P, Meir IA, Isaac S (2021) The impact of thermal mass and insulation of building structure on energy efficiency. Energy Build 241. https://doi.org/10.1016/j.enbuild.2021.110954

# Chapter 77
# Implementation of GIS-AHP Framework for the Identification of Potential Landfill Sites in Bengaluru Metropolitan Region, India

**A. D. Aarthi, B. Mainali, D. Khatiwada, F. Golzar, and K. Mahapatra**

**Abstract** Uncontrolled open dumping and burning of municipality solid waste (MSW) has resulted in soil, water, and air pollution in many urban cities in India. Landfills are the most common cost-effective solution for MSW management in many developing countries like India. However, the identification of suitable landfill sites always remains a challenging task as it involves selection of several environmental criteria set by the local authorities. The objective of this study is to identify the most potential landfill sites proposed by the Government in Bengaluru Metropolitan Region, Karnataka state, India using Geographic Information System enabled Analytical Hierarchy Process based multi-criteria evaluation technique. Several criteria and constraints as recommended by the local authorities along with the proximity to the solid waste processing plants are used to identify the potential landfill sites in the study region. The study identified three highly suitable sites (Neraluru, Gudhatti, Madivala) for landfills which are not only environmentally sustainable but also economically attractive as they are closer to the solid waste processing plants minimizing the transportation cost involved in the disposal of solid waste from the source to the final disposal sites in the study region.

**Keywords** Municipal solid waste · Landfill site selection · Circular economy · GIS enabled AHP technique · Bengaluru Metropolitan region

A. D. Aarthi (✉)
LKAB, Malmberget, Sweden
e-mail: aarthi.aishwarya.devendran@lkab.com

B. Mainali · K. Mahapatra
Department of Built Environment and Energy Technology, Linnaeus University, Växjö, Sweden

D. Khatiwada · F. Golzar
KTH Royal Institute of Technology, Stockholm, Sweden



N. S. Caetano and M. C. Felgueiras (eds.), *The 9th International Conference on Energy and Environment Research*, Environmental Science and Engineering,
https://doi.org/10.1007/978-3-031-43559-1_77

## 77.1 Introduction

Management of Municipal Solid Waste (MSW) has been a major challenge and is receiving growing attention in many urban cities because of its environmental and economic impact (Sharma and Chandel 2021). MSW can be converted into energy using advanced conversion technologies, known as Waste to Energy (WtE) systems. However, uncontrolled open dumping and burning of MSW result in soil, water, and air pollution and thus landfills are one of the most preferred waste management techniques due to its least-cost solution in developing countries like India (Ali et al. 2021). At present, Bengaluru Metropolitan Region (BMR) generates around 3056 tonnes of MSW per day with a per capita waste generation of 363 g. By 2031, BMR is expected to generate 13,911 tonnes of MSW per day. Thus, there is a need for additional landfill sites for the proper disposal of the generated waste considering the population growth and city expansion plan (Revised Structure Plan of BMR 2031). The current study identifies the most suitable sites for potential engineered[1] landfill sites closer to MSW processing plants in the disposal of solid waste in BMR considering the environmental criteria through the methodological framework based on Geographic Information System (GIS) and Analytical Hierarchy Process (AHP) based Multi Criteria Evaluation (MCE) technique.

## 77.2 Literature Review

MSW is considered a source of energy (Kaur et al. 2023) and there are several studies in the estimation of energy recovery potential through different WtE technologies. Despite modern WtE conversion technologies, engineered or sanitary landfills would play a key role in the management of MSW especially in Indian cities. AHP, a MCE technique, decomposes a complex multi criteria decision making process into several simple alternatives in the form of a hierarchy (Saaty 1990). Integrated GIS and AHP approach have been implemented for landfill site selection in various research (Kamdar et al. 2019). However, limited studies had been reported that utilize combined GIS and AHP based MCE approach in the identification of landfills in India. This current study will fill in the identified knowledge gap, thereby contributing to the optimal landfill selection through economic, environmental, and social criteria.

The Bruhat Bengaluru Mahanagara Palike (BBMP) is the administrative body responsible for civic amenities and some infrastructural assets of BMR (https://bbmp.gov.in/indexenglish.html). Solid waste management is an essential municipal service for which 50–60% of the BMR's budget is allocated. Open dumping and open burning of MSW, non-availability of source level waste segregation, and lack of information on the tracking system of the vehicles are some of the major challenges faced by the planning authorities in BMR (Surendra et al. 2021). Currently, BMR

---

[1] An engineered landfill is a site where controlled disposal and scientific treatment of solid waste is done.

does not have any appropriate scientific treatment techniques for treating the waste before sending them to landfill. This has led to the development of various unauthorized dumpsites. At present, landfills exist at Mavallipura, Mandur, Doddaballapur and Bommanahalli in BMR. The municipality needs more landfill sites to manage the increasing MSW generation in the city. The government has proposed potential landfill sites in Gudhatti, Neralur, Bendiganahalli of Attibele hobli, Madivala, Gowrenahalli and Samanduru in Anekal hobli, Galipuje, Honnaghatta of Doddaballapur Taluk, Chikkabonahalli of Devanahalli Taluk and Kalarikaval village in BMR.

## 77.3 Data and Methods

### *77.3.1 Study Area*

Bengaluru is the capital city of Karnataka state and the fifth largest city in India and is called the IT (Information Technology) capital of India. BMR located in southern India (Fig. 77.1) on the Deccan Plateau covers an area of 8005 $km^2$ and is the second largest metropolitan in India with a population of 12,765,000 as of 2021. BMR includes the districts of Bengaluru Urban, Bengaluru Rural and Ramanagara and is considered one of the fastest developing metropolitans in Asia (Dittrich 2004). With an increased population level, rapid industrialization, increase in the number of IT parks, and a rise in the standard of living of people, the generation rate of MSW accelerates (Naveen and Sivapullaiah 2020) making BMR one of the filthiest metropolitans in India.

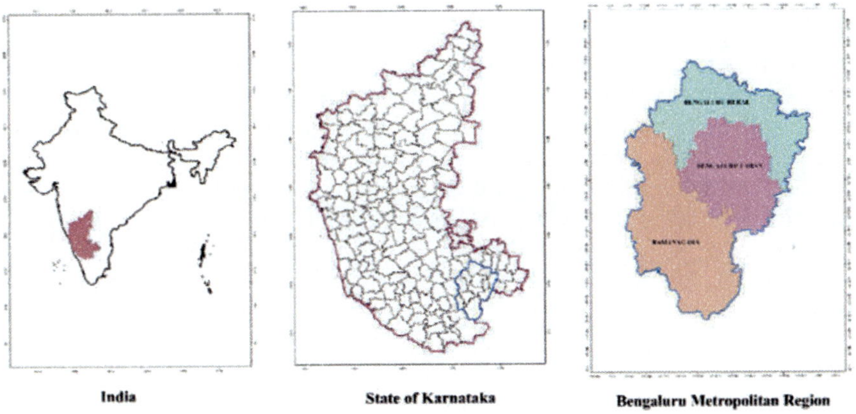


**Fig. 77.1** Location map of the study region

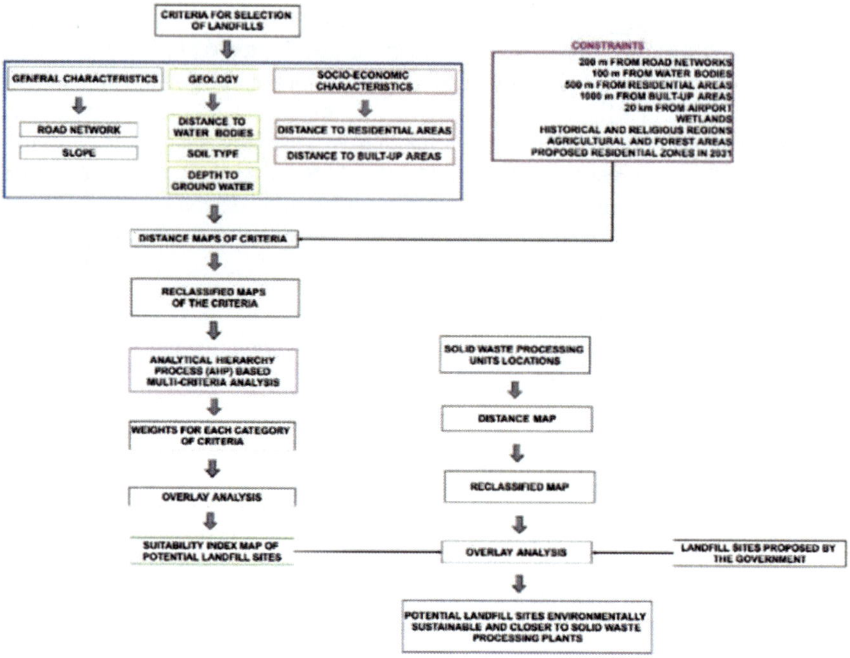


**Fig. 77.2** GIS enabled AHP technique adopted for the identification of potential landfill sites in BMR

## *77.3.2 Methodology*

### 77.3.2.1 Selection of Potential Criteria and Constraints

To identify potential landfill sites in BMR, general, geological, and socio-economic criteria as recommended by Health et al. (2023) are considered including (i) Road network, (ii) Soil types, (iii) Water bodies, (iv) Depth to Groundwater, (v) Residential Areas, (vi) Built-Up areas available from OpenStreet Map[2] and (vii) Slope derived from SRTM (Shuttle Radar Topography Mission) 1 Arc-second Global data available from U.S. Geological Survey (USGS)[3] as described in Fig. 77.2.

---

[2] https://www.openstreetmap.org/#map=4/62.99/17.64 (as accessed on <25 May 2021>).

[3] EarthExplorer (https://earthexplorer.usgs.gov/).

#### 77.3.2.2 AHP Based Multi-Criteria Technique for Landfill Site Selection

AHP was developed in 1970s (Saaty 1977) to solve various multi-criteria decision-making problems (Mardani et al. 2015). In the current study, seven criteria as mentioned in Sect. 77.3.2.1 are used to prepare the distance maps through proximity analysis in an ArcGIS environment. The distance maps are further reclassified into 5 categories including 5(Excellent), 4(Good), 3(Average), 2(Poor), and 1(Bad) based on their importance in the identification of landfill sites. These reclassified maps of the criteria are used to derive the weights through AHP analysis.

#### 77.3.2.3 Preparation of Landfill Suitability Index Map

The weight maps of the criteria obtained through AHP technique are used to prepare the suitability index map through overlay analysis (Vázquez-Quintero et al. 2020) in an ArcGIS environment. The suitability index map is categorized into 5 classes including excluded, unsuitable, moderately suitable, suitable, and highly suitable based on the suitability index values of 0–25, 25–50, 50–60, 60–70 and 70–100 respectively (Ohri et al. 2015).

#### 77.3.2.4 Identification of Potential Landfill Sites Based on the Distance Between Collection Points and Suitable Sites

The logistics cost involved in carrying the solid waste from the source points to the final disposal locations is huge (Martinez et al. 2019) and finding cost-effective and environmentally sustainable landfill sites is very essential for an efficient management of solid waste. In BMR, there are 201 number of solid waste processing plants and the potential landfill sites are selected from the landfill suitability map identified through AHP technique based on its proximity to the solid waste processing plants.

## 77.4 Results and Discussions

### *77.4.1 Preparation of Input Parameters for AHP Based Multi-Criteria Analysis*

The proximity and reclassified maps of seven criteria (Sect. 77.3.2.1) are prepared. The reclassified maps of the criteria (Fig. 77.3a–g) are used as input parameters for AHP analysis for the weight generation for the categories of the criteria. Figure 77.3h

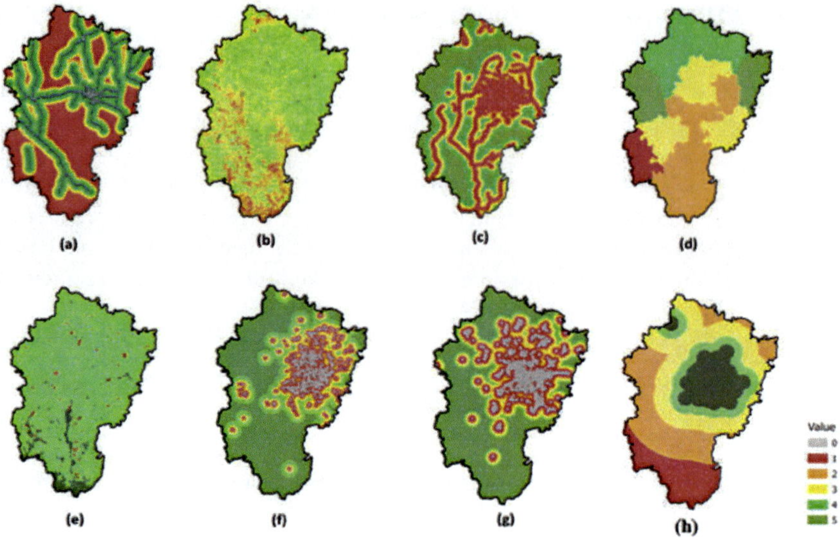


**Fig. 77.3** Reclassified Maps of **a** distance to road network; **b** slope; **c** distance to waterbodies; **d** depth to groundwater; **e** soil types; **f** distance to residential areas; **g** distance to built-up areas; **h** distance to solid waste processing plants location

represents the reclassified map of solid waste processing plants and contains 5 classes including 5, 4, 3, 2 and 1 indicating a distance of 5 km, 10 km, 20 km, 40 km and greater than 40 km respectively from the solid waste processing plants locations. Figure 77.3h is used in the identification of potential landfill sites based on its proximity to the waste processing plants based on the landfill suitability map prepared through AHP technique.

### *77.4.2 Preparation of Suitability Index Map for Landfill Site Selection in BMR*

The weights of each category of the criteria derived through AHP technique are given in Fig. 77.4. The numbers inside the parentheses in Fig. 77.4 correspond to the reclassified values of the criteria as described in Sect. 77.3.2.2 and Fig. 77.3. Each criterion is reclassified into 5 classes as suggested by Health et al. (2023) and is shown as the category values of each criterion in Fig. 77.4. The weight maps of the criteria obtained through AHP technique thereby are used to prepare the suitability index map (Fig. 77.5a) through overlay analysis in ArcGIS environment.

Based on the analysis, regions around Kudur, Gummanahalli, Karahalli, Sonnanayakanapura, Channapatna and Uyyamballi are considered highly suitable for the establishment of landfills in BMR. According to Revised Structure Plan of

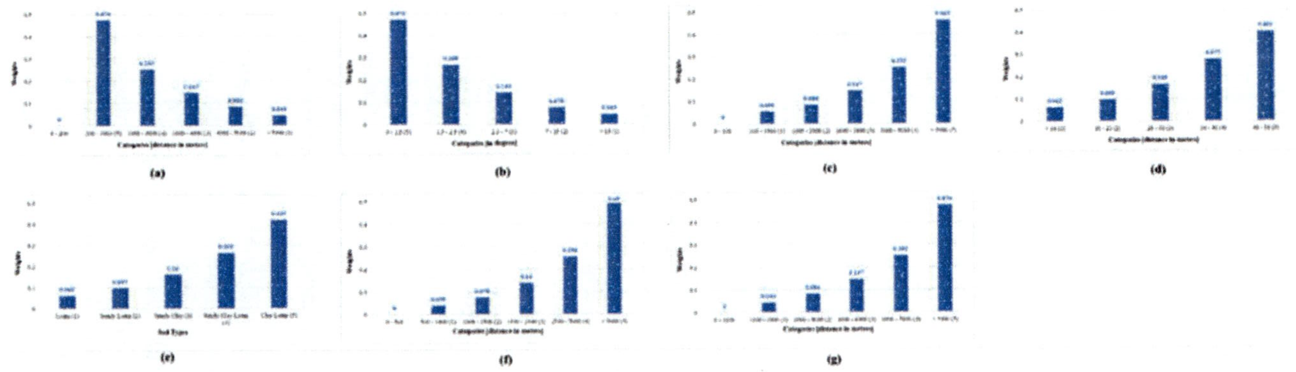


**Fig. 77.4** Weights assigned to the categories of criteria for the identification of potential landfill sites in BMR. **a** Road network; **b** slope; **c** distance to waterbody; **d** depth to groundwater; **e** soil types; **f** distance to residential areas; **g** distance to built-up areas

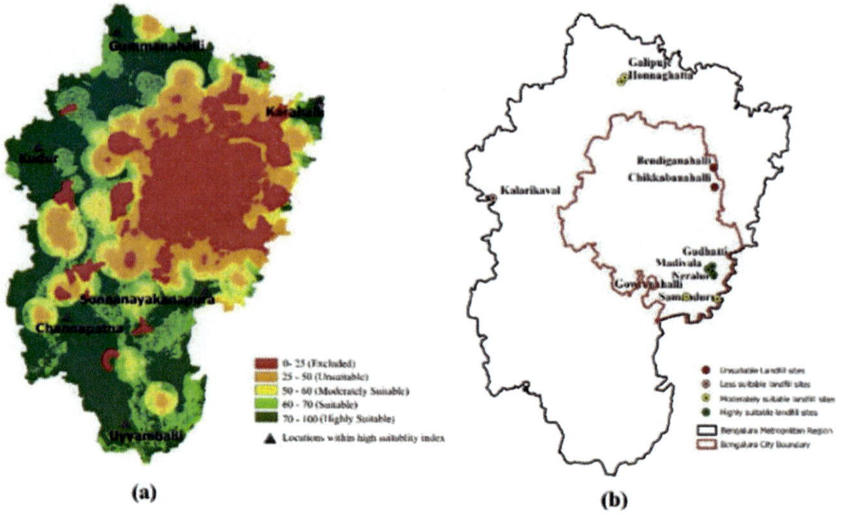


**Fig. 77.5** Suitability index map for landfill in BMR. **a** Landfill suitability map based on AHP technique; **b** potential landfill sites based on (**a**) and closer to solid waste processing plants

BMR (2031), the Government has proposed to set up landfill sites in BMR by 2031 as described in Sect. 77.2. Based on the suitability index map (Fig. 77.5a) prepared through AHP analysis along with the reclassified map of solid waste processing plants (Fig. 77.3h) the proposed landfill sites are categorized as highly suitable, moderately suitable, less suitable and unsuitable as shown in Fig. 77.5b. The proposed sites at Madivala, Neraluru, Gudhatti are considered highly suitable for landfill establishment as they fall within the high suitability index for landfill selection and lie within 10 km from the solid waste processing plants. Hence, when setting up the landfills, the Government may give preference to these three sites for setting up of landfills in BMR. Galipuje, Honnaghatta, Samanduru, Gowrenahalli sites fall under moderate landfill suitability index and lie 20 km away from the processing plants making them moderately suitable landfill sites in BMR. Though the landfill site at Kalarikaval comes under high suitable landfill index, since it is located at a distance of 40 km from the solid waste processing plants which may make the transportation of the MSW to this landfill expensive, this landfill site is categorized as less suitable. If there are any solid waste processing plants established closer to this landfill sites, then it may serve as a potential site in the future. Bendiganahalli and Chikkabanahalli sites lie closer to the solid waste processing plants (within 5 km). However, these regions are expected to undergo urban developments in the future (Revised Structure Plan of BMR 2031) and are not recommended for the setting up of landfill sites considering the negative impacts of a landfill on the health of the people living nearby them. Probably the Government might have considered these landfill sites as they lie closer to the processing plants which may reduce the transportation cost of the MSW which contributes a huge share in the budget of BMR. The results of the analysis would be useful to the local planning authorities to decide upon giving preference to sites while establishing the landfills in BMR.

## 77.5 Conclusions

In the current study, to identify potential landfill sites in BMR, GIS enabled AHP technique was implemented based on general, socio-economic, geological criteria and constraints. The landfill suitability map prepared through AHP analysis categorized the study region into five categories as excluded, unsuitable, suitable, moderately suitable, and highly suitable. Considering the increasing solid waste generation, the Government has proposed to set up new landfill sites in the study region by 2031. Based on the landfill suitability map, the study categorized the proposed landfill sites as highly suitable, moderately suitable, less suitable and unsuitable based on its location to the solid waste processing plants. Results of this study identified three potential sites at Gudhatti, Madivala and Neraluru as highly suitable which would help the local authorities while planning to set up the landfill sites in the future. The metropolitan region is thriving to achieve sustainable waste management in the future through reusing, recycling, and energy conversion of solid waste which might reduce the amount of waste disposed in the landfills thereby increasing the lifespan of the existing landfills and may possibly reduce the need for new landfills in the future. This needs further attention in the research.

**Acknowledgements** The research was financed by the Swedish Research Council FORMAS through project grant number 2017-00266.

## References

Ali SA, Parvin F, Al-Ansari N, Pham QB, Ahmad A, Raj MS, Anh DT, Ba LH, Thai VN (2021) Sanitary landfill site selection by integrating AHP and FTOPSIS with GIS: a case study of Memari Municipality, India. Environ Sci Pollut Res 28(6):7528–7550. https://doi.org/10.1007/s11356-020-11004-7

Central Public Health and Environmental Engineering Organisation (CPHEEO) (2023) Municipal solid waste management manual. Part II: the manual. In: Ministry of urban development, vol part II. http://cpheeo.gov.in/upload/uploadfiles/files/Part2.pdf. Last accessed 27 June 2023

Dittrich C (2004) Bangalore: globalization and securing survival in India's high-tech capital. ASIEN 103:45–58

Kamdar I, Ali S, Bennui A, Kuaanan T, Jutidamrongphan W (2019) Municipal solid waste landfill siting using an integrated GIS-AHP approach: a case study from Songkhla, Thailand. Resour Conserv Recycl 149(4):220–235

Kaur A, Bharti R, Sharma R (2023) Municipal solid waste as a source of energy. Mater Today Proc 81(2):904–915. https://doi.org/10.1016/j.matpr.2021.04.286

Mardani A, Jusoh A, Nor KM, Khalifah Z, Zakwan N, Valipour A (2015) Multiple criteria decision-making techniques and their applications—a review of the literature from 2000 to 2014. Econ Res-Ekonomska Istraživanja 28(1):516–571. https://doi.org/10.1080/1331677X.2015.1075139

Martinez JAS, Mendoza A, Vazquez MDRA (2019) Collection of solid waste in municipal areas: urban logistics. Sustainability 11(19):1–15. https://doi.org/10.3390/su11195442

Naveen BP, Sivapullaiah PV (2020) Solid waste management: current scenario and challenges in Bengaluru. In: Sustainable sewage sludge management. IntechOpen, London, pp 1–23. https://doi.org/10.5772/intechopen.90837

Ohri A, Singh PK, Maurya SP, Mishra S (2015) Sanitary landfill site selection by using geographic information system. In: Proceedings of national conference on open source GIS: opportunities and challenges, 9–10 Oct 2015, pp 170–180

Revised Structure Plan of BMR 2031 (2023) Bangalore Metropolitan region—draft report. https://data.opencity.in/Documents/Recent/Revised-Structure-Plan-2031-Draft-Report.pdf. Last accessed 27 June 2023

Saaty TL (1977) A scaling method for priorities in hierarchical structures. J Math Psychol 15(3):234–281. https://doi.org/10.1016/0022-2496(77)90033-5

Saaty TL (1990) How to make a decision: the analytic hierarchy process. Eur J Oper Res 48(1):9–26. https://doi.org/10.1016/0377-2217(90)90057-I

Sharma BK, Chandel MK (2021) Life cycle cost analysis of municipal solid waste management scenarios for Mumbai, India. Waste Manage 124:293–302. https://doi.org/10.1016/j.wasman.2021.02.002

Surendra HJ, Hema HC, Nagasahadevareddy K (2021) Analysis an assessment of solid waste management through field approach, Halasuru Ward, Bangalore. Int J Innov Res Sci Eng Technol 8(2):978–984. https://doi.org/10.15680/IJIRSET.2019.0802054

Vázquez-Quintero G, Prieto-Amparán JA, Pinedo-Alvarez A, Valles-Aragón MC, Morales-Nieto CR, Villarreal-Guerrero F (2020) GIS-based multicriteria evaluation of land suitability for grasslands conservation in Chihuahua, Mexico. Sustainability 12(1). https://doi.org/10.3390/SU12010185

# Chapter 78
# Role of Energy Sources in Achieving Carbon Neutrality Under the Condition of Economic Growth

**I. V. Filimonova, A. V. Komarova, K. D. Gladkikh, and A. Y. Novikov**

**Abstract** The economic growth of the countries is mainly associated with either the introduction of new technologies or involvement of the extra resources into production. However, industrial activities are still based on the combustion of fossil fuels, which poses implications for the climate change. Russia has been one of the five largest emitters of greenhouse gases for the past decade. Therefore, the achievement of carbon neutrality should be accompanied by the transition to the trajectory of sustainable development. This research considers the calculation of the greenhouse gas emission in the energy sector carried out on the basis of the guidelines of the Ministry of Natural Resources and Ecology of Russia, which was compiled on the basis of the IPCC methodology. The results showed that there is a general decrease in carbon dioxide emissions from stationary sources of fuel combustion in Russia, and the main source of emissions is combustions of fossil fuels in boilers and furnaces. The reduction of the emission in the Urals and Siberian Federal Districts is the result of the gasification process. However, despite the favorable trend, it is advisable to pursue an active climate regulation policy in these regions. Policy should pay attention to reducing emissions from coal, gasoline, diesel fuel and other petroleum products, which are the main sources of the greenhouse gases in these regions.

**Keywords** Energy sources · Greenhouse gases · Spatial organization · Fuel combustion · Carbon neutrality · Economic development

---

I. V. Filimonova (✉) · A. V. Komarova · K. D. Gladkikh · A. Y. Novikov
Novosibirsk State University, 1, Pirogov St., 630090 Novosibirsk, Russia
e-mail: FilimonovaIV@list.ru

Trofimuk Institute of Petroleum Geology and Geophysics, 2, Koptug Av., 630090 Novosibirsk, Russia

 
N. S. Caetano and M. C. Felgueiras (eds.), *The 9th International Conference on Energy and Environment Research*, Environmental Science and Engineering,
https://doi.org/10.1007/978-3-031-43559-1_78

## 78.1 Introduction

Climate problems are the centre of attention for the public, politicians and researchers around the world for the last decades. As a result of the active economic development of many countries, industrialization and greatly accelerated human activity, higher amount of the greenhouse gases has been produced, which could lead to irreversible climate change and global warming (Bennett 2021). The problems associated with lowering water levels, melting glaciers, deteriorating public health have intensified, which led to the adoption of the Kyoto Protocol and the Paris Agreement, which have main goals of the reduction of the greenhouse gas emissions. One of the main sources of greenhouse gases is the combustion of fossil fuel and energy sources in order to obtain energy (Kontorovich et al. 2017). Thus, it is necessary to adequately and timely adjust the permissible level of greenhouse gas emissions, including carbon dioxide, methane, nitrous oxide, etc. That is why at the moment there are a lot of studies related to the accurate assessment of the emissions into the atmosphere. Mostly they are based on calculations in accordance with the Intergovernmental Panel on Climate Change (IPCC) guidelines, which are fundamental for the formation of assessment methodology in different countries. Thus most of the studies consider country or regional specifics of the emission estimation. Sununta et al. (2019) carried out emission assessment and mitigation planning in Dai Sai Municipality. The authors concluded that a model of a city that actively introduces an increase in the number of solar roofs, the use of household LED lamps, and improved waste management using RDF technology is the best, because such activities decrease $CO_2$ emissions (Sununta et al. 2019).

Aliyu estimated nitrous oxide emissions from arable land in China based on regional and crop-specific emission factors. The authors concluded that emissions calculated based on regional coefficients differ from emissions calculated using IPCC indicators, which indicates the need to use them. Measuring local emission factors is one of the key ways to obtain consistent emission estimates and reduce uncertainty (Aliyu et al. 2006). Fearnside (2015) estimated emissions from tropical hydroelectric power plants. The author concluded that the considered emissions also have some weight in the total greenhouse gas emissions. Current values for this type of emission are grossly underestimated. According to the authors, the IPCC guidelines should be revised, so that the required level of reporting for dams reflects the full emissions of all greenhouse gases (Fearnside 2015). Kalt et al. (2016) provided analysis of the potential benefits of replacing carbon-intensive products compared to replacing fuel. In order to include the replacement of carbon-intensive products in national climate strategies, the shortcomings of existing default accounting methods need to be addressed. In this case, there is also a problem with the selection of approaches to the measurement (Kalt et al. 2016). Vásquez et al. (2015) carried out an assessment of greenhouse gas emissions from a university campus in Chile. This study discusses methods for measuring greenhouse gas emissions on a local scale. The main sources of greenhouse gas emissions are student and staff transportation and electricity consumption. In addition, the authors presented four scenarios for reducing greenhouse gas emissions (Vásquez et al. 2015). The purpose of this work is to provide quantitative and factor analysis of the carbon dioxide emissions from fuel combustion in the regions of the Russian Federation. The assessment is based on the guidelines of the IPCC (Soegoto and Utomo 2019). This result will reveal the potential for reducing the greenhouse effect to achieve carbon neutrality.

## 78.2 Materials

The object of research is the emission of carbon dioxide from fuel combustion. Over the past decades, most greenhouse gas emissions have traditionally come from the energy sector, amounting to the average level about 78.9%. In the Russian Federation, at the legislative level, a list of greenhouse gases subject to mandatory accounting is defined as follows: carbon dioxide ($CO_2$), methane ($CH_4$), nitrous oxide ($N_2O$), sulfur hexafluoride ($SF_6$), nitrogen trifluoride ($NF_3$), hydrofluorocarbons (HFCs) and perfluorocarbons (PFCs). However, in emissions from fuel combustion, more than 90% comes from carbon dioxide. In this regard, this study focused on estimating carbon dioxide emissions in the Energy sector. In accordance with the selected categories, guidelines establish the use of a calculation method based on information about the activities of the enterprise and certain emission factors. To conduct the study, an initial dataset was collected based on information from official statistical source—Rosstat (Federal State Statistics Service). For the analysis, the following indicators were selected: the consumption of fuel and energy resources in the country as a whole, as well as in the following federal districts: Central, Ural, North Caucasian, Far Eastern, Siberian, Southern, Volga (Privolzhsky), Northwestern. The collected data characterize the period 2017–2020 and reflect the total consumption of different types of fuel by main types of associated economic activities. The data is broken down by the following categories of consumption: boiler and furnace fuel, motor fuel supplied to the population, lubricating fuel and fuel consumed for non-fuel needs. The last two categories were not taken into account when estimating emissions, since they are used for non-energy needs. Finally, it is worth noting that the data was aggregated by statistical services on the basis of the form 4-TER «Information on the use of fuel and energy resources». This form of federal statistical observation is submitted by all organizations (except for small businesses) that consume fuel and energy, secondary resources, and also sell them to the population and other legal entities and individuals. Thus, it is possible to underestimate emissions due to not including small enterprises, but it is considered insignificant.

## 78.3 Methods

Fundamental for calculating greenhouse gas emissions in individual countries is methodology developed by the IPCC. Usually, countries develop their own methodological guidelines that take into account the climatic features of the territory on the basis of international methods. In this work, the calculation was carried out on the basis of the guidelines of the Ministry of Natural Resources and Ecology of Russia dated June 30, 2015 No. 300, compiled on the basis of the IPCC methodology. Before directly determining carbon dioxide emissions, the initial data presented in natural units, according to the methodology, must be converted into energy units using the formula 78.1.

$$FC_{j,y} = FC_{j,y\prime} * NVC_{j,y} * 10^{-3} \tag{78.1}$$

where $FC_{j,y}$—fuel consumption j in energy equivalent for period t in region i; $FC_{j,y\prime}$—fuel consumption j in physical terms for a period t in region i, tons or thousand m3; $NVC_{j,y}$—net calorific value of fuel j, MJ/kg or MJ/m$^3$.

In accordance with the recommendations of the IPCC, the recalculation should be carried out according to the formula 78.2, presented below. Appropriate actions need to be taken for each emission source.

$$E_{CO_2,y} = \sum_{j=1}^{n}(FC_{J,Y} * EF_{CO_2,j.y} * OF_{j.y}) \tag{78.2}$$

where $E_{CO_2,y}$—$CO_2$ emissions from stationary fuel combustion for period t in region i, ton $CO_2$; $FC_{J,Y}$—fuel consumption j in energy equivalent for period t in region i, TJ; $EF_{CO_2,j.y}$—emission factor $CO_2$ from stationary fuel combustion j, ton $CO_2$/unit; $OF_{j.y}$—fuel oxidation coefficient j, share; n—number of fuels used in period t (Eggleston et al. 2006).

Main types of fossil fuels have different properties, which affects their calorific value. Therefore, conversion coefficients to energy units and emission coefficients differ as well. The variety of the values of these coefficients for the four aggregate categories of fuels are presented in Table 78.1.

**Table 78.1** Conversion coefficients to energy units and emission coefficients by type of fuel used

| Fuel types | Conversion coefficient to energy units ($NVC_j$) Unit of measurement[TJ/th t (million m$^3$)] | Emission coefficient ($EF_{CO_2,j}$) [t $CO_2$/TJ] |
|---|---|---|
| Gas | [Thousand m$^3$] 33.8 | 54.4 |
| Oil products | [Ton] 42.5–43.7 | 69.3–74.1 |
| Coal | [Ton] 13.7–22.5 | 94.6–101.0 |
| Fuel wood | [Thousand m$^3$] 7.8 | 112.0 |

Further, the calculations and discussion of the obtained results were carried out. Additionally, set of indicators such as emissions per capita; emissions per unit of GDP; sensitivity of emissions to changes in the parameters of energy carriers were analyzed.

## 78.4 Results and Discussion

At the first stage of calculations estimates of carbon dioxide emissions into the atmosphere from the Energy sector in each federal district and in Russia as a whole were obtained. The information obtained allows us to consider emissions when separated by different types of fuel, by sources, by different purposes, as well as by territorial location. Data on total carbon dioxide emissions in the regions under consideration are presented in Table 78.2. The total emissions from stationary fuel combustion in Russia in 2020 amounted to 1097 million tons of $CO_2$. Emissions are not evenly distributed across the federal districts (Table 78.2).

Thus, the analysis of data on the usage of fuel and energy resources showed that in general in Russia a reduction in greenhouse emissions from stationary fuel sources is observed. However, the situation in the regions and districts differs a lot. The most significant part of the calculated emissions falls on the Central and Ural Federal Districts, which may indicate a strong influence of changes in the situation in these regions on the overall environmental situation in the country.

Distribution of total emissions $CO_2$ of the Russian Federation by intended use in 2019 shown in Fig. 78.1. The largest share of emissions from fuel use is formed when fuel is used for boilers and furnaces, which is associated with a large amount of resources spent on centralized heat and energy production in the Energy sector. The second largest category of emissions in the Central, Southern and North Caucasian Federal Districts are emissions from fuel supplied to the population. All other regions

**Table 78.2** Carbon dioxide emissions in different federal districts of the Russian Federation, million tons

| Area | 2017 | 2018 | 2019 | 2020 |
|---|---|---|---|---|
| Russian Federation | 1189.55 | 1162.09 | 1141.85 | 1097.93 |
| Central Federal District | 224.02 | 204.41 | 200.09 | 197.4 |
| Northwestern Federal District | 133.28 | 131.06 | 128.23 | 121.34 |
| Southern Federal District | 73.11 | 67.91 | 64.5 | 61.67 |
| North Caucasian Federal District | 27.26 | 26.17 | 27.09 | 26.55 |
| Volga Federal District | 202.56 | 196.28 | 191.19 | 176.69 |
| Ural federal district | 248.03 | 257.19 | 253.8 | 248.21 |
| Siberian Federal District | 210.49 | 188.35 | 186.35 | 175.14 |
| Far Eastern Federal District | 70.49 | 90.72 | 90.67 | 90.62 |

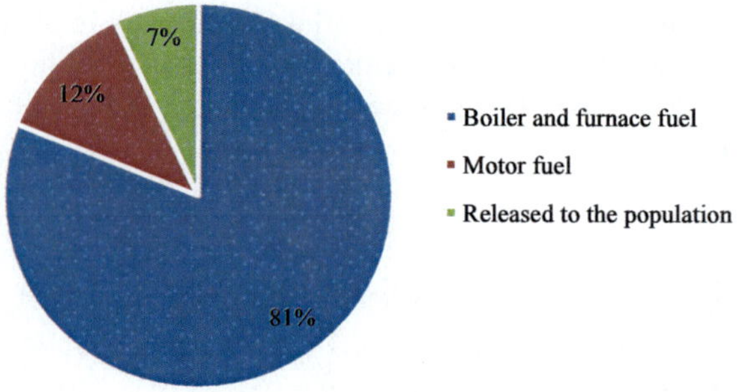


**Fig. 78.1** Distribution of total emissions $CO_2$ from Energy sector in the Russian Federation by intended use in 2019

spend most of their resources on the use of fuel as motor fuel, which is confirmed by the obtained results (Fig. 78.1).

Figure 78.2 shows the distribution of $CO_2$ emission in the Energy sector by fuel type in the Russian Federation in 2019. It can be seen, that most of the emissions are generated by gas combustion. Additional calculations showed that the largest share of $CO_2$ emissions is generated from gas usage in all federal districts, with the exception of the Siberian Federal District and Far East Federal District, which is due to the specifics of activities in this area.

In addition, when analyzing the obtained results using a system of indicators, the authors made the following important conclusions, which are consistent with the results of other authors.

- There is a noticeable tendency in the Russian Energy sector to switch to gas when using fuel as a source for boilers and furnaces (Eder et al. 2019).
- According to the analysis of the emissions from energy sources used as a motor fuel, the largest share is occupied by emissions from petroleum products, which is confirmed by practical results and research (Lao et al. 2023).
- According to the calculations of the carbon dioxide emissions per unit of gross product, the highest value of this indicator is observed in the Siberian and Ural

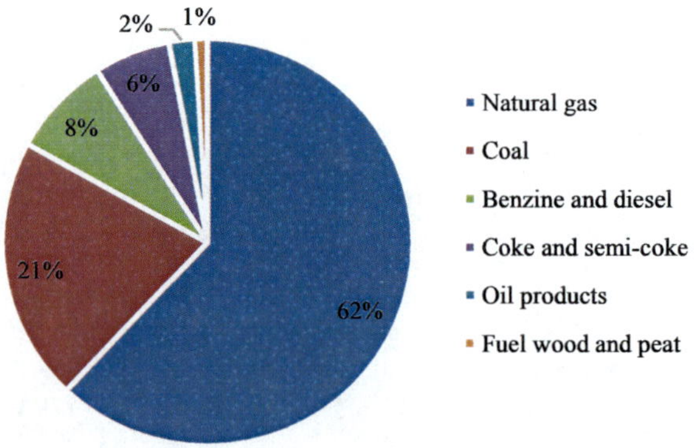


**Fig. 78.2** Distribution of $CO_2$ emission from energy sector in the Russian Federation by fuel type in 2019

Federal Districts, which can be directly related to unfavorable climate conditions in the regions.

- Analysis of emissions per capita confirms the previously noted trend: the highest value of the parameter falls on the Ural Federal District, which in this case is again associated with the presence of climate problems and the unfavorable demographic situation in the region.
- Sensitivity analysis showed that in the case of a change in the proportion of fuel used, gas is the most sensitive to emissions, due to the large amount of its use. In the case of sensitivity assessment when the initial parameter is changed by a fixed number, coal, gasoline and diesel fuel turned out to be the most different from all.
- In five regions out of eight studied, coal turned out to be the most polluting fuel, namely in the Central, Southern, Far Eastern, Siberian and Ural Federal Districts.
- The findings from the analysis suggest that the government needs to promote switch from coal, gasoline, diesel and other refined petroleum products to alternative fuel sources that produce less emissions. However, measures to achieve carbon neutrality are very costly for the country (Gren and Tirkaso 2021).

## 78.5 Conclusion

In the study the quantitative assessment of carbon dioxide emissions in the Energy sector of Russia with respect of the territorial division and types of fuels was carried out. When analyzing the obtained results, it was revealed that there is a general decrease in carbon dioxide emissions from stationary fuel combustion in Russia. However, this is not only the result of the ongoing climate policy, but also the outbreak of the coronavirus, which has negatively affected overall economic activity and provided the reduction in electricity generation.

Another important trend is the growth of the natural gas usage as the boiler and furnace fuel. It is mainly the result of the transition to gas in the Urals and Siberian Federal Districts, which occupy significant shares in total carbon dioxide emissions. However, despite this, these districts have large emissions per capita and per unit of gross regional product, which indicates the feasibility of an enhanced climate policy to achieve carbon neutrality in these regions. Moreover, according to the calculations, it is necessary to reduce emissions from the use of coal, gasoline, diesel fuel and other petroleum products, which produce the most pollutants for the atmosphere.

Thus, despite the positive dynamics in reducing emissions in Russia as a whole, as well as data indicating the success of the ongoing gasification policy in some regions, the state needs to make sufficient efforts in the form of implementing measures to reduce carbon dioxide emissions. The actions of the authorities should involve the introduction of new measures and preferably correspond to an intensive development scenario that allows them to lead to a positive effect in a timely manner.

**Acknowledgements** This study was carried out with the financial support of the Russian Science Foundation within the framework of the project №. 22-18-00424.

## References

Aliyu G, Luo J, Di HJ, Lindsey S, Liu D, Yuan J, Chen Z, Lin Y, He T, Zaman M, Ding W (2019) Nitrous oxide emissions from China's croplands based on regional and crop-specific emission factors deviate from IPCC 2006 estimates. Sci Total Environ 669:547–558

Bennett J (2021) Healthcare industry: a movement towards a reduction in greenhouse gas emissions. J Vascular Nursing Official Publ Soc Peripheral Vascular Nursing 39(4):89–90

Eder LV, Provornaya IV, Filimonova IV (2019) Problems of rational use of associated petroleum gas in Russia. Geogr Nat Resour 40:9–14

Eggleston HS, Buendia L, Miwa K, Ngara T, Tanabe K (2006) 2006 IPCC guidelines for national greenhouse gas inventories

Fearnside PM (2015) Emissions from tropical hydropower and the IPCC. Environ Sci Policy 50:225–239

Gren M, Tirkaso W (2021) Costs and equity of uncertain greenhouse gas reductions—fuel, food and negative emissions in Sweden. Energy Econ 104:105638

Kalt G, Höher M, Lauk C, Schipfer F, Kranzl L (2016) Carbon accounting of material substitution with biomass: case studies for Austria investigated with IPCC default and alternative approaches. Environ Sci Policy 64:155–163

Kontorovich AE, Eder LV, Filimonova IV (2017) Paradigm oil and gas complex of Russia at the present stage. IOP Conf Ser Earth Environ Sci 1:012010

Lao J, Song H, Wang C, Zhou Y (2023) Research on atmospheric pollutant and greenhouse gas emission reductions of trucks by substituting fuel oil with green hydrogen: a case study. Int J Hydrogen Energy 48(30):11555–11566

Soegoto ES, Utomo AT (2019) Marketing strategy through social media. IOP Conf Ser Mater Sci Eng 3:032040

Sununta N, Kongboon R, Sampattagul S (2019) GHG evaluation and mitigation planning for low carbon city case study: Dan Sai Municipality. J Clean Prod 228:1345–1353

Vásquez L, Iriarte A, Almeida M, Villalobos P (2015) Evaluation of greenhouse gas emissions and proposals for their reduction at a university campus in Chile. J Clean Prod 108:924–930